SOLAR MAGNETIC FIELDS

INTERNATIONAL ASTRONOMICAL UNION
UNION ASTRONOMIQUE INTERNATIONALE

SYMPOSIUM No. 43
HELD AT THE COLLÈGE DE FRANCE, PARIS, FRANCE
AUGUST 31 TO SEPTEMBER 4, 1970

SOLAR MAGNETIC FIELDS

EDITED BY

ROBERT HOWARD

Hale Observatories, Carnegie Institution of Washington,
California Institute of Technology, Pasadena, Calif., U.S.A.

D. REIDEL PUBLISHING COMPANY

DORDRECHT-HOLLAND

1971

Published on behalf of
the International Astronomical Union
by
D. Reidel Publishing Company, Dordrecht, Holland

Library of Congress Catalog Card Number 78–159656

ISBN-13: 978-94-010-3119-6   e-ISBN-13: 978-94-010-3117-2
DOI: 10.1007/978-94-010-3117-2

PREFACE AND INTRODUCTION

This symposium was held at the Collège de France in Paris from August 31 to September 4, 1970. The Organizing Committee consisted of V. Bumba, R. Howard (Chairman), K. O. Kiepenheuer, R. Michard, E. N. Parker, A. B. Severny, V. E. Stepanov, and T. Takakura. The Local Organizing Committee consisted of Miss G. Drouin (Secretary), R. Michard (Chairman), J. -C. Pecker, and J. Rayrole.

We are indebted to the Collège de France for their kind hospitality. I wish to express my gratitude to members of the Organizing Committee for advice and assistance and to R. Michard and the Local Organizing Committee, who were responsible for the smooth running of the sessions, the distribution and collection of the discussion sheets, and for a delightful Wednesday afternoon excursion to Meudon. It is a pleasure to thank J. W. Evans, V. E. Stepanov, K. O. Kiepenheuer, R. G. Giovanelli, T. G. Cowling, V. Bumba, W. C. Livingston, and J. M. Wilcox who kindly served as session chairmen. I also wish to thank Miss Judy Harstine and John M. Adkins of the Hale Observatories, for invaluable assistance in editing the proceedings. This Symposium has been supported financially by the International Astronomical Union.

An IAU Symposium on *Stellar and Solar Magnetic Fields* (No. 22) was held in Rottach-Egern near Munich, Germany just seven years ago, in 1963. At that time we could devote a symposium to both stellar and solar magnetic fields and still maintain a rather comfortable and uncrowded program. Now the subject of solar magnetic fields has grown so much that it is too big for one symposium. More than 160 persons attended IAU Symposium No. 43 in Paris, and there were 94 papers of various lengths. A number of papers which authors wished at the last moment to present had to be refused because of lack of time. If work in the field continues at this rate, future symposia will have to cover only part of the subject of solar magnetic fields.

An interesting measure of activity in the field is the number of photoelectric magnetographs which contributed results to the symposium. In the case of the Rottach-Egern meeting this number was three. In the case of IAU Symposium No. 43, I count at least 15 such instruments.

Growth in the subject of solar magnetic fields has proceeded in several subdisciplines. On the smallest scale a number of investigators have looked at the smallest known magnetic features on the Sun. Results have been somewhat disparate, but all agree that outside sunspots at least a large fraction of the magnetic flux is contained in small (~ 1000 km) bundles of lines of force. The magnetic field within such features ('gaps' or 'knots') may be expected to reach values of at least a few hundred gauss. High temperatures within such features tend to weaken many Fraunhofer lines commonly used to measure magnetic fields. The resulting line-profile changes adversely affect the signals one gets from most photoelectric or photographic magnetic field measuring techniques. The physics of these 'gaps' and their relation to the short-lived pores and the larger structure of the supergranular network and active regions are areas of intensive investigation at a number of observatories. The recent suggestion that magnetic fields in 'gaps' may be quantized is an intriguing one, and no doubt will be followed up by many investigators.

On a somewhat larger scale the magnetic structure of the supergranular network and active regions continues to be an active field for research. Several investigators in recent years have accurately measured the supergranular velocity fields and have associated magnetic field concentrations with the boundaries of the calcium network. The problem of exactly what association exists, if any, between flares and magnetic field changes in active regions continues after many years essentially unanswered. Many astronomers at several observatories continue to pursue this topic. Part of the problem has been a disappointingly inactive cycle for flares. Some high-resolution magnetic cine films obtained within the last year or so give promise of solving this problem in the near future – although in the next few years these beautiful observations will probably raise more questions than they will solve.

On the large scale, advances have been made in recent years in the association of photospheric magnetic fields extended above the photosphere with coronal features and with the interplanetary magnetic field in the neighborhood of the Earth. The large-scale distribution of magnetic fields on the solar surface has been studied extensively in recent years. The dynamics of the large-scale fields appear to be explainable at least in part by a combination of the eroding effects of supergranular motions on the active region magnetic fields and the shearing effects of differential rotation. The failure of the polar magnetic fields to reverse their polarities so far in this cycle has provided an interesting new factor to consider in connection with theories of the solar cycle.

The subject of solar magnetic fields has developed rapidly in recent years, and we can look forward to exciting advances in the years to come.

Hale Observatories, Carnegie Institution of Washington,   ROBERT HOWARD
California Institute of Technology, Pasadena, Calif., U.S.A.

TABLE OF CONTENTS

PART III / OBSERVATIONS OF SUNSPOT AND ACTIVE REGION MAGNETIC FIELDS

PART IV / OBSERVATIONS OF MAGNETIC FIELDS ASSOCIATED WITH FLARES AND OTHER TRANSITORY PHENOMENA

PART VII / THE POLAR FIELDS OF THE SUN AND THE MAGNETIC ACTIVITY CYCLE

PART VIII / THEORIES OF LARGE SCALE FIELDS AND THE ACTIVITY CYCLE

LIST OF PARTICIPANTS

Adam, M. G., University Observatory, Oxford, England
Altrock, R. C., Sacramento Peak Observatory, Sunspot, New Mex., U.S.A.
Altschuler, M. D., High Altitude Observatory, Boulder, Colo., U.S.A.
Andelin, J., Harvard College Observatory, Cambridge, Mass., U.S.A.
Anzer, U., Max-Planck-Institute für Physik und Astrophysik, Munich, Germany
Arnaud, J., Observatoire de Paris, Meudon, France
Athay, R. G., High Altitude Observatory, Boulder, Colo., U.S.A.
Avignon, Y., Observatoire de Paris, Meudon, France
Ballario, M. C., Osservatorio Astrofisico di Arcetri, Florence, Italy
Banos, G. J., Astronomical Institute – National Observatory, Athens, Greece
Bappu, M. K. V., Astrophysical Observatory, Kodaikanal, India
Beckers, J. M., Sacramento Peak Observatory, Sunspot, New Mex., U.S.A.
Blondel, M., Observatoire de Paris, Meudon, France
Boischot, A., Observatoire de Paris, Meudon, France
Bray, R. J., National Standards Laboratory, Sydney, Australia
Brueckner, G. E., University of Maryland, College Park, Md., U.S.A.
Bruzek, A., Fraunhofer-Institut, Freiburg, Germany
Bumba, V., Observatory Ondřejov, Ondřejov, Czechoslovakia
Cacciani, A., Osservatorio Astronomico su Monte Mario, Rome, Italy
Caroubalos, C., Observatoire de Paris, Meudon, France
Casanovas S.J., J., University of Tenerife, Canary Islands, Spain
Castellanos, J. M., NASA Station, Canary Islands, Spain
Chambe, G., Observatoire de Paris, Meudon, France
Charvin, P., Observatoire de Paris, Meudon, France
Cimino, M., Osservatorio Astronomico su Monte Mario, Rome, Italy
Cowling, T. G., The University, Leeds, England
Csada, I. K., Konkoly Observatory, Budapest, Hungary
Deubner, F. L., Fraunhofer Institute, Freiburg, Germany
Dodson-Prince, H., McMath-Hulbert Observatory, Pontiac, Mich., U.S.A.
Dollfus, A., Observatoire de Paris, Meudon, France
Dulk, G. A., Radiophysics Laboratory, Chippendale N.S.W., Australia
Dunn, R., Sacramento Peak Observatory, Sunspot, New Mex., U.S.A.
Durney, B., National Center for Atmospheric Research, Boulder, Colo., U.S.A.
Durrant, C. J., The Observatories, Cambridge, England
Elliott, I., Dunsink Observatory, Dublin, Ireland
Evans, J. W., Sacramento Peak Observatory, Sunspot, New Mex., U.S.A.
Fofi, M., Osservatorio Astronomico su Monte Mario, Rome, Italy

Fossat, E., Observatoire de Nice, Nice, France
Foukal, P., Hale Observatories, Pasadena, Calif., U.S.A.
Frazier, E. N., Aerospace Solar Observatory, Los Angeles, Calif., U.S.A.
Gaizaukas, V., Dominion Observatory, Ontario, Canada
de Genouillac, G., Observatoire de Paris, Meudon, France
Gilman, P. A., National Center for Atmospheric Research, Boulder, Colo., U.S.A.
Giovanelli, R. G., National Standards Laboratory, Sydney, Australia
Godoli, G., Osservatorio Astrofisico, Citta Universitaria, Catania, Italy
Grigoryev, V. M., SibIZMIRAN, Irkutsk, U.S.S.R.
Grossmann-Doerth, U., Fraunhofer-Institut, Freiburg, Germany
Hall, D. N. B., Kitt Peak National Observatory, Tucson, Ariz., U.S.A.
Harvey, J. W., Kitt Peak National Observatory, Tucson, Ariz., U.S.A.
Harvey, K. L., Lockheed Solar Observatory, Burbank, Calif., U.S.A.
Heckman, G., NASA Station, Canary Islands, Spain
Hedeman, E. R., McMath-Hulbert Observatory, Pontiac, Mich., U.S.A.
Herrera, F., NASA Station, Canary Islands, Spain
House, L. L., High Altitude Observatory, Boulder, Colo., U.S.A.
Howard, R., Hale Observatories, Pasadena, Calif., U.S.A.
Hudson, H., University of California at San Diego, La Jolla, Calif., U.S.A.
Jäger, F. W., Astrophysikalisches Observatorium, Potsdam, Germany
Jakimiec, J., Wroclaw University, Astronomical Institute, Wroclaw, Poland
Janssens, T. J., Aerospace Solar Observatory, Los Angeles, Calif., U.S.A.
Jensen, E., Institute of Theoretical Astrophysics, Oslo, Norway
Jordan, C., Culham Laboratory, Berkshire, England
Kiepenheuer, K. O., Fraunhofer Institute, Freiburg, Germany
Koeckelenbergh, A., Observatoire de Belgique, Brussels, Belgium
Köhler, H., Universitäts-Sternwarte, Göttingen, Germany
Kompfner, P., Oxford University, Oxford, England
Kopecký, M., Observatory Ondřejov, Ondřejov, Czechoslovakia
Krause, F., Zentralinstitut für Astrophysik, Berlin, Germany
Krieger, A., American Science and Engineering, Cambridge, Mass., U.S.A.
Kuklin, G. V., SibIZMIRAN, Irkutsk, U.S.S.R.
Kundu, M. R., University of Maryland, College Park, Md., U.S.A.
Kuperus, M., Astronomical Observatory 'Sonnenborgh', Utrecht, The Netherlands
Kuypers, J., Astronomical Institute at Utrecht, Utrecht, The Netherlands
Lacombe, C., Observatoire de Paris, Meudon, France
Lamb, F. K., Oxford University, Oxford, England
Lantos, P., Observatoire de Paris, Meudon, France
Lantos-Jarry, M. F., Observatoire de Paris, Meudon, France
Leblanc, Y., Observatoire de Paris, Meudon, France
Leighton, R. B., Hale Observatories, Pasadena, Calif., U.S.A.
Leroy, J., Observatoire du Pic du Midi, Bagnères de Bigorre, France
Lilliequist, C., High Altitude Observatory, Boulder, Colo., U.S.A.

Livingston, W. C., Kitt Peak National Observatory, Tucson, Ariz., U.S.A.
Lufkin, D., Solar Forecast Facility, Ent A. F. B., Colorado Springs, Colo., U.S.A.
Magnant, F., Observatoire de Paris, Meudon, France
Maltby, P., University of Oslo, Oslo, Norway
Martres, M. J., Observatoire de Paris, Meudon, France
Mattig, W., Fraunhofer Institut, Freiburg, Germany
Mayfield, E. B., Aerospace Solar Observatory, Los Angeles, Calif., U.S.A.
Mein, P., Observatoire de Paris, Meudon, France
Mein, N., Observatoire de Paris, Meudon, France
Mendis, D. A., University of California at San Diego, La Jolla, Calif., U.S.A.
Meyer, F., Max-Planck-Institut für Physik und Astrophysik, Munich, Germany
Michard, R., Observatoire de Paris, Meudon, France
Mogilevsky, E. I., IZMIRAN, Moscow Region, U.S.S.R.
Molnar, J., University of Göttingen Observatory, Göttingen, Germany
Mouradian, Z., Observatoire de Paris, Meudon, France
Muller, R., Observatoire du Pic du Midi, Bagnères de Bigorre, France
Musman, S., Sacramento Peak Observatory, Sunspot, New Mex., U.S.A.
Nagarajan, S., Institute of Mathematical Sciences, Madras, India
Nakagawa, Y., High Altitude Observatory, Boulder, Colo., U.S.A.
Namba, O., The Astronomical Institute at Utrecht, Utrecht, The Netherlands
Nardi, V., Stevens Institute of Technology, Hoboken, N. J., U.S.A.
Newkirk, Jr., G., High Altitude Observatory, Boulder, Colo., U.S.A.
Oertel, G. K., NASA, Washington, D.C., U.S.A.
Orrall, F. Q., University of Hawaii, Honolulu, Hawaii, U.S.A.
Pande, M. C., Uttar Pradesh State Observatory, Naini Tal, India
Pasachoff, J. M., Harvard College Observatory, Cambridge, Mass., U.S.A.
Pecker, J. C., Observatoire de Paris, Meudon, France
Pick, M., Observatoire de Paris, Meudon, France
Pierce, A. K., Kitt Peak National Observatory, Tucson, Ariz., U.S.A.
Pneuman, G. W., High Altitude Observatory, Boulder, Colo., U.S.A.
Rayrole, J., Observatoire de Paris, Meudon, France
Roddier, F., Faculté des Sciences de Nice, Nice, France
Rösch, J., Pic du Midi Observatory, Bagnères de Bigorre, France
Rosenberg, J., Astronomical Observatory 'Sonnenborgh', Utrecht, The Netherlands
Rust, D. M., Sacramento Peak Observatory, Sunspot, New Mex., U.S.A.
Sawyer, C., U.S. Department of Commerce, ESSA, Boulder, Colo., U.S.A.
Schanda, E., Institut für Angewandte Physik, Universität Bern, Bern, Switzerland
Schatten, K., NASA Goddard Space Flight Center, Greenbelt, Md., U.S.A.
Schatz, D., National Standards Laboratory, Sydney, Australia
Schmalberger, D. C., State University, Albany, New York, U.S.A.
Schmieder, B., Observatoire de Paris, Meudon, France
Schröter, E. H., Universitäts-Sternwarte, Göttingen, Germany
Semel, M., Observatoire de Paris, Meudon, France

Severny, A. B., Crimean Astrophysical Observatory, Nauchny, Crimean, U.S.S.R.
Sheeley, Jr., N. R., Kitt Peak National Observatory, Tucson, Ariz., U.S.A.
Simon, G. W., Sacramento Peak Observatory, Sunspot, New Mex., U.S.A.
Simon, M., State University of New York, Stony Brook, New York, U.S.A.
Simon, P., Observatoire de Paris, Meudon, France
Smerd, S. F., CSIRO, Chippendale N.S.W., Australia
Smith, E. v. P., University of Maryland, College Park, Md., U.S.A.
Smith, S. F., Lockheed Solar Observatory, Burbank, Calif., U.S.A.
Soboleva, N. S., Pulkovo Observatory, Leningrad, U.S.S.R.
Soru-Iscovici, I., Observatoire de Paris, Meudon, France
Sreenivasan, S. R., The University of Calgary, Calgary, Alberta, Canada
Steinberg, J. L., Observatoire de Paris, Meudon, France
Steinitz, R., Tel-Aviv University, Tel-Aviv, Israel
Stenflo, J. O., Lunds Universitet, Lund, Sweden
Stepanov, V. E., SibIZMIRAN, Irkutsk, U.S.S.R.
Steshenko, N. V., Crimean Astrophysical Observatory, Nauchny, Crimea, U.S.S.R.
Sweet, P. A., The University of Glasgow, Glasgow, Scotland
Sýkora, J., Skalante Pleso Observatory, Tatranska Lomnica, Czechoslovakia
Takakura, T., Tokyo Astronomical Observatory, Tokyo, Japan
Tandberg-Hanssen, E., High Altitude Observatory, Boulder, Colo., U.S.A.
Tanenbaum, A., University of California at Berkeley, Berkeley, Calif., U.S.A.
Tavares, J. C. T. L., Universidade de Lourenço Marques, Lourenço Marques,
    Mozambique
Title, A., Harvard College Observatory, Cambridge, Mass., U.S.A.
Tlamicha, A., Astronomical Observatory, Ondřejov, Czechoslovakia
Tuominen, J. V., Astrophysical Laboratory, Helsinki, Finland
Turon, P., Observatoire de Paris, Meudon, France
Ubarz, H., Astronomisches Institut der Universität Tübingen, Weissenau, Germany
Vaiana, G. S., American Science and Engineering, Cambridge, Mass., U.S.A.
Vrabec, D., Aerospace Corporation, Los Angeles, Calif., U.S.A.
Waldmeier, M., Swiss Federal Observatory, Zürich, Switzerland
Ward, F. W., AFCRL, Bedford, Mass., U.S.A.
Weiss, N. O., Cambridge University, Cambridge, England
Wiehr, E., University Observatory, Göttingen, Germany
Wilcox, J. M., University of California at Berkeley, Berkeley, Calif., U.S.A.
Wild, J. P., CSIRO, Chippendale N.S.W., Australia
Wilson, P. R., University of Sydney, Sydney, Australia
Wittman, A., Universitäts-Sternwarte, Göttingen, Germany
Wyller, A. A., Bartol Research Foundation, Swarthmore, Penn., U.S.A.
Zhugzhda, Y. D., IZMIRAN, Moscow Region, U.S.S.R.
Zirker, J. B., University of Hawaii, Honolulu, Hawaii, U.S.A.
Zwaan, C., Astronomical Observatory 'Sonnenborgh', Utrecht, The Netherlands

PART I

INSTRUMENTATION – MEASUREMENT OF
MAGNETIC FIELDS IN THE SOLAR ATMOSPHERE

THE MEASUREMENT OF SOLAR MAGNETIC FIELDS

JACQUES M. BECKERS

*Sacramento Peak Observatory, Air Force Cambridge Research Laboratories,
Sunspot, New Mexico 88349, U.S.A.*

Abstract. The different methods which have been used, or which may be used in the future, to measure solar magnetic fields are described and discussed. Roughly these can be divided into three groups (a) those which use the influence of the magnetic field on the electromagnetic radiation, (b) those which use the influence of the field on the structure of the solar atmosphere (MHD effects), and (c) those which use theoretical arguments. The former include the Zeeman effect, the Hanle effect, the gyro and synchrotron radiations and the Faraday rotation of radiowaves. The second includes the alignment of details at all levels of the solar atmosphere, and the calcium network, and the third makes use, for example, of the assumption of equipartition between magnetic and kinetic energy density.

1. Introduction

In preparing this discussion of the methods used to measure solar magnetic fields, I was tempted to describe in detail the refinements used today in the measurement of the field by means of the Zeeman effect. Other papers on this topic do, however, already exist and it seemed therefore wasteful. I, therefore, decided to try to review and discuss all the known different ways of magnetic field determination on the Sun.

I will divide these ways into three groups. The first utilizes the influence of the magnetic field on the solar electromagnetic radiation. It includes measurements made by means of the Zeeman effect, the Hanle effect (or resonance scattering), the gyro-resonance radiation and synchrotron radiation in the radio region, and the Faraday rotation of radio waves. The second group makes use of the influence of the magnetic field on the temperature and density structure of the solar atmosphere. In it fall the relation of the field with the Hα fibril structure and the calcium emission network. The third group utilizes theoretical arguments for the magnetic field determination. It includes, for example, the potential field calculations of coronal magnetic fields, and the equipartition of magnetic and kinetic energy.

2. Magnetic Fields as Determined by their Influence on Electromagnetic Radiation

2.1. THE ZEEMAN EFFECT

Solar Magnetographs utilizing the Zeeman effect have been described by Zhulin *et al.* (1962), Evans (1966), Beckers (1968b), and others. The Zeeman effect refers to the splitting of spectral lines in the presence of a magnetic field. For not too strong fields, where the Paschen-Back effect is negligible (and this is the case for all lines, but the Lithium lines, for magnetic field strengths encountered on the Sun), this splitting is proportional to the field strength. The splitting pattern is dependent on the details of the atomic transition (Beckers 1969); the strengths and the polarization of the components is dependent on the direction of the magnetic field. For strong fields

Howard (ed.), Solar Magnetic Fields, 3–23. All Rights Reserved.
Copyright © 1971 by the IAU.

$(B > 2000\,\mathrm{G})$ and for favorable lines (Zeeman triplets, $g = 3$) the splitting is large enough to be directly measurable. In that case, it is rather simple to measure both the amount and the direction of the magnetic field (see e.g. Beckers and Schröter (1969)). Most magnetographs are, however, built to measure very weak fields for which the Zeeman splitting $\Delta\lambda_B$ is much smaller than the line width. In this case, the splitting manifests itself only in a polarization of the line profile. The relation between this polarization and the magnetic field will be discussed in detail in Dr. Stenflo's review article. It is sufficient for the present discussion to give an approximate expression for this polarization based on the Seares formulae (Seares (1913)). One easily derives the following four Stokes parameters:

$$I = \text{Intensity} = \tfrac{1}{4}(1 + \cos^2\gamma)\left[p(\lambda - \Delta\lambda_B) + p(\lambda + \Delta\lambda_B)\right] + \tfrac{1}{2}\sin^2\gamma\, p(\lambda) \tag{1}$$

$$Q = \text{Linear polarization} = \tfrac{1}{4}\sin^2\gamma\left[2p(\lambda) - p(\lambda + \Delta\lambda_B)\right] \tag{1'}$$

$$U = 0 \tag{1''}$$

$$V = \text{Circular polarization} = \tfrac{1}{2}\cos\gamma\left[p(\lambda + \Delta\lambda_B) - p(\lambda - \Delta\lambda_B)\right]. \tag{1'''}$$

In Equations (1) the azimuth of the magnetic field coincides with the reference direction for the Q-Stokes parameter, $p(\lambda) = $ line profile and $\gamma = $ angle between $\mathbf{B}$ and the line of sight. For small $\Delta\lambda_B$ a Taylor expansion of (1) gives:

$$I \approx p(\lambda) \tag{2}$$

$$Q \approx -0.25\,\Delta\lambda_B^2\,\sin^2\gamma\,\mathrm{d}^2p/\mathrm{d}\lambda^2 \qquad (:)\ B_\perp^2 \tag{2'}$$

$$U \approx 0 \tag{2''}$$

$$V \approx \Delta\lambda_B\cos\gamma\,\mathrm{d}p/\mathrm{d}\lambda \qquad (:)\ B_\parallel. \tag{2'''}$$

By measuring V one therefore can determine the longitudinal magnetic field $B_\parallel$ and by measuring the direction and the amount Q of linear polarization one can measure the square of the transverse field $B_\perp^2$. Because for weak fields $Q \ll V$ it is very hard to measure transverse fields. Zeeman magnetographs are in fact polarimeters which measure either V (Longitudinal Magnetographs) or V, Q and U (Vector Magnetographs) and which through some kind of calibration procedure interpret these in terms of $B_\parallel$ and $B_\perp$ and through these in $\mathbf{B}$. The measurement of the polarization is generally done by means of electro-optical light modulators combined with polarizers and retardation plates. The number of possible combinations of these is virtually inexhaustible as the compilation of existing magnetographs in Table I shows. Those instruments in Table I which are built specifically to measure magnetic fields are called magnetographs. The two polarimeters are general purpose instruments which among others can measure the polarization due to the Zeeman effect. Figure 1 shows the optical diagram of the HAO polarimeter which is presently under construction. The two electro-optical light modulators cause the light intensity to vary with certain frequencies, each of the four Stokes parameters I, Q, U and V being connected with the amplitude of a specific frequency. This particular instrument will be connected with a

TABLE I

Summary of Zeeman magnetographs

Location	Type*	Lay-out**	Reference(s)
Aerospace	L	Photographic by image subtraction	–
Aerospace	L	Video system (under construction)	–
Cambridge (Malta)	L	EOLM ($\lambda/4,45°$), P	Beggs et al. (1964)
Capri	V	$\lambda/4$ (rot), EOLM ($\lambda/4,45°$), P	Deubner et al. (1969)
Crimea	V	$\lambda/4$ (0° and 45°), EOLM ($\lambda/4,45°$), P	Stepanov et al. (1962)
Hawaii	P	$\lambda/4$ (rot), P	–
HAO	L	EOLM ($\lambda/4,45°$), P (automatic calibration)	Lee et al. (1965)
HAO	P	EOLM ($\lambda/2.61,0°$), EOLM ($\lambda/6.50,45°$), $\lambda/4$ (45°), P (under construction)	–
Huntsville	V	Video System (under construction)	–
Izmiran	V	$\lambda/4$ (15°), EOLM ($\lambda/4,45°$), P	Ioshpa et al. (1962, 1965)
Kitt Peak	V	{$\lambda/2$ (22.5°) and none}, EOLM ($\lambda/2,45°$), P	Livingston et al. (1970)
Kitt Peak	L	EOLM ($\lambda/4,45°$), P (40 channel)	Livingston et al. (1970)
Kodaikanal	L	EOLM ($\lambda/4,45°$), P	
Locarno	V	$\lambda/13.7$ (0°), EOLM ($\lambda/4,45°$), P	Wiehr (1969)
Meudon	V	Photographic with lambdameter	Rayrole (1967)
Mt. Wilson	L	EOLM ($\lambda/4,45°$), P	Babcock (1953)
Ondrejov	V	$\lambda/4$ (13°), $\lambda/4$ (45°), EOLM ($\lambda/4,45°$), P	Kuznetzov et al. (1966)
Pulkovo	V	$\lambda/2$ (rot), EOLM ($\lambda/4,45°$), P	–
Rome	L	EOLM ($\lambda/4,45°$), P	–
Sac Peak	L	$\lambda/4$ (45°), P (measures spectrum separation)	Evans (1966)
Sibizmir	V	$\lambda/4$ (13°), $\lambda/4$ (45°), EOLM ($\lambda/4,45°$), P	Kuznetzov et al. (1966)
Sydney	L	Video System (under construction)	–

* L = Longitudinal Magnetograph, V = Vector Magnetograph, P = Polarimeter.
Most vector magnetographs work also in longitudinal mode only.
** Explanation: $\lambda/4$ (20°) = quarter wave plate with optical axis at 20° with respect to the axis of polarizer P.
EOLM ($\lambda/4,45°$) = Electro-optical light modulator, modulated at $\pm \lambda/4$ axis makes 45° with P.

rapidly scanning spectrometer so that the line profile can be measured in all Stokes parameters.

The Zeeman magnetograph is the best instrument so far to measure solar fields. There are, however, a number of serious difficulties in deriving the magnetic field from the magnetograph signal. I want to discuss these shortly.

(a) The calibration of the polarization in terms of $B_{\parallel}$ or $B_{\perp}$ is often very difficult. Partly this is due to a variation of the quantities $dp/d\lambda$ and $d^2p/d\lambda^2$ across the solar surface partly this is due to as yet poorly understood differences between calibration curves which are derived in different ways (Deubner (1969), Severny (1967)). Some longitudinal magnetographs solved much of these problems by either measuring the

JACQUES M. BECKERS

THE HIGH ALTITUDE OBSERVATORY
STOKES POLARIMETER

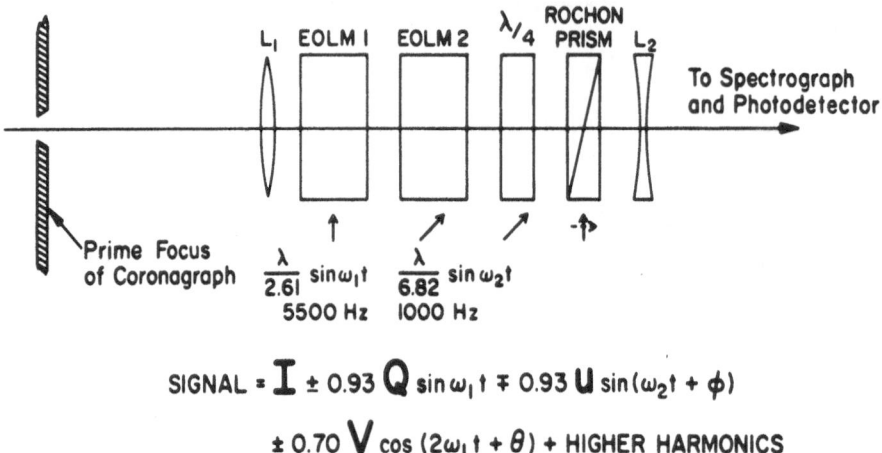

Fig. 1.   Diagram of the HAO polarimeter. The polarimeter is mounted in the prime focus of the
40 cm coronograph before any reflections take place. EOLM 1 and 2 are two *KD*P* electro-optical
light modulators and $\lambda/4$ is an achromatic quarter wave plate.

$dp/d\lambda$ (HAO magnetograph) or by measuring the difference in position of the line profiles as seen in opposite circular polarizations (Sac Peak magnetograph).

(b) In some vector magnetographs the Q and V Stokes parameters are detected at the same frequency. They are separated by respectively adding and subtracting the blue and red wing signals since presumably $dp/d\lambda$ (Equations (2)) is opposite in sign in the two wings whereas $d^2p/d\lambda^2$ is equal in sign. There is, however, no reason that the $dp/d\lambda$ is also equal in amplitude since asymmetric line profiles are quite common on the Sun so that this addition and subtraction may not separate the circular and linear polarization thus creating the danger of a serious mixing of the signals.

(c) The horizontal fine structure of the magnetic fields can influence strongly the measurements. One measures for example in a longitudinal magnetograph the circular polarization averaged over the scanning aperture or

$$V = \langle V(x, y)\rangle = \left\langle \Delta\lambda_B \cos\gamma \frac{dp}{d\lambda} I_{cont} \right\rangle. \qquad (3)$$

One therefore measures the average field across the aperture only if the product $dp/d\lambda \cdot I_{cont}$ is constant. Generally, this is not the case however. This is believed to be the principle cause for the large differences (as much as a factor of 3) in the magnetic field measurements made in different lines as shown in Figure 2. Similarly one also only measures the average of the square of the transverse field $B_\perp^2$ if $d^2p/d\lambda^2 \cdot I_{cont}$ is constant. Even if one eventually manages to measure the proper averages $\langle B_{||}\rangle$ and $\langle B_\perp^2\rangle$ one does not know yet the average vector field strength and direction because the averaging of the transverse field is done over the second power. One can measure,

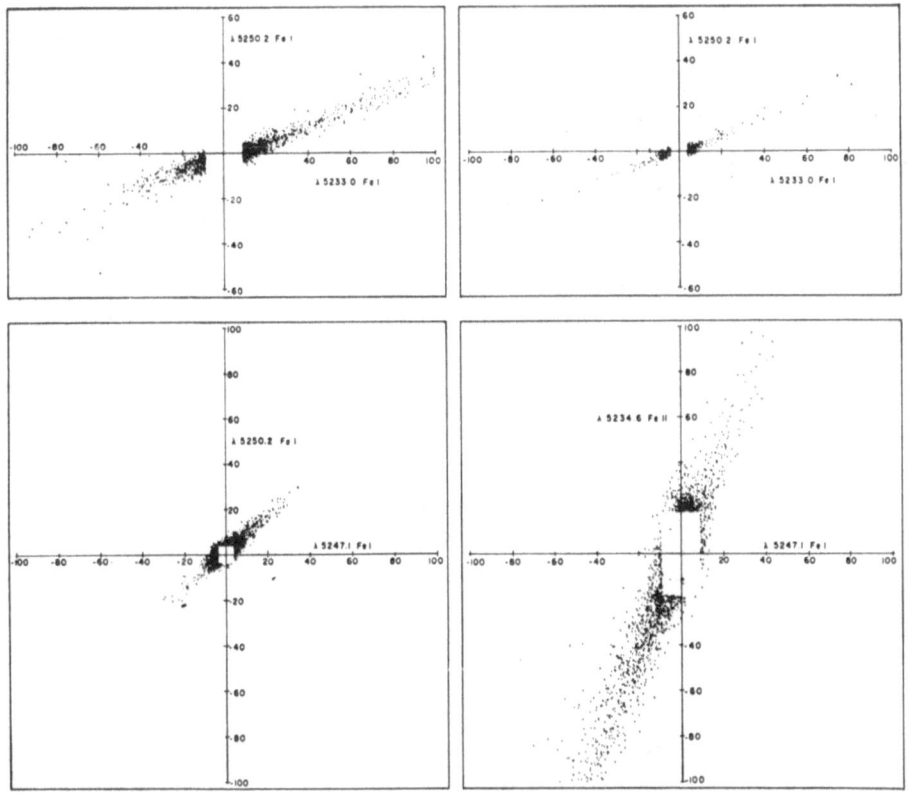

Fig. 2.   Comparison of the longitudinal magnetic field as measured in different lines by
Harvey and Livingston (1969).

for example, the wrong direction of the magnetic field vector even if this direction is
constant across the aperture and only the field amplitude changes.

(d) Vertical variations of the magnetic field cause other problems especially in
strong lines like Hα which are formed over a large range in depth over which these
variations may be very large. One may in these lines measure magnetic fields which
are very different in size and even opposite in sign, as compared to the real fields
(similar effects for Doppler shifts have been discussed by Athay (1970) and Beckers
(1968a)). In addition, one has to consider in this case magneto-optical effects like
Faraday rotation as will be discussed by Dr. Stenflo.

These difficulties in the interpretation of the magnetograph signals have fortunately
not withheld people from measuring solar fields. I mention them only to warn against
relying too strongly on the quantitative values of the field. For weak fields deviations
of 50% and more could easily occur.

Recent developments in solar magnetographs include adaptation of digital techniques, as well as computer reduction and visual display. Figure 3 shows for example
an isogauss map derived by computer from the Sac Peak magnetometer signal. Figure
4 shows an example of the Kitt Peak vector magnetograph results. Instead of drawing

isogauss lines this form of display generates an intensity picture for the signals of interest. The same is done for the full solar disk longitudinal field measurements made with the 40 channel magnetometer at Kitt Peak as displayed in Figure 5. Only by using simultaneous measurements in 40 channels does the photoelectric magnetograph compare in sensitivity with photographic ways measuring magnetic field (Beckers, 1968b). Figure 6 shows a magnetogram obtained at the Aerospace Observatory by direct photographic subtraction of images photographed in opposite circular polarizations.

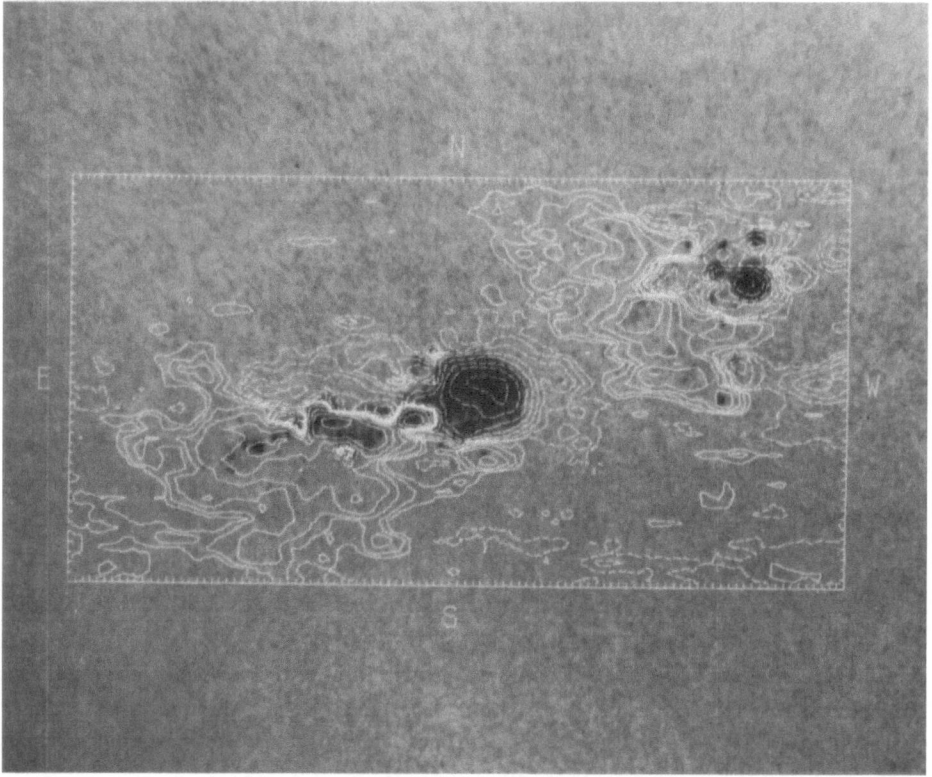

Fig. 3.   Isogauss map as produced by digital recording and by computer reduction of the Sacramento Peak Magnetometer signal (courtesy D.M. Rust).

In the future we look forward to many refinements of the photoelectric magnetograph as for example the television and digital image tube applications being undertaken in Sydney, Aerospace and Huntsville. We could also start to make use of the fact that the Zeeman splitting varies as the square of the wavelength so that the effect of the fields are very much larger in the infrared.

In the near infrared ($\sim 2\mu$) there are a number of photospheric lines with large Zeeman splitting which might profitably be used. The disadvantage in this wavelength region is of course the poor performance of the detectors. Fourier spectroscopy is of

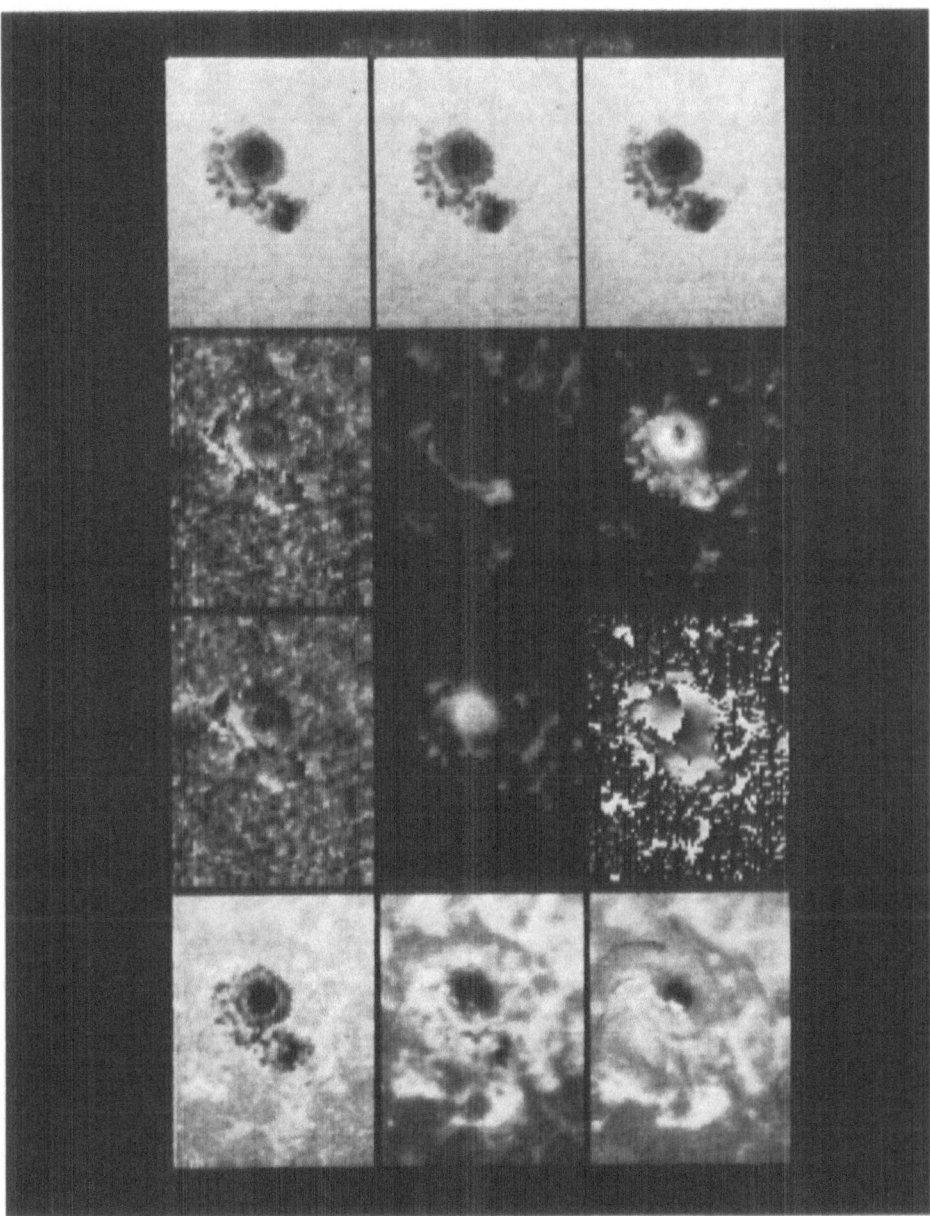

Fig. 4. Example of a display generated by computer of the Kitt Peak vector magnetograph signals. From left to right, top to bottom, the twelve displays represent: (1) the continuum intensity, (2) the wing brightness of $\lambda 5233$, (3) the wing brightness of $\lambda 5250$, (4) the Doppler velocity in $\lambda 5233$, (5) the positive longitudinal field, (6) the strength of the transverse field, (7) the Doppler velocity in $\lambda 5250$, (8) the negative longitudinal field, (9) the direction of the transverse field, (10) the brightness in the core of $\lambda 5250$, (11) the K232 brightness, and (12) the Hα brightness (courtesy W. Livingston and J. Harvey).

Fig. 5.   Example of a display generated by computer of the longitudinal magnetic field as measured with the 40 channel Kitt Peak magnetograph (courtesy W. Livingston, J. Harvey and C. Slaughter).

little help since the magnetograph uses only one or two distinct wavelengths. One might exploit however the possibility of making a spatial Fourier analysis along the spectrograph slit. This should give the same signal to noise ratio multiplex advantage as Fourier analysis of the spectrum gives so that it might be possible to achieve a good sensitivity. Spatial Fourier analysis can for example be made by means of Savart plate interferometers presently used to study modulation transfer functions (Steel (1969)) or by some direct mechanical device.

   In the extreme infrared or microwave region Dupree (1968) predicted the existence of recombination lines originating in high levels of hydrogenic coronal ions. These lines are very numerous at wavelengths longer than $700\,\mu$ which can only be observed from high altitude stations. Their Landé factor is about one resulting in a Zeeman

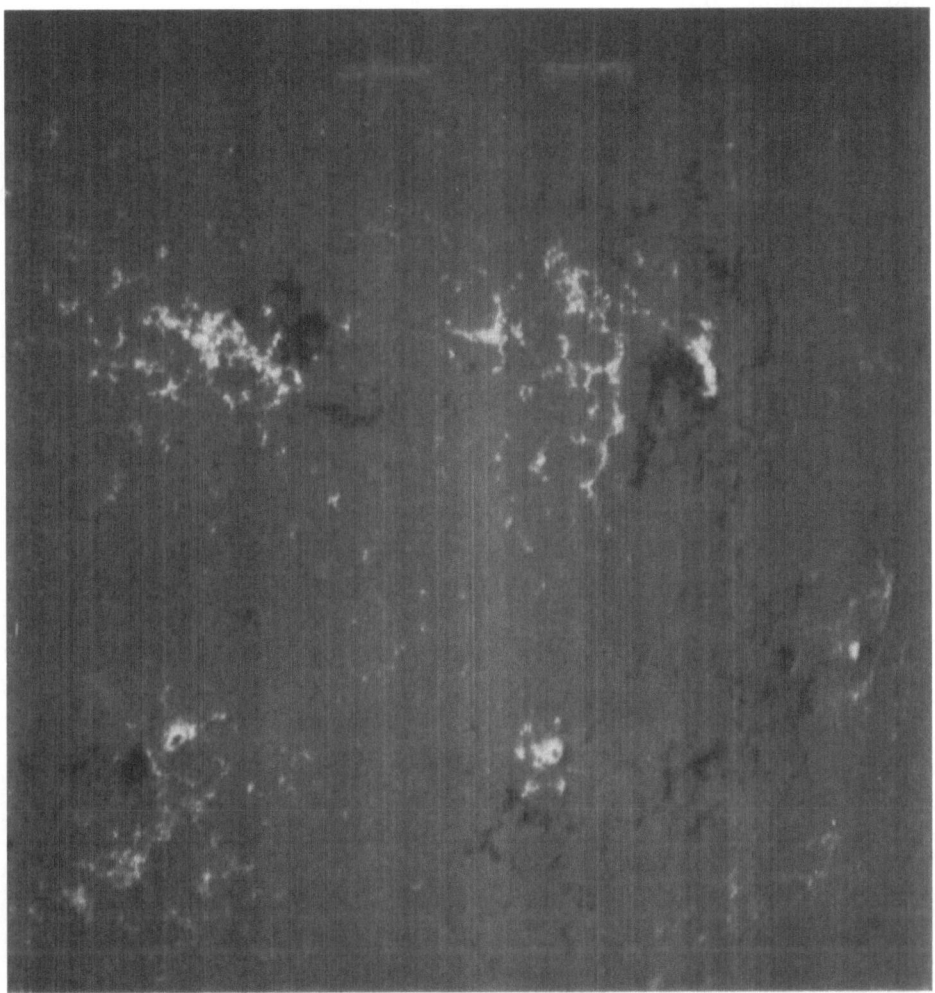

Fig. 6. Magnetogram obtained by photographic subtraction (Leighton's method) (courtesy D.
Vrabec, Aerospace Corporation).

splitting of $0.08\,\mu$, which equals the thermal Doppler width of the line ($T = 2 \times 10^6$ K),
for only 17 G for a line at 1 mm. This could become a very powerful tool for studying
coronal magnetic fields after the existence of these lines has been confirmed. It may be
that in this case the observation can best be done by Fourier spectroscopy. It is in this
context of interest therefore to describe the Fourier Transform of a Zeeman split line.
If $P(s)$ equals the Fourier transform of $p(\lambda)$ such that

$$P(s) = \int_{-\infty}^{+\infty} p(\lambda)\, e^{-2\pi i s\lambda}\, d\lambda \qquad (4)$$

it follows from Equations (1) that the Fourier transforms $\{P_I, P_Q, P_U, P_V\}$ of the four

Stokes parameters are:

$$P_I(s) = \left[\cos^2(\pi s \Delta \lambda_B) - \cos^2 \gamma \sin^2(\pi s \Delta \lambda_B)\right] P(s) \approx P(s) \qquad (5)$$

$$P_Q(s) = \sin^2 \gamma \sin^2(\pi s \Delta \lambda_B) P(s) \approx \pi^2 s^2 \Delta \lambda_B^2 \sin^2 \gamma P(s) \qquad (:) \ B_\perp^2 \qquad (5')$$

$$P_U(s) = 0 \qquad (5'')$$

$$P_V(s) = i \cos \gamma \sin(2\pi s \Delta \lambda_B) P(s) \approx 2\pi i s \Delta \lambda_B \cos \gamma P(s) \qquad (:) \ B_\parallel . \qquad (5''')$$

Where the $\approx$ signs refer to small $\Delta \lambda_B$. The P_Q and P_V becomes zero for $s = \frac{1}{2}\Delta \lambda_B$ and $s = 1/\Delta \lambda_B$ respectively so that $\Delta \lambda_B$ or $|B|$ can be determined much more easily from the P_Q and P_V profiles than from the Q or V profiles. For small $\Delta \lambda_B$ this zero point moves towards infinity and becomes hard to determine because of the small amplitude of $P(s)$. Then the P_Q and P_V contain only information on $B_\perp^2$ and $B_\parallel$ respectively.

2.2. THE HANLE EFFECT

Observations of prominences and of the corona outside the solar limb show that the emission lines often have a significant amount of linear polarization with the direction of polarization making some rather arbitrary angle with respect to the solar limb. This polarization is due to the so-called Hanle effect. The Hanle effect refers to the resonance scattering of bound electrons in the presence of a magnetic field (Mitchell and Zemansky (1961)). For a zero magnetic field and for electric dipole transitions this scattering results in linear polarization with the dominant electric vector parallel to the limb. For weak fields the excited electron gyroprecesses in the field so that the direction of polarization of the emitted radiation, as well as the degree of polarization, is changed. The degree and direction of linear polarization is therefore a function of the magnetic field. In addition to being related to the field the polarization is also dependent on the following factors:

(a) The ratio of collisional to radiative excitations. The fraction of the excitations which are collisional does of course not give rise to polarized emission.

(b) The type of Zeeman splittings of the upper and lower levels. Some Zeeman splittings (e.g. $J=1$ upper level, $J=0$ lower level) give a maximum polarization (100% for zero field and extreme anisotropic radiation field). Others (e.g. $J=0$ for upper level, $J=1$ for lower level) give always zero polarization.

(c) The ratio δ of Larmor frequency $\omega_L = eB/m_e c$ to the damping constant of the excited state. This determines the amount the electron can gyroprecess before emitting. Generally this damping constant is taken equal to the radiative damping constant. For dense media the collisional damping constant has to be taken into account also.

(d) The properties of the incident radiation field.

(e) The optical thickness of the object under study. Multiple scattering may become significant for large thicknesses.

(f) The interlocking with other atomic levels and other radiation transfer phenomena like frequency dependent source functions.

If all of these effects are known and taken into account one can calculate, for a known field, the polarization due to the Hanle effect. This has been done to various degrees of completeness for example by Hyder (1964) and House (1970). Figure 7 shows some of the results obtained by House. In Figure 7 the degree p_{max} and the angle α'_{max} of linear polarization is shown for a $J_{upper} = 1$ and $J_{lower} = 0$ transition like the Ca I $\lambda4227$ line or the Fe XIII $\lambda10747$ line. The magnetic field vector lies in a plane parallel to the solar limb and makes an angle θ' with the direction to the line of sight. The quantity Δ is directly proportional to the magnetic field. It is equal to 1 for 70 G in the permitted $\lambda4227$ line and 0.3 microgauss in the forbidden $\lambda10747$ line. For zero

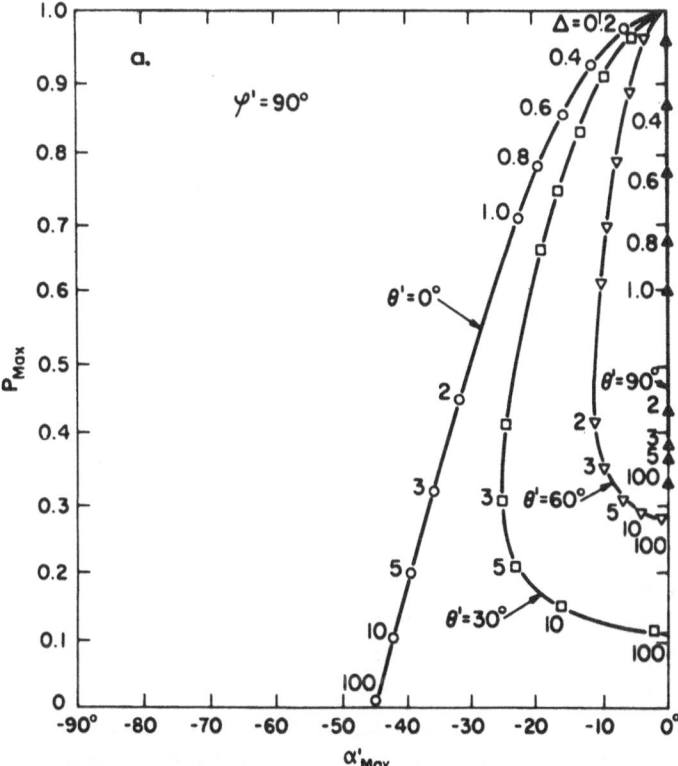

Fig. 7. Degree p_{max} and direction α'_{max} of linear polarization of a $J_{lower} = 0$ and $J_{upper} = 1$ line (e.g. $\lambda4227$). Illumination is unidirectional at right angle to the line of sight. The magnetic field vector lies in the plane at right angles to the illumination direction and makes an angle θ' with the line of sight. Δ is a measure of the field strength (see text). The collisional excitations and multiple scattering have been neglected (courtesy L. House).

field one has maximum polarization at $\alpha' = 0$ or parallel to the limb for the permitted electric dipole radiation like $\lambda4227$ and at right angles to the limb for the forbidden magnetic dipole radiation like $\lambda10747$. With increasing field strength the amount of polarization decreases and the direction changes. There is no simple relationship between the amount and direction of linear polarization and the strength and direction of the magnetic field. Only with very large approximation can one say that the degree of polarization is related to the total field strength and the direction to the longitudinal magnetic field strength.

Apart from the Hanle effect there are two other effects which may give rise to linear polarization. These are (a) the transverse Zeeman effect, and (b) the so-called impact polarization resulting from collisional excitation by anisotropic particle (electron) streams as occur during solar flares. Generally these two effects can be neglected.

The Hanle effect has been used for magnetic field determinations in prominences (e.g. Hyder, 1964, 1965, 1966; Brückner, 1966; Nikolskii *et al.*, 1970) and in the corona (e.g. Hyder, 1965; Charvin, 1965). It may also complicate transverse magnetic field observations on the solar disk by Zeeman magnetographs as discussed by Hyder (1968) and Lamb (1970).

2.3. THE GYRO-SYNCHROTRON RADIATION

Free electrons spiraling in a magnetic field also emit electromagnetic radiation whose frequency is directly related to the Larmor, or gyro frequency, $\omega_L = eB/m_e c = 1.76 \times 10^7 B \sec^{-1}$. For electrons with small kinetic energy the radiation occurs only at ω_L. This so called gyro-resonance radiation or magneto bremsstrahlung is strongest and circularly polarized in the so-called extraordinary mode when viewed along the field lines. Since on the Sun $B < 3000 \, G$ this radiation occurs at frequencies $< 10 \, GHz$ or at wavelengths $> 3 \, cm$ so that the upper chromosphere and corona are the regions accessible by this phenomenon. When the electron velocities are no longer small but still not relativistic, and this is the case for thermal electrons in the quiet solar corona ($v/c \approx 0.02$) and for some radio bursts, radiation occurs also at the lower harmonics of the gyro frequency. For relativistic electrons, the gyro-resonance radiation turns into synchrotron radiation. In the synchrotron radiation the emitted energy is concentrated mainly in the plane of gyration, it has a very wide continuum-like spectral distribution, and it is polarized in this plane of gyration. Very good descriptions of these types of radiation are given by Takakura (1967).

The gyro-synchrotron radiation is strongly modified by propagation effects before it reaches the Earth. These are both absorption and refraction effects. Absorption is the inverse of the emission process just described. It is most efficient when the emission is most efficient and it is therefore much stronger for the extraordinary mode of circular polarization than for the ordinary. The magnetic field generally decreases with increasing height in the solar atmosphere so that the Larmor frequency decreases. The gyro-absorption occurs therefore at lower frequencies than the gyro-emission so that this effect tends to suppress the observed emission at low frequencies. Because it absorbs the extraordinary mode of polarization it may also make the polarization of the transmitted radiation ordinary. Refractive effects also prohibit the extraordinary ray at the gyro-frequency to escape. Only the higher harmonics in the extraordinary mode can therefore escape. The ordinary mode on the other hand can escape unhindered at the gyro frequency.

These absorption and refraction effects make the interpretation of the radio emission observations very difficult. Radiative transfer theories have been given by Kawabata (1954) and Kai (1965). These theories permit a fairly good derivation of the observable radiation for a given magnetic field configuration and a given thermal and non-thermal electron distribution. To derive the magnetic field from the observation is however very difficult also because the observations are of low spatial resolution (worse than 1′) so that one probably averages over many different features. None-theless, many of the direct magnetic field measurements in the solar corona have come from the interpretation of gyro-resonance and synchrotron radiation effects.

Microwave type IV bursts ($IV \mu$) and impulsive microwave bursts with brightness tempeiatures of 10^9 K are the most likely candidates for gyro-synchrotron emission by non-thermal electrons. These bursts are partially circularly polarized in the ordinary mode at low frequencies (< 1000 MHz) and in the extraordinary mode at high fre-quencies (> 10000 MHz). The emission reaches a maximum at ≈ 2000 MHz. Takakura (1966, 1967) finds from a quantitative evaluation of the emission and absorption processes that this maximum for gyro-synchrotron radiation should occur at about four times the gyro frequency $f_H = \omega_L/2\pi = 2.8 \times B$ MHz which results in a field strength

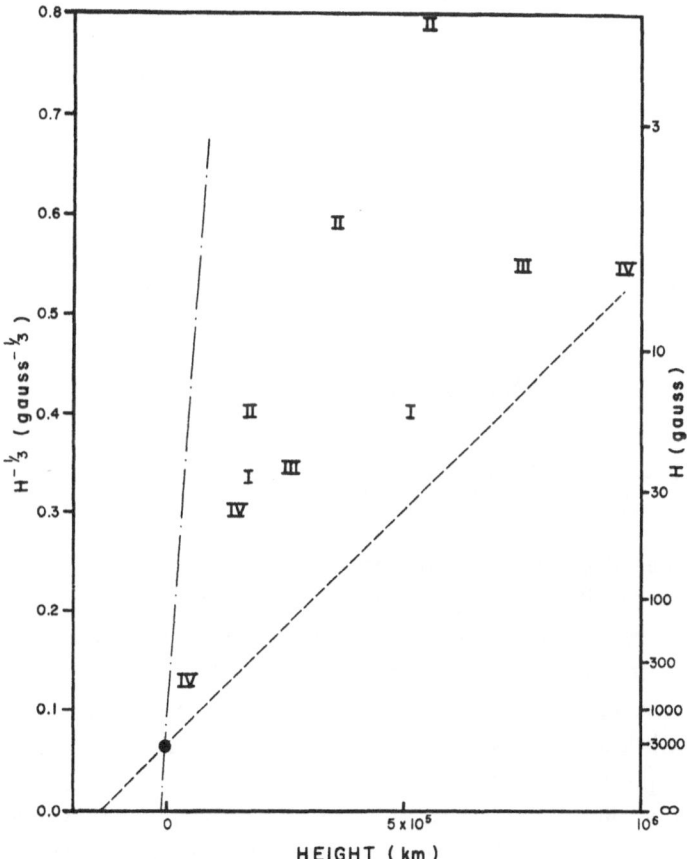

Fig. 8.   Magnetic field determinations in the corona above sunspots as derived from radiobursts. The roman numeral indicates the type of burst. In the plot $B^{-1/3}$ to height a magnetic dipole represents a straight line. The dashed line represents the magnetic dipole distribution by Correll et al. (1956) from the paths of 'rain' type prominences near sunspots. The dash-dot line represents the dipole distribution determined by Ioshpa et al. from the magnetic field variation across the sunspot.

of 200 G which is a reasonable value since the bursts occur ≈ 30000 km above sunspots. Similar arguments can be used to derive the field at greater heights from decimeter and meter type IV bursts (Figure 8). The interpretation of the polarization reversal is unclear. Takakura (1967) attributes it to a double source resulting from the bi-polar character of sunspots groups the preceding dominant spot giving the extraordinary mode at high frequencies and the following, weaker spot giving the ordinary mode at low frequencies. Earlier Takakura (1960) interpreted it as a single source phenomenon with the ordinary low frequency polarization caused by gyro-absorption. The latter interpretation results in a polarization reversal at 2–3 f_H which gives magnetic fields in agreement with the ones determined from the emission maximum.

Type I bursts, or the so-called noise storms, present a different example of magnetic field determination (Takakura, 1966). Their emission is thought to be due to plasma waves, their very high degree of circular polarization in the ordinary mode is interpreted as due to the inability of the extraordinary ray to escape because of refraction effects. This condition gives at the source of the plasma waves magnetic fields which are in agreement with those derived from the type IV bursts (Figure 8).

An example of gyro-resonance radiation from thermal electrons can be found in the radio emission occurring above sunspot umbrae. The properties of this emission have been discussed by Livshits et al. (1967). The brightness temperature of sunspot ɛ mounts to as much as 10^6 in the cm wavelength region, the radiation being polarized in the extraordinary mode. Livshits et al. derive a temperature model of the sunspots chromosphere and corona from these data with an assumed magnetic field distribution and a hydrostatic density profile.

2.4. THE FARADAY ROTATION

The amount of Faraday rotation is given by

$$\phi = 2.36 \times 10^{-3} f_{\mathrm{MHz}}^{-2} \int N_e B_z \, d_z$$

where the integral is taken over the path between source and observer and where B_z equals the longitudinal magnetic field. In principle it is therefore possible to estimate B_z if the amount of Faraday rotation is known. The latter can be derived for example from the frequency dependence of ϕ. The amount of rotation is however so large that ϕ changes significantly within the bandwidth so that the linear polarization of the source is lost. The requirements to observe the Faraday rotation therefore are: (a) a linearly polarized source. This could be, for example, a synchrotron radiation source or also a linearly polarized radar signal. (b) Narrow bandwidth so that the rapid frequency variation of ϕ can be studied (Akabane et al., 1961). The Faraday effect as a means of magnetic field determination in the solar corona has been used very little. Golnev et al. (1964) determined an upper limit to the field of 10^{-2} G at $5 R_\odot$ by using the Crab Nebula as the source of linear polarization while Bhonsle et al. (1964) determined relative values of the coronal field from a type III burst source.

3. Magnetic Fields as Studied by their Influence on the Solar Atmosphere

In the regions of the solar atmosphere where the magnetic pressure $P_B = B^2/8\pi$ exceeds the gas or kinetic pressure (resp. P_g and $P_k = \frac{1}{2}\varrho v^2$) one expects a strong influence of the magnetic field on the physical conditions. Such regions are: parts of the photosphere such as sunspots, pores and faculae, all of the chromosphere and all of the corona except the solar wind. From the study of the physical conditions of these regions it is therefore often possible to infer some of the properties of the magnetic field.

Fig. 9. Hα filtergram of an active region (courtesy R. B. Dunn).

3.1. The alignment of structures

Plasma motions generally follow magnetic field lines so that there tends to be an alignment of matter in the solar atmosphere which coincides with the direction of the magnetic field. The study of these alignments is, in my opinion, by far the best way of determining the projected direction or azimuth of the field lines. In this way one can determine the field direction in sunspot penumbrae from white light photography, and in the chromosphere from spectroheliograms. In the corona one can use limb observations made during an eclipse or with a coronagraph and disk observations made with high resolution X-ray telescopes. Also to be included in this category of measuring magnetic fields are the directions of the filamentary structures in the outer corona as inferred from the observation of anisotropic scattering of the crab nebula radio radiation. Figure 9 shows as an example an Hα filtergram of a sunspot region. Here, as in quiet regions, the alignment of the fibrils and elongated fine mottles shows the structure of the magnetic field in such detail as is virtually impossible to obtain by transverse magnetographs.

3.2. Local changes in the atmosphere model

When $P_B > P_g$ one expects significant changes of the temperature and density structure of the solar atmosphere. In the photosphere this condition occurs in sunspots and pores; wherever we see such a structure we are therefore sure that we also see a region with a magnetic field $>1500\,\mathrm{G}$. Because of the much lower densities, this condition holds over almost all of the chromosphere and the corona and over much of the upper photosphere. Figure 10 shows as example a spectroheliogram in the CN bands at $3888\,\text{Å}$. Chapman and Sheeley (1968) find that these CN brightness structures in the upper photosphere coincide with magnetic fields of the order of $300\,\mathrm{G}$. The same has been shown to be the case for the chromospheric H and K line network except that there the brightening occurs at weaker fields. We do not know yet however whether the amount of brightening is uniquely related to the absolute field strength, to the vertical field strength or to some other parameter related to the field. We also cannot determine the direction of the field, although, together with the alignment information, one can take a good guess at this.

3.3. The propagation of rapid disturbances

One often sees disturbances propagate with velocities greater than the sound velocities. One of the explanations for these high velocities is that the disturbance is of a magneto-hydrodynamic type which propagates with the Alfvén speed $V_A = B\sqrt{4\pi\varrho}$. Such explanations have been made plausible in a number of cases which therefore have resulted in an estimate of B. Takakura (1966) derives in this way the magnetic fields above a sunspot from the propagation in frequency, and therefore in height, of type II bursts (see Figure 8). Meyer (1968) inferred in this way from the propagation of the Moreton waves an average coronal magnetic field of $6\,\mathrm{G}$.

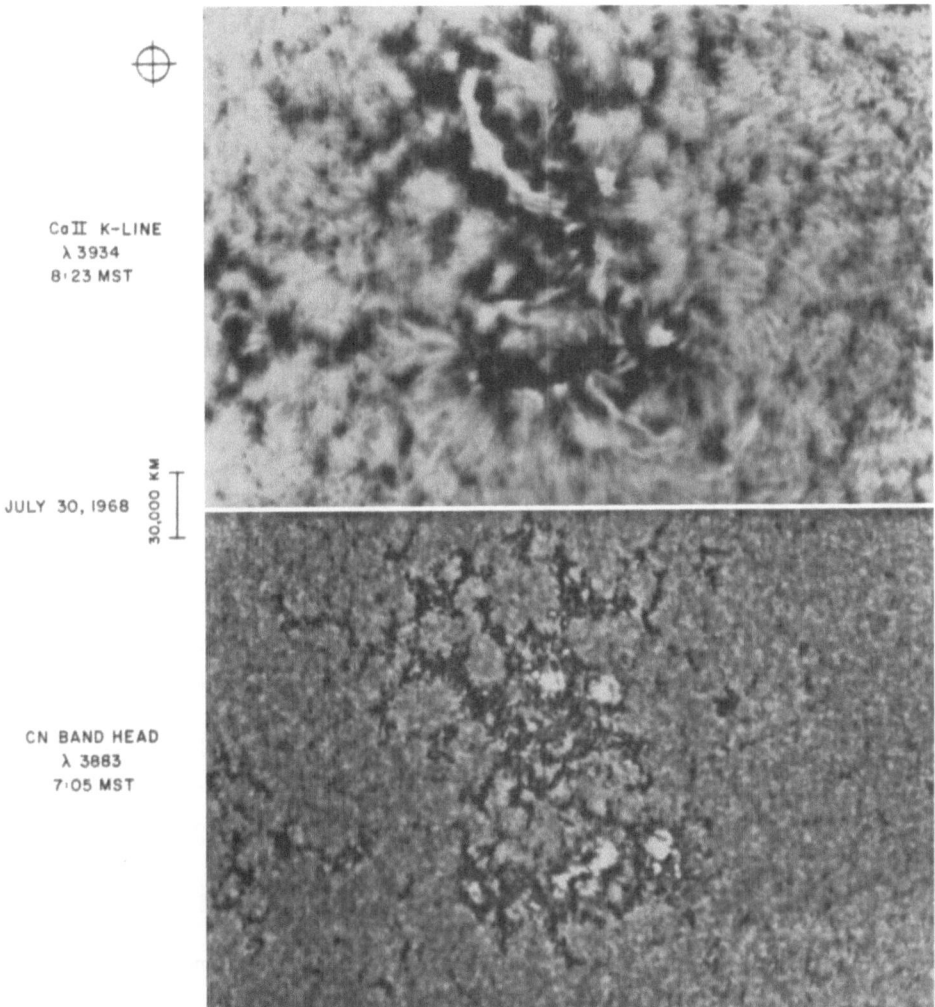

Fig. 10.   Spectroheliograms taken in the K line and in the CN bandhead. The latter shows the bright 'photospheric' network which coincides with the photospheric magnetic field structure (courtesy N. R. Sheeley, Jr.).

3.4. Oscillations of Prominences

After large flares filaments up to 40° away are occasionally seen to have a velocity oscillation which manifests itself in Doppler shift oscillations in the Hα line. Hyder (1966) and Kleczek *et al.* (1969) explain this phenomenon by introducing a restoring force resulting from the magnetic tension. The difference between the two investigations results from the interpretation of the oscillation to be respectively vertical or horizontal. This results in a difference in the interpretation of the damping mechanism of the oscillation. The period of oscillations in both cases comes out within the same order of magnitude. However, it is related to the magnetic field by: $B = 4\pi Hf\sqrt{\pi\varrho}$ where

H=height of the filament ($\approx 3 \times 10^4$ km) and $f=$ the frequency of oscillation ($\approx 10^{-3}$ s^{-1}). This results in a field strength of about 10 G which is indeed consistent with measurements made using the Zeeman and Hanle effects.

4. Magnetic Fields as Inferred from Theoretical Consideration

There are some ways in which the magnetic field on the sun can be inferred from purely theoretical arguments.

4.1. POTENTIAL OR CURRENT FREE FIELD CALCULATIONS

These calculations take the measurements of the longitudinal photospheric magnetic field as determined with Zeeman magnetographs. The assumption that there are no electric currents above the photosphere, or that curl **B**=0 gives then the magnetic

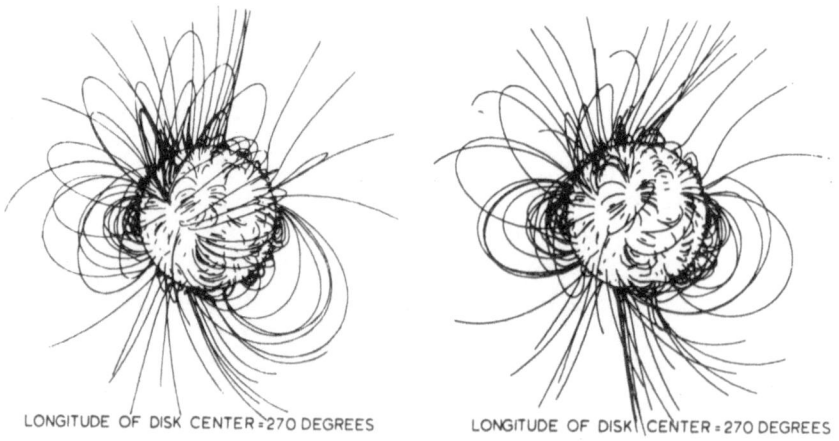

LONGITUDE OF DISK CENTER = 270 DEGREES   LONGITUDE OF DISK CENTER = 270 DEGREES

Fig. 11.   Magnetic field distribution in the corona as determined from the potential field assumption
(courtesy M. D. Altschuler and G. A. Newkirk).

field configuration in the corona. The general solution of this potential field approximation in spherical coordinates has been given by Altschuler *et al.* (1969). It uses an expansion of the magnetic potential in Legendre polynomials, the coefficients of this expansion being determined by a least squares fitting of the observed longitudinal photospheric magnetic field. Figure 11 gives an example of the field configuration in the corona obtained in this way. A refinement of the method permits the inclusion of the solar wind so that the field lines in the outer corona are made approximately radial.

4.2. EQUIPARTITION OF MAGNETIC AND KINETIC ENERGIES

In a convective medium magnetic fields tend to be concentrated towards the edges of the convection cells. According to Weiss (1966) it takes typically 3 turnover times to reach the maximum field strength whose magnitude is such that the kinetic and magnetic energy densities are comparable although the latter can never exceed the former.

For the supergranulation one derives in this way magnetic fields which are of the same magnitude as have been observed (Parker, 1963; Clark *et al.*, 1967). For the solar core this procedure gives magnetic fields of the order of 10^9 G and for the deep convective zone 10^5 G.

5. Conclusion

In Table II, I summarize the methods which can be used to determine the magnetic fields in the various zones on the Sun. Each of these methods has its own merits which makes it occasionally preferable over the other methods. None of the methods permits as yet an accurate determination of the complete field vector. The closest to this comes the Zeeman Vector Magnetograph. It can however only be used well in the solar photosphere. The interpretation of its signals in terms of the magnetic field vector is sufficiently complex and difficult that a reliability of a factor of 2 seems almost optimistic.

The magnetic field is however one of the, if not *the*, most important physical quantity in the observable solar atmosphere. Its measurement by any available method is therefore of the greatest importance. By future refinements of the measurements and their interpretation, by improvements in the theories of the solar radiation in the presence of a magnetic field and by exploring new ways in which to measure magnetic fields we may therefore expect to significantly improve our understanding of the physics of the solar atmosphere.

TABLE II

Summary of ways of measuring solar magnetic fields

Effect	Region on the Sun			
	Int.	Phot.	Chrom.	Cor.
Electromagnetic radiation				
Zeeman Effect Visual		X	X	(X)
Near IR ($\sim 2\mu$)		X	X	(X)
Far IR ($\sim 700\mu$)				X
UV			X	X
Hanle Effect (Hα, D$_3$, etc., Lyman α?)			X	X
Gyro-Synchrotron radiation (radio)				X
Faraday rotation				X
MHD Effects				
Alignments of structures (Hα, Corona, penumbral filaments)		(X)	X	X
Influence on T, P structure (K-brightness CN network)		(X)	X	X
Alfvén velocity $V_A = B/\sqrt{4\pi\varrho}$			X	X
Prominence Oscillations				X
Theory				
$P_g \approx P_B$	(X)	(X)		
Force free/Potential field			X	X

References

Akabane, K. and Cohen, M. H.: 1961, *Astrophys. J.* **133**, 258.
Altschuler, M. D. and Newkirk, G.: 1969, *Solar Phys.* **9**, 131.
Athay, R. G.: 1970, *Solar Phys.* **12**, 175.
Babcock, H. W.: 1953, *Astrophys. J.* **118**, 387.
Beckers, J. M.: 1968a, *Solar Phys.* **3**, 367.
Beckers, J. M.: 1968b, *Solar Phys.* **5**, 15.
Beckers, J. M.: 1969, AFCRL-Physical Sciences Research Papers, No. 371.
Beckers, J. M. and Schröter, E. H.: 1969, *Solar Phys.* **10**, 384.
Beggs, D. W. and Von Klüber, H.: 1969, *Monthly Notices Roy. Astron. Soc.* **127**, 133.
Bhonsle, R. V. and McNarry, L. R.: 1969, *Astrophys. J.* **139**, 1312.
Brückner, G.: 1966, in *Atti del Convegno sui Campi Magnetici Solari* (ed. by M. Cimino), G. Barbèra,
    Firenze, p.101.
Chapman, G. A. and Sheeley, N. R.: 1968, *Solar Phys.* **5**, 442.
Charvin, P.: 1965, *Ann. Astrophys.* **28**, 877.
Clark, A. and Johnson, H. K.: 1967, *Solar Phys.* **2**, 433.
Correll, M., Hazen, M., and Bahng, J.: 1956, *Astrophys. J.* **124**, 597.
Deubner, F. L. and Liedler, R.: 1969, *Solar Phys.* **7**, 87.
Dupree, A. L.: 1969, *Astrophys. J.* **152**, L125; and Thesis, Harvard Univ., 1968.
Evans, J. W.: 1966 in *Atti del Convegno sui Campi Magnetici Solari* (ed. by M. Cimino), G. Barbèra,
    Firenze, p.123.
Golnev, V. J., Parijsky, Y. N., and Soboleva, N. S.: 1969, *Izv. Pulk. Obs.* **23**, 22.
Harvey, J. and Livingston, W.: 1969, *Solar Phys.* **10**, 283.
House, L. L.: *J. Quant. Spectr. Radiative Transfer*, submitted 1970.
Hyder, C. L.: 1964, Thesis, University of Colorado.
Hyder, C. L.: 1965a, *Astrophys. J.* **141**, 1374.
Hyder, C. L., 1965b, *Astrophys. J.* **141**, 1382.
Hyder, C. L.: 1966, *Z. Astrophys.* **63**, 78.
Hyder, C. L.: 1968, *Solar Phys.* **5**, 29.
Hyder, C. L.: 1969, in *Atti del Convegno sui Campi Magnetici Solari* (ed. by M. Cimino), G. Barbèra,
    Firenze, p.110.
Ioshpa, B. A. and Obridko, V. N.: 1964, *Geomagnetism Aeronomy* (Engl. ed.) **4**, 12.
Ioshpa, B. A. and Obridko, V. N.: 1965, *Soln. Aktivnost* **2**, 131.
Ioshpa, B. A. and Obridko, V. N.: 1965, *Soln. Dann.* **3**, 54.
Kai, K.: 1965, *Publ. Astron. Soc. Japan* **17**, 309.
Kawabata, K.: 1964, *Publ. Astron. Soc. Japan* **16**, 30.
Kleczek, J. and Kuperus, M.: 1969, *Solar Phys.* **6**, 72.
Kuznetzov, D. A., Kuklin, G. V., and Stepanov, V. E.: 1966, *Results IQSY* **1**, 80.
Lamb, F. K.: 1970, *Solar Phys.* **12**, 186.
Lee, R. H., Rust, D. M. and Zirin, H.: 1965, *Appl. Opt.* **4**, 1081.
Livingston, W. and Harvey, J.: 1970, Preprint.
Livshits, M. A., Obridko, V. N., Pikel'ner, S. P.: 1967, *Soviet Astron.* **10**, 909.
Meyer, F.: 1968, in K. O. Kiepenheuer (ed.), 'Structure and Development of Solar Active Regions',
    IAU Symp. **35**, 485.
Mitchell, A. G. and Zemansky, M. W.: 1961, *Resonance Radiation and Excited Atoms*, Cambridge
    University Press, 2d Ed.
Nikolskii, G. M. and Kretsuriani, Ts. S.: 1970, *Soviet Astron.* **13**, 815.
Parker, E.: 1963, *Astrophys. J.* **138**, 552.
Rayrole, J.: 1967, *Ann. Astrophys.* **30**, 257.
Seares, F. H.: 1913, *Astrophys. J.* **38**, 99.
Severny, A. B.: 1967, *Izv. Krymsk. Astrofiz. Obs.* **36**, 22.
Steel, W.: 1964, *Opt. Acta* **11**, 9.
Stepanov, V. E. and Severny, A. B.: 1962, *Izv. Krymsk. Astrofiz. Obs.* **28**, 166.
Takakura, T.: 1960, *Publ. Astron. Soc. Japan* **12**, 325, 352.
Takakura, T.: 1966, *Space Sci. Rev.* **5**, 80.
Takakura, T.: 1967, *Solar Phys.* **1**, 304.

Weiss, N.: 1966, in *Atti del Convegno sui Campi Magnetici Solari* (ed. by M. Cimino), G. Barbèra, Firenze, p.299.
Wiehr, E.: 1969, *Solar Phys.* **9**, 225.
Zhulin, I. A., Ioshpa, B. A., and Mogilevsky, E. I.: 1962, *Geomagnetism Aeronomy* (Engl. ed.) **2**, 489.

Discussion

Semel: Are the given expressions for the Fourier transforms of the Stokes parameters valid for any model?

Beckers: The expressions refer to the Seares approximation for a Zeeman triplet. The detailed theory of line formation to be discussed by Dr. Stenflo tomorrow will be able to give the Fourier transforms for more general cases. The properties of the Fourier Stokes Parameters will be similar, however, to those discussed for the Seares approximation.

Maltby: In connection with the Zeeman measurements could you comment on the use of electro-optical polarization optics as compared to rotation of polarization optics. By rotating a linear polarizer at a much higher speed than the rotation frequency of a quarter wave plate all 4 Stokes parameters may be determined. Will the rotating system be too slow?

Beckers: There are two disadvantages in using rotating polarization optics: (1) The speed of rotation can generally not be made very high. This results in modulation frequencies which fall in the range of the frequencies associated with the atmospheric seeing. The seeing becomes in this case a serious source of noise. This is not the case for electro-optical devices which can be operated in the kilohertz range (2) Rotation of optics causes slight changes in the optical paths (dust, motion of light on photocathode etc.) resulting in spurious signals.

Foukal: Which infra-red lines around 2 microns do you consider best suited to Zeeman effect measurements?

Beckers: Dr. Hall has a list of these lines. I remember specifically a Titanium line at $2.3\,\mu$.

Simon, M.: I would like to add that there is another effect which may be used at radio wavelengths in addition to those discussed by Dr. Beckers. This is the suppression of gyro-synchrotron radiation in the presence of a plasma - the Razin effect. It results in a sharp low frequency cut-off, and was first observed in a moving Type IV burst by Boischot and his co-workers. The burst was analyzed by them, Ramaty, and also Bohlin and myself. Bohlin and I determined the field in the streamer in which the burst took place to be $\sim \frac{1}{4}$ G at a height $\sim 1 R_\odot$.

Beckers: This effect should indeed be added to my list.

Pasachoff: Pulsars are also sources of both circularly and linearly polarized radiation behind the solar corona, and so those that are occulted could conceivably be used as probes for Faraday rotation measurement, although the solar contribution is small. Frequency dispersions of pulses from pulsars is already being used to assess electron densities in the outer corona. Care must be taken to calibrate the polarization in different parts of the pulses.

For all rotation studies, one must independently monitor the considerable rotation introduced by the Earth's ionosphere. This contribution varies both rapidly and diurnally by large factors.

Beckers: This is indeed another possibility.

Brueckner: When extrapolation of measured photospheric fields into the chromosphere are made, the assumption of current-free fields is made from a certain altitude. At which optical depth is this assumption allowed?

Beckers: We do not know yet. There are conflicting opinions on whether the field above the photosphere, where the Zeeman magnetograph measurements are made, is current free.

THE CULGOORA MAGNETOGRAPH

J. V. RAMSAY, R. G. GIOVANELLI and H. R. GILLETT

CSIRO Division of Physics, National Standards Laboratory, Sydney, Australia 2008

Abstract. The magnetograph is based on a high-resolution filter which serves in place of a spectrograph, except that a reasonably large field of view (one-quarter of the Sun's diameter) can be observed at the one instant. Observations are made by obtaining filtergrams of opposite circular polarizations simultaneously in the wing of a magnetically sensitive line. Exposure times are about 0.3 s, the angular resolution of the magnetic field is about 2 arc s, closest frame repetition rates about 8 s. The filtergrams are processed subsequently by photographic or television subtraction. Semi-automatic photographic and/or TV subtractions yield magnetograms suitable for cinematographic projection though the subtractions are not yet as perfect as those obtained by individual subtraction.

1. Introduction

This instrument has been developed with the intention of producing magnetograms of spatial resolution approaching or surpassing 1 arc s with a fast repetition rate. The spatial resolution achieved at present is about 2 arc s, and observations may be obtained over a circular field of one-quarter the solar diameter every 8 s.

The magnetograph consists of a telescope with which filtergrams are obtained simultaneously in opposite circular polarizations in the wing of a magnetically sensitive line, and a subtractor which displays the image differences due to the longitudinal magnetic field. Subtractions can be obtained semi-automatically, yielding cinefilms with which rapid changes in field may be studied.

2. The Telescope

Most photoelectric magnetographs suffer from the disadvantage that light may be recorded at only one resolved image point at a time. Spectroheliograph methods of observing magnetic fields allow observations to be made at all points along the length of the spectrograph slit simultaneously, but considerable time is still taken in scanning to yield an adequate field of view. To overcome this disadvantage, the Culgoora magnetograph incorporates a narrow-band filter (Ramsay *et al.*, 1970), consisting of a series of Fabry-Perot interferometers and a prefilter to isolate light from the wing of a suitable spectral line, the filter replacing the more conventional spectrograph.

The filter has a clear diameter of 50 mm and may be adjusted to any spectral line in the range 4000–6600 Å by coarse tuning provided there is a suitable interference prefilter of about 20 Å pass band available to isolate the required spectral region. Fast, smooth fine-tuning is available over a range of about 5 Å. The pass band varies irregularly in width throughout the spectrum, being about 0.05 Å at 6100 Å and nowhere exceeding 0.1 Å. As in any interferometer, the peak wavelength is shifted to the blue for rays incident at increasing angles θ to the axis, the variation being proportional to θ^2; at 6100 Å it amounts to 0.05 Å for $\theta = 15$ arc min. We regard 0.05 Å as the

Howard (ed.), Solar Magnetic Fields, 24–29. All Rights Reserved.

tolerable limit of variation across the field, so the field diameter 2θ is about 30 arc min.

Interreflections between the three interferometers are quite strong. They preclude the filter from being used near a focal plane, for multiple reflections would then overlap the main image. To overcome this disadvantage the filter is used in a collimated beam, the individual interferometers being tilted relative to each other so that the multiple images produced by the final imaging lens (c, Figure 1) fall outside the useful field. Tilting the interferometers makes the transmission slightly asymmetric around the centre of the field, but this is generally negligible or tolerable.

The Airy disk of a lens of diameter 50 mm has a radius of about 3.1 arc s at 6100 Å, and it might have been hoped that the above filter-telescope lens combination would have an angular resolution of this order. In practice wave-front deformations on passing through the filter reduce the angular resolution to about 5 arc s. In the

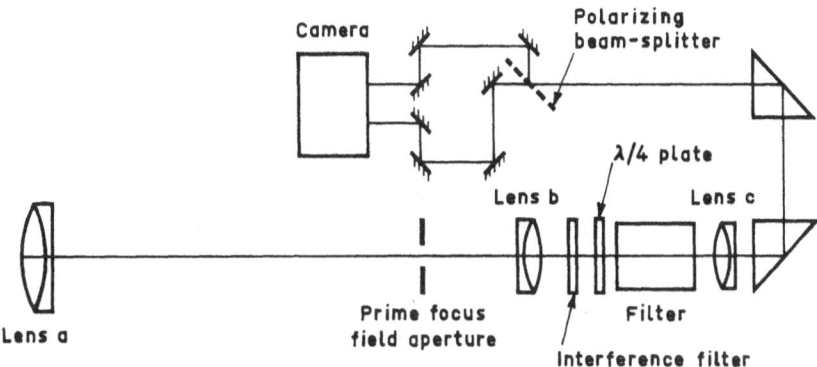

Fig. 1.   Schematic diagram of the magnetograph optical system.

present magnetograph we use a telescopic magnifying system (lenses a and b, Figure 1) before the filter, sacrificing angular field of view for improved spatial resolution on the Sun.

To decide on optimum focal lengths, we note that f_C should be short in order to keep exposure times short, however subject to the condition that the film does not limit the resolution. For f_C we have selected 180 cm, in the focal plane of which 5 arc s represents $\frac{1}{24}$ mm. This is to be compared with the resolution data for the most suitable film we have found, Kodalith Pan; manufacturers' data indicate that this has a one-dimensional transfer function of 0.8 for 30 line-pairs per mm. For the telescopic magnifier we have chosen a ×4 system, the prime objective lens a, made by Grubb Parsons, having a diameter of 20 cm and, for compactness, a focal length of 170 cm. Lens b has focal length one-quarter of this and a clear diameter rather larger than the minimum required. A chrome-plated stop at the focus of lens a restricts the angular field on the Sun to 7.5 arc min to match that of the filter. This stop is perforated, the air heated by the stop being sucked away through the perforations to avoid internal seeing degradation.

The telescope is mounted on an equatorial spar 305 cm long. To contain the total length of the optical system it has been necessary to fold the beam using two 90°–45° prisms as shown in Figure 1.

Pointing at the required solar region is achieved by fine adjustments incorporated in a separate photoelectric guider mounted on the spar.

The polarizing system consists of a $\lambda/4$ mica plate cemented between cover glasses and mounted before the filter, together with a polarizing beam splitter (Steel *et al.*, 1961) immediately before a 35 mm cine camera. The gate of the camera has been opened out to accommodate two frames, and a pair of rhomboidal prisms transfers the two images from the beam splitter to the camera. The axis of the beam splitter must be such as to sort out components polarized in and at right angles to the plane containing the incident and reflected rays in the two 90°–45° prisms, since the states of polarization of these rays are unaffected by the 45° reflections. This decides the orientations of the $\lambda/4$ plate axes, which must be at 45° to the face of the spar.

The solar image may be checked visually for alignment and focus via a reflex mirror immediately preceding the beam splitter. A second reflex mirror may direct light onto a photomultiplier for monitoring telescope transmission and thus establishing the wavelength.

In operation, a pair of photographs in opposite circularly polarized components may be obtained simultaneously in any desired position in a spectral line, a second pair being obtained 8 s later in the same position or, by automatic tuning, in any other wavelength out to ± 0.5 Å from the first. This sequence may be repeated automatically at intervals of approximately 16, 32, 64, 128 or 256 s. By making the wavelengths symmetrical about the line centre we may obtain Doppler records as well as magnetograms in opposite wings, though the Doppler pairs are not obtained simultaneously and so suffer from seeing differences.

Initial observations have been concentrated on the 6102.7 Å line of Ca I. Exposure times for a density of 1.00 (about the bottom of the straight-line portion of the Kodalith Pan film characteristic curve when processed in Kodak D19 developer for 5 min at 20 °C) are about 0.1–0.2 s in the neighbouring continuum and 0.3 s in the wings where magnetograms are normally obtained. There is a slight variation in density across the field which depends on the location in the line and is due to the variation of filter peak wavelength with field angle. This has not been found disadvantageous in obtaining magnetograms, and can be turned to advantage in the ready identification of the part of the line being recorded. However Doppler subtractions show a variation of background density across the field.

The angular resolution of the magnetograph could be improved by replacing lens *a* by one of larger diameter but of identical *f*/ratio, though this implies a corresponding reduction in angular field, and by improvement in the spacing uniformity of the interferometers by the use of different high-reflectance coatings. These improvements are currently being undertaken.

3. Subtraction

To obtain magnetograms three different methods of subtraction have been used. The first, the well-known technique of photographic subtraction (Leighton, 1959) yields the best results, though this method is too slow for processing large numbers of pairs of frames (500 upwards) obtainable in a day. The emulsion used for printing is Kodak Gravure positive film, and is developed in Ilford PQU developer, diluted 40 cc to a litre, at 25°C for approximately 4 min, the precise time being selected by trial and error to give $\gamma = 1 \pm 0.02$.

For fast subtraction, a semi-automatic photographic or a TV subtraction system is used. The photographic process has been developed in collaboration with Supreme Sound Studios, who carry out subtractions for us. A 35 mm positive print with $\gamma = 1$ is made on Eastman fine-grain positive film in an automatic optical printer at unit magnification. This positive is then used as a focal plane filter in the optical printer. The positive is placed in contact with unexposed film and is transported at the same rate through the printer as the unexposed film. The negative, displaced one frame with respect to the positive, is imaged on to the positive/unexposed film combination and an exposure taken, thus providing a magnetogram. Once the original alignment between the image of the negative and the positive has been established, the remainder of the process is automatic since the separation between each pair of frames to be subtracted is fixed by the beam splitter. Since a subtraction of the correct sense is obtained on only every second frame, every alternate frame must be removed in a subsequent stage of editing.

The biggest disadvantage of this method is difficulty of controlling the value of γ which, while very uniform along the film, is not as good as under laboratory conditions. Thus there is usually background 'noise' across the subtraction.

An alternative method for fast subtraction uses a modified TV subtractor developed by Amalgamated Wireless (Aust.) Ltd. This consists of two 'Vidicon' television cameras, one viewing the filtergram obtained in one circular polarization while the other views the oppositely polarized filtergram. The output of the two cameras is fed into a differential amplifier, the difference being displayed on a monitor screen which is photographed by a 35-mm cine camera. The filtergrams are presented to the Vidicons by means of two nominally identical optical systems. We are currently using diffuse illumination from fluorescent lamps to illuminate the filtergrams, and have found no disadvantage due to flicker. Provision has been incorporated in the subtractor for independent adjustment of image magnification, position and rotation in the two cameras, in order to achieve optimum superposition, while adjustable sawtooth and parabolic 'shading' is available in x and y coordinates to compensate in part for systematic density variations across the filtergrams. The film transport mechanism is operated manually. There is some drift in the system, particularly in the maintenance of image coincidence, so that occasional slight adjustments are required. However, a time of about 5 to 10 s is usually adequate for each magnetogram, so that a day's observations can be subtracted in about the same time.

Figure 2 shows two original filtergrams together with magnetograms obtained by photographic subtraction of the two frames, by the semi-automatic photographic method, and by TV subtraction. The TV subtraction is inferior in part to the photo-

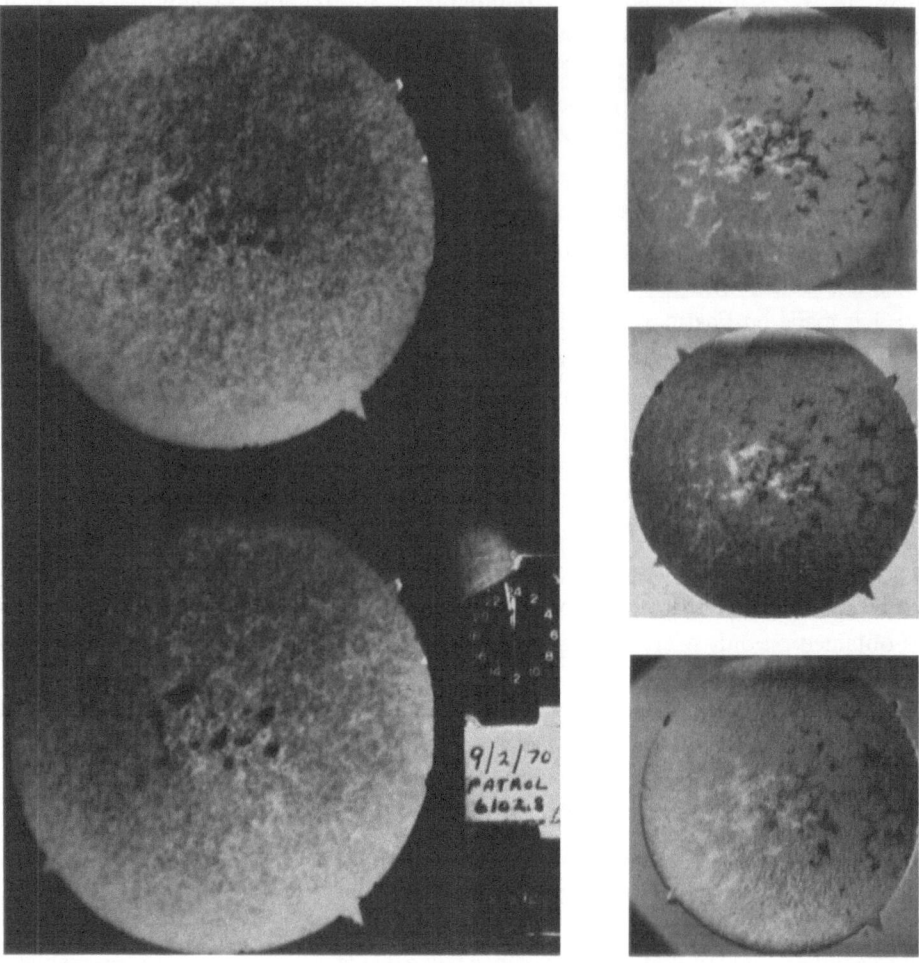

Fig. 2.   The two frames on the left hand side are simultaneous filtergrams in opposite circular polarizations in the blue wing of the Ca I absorption line at 6102.7 Å. The resulting subtractions of the two filtergrams are shown on the right hand side; the upper is by individual photographic process, the middle by a semi-automatic photographic procedure and the lower by a TV subtraction technique.

graphic methods because the Vidicons and the yokes used at present are not matched, each camera having its individual time base synchronized to the mains frequency, and the superposition is perfect over only part of the frame. Plans are in hand to improve on these drawbacks.

In the meantime, the semi-automatic photographic method and the TV technique provide acceptable solutions to the problem of producing cine films showing changes in magnetic fields, whilst the usual slower photographic subtraction of individual pairs of frames is used for more critical studies.

Acknowledgements

Many people have been involved in this project, amongst whom are A. F. Young, R. Riches, and D. Neill. We thank T. Nurse of Supreme Sound Studies for his cooperation in the development of the semi-automatic photographic methods. This work has been supported in part by the National Aeronautics and Space Administration under Contract NASr-208.

References

Leighton, R. B.: 1959, *Astrophys. J.* **130**, 366.
Ramsay, J. V., Kobler, H., and Mugridge, E. G. V.: 1970, *Solar Phys.* **12**, 492.
Steel, W. H., Giovanelli, R. G., and Smartt, R. N.: 1961, *Australian J. Phys.* **14**, 201.

A COMPLETE STOKES VECTOR POLARIMETER

FRANK Q. ORRALL

Institute for Astronomy, University of Hawaii, Honolulu, Hawaii 96822, U.S.A.

Abstract. A scanning photoelectric polarimeter of 16 cm aperture that can measure all four Stokes parameters of the visible radiation of the Sun's disk, the Corona, Moon and planets, has been constructed at the Institute for Astronomy and is installed on Mt. Haleakala. It is a two (orthogonal) channel system and uses a rotating $\lambda/4$ plate modulator. Photon counting is done by a digital computer that also Fourier analyzes the modulated output of the photomultipliers, and, from the Fourier components, computes the Stokes parameters in real time.

1. Introduction

Most of our knowledge of the physical nature of astronomical objects has been inferred from the intensity of the radiation $I(\lambda)$ received from them. But of course a complete description of the radiation (complete except for absolute phase) requires a measurement of the complete Stokes' vector $\{I, Q, U, V\}$. The parameters $Q(\lambda)$, $U(\lambda)$ and $V(\lambda)$ contain fundamental information about the magnetic fields, non-thermal processes, anisotropies in the source and the process of radiative transfer that can be found in no other way. Although a number of polarization measurements have been made on the Sun, few have measured all four Stokes' parameters except for some special purpose as in the vector magnetograph. At the Institute for Astronomy we have recently designed and constructed a complete Stokes' vector polarimeter-photometer that can be used to scan the Sun's disk and corona as well as the Moon and planets. At present, it uses broad or narrow band filters to isolate regions of the spectrum. Later, a high resolution scanning monochromer will be added so that the complete Stokes' four-vector can be measured as a function of wavelength.

2. The Instrument

A schematic drawing of the optics and electronics is shown in Figure 1. The telescope is a coronagraph of 16 cm aperture and 250 cm focal length ($f/15$). Two interchangeable telescope objectives are provided; one a singlet lens of coronagraph quality, the other an achromat. A 2.5 cm image of the Sun is formed on a conical surface with a small scanning aperture at its apex, and this conical surface reflects all of the light that does not pass through the aperture into baffles along the side of the telescope tube. It thus serves in place of an occulting disk. A second objective behind the aperture, collimates the light into a beam 2 cm in diameter. Immediately behind this is a modulator consisting of a rotating $\lambda/4$ plate and a polarizing beamsplitter (Wollaston prism). The two collimated beams from the beamsplitter fall on separate photomultipliers, which can be cooled by dry ice to reduce thermal noise. Suitable glass, gelatin or interference filters can be placed in front of the photomultiplier windows to isolate broad or narrow regions of the spectrum.

Howard (ed.), Solar Magnetic Fields, 30–36. All Rights Reserved.

The entire optical system fits in a cylindrical tube 22.5 cm in diameter and 4.5 meters long (including the dust tube in front of the objective). This tube is mounted on the 12-ft solar spar on Mt. Haleakala, Maui, at an elevation of 3054 m. The spar is pointed at the center of the Sun, but the entire polarimeter moves relative to the spar to scan the Sun or corona. Thus the polarimeter is always used on-axis. An aperture stop and mask are suitably placed in front of the collimator to make the instrument in all respects a Lyot coronagraph, provided that the singlet, low-scatter objective is used. If the scanning aperture used in the primary focal plane is small, a field lens (usually required in the Lyot coronagraph) is not needed.

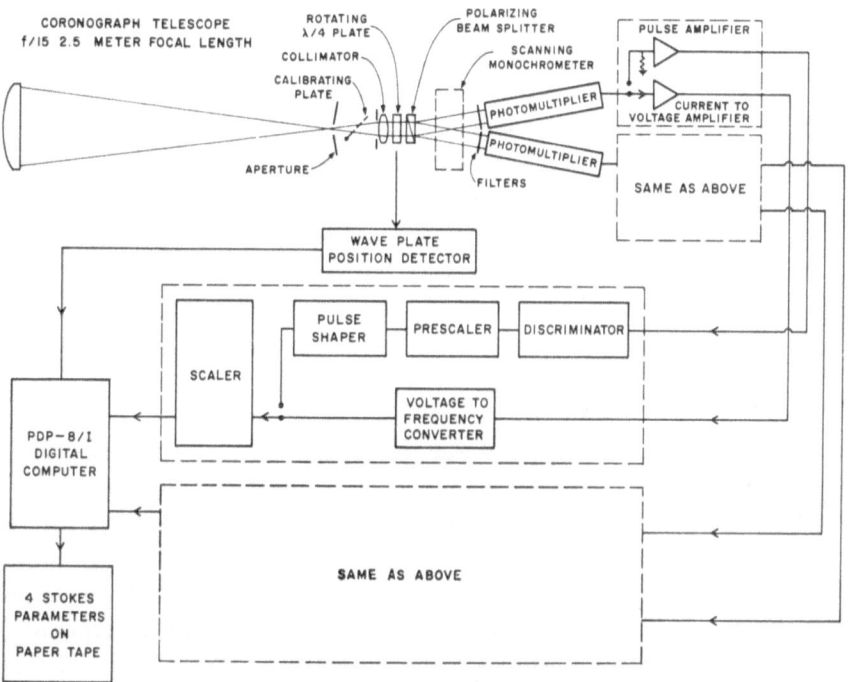

Fig. 1.   A schematic diagram of the optics and electronics of the complete Stokes vector polarimeter of the Institute for Astronomy.

A PDP-8/I digital computer points and controls the instrument, receives and processes data from it, and prints out the processed data on a typewriter or on high speed punched paper tape. The polarimeter will share time on the computer with other observatory instruments.

A. Scanning in the Plane of the Sky

The 12-ft Solar Spar on which the polarimeter is mounted remains photoelectrically pointed at the center of the Sun's disk to accommodate the other instruments mounted on it. Since the polarimeter scanning aperture is always on-axis in order to minimize instrumental polarization, the polarimeter must move relative to the spar in order to scan on the plane of the sky. This is accomplished by mounting the polarimeter on gimbals near its center of mass, and then driving its forward end in two coordinates by means of two precision screws driven by stepping motors. These stepping motors are under direct computer control, so that a raster scan, or any other type of scan can be performed on the plane of the sky by suitably programming the computer.

B. THE MODULATOR

The modulator is driven by a stepping motor geared so that 160 steps of the motor produces one full rotation of the $\lambda/4$ plate. The pulses for the stepping motor are provided (through a translator) by the kilocycle time base of the computer. Thus at 10^3 steps/sec the $\lambda/4$ plate rotates 6.25 times per second. Most polarimeters or magnetometers have modulators based on either the Pockels, Kerr or Piezo-optical effects and can be driven at frequencies from 10^3 to 10^6 Hz. The low modulation frequency (6.25 Hz) of our polarimeter has some disadvantages, but it enables us to use digital methods throughout. This makes it possible to detect and correct for spurious modulation which would otherwise be interpreted as polarization.

The computer keeps track of the angular position of the wave plate by counting steps. To insure that this count has not been lost, a small photoelectric sensor detects a fiducial mark on the edge of the wave-plate at each full rotation of the plate and the computer produces an error alarm if this does not come at the expected time.

C. ELECTRONICS

The output of the photomultiplier can be recorded in two alternate ways: the pulses produced by individual photons can be counted; or the photocurrent produced by a beam of photons can be measured. The pulsewidth is such that if more than about 5×10^6 pulses arrive each second, enough coincidences occur that the counting rate is no longer proportional to the photon arrival rate. Thus if the counting frequency exceeds about 5 megacycles, it becomes necessary to measure photocurrent instead of counting photons. Since this instrument is intended to be used over a wide range in intensities, both types of measuring systems are required.

As shown in Figure 1, each of the two photomultipliers has a separate channel of electronics, and each of these consists of two parallel (alternate) systems. Each pulse counting system has a pulse amplifier, and a discriminator which must be set at an appropriate level to distinguish between noise and true pulses due to photons. The current measuring systems consist of a current to voltage amplifier followed by a voltage to frequency converter. Thus each system produces counting rates that are proportional to the intensity of the light beam falling on the photocathode. In both systems the actual counting is done in the PDP-8/I computer.

3. Theory

Let $S \equiv \{I, Q, U, V\}$ be the Stokes vector of the incident beam of light to be measured, and let $S' = \{I', Q', U', V'\}$ represent the beam as modified by passing through the linear retarder (wave plate) and the linear polarizer (Wollaston prism). The transformation from S to S' can be expressed by the Mueller calculus:

$$S' = [M_P] [M_R] S. \tag{1}$$

Here $[M_P]$ and $[M_R]$ are the Mueller matrices of the polarizer and retarder respectively. They have a simple well known form if the retarder and polarizer are linear and ideal (Shurcliff, 1962).

The matrix $[M_R]$ will depend on the retardance δ, and on the instantaneous angle of orientation ψ of the fast axis of the wave plate. If the plate rotates at circular frequency ω then $\psi = \omega t$, and $[M_R]$ will be a function of time. We choose for simplicity the convention that $\psi = 0$ when the normalized fast eigenvector of the wave plate is $\{1, 1, 0, 0\}$. We require only the intensity of the light that falls on the photocathode – that is, the first component or parameter of S', namely I'. From Equation (1) we can readily find that

$$2I' = I \pm [(a + b \cos 4\omega t) Q + (b \sin 4\omega t) U + (c \sin 2\omega t) V] \tag{2}$$

where

$$a \equiv \tfrac{1}{2}(1 + \cos \delta)$$
$$b \equiv \tfrac{1}{2}(1 - \cos \delta) \tag{3}$$
$$c \equiv \sin \delta.$$

In Equation (2) and in what follows, the $+$ sign applies to the beam from the Wollaston prism whose normalized eigenvector is $\{1, 1, 0, 0\}$ and the $-$ sign to the other beam with the eigenvector $\{1, -1, 0, 0\}$. The retardance δ of the wave plate will only nominally be $90°$, and a, b and c can be considered constants of the instrument that will vary with the wavelength distribution of the light being measured. We note from Equation (2) that if δ approaches $180°$, the parameter V will be indeterminate.

After correcting the observed photocurrents for dark current and multiplying them by suitable calibration constants, their time dependent value $g(\psi = \omega t)$ should equal the right hand side of Equation (2). In practice we fit $g(t)$ to a Fourier series

$$g(t) = A_0 + \sum_{n=1} [A_i \cos n\omega t + B_i \sin n\omega t], \tag{4}$$

with $A_3 = B_3 = A_n = B_n = 0$ for $n \geqslant 5$. Comparing (4) with (2) we should expect (ideally):

$$A_0 = I + aQ$$
$$A_1 = B_1 = 0$$
$$A_4 = \pm bQ \tag{5}$$
$$B_4 = \pm bU$$
$$B_2 = \pm cV.$$

The Stokes' parameters I, Q, U, V can be inferred from the observed values of A_0, B_2, A_4, B_4, using either photomultiplier channel.

In practice, of course, the other Fourier coefficients are not equal to zero by amounts that cannot be explained by noise, indicating that there are spurious sources of modulation (besides instrumental polarization) which must also contribute to A_0, A_4, B_4 and B_2. Our experience suggests that these sources become important when one attempts to detect polarizations of less than about 0.1% and thereafter become increasingly important. We shall mention several possible mechanisms for this spurious modulation but shall postpone a discussion of their magnitude until the instrument has been more fully tested.

3.1. THE RETARDER OR POLARIZER MAY BE NON-LINEAR

That is, the fourth Stokes' parameter of its eigenvectors may be significantly different from zero. Thus, for example, a wave plate made of quartz will in general be an elliptical retarder unless it is made from a section cut parallel to the optic axis of the crystal. Proper construction of the optics can minimize this.

3.2. THE RETARDER OR POLARIZER MAY BE NON-PERFECT

An ideal linear polarizer would have only one eigenvector. In general, however, the minor eigenvector is non-zero. Further, the retarder may be dichroic, that is, its optical transmission may depend on the form of the incident polarized light. Mica wave plates have appreciable linear dichroism, and quartz shows circular dichroism. Sekera (1956) and Hodgdon (1965) have considered some effects of non-perfect components on the modulation of a rotating wave-plate polarimeter. Compound retarders can be constructed having very little dichroism (Evans, 1970).

3.3. THE ROTATION OF THE WAVE PLATE MAY INTRODUCE MODULATION BY EFFECTS OF GEOMETRIC OPTICS

If the two surfaces of the wave plate are not flat and parallel, or if the mechanical axis of rotation is not parallel to the optic axis, the collimated beam will move about on the photocathode. Since the transmission of the photomultiplier window, or the sensitivity of the photocathode, may be non-uniform, this will produce a modulation. Or, if there is a bit of dirt on wave-plate its shadow will move about on the photosurface. These effects will produce a modulation not only at the fundamental frequency ω (so that A_1 and B_1 will be non-zero) but also at higher harmonics. This modulation can be minimized by careful construction and alignment of the wave plate but is difficult to reduce it to a negligible value when measuring small amounts of polarization.

It is possible to modify Equation (2) to include all of the above effects. This will lead to a more complicated set of relations between the observed Fourier coefficients and the Stokes parameters than in Equation (5), and will include several instrumental calibration constants. These constants can be determined by calibration, and with them, the Stokes parameters can be determined in real time in the computer. We shall postpone a more complete discussion until we have better measurements of the magnitudes of these effects.

The complete vector **S** can be derived from either channel as we have done above, but this does not make use of the fact that the two channels contain orthogonal information. By adding and subtracting the observed values of I' from the two channels, one can derive each Stokes parameter using data obtained simultaneously. As Lyot showed, such two channel observations can reduce the effects of seeing and certain other types of noise. This can be accomplished using computer software.

4. Sensitivity

The accuracy or limiting polarizations that this instrument can attain will depend on the magnitudes of the photon noise, instrumental polarization, effects due to the Earth's atmosphere, spurious sources of modulation discussed above, and (for low counting rates) on the dark current. Only the photon noise can be predicted with confidence. As an example, consider the Sun's disk. Let N_λ be the number of photons counted each second with an average energy hc/λ within an effective band pass $\Delta\lambda$ by one of the photomultipliers. Then evidently

$$N_\lambda = (f_\lambda Aa\Delta\lambda\lambda e)/(a_0 hc), \tag{6}$$

where f_λ is the flux from the Sun's disk falling on the outside of the Earth's atmosphere; $A \simeq 176$ cm^2 is the effective area of the telescope objective; a and $a_0 = 2.9 \times 10^6$ are the areas of the scanning aperture and of the Sun's disk respectively, given in (arc s)^2. The efficiency factor $e \approx 0.02$ includes the detector efficiency, the transmission of the Earth's atmosphere and optical losses at $\lambda 5000$, as well as a factor $\frac{1}{2}$ for the beam splitter. At $\lambda 5000$, $f_\lambda = 200$ erg cm^{-2} A^{-1} s^{-1} and

$$N_{\lambda 5000} = 6 \times 10^7 \ a \ \Delta\lambda \text{ photons s}^{-1}. \tag{7}$$

Thus with $\Delta\lambda = 10$ Å and a 3×3 arc s scanning aperture, the counting rate will be $\sim 5 \times 10^9$ photons s^{-1} which will produce about the maximum photocurrent (50 μamp) that the photomultipliers can tolerate.

In this case, if photon noise were the limiting factor, one could detect polarization of $\sim 10^{-5}$ in one second. Laboratory tests show that instrumental or modulation effects are not important if the polarization is greater than $\sim 10^{-3}$, but the lower limit set by these effects can only be determined by extended tests on the Sun's disk.

Acknowledgements

This instrument was constructed as part of a program of J. T. Jefferies, J. B. Zirker and the author to study magnetic fields and fast particles in the Sun. This work has been supported by Project Themis through ONR grant N00014-68-A0149. R. Zane designed the electronic control and data gathering systems, supervised their construction, and with J. Harwood and H. Yee wrote the necessary computer programs. J. Barclay and H. Boesgaard designed the mechanical system which was constructed under the supervision of L. Emarine and K. Miller.

References

Evans, J. W.: 1970, (Private Communication).
Hodgdon, E. B.: 1965, *Appl. Opt.* **4**, 1479.
Sekera, Z.: 1956, *Adv. Geophys.* **3**, 43.
Shurcliff, W. A.: 1962, *Polarized Light*, Harvard University Press, Cambridge, Mass.

Discussion

Severny: (1) What is the minimal degree of polarization you could measure?

(2) Your scheme seems to be very similar to that described by Landestreet in *Astrophys. J.*, June 1970 and used for the measurements of the magnetic field in white dwarfs.

(3) What is the difference between your scheme and the scheme proposed by Lyot and used by Dr. Dollfus?

(4) How can you discriminate polarization due to magnetic fields from that due to scattering in the process of line formation?

Orrall: To answer your questions in order:

(1) We hope that when sufficient photons can be counted the smallest measurable polarization will be between 10^{-4} and 10^{-5}, a limit set by instrumental and sky polarization. But the instrument is not fully tested and we shall see.

(2) I think that his scheme uses also a rotating wave plate but I am not sure.

(3) Dr. Dollfus' instrument is also a two-channel polarimeter using a refractor on axis. It differs from ours in that it is designed to measure linear polarization and rotates about its optical axis to determine the direction of polarization. It uses a different modulation scheme from ours, and treats the output of the photomultipliers in an entirely different way.

(4) The problem of inferring the magnetic field from measured polarization is a difficult one that is not yet satisfactorily solved. We hope that this instrument will contribute to the solution of that problem. When it is we will be able to call our polarimeter a magnetometer!

Gaizauskas: It does not necessarily follow that the use of a much higher frequency of modulation of an EOLM in a photoelectric magnetograph produces a better field measurement. With modern systems, the major source of signal noise is produced by photon noise in the photomultipliers and by image motion. Photon noise can be quickly filtered out, but the power spectrum of image motion rises rapidly towards low frequencies (10 Hz and less). In areas of rapidly changing field strength, the major noise component is therefore produced by image motion. If image motion is monitored at the same time as the magnetic field is rapidly scanned, it becomes possible to reject the measurements made during periods of poor seeing to improve the data.

Orrall It is certainly true that image motion will be a major source of noise when measuring regions where the field is changing rapidly. Monitoring the seeing as you suggest is certainly a good idea. But I think that a high modulation frequency is still valuable

MEASUREMENTS OF MAGNETIC FIELDS

M. SEMEL

Observatoire de Meudon, Meudon, France

Abstract. It is demonstrated that certain interrelations between the Stokes parameters are much less dependent on line formation than generally expected. Thus the dependence of these parameters on the magnetic field may be made more explicit and may lead to reliable calibration of polarimetric measurements in terms of magnetic field.

1. Introduction

Three preliminary hints indicate that such an independence on line formation is very likely to occur.

First, the Sears formula, for the case of weak fields, is easily generalized by treating the transfer equations (Semel, 1967). That means that only broadening mechanisms should be estimated to make this formula valid.

Second, for strong fields, that is when all Zeeman components are separated, the polarization in each component is a function of the magnetic field only.

Third, in the relation $V/Q = \eta_V/\eta_Q$ in the Unno theory (1956) only the broadening mechanism and magnetic field should be considered. This relation can be easily generalized by treating the transfer equations. The only condition necessary is that the broadening mechanism is kept constant in the observed element of the Sun. (However neglect is made of all parasitic effects as depolarization etc...).

The critical factor in all these cases is the broadening mechanism.

2. Numerical Computation

Now we shall extend our treatment with the help of numerical computations using the Schuster-Schwarzschild model (Michard, 1961) which for our purpose is quite equivalent to the Milne-Eddington model (Unno, 1956). For each model three parameters of line formation should be specified:

(1) For Doppler broadening, ξ the half Doppler width;

(2) τ_0 for the $S-S$ model or η_0 for $M-E$ model;

(3) $A_{SS} = (I_0 - S)/I_0$ for $S-S$ model or $A_{ME} = \beta_0 \cos\theta/(1 + \beta_0 \cos\theta)$ for $M-E$ model.

In the proposed mathematical method A_{SS} and A_{ME} are eliminated. The dependence on ξ and on τ_0 or η_0 is shown by the computed curves. The terms r_I, r_V and r_Q are used to represent the Stokes parameters I, V and Q in the depression representation.

The case of a Zeeman triplet observed with circular analysis is considered first. The quantity calculated is

$$\Delta\lambda_G = \frac{\int \Delta\lambda r_V \, d\lambda}{\int r_I \, d\lambda}$$

Howard (ed.), Solar Magnetic Fields, 37–43. All Rights Reserved.
Copyright © 1971 by the IAU.

$\Delta\lambda = \lambda - \lambda_0$; λ is the wave-length and λ_0 is the origin (the wave-length of the undisturbed line).

The first results are given in Figure 1 where $\Delta\lambda_G$ is plotted against $\Delta\lambda_H$ both in units of ξ_0 the half Doppler width. $\Delta\lambda_H$ is related to the magnetic field H by the well known expression:

$$\Delta\lambda_H = 4.67 \times 10^{-13} \, g\lambda^2 H.$$

For practical reasons the values $-5\xi_0$ and $5\xi_0$ were taken as the lower and upper limits respectively in the integrals giving $\Delta\lambda_G$.

$\Delta\lambda_G$ approximates the longitudinal component ($\Delta\lambda_G \simeq \Delta\lambda_H \cos\psi$). The differences between the curves for $\tau_0 = 1.45$ and $\tau_0 = 3$ are very small. The curves are practically

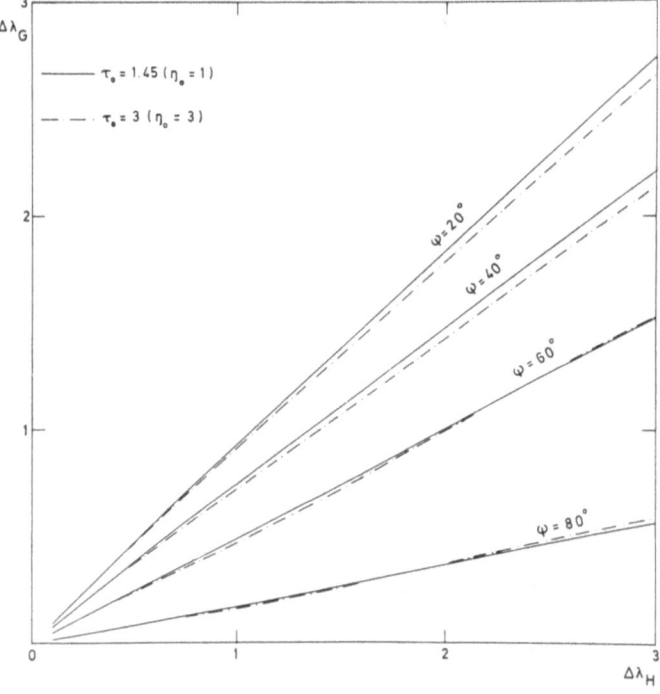

Fig. 1. For the case of Zeeman triplet observed with circular polarization. $\Delta\lambda_G$ is plotted against $\Delta\lambda_H$ in units of ξ_0 the half Doppler width.

$$\Delta\lambda_G = \frac{\displaystyle\int_{-5\xi_0}^{5\xi_0} \Delta\lambda \, r_V \, d\lambda}{\displaystyle\int_{-5\xi_0}^{5\xi_0} r_I \, d\lambda}$$

ψ is the angle between the magnetic field and the line of sight. $\Delta\lambda_G \approx \Delta\lambda_H \cos\psi$.

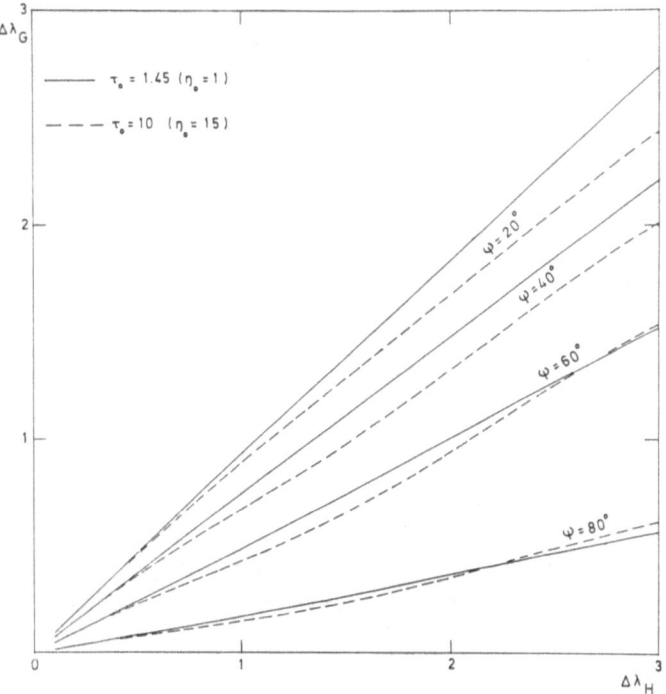

Fig. 2.   The same as in Figure 1. Comparison between high saturation and low saturation.

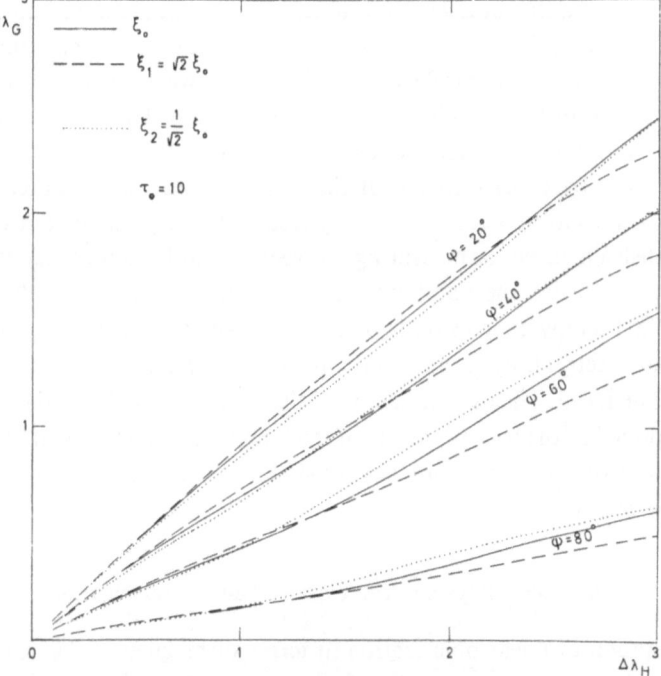

Fig. 3.   The same as in Figure 1. $\Delta\lambda_G$ is plotted against $\Delta\lambda_H$ in units of ξ_0, but three different values for the Doppler broadening are considered.

straight lines passing through the origin of the axes and consequently are independent of ξ. Figure 2 compares the curves for $\tau_0 = 1.45$ and $\tau_0 = 10$ (high saturation). The differences are of the order of only 10%. In Figure 3 a comparison is made between curves all corresponding to the same saturation $\tau_0 = 10$ but with different values for ξ: ξ_0, $\xi_1 = \sqrt{2}\xi_0$ and $\xi_2 = \xi_0/\sqrt{2}$. The dependence on ξ seems to be small. The deviation for ξ_1 is due to the fact that integrations were limited to the range $-5\xi_0 < \Delta\lambda < 5\xi_0$.

$\Delta\lambda_G$ is the displacement of the center of gravity of a line profile observed with circular polarization, and can be measured by a magnetograph. A proposal for such a method is described by Semel (1970a).

3. Observations

I would like now to sketch very briefly how progress in this direction was achieved with photographic observations. At the beginning we knew only the Sears formula for weak field. For the purpose of quantitative analysis Michard established a method which was the analog of a standard magnetograph. We soon felt the response is saturated for strong fields. Then Michard had the project of the Lambdameter (Rayrole et al., 1962). The advantages were evident. However we had as a first problem: the choice of slits for the Lambdameter, we made the most intuitive choice of slits from technical point of view only, Rayrole established a programme for computing the response of the Lambdameter using the Unno theory. He expected to measure the vector field by using different lines (Rayrole, 1964).

When I proceeded to test the longitudinal component approximation by comparison of observation with different lines, I was able to show that the approximation is fairly good even for strong fields. I was more and more convinced that the longitudinal component approximation should not be limited to weak fields and I had the conviction that this approximation is not due to pure hazard. This conviction led Rayrole and me to have many animated discussions and also to many tests by numerical computation and by experiments with the Lambdameter using different slits. For this purpose the photographic method proved to be very useful. Thus Rayrole was able to establish a method for measuring the vector field by simultaneously observing different lines and using the Lambdameter with different slits (Rayrole, 1967).

Now our first happy conclusion is that: By the analysis of the circularly polarized component of spectral lines formed in the presence of a magnetic field, magnetic data can easily be deduced. The displacements of the center of gravity of the profiles $r_I \pm r_V$ can be interpreted as the measurements of the longitudinal component of the magnetic field. In many cases this procedure is very slightly dependent on the unknown parameters of the line formation.

4. The Analysis in the Case of Linear Polarization

The interpretation of linear polarization in terms of magnetic fields cannot be made independent of the parameters of line formation. A dependence on the broadening mechanism is expected. However for many lines the broadening mechanisms are likely to be identical. Therefore it would be interesting to develop a method where the calibration of polarimetric measurements depends only on the broadening parameters. In such a case the disadvantage due to this dependence might have its counterpart, namely the determination of the broadening mechanisms, which has an important physical meaning.

Using the $S-S$ model I calculated:

$$\Delta\lambda_G = \frac{\displaystyle\int_0^{5\xi_0} \Delta\lambda r_Q \, d\lambda}{\displaystyle\int_0^{5\xi_0} r_I \, d\lambda} \, .$$

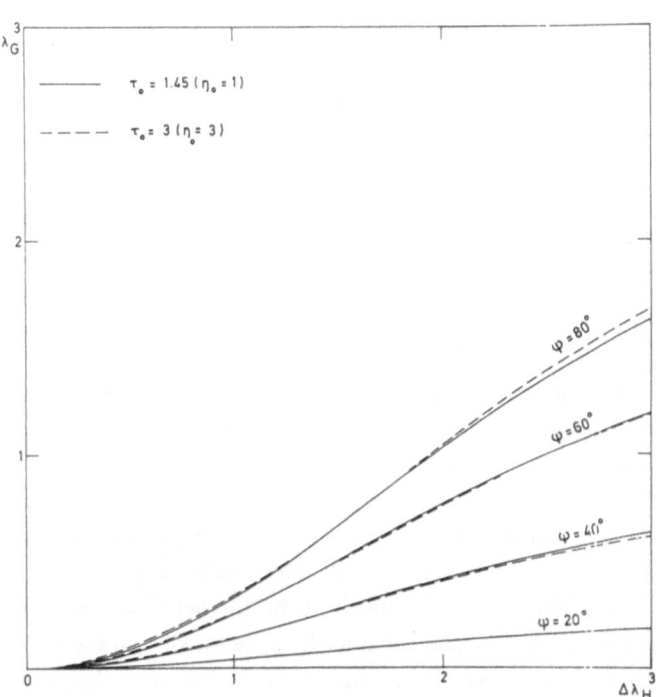

Fig. 4.   For the case of Zeeman triplet observed with linear polarization. $\Delta\lambda_G$ is plotted against $\Delta\lambda_H$ in units of ξ_0 the half Doppler width.

$$\Delta\lambda_G = \frac{\displaystyle\int_0^{5\,\xi_0} \Delta\lambda \, r_Q \, d\lambda}{\displaystyle\int_0^{5\,\xi_0} r_I \, d\lambda} \, .$$

The results are given in Figure 4. Again the variations of τ_0 are unimportant.

The choice of ξ is critical as expected. Thus the reduction of linear polarization measurements to magnetic data should simultaneously incorporate a determination of ξ. An attempt in this direction can be made by simultaneously observing several lines having different values of the Landé factor g. Assuming that ξ is the same for all lines used, then, the simultaneously measured values for $\Delta\lambda_G$ will only lie on the same curve of calibration if ζ is correctly evaluated.

A proposal for the 'realization' of this approach by a magnetograph was described in another paper and cannot be discussed in more detail here (Semel, 1970b).

5. Computation of Line Profiles

We now proceed to another aspect of the problem that is the shape of spectral line profiles, because in last analysis any observation is the analysis of line profiles either explicitly or implicitly. The following is the result of numerical computation made by Rayrole and myself.

Our first question was whether an observed profile in absence of magnetic field may allow to determine the model to be used and its parameters. We found that numerically speaking the $S-S$ and $M-E$ models are equivalent. That is we found that for each of the three parameters of one model we can find a choice of three parameters for the other model to obtain practically the same profile. Then we introduced the magnetic field but we find again that the two models give similar profiles when using for each model the parameters adjusted as above for the zero field profile. Thus we could continue our numerical exercises with one model only, the $M-E$ one. It became evident that the choice of the parameters might be critical. In the presence of a magnetic field the profiles may be quite different. As a result of high saturation, $\eta_0 > 4$, a new peak or a pseudo Zeeman component appears (Henoux, 1968; Göhring, 1969). We expected that such a peak would be easily distinguished and could be used as a criterion for the choice of η_0 [*]. However we never observed such a pseudo component in our films. We were not content to conclude that η_0 should always be small for all lines. We realized soon that the occurrence of such a pseudo component is highly related to the choice of ξ and that in fact ξ is likely to be smaller than was usually considered ($\xi = 40$ mÅ for Fe I). If ξ is made smaller by a factor of two the pseudo component may appear again (for weaker fields) but might be no more distinguished. Thus we came to the conclusion that the introduction of microturbulence to render high values for ξ should be highly criticized and that the analysis of line profiles in the presence of a magnetic field may be a key to the problem. High values for η_0 ($\eta_0 > 4$) are permissible.

Our next step was a demonstration of the equivalence of different arithmetic expressions to describe the profile of a bell and eventually also a spectral line profile. The $S-S$ and the $M-E$ models are examples. By considering quite different values

---

[*] The pseudo component should not be confused with other peculiarities. See Beckers and Schröter (1969).

for microturbulence, macroturbulence, finite width slits, instrumental profiles etc., we could get practically indistinguishable profiles with quite different values for the source function, η_0, and for ξ.

Finally we could confirm the results obtained from the simplified models by a series of computations I had made with empirical models: the BCA for the photosphere and the Henoux model for sunspots. The appearance of a pseudo component is highly related to the saturation and to the choice of the microturbulence in each model. Thus we conclude that our simple analysis with the Unno model is much more general than can be expected from such a simplified model.

We may conclude that the 'undisturbed profile' should not be considered for calibration of magnetic field measurements. But the Unno theory can be used when it is considered that:

(1) β_0 is unknown; (2) η_0 is unknown; and last but not least (3) ξ is unknown.

This theory can be used because the Stokes parameters are correctly introduced, and calibration of magnetic field measurements can be made without ad hoc fixing the line formation parameters.

A tentative attempt in this direction, observing simultaneously different lines, is described by Rayrole in this Symposium.

References

Beckers, J. M. and Schröter, E. H.: 1969, *Solar Phys.* **7**, 222.
Charvin, P., Rayrole, J., and Semel, M.: 1962, *Compt. Rend. Acad. Sci.* **254**, 2289.
Göhring, R.: 1969, *Solar Phys.* **8**, 271.
Henoux, J. C.: 1968, *Solar Phys.* **4**, 315.
Michard, R.: 1961, *Compt. Rend. Acad. Sci.* **253**, 2857.
Rayrole, J.: 1964, *Compt. Rend. Acad. Sci.* **258**, 1161.
Rayrole, J.: 1967, *Ann. Astrophys.* **30**, 257.
Semel, M.: 1967, *Ann. Astrophys.* **30**, 513.
Semel, M.: 1970a, *Astron. Astrophys.* **5**, 330.
Semel, M.: 1970b, *Astron. Astrophys.* **9**, 356.
Unno, W.: 1956, *Publ. Astron. Soc. Japan* **8**, 108.

Discussion

Brueckner: The 5250 line can be calculated with Henoux' spot model in a sunspot without changing η_0, ξ or g. Calculations by Olaf Moe at the Naval Research Laboratory have shown, that it is necessary to take into account the numerous molecular lines and include them into the continuous absorption coefficient. If one does so, the calculated line profile matches fairly well the observed one.

Semel: The necessity to introduce the molecular line in the computation of the line 5250.2 is evident. However your success is not in disagreement with my paper. In this paper the problem is not the possibility of a solution but the uniqueness of the solution. The profile of 5250.2 presented in this paper is merely a part of the 'mathematical exercise'. I have never tried to match observed profiles with calculated ones.

DIGITAL VIDEOMAGNETOGRAMS IN REAL TIME

THOMAS J. JANSSENS and NEAL K. BAKER

The Aerospace Corporation, San Fernando Observatory, Los Angeles, California, U.S.A.

Abstract. The Aerospace – NASA Videomagnetograph began operation one month ago, two years after components were ordered and construction began. The design grew out of a desire to obtain magnetic fields in real time using an optical filter. The aim was to study and analyze magnetic configurations and changes, quantitatively if possible, with high spatial and temporal resolution and as much sensitivity as possible. This instrument is restricted to the line-of-sight component of the magnetic field and is primarily intended for high resolution studies of selected regions of the sun. The rationale behind our approach is shown in the next section and the design details in the following.

1. Analysis of An Idealized Magnetograph

In most solar magnetographs the measurement of the line-of-sight component requires the same processes, namely: light collection, spectral isolation, polarization selection, detection, subtraction, scanning, and data presentation.

The performance characteristics include the sensitivity, ΔB; the area of a resolution element, a; the time resolution, T; and the number of resolution elements per scan, n. If we assume a given magnetic line and noise determined solely by the photon limit then we find that the product $n (\Delta B)^{-2} T^{-1} a^{-1}$ is invariant such that if n, T and a are changed, ΔB varies to hold the product constant. In theory this allows any desired trade off between spatial resolution, temporal resolution, and field of view but in practice such versatility is rather limited in actual magnetographs. Thus this product can perhaps be used then as a kind of performance figure of merit. It is proportional to the rate of photon detection and depends on the product εA (or εAN for a multi-channel detector) where ε is the overall efficiency of the system, A is the area of the objective lens, and N is the effective number of data channels. The results are not surprising: that someone with an ideal detector would want a large aperture telescope, a high efficiency, and a multichannel detector.

Our approach has been to concentrate on increasing the effective number of parallel data channels, while using a rather small aperture. The number of data channels and mode of scanning are interrelated. Three types of scanning might be called a point-by-point scan, a slit scan, and a simultaneous exposure. In the point-by-point scan light from one element passes through a spectrograph and is analysed with the procedure repeated for each element in turn. This gives an effective $N=1$. In a slit scanning system such as the photographic subtraction technique developed by Leighton, photons along the exit slit of a spectroheliograph are simultaneously detected. This typically gives an effective N of 200 to 1000. The third method of simultaneous exposure requires the detection of photons from the entire image at once. This could have an effective $N=40000$ for a 200×200 array.

Simultaneous exposure requires use of a filter for spectral isolation rather than a spectrograph. This generally restricts one's versatility in the narrowness of the spec-

Howard (ed.), Solar Magnetic Fields, 44–50. All Rights Reserved.
Copyright © 1971 by the IAU.

trum selected and the number of lines available for use, though filter technology has been steadily improving.

The two possible sensors for simultaneous exposure are film and video. With film one is restricted to a low duty cycle or a large expenditure of film. For example to take 0.2 S exposures each 10 S would decrease the effective N by a factor of 50 while to take exposures much more frequently would consume an enormous amount of film. In addition, photographic subtraction can be time consuming, difficult, and suitable from only single or double cancellation. Certainly photographic subtraction would not yield real time results.

A television camera is not without difficulties either. They usually cannot match film in resolution and signal-to-noise ratio at present. Also, the target of a video camera can hold only a given number of photoelectrons before being read off, so if one is to maximize the rate of photon detection the camera must be scanned at a fast rate, not operated in a slow scan mode. In a fast scan mode, the rates and capacity necessary to store and subtract video images in real time are quite a problem.

We decided to store, subtract, and average (enhance) the video images digitally rather than by analog means but found that the usual method of digitizing and storing the data for later analysis does not work because the video data is 4 million words per second, about 100 times faster than the usual magnetic tape recorder. A thousand reels of magnetic tape would be filled in only one hour at video rates.

Computing with the data is much faster than storing it but for our task the required speed and storage capacity would require a very large and expensive general purpose computer which was not available. To solve the problem we designed our own special purpose computer which would operate at these high video rates. The heart of the video computer is a high speed adder with a memory unit consisting of a digital magnetic disc which is used with sequential addressing instead of random accessing. The disc has 72 tracks, 130000 bits per track and rotates at 30 Hz. The advantages of a disc memory over core memory are, a much lower cost, a greater capacity, and higher access times. In addition, we can simultaneously use different parts of the disc for different tasks.

2. Description of the Videomagnetograph

The system shown in block diagram form in Figure 1 has four organization parts: (i) optical and video, (ii) data gathering and preprocessing section, (iii) display section, and (iv) analytic section.

The optical section consists of a telescope with a 15 cm diameter lens which forms a real image which is further enlarged onto the face of an SEC Vidicon. Spectral isolation of one wing of the 5324Å line of Fe I is achieved by a hybrid filter with a bandpass of 0.11Å similar to the filter developed by H. Ramsey of Lockheed. An electro optic crystal or a quarter wave plate is switched to alternately admit right and left handed circularly polarized light. One TV line is 1.5 arc s and the field of view is 300 arc s on a side.

The data gathering and preprocessing section includes the analog-to-digital converter, high speed adder, and 32 tracks of the disc divided into bands 1 and 2.

The incoming video signal is digitized into 8 bits (256 levels) by a 4 megacycle converter which is synchronized with the data flow of the disk. The output of the analog-to-digital converter can be added to or subtracted from a previously stored array in the data gathering area of the disk and then stored, thereby erasing the previous image array. This operation is controlled by the disk interface controller

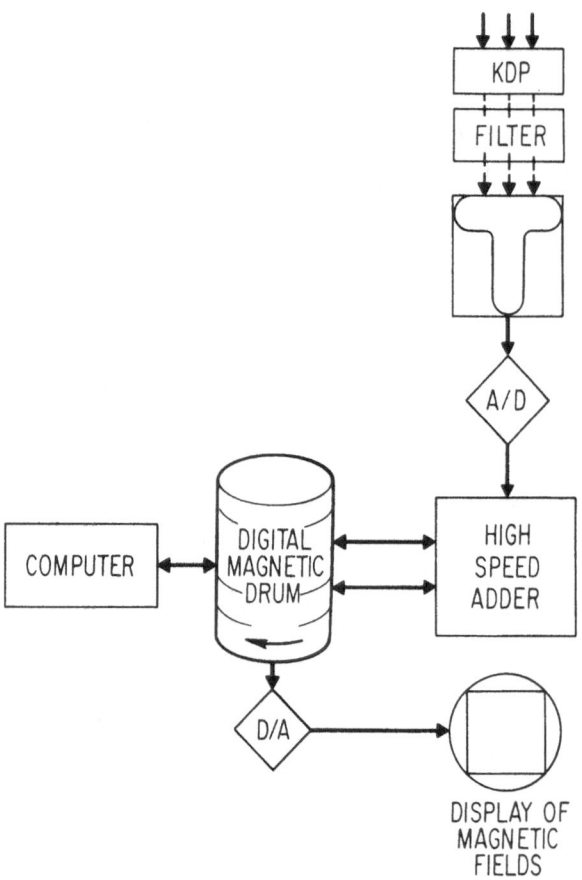

Fig. 1.   Block diagram of the Aerospace – NASA Videomagnetograph.

which requires one instruction per frame whether to add or subtract the incoming image. It is operated on an interrupt basis by a small general purpose computer attached to the system.

After the completion of the data gathering mode, the system can be interrogated for information on the entire frame. At present, we can get the number of elements greater than a certain value, value of the nth largest element, and mth smallest element, or the average value. The position of the element can not be found in this interrogation mode.

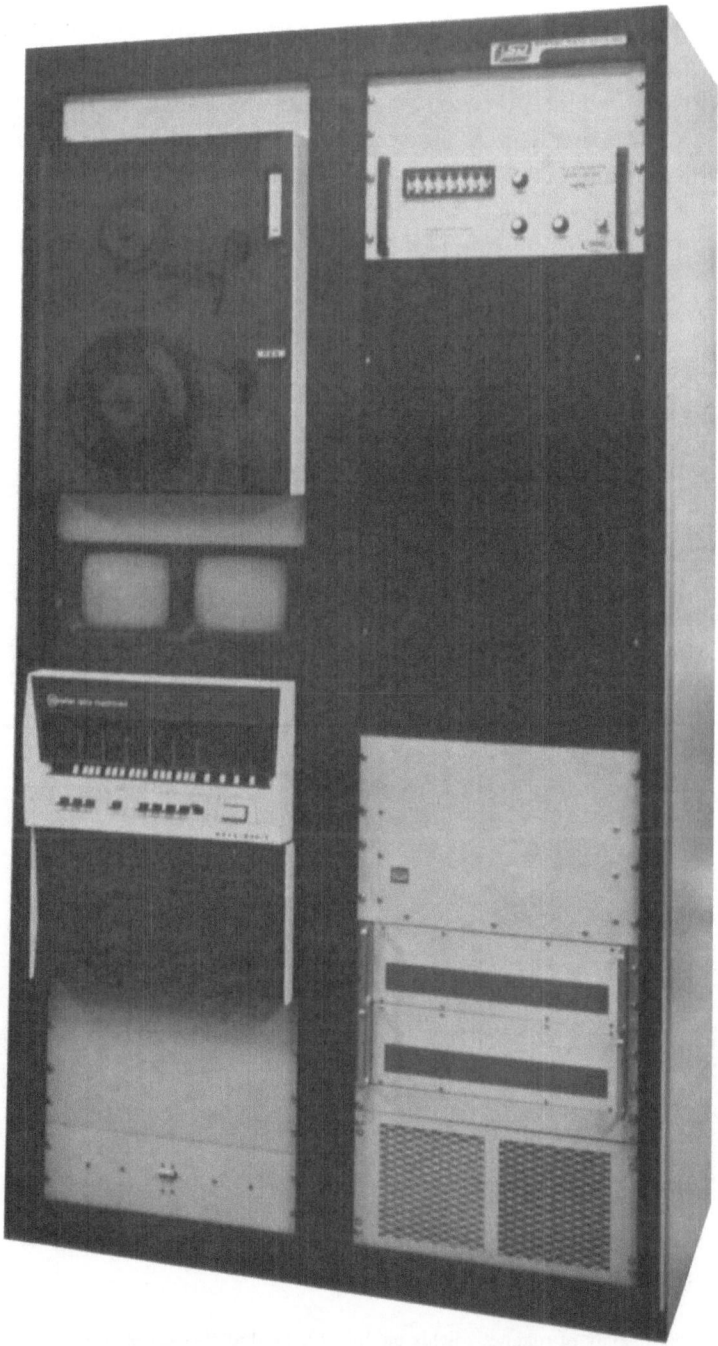

Fig. 2. Digital Processor for the Videomagnetograph.

For instance, the time to obtain the nth largest element in the array of 100000 points is 0.5 s. This information is mainly used to set the levels for the display and to make distribution plots.

In the display section after a number of magnetic frames have been averaged to enhance sensitivity and achieve data compression, the results are displayed on a TV monitor for visual inspection and recording photographically. The magnetic field can be displayed with each brightness made to correspond to any desired magnetic strength in a number of ways, ranging from arbitrary selection of 8 to 16 gray levels to automatic selection by the computer. The display can be either in a positive or negative mode, logarithmic or linear scaling. This display data is transferred to the display section of the disk and displayed to the operator. The 4 bit (16 gray levels) data is

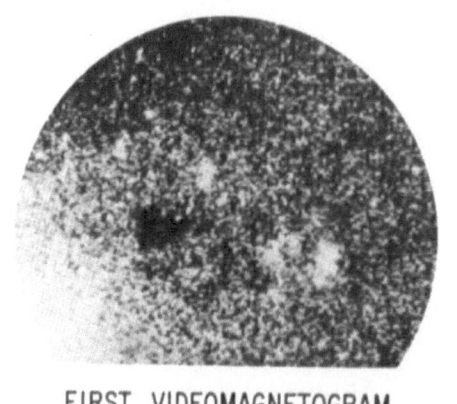

FIRST VIDEOMAGNETOGRAM
BY AEROSPACE-NASA VIDEOMAGNETOGRAPH

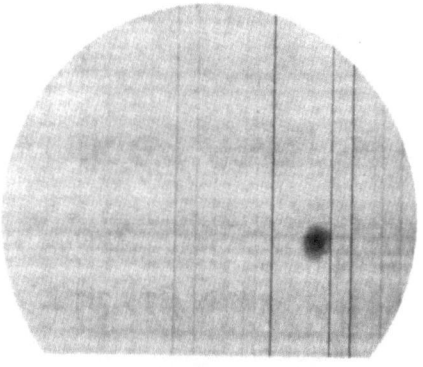

CONTINUUM
SPECTROHELIOGRAM

⊢⎯⊣
20 arc sec

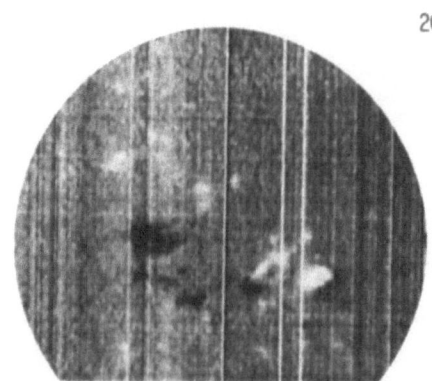

MAGNETIC FIELDS BY
PHOTOGRAPHIC SUBTRACTION

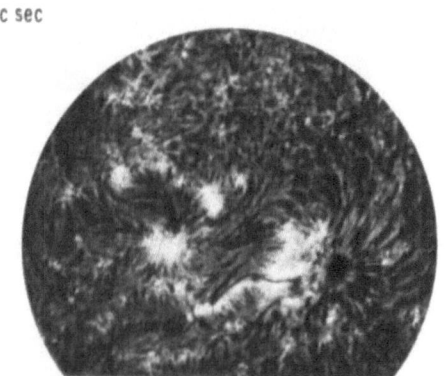

Hα FILTERGRAM

Fig. 3.   First real time display of magnetic fields on July 17, 1970 at the Aerospace San Fernando
Observatory. 40 s were required for 500 subtractions. Other pictures are for
comparison and confirmation.

connected to a digital-to-analog convertor and fed to two TV monitors which are synchronized to the disk. The operator can use a light pen on the monitor to select points or areas on which he wants more information, such as the numerical value of a specific point or position of the point. For areas, the operator can define rectangles or circles of various sizes in which he wants average values, maximum value, or minimum value.

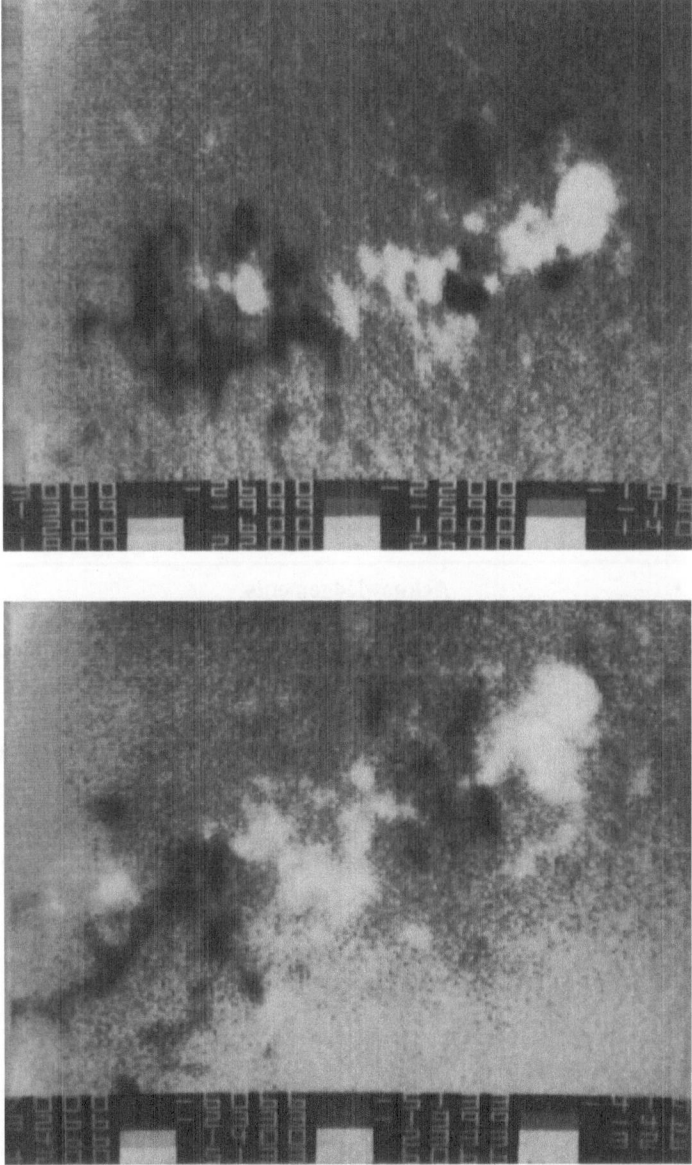

Fig. 4. Videomagnetograms showing the development of an active region during three days from October 19 to 22, 1970. The numbers next to the density bars are not calibrated in Gauss.

The display section can also be used for graphic and alphanumeric display of the data, ranging from intensity plots on arbitrary cross sections of the picture, to distribution plots of the entire frame.

The analytic section consists of a Varian 620i computer and forty-eight tracks on the disc.

This data storage area is used as an intermediate storage to free the data gathering area. From this area, the data can be transferred to permanent storage on tape. It can also be used to hold image correction data which the computer could use to correct the raw data for imperfections or shading in the TV camera. The Varian computer controls the system and is available for further analysis of the data either in real time or later. The digital equipment is shown in Figure 2.

3. Results and Present Status

The initial results from the Videomagnetograph are shown in Figure 3. The video-magnetogram shown is a polaroid picture of the display monitor. Several days later the magnetic fields from photographic subtraction were available for confirmation of the fields and their polarities.

The initial result is very noisy. Later and somewhat less noisy magnetograms made later do not have matching photographic magnetograms for comparison. Improvements in the video preamplifier is expected to greatly improve the sensitivity but this and other improvements are not yet complete. The present effort is mostly directed toward greater reliability, higher sensitivity and calibration.

Acknowledgements

The authors wish to express their appreciation to E. B. Mayfield, G. A. Paulikas, and R. A. Becker of the Aerospace Corporation for their continued support of this project, to Greg Kozlowski of the Aerospace Corporation for technical support in all areas, to Donald Robbins of NASA MSFC for his continual collaboration, to Dave Rutland of Spatial Data for design and construction of the high speed digital electronics, to Dale Vrabec of the Aerospace Corporation for many helpful discussions, to Gary Chapman of the Aerospace Corporation for taking simultaneous photographic magnetograms and to Mary Gates of the Aerospace Corporation for her aid in assembly of reports.

Discussion

Deubner: What was the exposure time for the photographic magnetogram, as compared with that for the video-magnetogram taking 40 s?

Janssens: The scan time for the photographic magnetic exposure was two minutes.

THE KITT PEAK MAGNETOGRAPH

IV: *40-Channel Probe and the Detection of Weak Photospheric Fields*

W. LIVINGSTON and J. HARVEY

Kitt Peak National Observatory*, Tucson, Arizona, U.S.A.

Abstract. The 40-channel magnetograph is described. Special attention is given to details of the fiber-optic probe which dissects the Fraunhofer line image and provides for a choice of spectral and spatial resolution. Examples of magnetograms taken with the instrument are shown. Maps of quiet areas, made with long integration times to achieve a noise of 0.4 G, yet with 5 arc-sec resolution, indicate the presence everywhere of a minimum background field of average strength 2–3 G.

1. Introduction

We describe here a new instrument of the Babcock-type, intended to achieve high resolution in magnetic observations without a sacrifice of limiting sensitivity. The number of spatial elements that are simultaneously observed has been increased 40-fold. Among other problems, the instrument is being applied to the study of the fine structure of weak fields.

The resolution realized with the photoelectric magnetograph depends on the size of the exploring aperture, the seeing, and the signal/noise ratio (which in turn is proportional to the square root of the light flux). As the aperture is made small, the order of one arc s or less, the flux is reduced and to maintain a given S/N the scan speed must be slower, a circumstance unfavorable for good resolution because of atmospheric seeing. Beckers (1968) has reviewed these conditions and concluded that for high resolution, photographic techniques are superior to single-channel photoelectric magnetographs. We agree with Beckers but have sought to improve our resolution by expanding our magnetograph from the single aperture to a linear array of 40, thus creating something like a photoelectric spectroheliograph.

2. The Fiber-Optic Probe

Figure 1 diagrams the essential optical components as positioned on the spectrograph. In front of the entrance slit is the analyzer for circularly polarized light; a KDP crystal (ISOMET, Inc.) having a clear aperture of 3×50 mm followed by a polaroid. The spectrograph slit is set in width to 1, $\frac{1}{2}$, or $\frac{1}{4}$ mm corresponding to approximately 2, 1, or $\frac{1}{2}$ arc s on the incident solar image.

In the camera plane of the 13.5 m Czerny-Turner spectrograph (FL collimator = FL camera) are two fiber-optic probes which serve as double exit slits along the direction of dispersion. One of the probes is equivalent to a rather broad set of exit slits (0.175 – 0.150 – 0.175 Å in our grating 5th order) and provides a match to the

---

* Operated by the Association of Universities for Research in Astronomy, Inc., under contract with the National Science Foundation.

Howard (ed.), Solar Magnetic Fields, 51–61. All Rights Reserved.

wings of the strong line of Fe I $\lambda 5233$. Harvey and Livingston (1970) have calculated the magnetograph response signal resulting from this line and slit arrangement to be linear ($\pm 1\%$) up to fields of 4000 G (paper II). Fortunately, the wings of Fe I $\lambda 5233$ are relatively insensitive to temperature changes so this line and probe provide wide dynamic range and, we believe, an accurate indication of the line of sight field.

The second probe is appropriate for a much narrower line such as Fe I $\lambda 5250$

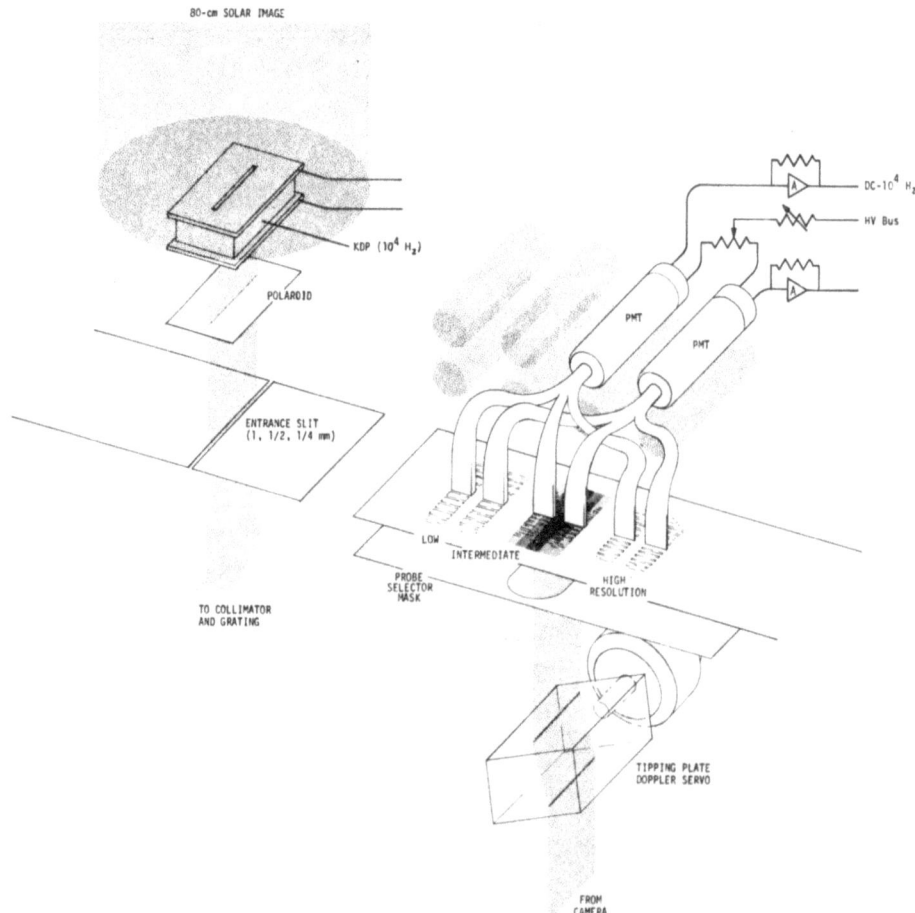

Fig. 1.   Arrangement on the spectrograph. Not shown is a second probe with parallel light feed to the PMTs. The tipping plate Doppler compensating servo is controlled by summed signals from all 40 (20) channels.

($0.067 - 0.028 - 0.067$ Å). For the detection of weak fields this line and probe provide a S/N about twice that obtained with Fe I $\lambda 5233$, but there is some uncertainty in the measurements because of the variable profile of Fe I $\lambda 5250$ (Chapman and Sheeley, 1968; Harvey and Livingston, 1969).

Each of the above spectral probes is subdivided into three sizes of spatial sections,

Fig. 2.   Pre-assembly photo of the fiber optic probe unit. The knob with chain drive to a counter translates the Fe I 5233 probe with respect to Fe I 5250. The right hand control moves both probes together over the 'probe selector mask'.

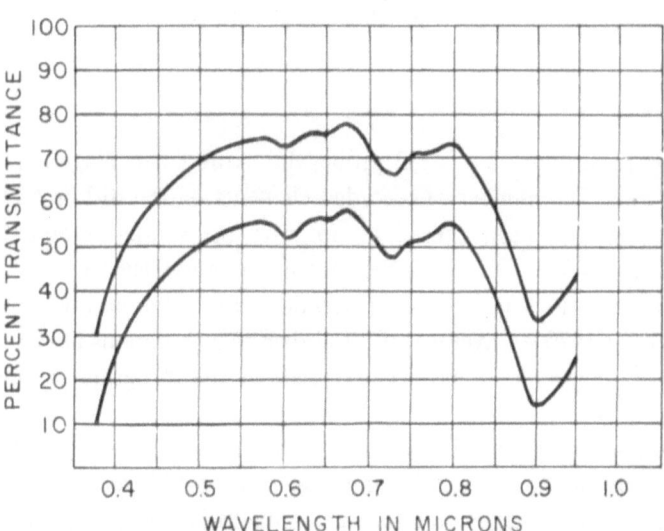

Fig. 3.   Transmission of 30 cm length of plastic fiber (length employed ≤ 30 cm). Upper curve assumed no end losses, *i.e.* optical contact to source-receiver. Lower curve includes end losses.

perpendicular to dispersion, providing 2, 1, and $\frac{1}{2}$ arc s resolution. Figure 1 indicates the forty channel parallel feed for the $\lambda 5233$ probe. There are a total of six possible groups of fibers converging on each PMT. The group in use is controlled by the probe selector mask. Additionally, there is provision for masking off half the groups of a given resolution, translating the $\lambda 5233$ probe sideways into spatial alignment with the $\lambda 5250$ probe, and thus obtain simultaneous 2 spectral channel and 20 spatial channel measurements. Figure 2 shows the probes, which were custom manufactured (Poly-Optics, Inc.). They are made from a plastic material having the transmission given in Figure 3. This transmission, combined with the S-10 spectral response of our inexpensive photomultiplier tubes (EMI 9524C), limit spectral coverage from about 4500 to 6000 Å.

The anode signal from each PMT is passed through a miniature preamplifier (bandwidth DC-10^4 Hz) located on the tube socket, then routed over a 10 m cable to an electronics rack. The DC components from each pair of tubes are converted to brightness and velocity signals. The AC component is phase-detected at 10^4 Hz to produce the magnetic signals. These three signals are sequentially sampled, converted to digital information, and recorded on magnetic tape. This data tape is used to prepare cathode-ray tube pictures, contour maps or other representations of the mapped fields. Further details concerning the instrument as a polarimeter (paper I), data reduction, calibration procedures (II), and display techniques (III) are given in papers by Livingston *et al.*, 1970.

3. Preliminary Results on Solar Fields

Two kinds of observations are being taken with the 40-channel system: Full Disk Maps of low resolution which are taken on a semi-routine basis, and Area Scans which cover smaller regions for special projects with possibly higher resolution and sensitivity.

FULL DISK MAPS

Using the low resolution Fe I $\lambda 5233$ probe, the entire solar image is mapped in 21 swaths in 40 min. The number of recorded elements on the disk is 6 (10^5). A time constant of 33 ms has been chosen to optimize the spatial resolution. The line-of-sight field strength measured ranges from the noise level (~ 15 G) to 3000 G. The spatial resolution is most often determined by image motion or 'seeing'. Beginning February 1970 an average of 3 full disk magnetograms have been obtained each month. We expect to continue and enlarge this program as telescope time allows.

A great utility to these full disk records is that the region of interest need not be anticipated in advance. For example, on the record for 7 March, 1970 (Figure 4) the rather insignificant region just off the far south meridian has proven interesting in the interpretation of nearly simultaneous X-ray pictures (Van Speybroeck *et al.*, 1970).

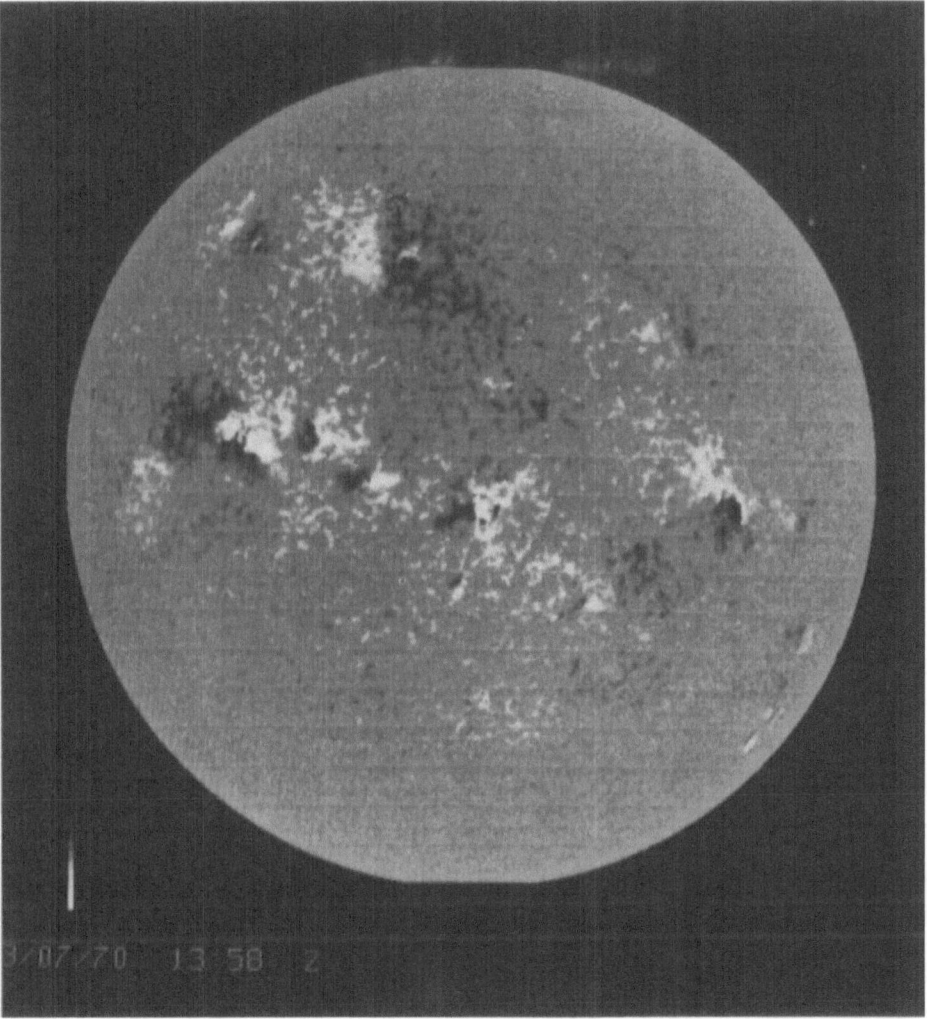

Fig. 4.   Full disk magnetogram of 7 March, 1970 taken in Fe I 5233. Fields of positive polarity are
indicated dark, negative light.

AREA SCANS

In this observing mode the image is step scanned. For each step the image is held
fixed while the computer samples and integrates the signals from the 40 channels.
After this integration period, preselected from 0ˢ015 to 16ˢ0, the image is stepped one
resolution element and the process repeated.

Figures 5 through 7 afford a comparison of the full disk and area scan results.
Figure 5 is a full disk brightness map made in Fe I λ5250 under relatively poor seeing.
Figure 6 is an area scan of the region west of the central meridian taken the following
morning under good seeing conditions. The low resolution probe was employed for

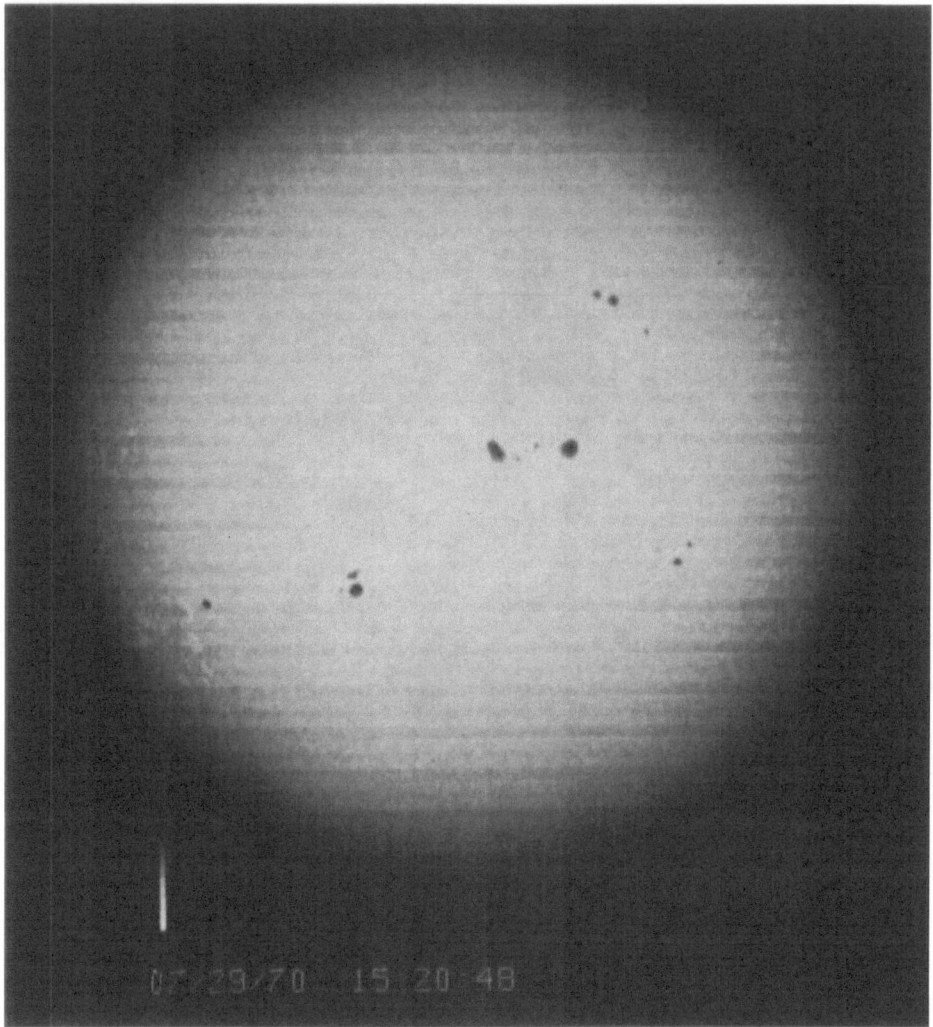

Fig. 5. Brightness record for the full disk map of 29 July, 1970 at 15:20 local time. North is top, West is right. Resolution is seeing limited.

both the observations but while the full disk observation is seeing limited, the area scan record is probe limited. Finally, Figure 7 is a four-swath scan of this same region with the $\frac{1}{2}$ arc s probe demonstrating an achieved resolution comparable to that obtained with photographic methods.

4. Measurements of Weak Fields

Compared with photographic techniques the 40-channel system can detect much weaker fields. Fe I $\lambda 5250$ area scans made with a long integration time of 16 sec yield

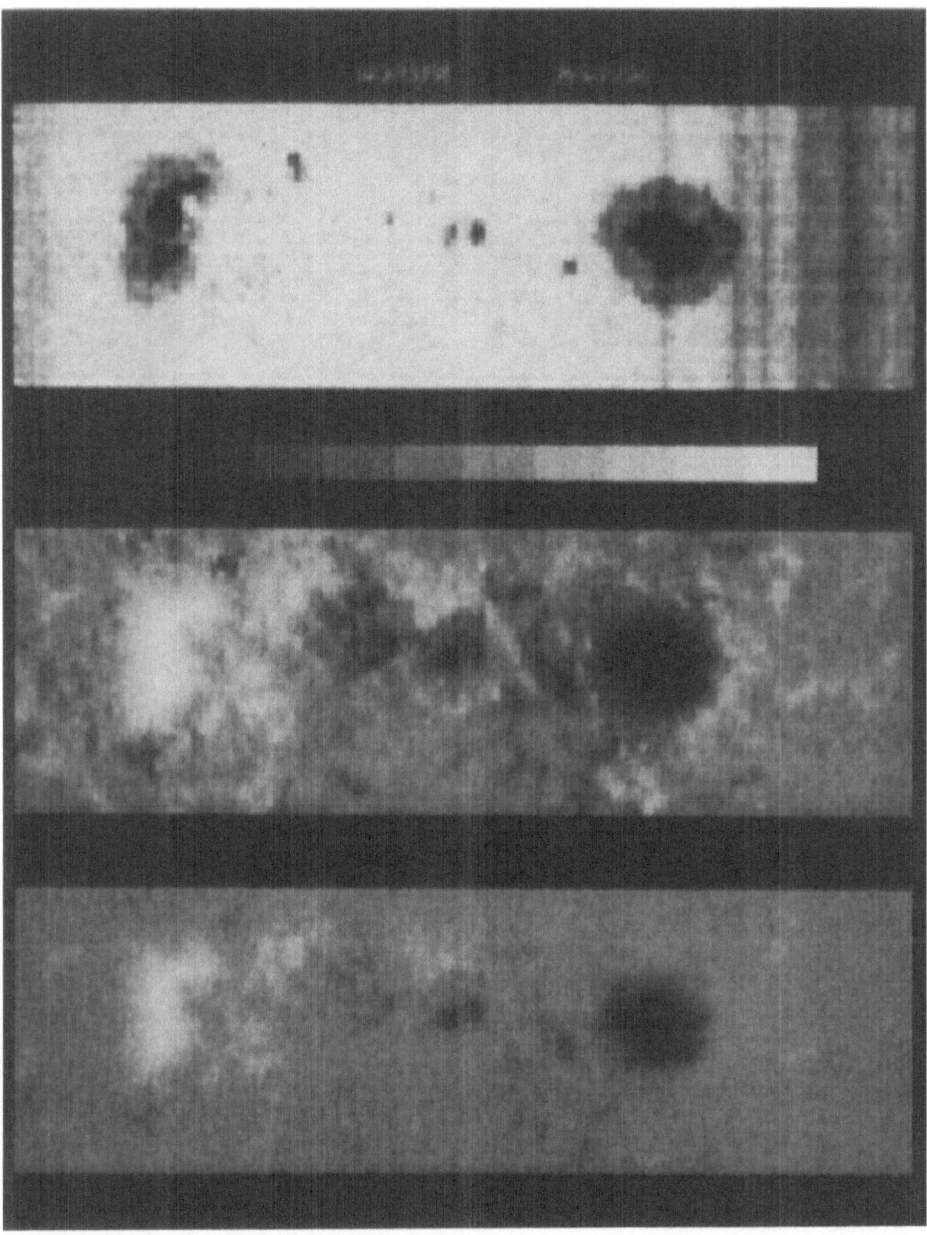

Fig. 6. Area scan of the region near the center of Figure 5 taken on 30 July, 1970 at 07:35 local time under good seeing conditions. Resolution is limited by the 2 arc s probe. Top frame is the brightness; a thin cloud produced the streaks near the right edge. In the bottom frame the departure from grey is proportional to the field strength and in the middle frame to the square root of the field strength to emphasize weak fields.

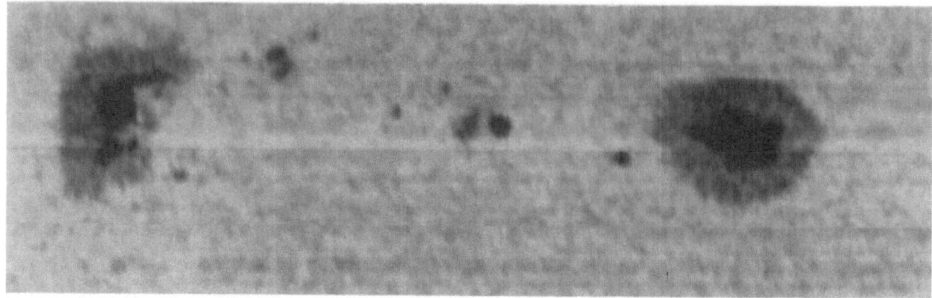

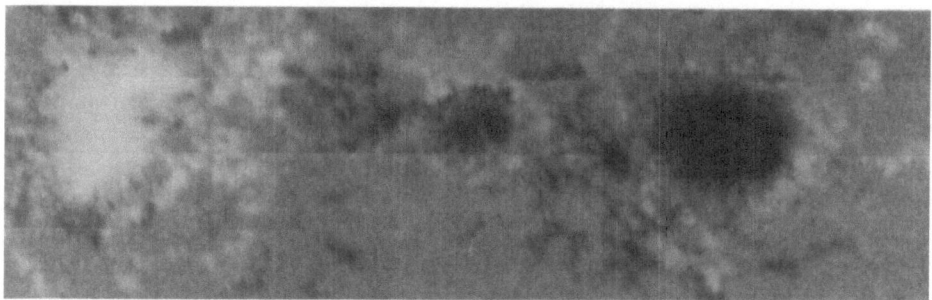

Fig. 7.   Same as Figure 6 except with $\frac{1}{2}$ arc s probe. Now the spatial resolution
is again seeing limited.

maps of good resoution ($\sim$5 arc s) yet with a noise of about 0.4 G. The combination
of good resolution and high sensitivity permits us to discriminate between the well
known small regions with strong magnetic fields and possible large-scale areas with
weak magnetic fields. Preliminary observations indicate that between the strong
fields a *weak field is found everywhere*. Assuming that the normal disk profile of Fe I
λ5250 holds for these weak field regions, the average field strength is 2–3 G.

Evidence for the weak fields is contained in Figures 8–10. We have plotted the isogauss contours for fields stronger than 20 G which are concentrated in small areas. Below 20 G only the polarity is displayed as light and dark shading. The pattern of data elements is visible in this shading. Except for a very local coherence extending over about 5 arc s caused by the seeing, the data sample elements are independent. The areas of like polarity are seen invariably to extend over many data elements. We take this coherence of polarity over extended areas as evidence for the weak but

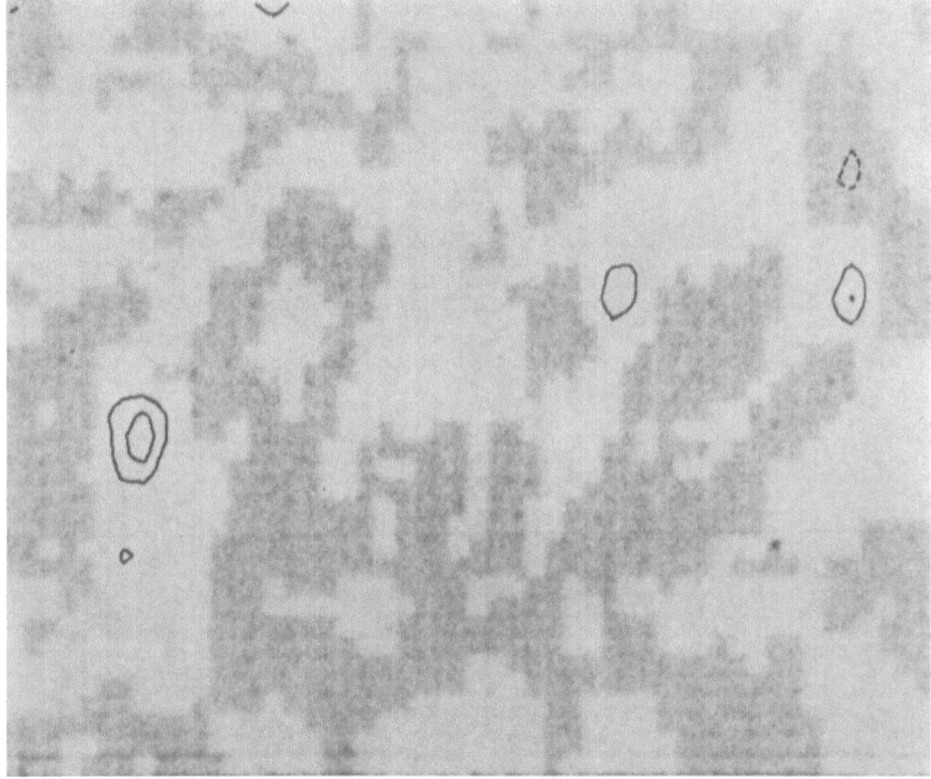

Fig. 8. Area scan with 16 s time constant, Fe I 5250, 2 arc s probe, of high latitude quiet region. Contour intervals are 20, 40,... G. Noise ~ 0.4 G. Shading indicates polarity of weaker, non-contoured fields. Contiguous character of these polarity indicators is evidence for a large scale weak field.

significant background field. Figure 10 is a recording identical in every way to the preceding except taken with the KDP off. The pattern produced is essentially random. However, it is well to note that this does not prove that the pattern seen when the modulation is on is actually due to a weak background field. Whenever one measures weak magnetic fields with a Babcock magnetograph it is important to make sure that there is no instrumental elliptical polarization. When such polarization is present, a small Doppler shift can masquerade as a weak magnetic field. Since we can only

W. LIVINGSTON AND J. HARVEY

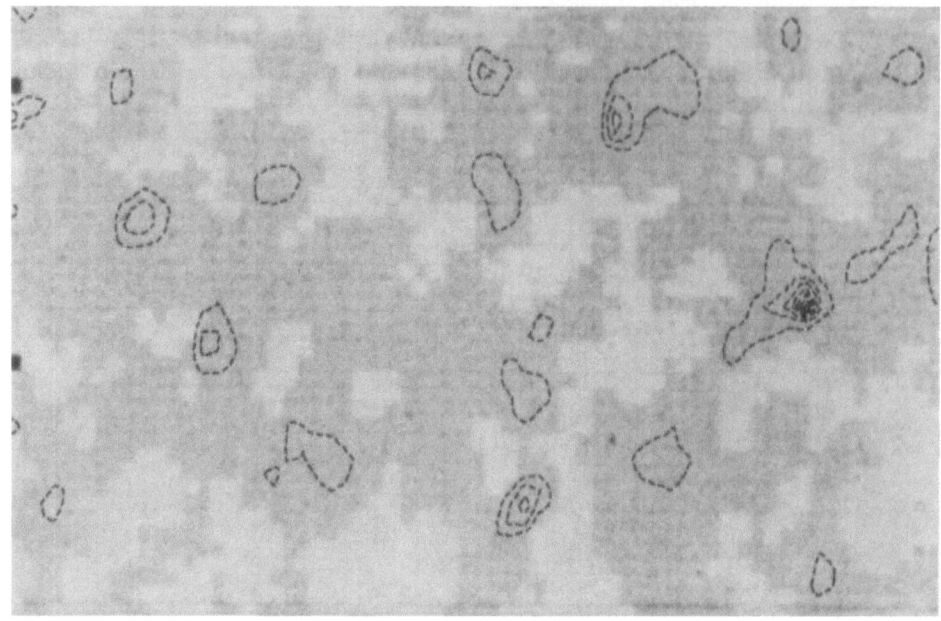

Fig. 9. Similar to Figure 8 but in opposite hemisphere.

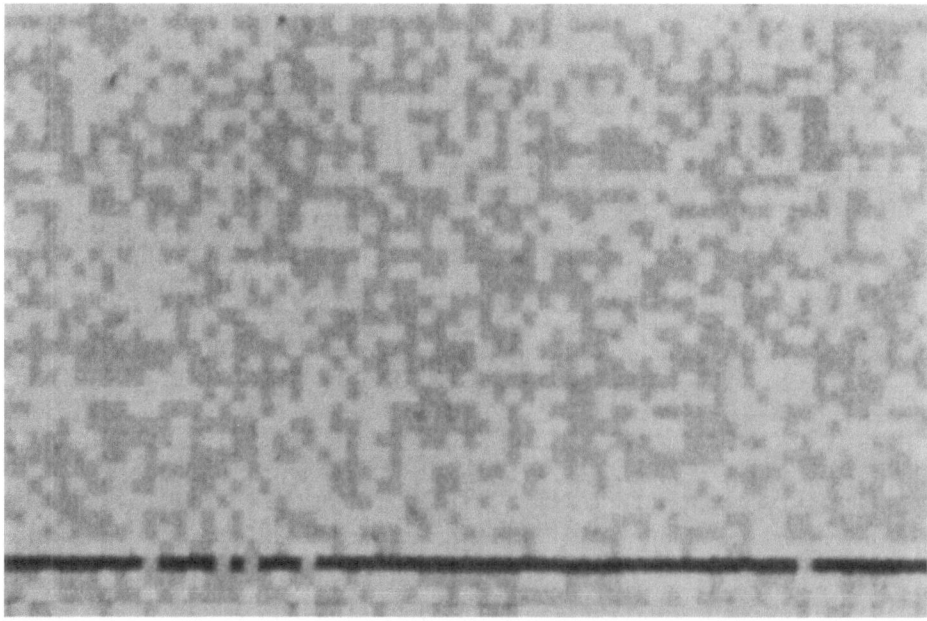

Fig. 10. Same as Figure 9 except with KDP off, showing random polarity pattern for comparison. One channel was erratic and has been deleted from the diagram.

correct the average Doppler shift of the 40 spatial channels it is possible that the weak field we see is just the pattern of velocity fields present. Further observations are in progress.

References

Beckers, J. M.: 1968, *Solar Phys.* **5**, 15.
Chapman, G. A. and Sheeley, N. R. Jr.: 1968, *Solar Phys.* **5**, 442.
Harvey, J. and Livingston, W.: 1969, *Solar Phys.* **10**, 283.
Livingston, W. and Harvey, J.: 1971, (paper I) submitted to *Solar Phys.*
Harvey, J. and Livingston, W.: 1971, (paper II) submitted to *Solar Phys.*
Livingston, W., Harvey, J., and Slaughter, C.: 1970, (paper III) IAU Colloquium No. 11, *Automation in Optical Astrophysics*, Edinburgh, 12–14 August, 1970.
Van Speybroeck, L., Krieger, A., and Vaiana, G.: 1970, *Nature* **227**, 818.

Discussion

N. R. Sheeley, Jr.: This is going to be a dirty question. Would you care to comment on the existence of magnetic quanta?

W. Livingston: It seems likely we are observing three rather different kinds of fields on the Sun: sunspot associated fields, strong but non-sunspot small-scale fields (your 'gaps'), and now these diffuse very weak fields. Any notion of a single-valued or quantized characteristic would apply only to the gap fields. But I would rather not expand on this here.

G. W. Simon: What are the typical field strengths of the contoured strong fields when you integrate for 16^s and find a weak background field 2–3 G? (Figures 8–10).

W. Livingston: They range 20 to 140 G in the sample studied. One must realize that the achieved resolution in these long integration observations is not nearly as good as that shown in Figures 6–7 because of seeing, so I would avoid a quantitative interpretation. Presumably they belong to the "magnetic knot" or gap category.

A PRESSURE SCANNING FABRY-PEROT MAGNETOMETER*

T. D. FAY and A. A. WYLLER

Bartol Research Foundation of The Franklin Institute, Swarthmore, Pennsylvania, U.S.A.

Abstract. An oscillating magnetic analyzer (KDP crystal plus Glan-Thompson prism) is coupled to an echelle-interferometer spectrograph. The single slit magnetometer by pressure variations can be made to scan the entire profiles of the circularly and linearly polarized Zeeman components. Freon gas is used as the scanner gas with wavelength displacements of 0.02 Å per 0.1 in. Hg pressure change at the NaD lines. The available scan range is 15 Å in the visual spectral region.

In order to understand more fully the operation of our magnetometer, it is necessary to describe in some detail the construction of the spectrometer to which the magnetometer is attached. The Bartol spectrometer has been planned with the view in mind to provide a completely self-contained, mobile, high resolution spectrometer equivalent in performance to several conventional coudé spectrograph camera assemblies with a variety of spectral resolutions and wavelength coverages. Briefly speaking the design goals have been the full utilization of the light of solar surface features or stellar images (diam 2″–4″) anywhere in the wavelength range 4000 to 12000 Å with spectral resolutions continuously variable from 10^3 to 10^6 and dual channel pulsecounting facilities.

The integrated Fabry-Perot spectrometer which has been developed at Bartol to meet these demands consists of four subassemblies: (a) a Fabry-Perot interferometer; (b) a predispersing echelle spectrograph; (c) a combined pressure chamber and pressure control system; (d) a dual channel pulse counting system.

The servo-controlled Fabry-Perot interferometer of Ramsay's design uniquely provides the required versatility in spectral resolution and wavelength coverage with one and the same instrument. Two pairs of interferometer plates cover the spectral range $\lambda\lambda 4000$–12000 with an effective finesse of 30. The plate separations can be varied continuously from one tenth to one hundred millimeters and the plates are maintained parallel to $\lambda/80$ by electronic servo controls. The geometrical plate separation is maintained to an accuracy of 10 Å also by servo controls.

The Fabry-Perot interferometer necessitates a predispersing unit to filter out adjacent passbands. To this end a Hilger-Engis monochromator (Model 600) has been modified at the factory to permit installation of a Bausch and Lomb echelle grating (300 lines/mm, blazed at 63°26′). This eliminates the need for grating changes in covering the above broad wavelength interval. Overlapping orders are separated by suitable broadband interference filters.

The pressure chamber (cylindrical with 2 ft diam and 4 ft length) is mounted on a heavy table, which rolls easily but can be firmly jacked up from the floor for steady coupling by laser means to a given telescope coudé configuration. The installation of remote controls (externally on the pressure chamber) for the parallelism adjustment

* Work supported by NASA Grant NGR-39-005-066.

of the interferometer plates and the tilt variations of the grating have been carried out. This eliminates the need for opening the pressure chamber every time readjustments of the plates and changes to other wavelengths or spectral resolutions are required. A system of 6 entrance and exit pinholes (2, 0.4, 0.2, 0.1, 0.05 and 0.025 mm) movable by micrometer screws have also been installed and are remotely controlled.

The pressure generator and controller has been acquired from the Texas Instrument Co. Its unique precision in control and setting (to a few hundredths of a mm Hg) enables one to use freon gas with five times more sensitive variations in the index of refraction to pressure variations than air, which has been most commonly used in astronomical applications. The pressure chamber has been designed for pressure scanning over 4 atm, which yields in the visual a spectral scan width of about 20 Å. The servo-controlled pressure generator does not involve continuous leakage but enables one to remain at a given pressure level and pulse count to the required precision. At the NaD-line a pressure change of 0.1 in. Hg moves the spectrometer passband 0.02 Å.

Pulse counting facilities have been developed for counting at low light levels (10–100 counts per second). The EMI9529 photomultiplier tube records the intensity of the line element under consideration and is dry-ice cooled. The echelle spectrograph, by the use of a beam splitter, permits one to monitor a 50 Å wide continuum strip inside the pressure chamber while having it centered on the line feature wherever in the spectrum one may wish to operate. This second channel serves as a sunspot guiding monitor. A third broadband channel outside the entrance pinholes is utilized for the monitoring of transparency changes.

The Bartol magnetometer is a one slit system which has the advantage of simplicity and reduces the difficulty of the 'Doppler error' in multi-slit-systems. The penalty for this choice of system lies in the problem of eliminating the instrumental polarization, which however seems possible to overcome in simple mirror-lens systems.

Furthermore, the double-slit method only eliminates circular polarization. If the linear polarization component is to be measured, elimination techniques similar to those for the one-slit-system have also to be employed, which can be achieved either through a phase compensator or by a direct measurement of the instrumental linear polarization. Since the Bartol system uses a 15-in. siderostat (diaphragmed down to 4 in.) with a single plane mirror feeding the objective lens-doublet, the instrumental polarization characteristics are rather simpler than in the multi-mirror systems of tower telescopes or conventional stellar coudé-configurations.

An unusual advantage of the present spectrometer, which has overridingly dictated the choice of a single slit-system, is the capability it gives for pressure scanning through the entire profiles of all the Zeeman components. This will yield information on the variation of the magnetic field with optical depth, i.e. the magnetic field gradients, which are so important for the construction of realistic spot models. This is very difficult if not impossible with a two-slit system which requires perfect symmetry about line center in the placement of the two slits. In our simple system with a bandpass of 0.02 Å for the Fabry-Perot echelle combination we typically record for the spot continuum 1000 counts/s with the magnetic analyzer in position and an entrance aperture of 2 s of arc (50μ entrance and exit pinhole apertures).

The magnetic analyzer consists of a KD*P (KD_2PO_4) crystal with 80% transmission over the wavelength range 4000 to 13000 Å and a half-wave voltage of 4.0 kV at $\lambda 5000$. It thus can act as a quarter-wave or half-wave plate over a wide wavelength region. It is followed by a Glan-Thompson prism, which acts as an analyzer at 45° to the KD*P plate. The complete magnetometer transmits about 25% of the sunspot beam intensity. Both are placed immediately in front of the spectrograph optics, just behind the entrance pinhole, to eliminate the instrumental polarization due to the spectrograph itself.

For the measurement of circularly polarized components the voltage to the KDP crystal will oscillate at a frequency of about 15 c/s to minimize the effects of transparency fluctuations. The difference count between lefthand and righthand circularly polarized light will be recorded at each wavelength passband.

In order to measure the linearly polarized light, the Glan-Thompson prism will rotate also at about 15 c/s and the resulting ac-signal will be recorded by a pulse-counter. Under consideration are also broad and narrow band depolarizers to eliminate residual polarizations in the glass surfaces of the photomultiplier tubes.

A complete description of the general instrument will appear in *Applied Optics*.

Discussion

Foukal: What is the effective finesse of the Fabry-Perot alone, without stepping the echelle?
    Wyller: It is 25 with up to 20% variation over the wavelength region $\lambda\lambda$ 4000–6600.

SACRAMENTO PEAK MAGNETOGRAPH

RICHARD B. DUNN

*Sacramento Peak Observatory, Air Force Cambridge Research Laboratories,
Sunspot, New Mexico, U.S.A.*

Abstract. The Sac Peak magnetograph (DZA) has been modified from Evans' original scheme so that it measures the displacement of the right and left hand circularly polarized lines separately. The computer reduction calculates the Zeeman and radial velocity signals. A grating servo system has been added to correct for slow temperature drifts in the spectrograph. A paper-tape reader controls the raster scan and the formatting of data on to magnetic tape.

The Sacramento Peak Magnetograph has been briefly described by Evans (1966) in his summary of Solar Magnetographs. He refers to this device as the DZA, which is an abbreviation for Doppler-Zeeman Analyzer. The DZA contains servos that continually re-center the spectral lines on analyzing slits of fixed separation. When operating on strongly split lines, it overcomes the saturation problems of a Babcock magnetograph. The DZA operates as a longitudinal magnetograph in the presence of both σ and π components unless the components are completely separated, in which case it measures total field.

This paper describes a number of significant modifications to the instrument and

COMPACT SPECTROGRAPH

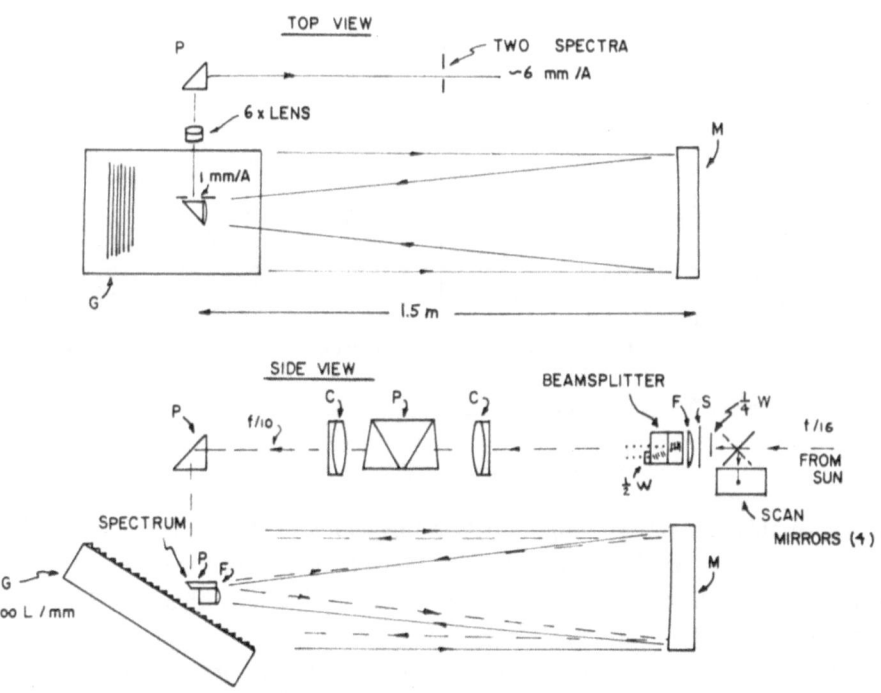

Fig. 1.   Compact spectrograph.

Howard (ed.), Solar Magnetic Fields, 65–70. All Rights Reserved.
Copyright © 1971 by the IAU.

the addition of a data collection system. Utilizing a compact spectrograph designed by himself, and shown schematically in Figure 1, Evans formed two spectra simultaneously, one above the other. The top one is analyzed for the right-handed circularly polarized component, the lower for the left-hand component. The analyzer consists of a $\frac{1}{4}$ wave plate in front of the spectrograph slit and a polarizing beamsplitter immediately after the slit. The beamsplitter is made from two pieces of cleaved calcite with a $\frac{1}{2}$ wave plate in between to equalize the optical path length between the two images. Another $\frac{1}{2}$ wave plate cemented to one-half of the exit face of the beamsplitter rotates the plane of polarization of one of the beams so that it coincides with the other and so that they both coincide with the azimuth of highest reflectance of the grating.

There is no chopping between the two circularly polarized components because there is no KD*P crystal in the analyzer. The DZA simply measures the displacement of the spectral line relative to its average position by balancing the light passing through a slit on each side of the line as shown in Figure 2. An unbalance causes an error signal

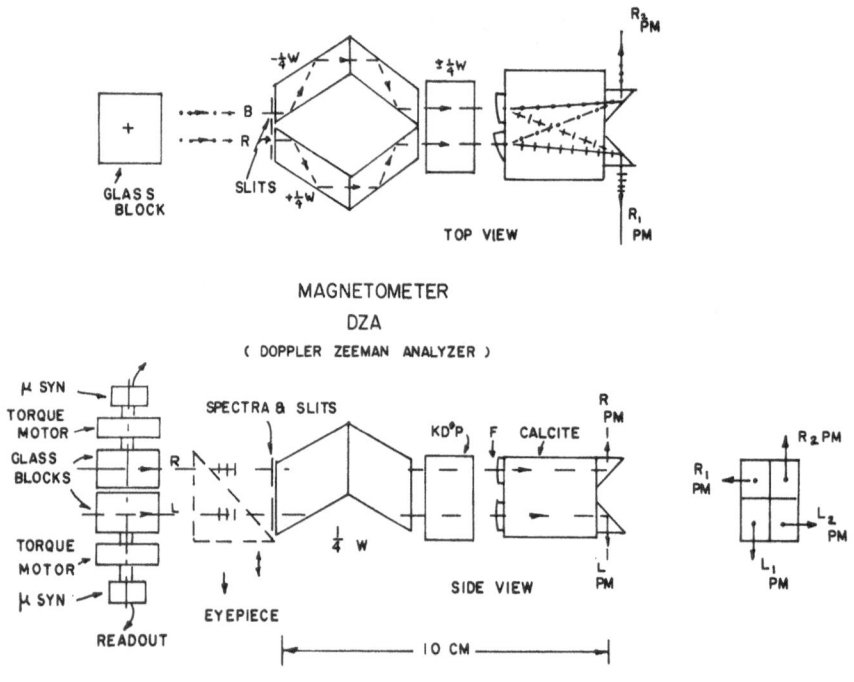

Fig. 2.   Sacramento Peak magnetograph.

that is amplified in a servo loop to drive a torque motor that, in turn, rotates a cube of glass until the signal is nulled. The amount of rotation of the cube, and thus the displacement of the line, is measured by a small transformer with a rotating armature called a 'Microsyn'.

The $\frac{1}{4}$ wave plate, KD*P crystal and calcite polarizer are simply a convenient means devised by Evans for chopping between the two slits to develop the error signal for the servo system. The two spectra are already polarized by the analyzer at the slit of the spectrograph. The chopping is accomplished by adding a fixed plus quarter wave to one slit and a minus quarter wave to the other. Evans used achromatic wave plates consisting of Fresnel rhombs. During one portion of the chop the KD*P adds a plus $\frac{1}{4}$ wave to the light to linearly polarize it again. The subsequent calcite beamsplitter sends the light from the red slit to the R_1 photomultiplier and the light from the blue slit to the R_2 photomultiplier. On the other portion of the chop the KD*P crystal adds a minus $\frac{1}{4}$ wave and the light from the blue slit is sent to the R_1 photomultiplier and that from the red slit is sent to the R_2 photomultiplier. All the light is utilized in this arrangement. The beamsplitters are made from calcite and the surfaces are coated so that the loss of light due to absorption is small. The signals from R_1 and R_2 are added to form the R error signal. An identical arrangement is used in the L channel.

The original DZA utilized two glass cubes in series so that the Microsyns measured line displacements proportional to Zeeman and Doppler signals instead of to R and L. In adapting the DZA to digital recording we decided to eliminate this feature in order to allow corrections to be made by the computer to the individual R and L

MAGNETOGRAPH   ELECTRONICS

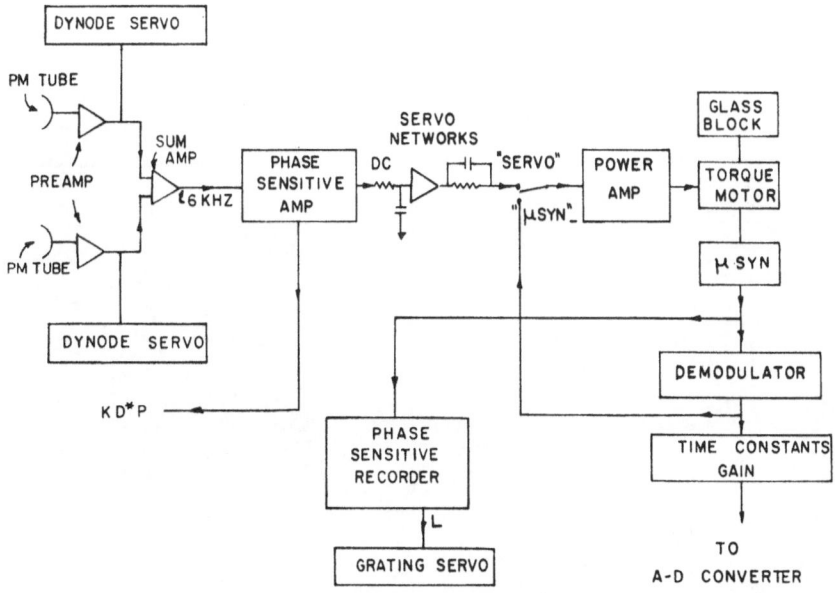

Fig. 3.   Magnetograph electronics.

channels. The computer then adds and substracts the R and L Microsyn signals to form the Doppler and Zeeman signals.

The block diagram for the electronics is shown in Figure 3. The output of the photomultiplier preamps is summed in an operational amplifier before it is phase-detected by a phase-sensitive amplifier. The output of the phase-sensitive amplifier is compensated to correct the frequency response of the servo system and drives the torque motor that rotates the glass block. The output of the Microsyn is demodulated and sent to a digital tape system consisting of a multiplexer, an analog-to-digital converter, a tape format controller and a 7-track tape drive. An auxiliary analog recorder traces the R and L signals and by means of a circuit that adds and subtracts the R and L signals, traces the Zeeman and Doppler signals.

The dynodes of the photomultiplier tubes are servoed to hold the DC output of each photomultiplier constant so that the servos operate at constant gain even while measuring the fields of sunspots.

Two of the four dynode signals, the coordinates on the Sun, and the R and L signals are multiplexed and recorded on the magnetic tape.

The format of the scan is predetermined by a paper tape and machine-tool controller that programs Right Ascension and Declination stepping motors to drive a scanner in front of the slit. One step corresponds to 0.3 s of arc. Auxiliary control characters on the paper tape cause the clock, record and end-of-file gaps, and the information set manually on digital switches to be recorded on the magnetic tape.

In order to compensate for thermal drifts, an additional servo very slowly rotates the grating to keep the spectral lines centered on the DZA. It derives its signal from the recorder output.

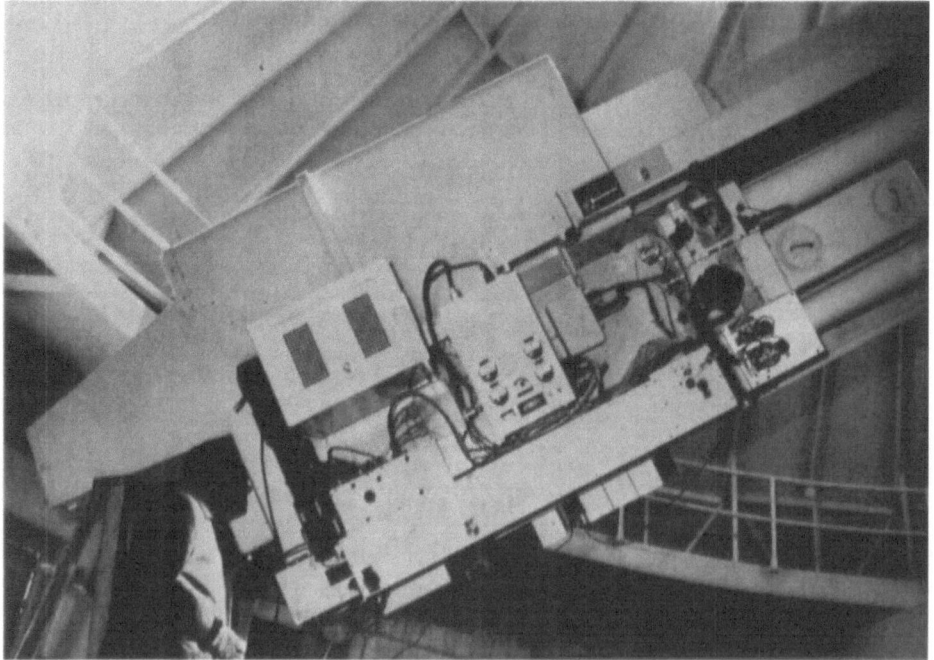

Fig. 4.   Photograph of magnetograph. The DZA is on top of the compact spectrograph. Light comes into the system from the right hand side. The scanner is immediately to the right of the compact spectrograph. A small rack mounted above the DZA contains the power supplies for the stepping motors. To the right is a second box of electronics containing the dynode servo systems.

The system has been used to record data for several thousand scan series. The tape-recorded records are analyzed in a digital computer. Currently it is possible to produce three (3) contoured maps – of magnetic field, velocity field and brightness – in about two minutes. We record the maps on microfilm or on drawing paper for detailed analysis. Because the instrument measures only relative displacements of the L and R components of the line, a substantial portion of the reduction process involves the location and removal of the relative zeros of magnetic and velocity field. Magnetic field measurements have been restricted mainly to active regions, since it is possible, in that case, to assume that the regions of spatially smooth field surrounding the steep and irregular fields of active centers are regions of approximately zero field. That is, we find the zero by fitting the flat field regions – not by equating mean measured field with zero field.

The DZA is still undergoing changes. At present, we are adding tachometer feedbacks to the torque motor servos to improve the low frequency response and we are changing the 1P21 photomultiplier tubes to an extended red-sensitive tube. We are in the final stages of installation of a data-link between the DZA and our medium-size XDS computer. We want to send the data directly to the computer during observations because data can be recorded more rapidly that way than with

Fig. 5. Photograph of magnetograph electronics and data collection system. The phase sensitive amplifiers are in the left rack together with servo control electronics. The center rack contains the machine-control system with the paper-tape reader and servo power amplifiers. The third rack contains the digital data collection system including the clock, multiplexer, analog-to-digital converter, controller, and magnetic tape unit.

the currently-used incremental tape recorder. Also, we will be able to produce nearly real-time maps of the velocity, magnetic and brightness fields being observed. The maps will be drawn on a storage-type oscilloscope and transmitted via television to the observer at the telescope. We feel that this feedback to the observer will significantly improve the quality of the data obtained.

Photographs of the instrument and its electronics are shown in Figure 4 and 5. The DZA presently operates with a 40 cm telescope. Maps made from the instrument will be shown in other papers presented by Sac Peak staff at this meeting.

Reference

Evans, J. W.: 1966, *Atti Del Convegno Sui Campi Magnetici Solari,* (ed. by M. Cimino), G. Barbèra, Firenze.

SYSTEMATIC ERRORS OF THE CRIMEAN VECTOR
MAGNETOGRAPH

V. A. KOTOV

Crimean Astrophysical Observatory, Crimea, U.S.S.R.

Abstract. The influence of some 'effective' asymmetry (miscentering) of spectral line on the accuracy of the complete field vector is considered for the measurements with the Crimean magnetograph. This miscentering appears in the device for $H_\perp$ records if the longitudinal field is strong.

In several papers (Kuznecov *et al.*, 1966; Beckers, 1968) it has been pointed out that the transversal field-mode of magnetographs similar to the Crimean one possesses a systematic error connected with the influence of the longitudinal field on the line of sight velocity signal. Namely when the phase-difference produced by the *ADP* crystal vanishes, the electro-optic light modulator (EOM) works as an analyser of circular-polarized light. Recently Wiehr (1969) has considered this effect for the Locarno magnetograph and for one of the modes of the Crimean magnetograph (when the optical axes of $\lambda/4$-plate and *ADP* coincide). But the detailed analysis of possible errors in the determination of $H_\parallel$, $H_\perp$ field components, inclination γ and azimuth χ

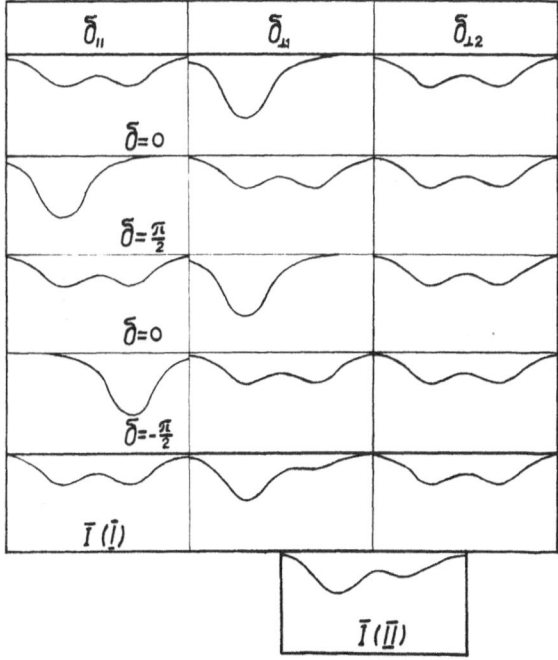

Fig. 1.   The line profiles produced by the EOM at different phase-differences δ of crystal in different types of light modulation ($\delta_\parallel$, $\delta_{\perp 1}$, $\delta_{\perp 2}$). The line profiles I average in time for modes I and II are shown at the bottom of the picture.

Howard (ed.), Solar Magnetic Fields, 71–75. All Rights Reserved.

has not been performed. We have also estimated these errors in the new variant of the
Crimean magnetograph (Nikulin, 1967).

The Crimean magnetograph gives two possibilities to measure the total magnetic
vector: the mode (I) for successive separate recording of three field components, described by Stepanov and Severny (1962), and another mode (II) for simultaneous
recording (Nikulin, 1967). Figure 1 presents schematically the line profiles produced
by the EOM at different phase-differences δ produced by the ADP crystal. We see
asymmetry of the line profile I average in time (at the bottom of the picture) for the
second mode; in mode I the asymmetry appears only for $\delta_{\perp 1}$ recording.

To examine this effect we have calculated a fictitious velocity, due to asymmetry,
and magnetic signals for Fe I $\lambda 5250$ line using Unno (1956) theory. This velocity signal
upon the field strength H and inclination γ is plotted in Figure 2. The dotted and solid
lines refer to modes I and II respectively. We see that fictitious velocities can reach
3–5 km/s for $H \gtrsim 1500$ G, especially for mode I (separate recording of $\delta_{\perp 1}$ signal). In
this case the line-shift is twice as much as it is in mode II for $H \lesssim 1000$ G and for small
values of γ.

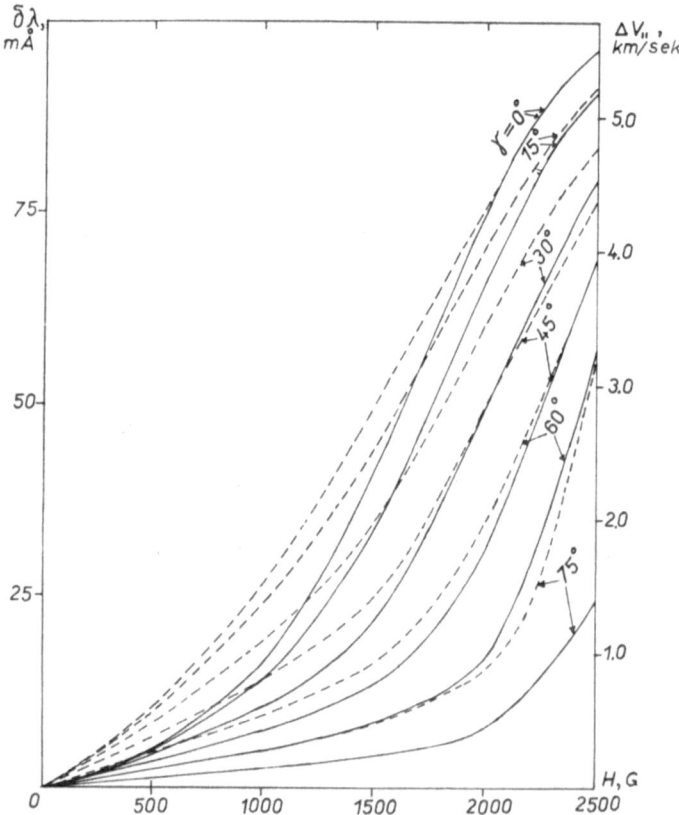

Fig. 2.   The dependence of the line shift produced by asymmetry of the line profile upon field strength
H and inclination γ.

The dependence of the ratio $\delta_\perp/\delta_\perp^0$ ($\delta_\perp^0 = \sqrt{\{(\delta_{\perp 1}^0)^2 + (\delta_{\perp 2}^0)^2\}}$ is the undistorted value) on H and γ for mode II is illustrated in Figure 3. The errors may be quite appreciable for strong $H_\parallel$, but when $\gamma \gtrsim 45°$ and $H \lesssim 1500\,\mathrm{G}$ the errors do not exceed 12%; for $H \lesssim 1000\,\mathrm{G}$ they are less than 18% for any value of γ. These errors are well within the limits of the usual accuracy of magnetograph measurements. As may be realized

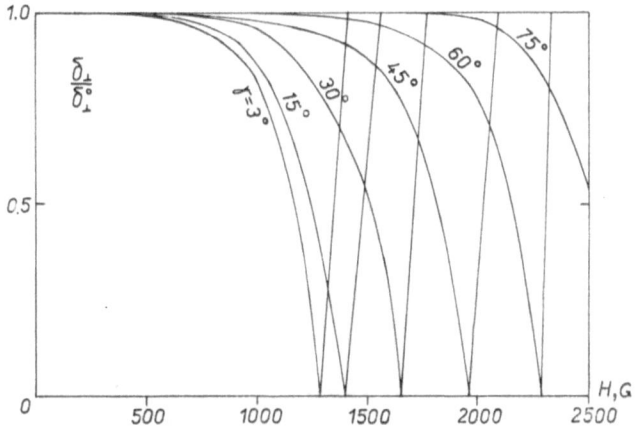

Fig. 3.   The ratio $\delta_\perp/\delta_\perp^0$ upon H and γ in mode II.

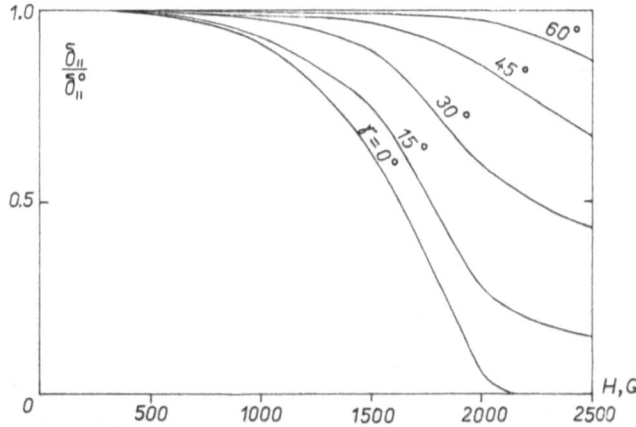

Fig. 4.   The ratio $\delta_\parallel/\delta_\parallel^0$ as a function of field strength H and inclination γ (mode II).

from Figure 4, the relative errors of $\delta_\parallel$, in general, are smaller than those of $\delta_\perp$. When $H \approx 1500\,\mathrm{G}$ errors do not exceed 10% for $\gamma > 30°$.

The calculation shows that the errors in azimuth can appear only in the first mode and they do not exceed 8° for $H \approx 1500\,\mathrm{G}$ and $\gamma > 45°$; when $H \lesssim 1000\,\mathrm{G}$ errors are less than 8° for arbitrary γ.

To summarize the results we conclude that the total vector measurements with a Crimean-type magnetograph are quite satisfactory for $H \lesssim 1000\,\mathrm{G}$. For stronger fields the errors are acceptable if the inclination $\gamma \gtrsim 45°$. When $H > 1500\,G$ and $\gamma < 45°$ the miscentering effect leads to considerable errors, but for such strong fields magneto-graph results, in general, are inadequate due to saturation. So the miscentering is only of importance when the longitudinal field $H_{\parallel} > 1500\,G$. The second mode of the Crimean magnetograph has a smaller error than that of mode I.

One should note that all calculations of the magnetograph signal are usually based on some theory of the line formation in the presence of magnetic fields, and thus they

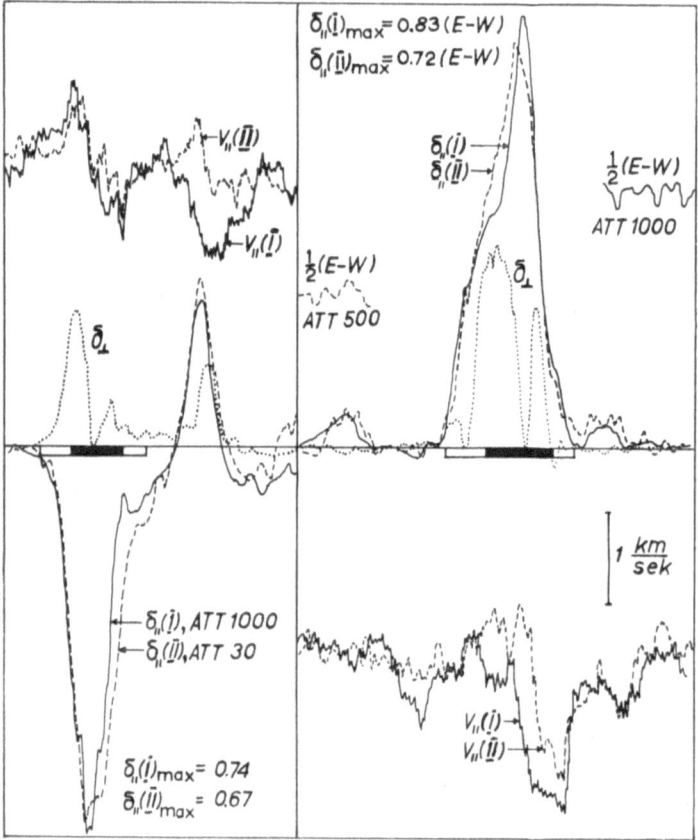

Fig. 5.   The recordings of $\delta_{\parallel}$, $\delta_{\perp}$ and velocity signals obtained in a spot by both modes of the magneto-graph. The value E-W means the calibration signal.

are scarcely consistent in view of the simplifications inherent in theoretical models. The values of systematic errors found here and earlier by Wiehr (1969) are much in excess, because the numerous observations with our magnetograph give clear evidence that the fictitious velocities turned out to be considerably smaller than the calculated values. The most surprising result from Figure 5 is that the difference between maxi-mum signals $\delta_{\parallel}$ obtained in both modes is found to be smaller than 14% and the maximum fictitious velocity $\Delta V_{\parallel} \lesssim 1$ km/s, while the field $H_{\parallel} \gtrsim 2000\,G$ in the spot.

A number of investigations made with the magnetograph and using the photographic technique have shown that the field in the penumbra is mainly transversal ($\gamma > 50°$); in the umbra often $\gamma > 30°$ and only in a very narrow region is the field purely longitudinal. These facts together with our calculations give a reason to consider the measurements with the Crimean vector magnetograph to be quite reliable up to field-strength $H_{\parallel} \approx 1500\,G$. Probably this value of $H_{\parallel}$ can be much greater when the empirical calibration is applied. This follows also from the fact that results on the field structure in spots obtained in the $\lambda 5250$ line are confirmed by similar data obtained in spectral lines of small magnetic sensitivity, for which errors introduced by miscentering are almost negligible.

It should be stressed here that the miscentering effect cannot explain the discrepancy between theoretical and empirical calibration curves originally found by Severny (1967). This discrepancy is still not quite clear and should be considered in future research. It is possible that this discrepancy has the same origin as the discrepancy between calculated and observed miscentering effects that we have just described.

References

Beckers, J. M.: 1968, *Solar Phys.* **5**, 15.
Kuznecov, D. A., Kuklin, G. V., and Stepanov, V. E.: 1966, *Rezultaty nabljud. i issled. v period MGSS* **1**, 80.
Nikulin, N. S.: 1967, *Izv. Krymsk. Astrofiz. Obs.* **36**, 76.
Severny, A. B.: 1967, *Izv. Krymsk. Astrofiz. Obs.* **36**, 22.
Stepanov, V. E. and Severny, A. B.: 1962, *Izv. Krymsk. Astrofiz. Obs.* **28**, 166.
Unno, W.: 1956, *Publ. Astron. Soc. Japan* **8**, 108.
Wiehr, E.: 1969, *Solar Phys.* **9**, 225.

Discussion

Schröter: Did you ever consider light retardation in your instrument, since if such an instrumental retardation exists you should in reality not observe in mode 1 and in mode 2 (the two positions of the $\lambda/4$ plate) but somewhere in between. This very probably can explain why you are not observing the miscenterings as expected.

Severny: The 'retardation' properties of our solar tower (or better to say the instrumental *elliptical* polarization) have been considered in several papers: by Stepanov and Severny (1962, *Publ. Crim. Obs.*), by Severny (1964, *Publ. Crim. Obs.*). If the coatings of mirrors are not fresh and proper and we avoid oblique incidence the effect of retardation (due to reflection from aluminized surfaces) is not more than 0.2 %. The elliptical polarization due to scattered light inside a sunspot can easily be measured by using a magnetically insensitive line e.g. $\lambda 5123$, and this effect brings no more than 200 G inside the umbrae of sunspots (with a field ≥ 1500 G). The effects of miscentering and the instrumental polarization are probably most appreciable in such cases as Coudé spectrographs like Locarno (and Oxford) because of the strong influence of changing the azimuth of polarization (rotation of the field of view).

ANALOG VIDEO MAGNETOGRAMS IN REAL TIME

R. C. SMITHSON and R. B. LEIGHTON

California Institute of Technology, Pasadena, Calif., U.S.A.

For many years solar magnetic fields have been measured by a variety of techniques, all of which exploit the Zeeman splitting of lines in the solar spectrum. One of these techniques (Leighton, 1959) involves a photographic subtraction of two monochromatic images to produce a picture of the Sun in which the line-of-sight component of the solar magnetic field appears as various shades of gray. In a magnetogram made by this method, zero field strength appears as neutral gray, while magnetic fields of one

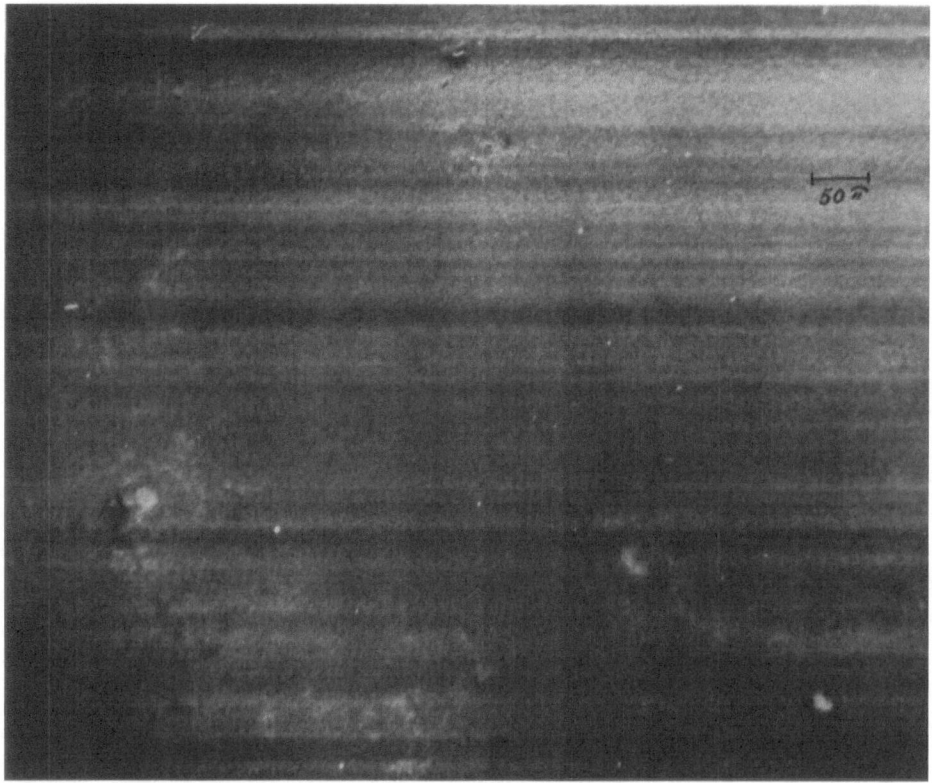

Fig. 1.   An early magnetogram produced by the photographic subtraction technique. The monochromatic images were made using the spectroheliograph at the 60 ft tower at Mount Wilson, September 11, 1960.

polarity or the other appear as lighter or darker areas, respectively. Figure 1 shows such a magnetogram.

The photographic subtraction method consists of using a beamsplitter and spectro-

Howard (ed.), Solar Magnetic Fields, 76–83. All Rights Reserved.
Copyright © 1971 by the IAU.

heliograph (or filter) to make two simultaneous monochromatic pictures of the Sun, one of the pictures being made in right-hand circularly polarized light and the other in left-hand circularly polarized light. Both pictures are taken in the same wavelength – a wavelength that falls in the wing of a solar line susceptible to the Zeeman effect and having a simple triplet Zeeman pattern. In the presence of a magnetic field, one component of the triplet shifts above the zero-field frequency, and the other shifts an equal amount below, and these outer components are oppositely circularly polarized if the field is directed toward or away from the observer. If the frequency at which the two pictures are taken is in the wing of the absorption line where intensity is varying rapidly as a function of frequency, a small frequency shift will be translated into a difference in intensity between the two pictures (see Figure 2).

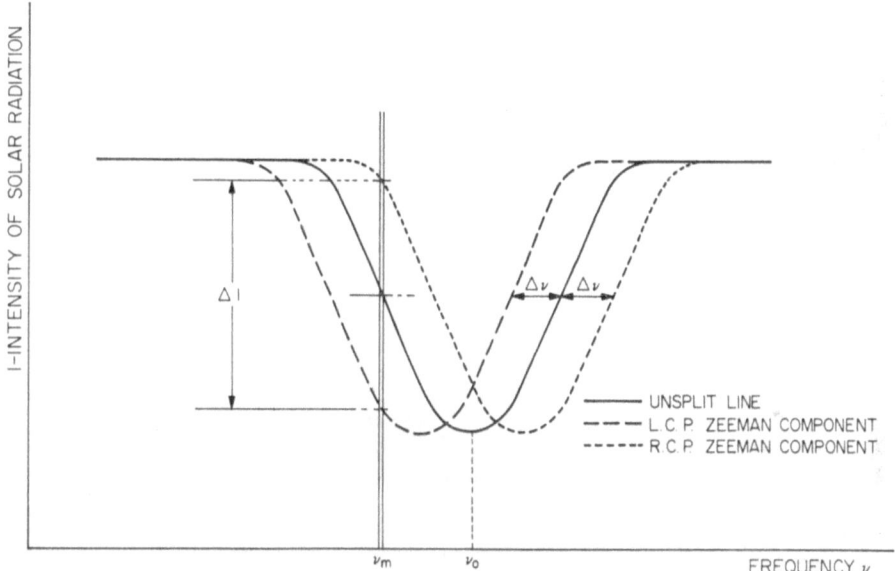

Fig. 2. The wing of the absorption line is used to translate Zeeman splitting into a difference in image intensity. v_0 is the frequency of the unsplit line, and v_m is the frequency at which the monochromatic images are made. A shift Δv in the frequency of the Zeeman components produces an intensity difference ΔI.

Once the two pictures have been made, a unit contrast, positive copy is made of one of them in such a way that if it is superimposed upon its own negative, a uniform gray results. This positive is then superimposed upon the other negative, producing a picture in which all intensity variations *common* to both original pictures are cancelled, leaving only the *difference* between the two images as the final magnetogram (see Figure 3).

The photographic subtraction method gives magnetograms of high spatial resolution compared to those made by scanning magnetographs which use photomultiplier tubes as the light-sensing element, but is only about 1/10 as sensitive in terms of field strength. The chief drawback of the method is the complexity of the photographic processes involved. Making suitable positive copies is an exacting and time consuming process and, typically, several days elapse before the final magnetograms are prepared. Furthermore, this complexity discourages one from preparing

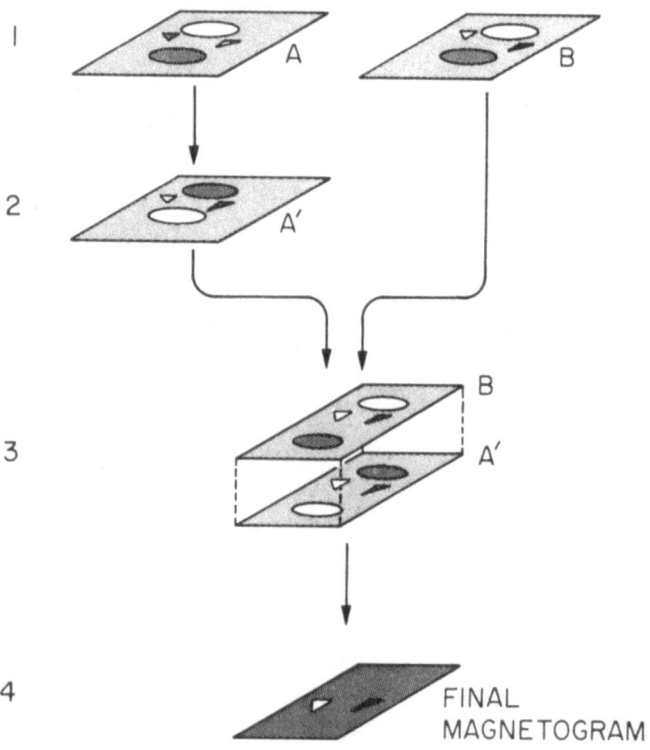

Fig. 3.   The photographic subtraction method. In step 1, the two monochromatic images A and B have been made. In step 2, a unit contrast, positive copy of A has been made (A'). In step 3, A' and B are sandwiched together to form the final magnetogram (step 4).

large numbers of magnetograms such as would be required for time-lapse movies. At Caltech, the photographic cancellation process has been automated for greater ease in processing large numbers of pictures (Roberts, 1970), but it remains time-consuming and limited in sensitivity.

The video differential photometer described here is the result of an attempt to adapt television technology to the task of producing an electronic equivalent of the photographic subtraction technique. Advantages of *electronic* subtraction would be: potentially greater field sensitivity, immediate availability of the subtracted magnetograms (virtually in real time), and greater versatility in combining and comparing image data.

A definite factor in the design of the present system was the desire to keep the cost as low as possible. From the standpoint of cost, it seemed most desirable to use standard 525 line, 30 frame per second, commercial television equipment. Because of the large commercial market, prices are generally much lower and a much wider range of equipment is available than is available in other possible formats. Of course, a slow-scan system would have the advantages of better intrinsic signal-to-noise ratio and greater ease of digitization for computer processing of pictures. It would, however, have the serious disadvantages of higher cost and less readily available component parts. Nevertheless, it remains a promising area for future investigation.

Once the commercial format was decided upon, a signal-to-noise ratio of the camera output of about 300:1 could be assumed: this is a typical value for a moderately priced commercial television camera. Since an ability to detect intensity differences between two pictures of 1:1000 (0.1%) was considered desirable, a method of improving the signal-to-noise ratio had to be found. It was also necessary to find a means of storing the finished magnetograms.

The means used to fulfill both these functions is through a many-track video disc recorder. The system signal-to-noise ratio can then be improved by averaging several magnetograms. (The signal-to-noise ratio should be improved by a factor of nearly $N^{1/2}$, if N magnetograms are averaged.) The recorder can also be used to store sequential pictures in right and left circularly polarized light so that the recorded pictures can be subtracted later. This eliminates the need for a beam splitter in cases when the image does not move appreciably between the times when the two pictures are recorded. A disc recorder was chosen over a video tape recorder because of its generally better signal-to-noise ratio and its much faster and easier access to individual stored pictures for subtraction or averaging. The recorder used has two movable video read/write heads with 153 tracks available to each, and two fixed video read/write heads. It is possible to read from two heads and write on a third simultaneously, a feature which is necessary for subtracting and averaging recorded pictures, and which would be almost impossible using video tape techniques.

All video subtraction, averaging, and other processing is done in analog form, since real-time digitization at commercial picture rates is difficult and storage of the digital information requires costly high-speed memory systems.

The system is under the control of a patchcord-programmable controller designed and built at Caltech. All video recording and processing, as well as control of peripheral equipment is handled automatically.

The final magnetogram is displayed on an ordinary television monitor, where it may be photographed frame-by-frame by an automatic cine camera under the control of the system automatic controller.

A functional block diagram of the differential video photometer is shown in Figure 4. The system can operate either as a two-camera system in which the two pictures to be subtracted are viewed simultaneously through a beamsplitter, or as a single-camera system in which the two pictures are recorded sequentially. In the case of the dual camera system, differences in response of the two cameras are easily averaged out by recording a (camera A-camera B) picture, then reversing the role of the two cameras and recording a (camera B-camera A) picture. The two pictures are then averaged, retaining the desired information but cancelling out systematic differences between the two cameras.

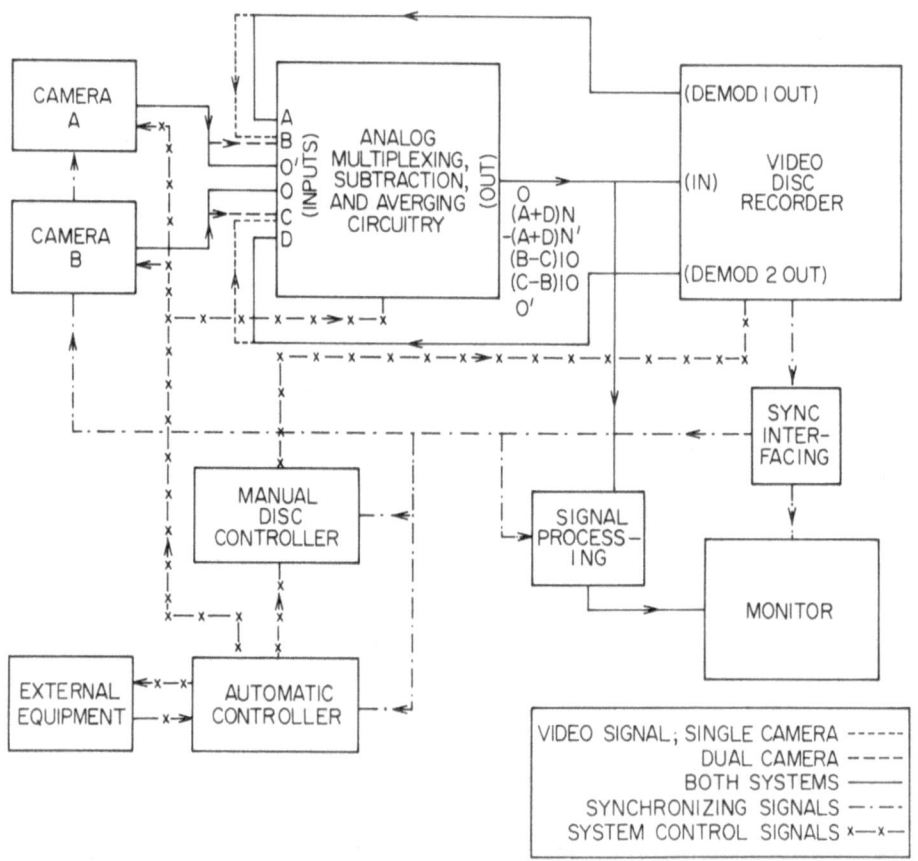

Fig. 4.   A functional block diagram of the differential video photometer. Direction of flow of the various signals is indicated by arrows. A list of the outputs which may be selected is given near the output of the multiplexing circuit. The video disc recorder has two independent outputs for recorded video, labeled 'demod 1' and 'demod 2'.

In order to produce the desired 0.1% sensitivity, it is necessary to average about 100 subtracted pairs. (The actual number is taken as 128 for convenience.) An effective signal-to-noise ratio of 1000:1 can be produced by averaging, even though the disc recorder has a signal-to-noise ratio of only 100:1, by contrast-enhancing the sub-tracted pictures by a factor of about 10, so that the voltage level of the noise to be averaged out is large compared with that of the disc recorder noise. This technique causes large difference signals to saturate, however, and sensitivities of 0.1% cannot be realized if the maximum differences to be faithfully recorded are much greater than two or three percent. Systematic differences in the recorder heads are averaged out in

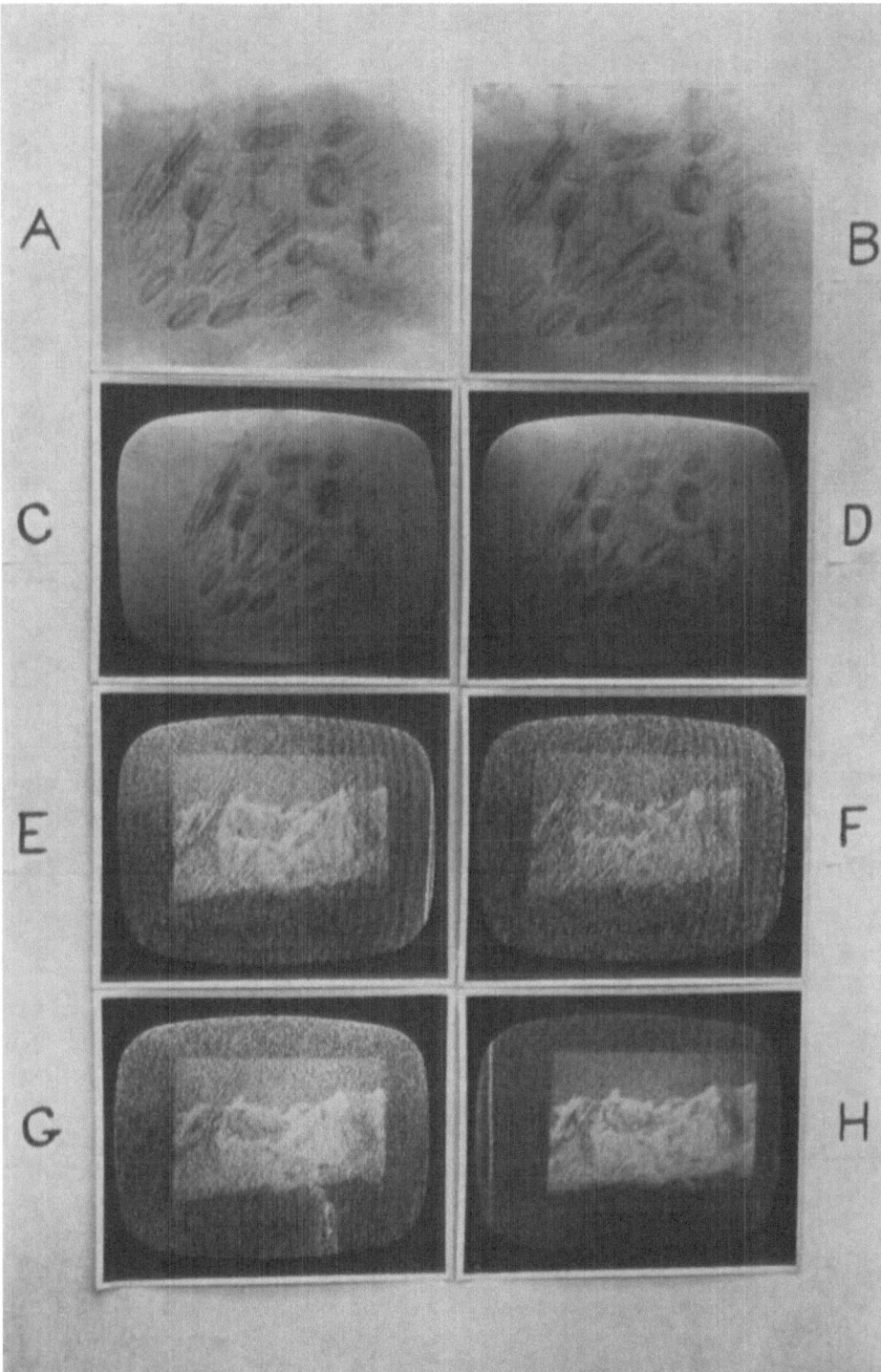

Fig. 5. A laboratory test of the system in the single camera mode. *A* shows a test target. *B* shows the same test target with a slide projected on it. Highlight intensity of the slide is about $2\frac{1}{2}\%$ of the total light incident on the target. *C* and *D* show a television view of the target without and with the projected slide. *E* and *F* show the result of a simple single subtraction with a contrast enhancement of 10. *G* is an average of *E* and *F*. Notice the improvement in the signal-to-noise ratio and the cancelling out of systematic recording errors. *H* shows an average of 128 subtracted pairs. The intensity of the sky in the mountain scene is about 0.1% of the total light intensity incident on the target.

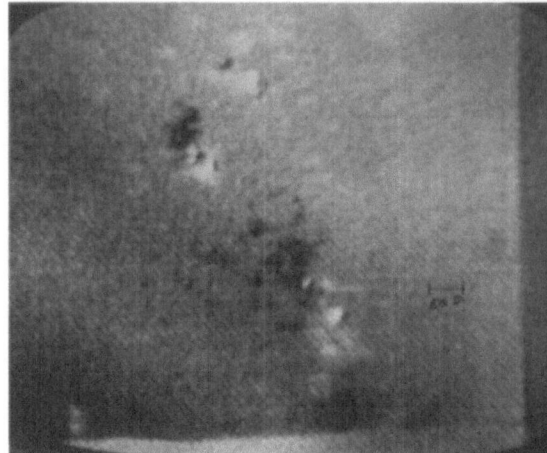

Fig. 6a.

Fig. 6b.

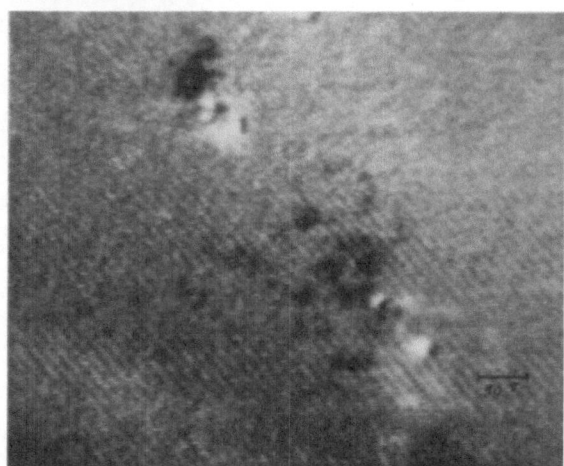

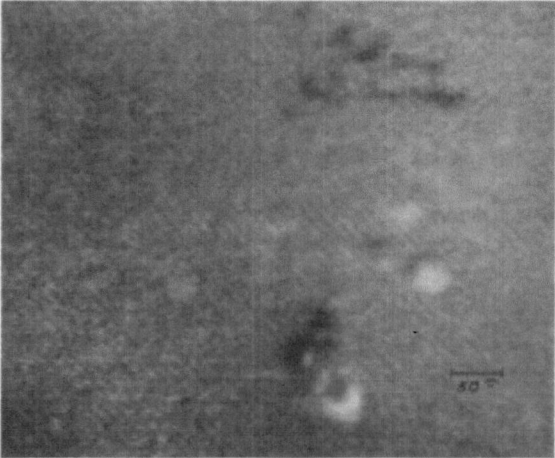

Fig. 6c.

Figs. 6a-c. Three videomagnetograms made at Caltech on October 17, 1970. The monochromatic images were made using a $\frac{1}{8}$ Å bandpass Lyot filter kindly loaned to the authors by the Lockheed Solar Observatory.

a manner similar to that used in the dual camera system. Figure 5 shows the results of a laboratory test of the system. Figure 6 shows some typical magnetograms. Both were taken using the photometer in the single-camera mode. As yet, little experience has been gained with the double-camera mode of operation. In the single-camera mode, about ten seconds are required to record the initial pictures and about 45 sec to subtract and average them. Thus, about one magnetogram per minute can be made. The system is presently in use at Downs Laboratory at Caltech, in a study of magnetic fields in developing solar active regions.

Acknowledgement

This work was supported in part by the Office of Naval Research (ONR) under grant NONR-220(55) and the National Aeronautics and Space Administration (NASA) under grant NGR-05-002-142.

References

Leighton, R. B.: 1957, *Astrophys. J.* **130**, 366.
Roberts, P. H.: 1970, Ph.D. Thesis, California Institute of Technology (unpublished).

A NEW COMPLETELY DIGITIZED FILTER MAGNETOGRAPH

G. E. BRUECKNER

Naval Research Laboratory, Washington, D.C., U.S.A.

Abstract. The optics and electronics of a new filter magnetograph will be described. The instrument uses a Zeiss 0.13 Å birefringent filter to isolate magnetic sensitive lines. All four Stokes parameters can be measured. A Westinghouse SEC vidicon WX 30 654 serves as the detector. The data are completely digitized and transmitted in real time into a Univac 1108 computer.

The instrument is under construction as a joint project of the Naval Research Laboratory, Washington, D.C. and the Marshall Space Flight Center, Huntsville, Alabama. It has been designed to obtain magnetograms of active areas on the Sun with a time resolution of 10 s, a spatial resolution of 1 arc s and a magnetic accuracy of 20 G. The instrument should be capable of accumulating as much quantitative circular and linear polarization measurements as possible to search for short time field changes. The data handling therefore was an equally important part of the design as the outline of the magnetograph itself.

Only a magnetograph using a television type sensor and a filter can meet these requirements, if attached to a small size solar telescope.

Similar magnetographs have been constructed or are under construction by Giovanelli and Ramsay (1971), Janssens (1971) and Smithson and Leighton (1971).

Livingston's approach to increase the information collecting capacity of a Babcock type magnetograph by placing 40 pairs of detector along the slit is not practical for our application, because it would require permanent access to a large telescope and a large spectrograph.

1. The Optical System

Figure 1 shows a schematic diagram of the optics. A 30 cm cassegrain telescope (2,3) serves as an image forming system. A heat reflection interference filter (1, band-width 300 Å) in front of the cassegrain telescope avoids excessive heating of the secondary mirror. In order to reduce the instrumental polarization and avoid any long term changes of it, a cassegrain telescope has been chosen. The polarimeter is attached straight to the telescope to eliminate any reflections other than normal incidence in front of the polarization analyzer (10). A tilted glass plate allows compensating of the residual instrumental polarization. Accurate calibration for circular and linear polarization can be obtained by introducing a tilted glass plate (7) and a quarter wave-plate (8). The polarization analyzer (10) consists of two quarter wave-plates and two KD*P crystals to allow measurements of all three Stokes parameters. A Zeiss birefringent filter (half-width 0.13 Å) separates the magneticly sensitive line Fe I 5250 Å. The filter is tunable over a range of ± 8 Å with an accuracy of better than 0.02 Å to

Howard (ed.), Solar Magnetic Fields, 84–88. All Rights Reserved.
Copyright © 1971 by the IAU.

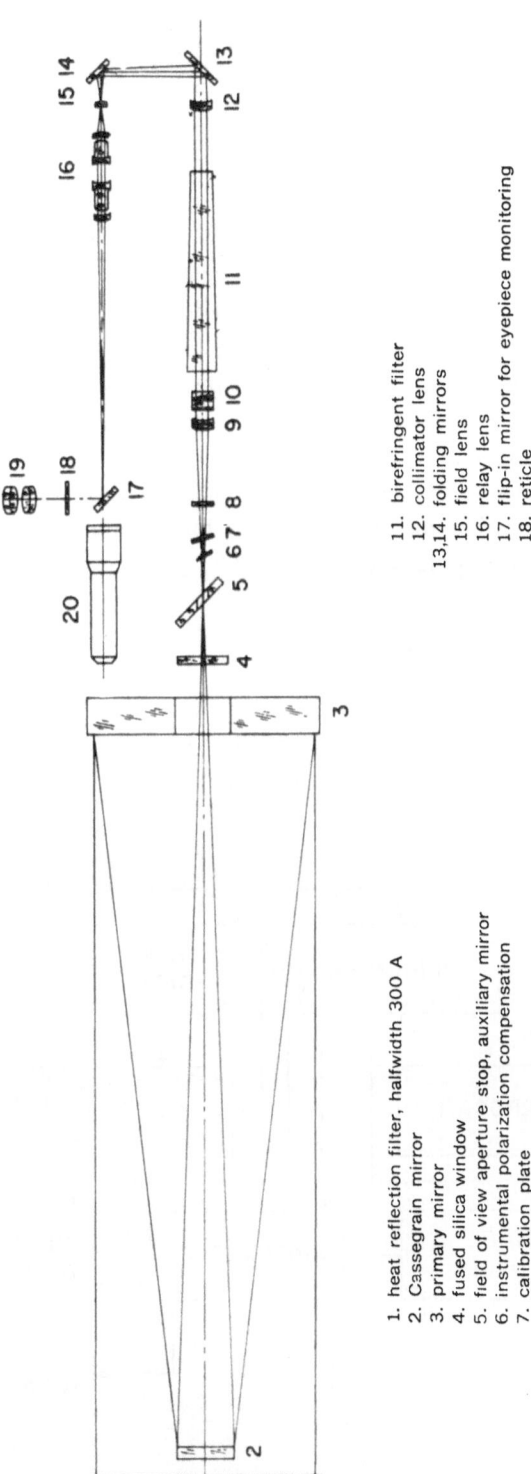

1. heat reflection filter, halfwidth 300 A
2. Cassegrain mirror
3. primary mirror
4. fused silica window
5. field of view aperture stop, auxiliary mirror
6. instrumental polarization compensation
7. calibration plate
8. /4 plate for L. Z. E. calibration
9. collimator lens
10. polarization analyser

11. birefringent filter
12. collimator lens
13,14. folding mirrors
15. field lens
16. relay lens
17. flip-in mirror for eyepiece monitoring
18. reticle
19. eyepiece
20. SEC vidicon

Fig. 1.   Real time solar magnetograph, optics schematic.

allow polarization measurements in different parts of a line and also in different lines in the neighborhood of 5250 Å. The filter and the polarization analyzer are used in a collimated beam. Two different relay lenses at (16) can be used to project the solar image onto the faceplate of the TV tube. The field of view of the instrument is 5×5 arc min^2 or 2×2 arc min^2.

2. The Sensor

In order to obtain a maximum signal-to-noise ratio, a Westinghouse SEC vidicon WX 30654 has been chosen as receiver. This tube has a high target storage capacity. Furthermore, the line scanning density has been reduced to 45 TV lines cm^{-1}. (This corresponds to one picture element per arc second in the case of the 2×2 arc min field of view.) We should obtain a rms signal-to-noise ratio of 200:1 for each picture element. It has been shown theoretically that this signal-to-noise ratio can be achieved. Laboratory measurements using a similar but smaller SEC vidicon WL 30691 have confirmed the calculated values. We also obtain a nearly 100% transfer modulation function by operating the tube with this low scanning density. The magnetic resolution in one pair of polarized images will be ± 20 G.

3. The Electronic System

Figure 2 and Figure 3 show a very simplified electronic block diagram. The camera tube is a part of the data system rather than an independent variable. A particular part of the image is addressed by the programmer via sweep circuits. At the appropriate

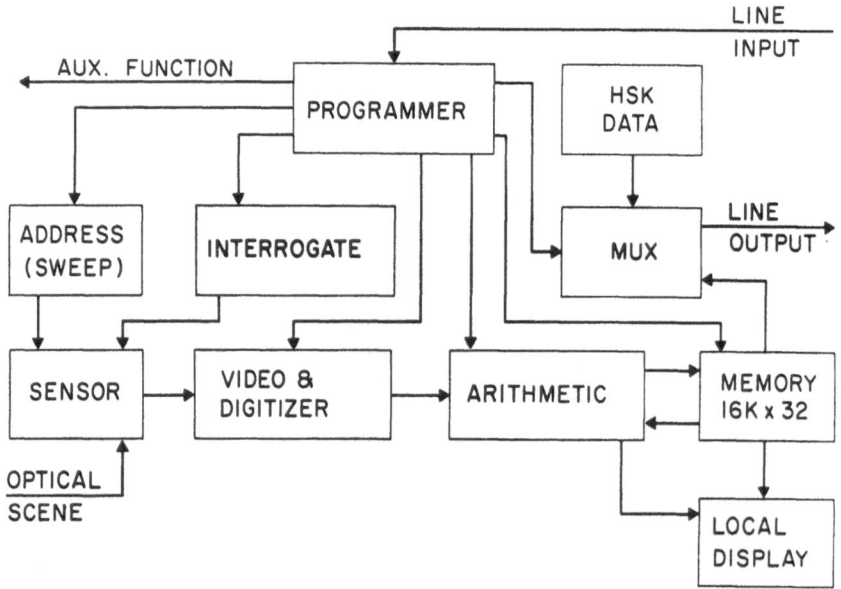

Fig. 2.   Real time solar magnetograph, block diagram of electronics, located near the telescope.

time, the SEC target is interrogated, and the stored information read, digitized, and held in a register. At the completion of this function, the programmer calls up from memory past history of the address in question, adds the new information and places the new number back at its assigned location in the core memory. This process repeats

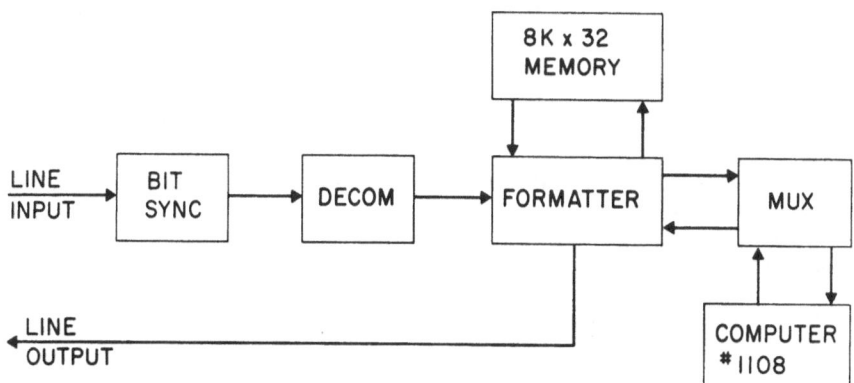

Fig. 3.   Real time solar magnetograph, block diagram of electronics, located near the computer.

until the entire data frame is read into the core memory, at which time several options are available. Some of these options are: (1) the data field may re-cycle with the sensor input either on or off, (2) a sensor polarizing data field may be executed, (3) this portion of the data system may stop awaiting some external event, such as changing of the KDP voltage, indexing the birefringent filter, data dump, etc.

Two KDP crystals in front of the birefringent filter act as polarization analyzer. Discrete voltage level combination applied to those crystals result in different phase shifts of the analyzer combination, which are $-\lambda/4, 0, +\lambda/4$ and $+\lambda/2$. The target of the SEC vidicon is exposed to light of one of the specific phase shifts of the analyzer, which is controlled by the programmer. After readout, digitization and dump into one half of the memory, the programmer switches the analyzer into another phase shift. The target is then exposed to this opposite polarized scene which is digitized and dump-ed into the other half of the core memory. Subsequent images of the same analyzer phase shift can be added in the memory to enhance the signal-to-noise ratio.

The contents of the memory can be displayed as an analog image in rapid sequence on an oscilloscope screen and reloaded into the memory. This allows a quick visual evaluation of the image quality, which is determined by the seeing characteristics of the atmosphere during target exposure. Unwanted images can be disposed immediately and replaced by new exposures.

At convenient intervals, housekeeping data such as filter settings, local time, etc. must be added to the data stream so that identification can be made. This is accomp-lished by a conventional digital multiplexer.

A Receiver/Formatter unit is located at the computer central, located some 10 000 ft away from the observatory tower. Since the signals from the tower have been serialized for transmission through a coaxial cable, a means must be provided for recovering the original reference. This is accomplished by use of a bit synchronizer. A decommutator is used to group convenient numbers of bits for parallel presentation to the Formatter. The Formatter performs the function of controlling the 8KX32 memory where a complete data frame may be placed while awaiting time on the computer. This step is necessary because the computer is time shared via the multiplexer, and entry cannot be guaranteed at any specific time.

References

Brueckner, G. E. and Tucker, B. J.: 1970, in 'Astronomical Use of Television Type Image Sensors', Princeton Univ. Press, in press.

Giovanelli, R. G. and Ramsay, J. V.: this volume, p.293.

Janssens, T. J.: 1971, in 'Astronomical Use of Television Type Image Sensors', Princeton Univ. Press, in press.

Smithson, R. C. and Leighton, R. B.: this volume, p.76.

Discussion

Severny: Would you think that magnetograph records of Hα are not reliable for measurements of the chromospheric field? It gives still reasonable data when used in a proper way.

Brueckner: Yes, we are planning to have besides the 5250 filter an Hα filter.

Dunn: Do you think it will be necessary to make a point-by-point correction? If you make such a correction what signal-to-noise do you expect.

Brueckner: We think that a correction is only necessary for the large scale change in sensitivity of the tube. The small scale sensitivity changes are less than 2% (from one to the next image element). A great deal of effort has been invested into the circuits to make sure that the reading beam of the tube scans with very high accuracy in order to eliminate the small scale fluctuations by subtraction of the opposite polarized images.

DIFFICULTIES IN THE SIMULTANEOUS MEASUREMENT
OF ALL STOKES PARAMETERS

E. WIEHR

Universitäts-Sternwarte, Göttingen, Germany

Telescopic phase retardation in connection with the polarizer behind a polarimeter's KDP-crystal strongly influences the Zeeman pattern, since the circular polarized parts of the two σ-components are strengthened and weakened respectively. This influence depends on the hour angle if the solar image rotates with respect to the polarizer (e.g. for Coudé telescopes).

Figure 1 schematically shows the behaviour of the Zeeman triplet Fe $\lambda 6302.5$ in

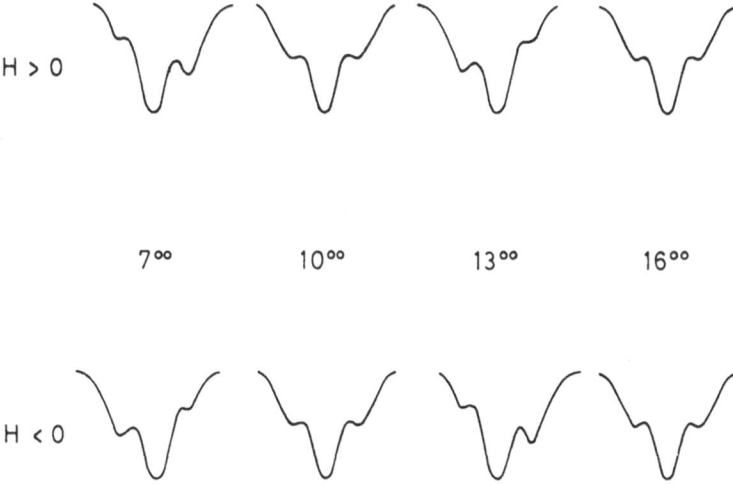

Fig. 1. Influence of the telescopic phase retardation on the Zeeman triplet Fe λ 6302.5 as measured in two sunspot umbrae of opposite polarity (schematic representation).

two sunspot umbrae of opposite polarity as measured at the Locarno observatory during one day. Since the asymmetric profiles are miscentered by the Doppler compensator, all polarimeters which separate the Stokes parameters V and U (or Q) using their different distribution along the line profile, cannot be used without special precautions.

In the Locarno polarimeter, for instance, which uses a three-slit arrangement, U and Q are measured in the π-component (Wiehr, 1969). Since U and V both give signals with the KDP's frequency, they can only be separated if V vanishes in the central exit slit. This, however, is only assured if the inversion point $V(\lambda)=0$ coincides exactly with the center of the central slit. The above mentioned miscentering (and naturally each other miscentering!) yields a cross-talk between U and V which completely falsifies the sensitive U-signal.

Howard (ed.), Solar Magnetic Fields, 89–90. All Rights Reserved.
Copyright © 1971 by the IAU.

Another effect results from the decrease of circular polarization caused by the telescope's phase retardation which partially converts the circular into linear polarized light. Consequently a too small value for the quantity V is measured. A comparison of measurements at 7^h and 10^h (see Figure 1) yields a maximal error of a factor two.

A similar error occurs for the linear polarization which is diminished by the phase retardation. One thus obtains a too small value for the Stokes parameters U and Q.

Up to now, quantitative investigations about the telescopic phase retardation have only been carried out by Jäger and Oetken (1963) for a Coelostat and for a Cassegain telescope as well as by Wiehr (1971) for the Grégory-Coudé telescope at Locarno. Similar effects are to be expected for all telescopes.

References

Jäger, F. W. and Oetken, L.: 1963, *Publ. Astron. Obs. Potsdam* No. 103, Vol. **31**, p. 30.
Wiehr, E.: 1969, *Solar Phys.* **9**, 225.
Wiehr, E.: 1971, *Solar Phys.* **18**, 226.

REDUCTION OF THE PARASITICAL SIGNAL OF CIRCULAR POLARIZATION ON AN ANTENNA OF VARIABLE PROFILE WITH THE HELP OF A GRATING

N. A. ESEPKINA, V. Y. PETRUNKIN, N. S. SOBOLEVA, G. M. TIMOFEEVA,
and A. V. REINER

Pulkovo Observatory, Leningrad, U.S.S.R.

Abstract. A method of reducing circular parasitical polarization in the antennas of variable profile with the help of a grating of curved wires is reported. The experimental verification of the method shows that the parasitical signal practically vanishes. It was established that using this method, simultaneous observations at three or even five wavelengths can be made.

It is well known that the diagram of an antenna of variable profile has parasitical lobes of cross polarization (Esepkina *et al.*, 1961; Kuznezova and Soboleva, 1964). The maxima of these lobes are situated in one of the main planes. The cross-polarized lobes are displaced in phase by 90° with respect to the main polarization, therefore it is difficult to study the circular polarization of the radio sources. In particular it is important in the case when solar magnetic fields are studied by radio methods.

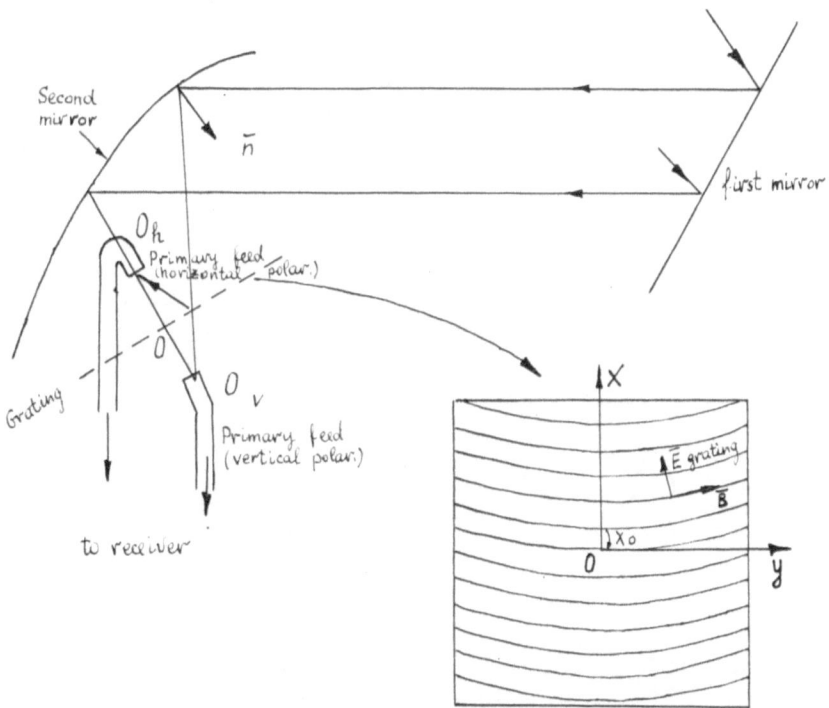

Fig. 1.   The method of measuring circular polarization by means of a grating.

Howard (ed.), Solar Magnetic Fields, 91–93. All Rights Reserved.
Copyright © 1971 by the IAU.

To overcome these difficulties in observations of circular polarization, the installation of a grating of curved wires was suggested (Esepkina *et al.*, 1969). This grating is placed in front of the primary feed. The method of measuring the circular polarization by means of a grating is shown in Figure 1.

The scheme (Figure 2) comprises two feeds of different polarization (horizontal and vertical), the grating being inserted between them. The signal with vertical polarization from the celestial source passes completely through the grating, however only the useful part of it (without cross polarization) is received by the feed. The same effect takes place when we deal with the horizontal polarization, which is reflected from the grating.

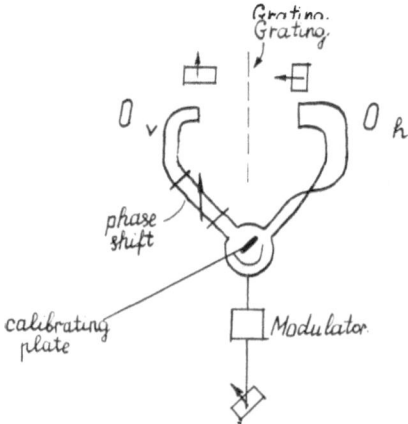

Fig. 2.   The scheme of the feeds.

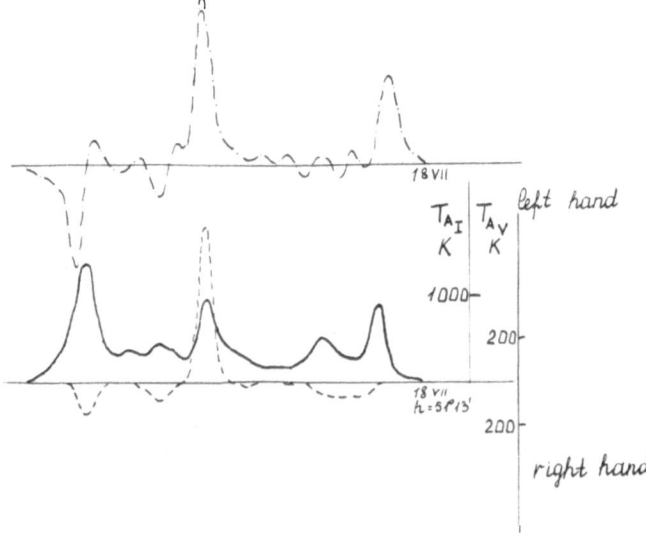

Fig. 3.   Scans of circularly polarized and unpolarized emission from the slowly varying component of the Sun. The upper curve is without compensation of the parasitical signal. The lower curve (dashed) is the same polarized signal with compensation.

One of the polarizations is then phase-shifted and added to another in a cylindrical waveguide. After this the signal is modulated with the ferrite modulator. The shape of the wires was calculated with the aid of a Minsk 22 computer.

To check this method, observations of radio source Tau-A were made. The source Tau-A has no circular polarization. After this experiment it was proved that a circular parasitical signal can be reduced by more than a factor of 10–20.

Observations of the circular polarization of the Sun by the proposed method were commenced with the Pulkovo large radio telescope in July 1968 (Esepkina *et al.*, 1968). The scans of circularly polarized and unpolarized emission from the slowly varying component of the Sun are shown in Figure 3. Circular polarization without compensation of parasitical signal is depicted by the upper curve. The next (dotted line) is the same polarized signal with compensation of parasitical signal using the grating.

Regular observations of circular polarization with the Pulkovo large radio telescope started in March 1969 simultaneously at three wavelengths. The results will be published in the bulletin *Solnechny Dannye* (Russian *Solar Data*).

References

Esepkina, N. A., Kaidanovsky, N. L., Kuznezov, B. G., Kuznezova, G. V., and Khaikin, S. E.: 1961, *Radiotechnica y Electronica* **6**, 1947.

Esepkina, N. A., Soboleva, N. S., and Timofeeva, G. M.: 1968, *Soln. Dann.* No. **8**, 86.

Esepkina, N. A., Petrunkin, V. J., Soboleva, N. S., and Timofeeva, G. M.: 1969, *Radiotechnica y Electronica* **10**, 1871.

Kuznezova, G. V. and Soboleva, N. S.: 1964, *Isvestia GAO* **23**, 122.

A SHORT REPORT ON THE MAGNETIC BEAM ABSORPTION FILTER RESEARCH AT THE ROME ASTRONOMICAL OBSERVATORY

A. CACCIANI, M. CIMINO, and M. FOFI

Rome Astronomical Observatory, Rome, Italy

Abstract. Some possible developments of magnetic beam absorption devices are described, in view of their very promising possibilities of practical application in the field of solar magnetic field observations.

1. The first apparatus described by Cimino *et al.* (1968a, b) essentially consists of two atomic (or vapor) beam filters in the schematic disposition of Figure 1.

   A chemical element (say sodium) is evaporated in the cells (Figure 1a), where a pressure of the order of 10^{-4}–10^{-5} mm of Hg is maintained. With the atomic beam at 90° with the incident light, the statistical Doppler broadening, which is the most important cause of line broadening, may be avoided. In our first experiment a sodium-vapor lamp, which has a large emission band simulating a continuous spectrum in a limited wavelength region, has been used (Figure 1b shows filtering steps and polarizations). If the laboratory magnetic field is parallel to the incident light, only the σ components are absorbed, and if the beam is between two crossed nicols, all continuous radiation is extinguished except the light in the absorption regions. In fact, the linear polarization introduced by the first nicol may be considered composed of light

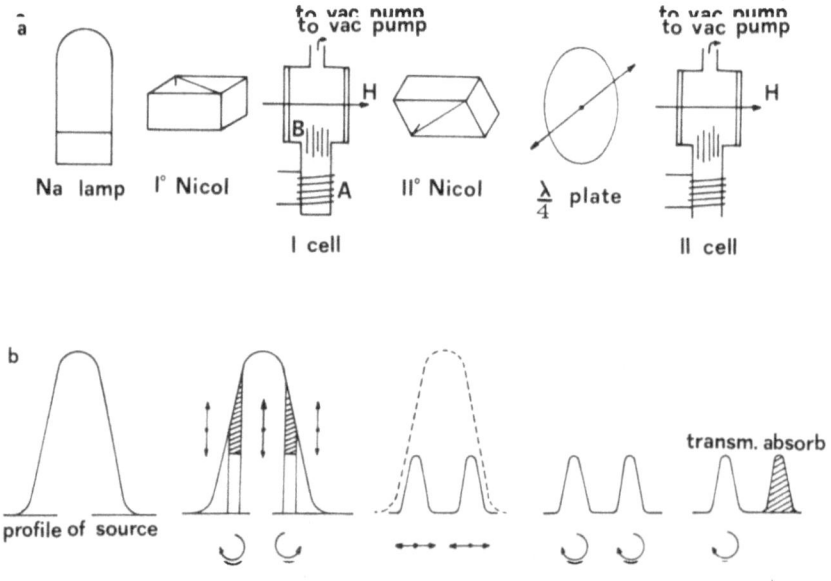

Fig. 1.   (a) Schematic experimental disposition; (b) Filtering steps and polarization.

Howard (ed.), Solar Magnetic Fields, 94–98. All Rights Reserved.
Copyright © 1971 by the IAU.

of two opposite circular polarizations, one of which is absorbed by the two lines of the beam and another half of which is allowed to pass through the second nicol (Klinkenberg, 1966, p. 77).

Of course, the real things go in a rather more complicated way than the schematic one. The physical effect we have pointed out in the schematic description is the *absorption effect* caused by the Zeeman levels of the atoms in the beam (inverse Zeeman effect). Such an effect must not be confused with the Maccaluso-Corbino effect (i.e. the magneto-optic Faraday effect in the neighborhood of an absorption line of the gas), though the two effects are intimately connected, and must be considered together for a complete understanding of the theory of the filter. In our apparatus the absorption effect plays a *basic role* and, as far as we know, we think that it has never been applied in a real optical instrument in astrophysics; on the contrary, the M.C. effect was considered in astrophysics by Öhman (1960), in different occasions and also in the theory of radiation transfer in the solar atmosphere.

2. An elementary theory of the magnetic beam absorption filter was developed by Cimino *et al.* (1970). The schematic explanation of the filter behavior given in our first note may be considered satisfactory, as far as the absorption effect is concerned; otherwise, as a consequence of the theory, it is important to consider not only the M. C. effect, but also the behavior of the cells in relation to the central optical depth of the line, and to the value and the uniformity of the magnetic field.

Let us consider the first cell of the apparatus, between the two crossed nicols; assuming the value of the magnetic field as constant along the cell, Figure 2 and Figure 3, calculated by Cacciani and Fofi, give the transmission of the cell both for the absorption and for the M. C. effect *versus* $z-g_i z_H$, which represents the distance from the center of the i-line of the Zeeman multiplet, expressed in units of half the natural line width $(\gamma/2)$.

It is of interest to note how the light transmitted by the absorption effect is restrained within a narrow band around the line center; but at the same time the M.C. effect may give a more important contribution also in the neighborhood of the center of the line. Indeed this is an advantage when using the magnetic filter as a spectrograph, since more light is allowed to pass in a narrow pass-band, which represents the real pass-band of the cell.

Let us now consider the second cell. Because of the $\lambda/4$ plate in the front, *all* the light is circularly polarized in one direction only. If the magnetic field around such a cell is *not* uniform (say, increasing of the order of 200 G), Figure 4 shows how well the pass-band of the first cell, and one of the two Zeeman components is therefore allowed to pass completely, while the other is completely cut down by absorption. Therefore we have called the second cell the 'absorption cell', while the first may be called the 'transmission one'. So we have a spectrographic exit slit, which may be shifted along a wing of a solar line by tuning the laboratory magnetic field.

3. More recently a new scheme was given by Cacciani (Cimino *et al.*, 1970) in order

to use magnetic beam filters as solar magnetographs with a very high resolution pass-
band (down to the order of 10^{-2} Å in practice and less in theory).

As shown in Figure 6, the transmission cell between two polarizers (in the experimental tests the cell was substituted by two identical ones, in order to reduce

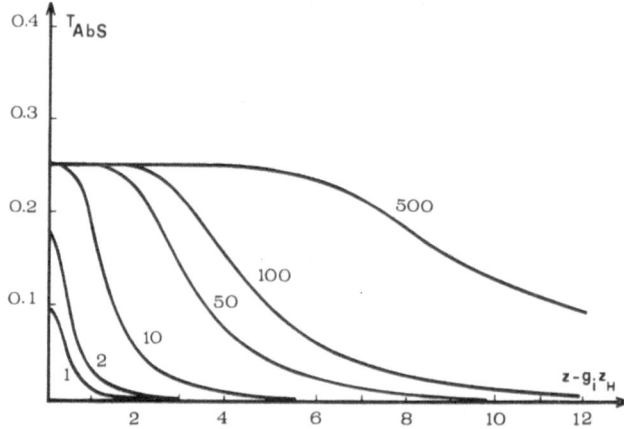

Fig. 2. T_{abs} = transmitted light for absorption effect, from different values of τ_0 ($\frac{1}{2}$ central optical
depth) in uniform magnetic field.

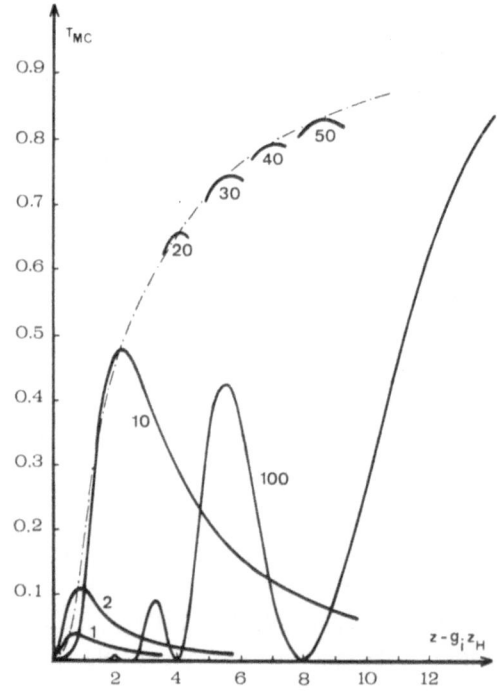

Fig. 3. T_{MC} from different values of τ_0 ($\frac{1}{2}$ central optical depth) in uniform magnetic field.

scattered light and trouble from the M.C. effect) is followed by the absorption cell
(in a non-uniform, increasing magnetic field), but the $\lambda/4$ plate was put *behind* the cell.
With such an arrangement, both the Zeeman components, as transmitted by the first
cells, completely pass through the second cell and with opposite circularly polarized
light; the calcite at 45° to the $\lambda/4$ plate discriminates them like a couple of separate
magnetographic exit slits. On the contrary, the residual light, both M. C. light as well

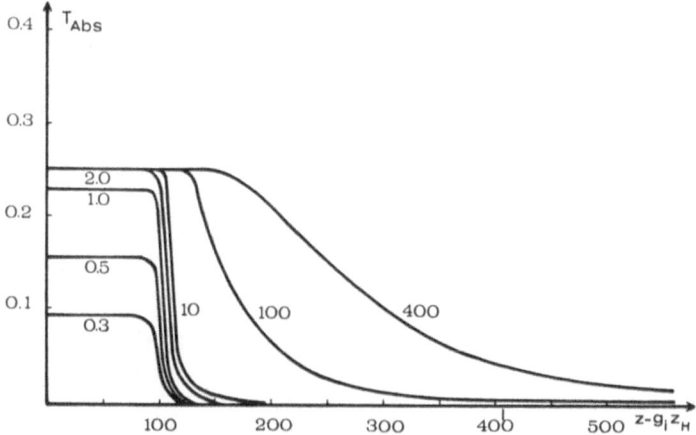

Fig. 4. T_{abs} from different values of τ_0 ($\frac{1}{2}$ central optical depth) in non-uniform magnetic field.

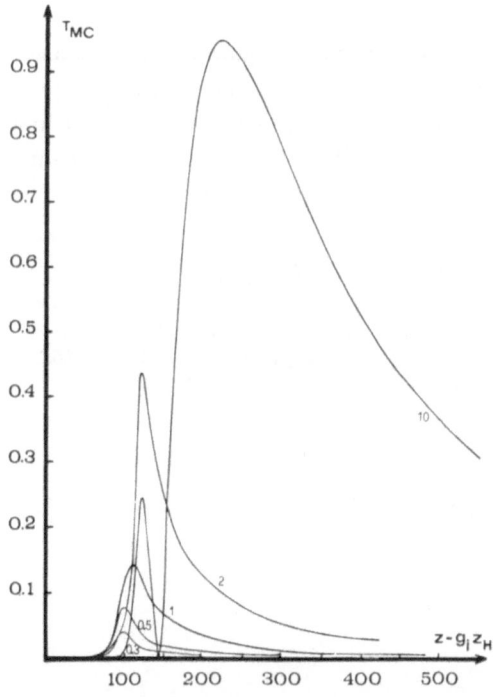

Fig. 5. T_{MC} from different values of τ_0 ($\frac{1}{2}$ central optical depth) in non-uniform magnetic field.

as continuum outside the absorption band, are linearly polarized, and will be changed by the $\lambda/4$ plate in elliptic light, with their axes parallel, in every case, to the $\lambda/4$ retardation axes. Therefore the calcite equally divides all the residual light between the ordinary and the extraordinary rays; the residual light only acts in increasing the statistical noise. Since the detector may be electronic tubes as well as photographic plates, we can use also the Leighton method.

Note that we have a true zero level in the velocity field measurements. If used as a magnetograph, the Doppler effect cannot be compensated. In this case it may be

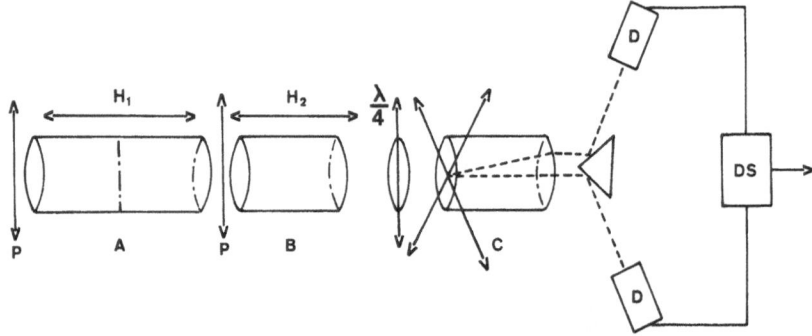

Fig. 6.   Scheme for a two window magnetograph.

P = polarizers; A = couple of identical cells separated by a linear polarizer at 90° with P; B = Absorption cell; C = calcite; DS = differential system; H_1 = Uniform magnetic field; H_2 = non-uniform magnetic field.

convenient to go back to the scheme of Figure 1, use the exit slit as a 'single' Babcock window, and correct the Doppler effect by tuning the magnetic fields around the cells.

Let us also call attention to what we have noted in our first note (Cimino *et al.*, 1968a, b) on the important possibility of calibrating the instrument without theoretical calculations, only by varying the laboratory magnetic field.

References

Cimino, M., Cacciani, A. and Sopranzi, N.: 1968a, *Solar Phys.* **3**, 618.
Cimino, M., Cacciani, A., and Sopranzi, N.: 1968b, *Appl. Opt.* **7**, 1654.
Cimino, M., Cacciani, A., and Fofi, M.: 1970, *Solar Phys.* **11**, 319.
Öhman, Y.: 1960, *Scientia* **54**, 1.

PART II

THE INTERPRETATION OF MAGNETOGRAPH RESULTS – THE FORMATION OF ABSORPTION LINES IN A MAGNETIC FIELD

THE INTERPRETATION OF MAGNETOGRAPH RESULTS: THE FORMATION OF ABSORPTION LINES IN A MAGNETIC FIELD

J. O. STENFLO

Astronomical Observatory, Lund, Sweden

Abstract. The theory of line formation in a magnetic field is reviewed. It is shown how the formulations by Unno, Beckers, Stepanov and Rachkovsky are related to each other. The general treatment of true absorption, anomalous dispersion, radiative scattering and level-crossing interference is discussed. Special attention is paid to the inhomogeneous nature of the solar atmosphere. The properties of magnetic filaments are reviewed. It is shown how a filamentary structure will drastically influence the interpretation of magnetograph observations. The treatment of line formation in a turbulent magnetic field is also discussed.

The fine structure of the solar atmosphere will influence different types of magnetographs in entirely different ways. The relation between the observed magnetic field and the resolution of the instrument is discussed for a Babcock-type, Evans-type and a transversal magnetograph. Finally some suggestions for future work in this field are listed.

1. Introduction

The most important method to determine solar magnetic fields is to observe the Zeeman effect in the solar atmosphere. The polarization in magnetic-sensitive spectral lines can be analysed by sophisticated methods, and from the relative intensities and displacements of the differently polarized line components one can draw conclusions about the magnetic field at the level in the solar atmosphere where the spectral line is formed.

The main trouble in the interpretation of magnetograph observations has not been that we do not know enough about line formation in a magnetic field; it has been our lack of knowledge of the fine structure of the field. The magnetograph readings have been interpreted in terms of a homogeneous magnetic field although it has been completely clear that most of the magnetic features have not been resolved.

There are however a great number of unsolved problems in the basic theory of line formation in a magnetic field. And this theory forms the general basis for all interpretations of what is measured with a magnetograph. Before we can present a general formulation of the radiative transfer problem in a magnetic field we have to start with a suitable representation of polarized light.

2. Representation of Polarized Light

The vibration of the electrical vector of polarized light propagating along the z-axis in an xyz coordinate system can be described by

$$\xi_x = \xi_1 \cos(\omega t - \varepsilon_1)$$

and

$$\xi_y = \xi_2 \cos(\omega t - \varepsilon_2),$$

(1)

Howard (ed.), Solar Magnetic Fields, 101–129. All Rights Reserved.
Copyright © 1971 by the IAU.

where ξ_x and ξ_y are the components of the vibration along the x and y axis, and ω is the circular frequency. The four constants ξ_1, ξ_2, ε_1, and ε_2 are sufficient to define the state of polarization.

It is convenient to describe the polarization by means of a complex vector $\mathbf{m}$. Let $\mathbf{e}$ and $\mathbf{e}'$ be complex orthogonal unit vectors. Then

$$\mathbf{e} \cdot \mathbf{e}^* = \mathbf{e}' \cdot \mathbf{e}'^* = 1$$

and   (2)

$$\mathbf{e} \cdot \mathbf{e}'^* = 0,$$

where $\mathbf{e}^*$ denotes the conjugate form of $\mathbf{e}$. The polarization vector $\mathbf{m}$ can be written

$$\mathbf{m} = E\mathbf{e} + E'\mathbf{e}',$$ (3)

where E and E' are the complex amplitudes of the electric vector. If we choose a system in which $\mathbf{e}$ and $\mathbf{e}'$ represent linear polarization, e.g. $\mathbf{e} = \hat{x}$ and $\mathbf{e}' = \hat{y}$, $\hat{x}$ and $\hat{y}$ being real unit vectors along the x and y axis, the vibration in (1) can be represented by (3) if we put $E = \xi_1 e^{-i\varepsilon_1}$ and $E' = \xi_2 e^{-i\varepsilon_2}$. The time-dependent factor $e^{i\omega t}$ is ignored as it is of no interest for the polarization. If $e^{i\omega t}$ is included, we obtain (1) by taking the real part of $\mathbf{m}$.

Polarized light can be completely described by the four Stokes parameters I_1, I_2, U and V.

$$\begin{aligned}
I_1 &= \overline{EE^*}, \\
I_2 &= \overline{E'E'^*}, \\
U &= \sqrt{2}\,\mathrm{Re}\,(\overline{EE'^*}), \\
V &= -\sqrt{2}\,\mathrm{Im}\,(\overline{EE'^*}).
\end{aligned}$$ (4)

A line above the symbols means a time average. Re and Im mean that the real and imaginary part of the quantity is taken, respectively.

In our system of linear polarization, (4) can be rewritten in the form

$$\begin{aligned}
I_1 &= \overline{\xi_1^2}, \\
I_2 &= \overline{\xi_2^2}, \\
U &= \sqrt{2}\,\overline{\xi_1\xi_2\cos(\varepsilon_1 - \varepsilon_2)}, \\
V &= \sqrt{2}\,\overline{\xi_1\xi_2\sin(\varepsilon_1 - \varepsilon_2)}.
\end{aligned}$$ (5)

The parameters U and V used by Chandrasekhar (1950) and Unno (1956) are larger than ours by a factor of $\sqrt{2}$. However, by defining the parameters as in (4) and (5), the matrices appearing in the theory will be symmetric or anti-symmetric.

In many papers the parameters I and Q are used instead of I_1 and I_2. We have the relations $I = I_1 + I_2$ and $Q = I_1 - I_2$.

A compact form of representation of polarized light is by means of a four-dimensional vector

$$\mathbf{I} = \begin{pmatrix} I_1 \\ I_2 \\ U \\ V \end{pmatrix}.$$ (6)

The values of the Stokes parameters depend on the choice of the vector system (e, e'). Suppose that e and e' as before represent linear polarization along the x and y axis, and that we make a rotation of the coordinate system by an angle φ in the anti-clockwise direction. The unit vectors in the new system are

$$e_1 = e \cos\varphi + e' \sin\varphi,$$
$$e_1' = -e \sin\varphi + e' \cos\varphi. \tag{7}$$

The Stokes parameters in the new system are obtained by a linear transformation

$$I^{(1)} = L_1 I. \tag{8}$$

$$L_1 = \begin{pmatrix} p & 1-p & q & 0 \\ 1-p & p & -q & 0 \\ -q & q & 2p-1 & 0 \\ 0 & 0 & 0 & 1 \end{pmatrix}, \tag{9}$$

where

$$p = \cos^2\varphi,$$
$$q = \frac{1}{\sqrt{2}} \sin 2\varphi. \tag{10}$$

The unit vectors describing mutually orthogonal elliptical polarization can be written

$$e_2 = e \cos\varphi + e' i \sin\varphi,$$
$$e_2' = e i \sin\varphi + e' \cos\varphi, \tag{11}$$

where e and e' still represent linear polarization. The principal axis of the ellipses are along e and e'. The ratio between the axis of the ellipses is $\tan\varphi$ or $\operatorname{ctn}\varphi$. The linear transformation between the two systems in (11) is

$$L_2 = \begin{pmatrix} p & 1-p & 0 & q \\ 1-p & p & 0 & -q \\ 0 & 0 & 1 & 0 \\ -q & q & 0 & 2p-1 \end{pmatrix}. \tag{12}$$

The inverse transformation is determined by

$$L_{1,2}^{-1} = L_{1,2}^T, \tag{13}$$

where L^T is the transpose of L.

3. Radiative Transfer in a Magnetic Field

3.1. TRUE ABSORPTION

This is the case for which the theory is most developed. The transfer equation can conveniently be written using the compact matrix formulation.

$$\cos\theta \frac{d}{d\tau} I = (1+\eta)(I-B). \tag{14}$$

θ is the heliocentric angle, and τ is the optical depth related to the continuous absorption coefficient. **1** is the unit matrix. The intensity vector **I** is defined by (6) and the source function vector by

$$\mathbf{B} = \begin{pmatrix} B/2 \\ B/2 \\ 0 \\ 0 \end{pmatrix}, \tag{15}$$

B being the Planck function.

In the system of Stokes parameters (5), the absorption matrix can be written

$$\eta = \begin{pmatrix} a_+ & 0 & b & c \\ 0 & a_- & b & c \\ b & b & (a_+ + a_-)/2 & 0 \\ c & c & 0 & (a_+ + a_-)/2 \end{pmatrix} \tag{16}$$

where

$$a_{\pm} = \tfrac{1}{2}[\eta_p - \tfrac{1}{2}(\eta_r + \eta_b)]\sin^2 \gamma\,(1 \pm \cos 2\chi) + \tfrac{1}{2}(\eta_r + \eta_b), \tag{17}$$

$$b = \frac{1}{2\sqrt{2}}[\eta_p - \tfrac{1}{2}(\eta_r + \eta_b)]\sin^2 \gamma \sin 2\chi, \tag{18}$$

$$c = \frac{1}{2\sqrt{2}}(\eta_r - \eta_b)\cos \gamma. \tag{19}$$

$\eta_{p,r,b}$ denotes the ratio between the coefficients of line absorption and continuous absorption for the three Zeeman components. Index p refers to transitions for which the change in magnetic quantum number $\Delta m = 0$. For r, $\Delta m = -1$, and for b, $\Delta m = 1$. We have

$$\eta_p = \eta_0 H(\alpha, v),$$
$$\eta_r = \eta_0 H(\alpha, v - v_H), \tag{20}$$
$$\eta_b = \eta_0 H(\alpha, v + v_H),$$

where η_0 is the ratio between the line absorption coefficient at the line centre and the continuous absorption coefficient. The function H giving the form of the absorption profile is usually assumed to be a Voigt function.

$$\alpha = \gamma_d/\Delta\lambda_D, \tag{21}$$

γ_d being the damping constant and $\Delta\lambda_D$ the Doppler width.

$$v = \frac{\Delta\lambda}{\Delta\lambda_D} \tag{22}$$

and

$$v_H = \frac{\Delta\lambda_H}{\Delta\lambda_D}. \tag{23}$$

The Zeeman splitting is determined by

$$\Delta\lambda_H = 4.67 \times 10^{-13} \lambda^2 g H. \tag{24}$$

H is the magnetic field strength in G and g the Landé factor. The wavelength should be given in Å. Doppler shifts can also easily be included in (20).

The geometry of the system to which $\boldsymbol{\eta}$ in (16) refers is described by Figure 1. γ is the angle between the magnetic field and the line of sight and χ the azimuth angle of the field vector.

With the formulation of the transfer equations given by (14)–(19) we are not restricted to the case of homogeneous magnetic fields. The field may exhibit any variation in magnitude and direction with depth in the solar atmosphere. The ab-

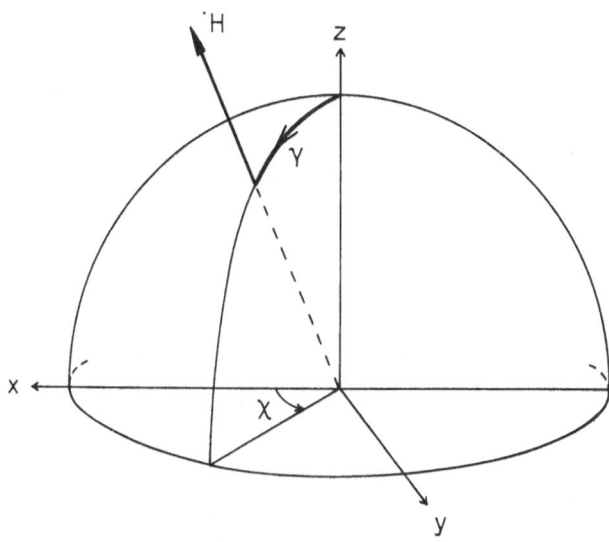

Fig. 1. Geometry for the case of true absorption. The z-axis is assumed to be directed along the line of sight. χ is the azimuth angle of the magnetic-field vector, γ the angle between the field and the line of sight.

sorption coefficients, line broadening mechanisms and radial velocities may also have any variation with depth.

With (14)–(19) one can calculate the polarization of the emerging radiation for inhomogeneous fields and different model atmospheres. This was first done by Beckers (1969a, b). His formulation of the transfer equations is equivalent to ours. Beckers integrated the four coupled differential equations by means of a Runge-Kutta technique.

The absorption matrix can be written in many different ways depending on what system of Stokes parameters has been used. Transformations from one system to another can be made by means of the matrices $\mathbf{L}_{1,2}$ given by (9) and (12). If we denote matrices in the new system with a prime, we have

$$\cos\theta\,\frac{d}{d\tau}\mathbf{I'} = \mathbf{L}_{1,\,2}\,(1+\mathbf{\eta})\,\mathbf{L}_{1,\,2}^{-1}(\mathbf{I'} - \mathbf{B'}). \tag{25}$$

It is easy to show that the source function $\mathbf{B}$ is not changed by the transformations $\mathbf{L}_{1,\,2}$. Hence (25) can be written

$$\cos\theta\,\frac{d}{d\tau}\mathbf{I'} = (1+\mathbf{\eta'})\,(\mathbf{I'} - \mathbf{B}), \tag{26}$$

where

$$\mathbf{\eta'} = \mathbf{L}_{1,\,2}\mathbf{\eta}\mathbf{L}_{1,\,2}^{T}, \tag{27}$$

and (13) has been used.

By applying the transformation $\mathbf{L}_1$ with φ in (10) equal to χ, we rotate the coordinate system in Figure 1 so that the magnetic field vector will be in the xz-plane. Then we obtain the absorption matrix in the same system as that used by Unno (1956) when he first developed the theory of line formation in a magnetic field. Of course we also obtain Unno's equations directly by putting $\chi=0$ in (17) and (18). The Unno formulation is unable to treat line formation in magnetic fields of variable azimuth.

If we in (27) let $\mathbf{\eta}$ be the Unno absorption matrix and use the $\mathbf{L}_2$ transformation (12) with φ determined by

$$\operatorname{ctn}\varphi = \{[\eta_p - \tfrac{1}{2}(\eta_r + \eta_b)]\sin^2\gamma$$
$$\pm\sqrt{[\eta_p - \tfrac{1}{2}(\eta_r + \eta_b)]^2\sin^4\gamma + (\eta_r - \eta_b)^2\cos^2\gamma}\}/(\eta_r - \eta_b)\cos\gamma, \tag{28}$$

the absorption matrix $\mathbf{\eta}$ will be diagonalized.

$$\mathbf{\eta'} = \begin{pmatrix} \eta_+ & 0 & 0 & 0 \\ 0 & \eta_- & 0 & 0 \\ 0 & 0 & (\eta_+ + \eta_-)/2 & 0 \\ 0 & 0 & 0 & (\eta_+ + \eta_-)/2 \end{pmatrix}. \tag{29}$$

$$\eta_\pm = \tfrac{1}{2}[\eta_p - \tfrac{1}{2}(\eta_r + \eta_b)]\sin^2\gamma + \tfrac{1}{2}(\eta_r + \eta_b)$$
$$\pm\tfrac{1}{2}\sqrt{[\eta_p - \tfrac{1}{2}(\eta_r + \eta_b)]^2\sin^4\gamma + (\eta_r - \eta_b)^2\cos^2\gamma}. \tag{30}$$

It is seen that

$$\eta_+ + \eta_- = a_+ + a_-, \tag{31}$$

where $a_\pm$ are given by (17) for $\chi=0$.

The diagonal form (29) is by far the most suitable for calculations of the polarization of the emerging radiation, since the four transfer equations are independent of each other and can be solved by means of all methods that have been developed for unpolarized radiation. Unfortunately the matrix does not diagonalize for arbitrary values of the azimuth χ.

If we assume that the radiation becomes unpolarized as $\tau\to\infty$ the parameters U'

and V' must be $\equiv 0$ throughout the atmosphere. This is so because the source function is zero for U' and V'. With this boundary condition the four transfer equations reduce to the two

$$\cos\theta \, \frac{d}{d\tau} I_{\pm} = (1 + \eta_{\pm})\left(I_{\pm} - \frac{B}{2}\right). \tag{32}$$

Here we have used the notation I_+ and I_- instead of I_1' and I_2'.

When Stepanov (1958a, b) independently of Unno first developed the theory of line formation in a magnetic field he started by calculating the line absorption coefficients from the classical magneto-optical theory. This yielded $\eta_{\pm}$ in (30). Stepanov (1958a) found that η_+ and η_- correspond to absorption of two mutually orthogonal beams $I_{\pm}$. An atmosphere that absorbs I_+ is completely transparent for I_- and vice versa. I_+ and I_- cannot interfere (Chandrasekhar, 1950). In this way Stepanov (1958b) arrived at the Equations (32).

Rachkovsky (1961a) showed that Stepanov's Equations (32) could be deduced from Unno's equations but that the two more equations in (29) had to be added to make the system fully equivalent to that of Unno (Rachkovsky, 1961b).

There is one important restriction which we must notice when we use (32). When solving (32) the system for the Stokes parameters must be the same for all τ, which means that the transformation determined by $\operatorname{ctn}\varphi$ in (28) must be independent of depth. This means in the general case that the magnetic field has to be homogeneous and that the Voigt functions in (20) are independent of τ. η_0 may however have any variation with τ.

In the case of a purely longitudinal ($\gamma = 0°$) or transversal ($\gamma = 90°$) field with constant azimuth, however, any variations in the field strength, radial velocity and line broadening parameters are allowed for.

Stepanov (1958b) gave analytical solutions of (32) and calculated the profile of the line Fe I 6173 Å in sunspots. Mattig (1966) suggested numerical solutions of (32) using different model atmospheres. Such numerical integrations of (32) were made by Moe (1968), Moe and Maltby (1968), Evans (1968, 1969a, b), and Evans and Dreiling (1969).

Having obtained a solution $\mathbf{I}'$ for the emerging radiation in Stepanov's system we can easily transform it to the common Stokes parameters $\mathbf{I}$ by

$$\mathbf{I} = \mathbf{L}_2^T \mathbf{I}'. \tag{33}$$

If $U' = V' = 0$ we find

$$\begin{aligned}
I &= I_+ + I_-, \\
Q &= (I_+ - I_-)\cos 2\varphi, \\
U &= 0, \\
V &= (I_+ - I_-)\, 2^{-1/2} \sin 2\varphi,
\end{aligned} \tag{34}$$

where φ is determined by (28). Note that U and V are defined as in (5) and are smaller than Chandrasekhar's (1950) U and V by a factor of $\sqrt{2}$.

3.2. ANOMALOUS DISPERSION

In the preceding section we have ignored the variation of the refractive index n within the line. In a magnetic field this effect may be of some importance, however, since the refractive index will be different for the differently polarized components of a Zeeman-split line. Hence we obtain different retardations of the components. In a longitudinal magnetic field there will be relative retardations between the left- and right-hand circular polarizations which causes a rotation of the polarization ellipse. This is the *Faraday rotation* or *Macaluso-Corbino* effect. In the case of a transversal magnetic field we have instead linear birefringence resulting in the so-called *Voigt effect* (Born, 1965). In the general case we have *elliptical birefringence*.

The refractive index n in the line in the absence of Doppler broadening is determined by (Beckers, 1969a)

$$n - 1 = \frac{\kappa_0 \lambda}{4\pi^{3/2} H(\alpha, 0)} \frac{v}{v^2 + \alpha^2},\tag{35}$$

where v and α are given by (22) and (21). κ_0 is the line absorption coefficient at the centre of the line, and $H(\alpha, 0)$ is the Voigt function for $v=0$. Including Doppler broadening, we obtain (Born, 1965; Beckers, 1969a; Rachkovsky, 1962a)

$$n - 1 = \frac{\kappa_0 \lambda}{2\pi H(\alpha, 0)} F(\alpha, v),\tag{36}$$

where

$$F(\alpha, v) = \frac{1}{2\pi} \int\limits_{-\infty}^{+\infty} \frac{u}{u^2 + \alpha^2} e^{-(u-v)^2} du.\tag{37}$$

Rachkovsky (1962a) expanded $F(\alpha, v)$ in a power series

$$F(\alpha, v) \approx \sum_{i=0}^{3} \alpha^i F_i(v).\tag{38}$$

The functions F_i have been tabulated by Beckers (1969a).

Anomalous dispersion can be included in the transfer equation (14) by writing the absorption matrix η as

$$\eta = \eta_a + \eta_\delta,\tag{39}$$

where η_a is our earlier absorption matrix given by (16)–(19), while

$$\eta_\delta = \begin{pmatrix} 0 & 0 & d & \frac{1}{\sqrt{2}} e \sin 2\chi \\ 0 & 0 & -d & -\frac{1}{\sqrt{2}} e \sin 2\chi \\ -d & d & 0 & -e \cos 2\chi \\ -\frac{1}{\sqrt{2}} e \sin 2\chi & \frac{1}{\sqrt{2}} e \sin 2\chi & e \cos 2\chi & 0 \end{pmatrix},\tag{40}$$

where

$$d = \frac{1}{\sqrt{2}} (\delta_r - \delta_b) \cos \gamma, \tag{41}$$

$$e = - [\delta_p - \tfrac{1}{2}(\delta_r + \delta_b)] \sin^2 \gamma, \tag{42}$$

$$\delta_p = \eta_0 \, F(\alpha, v),$$

$$\delta_r = \eta_0 \, F(\alpha, v - v_H), \tag{43}$$

$$\delta_b = \eta_0 \, F(\alpha, v + v_H).$$

The corresponding expressions by Beckers (1969a, b) are too small by a factor of $2 H(\alpha, 0)$.

By applying on η_δ the transformation L_1 as in (27) with φ in (11)$=\chi$, we obtain the same matrix as that found by Rachkovsky (1962a, b, 1967d) when he developed the theory of anomalous dispersion in stellar atmospheres. This matrix is also obtained by putting $\chi = 0$ in (40).

Generally the effect of elliptical birefringence in stellar atmospheres is small and can in most cases be neglected. The Faraday rotation is limited because the matrix coefficients η_i and δ_i in (20) and (43) are of the same order of magnitude. The rotation is proportional to the optical thickness $\Delta\tau$ of the line-forming layer. For $\Delta\tau = 1$ it is well below one radian. Radiation for which the rotation is larger than one radian is simply so much absorbed that it does not escape from the atmosphere. The effects of anomalous dispersion are thus much smaller than suggested by Kai (1968) and cannot explain the rapid changes in the field azimuth observed by Severny (1964, 1965).

The solution of the transfer equations for a homogeneous magnetic field with $\chi = 0$, assuming a Milne-Eddington atmosphere and a linear Planck function

$$B = B_0 \, (1 + \beta_0 \, \tau), \tag{44}$$

but including anomalous dispersion, was found by Rachkovsky (1962b, 1967d). If quantities with index zero refer to the continuum, the solution is

$$r_I(0, \theta) = \frac{I_0(0, \theta) - I(0, \theta)}{I_0(0, \theta)} = \frac{\beta_0 \cos \theta}{1 + \beta_0 \cos \theta}$$

$$\times \left[1 - \frac{\eta_I (\eta_I^2 + l^2)}{\eta_I^2 (\eta_I^2 + l^2 - m^2) - \alpha^2} \right], \tag{45}$$

$$r_Q(0, \theta) = \frac{Q_0(0, \theta) - Q(0, \theta)}{I_0(0, \theta)} = \frac{\beta_0 \cos \theta}{1 + \beta_0 \cos \theta} \frac{\eta_I^2 (a_+ - a_-)/2 - \alpha e}{\eta_I^2 (\eta_I^2 + l^2 - m^2) - \alpha^2},$$

$$r_U(0, \theta) = \frac{U_0(0, \theta) - U(0, \theta)}{I_0(0, \theta)} = - \frac{\beta_0 \cos \theta}{1 + \beta_0 \cos \theta} \frac{\eta_I \beta}{\eta_I^2 (\eta_I^2 + l^2 - m^2) - \alpha^2},$$

$$r_V(0, \theta) = \frac{V_0(0, \theta) - V(0, \theta)}{I_0(0, \theta)} = \frac{\beta_0 \cos \theta}{1 + \beta_0 \cos \theta} \frac{\eta_I^2 c + \alpha d}{\eta_I^2 (\eta_I^2 + l^2 - m^2) - \alpha^2},$$

where

$$\eta_I = 1 + \tfrac{1}{2}(a_+ + a_-),$$
$$l^2 = e^2 + 2d^2,$$
$$\alpha = 2cd - \tfrac{1}{2}(a_+ - a_-)e, \qquad\qquad (46)$$
$$\beta = -ce - \tfrac{1}{2}(a_+ - a_-)d,$$
$$m^2 = \tfrac{1}{4}(a_+ - a_-)^2 + 2c^2.$$

$a_\pm$, c, d and e have been defined earlier in (17), (19), (41) and (42). Note that the U and V parameters are here defined as in (5).

3.3. RADIATIVE SCATTERING

So far we have only treated the case of true absorption. Radiative scattering has mostly been neglected when discussing the polarization in commonly used spectral lines. This neglect is made simply because the theory of scattering in a magnetic field is so complicated, but there is no physical justification for leaving scattering out. Scattering plays an important role in most absorption lines used for magnetic-field measurements. Lines formed in true absorption weaken and disappear when we pass from the centre of the disc to the limb, whereas the profile of a commonly used line such as Fe I 5250.2 Å varies only little from centre to limb.

Due to its complication, the theory of scattering has only been developed for homogeneous magnetic fields. In its general form with true absorption and coherent scattering, the transfer equation reads

$$\cos\theta \, \frac{d}{d\tau} \mathbf{I} = (1+\eta)\mathbf{I} - (1-\varepsilon) \int_{4\pi} \mathbf{S}(\gamma, \varphi; \gamma', \varphi')\,\mathbf{I}(\gamma', \varphi') \frac{d\omega'}{4\pi}$$
$$- (1+\varepsilon\eta)\mathbf{B}. \qquad\qquad (47)$$

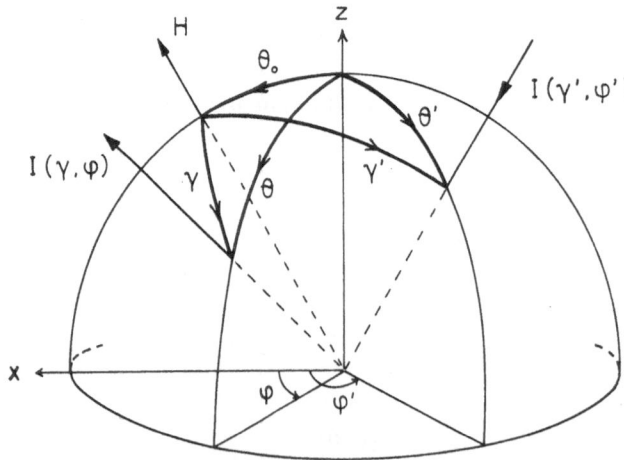

Fig. 2.   Geometry for the case of radiative scattering. The z-axis is assumed to be normal to the stellar atmosphere. The magnetic-field vector lies in the plane made up by the x- and z-axis. φ is the azimuth angle of the scattered ray, γ the angle between the field and the scattered ray. φ' and γ' are the corresponding quantities for the incident ray.

$S(\gamma, \varphi; \gamma', \varphi')$ is the scattering matrix, γ' the angle between the incident ray and the direction of the magnetic field, γ the angle between the scattered ray and the magnetic field, and φ' and φ the azimuths of the incident and scattered ray (see Figure 2). $I(\gamma', \varphi')$ describes the incident radiation. It is assumed that the fraction ε of the absorbed energy is transformed to thermal energy, while the rest is coherently scattered.

If the scattering is instead incoherent and we assume full redistribution of the absorbed quanta over the sublevels of the excited level, we should integrate the scattering term in (47) over all frequencies. Incoherent scattering of this kind will occur if the atom is subject to frequent collisions while it is in the excited state.

We have seen that in the case of true absorption the radiative transfer problem could be simplified considerably by diagonalizing the absorption matrix $\boldsymbol{\eta}$. When scattering is included the situation is no more that simple, because the matrices $\boldsymbol{\eta}$ and $S(\gamma, \varphi; \gamma', \varphi')$ do not diagonalize simultaneously.

Stepanov (1958b, 1960a, b, c, 1962) was the first who included scattering in a theory of line formation in a magnetic field. To be able to solve the problem Stepanov made several simplifying approximations which were not physically clear. A more rigorous theory of scattering was developed by Rachkovsky (1963a, b, 1965, 1967a, b, d). Rachkovsky (1965) used the rest intensities r_+ and r_- to compare the results from the different theories.

$$r_\pm = 2I_\pm/I_0, \tag{48}$$

where $I_\pm$ were defined in (32) and I_0 is the intensity of the continuous spectrum. The parameter η_0 was chosen independently for the different theories so that the theoretical line profile in the absence of a magnetic field at the centre of the disc should agree with the observed profile of the line Fe I 5250.2 Å. The results are shown in Figure 3 for a transversal field with $\chi=0$ and $v_H=1.0$. It appears that there is a quite good agreement between Rachkovsky's more rigorous and Stepanov's simplified theory, while Unno's theory gives strongly deviating results when we approach the limb. The discrepancies are as expected largest close to the line centre but become quite small far out in the wings. Hence it is necessary to use the theory of radiative scattering when calibrating solar magnetographs which use light from the centre of the line, e.g. the Locarno magnetograph.

For magnetographs working in the wings of the line it is generally not necessary to account for scattering. Unno's theory is a good approximation for $v > 1.5$ in the line wings and for the central part of the disc, i.e. for $0.7 < \cos\theta < 1$. Closer to the limb, however, it cannot be used.

In the wings for $v > 1.5$ Stepanov's theory can be used for practically the whole solar disc. For $v < 1.5$, however, the errors may be as large as 40%, and hence Rachkovsky's theory has to be used.

Rachkovsky (1963a, b, 1965) first developed the theory of scattering for atoms with a non-split upper level ($j_u=0$), i.e. for the transition $j_l=1$, $j_u=0$. Obridko (1965a) found the scattering matrix for the transition $j_l=0$, $j_u=1$ (which we have e.g. for the line FeI 5250.2 Å) and indicated how the scattering matrix could be found for any other transition. Rachkovsky (1967a) then developed a simple method to find the scattering matrix for arbitrary splitting of both the upper and lower levels and showed that it could be expressed as the sum of dyad products of matrices in such a way that one matrix only depends on the angle of incidence γ' while the other only depends on γ. The equations of radiative transfer could then be integrated. It was shown (Rach-

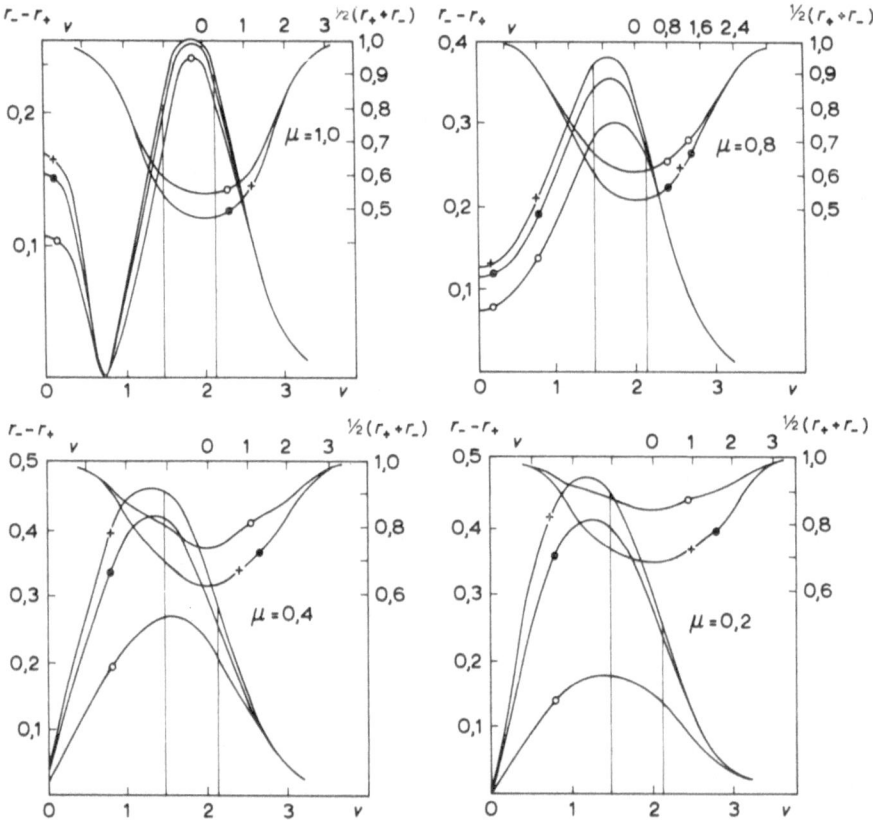

Fig. 3. Comparison according to Rachkovsky (1965) between calculations based on Rachkovsky's (filled circles) and Stepanov's (crosses) theories of scattering and Unno's (open circles) theory for true absorption. The magnetic field is assumed to be transversal with a Zeeman splitting $v_H = 1.0$. μ is the cosine of the heliocentric angle. The left and lower scales refer to $r_- - r_+$, the right and upper scales to $\frac{1}{2}(r_+ + r_-)$. The vertical lines show a typical position of an exit slit far out in the line wings. The function $r_- - r_+$ is symmetric around zero and is therefore only given for positive v.

kovsky, 1967d) that the problem of solving the transfer equations could be reduced to the problem of finding a solution to a system of integral equations for the generalized Ambartsumian φ function (Ambartsumian, 1943). Obridko (1965b) on the other hand used a method by Chandrasekhar to explore his equations for the transition $j_l = 0, j_u = 1$.

Rachkovsky (1967d) also succeeded to solve the transfer equations when both anomalous dispersion and coherent scattering for arbitrary splitting were taken into account. The case of incoherent scattering was discussed in an earlier work by Rachkovsky (1963b) and has recently been treated in greater detail by Domke (1969a, b) and Rees (1969).

3.4. THE HANLE EFFECT. LEVEL-CROSSING INTERFERENCE

Early laboratory experiments on resonance fluorescence (Wood and Ellett, 1924; Hanle, 1924) showed how the degree of polarization and the polarization angle of the scattered light varied with the magnetic field strength and the direction and polarization of the incident radiation. The theory of these polarization phenomena, usually called the *Hanle effect*, has been developed both from the classical theory (Breit, 1925) and the quantum theory of radiation (Dirac, 1927a, b; Weisskopf, 1931; Breit, 1932, 1933).

The theory shows that when the splitting in a magnetic field is not complete so that the different sublevels overlap, they are not independent of each other but there exist phase relations between their wave functions. This interference between the sublevels is often termed *level-crossing interference*.

This level-crossing interference has simply been neglected in the theories of radiative scattering we have discussed so far. The first calculation of the scattering matrix with the interference terms included was made by Obridko (1968) for the transition $j_l = 0, j_u = 1$. He used the methods developed by Hamilton (1947). I have transformed Obridko's matrix to our parameter system (5). The geometry is described by Figure 2 if we put $\theta_0 = 0$. The scattering matrix can be conveniently written in the form

$$\mathbf{S}(\gamma, \varphi; \gamma', \varphi') = \tfrac{3}{4}[\mathbf{S}_b^{(0)} + \tfrac{1}{2}\sqrt{\eta_p}(\sqrt{\eta_r} + \sqrt{\eta_b})\sin\gamma\sin\gamma'\mathbf{S}_i^{(1)}$$
$$+ \sqrt{\eta_r\eta_b}\mathbf{S}_i^{(2)} + \sqrt{\eta_p}(\sqrt{\eta_r} - \sqrt{\eta_b})\sin\gamma\sin\gamma'\mathbf{S}_i^{(3)}], \quad (49)$$

where indices b and i point out that the respective matrices describe pure blending or interference effects. In earlier works only the azimuthally symmetric blending matrix $\mathbf{S}_b^{(0)}$ has been taken into account but the interference matrices $\mathbf{S}_i^{(n)}$, $n = 1, 2, 3$, have been disregarded.

$$\mathbf{S}_b^{(0)} = \begin{pmatrix} 2\eta_p(1-\mu^2)(1-\mu'^2) + a_-\mu^2\mu'^2 & a_-\mu^2 & 0 & 2c'\mu^2 \\ a_-\mu'^2 & a_- & 0 & 2c' \\ 0 & 0 & 0 & 0 \\ 2c\mu'^2 & 2c & 0 & 2a_-\mu\mu' \end{pmatrix}, \quad (50)$$

where a_- is obtained from (17) with $\chi = 0$, i.e. $a_- = (\eta_r + \eta_b)/2$, while c is given by (19). c' is obtained by replacing γ by γ'. $\mu = \cos\gamma$ and $\mu' = \cos\gamma'$.

$$\mathbf{S}_i^{(1)} = \begin{pmatrix} 4\mu\mu'\cos(\varphi'-\varphi) & 0 & 2\sqrt{2}\,\mu\sin(\varphi'-\varphi) & 0 \\ 0 & 0 & 0 & 0 \\ -2\sqrt{2}\,\mu'\sin(\varphi'-\varphi) & 0 & 2\cos(\varphi'-\varphi) & 0 \\ 0 & 0 & 0 & 2\cos(\varphi'-\varphi) \end{pmatrix}, \quad (51)$$

$$
\mathbf{S}_i^{(2)} = \begin{pmatrix} \mu^2\mu'^2\cos 2(\varphi'-\varphi) & -\mu^2\cos 2(\varphi'-\varphi) & \sqrt{2}\,\mu^2\mu'\sin 2(\varphi'-\varphi) & 0 \\ -\mu'^2\cos 2(\varphi'-\varphi) & \cos 2(\varphi'-\varphi) & -\sqrt{2}\,\mu'\sin 2(\varphi'-\varphi) & 0 \\ -\sqrt{2}\,\mu\mu'^2\sin 2(\varphi'-\varphi) & \sqrt{2}\,\mu\sin 2(\varphi'-\varphi) & 2\mu\mu'\cos 2(\varphi'-\varphi) & 0 \\ 0 & 0 & 0 & 0 \end{pmatrix},
$$

$$\tag{52}$$

$$
\mathbf{S}_i^{(3)} = \begin{pmatrix} 0 & 0 & 0 & -\sqrt{2}\,\mu\cos(\varphi'-\varphi) \\ 0 & 0 & 0 & 0 \\ 0 & 0 & 0 & \sin(\varphi'-\varphi) \\ -\sqrt{2}\,\mu'\cos(\varphi'-\varphi) & 0 & -\sin(\varphi'-\varphi) & 0 \end{pmatrix}. \tag{53}
$$

It is readily seen that

$$
\mathbf{S}(\gamma, \varphi; \gamma', \varphi') = \mathbf{S}^T(\gamma', \varphi'; \gamma, \varphi). \tag{54}
$$

This equation expresses Helmholtz's principle of reciprocity for single scattering.

When $H \to 0$ we get $\eta_p = \eta_r = \eta_b = \eta$, $a_- = \eta$, and $c = c' = 0$. It can be directly verified that $\mathbf{S}(\gamma, \varphi; \gamma', \varphi')/\eta$ then becomes the well-known phase-matrix for Rayleigh scattering (Hamilton, 1947; Chandrasekhar, 1950).

The Doppler effect is not included in (49). Formally there is however no problem in including the Doppler effect. This is done by transforming the frequencies from a coordinate system following the atom to a system at rest and then integrate each term in (49) over the velocity distribution of the atoms. In the blending terms we obtain the well-known Voigt profiles if the velocity distribution is Maxwellian; the expressions for the interference terms will be more complicated.

If we consider a single scattering process, considerable errors may arise if the interference terms are neglected. In the transfer Equation (47), however, the light is integrated over all angles of incidence γ' and φ'. This means that the effects of interference will be small unless the radiation field is strongly anisotropic. The latter case is present in prominences which are illuminated by light from the solar disc.

It has been shown by Rachkovsky (1964) that for the transition $j_l = 1$, $j_u = 0$ there will be no influence of level-crossing interference on the scattering matrix. This is the transition for the lines FeI 6173.3 and FeI 6302.5 Å. Spectral lines corresponding to the transition $j_l = 0$, $j_u = 1$ that we have discussed are the commonly used line FeI 5250 Å, the resonance line CaI 4227 Å, and the forbidden coronal line FeXIII 10747 Å.

It was suggested by Hyder (1968) that the omission of the Hanle effect had caused serious errors in the interpretation of longitudinal magnetograph observations, and he calculated correction factors which should be used for the revision of the earlier measurements. These calculations were based on the assumption that the magnetic field was purely longitudinal ($\gamma = 0$). The signal from a longitudinal magnetograph is proportional to the Stokes parameter V. It is readily verified from (49)–(53) that the interference terms will have no contribution to the magnetograph signal. The scattered V is obtained by a pure blending of the red and blue Zeeman components, as was pointed out by Stenflo (1969).

The last term in (49) describes level-crossing interference which does not occur at zero magnetic field but only at intermediate field strengths. Due to the presence of this term the statement by Lamb (1970) that the interference effects are unobservable with longitudinal magnetographs unless the incident light is at least partially circularly polarized is not entirely correct; longitudinal magnetographs may be influenced somewhat also when the incident light is linearly polarized or unpolarized.

Extensive quantum mechanical calculations of the polarization of resonance radiation in a magnetic field have been made by Lamb (1970) and House (1970a, b, 1971). Such calculations are most important not only to give a deeper understanding of the Hanle effect but also to make possible the treatment of effects which cannot be adequately handled by the classical theory, e.g. the effect of collisions.

4. The Zeeman Effect for Inhomogeneous Fields

It is obvious that attempts to interpret Zeeman effect measurements in terms of a homogeneous magnetic field are unrealistic. Therefore the theories discussed so far have to be generalized to treat inhomogeneous fields before they can be successfully applied to the Sun.

The importance of accounting for inhomogeneities when interpreting magnetograph observations has been pointed out by Alfvén (1952, 1967) and Alfvén and Lehnert (1956). Only recently, however, these effects have been considered in detail (Stenflo, 1966a, b, 1968a, b).

Before we start to discuss the Zeeman effect in an inhomogeneous atmosphere, we will review some general properties of solar magnetic fields.

4.1. FILAMENTARY STRUCTURE: GENERAL PROPERTIES

A large part of the total magnetic flux on the Sun seems to emerge through small 'flux points' with high field strengths, while the field in between these points is quite weak. In the following we will call the regions with strong fields magnetic 'filaments'.

Although we are presently not able to treat the filaments in a satisfactory way from the point of view of theoretical plasma physics, our knowledge about the physical conditions in solar filaments has increased significantly in recent years. Sheeley (1967) found filaments with field strengths exceeding 300 G and diameters less than one second of arc far from active regions. The location of these high field strengths coincided with strong weakenings of many spectral lines and also showed good correlation with the location of regions of $Ca^+ K_{232}$ emission. The intensity of the continuous spectrum was smaller and the spectral lines were generally red-shifted in the filaments. Investigating the line weakenings further, Chapman and Sheeley (1968) found that they were mainly due to more excitation and ionization in the magnetic filaments than outside them. This was explained as caused by a higher temperature in the filaments. As a consequence of the line weakenings, there is a one-to-one correspondence between the magnetic filaments and the photospheric network recorded in the same spectral line as used for the magnetic-field observations. The chromospheric network on the other hand is much coarser and shows less correlation with the photospheric magnetic field.

In a very careful study of the active region around a big unipolar sunspot, Beckers and Schröter (1968a) found more than 2000 what they called magnetic knots around the sunspot. The field strengths in the knots varied between 600 and 1400 G, and the typical diameter of a knot was 1000 km. The total magnetic flux through the magnetic knots balanced the flux through the sunspot. The knots seemed to coincide with dark intergranular regions and Ca^+ plages, show a generally downward flow of matter and have lifetimes exceeding 30 min. Beckers and Schröter (1968c) also found striking inhomogeneities within the umbra of the sunspot.

The weakenings of spectral lines in magnetic filaments have a serious effect on the readings of a magnetograph. The main contribution to the average line profile, which is used for calibration, comes from the interfilamentary medium. The magnetograph signal, however, is proportional to the steepness of the wings of the weakened line. Harvey and Livingston (1969) made magnetograph recordings simultaneously in different spectral lines and found large discrepancies between the results obtained. These systematic differences between the apparent field strengths obtained with the different lines could be explained entirely in terms of line weakenings of the temperature-sensitive spectral lines. Harvey and Livingston also looked for non-magnetic line weakenings in the 5250 Å line but could not find any.

If the line weakenings are caused by a temperature increase in the magnetic filaments, one would expect that the weakening depends on the field strength. Harvey and Livingston (1969) found, however, that the weakenings seemed to be independent of the recorded field strength. There seemed to be mainly two alternative interpretations of this strange result: (1) The temperature increase does not depend on magnetic-field strength if the field exceeds some small threshold value. (2) Magnetic fields outside active regions do only occur in filaments having practically one and the same field strength.

Livingston and Harvey (1969) tested the second alternative and found clear indications on what they called "quantization in photospheric magnetic flux". They determined the magnetic flux in a single filament to be 2.8×10^{18} Mx. The diameter of a filament was not known. Assuming the cross-section to be $(1'')^2$, the field strength would be 525 G. The true cross-section might well be smaller, in which case the field strength would be correspondingly higher.

The strong temperature-sensitivity of the Fe I 5250 Å line has also been pointed out in works by Wiehr (1970) and Staude (1970b), who find that the effect is mainly caused by the low excitation potential of the lower level of the line. Wiehr (1970) suggests that the line Fe I 6302.5 Å should be suitable for magnetograph observations, since it does not change when one goes from the photosphere to the umbra of a sunspot.

Let us now consider how the line profiles are influenced by a temperature rise in the magnetic filaments. The line depth can be calculated from the equation (Gussman, 1968)

$$r_\lambda(\mu) = \int_0^\infty g(\tau, \mu) \left(1 - e^{-\tau\lambda/\mu}\right) d\tau. \qquad (55)$$

τ and $\tau + \tau_\lambda$ are the optical depths in the continuum and the line, respectively, and μ is the cosine of the heliocentric angle. The weight function is

$$g(\tau, \mu) = \frac{e^{-\tau/\mu}}{I_0(0, \mu)} \frac{dB(T)}{d\tau},$$ (56)

where $I_0(0, \mu)$ is the intensity of the emerging radiation in the continuum, and $B(T)$ is the Planck function.

For weak lines (55) becomes

$$r_\lambda(\mu) = \frac{1}{\mu I_0(0, \mu)} \int_0^\infty e^{-\tau/\mu} \frac{dB(T)}{d\tau} \tau_\lambda \, d\tau.$$ (57)

By far the most temperature-sensitive factor in the expression for τ_λ is the relative population number $n_{r,s}/\Sigma n_r$. This ratio between the number of atoms in excitation level s and ionization stage r to the total number of atoms of that element is determined by the Boltzmann and Saha equations. We can expand $\Delta n_{r,s}/\Sigma n_r$ in powers of $\Delta T/T$, but since $\Delta T/T$ is a small quantity, we need not consider higher orders than the first. If

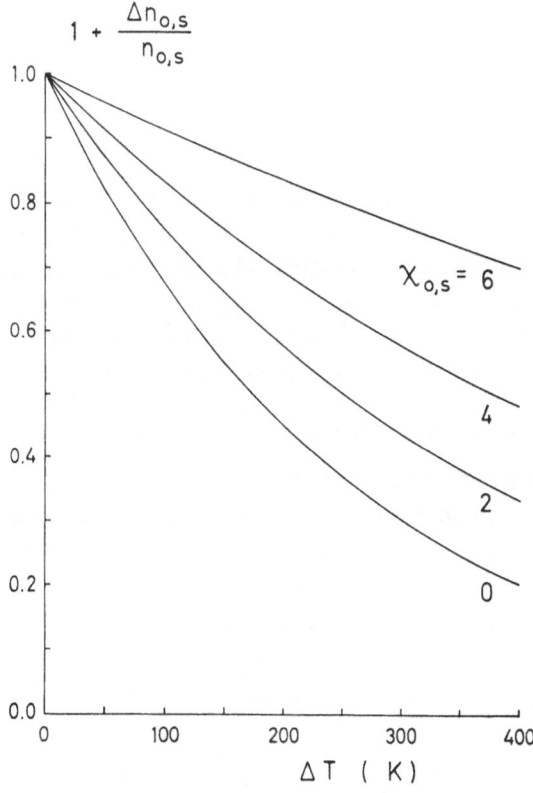

Fig. 4. The relative change in population number $n_{0,s}$ of level s for neutral iron atoms as a function of the temperature increase ΔT. The curves are drawn for different values of the excitation potential of level s (eV).

we also assume that the change in the electron density can be neglected, we obtain for neutral metal atoms

$$1 + \frac{\Delta n_{0,s}}{n_{0,s}} \approx \exp\left[-\left(\frac{3}{2} + \frac{\chi_i - \chi_{0,s}}{kT}\right)\frac{\Delta T}{T}\right].$$ (58)

In deriving (58) we have used the fact that metal atoms in the photosphere are predominantly singly ionized. χ_i and $\chi_{0,s}$ are the ionization and excitation potentials, respectively. Equation (58) is illustrated in Figure 4 for iron ($\chi_i = 7.90$ eV). If no other effect on the line profile were present, $1 + \Delta n_{0,s}/n_{0,s}$ would equal the ratio between the depths of the weakened and unweakened line, as seen from (57). The weakening is very large for low excitation potentials. The Fe I 5250 Å line has an excitation potential of only 0.12 eV, which to a large extent explains why it is so sensitive. A weakening of the same magnitude as observed by Harvey and Livingston (1969) would be obtained with $\Delta T \approx 300$ K in (58).

The proportionality between $1 + \Delta n_{0,s}/n_{0,s}$ and the line weakening is valid only in the case of weak lines. None of the lines commonly used in magnetograph observations can however be regarded as being weak, since they are not on the linear part of the curve of growth. When saturation plays a role, one has to use a larger ΔT to obtain the same weakening as for weak lines.

There is however another factor in the expression for the line depth in (55)–(57) that depends on the temperature, and that is the temperature gradient or $dB/d\tau$. If the temperature gradient is zero, the line disappears. The effect of $dT/d\tau$ depends on the level in the solar atmosphere at which the temperature rise takes place. As different lines are formed at different levels, this effect will vary from line to line.

It is possible to determine what part of the solar atmosphere that is heated by making numerical integrations of (55) for different filament models and compare the results with the observed weakenings in a great number of lines. Such a determination of the physical structure of a filament would provide an improved base for theoretical attacks on the very fundamental problem of the origin of the filaments.

4.2. FILAMENTARY STRUCTURE: INTERPRETATION OF THE MAGNETIC-FIELD OBSERVATIONS

Let us as a special case of a 'multi-stream' model of the solar atmosphere assume that we have two 'streams', one being the filaments (index f) and the other the interfilamentary medium (index i). Further we assume that the magnetograph signal is reduced in the filaments by a factor δ due to line weakening, saturation, Doppler-shift or reduced intensity. The longitudinal magnetic field in the filaments, which occupy a fraction A_f of the solar surface, is H_f. We have corresponding notations for the interfilamentary medium. It then follows that the true average longitudinal field is

$$H = A_f H_f + A_i H_i,$$ (59)

while the observed average longitudinal field is

$$H_{obs} = \delta A_f H_f + A_i H_i.$$ (60)

If the filaments occupy only a small fraction of the solar surface, so that $A_f \ll 1$ and $A_i \approx 1$, (59) and (60) give

$$H \approx \frac{1}{\delta} H_{obs} - \frac{1-\delta}{\delta} H_i.$$ (61)

This is the relation between the true field and the observed field with the inter-filamentary field as an unknown parameter, if we suppose that δ is known. For the mostly used line Fe I 5250.2 Å, δ has been provisionally determined by Harvey and Livingston (1969) for the central part of the solar disc, outside active regions. With the exit slits usually used in the Mt Wilson magnetograph (7–87 mÅ), δ would be 0.31, for Crimea (36–93 mÅ) $\delta \approx 0.43$.

If we use $\delta = \frac{1}{3}$, (61) becomes

$$H \approx 3H_{obs} - 2H_i.$$ (62)

(61) or (62) could possibly be used for the revision of all earlier magnetograph ob-servations in the 5250 Å line. (62) is illustrated in Figure 5. One difficulty is that we

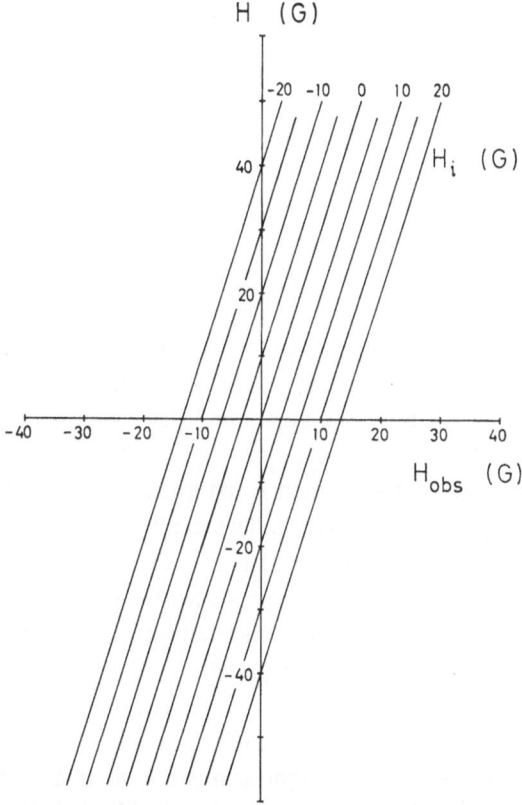

Fig. 5. Relation, given by (62), between the true average longitudinal field and the observed average longitudinal field (for a Babcock-type magnetograph) for a particular filamentary model with different values of the interfilamentary field strength H_i.

do not know H_i, which may have a sign opposite to the observed field. Furthermore, we do not know if δ varies across the solar disc or with the solar cycle.

Let us now turn to the case of transversal field measurements. In the observing scheme of the Crimean magnetograph, with which the pioneer work on transversal fields was done, two recordings $i_{\perp 1}$ and $i_{\perp 2}$ with different orientations of the polarization analyser are made. As was shown by Stepanov and Severny (1962), we can write

$$i_{\perp 1,\,2} \sim I_0 f_\perp (H) \sin^2 \gamma \begin{Bmatrix} \sin 2\chi \\ \cos 2\chi \end{Bmatrix}, \tag{63}$$

where χ is the azimuth angle of the field vector and $f_\perp (H)$ is a function that depends on the line profile and the magnetic field strength H. For weak fields (below a few hundred gauss), $f_\perp (H)$ is proportional to H^2, but it saturates and decreases again for strong fields (a few thousand gauss).

Neglecting line weakenings and fluctuations in γ and χ, the observed transversal field in the case of a multi-stream model is determined by

$$f_\perp (H_{\mathrm{obs}}) = \sum_k A_k f_\perp (H_k). \tag{64}$$

In discussing the filamentary structure for the transversal case we will allow A_f to assume any value from zero to one, but assume that the filaments carry all the flux, i.e. $H_i = 0$ and $H = A_f H_f$. We thus obtain from (64)

$$f_\perp (H_{\mathrm{obs}}) = A_f f_\perp (H_f), \tag{65}$$

or

$$f_\perp (H/A_f) = f_\perp (H_{\mathrm{obs}})/A_f. \tag{66}$$

For weak fields (66) reduces to

$$H = \sqrt{A_f} H_{\mathrm{obs}}. \tag{67}$$

Hence, if the filaments occupy only a small fraction of the solar surface, the observed field strength will be much *greater* than the true field strength. If the microspots occupy 1% of the solar surface, the apparent field will be 10 times the actual field according to (67).

For larger field strengths the function $f_\perp$ saturates very quickly. Particularly this is the case for $f_\perp (H/A_f)$, when A_f is small. This explains the strange behaviour of the curves in Figure 6, where we have plotted H as a function of H_{obs} for some different values of A_f. To do this the $f_\perp (H)$-curve presented by Rachkovsky (1967c) calculated according to his theory including radiative scattering has been used. All earlier transversal field measurements have been interpreted in terms of $A_f = 1$, but it is seen from Figure 6 that the interpretation is drastically changed when a filamentary structure is introduced. The observations of transversal fields exceeding 1000 G in active regions indicate however that our simplified approach is not valid there but that there must also exist, in and close to the sunspots, magnetic fields that fill up the whole space and are not merely inside narrow filaments. Nevertheless Figure 6 shows how extremely sensitive the observed transversal field is to the assumed model of the field.

One interesting feature of Figure 6 is that it opens a new possibility to test the idea of Livingston and Harvey (1969) that the magnetic flux outside active regions is what they call 'quantized'. This case coincides with the assumptions on which Figure 6 is based. The true average field H at high heliographic latitudes could be estimated from measurements of the average longitudinal field $H \cos \gamma$ assuming that the average direction of the field is along the solar radius. We assume that corrections for line weakenings are made for both the observed longitudinal and transversal fields. Close to the poles the average field will have an appreciable transversal component. Knowing

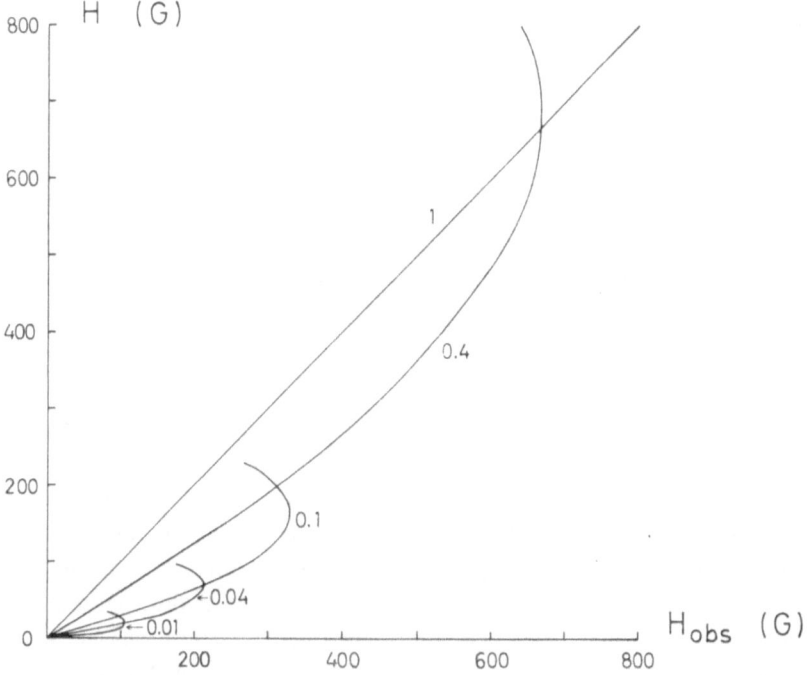

Fig. 6.   Relation between the true average transversal field and the observed average field for the filamentary model described in the text. The different curves refer to the cases that 1%, 4%, 10%, 40% and 100% of the solar surface is occupied by the filaments. The last case is a homogeneous field, represented by the straight line in the figure. All earlier transversal field measurements have been interpreted with the assumption of a homogeneous field inside the scanning aperture.

H the observed transversal field should be obtained from the intersection between a horizontal line in Figure 6 and the curve for the appropriate A_f. If $A_f = 10^{-2}$, which seems reasonable, $H = 5$ G would give $H_{obs} \approx 50$ G, and $H = 10$ G would give $H_{obs} \approx 90$ G. This is close to what should be possible to detect with transversal magnetographs by statistically reducing the noise and paying extreme attention to instrumental polarization and the position of the zero level.

4.3. Magnetic Microturbulence

From observations of the equivalent widths of spectral lines it can be concluded that there must exist a small-scale turbulence with eddies considerably smaller than 100 km having rms turbulent velocities of 1–2 km/s. As there is a coupling between the velocity field and the magnetic field in magnetohydrodynamics, one could expect that there should be fluctuations in the magnetic field on the same small scale. If there is equipartition between the magnetic and kinetic energy, the field strength should be several hundreds of gauss.

Measurements of the correlation between magnetic fields and the solar granulation have shown (Steshenko, 1960; Semel, 1962; Leighton, 1965; Livingston, 1968; Beckers and Schröter, 1968b; Howard and Bhatnagar, 1970) that if there exists any granular magnetic field at all, it cannot exceed a few tens of gauss.

The solar granulation, however, represents the long-wavelength part of the turbulence spectrum, if we at all can regard granulation as a kind of turbulence. The extremely complicated theory of hydromagnetic turbulence, which has only been developed for some idealized situations (Chandrasekhar, 1955; Kaplan, 1959; Kraichnan and Nagarajan, 1967; Nagarajan 1970, 1971), permits different solutions with different distributions of the magnetic and kinetic energies over the turbulence spectrum. Thus the magnetic energy density could be concentrated to the small eddies.

Attempts to determine the turbulent field strength of the optically thin eddies have been made (Unno, 1959; Fitremann and Frisch, 1969). The expected effects are, however, extremely small and difficult to observe, and the results that have been obtained have hardly been conclusive.

For the calculation of the polarization in spectral lines in a turbulent atmosphere we will here restrict ourselves to the special case that we only have to do with optically thin turbulent elements. In analogy with the terminology for velocity fields we will call these small-scale fluctuations in the magnetic field *magnetic microturbulence*.

An approach to a theory of magnetic microturbulence has been made by Stenflo (1968b). Staude (1970a) has also discussed the effect of microfluctuations. Ordinary microturbulence is treated by averaging the velocity-dependent terms (the absorption coefficients) in the equation of radiative transfer over a distribution of turbulent velocities. Analogously a microturbulent magnetic field may be treated by averaging the terms in the transfer equation that depend on the magnetic field over a distribution of field vectors. In the case of true absorption (14) becomes

$$\cos\theta \, \frac{d}{dt} \mathbf{I} = - \langle \kappa\varrho \, (1+\eta) \rangle \, \mathbf{I} + \langle \kappa\varrho \, (1+\eta) \, \mathbf{B} \rangle, \tag{68}$$

t being the geometrical depth, κ the coefficient of continuous absorption, and ϱ the density. The averagings should be made not only over a distribution of field vectors but also over distributions of radial velocities and temperatures, as all these parameters are generally correlated in a turbulent medium. The presence of a microturbulent magnetic field may substantially change the polarization of the emerging radiation. Although it is so important, no calculations of line formation for a realistic model of a turbulent medium have yet been made.

5. Relations Between the Observed Magnetic Field and the Resolution of the Instrument

When the scanning aperture of a magnetograph is made smaller so that more and more of the inhomogeneities in the solar atmosphere can be resolved, one would expect that the observed average magnetic field will change. The way in which the magnetograph reading is related to the resolution is, however, far from obvious and depends strongly on the magnetograph construction and on the true structure of the magnetic field.

The first attempt to investigate the relation between the observed average field far from active regions and the slit size was made in 1965 by Stenflo (1966a, b) with the Babcock-type magnetograph of the Crimean Astrophysical Observatory. A strong increase of the observed field was found for the smallest slit size that could possibly be used then, 7 (arc s)^2.

The preliminary observations of the dependence on slit size were repeated one year later (Stenflo, 1968a) with the Crimean magnetograph. With a somewhat better statistical material and improvements in the magnetograph, particularly in the control of the position of the zero line, a considerably smaller increase of only about 20% for the observed average field was found for the 7 (arc s)^2 aperture. The systematic displacement of the zero line due to infiltration of instrumental linear polarization was found to be larger for smaller slit sizes. This could to some extent have influenced the results from 1965, when the accurate position of the zero line could not be determined (the zero line obtained with the ADP crystal switched off does not represent the true zero line for $H_\parallel = 0$).

A theoretical investigation of the problem showed (Stenflo, 1968b) how an increase with resolution of the average field observed with a Babcock-type magnetograph is physically possible and may be caused by the better compensation of the Doppler shifts of small velocity structures. The increase with resolution will not exceed a few tens of per cent, however. It was also shown that if line weakenings occur in magnetic regions, the magnetograph will give incorrect values of the average field for any resolving power used.

The behaviour of the Evans-type magnetograph (Evans, 1964) is completely different. In contrast to the Babcock magnetograph, which has fixed exit slits and measures the intensity difference between right- and left-hand circular polarization, the Evans-type magnetograph is a servo instrument with the exit slits moving with the spectral line components, which directly determines the Zeeman splitting. If all magnetic features were resolved, the Evans-type magnetograph would give the correct field strength independently of any line weakenings that may occur.

Let us consider how the signal from an Evans-type magnetograph is influenced by the scanning aperture used in the case of a filamentary structure with the filaments evenly distributed (no clustering). We assume that the slit area is S, the cross-section of one filament is S_f, and that the filaments occupy a fraction A_f of the solar surface. Further we assume that the line depth is $r(v)$ with $v = \Delta\lambda/\Delta\lambda_D$ and that the Zeeman displacement in the filaments is $v_H = \Delta\lambda_H/\Delta\lambda_D$. The spectral line is assumed to be weakened by a factor δ in the filaments but has the same form as the unweakened line. The magnetic field is directed along the line of sight. When *one* filament is inside the slit the magnetograph will "see" the line profile in one circular polarization $r_{obs}(v)$ determined by

$$r_{obs}(v) = \frac{S - S_f}{S} r(v) + \frac{S_f}{S} \delta r(v - v_H). \tag{69}$$

The exit slits will move with this observed profile until the difference signal from the two photomultipliers is zero. The position of the exit slits then determines the observed Zeeman displacement.

The area on the sun that contains precisely one filament is S_f/A_f. If the slit area S

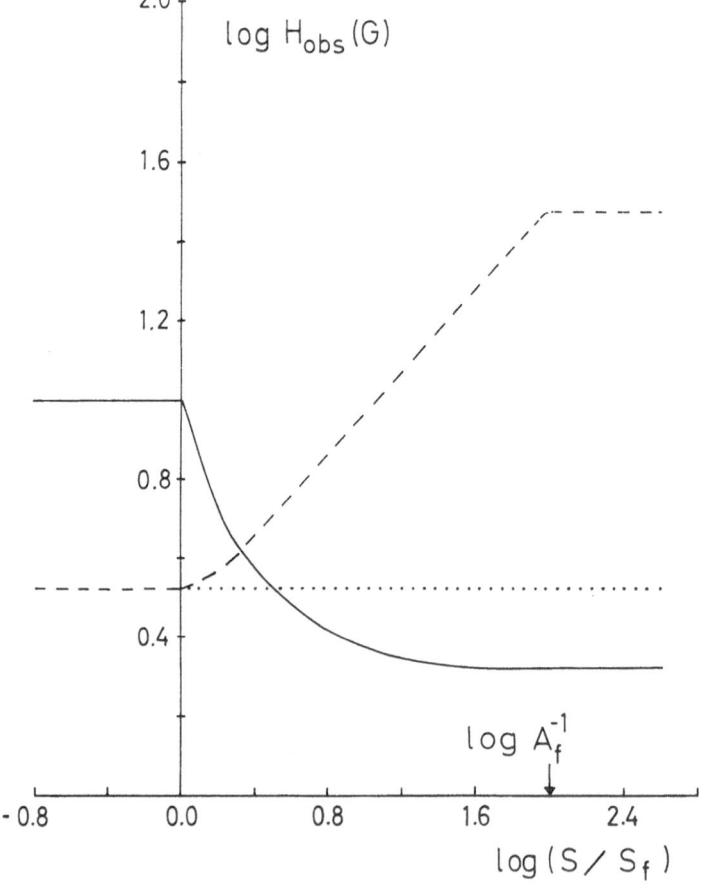

Fig. 7.   Dependence of the average observed field H_{obs} on the slit area S for a filamentary model of the field, in which the filaments occupy a fraction $A_f = 0.01$ of the solar surface, the cross-section of a filament is S_f, the filamentary magnetic field is 1000 G and the interfilamentary field zero. Solid line: Evans-type magnetograph. Dotted line: Babcock-type magnetograph. Dashed line: Transversal magnetograph of the Crimean type. For the transversal field measurements it is assumed that the field in the filaments is purely transversal, for the longitudinal field measurements it is assumed that it is purely longitudinal.

is smaller than S_f/A_f there will be $S_f/(SA_f)$ possible slit positions within the area S_f/A_f, and for all but one there is no magnetic filament within the slit. The resulting Zeeman displacement from (69) should accordingly be multiplied by $A_f S/S_f$ to give the average field strength observed with the magnetograph. The true field strength should according to this model be recorded for slits with $S \leqslant S_f$, while a considerably lower value of the field will be found for larger slits. The influence of the slit size will to some extent depend on the wavelength separation of the exit slits.

Numerical calculations of the effect have been made assuming that $r(v)$ can be described by a Gauss function and that the infinitely narrow exit slits are situated at a distance of 1.0 in the parameter v from each other. Further we assume that $v_H = 1$, $A_f = 0.01$ and $\delta = \frac{1}{3}$. To give the results in G, we assume that $v_H = 1$ corresponds to a field of 1000 G in the filaments. The true average field is thus $1000 \, A_f \, G = 10$ G. In Figure 7 we have given $\log H_{obs}$ as a function of $\log (S/S_f)$, where H_{obs} is the observed average field. The field observed with a longitudinal Babcock-type magnetograph is also illustrated (dotted line) for the same model of the field, assuming that the effects of Doppler and brightness compensations can be disregarded.

The influence of a filamentary structure on the interpretation of measurements with the transversal Crimean magnetograph was illustrated in Figure 6. The curves in this diagram have been used to calculate the dependence of the observed transversal magnetic field on the slit area S in our filamentary model. The result is presented in Figure 7.

It appears from Figure 7 that the observed average field shows an entirely different dependence on the slit area in the three cases. If such relations between H_{obs} and S could be determined empirically, it would be possible not only to test the assumption of a filamentary structure but also to obtain information on the important parameters S_f, A_f and H_f.

6. Conclusive Remarks. Suggestions for the Future

Obviously there is still much to be done before we can correctly interpret what is measured with a magnetograph. But our insight in the problems has increased significantly during the last years, and we are beginning to reveal the nature of the filamentary structure of the solar atmosphere.

As a kind of summary some recommendations and suggestions for future work in this field will be listed.

(1) The line Fe I 5250.2 Å should not be used further in magnetograph observations because it is so temperature-sensitive. The suitability of other lines should be thoroughly investigated.

(2) All earlier observations in the line Fe I 5250.2 Å should be revised. To do this we need to know the relation between the filamentary and interfilamentary fields. The variation of the line weakenings over the solar disc and with the solar cycle should also be determined.

(3) The filamentary nature of the field could be tested and studied by making recordings with different slit sizes with a Babcock-type, Evans-type and a transversal

magnetograph. This could give information on the size and distribution of the filaments.

(4) A physical model (mainly temperature stratification) of a filament could be established by recording the magnetic field in a great number of spectral lines, which are weakened differently in the filaments.

(5) Earlier interpretations of transversal field measurements should be reexamined by making recordings of the transversal field in *active regions* with different slit sizes.

(6) The theory of line formation in a magneto-turbulent medium should be further developed. The polarization in the lines should be calculated for a realistic turbulent model in which the different physical parameters are coupled to each other.

(7) Numerical calculations of the effect of level-crossing interference on measurements of prominence magnetic fields should be made.

Of course this list of suggestions is very incomplete, but nevertheless it illustrates how much there is to be done. All our knowledge of solar magnetic fields is based on the theory we use to interpret the observations, and it is no wonder that this theory has received so much attention during the last years.

References

Alfvén, H.: 1952, *Arkiv Fysik* **4**, 407.

Alfvén, H.: 1967, in *Magnetism and the Cosmos* (ed. by Hindmarsh, Lowes, Roberts, and Runcorn), Oliver & Boyd, Edinburgh/London, p. 246.

Alfvén, H. and Lehnert, B.: 1956, *Nature* **178**, 1339.

Ambartsumian, V. A.: 1943, *Dokl. Akad. Nauk S.S.S.R.* **38**, No. 8.

Beckers, J. M.: 1969a, *Solar Phys.* **9**, 372.

Beckers, J. M.: 1969b, *Solar Phys.* **10**, 262.

Beckers, J. M. and Schröter, E. H.: 1968a, *Solar Phys.* **4**, 142.

Beckers, J. M. and Schröter, E. H.: 1968b, *Solar Phys.* **4**, 165.

Beckers, J. M. and Schröter, E. H.: 1968c, *Solar Phys.* **4**, 303.

Born, M.: 1965, *Optik*, 2nd ed., Springer-Verlag, New York.

Breit, G.: 1925, *J. Opt. Soc. Am.* **10**, 439.

Breit, G.: 1932, *Rev. Mod. Phys.* **4**, 504.

Breit, G.: 1933, *Rev. Mod. Phys.* **5**, 91.

Chandrasekhar, S.: 1950, *Radiative Transfer*, Clarendon Press, Oxford.

Chandrasekhar, S.: 1955, *Proc. Roy. Soc. London* **233**, 322 and 390.

Chapman, G. A. and Sheeley, Jr., N. R.: 1968, *Solar Phys.* **5**, 442.

Dirac, P. A. M.: 1927a, *Proc. Roy. Soc. London* **114**, 710.

Dirac, P. A. M.: 1927b, *Proc. Roy. Soc. London* **114**, 243.

Domke, H.: 1969a, *Astrofiz.* **5**, 525.

Domke, H.: 1969b, *Monatsber. Deut. Akad. Wiss. Berlin* **11**, 269.

Evans, J. C.: 1968, Report at the MASUA Theoretical Physics Conference, University of Nebraska, Nov. 2–3, 1968.

Evans, J. C.: 1969a, Report at the Special Meeting on Solar Astronomy of the AAS, Pasadena, Febr. 18–21, 1969.

Evans, J. C.: 1969b, Report at the IAU Symposium on Laboratory Astrophysics, York University, Toronto, Ontario, Canada, Nov. 7–8, 1969.

Evans, J. C. and Dreiling, L. A.: 1969, Report at the 130th Meeting of the AAS, State University of New York, Albany, New York, Aug. 11–14, 1969.

Evans, J. W.: 1964, in *Atti del Convegno sui Campi Magnetici Solari* (ed. by G. Barbèra), Firenze.

Fitremann, M. and Frisch, U.: 1969, *Compt. Rend. Acad. Sci. Paris* **268**, 705.

Gussman, E. A.: 1967, *Z. Astrophys.* **65**, 456.

Hamilton, D. R.: 1947, *Astrophys. J.* **106**, 457.

Hanle, W.: 1924, *Z. Phys.* **30**, 93.

Harvey, J. and Livingston, W.: 1969, *Solar Phys.* **10**, 283.

House, L. L.: 1970a, *J. Quant. Spectr. Radiative Transfer* **10**, 909.

House, L. L.: 1970b, *J. Quant. Spectr. Radiative Transfer* **10**, 1171.

House, L. L.: 1971, *J. Quant. Spectr. Radiative Transfer* **11**, 367.

Howard, R. and Bhatnagar, A.: 1969, *Solar Phys.* **10**, 245.

Hyder, C. L.: 1968, *Solar Phys.* **5**, 29.

Kai, K.: 1968, *Publ. Astron. Soc. Japan* **20**, 154.

Kaplan, S. A.: 1958, in *Electromagnetic Phenomena in Cosmical Physics* (ed. by B. Lehnert), Cambridge University Press, p. 504.

Kraichnan, R. H. and Nagarajan, S.: 1967, *Phys. Fluids* **10**, 859.

Lamb, F. K.: 1970, *Solar Phys.* **12**, 186.

Leighton, R. B.: 1965, in R. Lüst (ed.), 'Stellar and Solar Magnetic Fields', *IAU Symp.* **22**, 158.

Livingston, W.: 1968, *Astrophys. J.* **153**, 929.

Livingston, W. and Harvey, J.: 1969, *Solar Phys.* **10**, 294.

Mattig, W.: 1966, in *Atti del Convegno sulle Macchie Solari* (ed. by G. Barbèra), Firenze.

Moe, O. K.: 1968, *Solar Phys.* **4**, 267.

Moe, O. K. and Maltby, P.: 1968, *Astrophys. Letters* **1**, 189.

Nagarajan, S.: 1970, *Phys. Fluids*, in press.

Nagarajan, S.: 1971, this volume, p. 487.

Obridko, V. N.: 1965a, *Astron. Zh.* **42**, 102.

Obridko, V. N.: 1965b, *Astron. Zh.* **42**, 502.

Obridko, V. N.: 1968, *Soln. Aktivnostj* No. 3, 64.

Rachkovsky, D. N.: 1961a, *Izv. Krymsk. Astrofiz. Obs.* **25**, 277.

Rachkovsky, D. N.: 1961b, *Izv. Krymsk. Astrofiz. Obs.* **26**, 63.

Rachkovsky, D. N.: 1962a, *Izv. Krymsk. Astrofiz. Obs.* **27**, 148.

Rachkovsky, D. N.: 1962b, *Izv. Krymsk. Astrofiz. Obs.* **28**, 259.

Rachkovsky, D. N.: 1963a, *Izv. Krymsk. Astrofiz. Obs.* **29**, 97.

Rachkovsky, D. N.: 1963b, *Izv. Krymsk. Astrofiz. Obs.* **30**, 267.

Rachkovsky, D. N.: 1965, *Izv. Krymsk. Astrofiz. Obs.* **33**, 111.

Rachkovsky, D. N.: 1967a, *Izv. Krymsk. Astrofiz. Obs.* **36**, 3.

Rachkovsky, D. N.: 1967b, *Izv. Krymsk. Astrofiz. Obs.* **36**, 9.

Rachkovsky, D. N.: 1967c, *Izv. Krymsk. Astrofiz. Obs.* **36**, 51.

Rachkovsky, D. N.: 1967d, *Izv. Krymsk. Astrofiz. Obs.* **37**, 56.

Rees, D. E.: 1969, *Solar Phys.* **10**, 268.

Semel, M.: 1962, *Compt. Rend. Acad. Sci. Paris.* **254**, 3978.

Severny, A. B.: 1964, *Izv. Krymsk. Astrofiz. Obs.* **31**, 126.

Severny, A. B.: 1965, *Izv. Krymsk. Astrofiz. Obs.* **33**, 3.

Sheeley, Jr., N. R.: 1967, *Solar Phys.* **1**, 171.

Staude, J.: 1970a, *Solar Phys.* **12**, 84.

Staude, J.: 1970b, *Solar Phys.* **15**, 102.

Stenflo, J. O.: 1966a, *Observatory* **86**, 73.

Stenflo, J. O.: 1966b, *Arkiv. Astron.* **4**, 173.

Stenflo, J. O.: 1968a, *Acta Univ. Lund.* II No. 1 (= *Medd. Lunds Astron. Obs.* Ser. II Nr: 152).

Stenflo, J. O.: 1968b, *Acta Univ. Lund.* II No. 2 (= *Medd. Lunds Astron. Obs.* Ser. II Nr: 153).

Stenflo, J. O.: 1969, *Solar Phys.* **8**, 260.

Stepanov, V. E.: 1958a, *Izv. Krymsk. Astrofiz. Obs.* **18**, 136.

Stepanov, V. E.: 1958b, *Izv. Krymsk. Astrofiz. Obs.* **19**, 20.

Stepanov, V. E.: 1960a, *Izv. Krymsk. Astrofiz. Obs.* **23**, 291.

Stepanov, V. E.: 1960b, *Izv. Krymsk. Astrofiz. Obs.* **24**, 293.

Stepanov, V. E.: 1960c, *Astron. Zh.* **37**, 631.

Stepanov, V. E.: 1962, *Izv. Krymsk. Astrofiz. Obs.* **27**, 140.

Stepanov, V. E. and Severny, A. B.: 1962, *Izv. Krymsk. Astrofiz. Obs.* **28**, 166.

Steshenko, N. V.: 1960, *Izv. Krymsk. Astrofiz. Obs.* **22**, 49.

Unno, W.: 1956, *Publ. Astron. Soc. Japan* **8**, 108.

Unno, W.: 1959, *Astrophys. J.* **129**, 375.
Weisskopf, V.: 1931, *Ann. Phys.* **9**, 23.
Wiehr, E.: 1970, *Solar Phys.* **11**, 399.
Wood, R. W. and Ellett, A.: 1924, *Phys. Rev.* **24**, 243.

Discussion

Pecker: (1) For *numerical* purposes, in small magnetic fields the Stokes parameters I and Q are more useful than the I_1 and I_2, which are convenient in the algebraic treatment. One actually needs to know accurately Q, not I (this being a comment valid for weak fields, of course!).

(2) It is clear that in *weak* fields, the coupling of magnetic effects and source function is essential; one cannot use a source function of 'fine scattering' when depolarizing collisions are quite essential!

(3) Would the *ionized iron* lines be a better choice than the neutral iron lines often in use, as they are less temperature dependant?

Stenflo: The main reason why the line Fe I 5250 Å is so temperature sensitive is the low excitation potential of the lower level, 0.12 eV. A small temperature change will cause a large change in the population number of that level. Ionized lines may be better, but are also affected by temperature changes. One should carefully investigate the behavior of different lines to find the most suitable one for magnetograph observations.

Athay: If I understand correctly, the scattering is treated coherently. This is incorrect, of course, since the scattering is actually incoherent. Could you comment on the effect of the incoherence. Secondly, the presence of velocity fields in the Sun generally makes the lines asymmetric. Could you comment on this effect.

Stenflo: The case of incoherent scattering has been treated by Rachkovsky and Domke, but as far as I know, nobody has made quantitative estimates of how much the results will differ from the case of coherent scattering. The effect of asymmetry of the line profile on the magnetograph recordings is always present but is very small compared with the other effects I have discussed.

Schatten: Observations of the interplanetary magnetic field have been compared with the photospheric magnetic field as measured by magnetographs by Wilcox and Ness, Schatten *et al.*, and Severny *et al.* They have shown that the 'mean' photospheric field agrees with the observed interplanetary field both in polarity and in *magnitude*. The agreement is probably accurate to about 80%. Thus it is difficult to reconcile the factor of 3 weakening in field strength in the magnetograph observations as you suggest. I would like to know your comments on this problem.

Stenflo: As I see it the really significant results you mentioned is the excellent correlation between the *polarities* of the solar and interplanetary magnetic fields. The polarity pattern on the Sun will probably not be changed when taking into account the line weakenings in the magnetic filaments. The establishment of the relation between the *magnitudes* of the solar and interplanetary magnetic fields is not so certain, and I think one has to reexamine this problem again.

Michard: I suggest that tests of the influence of an unresolved fine structure of magnetic fields on measured averages, can best be made by the simultaneous study of lines with different Zeeman patterns: these will be differently influenced by the fine structures.

Stenflo: There are many ways in which the influence of the fine structure may be studied. In addition to the suggestion by Dr Michard, one of the best methods would be to make magnetograph recordings in a great number of lines of different temperature-sensitivity to determine what layers of the solar atmosphere are heated.

Wiehr: Fe λ 5250 can *not* be replaced by an Fe+ line since the latter one shows the opposite behavior to Fe λ 5250. This was shown by Harvey and Livingston for the filamentary structures and by myself for sunspots. One should use a 3.5 eV line like Fe λ 6303.

Stenflo: Even the Fe λ 6303 Å line seems to be affected by the temperature increase in the filaments.

Lamb: I would like to make a brief comment concerning the effects of collisional depolarization. Although collisional depolarization is expected to be quite important in the lower regions of the solar atmosphere, there are strong theoretical reasons for expecting that the rate of collisional depolarization will not depend on magnetic field strength for fields of less than 5×10^3 G.

Deubner: Can you give theoretical arguments why the filamentary structure should be so pronounced as you mentioned, using values down to 1% of the total area being occupied by magnetic field strands in order to estimate the ratio of H_{obs} to the true field strength H? E.g. yesterday we learned from Livingston's high resolution magnetograms that there are magnetic fields evenly distributed all over the solar surface.

Stenflo: Unfortunately there is no theory that can satisfactorily explain why the filamentary structure is so pronounced. The observations by Livingston and Harvey indicate that most of the total magnetic flux is 'channelled' through these narrow filaments. At the same time they observe weak fields of more or less random distribution in between the filaments. The theoretical relation between the filamentary and interfilamentary magnetic fields is not clear, and we do not know how large a part of the total flux is in the interfilamentary medium.

Dunn: Let us not encourage people to duplicate the Evans type of magnetograph. You see it follows the center of gravity of one of the components and the π component. Thus in the presence of a π component it becomes a longitudinal type magnetograph. With no π component it measures total field. So in a weak sunspot it measures longitudinal field. As the spot strengthens the slits may jump between the π component and σ component. Years ago I tried to solve this problem by chopping between the continuum and the edge of the line. The noise went up but it was a successful solution, however it did not really appear practical for daily mapping. I think an equally good but simpler solution is a two channel Babcock magnetograph operating on a weakly and strongly split line, or perhaps one of the magnetographs described here that measure the Stokes' parameters, or one that records the entire line profile.

Stenflo: I think all types of magnetographs are most valuable, because from their different behavior in the presence of a filamentary structure, we can learn about the unresolved fine structure of the magnetic field.

Severny: We have not been able to find out outside of active regions such a drastic change in the line profile $\lambda\,5250$ as was found at Kitt Peak. Perhaps the resolution in our observations ($2'' \times 4''$) was not as good as at Kitt Peak, but it is still quite adequate for such a search of fine structure. Especially for polar regions we were not able to find results like that found at Kitt Peak regarding the line profile of $\lambda\,5250$. I have no doubt that inside active regions we have such an effect of decrease of the steepness of $\lambda\,5250$, but I am not sure that this takes place also for the regions of general field outside active regions. So that I think we should consider the statement that we should change all our measurements on $\lambda\,5250$ by a factor of 3 with some reserve.

COHERENCE PROPERTIES OF POLARIZED RADIATION
IN WEAK MAGNETIC FIELDS

LEWIS L. HOUSE

High Altitude Observatory, National Center for Atmospheric Research* Boulder, Colorado, U.S.A.

Abstract. The scattering of radiation in the presence of weak magnetic fields can give rise to coherence or interference phenomena that will profoundly affect the frequency, geometric, and polarization properties of the scattering event. In this paper we discuss and illustrate some of the features of the coherence phenomena associated with the scattering redistribution for the normal Zeeman triplet. The frequency dependent as well as the frequency independent scattering function is considered in a linear polarization basis. In addition we illustrate some properties of this redistribution function in the Stokes representation. Since the primary purpose of this paper is to demonstrate the nature of some of the properties of the coherence problems, that might be important in the interpretation of magnetic fields from polarization measurements of scattered radiation, it has been necessary in this initial work to neglect several features of the problem which are noted in the paper and are currently under investigation.

1. Introduction

To deduce the strength and direction of magnetic fields in solar prominences from the measurement of polarization requires that one utilize the scattering theory of radiation to interpret the measurement. In particular, for scatterings that take place in the presence of weak magnetic fields one must include in the scattering theory the quantum electrodynamic effects of coherence or correlation. This phenomena arises when the sublevels of the scattering atoms are weakly removed from degeneracy by the magnetic field and it manifests itself in the fact that a coherence or correlation is produced between the incoming and outgoing properties of the radiation field – that is, there is a coherence produced between the incoming and outgoing frequencies, directions of propagation, and polarizations. Thus if the scattering takes place under conditions that produce coherence, one may not interpret the measurement of polarization in terms of magnetic fields using the standard theory of scattering in normal Zeeman patterns. Hyder (1964), as well as others, have pointed this fact out in relation to scattering in prominences. This coherence phenomena was discussed in relation to laboratory experiments by Hanle (1924) after whom the effect is named.

In a series of papers (House 1970a, b, c) a general formulation has been given for the scattering redistribution function accounting for the frequency, geometric, and polarization properties of the event, applicable to any dipole transition. Beginning with the work of Weisskopf (1931), it has been possible to cast the general scattering redistribution function into a form that conveniently illustrates how the scattering in the standard Zeeman pattern is modified due to the coherence effects. The theory has been formulated in three polarization basis sets: linear, circular, and in terms of

* The National Center for Atmospheric Research is sponsored by the National Science Foundation.

Stokes parameters. A discussion of the coherence problem has also been given recently by Lamb (1970, 1971).

In this paper I should like to show the results of some calculations that illustrate the nature of the coherence phenomena associated with the general scattering function. First, however, let me briefly remind you in somewhat more detail of the origin of the coherence effects.

2. Origin of the Coherence

To see more explicitly how the coherence effects arise let us refer to Figure 1. Here we depict two bound levels of an atom, split into their magnetic sublevels. The com-

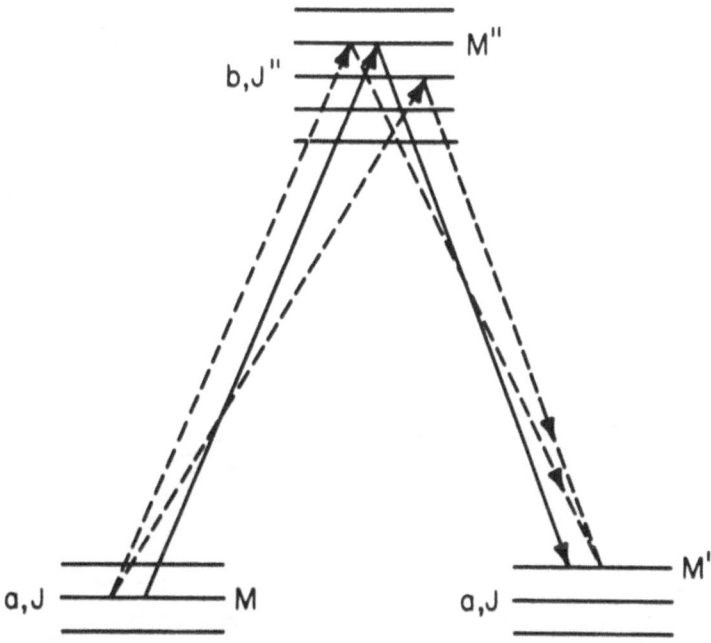

Fig. 1.   Schematic energy level diagram showing quantum number designations and typical transitions where coherency effects are not encountered (——) and where coherency does arise (----).

ponents for the ground state are shown a second time so that the difference between the initial and final states of the atom may be more easily seen.

Let us consider two scattering events. First, in the situation where the sublevels are far removed from degeneracy, there is no overlap of the levels, that is, each level is independent. If the excitation raises the atom to an excited state sublevel M'', the subsequent decay occurs from this level back to the ground state (sublevel M') – neglecting collisions and interlocking with other levels. However, if the sublevels are not far removed from degeneracy, that is, if their separation is comparable to or less than the level width, then, quantum mechanically these levels are virtually indistinguishable. In computing the scattering amplitude one must therefore sum over all the intermediate sublevels that could be excited by the incident radiation and could decay back to the ground state. This must be done, of course, accounting for the angular momentum selection rules. Because the summation over intermediate levels is carried out before one squares the amplitude to obtain the intensity of the scattered radiation, cross terms can arise and these in fact produce the coherence or correlation. After taking the modulus squared, sums over initial as well as final substates must also be carried out. In this situation of weak fields, the scattering event cannot be considered

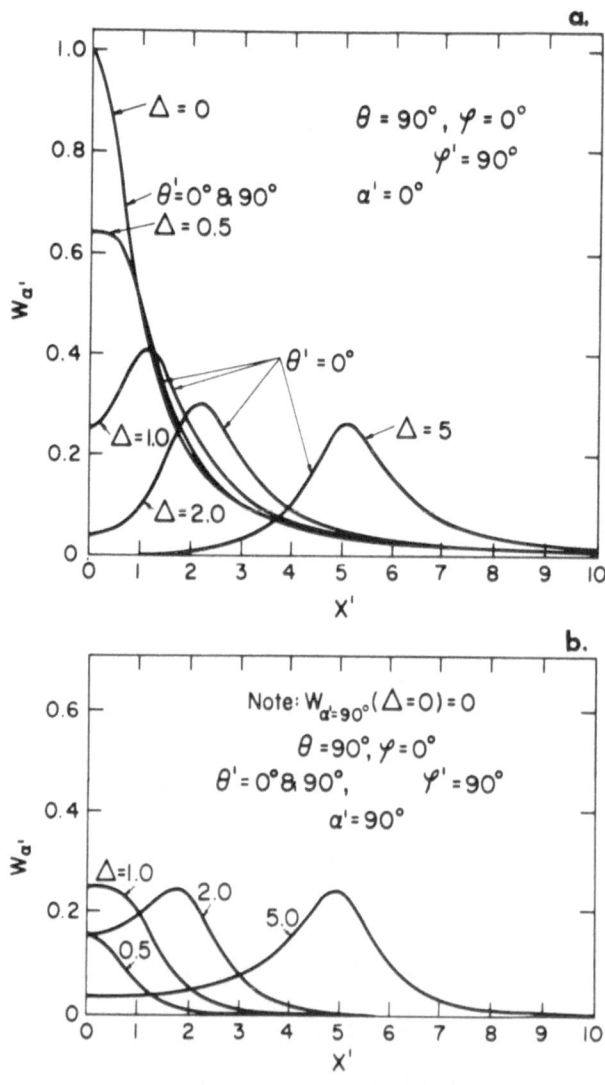

Fig. 2.   Emission probability as a function of frequency for various values of Δ and for $\theta' = 0°$ and $90°$; (a) applies for the polarization angle $\alpha' = 0°$ while (b) applies for the angle $\alpha' = 90°$. Note in (a) the emission probability is independent of Δ for the observing direction $\theta' = 90°$ and it is, in fact, given by the curve labeled $\Delta = 0$.

as a sequence of an absorption followed by an emission; it must be treated as a single event. Rather than discuss or illustrate the complicated equations that arise in this process, let me proceed to illustrate some calculations based upon the general scattering redistribution function.

3. The Scattering Redistribution Function in Linear Polarization

For the illustrations that I have selected, I have chosen the atomic transition $J=0$ to $J''=1$, that is the normal Zeeman triplet. In addition, for simplicity we will assume that the incident radiation field is unpolarized and independent of frequency over the width of the redistribution function. We shall treat the problem in a linear polarization basis, treat only single scatterings, and neglect collisions.

(a) *Frequency Dependent* – With these assumptions in mind, the next figure, Figure 2, shows a sample computation of the frequency dependent redistribution function where the magnitude of the function is on an arbitrary vertical scale and the horizontal axis is frequency measured from line center in units of the natural width. The splitting due to the magnetic field which is specified by Δ, is also measured in units of the natural width. (I might say for reference that for a transition having an Einstein transition probability of 10^8 s^{-1}, the value of $\Delta=1$, corresponds to a magnetic field of about 6G.) The incident radiation is taken to be perpendicular to the magnetic field, $\theta=90°$, and in part (a) of the slide, we have the redistribution function as seen in one state of linear polarization specified by the polarization angle $\alpha'=0°$ while in part (b) we have the orthogonal state of polarization, $\alpha'=90°$. The angle θ' is measured between the direction of the magnetic field and the line of sight. In part (a) of this figure we see that for viewing the radiation along the direction of the field lines, i.e. $\theta'=0°$, the redistribution function gradually decreases in amplitude and shifts outward as the field strength increases. However, if we look perpendicular to the field, the one function labeled $\Delta=0$ is in fact independent of the field strength. Thus in this state of polarization, as the field strength increases, it is only when we look along the magnetic field that the components shift outward. In the orthogonal state of polarization, as shown in part (b), however, these redistribution functions are obtained for directions both along and perpendicular to the magnetic field.

For the larger values of Δ, the scattering redistribution function is approaching that of the normal Zeeman pattern, for we can see that along the field the components in the two orthogonal states of polarization approach being equivalent which will correctly yield zero linear polarization at the position of the shifted component, i.e. at $x'=\Delta$. Perpendicular to the field, we will have a central unshifted component in one state of polarization and a shifted component in the orthogonal state of polarization, thus giving the usual Zeeman results. This figure therefore illustrates the transition from weak fields toward the normal Zeeman pattern.

The polarization resulting from these scattering functions is better illustrated in the next slide, Figure 3, where we can see the degree of linear polarization as a function of frequency for the scattering functions given on the previous slide. As one can see, for viewing both along the magnetic field as shown in part (a) and perpendicular to the field as shown in part (b), the degree of polarization at the position of the shifted component, $x' = \varDelta$, is very sensitive to the field strength; along the field varying from 100% to 0% as $\varDelta$ varies from 0 to 5 and perpendicular to the field varying from plus 100% to -70% at $\varDelta = 5$.

(b) *Frequency Independent* – Next we illustrate some of the more gross properties of the scattering function that have been obtained by integrating over both incoming and outgoing frequencies: that is, we shall look for a moment at some of the properties of the frequency independent scattering function.

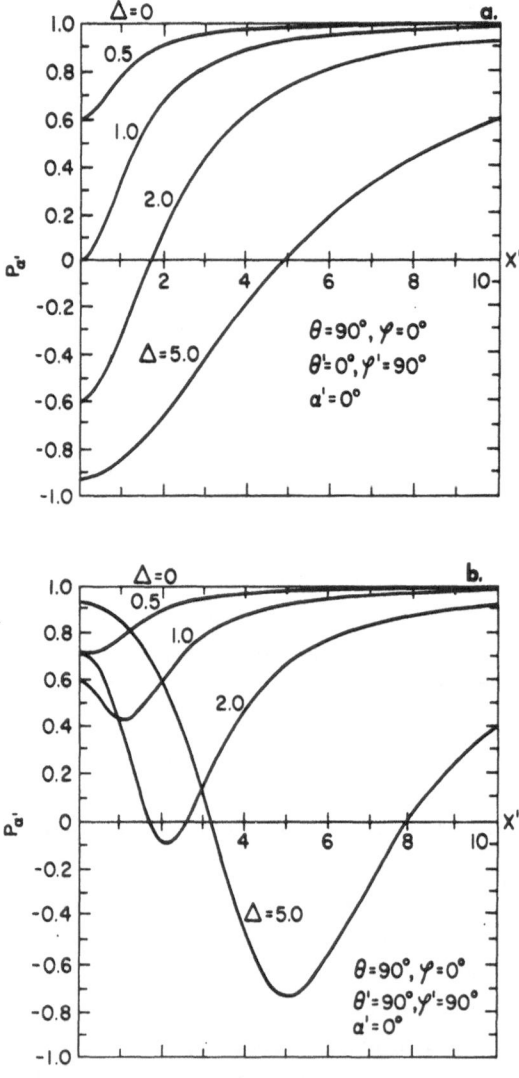

Fig. 3.   Degree of polarization for emission probabilities of Figure 2, as a function of frequency curves are given for various values of $\varDelta$ and for two viewing directions (a) $\theta' = 0°$, $\phi' = 90°$ and (b) $\theta' = \phi' = 90°$.

In Figure 4, scattering diagrams show the intensity of the radiation field again for unpolarized radiation that is incident perpendicular to the magnetic field and unpolarized and where the observer looks along the magnetic field. The length of the vector in these diagrams is proportional to the intensity of the radiation scattered into the state of linear polarization at an angle α, where α is measured counter clockwise from the axis projected vertically downward. These calculations, for various values of the field strength as specified by Δ, show the essential features of the Hanle effect. We see at

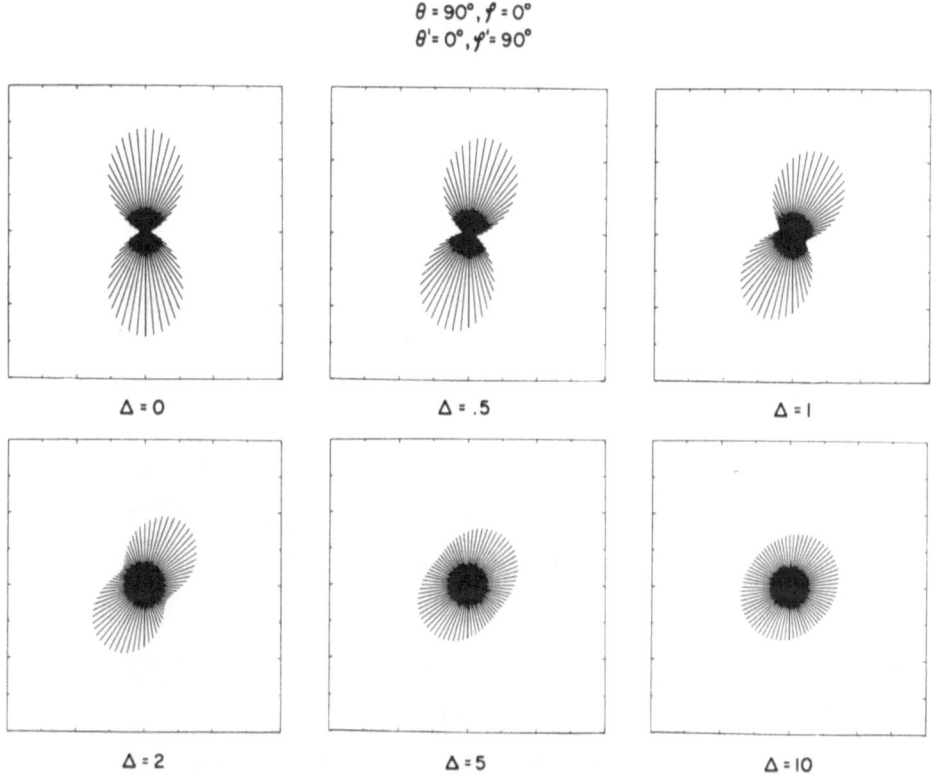

$$\theta = 90°, \mathscr{f} = 0°$$
$$\theta' = 0°, \mathscr{f}' = 90°$$

Fig. 4.   Frequency independent scattering probability for various values of α'. Vectors indicate magnitude of scattering probability as a function of α' where α' is measured counter-clockwise from the downward vertical direction.

zero magnetic field, we have the usual $\sin^2$ distribution of radiation and a maximum polarization of 100%. As the strength of the field is increased, the angle for the maximum degree of polarization rotates and at the same time, the maximum degree of polarization decreases because the radiation is filling in the direction orthogonal to the direction of maximum intensity. For large values of Δ, we see that unpolarized radiation would be obtained, that is, equal intensities at all polarization angles.

To see how the angle between the magnetic field and the observer influences the polarization, we refer to the next figure, Figure 5. Here I have plotted the maximum

polarization against the angle at which this polarization occurs. The different curves
are parameterized by the angle between the observer and the magnetic field, that is,
we have θ' varying from along the field, $\theta'=0°$, to $\theta'=90°$, which is perpendicular to
the magnetic field. The points distributed along the curves are for the indicated values
of the magnetic field strength in terms of Δ. The curve for $\theta'=0°$ shows the maximum
polarization which varies from 100% to 0%, as the angle of maximum polarization
varies from 0° to 45°. As one alters the viewing angle from along the field to 30°, 60°,
and finally to 90° to the field, we see that the range in maximum polarization, as well

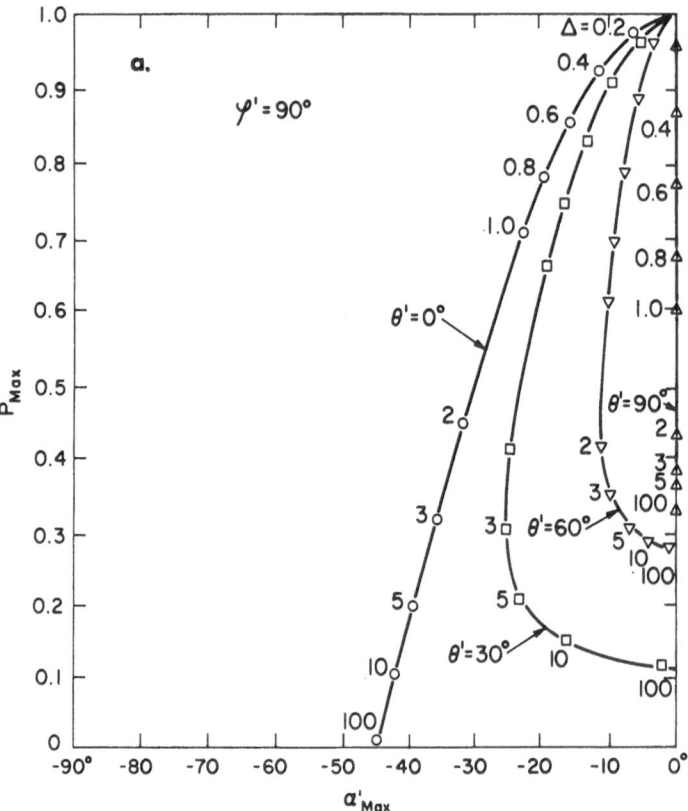

Fig. 5.   Degree of maximum polarization versus angle of maximum polarization for various values
of θ' and Δ. The different values of Δ are labeled at appropriate positions along the curves.

as the angle of maximum polarization, become somewhat restricted. Thus, if one
were to view a single scattering from a point perpendicular to the incident radiation,
then, from a measurement of the maximum polarization and the angle of maximum
polarization, one could in principle determine the strength of the field as well as the
angle between the line of sight and the direction of the field.

Naturally, in the solar atmosphere one does not have such a restricted direction for
the incident radiation field, as we have assumed in the previous calculations, because
from any single scattering point above the disk, for example in a prominence, one

must consider radiation incident from all angles subtended by the solar disk. The effects of such an integration over the disk of the Sun are shown on the next figure, Figure 6. Here we have plotted the maximum polarization as a function of height above the limb. Part (a) of the figure applies to the situation where the line of sight is

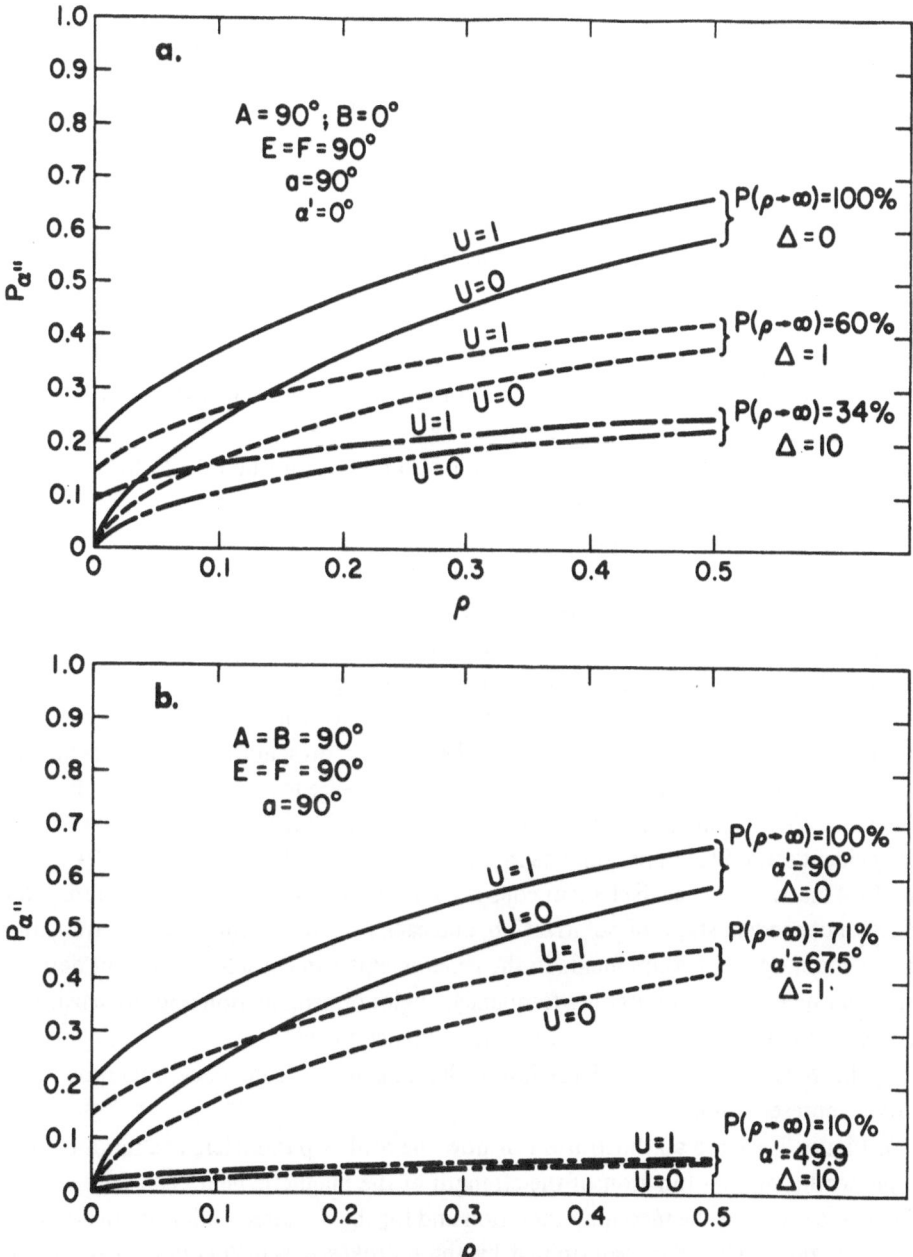

Fig. 6.   Degree of polarization as a function of height, ρ, for two values of the limb darkening coefficient ($U = 0$ and 1) and for three values of magnetic field ($\varDelta = 0$, 1, and 10). The magnetic field is tangent to the limb and the observer's direction is perpendicular to the field in (a) whereas in (b) the observer's direction is along the field.

perpendicular to the magnetic field and part (b) applies for the observer looking along the direction of the magnetic field. In both cases the line of sight and the magnetic field are tangent to the limb. Curves are given for three values of the strength of the field in terms of Δ and for each Δ, two curves are given, one for a limb darkening coefficient $U=1$ and the other for no limb darkening, or $U=0$. In this figure we see the extent to which the degree of anisotropy as well as the strength and direction of the magnetic field combine to control the degree of linear polarization. The asymptotic limits for the degree of polarization from a point source are given. We note that quite close to the surface of the disk, the polarization can in general vary from about 0 to 20% depending upon the combination of the various parameters.

4. Circular and Stokes Polarization Basis Sets

In the previous calculations a linear basis set for the polarization vector has been used. We may if we wish, however, expand the scattering function in terms of a circular polarization basis set. A series of calculations similar to those just shown in terms of circular polarization have been done. However, rather than show some of these calculations, I would prefer to say something about the formulation of the scattering redistribution function in the Stokes representation.

For a particle, such as a photon having two spin components, a very convenient approach to the Stokes representation is through the quantum mechanical density matrix formulation. In this approach, the fundamental quantum mechanical nature of the Stokes parameters becomes apparent and they are seen to be just the expansion coefficients of the density matrix in terms of the Pauli spin operators or equivalently the projections for the Pauli spin operators in a polarization space. In transforming the scattering redistribution function into the Stokes representation, one subsequently obtains a Mueller matrix for the resonance fluorescence process which includes coherence effects. Thus, the atom can be treated as an 'optical device' that transforms the incident set of Stokes parameters into an outgoing set. The general Mueller matrix is a 4×4 matrix where each element consists of four terms because of the combinations of two independent states of polarization, and each term depending upon the transition may contain up to 19 components in the general expansion; each component depending upon incoming and outgoing frequency, angles of propagation and polarizations, as well as the strength and direction of the magnetic field.

Again we show a sample calculation of the influence of coherency, this time in the Stokes representation.

In Figure 7, we see an illustration of how the Stokes parameters are modified in a single scattering as a function of the strength of the magnetic field. The Stokes parameters are listed for the incoming radiation and for the scattered radiation. In addition, the polarization ellipse as determined by these Stokes parameters is also shown. In this particular case, which is treated as frequency independent, we have chosen an arbitrary state of elliptic polarization for the photon which is incident along the direction of the magnetic field; in this example, the scattering takes place in the forward direction. (The rotation of the ellipse axis at zero magnetic field, relative to the incident radiation is due to a change in the local coordinate system and should be disregarded at this time.) For increasing magnetic field, the polarization ellipse is seen to rotate and at the same time the scattered photon becomes circularly polarized as indicated by S_3, while the degree of linear polarization, indicated by S_1 and S_2 decreases. This of course must be the case, since in the usual Zeeman theory, only

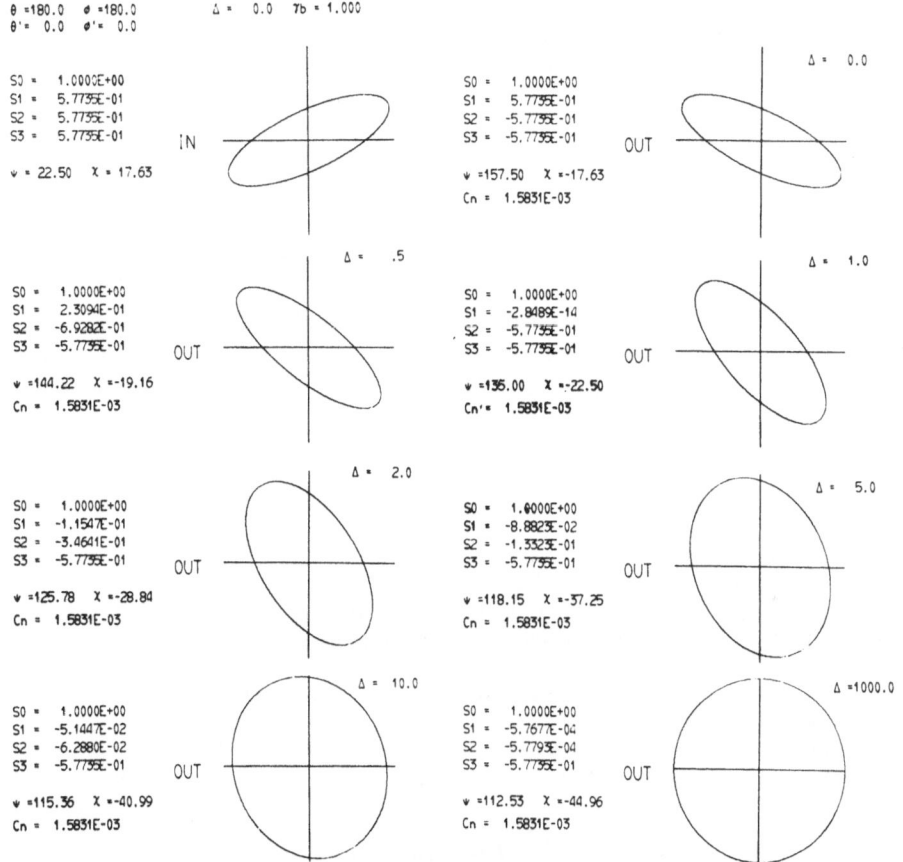

Fig. 7.   Stokes parameters and polarization ellipses for a single scattering as a function of Δ. The radiation incident (IN) along the direction of the magnetic field is scattered (OUT) in the forward direction. The four Stokes parameters as well as the parameters defining the polarization ellipse are shown. Rotation between IN and OUT: $\Delta = 0$ is a result of change of coordinate systems and is not to be regarded as real.

circular polarization is observed along the magnetic field if the excitation occurs through circular polarization. This illustration is again meant to point out the sensitivity of the scattering process to weak magnetic fields and how, for example, if one is to use a polarimeter to determine the Stokes parameters of the radiation field scattered by a prominence – the coherence or Hanle effect must be considered in detail.

5. Additional Features of the Scattering Problem

Three other attributes to the scattering problem must be treated before the theory can be applied to the interpretation of observational data, these are – the influence of the Doppler effect, collisions, and multiple scattering. The treatment of the Doppler effect is straight-forward, but because of the coherence features of the problem it requires an excessive amount of space and time even on a computer as large as the CDC-6600. These numerical calculations are however, in progress. Collisions at the present time can be incorporated by increasing the damping width of the states. To include collisions in detail, the statistical equilibrium equations must be treated and since in the radiation part of the problem, the scattering is considered a single event, this leads to difficult problems in including collisions.

Multiple scattering, that is setting up the transfer equation using the general redistribution function, is also complicated because of the coherence or correlation in all of the variables, each of which depends upon the strength and direction of the local magnetic field. Work is currently in progress on this part of the problem where the radiation transport is being done using the Monte Carlo technique. This technique is selected so that a wide range of geometric configurations may be handled. We have already reported some simplified preliminary studies on the influence that multiple scattering and the Hanle effect have upon the depolarization of scattered radiation (House and Cohen, 1969).

References

Hanle, W.: 1924, *Z. Phys.* **29**, 93.
House, L. L.: 1970a, *J. Quart. Spectr. Radiative Transfer.* **10**, 909.
House, L. L.: 1970b, *J. Quart. Spectr. Radiative Transfer.* **10**, 1171.
House, L. L.: 1970c, *J. Quart. Spectr. Radiative Transfer.* in press.
House, L. L. and Cohen, L. C.: 1969, *Astrophys. J.* **157**, 261.
Hyder, C. L.: 1964, *Astrophys. J.* **140**, 817 (For additional references see House, 1970).
Lamb, F. K.: 1970, *Solar Phys.* **12**, 186.
Lamb, F. K.: 1971, this volume, p. 149.
Weisskopf, V.: 1931, *Ann. Phys.* **9**, 23.

Discussion

Brueckner: (1) Did you use in your calculations the assumption of pure scattering?

(2) Are your calculations concentrated on optically thin lines (one scattering event)?

House: As stated in the introduction and conclusions, I treat at this time only a single scattering event. The main purpose of this presentation is to illustrate the properties of the scattering redistribution function. It is necessary to include effects of Doppler broadening, collisions, and multiple scattering before calculation may be compared with observations in prominences.

PASCHEN-BACK EFFECT OF THE LITHIUM RESONANCE
DOUBLET IN SUNSPOTS

P. MALTBY

Institute of Theoretical Astrophysics, University of Oslo, Norway

Abstract. The formation of the resonance doublet lines of Li^6 and Li^7 in sunspots is discussed. It is shown that the magnetic splitting of the lines must be determined according to the (partial) Paschen-Back pattern. As a first approximation to the problem a detailed calculation of the line profile is given for the case of pure absorption and local thermodynamic equilibrium.

1. Introduction

The Li I resonance lines at $\lambda 6708$ have been considered by several authors, in particular in connection with abundance determination.

In the case of penumbral/umbral spectrum the resonance lines have been treated as magnetically inactive (e.g. Greenstein and Richardson, 1951) or as affected by the Zeeman splitting (e.g. Wiehr *et al.*, 1968). However, the separation of the $^2P_{3/2}$ and $^2P_{1/2}$ levels of Li I is so small (0.151 Å) that the magnetic splitting in sunspot fields is comparable to the doublet separation. Thus, the intensity and wavelength position of the line components are determined according to the Paschen-Back pattern. Recently, Traub (1968) has taken the Paschen-Back effect into account – he has also considered the effect of the nuclear spin, but found that this additional effect may be neglected for magnetic field strengths above approximately 1000 G.

Let us consider the problem of the formation of the Li I $\lambda 6708$ line, consisting of the Li^7 doublet $\lambda\lambda 6707.761$, 6707.912 and the Li^6 doublet $\lambda\lambda 6707.921$, 6708.072. Each doublet is split into $6+4=10$ components in the magnetic field (neglecting 2 components associated with the nuclear spin). Thus, the first problem is to determine the intensity and wavelength position of each component. This information may be taken from Traub (1968) or several earlier authors. The second problem is the formation of the magnetically active lines in the spot atmosphere. Traub (1968) tried to solve this problem by establishing a linear approximation to the Planck function at each depth and assuming that Seares' formulae could be applied. We will follow another approach by using the line formation theory of Kjeldseth Moe (1968) and apply this theory to lines blended by other magnetically active lines.

As we are dealing with a resonance doublet, deviations from local thermodynamic equilibrium (LTE) as well as line scattering may be important. The calculations are based on the assumptions of pure absorption and LTE and should accordingly be regarded as a first, rough approximation to this line formation problem.

2. Theoretical Considerations and Results

The wave mechanical treatment of magnetic splitting at intermediate strengths of

Howard (ed.), Solar Magnetic Fields, 141–147. All Rights Reserved.

magnetic field was given by Darwin (1927). Numerical examples are given by Darwin (1928) for *s-p* and *p-d* doublets and the *s-p* triplet. In our case, the *s-p* doublet, the problem was considered already by Voigt (1913) using a classical approach. Sommerfeld (1922) gave the quantum theory modification of Voigt's work and Heisenberg and Jordan (1926) gave a quantum mechanical treatment for doublets. For our numerical work we will use the results given by Darwin (1928).

Let us consider the vector model for Li, in which only the outermost electron plays a role. As a result of the spin moment, **s**, the orbital plane precesses about the total angular momentum, **j**. We here regard **j** as composed of the orbital moment **l** and the spin moment, **s**. We may regard the precession frequency of **s** and **l** about **j** as a measure of the doublet separation (e.g. Pauling and Goudsmit, 1930). In dealing with the Zeeman effect it is often useful to regard the splitting of a level as a measure of the precession frequency of **j** about the external magnetic field, **H**. For the Zeeman effect we may assume that the precession of **s** and **l** about **j** is much faster than the Larmor precession. For strong magnetic fields and small doublet separation the precession frequencies become comparable and another approach must be used.

For a very strong magnetic field **l** and **s** wil lboth precess independently about the external field direction. For the Li doublet in sunspot spectra we are interested in the case where the external magnetic field is of the same order of magnitude as the internal magnetic field. In this case the precessions of **s** and **l** about **j** are not uniform, whereas the total projection of the mechanical moment on **H** is independent of magnetic field strength. The quantum mechanical calculations are done by taking into account simultaneously the magnetic field and the spin-orbit interaction. The perturbation theory used in quantum mechanics will not be given here. For a detailed discussion of the (partial) Paschen-Back effect we refer to Bethe and Salpeter (1957) and references given there.

Let us next consider the line formation of magnetically active lines. If we assume that the magnetic field is homogeneous and that the line is formed by pure absorption we may apply the solution given by Kjeldseth Moe (1968). A formal solution may also be

TABLE I

Data used in the numerical calculations

Parameter	Value/Source
Magnetic pattern	Darwin (1928)
f-value	Bickel *et al.* (1969)
Sunspot model	$\Delta\theta = 0.48$ model of Zwaan (1965)
Continuum absorption coefficient	Bode (1965)
Abundance of Li	$\log N$ (Li) = 0.80, Engvold *et al.* (1970)
Isotope ratio Li^6/Li^7	0.05 (assumed value)
Position on solar disc	Centre
Partition function	Bascheck *et al.* (1966)
Van der Waal damping	$a = 0.18$, see Aller (1963)
Turbulent velocity	zero (assumed value)

found in the case that scattering is important. However, the scattering matrix for the line must be derived and this is generally relatively difficult unless we restrict the discussion to a particular model atmosphere (as a Milne-Eddington atmosphere). Thus, as a first approximation we will regard the lithium line as formed by pure absorption. We will further assume that the sunspot atmosphere is in LTE and that the Doppler width of the line absorption coefficient may be regarded as constant through-

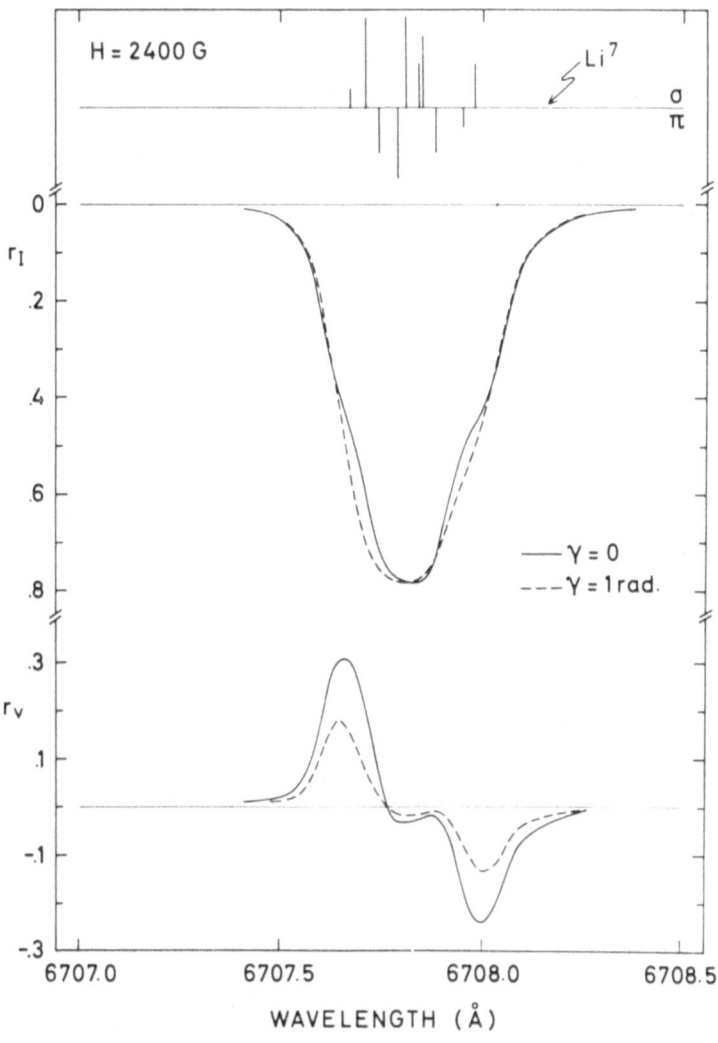

Fig. 1.

Figs. 1–4.   The line depth $r_I = (I_0-I)/I_0$ and the fraction of circular polarization $r_v = -V/I_0$ (V is the Stokes parameter) as function of wavelength for the Li^6 and Li^7 resonance doublets. The Paschen-Back pattern is indicated (top) for the Li^7 components – the pattern is identical for Li^6. The magnetic field strength, H, is given in the figure. The calculations are done for an angle $\gamma = 0$ and 1 radian between the magnetic field direction and the line of sight.

out the atmosphere. With these assumptions the same type of polarization will be absorbed and emitted at all depths at a fixed frequency within the line. The solution given by Kjeldseth Moe (1968) is obtained by regarding the light as composed of two independent light beams of elliptically polarized light in opposite states of polarization. In this way the problem may be transformed to a line formation problem similar to the non-magnetic case.

When two or more magnetically active lines blend with each other different types of polarization will generally be absorbed at different depths in the atmosphere. In the general case the type of solution outlined above is therefore not applicable, as the contribution functions for the line profiles are different. However, in the case of the lithium resonance doublet both lines start from the same atomic level and the contribution

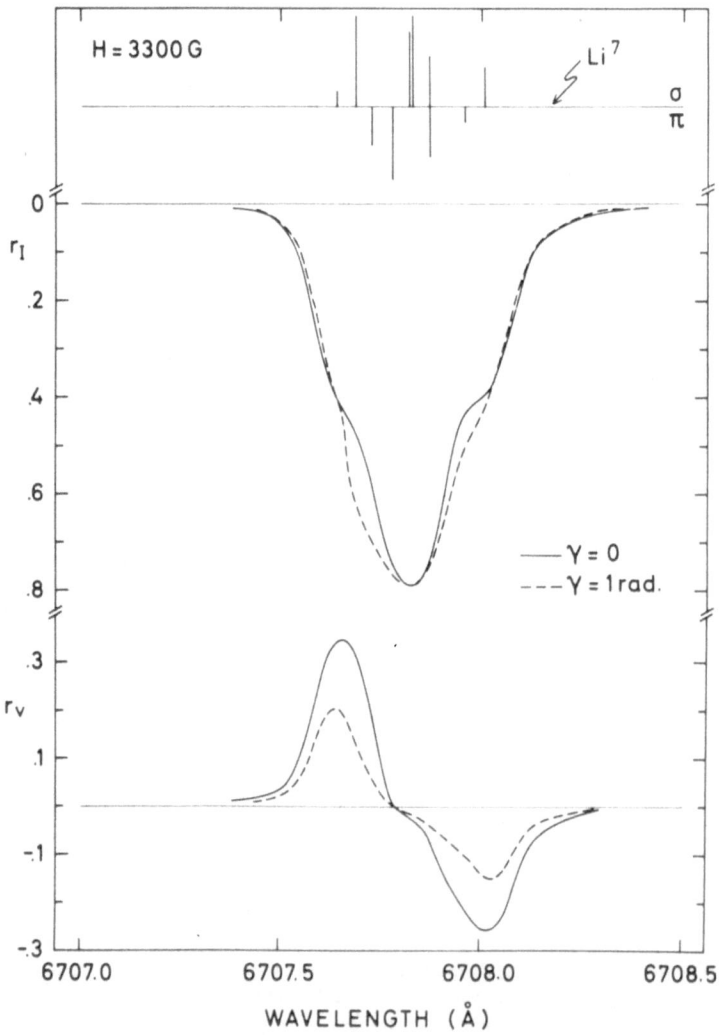

Fig. 2.

functions are identical except for a numerical factor. This numerical factor is determined by the oscillator strength and the Li^6/Li^7 isotope ratio. Accordingly, we have calculated the composite line profile for the two doublets using the input data given in Table I. A more detailed account of the line calculation is given by Engvold *et al.* (1970).

The results of the calculations are shown in Figures 1–4 for different values of the magnetic field strength. Two values of the angle γ between the line of sight and the magnetic field have been used in the calculations. Figures 1–4 give the line depth r_I for the total intensity as well as the quantity r_v for the circularly polarized intensity. These quantities are defined as $r_I = (I_0 - I)/I_0$ and $r_v = -V/I_0$, where I and I_0 are the intensities in the line and in the continuum, respectively and V is the Stokes parameter that determines the fraction of circularly polarized light.

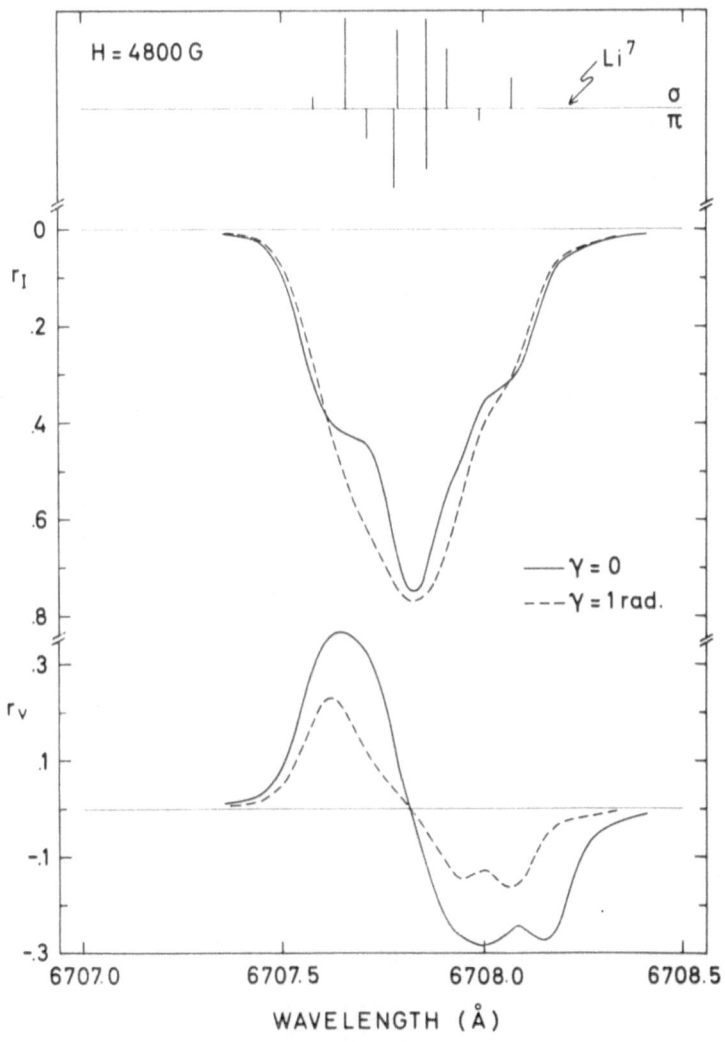

Fig. 3.

The calculations show that an error in the line profile is introduced if a Zeeman splitting pattern is used instead of the Paschen-Back pattern. The error, Δr_I, in the line depth r_I is of the order of 0.01 for a magnetic field strength of $H=2400\,$G, but increases rapidly to $\Delta r_I \approx 0.03$ for $H=3300\,$G and to $\Delta r_I \approx 0.10$ for $H=4800\,$G. The errors in r_v are comparable in magnitude. Thus, the Paschen-Back splitting must be taken into account for sunspot magnetic fields. For very strong fields $(12000\,$G$)$ we approach a complete Paschen-Back splitting pattern.

Figures 1-4 show considerable changes in the line profile and polarization characteristics as the strength and direction of the magnetic field are altered. Thus, although the Doppler-width of the line is considerable, detailed observations of the Li line may give valuable information about the magnetic field configuration in sunspots.

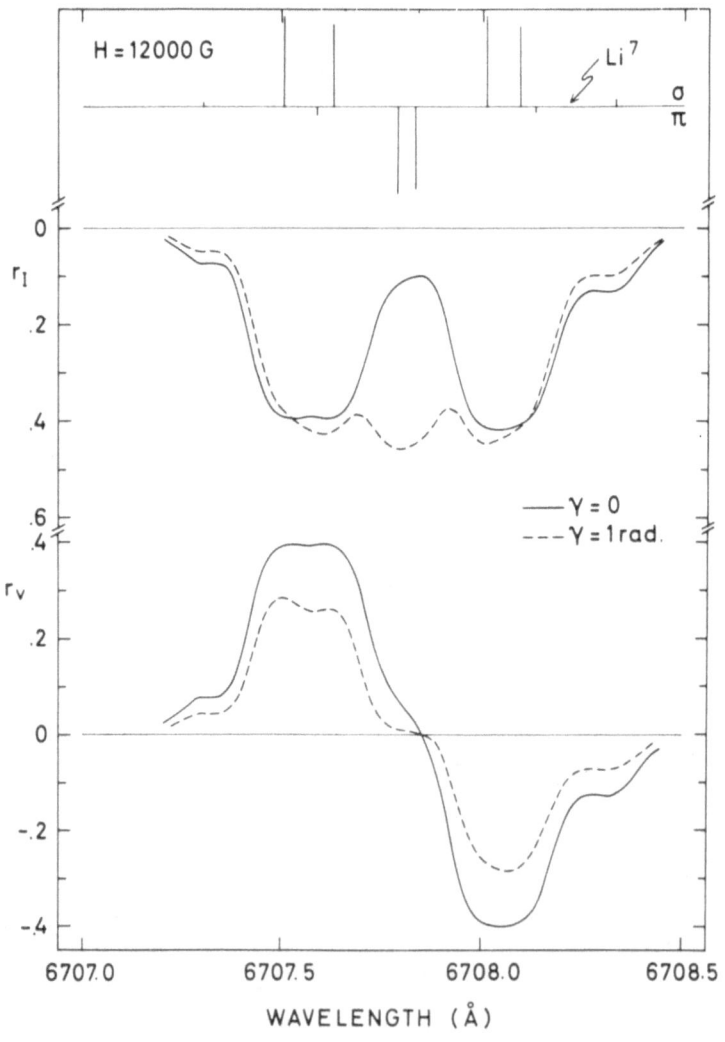

Fig. 4.

Maltby and Engvold (1970) have recently found that the Li resonance doublet show a faint intensity reversal in sunspot spectra. This probably means that part of the Li doublet is formed in relatively high layers of the sunspot atmosphere. Accordingly, we must emphasize that the calculations given here should be regarded as a first approximation to the line profile. On the other hand, as the Li doublet very probably is formed in relatively high layers of the spot atmosphere, better line formation theory and observations may give us detailed information about the magnetic field configuration in higher layers of the sunspot atmosphere.

Acknowledgement

The author wishes to thank O. Engvold for help with the numerical calculations.

References

Aller, L. H.: 1963, *The Atmospheres of the Sun and Stars*, Ronald Press, New York, p. 324.

Bethe, H. A. and Salpeter, E. E.: 1957, in *Handbuch der Physik* (ed. by S. Flügge), Springer-Verlag, Vol. XXXV, p. 291.

Bickel, W. S., Martinson, I., Lundin, L., Buchta, R., Bromander, J., and Bergström, I.: 1969, *J. Opt. Soc. Am.* **59**, 830.

Bode, G.: 1965, Thesis, Kiel.

Darwin, C. G.: 1927, *Proc. Roy. Soc.* **A115**, 1.

Darwin, K.: 1928, *Proc. Roy. Soc.* **A118**, 264.

Engvold, O., Kjeldseth Moe, O., and Maltby, P.: 1970, *Astron. Astrophys.* **9**, 179.

Greenstein, J. L. and Richardson, R. S.: 1951, *Astrophys. J.* **113**, 536.

Heisenberg, W. and Jordan, P.: 1926, *Z. Phys.* **37**, 270.

Kjeldseth Moe, O.: 1968, *Solar Phys.* **4**, 267.

Maltby, P. and Engvold, O.: 1970, *Solar Phys.* **14**, 129.

Pauling, L. and Goudsmit, S.: 1930, *The Structure of Line Spectra*, McGraw-Hill, New York, p.74.

Sommerfeld, A.: 1922, *Z. Phys.* **8**, 257.

Traub, W. A.: 1968, Thesis, University of Wisconsin.

Voigt, W.: 1913, *Ann. Phys.* **41**, 403.

Wiehr, E., Stellmacher, G., and Schröter, E. H.: 1968, *Astrophys. Letters* **1**, 181.

Zwaan, C.: 1965, *Rech. Astron. Obs. Utrecht* **17**, (4).

THE CROSSOVER AND MAGNETO-OPTICAL EFFECTS
IN SUNSPOT SPECTRA

V. M. GRIGORYEV and J. M. KATZ

Siberian Institute of Terrestrial Magnetism, Ionosphere and
Radio Propagation, Irkutsk, U.S.S.R.

Abstract. Two peculiarities of the magnetic splitting of a line in sunspot spectra have been investigated. The one is that in a rather small region of the penumbra near the umbra-penumbra boundary the π-component is absent in one circular polarization spectrum while both σ-components are present. In the spectrum of opposite circular polarization the σ-components are absent but the π-component is present. The second peculiarity consists of the anomalous splitting of the π-component of Zeeman triplets which are of the same and opposite signs in comparison with splitting of the σ-components. The nature of these effects is discussed.

THE EFFECT OF COLLISIONS ON SPECTRAL LINE
FORMATION IN SOLAR MAGNETIC REGIONS

F. K. LAMB

*Dept. of Theoretical Physics, Oxford University, Oxford, England**

Abstract. The present paper describes the results of a preliminary investigation of the effects on the absorption, emission, and scattering of polarized light caused by collisions between the atoms of interest and surrounding perturbers under physical conditions typical of solar magnetic regions. The description of these effects in terms of atomic level polarization is reviewed and the processes which can lead to atomic level polarization in solar magnetic regions and those which tend to reduce it are summarized. The effects associated with collisional relaxation of atomic level polarization are discussed and a method for estimating relaxation rates is described. Estimates of collisional relaxation rates are used to calculate upper limits on the degree of polarization of the levels involved in the formation of five magnetically-sensitive FeI absorption lines. The implications of the results for the form of radiative transfer equations used to describe the formation of these lines are discussed.

1. Introduction

The correct interpretation of those solar magnetic field measurements which are based on determination of the polarization of light in magnetically-sensitive solar absorption and emission lines depends on a correct choice of the models of absorption, emission, and scattering processes which are used to describe the formation of these lines. In order to choose models which best describe line formation in solar magnetic regions it is necessary to study the details of the interaction between atoms and polarized light under the physical conditions typical of these regions. Of particular importance for the physical conditions which exist in most solar magnetic regions are the effects on the absorption, emission, and scattering of polarized light due to collisions between the atoms of interest and surrounding perturbers. In spite of intense experimental and theoretical work in recent years (for a review of this work see Berman and Lamb, 1969), the problem of describing these effects remains only partially solved. Nevertheless, this recent work has shown that when collisional relaxation is important many features of the absorption, emission, and scattering of polarized light by atoms may be conveniently described in terms of the density matrices for each of the atomic levels involved in these processes (Cohen-Tannoudji, 1962; Omont, 1965). The atomic level density matrices alone can provide only a partial description of these processes, however, since they cannot give detailed information about the spectral features of the interaction between the atoms of interest and the polarized radiation field. In spite of this limitation, an investigation of the absorption, emission, and scattering of polarized radiation under the physical conditions typical of solar magnetic regions using the level density matrix formalism can still shed much light on the types of effects which are to be expected and can give some criteria for deciding on the importance of collisional relaxation.

* Present address: Department of Physics, University of Illinois at Urbana-Champaign, Urbana, Illinois, U.S.A.

Howard (ed.), Solar Magnetic Fields, 149–161. All Rights Reserved.
Copyright © 1971 by the IAU.

The present paper describes some of the results of such a preliminary investigation. The discussion in the following sections will be concerned exclusively with the state of an optically thin atomic assembly exposed to a given radiation environment and the conclusions which it is possible to draw concerning the details of absorption, emission, and scattering of polarized radiation by such an assembly. Although the implications of these conclusions for the form of the radiative transfer equations will be discussed, the problem of solving the radiative transfer equations will not be considered. The features of the interaction of polarized radiation with the atoms of interest which can be treated using the level density matrix formalism may be described by the concept of atomic level polarization: an atomic level will be said to be polarized if the reduced density matrix of the atom within the subspace of the level is not proportional to the unit matrix. In Section 2 we first recall briefly the ways in which atomic level polarization affects the polarization of light in solar absorption and emission lines and then review, in the context of solar magnetic regions, the processes which can lead to atomic level polarization and those which tend to reduce it. The influence of the magnetic field and the role of collisional relaxation are discussed in Section 3, and a way of estimating collisional relaxation rates is described in Section 4. Finally, in Section 5 some of the conclusions which have emerged from this investigation are discussed.

2. Atomic Level Polarization in Solar Magnetic Regions

2.1. THE ROLE OF ATOMIC LEVEL POLARIZATION IN THE POLARIZATION OF LIGHT IN A SPECTRAL LINE

The polarization of light in solar absorption and emission lines may result from the action of one or more of three distinct processes. First, in the presence of a magnetic field the light in absorption or emission lines may become partially polarized simply as the result of Zeeman splitting. Although the radiation absorbed or emitted at each frequency in the line by an assembly of atoms is then at least partially polarized, if atomic level polarization does not accompany the Zeeman splitting and if the assembly is optically thin in the line the total radiation absorbed or emitted, when integrated in frequency over the whole of the line, will be isotropic and unpolarized (one speaks of no 'net' polarization of the line). Second, the light in Zeeman split absorption or emission lines formed in an assembly which is not optically thin may show a different partial polarization at each frequency from that of an optically thin assembly. Even in the absence of atomic level polarization, this phenomenon usually leads to net polarization of the light in the line. Finally, light in absorption or emission lines may become polarized as a result of the polarization of one or both of the atomic levels involved in the formation of the line, since in this case the assembly will preferentially absorb, emit, and scatter radiation of a particular polarization and angular distribution in radiative processes beginning at each of the polarized levels.* Generally

---

* The general phenomenon of level polarization includes the Hanle effect and, more generally, any type of level-crossing interference as special cases.

speaking, atomic level polarization leads to net polarization of the light in the line, whether or not the line also undergoes Zeeman splitting or the assembly is optically thin. It is this last process with which we are concerned in the present paper.

Consider the absorption, emission, and scattering of polarized light in terms of the simplified three-level atomic model shown in Figure 1. In scattering light, the atom undergoes induced absorption from an initial level A (states $a, a', \ldots$) to an intermediate (excited) level B (states $b, b', \ldots$) and subsequently makes a transition via spontaneous emission to the final level C (states $c, c', \ldots$). The reduced atomic density matrix

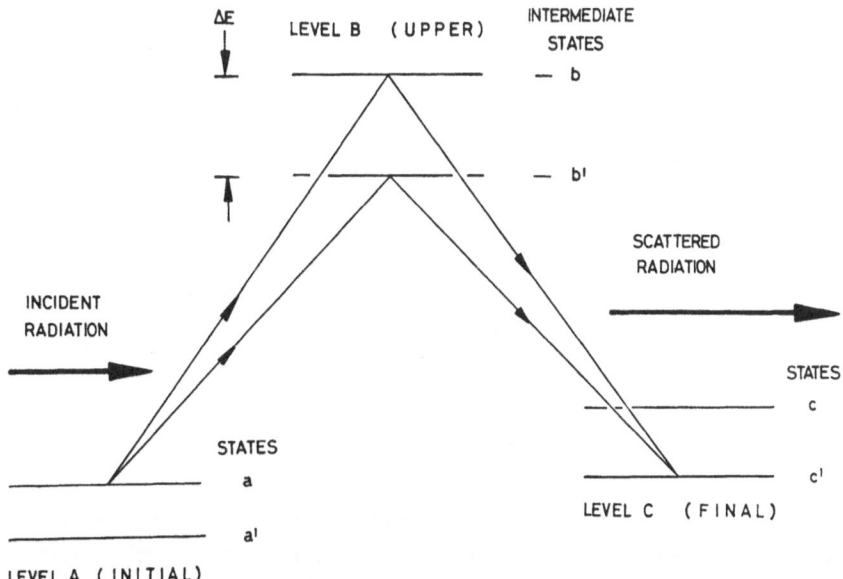

Fig. 1.   Simplified three-level atomic model used to illustrate level polarization effects in the absorption, emission, and scattering of polarized light. A typical scattering process is shown in which the atom undergoes induced absorption from an initial level A to an intermediate level B and subsequently makes a transition via spontaneous emission to the final level C. In the absence of collisional relaxation, the scattering of a stationary radiation field will produce coherence between the different energy eigenstates of the intermediate level provided the angular frequency separation $\Delta E/\hbar$ between such states is less than or of the order of the radiation damping rate Γ^R of the same states. If coherence exists between the energy eigenstates of the initial level, this must also be taken into account.

within the subspace of level A will be denoted ϱ^A, that within B by ϱ^B, etc. Generally speaking level C may be, but need not be, identical with A. However, in the present discussion we wish to focus on the way in which atomic level polarization affects the polarization of light in a particular spectral line involving the two levels A and B and hence the term 'scattering' will be used only to refer to scattering processes of the type $A \to B \to A$ in which the initial and final levels (but not necessarily the initial and final states) are identical. Simple absorption or emission in the line is then represented, respectively, by the single transitions $A \to B$ or $B \to A$. The polarization of light absorbed in the line will be determined by the polarization of the initial level, that is by ϱ^A, while the polarization of light emitted into the line will be determined by the polarization of the excited level, that is by ϱ^B. The polarization of light scattered into the line will be determined partially by the polarization of the initial level and partially by the polarization properties of the incident light which takes part in the scattering process; the result may be expressed as a contribution to the polarization of the excited level (and hence to ϱ^B) due to this scattering process.

All present theories of spectral line formation in solar magnetic regions have assumed both that the initial level A is unpolarized and that all processes which populate the excited level B other than direct radiative excitation from level A do not contribute to the polarization of level B. In this case the only level polarization which can occur is that of level B in the scattering process $A \rightarrow B \rightarrow A$.* If level A is populated to some extent by radiative excitation or if level B is populated to some extent by radiative excitation from levels other than A, these assumptions must be justified.

2.2. Processes Affecting the Extent of Level Polarization in the Solar Atmosphere

In an essentially isotropic environment (as, for example, deep in the solar interior) an assembly of atoms will exhibit no level polarization. If, however, the physical environment is not isotropic, the anisotropy will generally lead to atomic level polarization. In the solar atmosphere atomic level polarization might be produced (i) by impact polarization due to collisions with surrounding perturbers which have an anisotropic velocity distribution; (ii) by radiative excitation when the incident radiation is partially polarized or spatially anisotropic; or (iii) by the presence of magnetic fields in conjunction with excitation by radiation, the polarization and intensity of which vary with frequency. In this last case the variation with frequency may itself be produced by Zeeman splitting of atomic levels in the magnetic field. On the other hand atomic level polarization in the solar atmosphere might be reduced (iv) by the presence of a magnetic field; (v) by collisional mixing within the level when the perturber velocity distribution is isotropic; (vi) by collisional mixing with other levels when the perturber velocity distribution is isotropic; or (vii) by radiative decays or recombinations into the level.

Impact polarization is unlikely to be an important source of atomic level polarization, at least in the photosphere and lower chromosphere, since there are strong grounds for believing that the velocity distribution there is essentially isotropic (cf. Böhm, 1960). It may be significant at greater heights in the solar atmosphere, but the present evidence makes this, too, seem unlikely (Charvin, 1965). However, atoms in prominences and the corona will tend to exhibit level polarization because of exposure to anisotropies in the ambient radiation field which are due to the effects of

---

* Expressions for scattering operators which include this contribution to the polarization of level B have been obtained by Obridko (1965), Rachkovsky (1967), and Domke (1969) for the case when coherence between the different energy eigenstates in level B and collisional relaxation of the level polarization can be neglected. Expressions for scattering operators have also been derived by Stepanov (1962), Rachkovsky (1963), and Domke (1969) for the case when both levels A and B are completely unpolarized.

limb darkening and the finite angular size of the solar disk. Atoms in the lower chromosphere and photosphere will experience this same radiation anisotropy to a lesser extent. In addition, atoms in solar magnetic regions are subject to level polarization via mechanism (iii). On the other hand, in solar magnetic regions the type of level polarization which can occur will be limited to some extent by the strength of the magnetic field. Throughout the solar atmosphere the polarization of low-lying atomic levels will tend to be reduced by radiative decays into these levels. For atoms located in the lower chromosphere and the photosphere, the polarization of all atomic levels will tend to be reduced by collisional mixing within the level, while level depolarization due to collisional mixing *between* levels will generally be much less important. The reasons for these conclusions will be described in more detail in the sections which follow.

3. Collisional Relaxation Effects and the Influence of the Magnetic Field

As noted in the previous section, the nature and extent of atomic level polarization which occurs in a given solar magnetic region may be profoundly affected both by the magnetic field there and by collisional relaxation. The present section first reviews briefly the way in which level polarization depends on magnetic field strength in the absence of collisions and then discusses the effects of collisional relaxation, including the way in which the dependence on field strength is altered.

3.1. INFLUENCE OF THE MAGNETIC FIELD

The strength of the magnetic field in a given solar magnetic region partially determines the type of atomic level polarization which can occur there. The nature of this effect may perhaps be most clearly seen by considering the way in which the forms of the level density matrices depend on the strength of the magnetic field. For the sake of definiteness in the following discussion we shall consider a particular atomic level labelled S, but completely analogous considerations hold for other atomic levels.

For atoms without hyperfine structure,* there are two basic regimes which are characterized by the magnitude of the Larmor frequency $\omega_L(S)$ for the states in level S relative to the radiation damping rate $\Gamma^R(S)$ for these same states. When the strength of the magnetic field is such that $\omega_L(S)/\Gamma^R(S) \lesssim 1$, coherence between the different energy eigenstates within level S (represented by non-zero off-diagonal elements in the level density matrix ϱ^S expressed in the energy representation; cf. Cohen-Tannoudji, 1962) is possible and the particular type of level polarization often referred to as level-crossing interference (Franken, 1961) can occur. On the other hand, when the strength of the magnetic field is such that $\omega_L(S)/\Gamma^R(S) \gg 1$, the level density matrix ϱ^S is necessarily diagonal in the energy representation and there can be no coherence between the different energy eigenstates within level S. Only in this case can the different types of level polarization which are possible be described in terms of the populations

---

* Throughout the present paper hyperfine structure is assumed to be absent; the complications which arise when hyperfine structure is present are discussed in Lamb and ter Haar (1971).

of the energy eigenstates. The precise way in which the field strength H enters expressions for the operators describing absorption, emission, and scattering depends
on which of the irreducible tensor components of the atomic level density matrix are
involved. For example, expressions for the Stokes parameters I, Q, and U depend on
both of the two dimensionless parameters $(2\omega_L/\Gamma^R)$ and (ω_L/Γ^R), whereas expressions
for the Stokes parameter V depend only on (ω_L/Γ^R). In order to have a convenient
parameter by which to indicate which regime obtains in any given circumstance, we
shall refer to the 'critical field strength' H_c for which $2\omega_L/\Gamma^R = 1$. The value of H_c
depends on the particular atomic level in question and is given by the relation

$$H_c(S) = 5.66 \frac{\Gamma^R(S)}{g_J(S)},\qquad(1)$$

where H_c is in gauss, the radiation damping rate Γ^R is in units of $10^8 \, \mathrm{rad\,s^{-1}}$, and
g_J is the Landé g-factor for level S.

3.2. EFFECTS DUE TO COLLISIONAL RELAXATION

In the present subsection the influence on atomic level polarization of collisions between the atoms of interest and surrounding perturbers will be described within the
limitations of the level density matrix formalism. Attention will be focussed on the
effects of collision-induced transitions between states within the same level while the
effects of collision-induced transitions between states in different levels will be largely
neglected. The justification for this procedure will be discussed in Section 4.

At very low particle densities where collisional relaxation can be neglected, all
$2j+1$ Zeeman states of a given atomic level with total electronic angular momentum j
decay at the same rate, namely the radiation damping rate Γ^R. At higher particle
densities the Zeeman states can relax to each other through collisions so that, in
general, $(2j+1)^2$ parameters are necessary to describe the decay of the atomic level.
However, it has been shown on theoretical grounds that in an *isotropic* environment
the interaction with surrounding perturbers will lead to the independent relaxation of
each multipole moment of the atomic density matrix (Fano, 1963; Ben-Reuven,
1966a,b). For a single atomic level, this leads to a reduction in the number of different
level relaxation times, from $(2j+1)^2$ to $2j+1$, with one relaxation time for each 2^K-
pole moment of the level density matrix.

Even when the isotropy of the environment is destroyed by the presence of a magnetic field, it is expected on theoretical grounds that each multipole moment of the
density matrix for a single atomic level will relax independently provided the typical
frequencies ω_c occurring in the spectrum of the effective collision Hamiltonian (described in Section 4 below) satisfy

$$\omega_c \gg \omega_L.\qquad(2)$$

This result has been confirmed experimentally (see for example Happer and Saloman,
1967; and Omont and Meunier, 1968; the experimental data have been summarized by
Berman and Lamb, 1969). For collisions with either electrons or atomic hydrogen at
the temperatures typical of solar magnetic regions $\omega_c \gtrsim 10^{13} \, \mathrm{rad\,s^{-1}}$; since $\omega_L \lesssim 1.5 \times$

10^{11} rad s^{-1} even for the very strongest magnetic fields and largest g-factors of interest, (2) is well-satisfied in these regions.

With the assumption of the independent relaxation of each multipole moment of the density matrix for a single atomic level, the total relaxation rate for a given 2^K-pole moment including both radiation damping and collisional relaxation may be written

$$\Gamma^{(K)} = \Gamma^R + \Gamma_C^{(K)}, \tag{3}$$

where $\Gamma_C^{(K)}$ is the collisional relaxation rate of the Kth moment of the atomic level. For absorption, emission, and scattering processes which involve only electric dipole transitions, the only moments of the level density matrices (and hence the only $\Gamma_C^{(K)}$) of any significance are those with $K \leqslant 2$. It may be remarked that while $\Gamma_C^{(1)}$ and $\Gamma_C^{(2)}$, which represent the effects of collision-induced transitions between states within the *same* atomic level, are comparable to the ordinary collisional broadening rate, $\Gamma_C^{(0)}$ is given by the rate of collision-induced transitions between states in *different* levels and thus is ordinarily several orders of magnitude smaller at temperatures typical of the solar atmosphere. Both theory and experiment indicate $\Gamma_C^{(1)} \approx \Gamma_C^{(2)}$ (the exact numerical calculation for foreign-gas collisions in the dipole-dipole approximation gives $\Gamma_C^{(1)}/\Gamma_C^{(2)} = 1.12 \pm 0.02$; see Berman and Lamb, 1969). Since only very rough estimates of the absolute magnitudes of the $\Gamma_C^{(K)}$ are possible at the present time for the cases of interest to us, no distinction between $\Gamma_C^{(1)}$ and $\Gamma_C^{(2)}$ will be made in the discussion which follows.

Collisional relaxation of atomic level polarization affects the interaction of atoms with the radiation field in two different ways, which may be described intuitively as follows (for a more detailed discussion see Lamb and ter Haar, 1971): (i) the 2^K-pole moments of the atomic level density matrices with $K > 0$ relax faster than the total level populations (the $K = 0$ moments). This is apparent from the form of Equation (3) and the discussion immediately following it, and results in a reduction of the proportion of polarized radiation associated with absorption, emission, and scattering processes beginning at these levels. If we make the approximations*

$$\Gamma_C^{(1)} = \Gamma_C^{(2)} \equiv \Gamma_C, \tag{4}$$

and

$$\Gamma_C^{(0)} = 0, \tag{5}$$

then the ratio of polarized or anisotropic absorption, emission, or scattering associated with the $K > 0$ multipole moments of a given atomic level relative to the isotropic unpolarized absorption, emission, or scattering associated with the $K = 0$ moment of the same level is roughly given by

$$f_p = \frac{\Gamma^R}{(\Gamma^R + \Gamma_C)}, \tag{6}$$

when the level is populated entirely by completely polarized or completely collimated

---

* Approximation (4) has already been seen to be quite good; as noted earlier, the approximation represented by Equation (5) will be discussed in Section 4.

radiation in the absence of magnetic fields. If the exciting radiation is not completely collimated or completely polarized, if the level is populated to some extent by iso-tropic processes (which may include radiative decays, recapture, or collision-induced transitions into the level), or if there is a magnetic field present (see Subsection 3.1), then the importance of polarized absorption, emission, and scattering beginning at the given level will be correspondingly less. (ii) The energy distribution associated with each state in a given atomic level is broadened relative to the energy distribution in the absence of collisions. This makes possible coherence between different energy eigenstates within the same level at higher magnetic field strengths than in the absence of collisions by increasing the energy 'overlap' of these states. Consequently, in the presence of collisional relaxation the critical magnetic field strength H_c for a given level, say S, is given (within the approximations represented by Equations (4) and (5)) by a modified form of Equation (1), namely

$$H_c(S) = 5.66 \frac{[\Gamma^R(S) + \Gamma_C(S)]}{g_J(S)}, \tag{7}$$

where the radiation damping rate Γ^R and the multipole collisional relaxation rate Γ_C are in units of 10^8 rad s^{-1}. Both effects (i) and (ii) are clearly illustrated by the experimental results (see for example those of Barger, 1967; and Happer and Saloman, 1967, cited earlier).

From these results it seems clear that, except in the corona (see Charvin, 1965), the relative importance of collisional relaxation of atomic level polarization will have a decisive effect on the polarization of spectral lines formed in solar magnetic regions.

4. Estimating Collisional Relaxation Rates

4.1. QUALITATIVE CONCLUSIONS

In spite of intense work in recent years, the accurate calculation of collisional relax-ation rates for the cases of interest here remains beyond the scope of present techniques. Nevertheless, a number of important conclusions emerge from a qualitative analysis of atomic level depolarization by collisions with atomic hydrogen in solar magnetic regions: (i) the impact approximation is valid; (ii) the effects of collision-induced transitions between levels will generally be small; (iii) essentially all collisions are diabatic with respect to the states within the same atomic level (even those collisions with very small impact parameters which are essentially adiabatic with respect to the states within the same atomic level will partially reorient the atom); (iv) collisional relaxation rates are not expected to depend on the strength of the magnetic field for the field strengths and temperatures which exist in solar magnetic regions. The basis for these conclusions is as follows.

The impact approximation is valid (Baranger, 1958) if

$$n_H \pi b_c^3 \ll 1, \tag{8}$$

where n_H is the number density of hydrogen perturbers and b_c is the effective collision radius. For metal atoms perturbed by collisions with atomic hydrogen at $T \approx 5 \times 10^3$ K, b_c is expected to be of order 10^{-7} cm (a more detailed discussion of this point is given in Lamb and ter Haar, 1971). Since $n_H \lesssim 10^{17}$ cm^{-3} in the solar atmosphere, the validity condition (8) is certainly satisfied in this case.

Assuming all perturbers are in their ground state and that there are no accidental resonances in the atom-perturber interaction, an effective collision Hamiltonian of second order in the atom-perturber interaction may be introduced by summing over internal perturber variables (cf. Omont and Meunier, 1968; Berman and Lamb, 1969). This effective collision Hamiltonian will have non-vanishing matrix elements only between those atomic states with a separation in angular frequency units $\Delta\omega \lesssim \omega_c$, where ω_c is typical of the frequencies occurring in the Fourier transform of the effective collision Hamiltonian. Since the duration of a collision with impact parameter b is $\tau_d \approx b/v_r$, where v_r is the rms relative velocity between atom and perturber, the Fourier transform will contain frequency components up to $\omega_c = \tau_d^{-1} \approx v_r/b_c$. For the temperatures and atoms of interest, $b_c \approx 10^{-7}$ cm and $v_r \approx 10^6$ cm s^{-1} giving $\omega_c \approx 10^{13}$ rad s^{-1}. Since for most states lying in different atomic levels $\Delta\omega \gg \omega_c$, collision induced transitions between such states will be negligible. This condition may not be completely satisfied in some cases, particularly for the higher levels of the transition metals. For the temperatures of interest, however, we expect only a small contribution to the collisional relaxation rate due to transitions between levels. This is the reason for adopting the approximation $\Gamma_C^{(0)} = 0$ in Subsection 3.2.

On the other hand (cf. Spitzer, 1940; Rebane and Rebane, 1966) almost all collisions are diabatic with respect to states within the same atomic level. For Zeeman state separations $\Delta\omega \approx 10^{11}$ rad s^{-1}, which are the largest occurring in solar magnetic regions, distant collisions at impact parameters out to $b_{max} \approx 10^{-5}$ cm are still diabatic. Only collisions at very large impact parameters $b \gg b_{max}$ and at very small impact parameters $b \ll b_c$ (for which the frequency separation $\delta\omega$ caused by interaction with the perturber is much greater than ω_c) will be adiabatic for states within the same level. However, even close adiabatic collisions will partially reorient the atom. Because almost all collisions are diabatic in this sense, collisional depolarization rates are likely to be determined largely by the range of the atom-perturber interaction. For the same reason (i.e. because, even for the highest magnetic fields and largest g-factors of interest, the Zeeman state separations are much less than ω_c) collisional depolarization rates are not expected to depend on the strength of the magnetic field in solar magnetic regions.

4.2. QUANTITATIVE ESTIMATES

Rough estimates of the rate of collisional relaxation of level polarization have been calculated for a number of atomic levels which are of interest because they are involved in the formation of some of the spectral lines commonly used to study solar magnetic fields. The method of calculation used here is similar to that of Byron and Foley (1964) and Omont (1965), though the approximations used are somewhat more brutal. In addition to the classical path approximation, we have assumed that the atom-perturber interaction is adiabatic with respect to states in different atomic levels. It has further been assumed that the atom-perturber interaction is adequately described by the van der Waals term and that this term may be replaced by the London approximation for the static van der Waals interaction between two atoms. Because of these approximations a few words of caution concerning the accuracy of the results are in order. As noted earlier, the assumption of complete adiabaticity with respect to different atomic levels is not satisfied for some of the higher levels of the transition metals. There is also considerable evidence (see for example Behmenburg, 1964; Griem, 1964; Hindmarsh et al., 1967; Roueff and Van Regemorter, 1969) that the atom-perturber interaction is often poorly represented by the van der Waals term. Furthermore, the time dependence of the actual interaction between atom and perturber results in contributions to the effective collision Hamiltonian due to various virtual transitions of the atom and perturber which are different from those given by the London approximation for the static van der Waals interaction (cf. Berman and Lamb, 1969). In spite of these rather serious difficulties, the approximations used here are expected to provide order-of-magnitude estimates which may be useful in identifying cases where collisional level depolarization is either unimportant or essentially complete, but it is clearly not possible at the present time to resolve doubtful cases. On the whole we expect the present method to underestimate the depolarization rate.

5. Results and Conclusions

Using atomic level multipole collisional relaxation rates calculated according to the method outlined in Subsection 4.2., estimates of the maximum possible degree of level polarization f_p and the critical magnetic field strength H_c have been made for the levels involved in a number of magnetically-sensitive solar absorption lines. The results for five Fe I lines ($\lambda\lambda 5131$, 5250, 6173, 6302, and 8468) are presented in Table I. The radiative relaxation rates adopted are based on the f-values of Corliss and Warner (1964). These represent lower limits since f-values for some of the transitions which contribute were unavailable, but it is thought that the most important transitions have been included. The collisional relaxation rates were calculated for an atomic hydrogen density $n_H = 1.6 \times 10^{17}$ cm^{-3} and thermal temperature $T = 4500$ K, using the estimates of Fe I radii given by Warner (1969). The multipole collisional relaxation rate is expected to depend somewhat more sensitively on atomic hydrogen density than on temperature ($\Gamma_c \propto n_H T^{3/10}$ for the van der Waals interaction). When account is also taken of the much greater variation of hydrogen density with height in the solar atmosphere compared to that of the temperature, it is clear that the value of f_p for lines formed at heights in the solar atmosphere greater than those corresponding to these values of n_H and T will be greater than that given in Table I.

The results presented in Table I indicate that for the atomic levels involved in the formation of the lines listed there, collisional relaxation of atomic level polarization occurs faster than radiative decay at temperatures and neutral hydrogen densities typical of the photosphere. In all cases f_p is less than 0.1 and when account is also

TABLE I

Maximum possible level polarization and critical magnetic field strength for the levels involved in five magnetically-sensitive solar absorption lines

(1)	(2)	(3)	(4)	(5)	(6)
5131.478	Fe I	$\ll 1^a$	3.4×10^{-3c}	34	43
5250.218	Fe I	$\ll 1^a$	4.5×10^{-6d}	*	31
6173.348	Fe I	$\ll 1^a$	8.3×10^{-2e}	34	*
6302.508	Fe I	2.1×10^{-2b}	8.6×10^{-2f}	44	*
8468.417	Fe I	$\ll 1^a$	2.1×10^{-2b}	34	44

(1) Wavelength in Å;
(2) Chemical element;
(3) Maximum level polarization f_p for the lower level;
(4) Maximum level polarization f_p for the upper level;
(5) Critical magnetic field strength H_c in gauss for the lower level;
(6) Critical magnetic field strength H_c in gauss for the upper level.

[a] No radiative decays from these levels are listed by Moore (1945); the natural level lifetime is expected to be long.
[b] The computed radiative relaxation rate Γ^R includes the transitions 3443, 3465, 3476, 8327, 8468 8804.
[c] Γ^R includes 3397, 4090, 5079, 5131.
[d] Γ^R includes 5225, 5250.
[e] Γ^R includes 3850, 6173, 6862.
[f] Γ^R includes 5230, 5576, 6302.
* These levels have zero total electronic angular momentum ($j = 0$) and hence cannot be polarized.

taken of the other effects (described in Subsection 2.2) which lead to a reduction in level polarization, the actual degree of level polarization is likely to be an order of magnitude or more smaller than this. In such cases the usual neglect of level polarization in the treatment of absorption and emission processes will be a good approximation and transfer equations which treat scattering assuming complete depolarization may be expected to give a better description of the formation of such lines than those which neglect it. On the other hand, at greater heights in the solar atmosphere collisional relaxation will be less complete. For spectral lines formed primarily at these greater heights, an accurate description of line formation will require a treatment of scattering which includes the effects of partial level depolarization by collisions and an investigation of the level polarization effects described in Subsection 2.1 which appear in absorption and emission.

The results presented in Table I also indicate that in spite of the tendency for collisional relaxation to increase the critical magnetic field strength, H_c, at photospheric densities the energy 'overlap' of states within the same atomic level will not be significant for the lines listed there when the magnetic field strength is greater than 10^2 G. For this reason and also because of the depolarizing effect of collisions, level-crossing interference is unlikely to be important for these lines when formed in the lower regions of sunspot magnetic fields. However, level-crossing interference may well be important for lines formed in weaker magnetic fields at greater heights in the solar atmosphere.

The present preliminary investigation suggests that, except in the corona, collisional relaxation of atomic level polarization will have a decisive effect on the polarization of spectral lines formed in solar magnetic regions. Much work remains to be done, particularly to improve the reliability of calculated collisional relaxation rates and to explore the spectral features of the interaction between atoms and polarized radiation in the presence of collisions with surrounding perturbers, before an accurate description of the polarization of spectral lines formed in regions where collisional relaxation is only partially complete can be achieved.

References

Baranger, M.: 1958, *Phys. Rev.* **111**, 481.
Barger, R. L.: 1967, *Phys. Rev.* **154**, 94.
Behmenburg, W.: 1964, *J. Quant. Spectr. Radiative Transf.* **4**, 177.
Ben-Reuven, A.: 1966a, *Phys. Rev.* **141**, 34.
Ben-Reuven, A.: 1966b, *Phys. Rev.* **145**, 7.
Berman, P. R. and Lamb, W. E.: 1969, *Phys. Rev.* **187**, 221.
Böhm, K.-H.: 1960, *Stars and Stellar Systems* **6**, p. 88.
Byron, F. W. and Foley, H. M.: 1964, *Phys. Rev.* **134**, A625.
Charvin, P.: 1965, *Ann. Astrophys.* **28**, 877.
Cohen-Tannoudji, C.: 1962, *Ann. Phys. Paris* **7**, 423.
Corliss, C. H. and Warner, B.: 1964, *Astrophys. J. Suppl.* **8**, 395.
Domke, H.: 1969, *Astrofiz.* **5**, 525.
Fano, U.: 1963, *Phys. Rev.* **131**, 259.
Franken, P. A.: 1961, *Phys. Rev.* **121**, 508.
Griem, H. R.: 1964, *Plasma Spectroscopy*, McGraw-Hill Publishing Co., Inc., New York.
Happer, W. and Saloman, E. B.: 1967, *Phys. Rev.* **160**, 23.
Hindmarsh, W. R., Petford, A. D., and Smith, G.: 1967, *Proc. Roy. Soc. London* **A297**, 296.
Lamb, F. K. and ter Haar, D.: 1971, *Phys. Repts.*, in press.
Moore, C.: 1945, *Princeton Obs. Contr.*, No. 20.
Obridko, V. N.: 1965, *Astron. Zh.* **42**, 102 = *Soviet Astron.* **9**, 77.
Omont, A.: 1965, *J. Phys. Paris* **26**, 26.
Omont, A. and Meunier, J.: 1968, *Phys. Rev.* **169**, 92.
Rachkovsky, D. N.: 1963, *Izv. Krymsk. Astrofiz. Obs.* **29**, 97.
Rachkovsky, D. N.: 1967, *Izv. Krymsk. Astrofiz. Obs.* **36**, 3.
Rebane, V. N. and Rebane, T. K.: 1966, *Opt. Spectroskopiya* **20**, 185 = *Opt. Spectry.* **20**, 101.
Roueff, E. and Van Regemorter, H.: 1969, *Astron. Astrophys.* **1**, 69.
Spitzer, L.: 1940, *Phys. Rev.* **58**, 348.
Stepanov, V. E.: 1962, *Izv. Krymsk. Astrofiz. Obs.* **27**, 140.
Warner, B.: 1969, *Observatory* **89**, 11.

Discussion

Pecker: The collisions which enter Lamb's expression for f_p are by no means to be confused with the ones which affect the source function. The former concern only those collisions producing a change in the magnetic quantum number.

Lamb: One must distinguish between the different collisional relaxation rates, each describing the relaxation of a different atomic multipole moment, which may affect the form of the source function. The present investigation was primarily concerned with the rate of collisional mixing between states within the same atomic level, which is described by the collisional relaxation rates $\Gamma_c^{(K)}$ ($K \neq 0$) of the various higher multipole moments of the individual atomic levels. Collisions with surrounding perturbers will also broaden the absorption and emission line profiles of individual atoms; the broadening of the spectral profile associated with a given multipole transition is described by the collisional relaxation rate of the relevant multipole moment connecting the initial and final levels. Finally, collisions may induce transitions between different atomic levels; the effect of collisions on the total population of a given level may be described by the level monopole collisional relaxation rate $\Gamma_c^{(0)}$. In general all of these different relaxation rates will have an important effect on the form of the source function. Although the relaxation rate which enters the expression for the parameter f_p describes the relaxation of a different atomic moment than that described by the collisional broadening rate, it is usually of the same order of magnitude.

Pecker: The height dependence of the collisional relaxation rate might in practice bring many difficulties, the 'critical magnetic field' being height-dependent.

Lamb: The critical magnetic field strength H_c will vary with height due largely to the dependence of the collisional relaxation rate Γ_c on the atomic hydrogen density n_H, and this may complicate the description of line formation if the magnetic field in the line-forming region is of the order of H_c.

LINE FORMATION IN INHOMOGENEOUS MAGNETIC FIELDS

R. GÖHRING

Fraunhofer Institut, Freiburg, Germany

Until now, analytic solutions for the problem of line formation in magnetic fields have been given only under special assumptions concerning either the magnetic field or the atmospheric model. Conditions are given for which the system of differential equations can be solved analytically. For the general case a method is given to solve the system in approximation and analytically. LTE is generally assumed which means that we use the system given by Unno (1956) or that given by Beckers (1969) including the Faraday rotation.

1. The Separation of the System of Differential Equations

After transforming the system of differential equations such that the x-axis of the coordinate system is in the azimuth direction of the magnetic field for all depths, the matrix A of the system has a very simple form (Göhring, 1970)

$$\mu \frac{d\mathbf{I}}{d\tau} = A\mathbf{I} - \mathbf{S}; \quad \mathbf{I} = \{I, Q, U, V\}. \tag{1}$$

The elements of the matrix A are functions of the absorption coefficient η, the parameter ϱ describing the Faraday-rotation and $d\phi/d\tau$, the depth-dependence of the azimuth angle.

It is possible to solve the system analytically if we are able to transform the system (1) into a system with a diagonal matrix. Introducing a matrix L and the transformation

$$\mathbf{I} = L\mathbf{I}' \tag{2}$$

we get

$$\mu \frac{d\mathbf{I}'}{d\tau} = \left\{ L^{-1}AL - \mu L^{-1} \frac{dL}{d\tau} \right\} \mathbf{I}' - L^{-1}\mathbf{S}. \tag{3}$$

We have not systematically tried to find a matrix L such, that

$$D = \left\{ L^{-1}AL - \mu L^{-1} \frac{dL}{d\tau} \right\}$$

is a diagonal matrix – it is possible that there exists none – but we have determined L so that $L^{-1}\,AL$ is diagonal. It can be shown that this is possible. In this case the columns of L are the normalized eigenvectors of A. The components L_{ij} are only functions of the magnetic field strength H, the field inclination γ, the damping constant a, the Doppler width $\Delta\lambda_D$ and $d\phi/d\tau$.

Howard (ed.), Solar Magnetic Fields, 162–164. All Rights Reserved.
Copyright © 1971 by the IAU.

Now only the first term in D is diagonal and we have to ask for

$$\mu L^{-1} \frac{dL}{d\tau} \equiv 0. \qquad (4)$$

In the general case this term is neither diagonal nor zero. It can be shown that (4) is fulfilled only under the following conditions:

(a) pure longitudinal or transverse field; (b) homogeneous field: $dH/d\tau = d\gamma/d\tau = d\phi/d\tau = 0$ and depth independent Voigt-function $d\alpha/d\tau = d/d\tau(\varDelta\lambda_D) = 0$.

Assuming (a) Hubenet (1955) has given the general solution of (1), under the assumption (b) the problem was solved by Mattig (1966) and Kjeldseth Moe (1968), both Hubenet and Mattig, and Kjeldseth Moe neglected Faraday rotation. Special solutions with the second assumption are given by Unno (1956), Michard (1961) and Wiehr (1968). If the conditions (a) and (b) are not fulfilled, we have to look whether

$$L^{-1}AL \gg \mu L^{-1} \frac{dL}{d\tau}$$

or

$$L^{-1}AL \approx \mu L^{-1} \frac{dL}{d\tau}.$$

In the first case we are able to solve the system of differential Equations (1) by the method of the perturbation theory, in the second case we have to integrate the four linear differential equations numerically as Beckers (1969) has done.

2. The Faraday Rotation as a Perturbation

It can be shown that it is possible to consider the Faraday rotation as a perturbation. The zero order approximation is then given by

$$\mu \frac{d\mathbf{I}^{(0)}}{d\tau} = (1 + \eta_0)\, \mathbf{I}^{(0)} - \mathbf{S} \qquad (5)$$

the higher approximations can be written in the form

$$\mu \frac{d\mathbf{I}^{(i)}}{d\tau} = (1 + \eta_0)\, \mathbf{I}^{(i)} + \varphi \mathbf{I}^{(i-1)} \qquad (6)$$

η_0 is the absorption matrix, φ the matrix of the Faraday rotation. It can be shown that the solutions of (5) and (6) are members of a convergent series and that, with four iteration steps, the result is better than 1%.

3. The General Solution of the System

The possibility to take into account the Faraday rotation as a perturbation and to find a convergent series, the general solution of (1) is reduced to the solution of the zero order approximation (5), the system without Faraday rotation.

Using the transformation (2) we can write for (5) (with $\mathbf{I}^{(0)} = \mathbf{J}$):

$$\mu \frac{d\mathbf{J}}{d\tau} = \left\{ L^{-1}(1 + \eta_0) L - \mu L^{-1} \frac{dL}{d\tau} \right\} \mathbf{J} - L^{-1}\mathbf{S}. \tag{7}$$

Here again L is determined such that $L^{-1}(1+\eta_0)L$ is diagonal (this is possible!) and $\mu L^{-1}(dL/d\tau)$ is assumed to be a small perturbation. All calculations for the atmosphere from Holweger (1967) and Kneer (1970) and with a strong depth-dependence of both the magnetic field strength and the field inclination have shown that this assumption can be made.

With the perturbation theory we can write for the different approximations

$$\mu \frac{d\mathbf{J}^{(0)}}{d\tau} = L^{-1}(1 + \eta_0) L\mathbf{J}^{(0)} - L^{-1}\mathbf{S}. \tag{7}$$

$$\mu \frac{d\mathbf{J}^{(i)}}{d\tau} = L^{-1}(1 + \eta_0) L\mathbf{J}^{(i)} - \mu L^{-1} \frac{dL}{d\tau} \mathbf{J}^{(i-1)}. \tag{8}$$

The solution of (7) can be given analytically (the matrix is diagonal!). It has a form like that given by Mattig (1966) and Kjeldseth Moe (1968). The higher approximations (8) can be derived recursively. The series converges relatively fast and the computing time needed is shorter by a factor of 20 using the perturbation theory compared with the pure numerical integration.

References

Beckers, J. M.: 1969, *Solar Phys.* **9**, 372.
Göhring, R.: 1970, Dissertation, Freiburg.
Holweger, H.: 1967, *Z. Astrophys.* **65**, 365.
Hubenet, H.: 1955, *Z. Astrophys.* **35**, 245.
Kjeldseth Moe, O.: 1968, *Solar Phys.* **4**, 267.
Kneer, F.: 1970, Dissertation, Freiburg.
Mattig, W.: 1966, in *Atti del Convegno Sui Campi Magnetici Solari* (ed. by M Cimino), G. Barbèra. Firenze = *Mitt. Fraunhofer Inst.* Nr. 62.
Michard, R.: 1961, *Compt. Rend.* **253**, 2857.
Unno, W.: 1956, *Publ. Astron. Soc. Japan* **8**, 108.
Wiehr, E.: 1968, Dissertation, Göttingen.

PART III

OBSERVATIONS OF SUNSPOT AND
ACTIVE REGION MAGNETIC FIELDS

ON MAGNETIC FIELDS IN SUNSPOTS AND ACTIVE REGIONS

E. H. SCHRÖTER

University Observatory, Göttingen, Germany

Abstract. This review is restricted to a few topics of the extensive problem under discussion; it refers to the literature since 1967 only. The literature prior to 1967 is reviewed in Zwaan, 1968; Howard, 1967; Lüst, 1965; Kiepenheuer, 1968; Jäger, 1964; and Bray and Loughead, 1964.

1. Comments on the Thermodynamical Structure of Sunspots

1.1. RECENT UMBRAL AND PENUMBRAL MODELS

The homogeneous hydrostatic equilibrium models for the umbra of a so-called 'typical sunspot' derived recently by Hénoux (1969), Hong Sik Yun (1971), Stellmacher and Wiehr (1970) agree rather well with one another (Figure 1). Hénoux's model has been derived from continuous radiation data and checked with the equivalent widths of medium strong lines. The model of Hong Sik Yun reproduces the wavelength dependence of the spot contrast up to 1.6 μ, its center to limb variation (see Mattig (1969a) and Wittmann and Schröter (1969)) and the wings of the NaD-lines. The Stellmacher-Wiehr model is a modified Hénoux model and capable of reproducing the line profiles of magnetically unsplit lines, as observed by the authors. A working model in the range of validity of these models seems capable of describing most of the available continuous radiation and Franhofer line data within observational errors. Deviations from hydrostatic equilibrium, at least in the upper layers where most of the lines originate, do not seem serious. These models refer to the

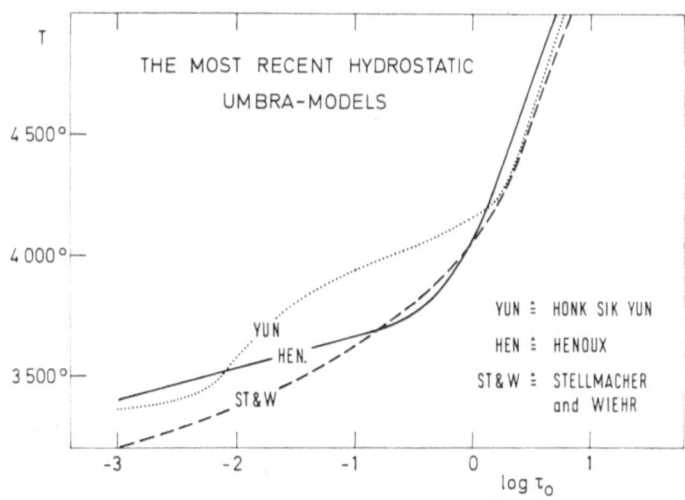

Fig. 1.   Comparison of the three most recent homogeneous umbral models (hydrostatic equilibrium models).

'mean averaged umbra' and not to the umbral region between the umbral dots. The relatively low umbral temperature gradient with its consequent departures from radiative equilibrium may be explained in terms of a lateral influx from the photo-sphere (cf. Zwaan, 1965 and Hong Sik Yun, 1971), in terms of a strong blanketing effect, caused by the thousands of molecular lines originating in the upper layers of umbrae, or is simulated by the umbral dots. Then the later have to be not of convective origin (Wilson, 1969) but non-radiative heated phenomena of the upper layers. At-tempts to incorporate umbral inhomogeneities in inhomogeneous umbral models (Makita, 1963 and Obridko, 1968) are premature. High resolution spectral atlases of umbrae are under preparation at Kitt Peak and Göttingen (Wöhl *et al.*, 1970).

The penumbral homogeneous hydrostatic equilibrium model of Kjeldseth Moe and Maltby is also capable of well reproducing all available observational data (see e.g. Moe and Maltby, 1969 and Schleicher and Schröter, 1971). Its physical significance, however, is rather limited because of the strong inhomogeneity of the penumbra.

1.2. THE STRUCTURE OF A 'TYPICAL SUNSPOT'

So far spectroscopic observations of umbrae seem not to contradict the assumption of hydrostatic equilibrium (see also Tepliskaja and Turchina, 1969). A vertical section through a typical sunspot can be derived (see Figure 2) by taking the mean of the results from the investigations (Wittmann and Schröter, 1969; Wilson and Cannon, 1968; Wilson, 1968; Wilson and McIntosh, 1969; Jensen *et al.*, 1969; Mattig, 1969b; and Ruhm, 1969). In these investigations either the center to limb variation of the sunspot intensity profile (Wilson-effect) or the density scale height have been discussed. There is some controversy whether or not the depression vanishes for small optical depths. An almost force-free magnetic field in the upper layers and

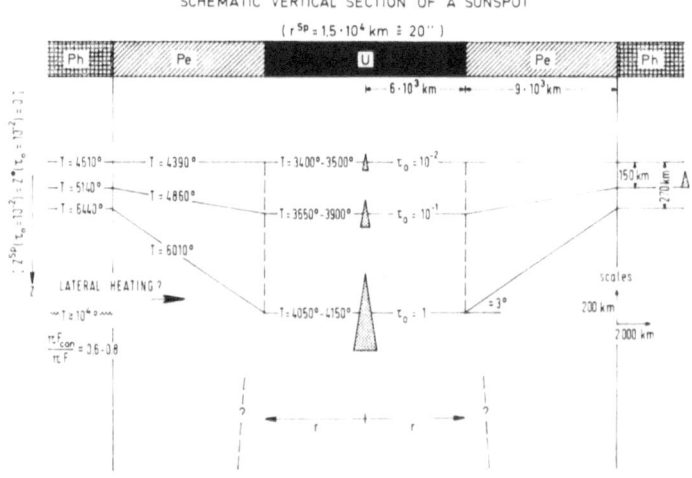

Fig. 2.   A vertical section through a typical sunspot derived from Wittmann and Schröter (1969), Wilson and Cannon (1968), Wilson (1968), Wilson and McIntosh (1969), Jensen *et al.* (1969), Mattig (1969), and Ruhm (1969). ▲ indicates the density scale height.

an increasing prevalence of magnetic forces in deeper layers of umbrae are the consequences of this model. Wilson's conclusion that the diameters of umbrae increase with depth should be checked by more sophisticated observations.

2. The Average Magnetic Field of a Sunspot

It is well known that the magnetic field structure may vary in a complicated manner from one spot to another, especially when complex sunspot groups are considered (see e.g. Jayanthan, 1970). In order to understand some general aspects of the sunspot magnetic field the following discussion is restricted to single, symmetric, long lived sunspots.

2.1. THE RADIAL VARIATION OF THE MAGNETIC FIELD STRENGTH

Redeterminations of $H(\varrho)$ (where ϱ is the distance from the spot center in units of its radius) generally confirm previous results (Adam, 1969; Rayrole and Semel, 1967, 1970; Beckers and Schröter, 1969; and Deubner and Göhring, 1970) resulting in the conventional axial magnetic field pattern. Beckers and Schröter showed that the observed $H(\varrho)$ curve may be well represented by

$$H(\varrho) = H(0) \left[1 + \varrho^2\right]^{-1} \qquad 0 \le \varrho \le 1 \tag{1}$$

which leads to a magnetic field strength at the outer sunspot border one half of that at the center. This formula is to be compared with Broxon's and Mattig's formulae

$$H(\varrho) = H(0) \left[1 - \varrho^2\right] \qquad H(\varrho) = H(0) \left[(1 - \varrho^4) e^{-2\varrho^2}\right] \tag{2}$$
$$\text{(Broxon)} \qquad\qquad\qquad \text{(Mattig)}$$

$$0 \le \varrho \le 1.$$

The $H(\varrho)$ curve as derived by Beckers and Schröter (1969) differs from previous results in two aspects: (a) the penumbral field was found to be considerably larger than assumed up to the present, (b) the magnetic field was found to tend sharply to zero at the outer penumbral border. Altrock (1969) and Mallia (1970) confirmed the high field strength in the penumbra, whereas Deubner and Göhring (1970) measured considerably smaller fields at the penumbral-photospheric border. Kossinskij (1967) found the magnetic sunspot radius to be larger than the spot intensity radius in contradiction to the measurements (b) by Beckers and Schröter (1969).

The saturation effect in Zeeman split lines as discussed by Maltby and Kjeldseth (1968), Hénoux (1968), Beckers and Schröter (1968a), Göhring (1969) and Staude (1970) may lead to an incorrect interpretation of the $H(\varrho)$-curve and of the height gradient when strongly saturated lines (such as $\lambda 5250$ Å) are used for the observations. This effect leads to a much larger separation of the maxima of the Stokes parameter V when compared with the true field if $\gamma \neq 0$ (γ is the inclination angle of the field to the line of sight). Since no discontinuous changes in $H(\varrho)$ at the umbra-penumbra border are observed, one concludes that either the umbral depression at those depths where the lines originate is negligible or that the model of an inclined depression within the penumbra (see Figure 2) is correct.

2.2. THE INCLINATION OF THE AVERAGE SUNSPOT MAGNETIC FIELD

Redeterminations of the zenith angle $\alpha(\varrho)$ of the magnetic field across a symmetrical sunspot (Adam, 1969; Beckers and Schröter, 1969; and Deubner and Göhring, 1970) generally confirm the classical Hale and Nicholson law $\alpha = \pi/2 \times \varrho$.

However, when comparing the results from different observers in detail (see e.g. Figure 9 in Beckers and Schröter, 1969) one finds: (a) a very large scatter in the various determinations, (b) only those observations based on the classical 'neutral line' method, which are observations of the location of the transversal effect within the sunspot during its disc passage, lead to zero values of α in the spot center; all other methods (intensity ratio of the σ-to-π-component, magnetographic and lambda-meter measurements) result in non zero values of α in the spot center, (c) the zenith angles measured by the latter methods are systematically larger than those obtained from the 'neutral line' method.

Retardation in solar instruments, generally neglected, may well be responsible for these effects. Almost all solar telescopes use mirrors at high angles of incidence (coe-lostats or Coudé-mirrors) which produce not only linear polarization but also light retardation caused by phase changes.* Only a few observers have considered the effect of instrumental retardation on their magnetic field measurements (e.g. Adam at Oxford, Jäger and Oetken at Potsdam).

Assume that this retardation makes the instrument behave like a $\lambda/4$ plate for a small fraction p of the incident light and that it does not change the polarization state of the remaining fraction $(1-p)$. The axis of the instrumental polarization ellipse will rotate during the day. For simplicity consider this axis to be parallel to the analyzing plate.

Then a fraction $(1-p)$ of the Zeeman light-pattern is analyzed with a $\lambda/4$ plate and a polaroid and the remaining fraction p through a $\lambda/2$ plate and a polaroid.

The simple Unno formulae for large splitting are a reasonable approximation for umbral magnetic fields. Using these formulae (multiplied by p) for a $\lambda/2$ plate + po-laroid arrangement and multiplied by $(1-p)$ for a $\lambda/4$ plate + polaroid arrangement one obtains:

$$r_{\sigma_1,\text{obs}} = f(\theta) \frac{\frac{1}{4}\eta_\lambda(1 - \cos\gamma)^2}{1 + \frac{1}{2}\eta_\lambda(1 + \cos^2\gamma)} \times$$

$$\times \left\{1 - p\frac{\sin^2\gamma\sin 2\varphi - 2\cos\gamma}{(1 - \cos\gamma)^2}\right\} \triangleq f(\theta)\frac{\frac{1}{4}\eta_\lambda(1 - \cos\hat{\gamma})^2}{1 + \frac{1}{2}\eta_\lambda(1 + \cos^2\hat{\gamma})}$$

$$r_{\sigma_2,\text{obs}} = f(\theta)\frac{\frac{1}{4}\eta_\lambda(1 + \cos\gamma)^2}{1 + \frac{1}{2}(1 + \cos^2\gamma)} \times$$

$$\times \left\{1 - p\frac{\sin^2\gamma\sin 2\varphi + 2\cos\gamma}{(1 + \cos\gamma)^2}\right\} \triangleq f(\theta)\frac{\frac{1}{4}\eta_\lambda(1 + \cos\hat{\gamma})^2}{1 + \frac{1}{2}\eta_\lambda(1 + \cos^2\hat{\gamma})}$$

* If the incident light is linearly polarized, one will generally measure elliptically polarized light at the exit.

$$r_{\pi, \text{obs}} = f(\theta) \frac{\frac{1}{2}\eta_\lambda \sin^2\gamma}{1 + \eta_\lambda \sin^2\gamma} \times$$

$$\times \{1 + p \sin 2\varphi\} \triangleq f(\theta) \frac{\frac{1}{2}\eta_\lambda \sin^2\hat{\gamma}}{1 + \eta_\lambda \sin^2\hat{\gamma}}$$

where $f(\theta) = (1 + \beta_0 \cos \vartheta / 1 + \beta_0)$, γ the true and $\hat{\gamma}$ the 'observed' inclination angle of the magnetic field to the line of sight and φ the azimuth angle of the field. Since $p \sin 2\varphi > 0$ in the range $0 < \varphi < 90°$, the π-component is strengthened by instrumental retardation (it is unchanged for $\varphi = 0°, 90°$). For small angles γ:

$$\sin^2\gamma \sin 2\varphi + 2 \cos \gamma > 0 \quad \text{and} \quad \sin^2\gamma \sin 2\varphi - 2 \cos \gamma < 0.$$

Hence, the stronger σ_2-component is weakened and the weaker σ_1-component is strenghtened considerably ($(1 - \cos^2\gamma)$ is almost zero) by instrumental retardation.

The effect is least for $\gamma = 90°$; generally the Zeeman components will show 'inconsistent' intensity ratios.

Therefore, when deriving $\hat{\gamma}$ from the intensity ratios σ_1/π, σ_2/π, σ_1/σ_2 and neglecting retardation in the instrument, one will obtain $\hat{\gamma} > \gamma$.

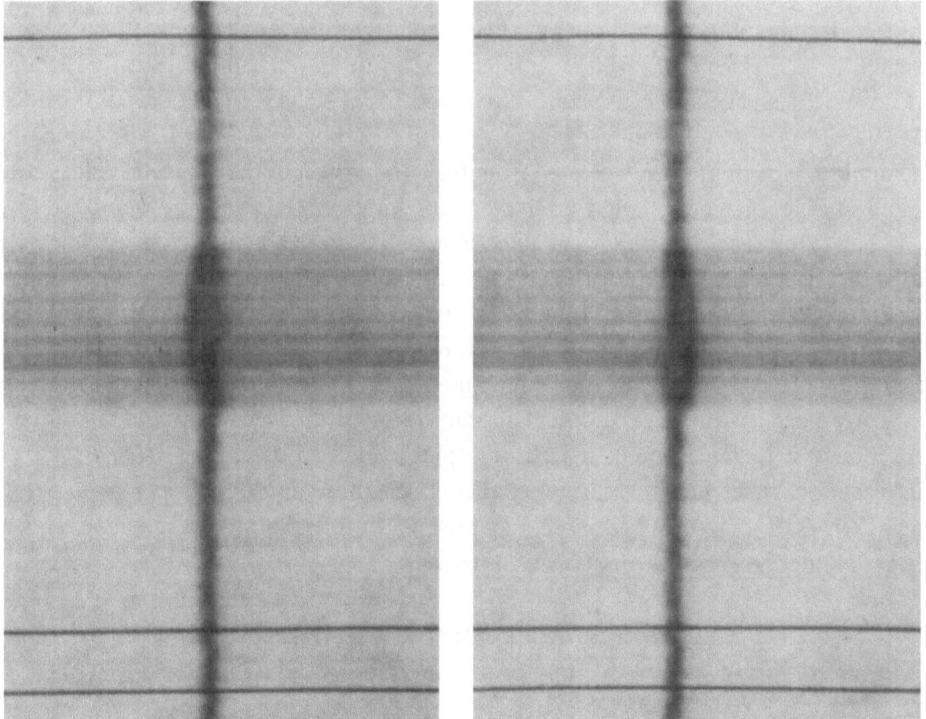

Fig. 3. The magnetically sensitive Fe-line $\lambda 6173.3$ Å in a sunspot penumbra, observed through a Wollaston prism and polaroid (Sacramento Peak Big Dome coelostat and spectrograph). Note the inconsistency of the Zeeman pattern in the two spectra due to retardation in the instrument; e.g. the π-component does not vanish (as it should) when both spectra are subtracted from each other.

Computations show, that if $\gamma = 10°$ (the smallest observable true inclination angle in the spot center because of the spot's disc position), $\hat{\gamma} \sim 25° - 35°$ when $p = 0.1$ and $\eta_0 = 10$. The retardation parameter may vary from instrument to instrument; Adam (1969) measured $p = 0.07$; at Locarno $p \sim 0.05$ was found.

The instrumental retardation (or the inconsistency of the Zeeman pattern) will generally escape detection in the lambdameter and magnetographic techniques and can best be recognized when the Stokes parameters of the entire line profile are measured. The scatter of the various determinations of $\alpha(\varrho)$, especially for small α, has to be interpreted in terms of (a) different values of p, (b) different positions of the instrumental retardation axis (at the time of observations) for the various instruments used, and (c) the dissimilar influence of instrumental retardation on the many analysing techniques employed. Since the effect of instrumental retardation is negligible for $\gamma = 90°$ and $\hat{\gamma} > \gamma$ when $\gamma < 90°$, the larger values for α obtained from methods other than the 'neutral line' method are well understood. Since retardation is present in almost all solar telescopes and has almost been ignored, earlier observations leading to a non vertical field in the spot center (Bumba, 1962; Bray and Loughead, 1964) are inconclusive.

2.3. THE HEIGHT GRADIENT OF THE SUNSPOT MAGNETIC FIELD

All redeterminations of the magnetic field height gradient in sunspots (Rayrole and Semel, 1970; Beckers and Schröter, 1969; Ikhsanov, 1968; Dubov, 1965, Guseynov, 1970; Wiehr, 1969; and Kusnezov, 1968) lead to 1.0–0.5 G/km for the spot center, 0.3–0.2 G/km for the umbra-penumbra border and to 0.2–0.05 G/km within the penumbra. Height gradients derived from the comparison of the magnetic fields, measured in different lines, should be treated with caution. This method is subject to a number of misleading effects and uncertainties (scattered light if ionized lines are used, LTE and NLTE; the problem of contribution curves and heights of origin; saturation effects as discussed by Maltby and Kjeldseth Moe (1968), Hénoux (1968), Beckers and Schröter (1968), Göhring (1969), and Staude (1970). Because of this, and the large scatter in Guseynov's measurements (1970) his result that occasionally a change of sign in the height gradient appears, is a typical 'over-interpretation' of data. At present it is unfeasible to use chromospheric lines to determine $\partial H_z / \partial z$ since no information of the height of origin of these lines above sunspots is available. Moreover, errors of a factor 2 and more may arise when interpreting the measurements in terms of line formation theory (Wiehr, 1969).

3. The Fine Structure of the Sunspot Magnetic Field

3.1. THE UMBRAL FINE STRUCTURES

A number of properties of the bright umbral fine structures, the umbral dots, have been studied by Beckers and Schröter (1968b). Many questions regarding the relationship of these features to the umbral magnetic field remain. Are umbral dots permanent phenomena in umbrae or do they occur only in certain stages of sunspot evolution?

Do they have the same magnetic field as the umbra or a zero field or even perhaps one of opposite sign? This is unknown, since all attempts to measure the magnetic field in umbral dots have failed. Are umbral dots of convective origin? If so, how is one then to explain their life time being considerably longer than their radiative cooling time? Are they non-radiative heated phenomena of the umbral upper layers (Wilson, 1969)? Do they occur in all spots with the same spatial density and the same intensity contrast, or do these vary from spot to spot?

The existence of a fine scale structure of the umbral magnetic field had been suggested by the Crimean observers several years ago (see e.g. Severny, 1965) who concluded this from the observed large rotation of the polarization plane with depth.

Mogilevskij *et al.* (1967) and later Beckers and Schröter (1968a) ascribed the splitting of the π-component of a Zeeman triplet into two opposite circularly polarized components to the magnetic field in umbral fine structures (Figure 4). The π-component splitting has also been observed and measured recently by Deubner and Liedler (1969) and Mehltretter (1969). This splitting leads to an opposite polarity (if interpreted in terms of magnetic splitting) when compared with the ordinary splitting and corresponds to 200–500 G. It should be emphasized that this effect which appears in

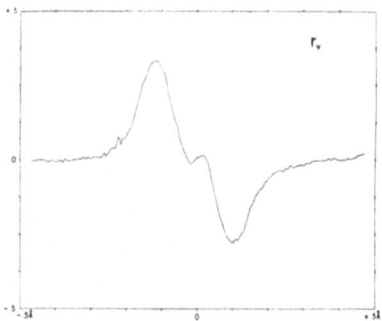

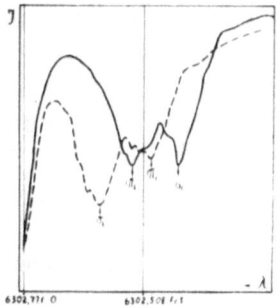

Fig. 4. The splitting of the π-component of $\lambda 6173.3$ Å and $\lambda 6302.5$ Å into two opposite circularly polarized components (a) after Beckers and Schröter (1968a), (b) after Mogilievsky *et al.* (1967).

the Stokes parameter V has no connection with the saturation effect appearing in the Stokes parameter I as studied by Maltby and Kjeldseth Moe (1968), Hénoux (1968), Beckers and Schröter (1968a), Göhring (1969) and Staude (1970). Surkov (1969), observing Doppler-shifts of the split π-components, came to the conclusion that this effect could be explained well in terms of Faraday-rotation. Beckers (1969) found this explanation unlikely, though possible. It has not yet been checked whether or not the π-component splitting occurs in all spots or only in some umbrae. It is very unlikely that this effect is caused by light scattered into the umbra from adjacent magnetic knots of opposite polarity. At the present there are two alternate explanations:

(1) In certain umbrae, the magnetic field of 2000–4000 G has a fine structure of 200–500 G of opposite polarity embedded in it. (Note: the intensity of the split π-components has been estimated to be ~ 0.1 of the σ-components and this is of the same order of magnitude as the contribution of the umbral dots to the composed umbral spectrum.)

(2) The effect is caused by a strong variation of the magnetic field azimuth with depth (Rachkovsky and Beckers, 1969). This, however, leads to the strange conclusion that, independent of the hemisphere, all spots with north polarity must have an anti-clockwise rotation of the field with increasing depth whereas all spots with south polarity must have magnetic fields rotating clockwise.

Steshenko (1967) gave direct evidence of the umbral magnetic field inhomogeneity. He observed a small umbral element ($\lesssim 2''$) showing a magnetic field strength of > 5000 G, embedded in an umbral field of 2400 G.

A new fine structure phenomenon of the upper-most layers of umbrae, the umbral flashes', has been detected by Beckers and Tallant (1969) (see also Wittman, 1969). The correlation of these Ca^+ K-line features with $H\alpha$ umbral fine structures and with umbral dots is unknown. Havnes (1970) succeeded recently in interpreting most of the properties of umbral flashes in terms of a magnetic-acoustic wave model.

3.2. THE FINE STRUCTURE OF THE PENUMBRAL MAGNETIC FIELD

Recent observations regarding the penumbral magnetic field fine structures lead to:

(1) a downward motion in dark interfilamentary regions with respect to the bright filaments (Mattig and Mehltretter, 1967 and Beckers and Schröter 1969,);

(2) a weaker longitudinal field in bright filaments according to Beckers and Schröter (1969) and Mamadazimov (1969), but a stronger one according to Mattig and Mehltretter (1967). This controversial result is apparently caused by insufficient spatial resolution achieved up to now;

(3) the observed local magnetic field fluctuations are $\sim 30\%$ of the average field and may in reality be considerably larger. (Do observations of the average penumbral field and its fluctuations repudiate a penumbral fine structure model in which small field-free regions exist?);

(4) the field in dark interfilamentary regions being more horizontal as compared with the surroundings (Beckers and Schröter, 1969);

(5) extensions of the penumbral filaments into the umbra are showing considerably weaker and more horizontal fields compared to the umbral surroundings;

(6) umbral light bridges definitely showing a much smaller magnetic field ($\Delta H > 300$ G) than their surrounding umbra (Beckers and Schröter, 1969; Abdussamatov, 1970; and Bumba, 1967a). A larger zenith angle of the field in light bridges is indicated.

Grigorjev (1969a) showed that the finite resolution falsifies local fluctuations in $H_{\parallel}$ more than those in $H_{\perp}$. Stenflo (1968) made the first attempt to interpret the penumbral magnetic field observations in terms of a filamentary magnetic field model. Several observers report to have found the $H_{\parallel} = 0$ line not to coincide with the neutral line of Doppler shifts when observing the penumbral magnetic and velocity field during the sunspot's passage across the disc. They attributed this discrepancy to the fact that both magnetic fields and motions are strongly inhomogeneous within the penumbra.

4. The Magnetic Field in Pores and in the Photosphere around Sunspots

4.1. The magnetic field in pores

Observations by Steshenko (1967), Bumba (1967), Beckers and Schröter (1968c) showed the magnetic field in pores and small spots to be never less than 1200 G. A rather flat distribution of the magnetic field strength across pores has been observed (Steshenko, 1967) which, in fact when corrected for seeing, may lead to an almost constant field strength across a pore.

4.2. Fine scale magnetic fields around sunspots

Strong magnetic fields (500–1400 G) in tiny photospheric areas ($\lesssim 2''$) around sunspots have been observed recently by a number of observers (Steshenko, 1967; Bumba, 1967a; Sheeley, 1967; Beckers and Schröter, 1968c; Grigorjev, 1969b; and Abdussamatov and Krat, 1969). All observers agree that generally these features ('gaps', 'magnetic knots', 'micro-pores') coincide with photospheric long lived dark intergranular regions, bright Hα flocculi and the Ca^+ K-network. It may be suspected, but has not yet been proved definitely, that they occur at the boundaries of supergranules (see e.g. Grigorjev, 1969b). Roughly 10 knots per 100 granules occur near spots and 1 knot per 100 granules far away from active regions (Beckers and Schröter, 1968c and Sheeley, 1967). Since their number in an active region is quite large, they play an important role in the balance of the magnetic flux and it can be concluded that magnetic knots are an essential part of a solar active region and not occasional phenomena around single spots. According to Grigorjev (1969b) these features appear in continuous dark regions when the field is almost vertical (magnetic knots?) and in continuous bright points when the field is almost horizontal (constituents of faculae?). The hydromagnetodynamic stability of these 'micro-pores' has been studied by Zwaan (1967). Livingston and Harvey (1969) claimed to have found evidence for a quantization in the fine scale photospheric magnetic flux. In view of the finite resolution achieved this claim is premature.

5. Dependence of the Spot Magnetic Field on Area and Temperature

Recent results regarding magnetic fields in knots, pores and small spots suggest that there may exist a threshold value of 1200–1500 G for the magnetic field required to form a visible (dark) pore or sunspot like phenomenon. (Note, that $H^2/8\pi$ is of the same order of magnitude as the photospheric gas pressure at $\tau \approx 1$). Until a few years ago one could argue that the magnetic field increases with area (Ringness and Jensen, 1960; Ringness, 1965), the spot intensity decreases with area, and hence the spot temperature depends on the magnetic field strength. In fact, such a relation $T_{\mathrm{eff}} = f(H)$ has been used in the theoretical sunspot models of Yun (1968) and Stankiewicz (1967) based on the idea that the inhibition of convection increases continuously with H. Both relations $I = f(A)$ (Rossbach and Schröter, 1970; Makarov, 1968) and $H = f(A)$ where A is the spot area, are now subject to strong doubt. Kopecký (1969) tried to explain both the recent and earlier observations in terms of a rather artificial two-component sunspot magnetic field.

5.1. Temporal variations of the sunspot magnetic field

Several attempts have been undertaken to remeasure time variations of the magnetic field in sunspots in connection with the latter's development and with the occurrence of flares (Kolpakov, 1968; Künzel, 1967; Ikhsanov, 1967a, b). Image motion and image blurring influence strongly magnetic field measurements (see Severny and Deubner, 1968); changing seeing conditions may well simulate temporal magnetic field variations. One should therefore consider the above results with caution. A typical number for dH/dt in a normally developing active region is found to be ~ 10 G/h in good agreement with the values given years ago by Cowling. Occasionally, changes up to 200 G/h can be observed in certain very active parts of a region, often associated with flares. In order to avoid interpretation difficulties due to seeing, simultaneous observations of temporal field changes in the same active region with magnetographs at three (or more) sites should be carried out. Only those changes which have been observed by at least two observers simultaneously should be accepted. Such an attempt has been undertaken recently by the observatories of Pulkovo, Potsdam and Schemecha.

6. Some Comments on the Development of Solar Active Regions

The following aspects regarding the development of solar active regions (SAR) have come to the forefront during the last few years:

(1) recent investigations by Bumba and Howard, 1965; Bumba and Howard, 1969; Ness and Wilcox, 1966; and Bumba et al., 1969, show that SAR preferably occur in certain long-lived zones of about 10° in longitude. These zones rotate with a synodic period of 27 days; their magnetic field structure determines the structure of the interplanetary magnetic field (see the numerous papers by Wilcox, Schatten and others);

(2) new activity is caused by the interaction of new ascending magnetic fields with

the old background field. Bappu *et al.*, (1968) found the new active region to occur near the border of the old magnetic region;

(3) it is well established that supergranulation plays an essential role in the interaction between the new and old magnetic fields. Motions in granules and supergranules redistribute the magnetic flux of a SAR by a random walk over a large area (endproduct: background field). At the boundaries of the supergranulation cells the first disturbance of this background field by the new fields occurs. Observations indicate that even spots may occur only at the boundaries of supergranules;

(4) changes in the background field, indicating new activity, may occur two or three days prior to the first appearance of sunspots (Bappu *et al.*, 1968). The onset of new activity is always indicated by a brightening of the chromospheric network (especially the K_{232}-network);

(5) almost all observers find small and large scale downward motions in regions of spot formation. Bhatnagar (1970) observed a conspicuous descending motion over all active regions of 0.6–0.8 km/s. Hence, the appearance of new magnetic flux is generally accompanied with descending material in layers where the lines originate;

(6) Bappu *et al.* (1968) observed the first changes in the old background field pattern to consist of the appearance of 'magnetic hills' of the longitudinal component, coinciding with the K_{232} emission regions. One is immediately reminded of the photospheric small scale magnetic fields as knots, gaps, micropores (Sheeley, 1967; Beckers and Schröter, 1968c; Grigorjev, 1969b; and Abdussamatov and Krat, 1969). That the magnetic hills of Bappu *et al.*, 1968, are of a larger size and show a smaller maximum field is not necessarily an implication against such a comparison, since the spatial resolution for magnetographic observations is worse than that achieved by Sheeley (1967), Beckers and Schröter (1968c), Grigorjev (1969a), and Abdussamatov and Krat (1969).

The new activity may be suspected to consist of the penetration of the new strong magnetic field (within knot-like features) into the old weak background field at the boundaries of adjacent supergranules. Bappu *et al.* observed the appearance of new small scale fields of opposite polarity within one supergranule boundary in the prespot phase of the SAR and this is in favour of the above interpretation.

Subsequently, more magnetic knot like features appear within the background field lifting the constituents of a large subphotospheric magnetic flux tube into the photospheric and chromospheric layers. In chromospheric layers the K_{232}-prespot network is built. If the spatial density of the new penetrated small-scale fields reaches a critical value, the individual small-scale flux tubes are merged into a larger magnetic pattern, thus forming a spot.

A rough estimate strengthens the presumed close correlation between magnetic knot-like photospheric field regions and the bright Ca^+K emission areas as observed in the prespot phase of a SAR.

Assume the diameter of a photospheric magnetic knot to be 800 km, its magnetic field to be 1000 G. The magnetic flux is then 5×10^{18} Mx. The size of the bright Ca^+K emission features in the prespot phase is roughly 3 s of arc (or ~ 2000 km). Attributing the photospheric flux to this area one obtains for the average magnetic field in these features $H \sim 40$ G, in accordance with observations. Assuming further that this Ca^+K emission occurs 2000–3000 km above the photosphere one arrives at a height gradient of the magnetic field of 0.5–0.3 G/km, again in agreement with the results of Section 3.3.

More observations along the lines of Bappu *et al.* and extended by additional simultaneous high resolution magnetic field measurements (Sheeley, 1967; Beckers and Schröter, 1968; Grigorjev, 1969a; and Abdussamatov and Krat, 1969) are needed to test whether or not fine-scale magnetic structures like 'gaps', 'magnetic knots', and 'micro-pores' are the first agents in the formation of a new solar active region.

References

Abdussamatov, H. J.: 1970, *Astron. Zh.* **47**, 82.
Abdussamatov, H. I. and Krat, V. A.: 1969, *Solar Phys.* **9**, 420.
Adam, M. G.: 1969, *Monthly Notices Roy. Astron. Soc.* **145**, 1.
Altrock, R. C.: 1969, *Solar Phys.* **7**, 343.
Bappu, M. K. V., Grigorjev, V. M., and Stepanov, V. E.: 1968, *Solar Phys.* **4**, 409.
Beckers, J. M.: 1969, *Solar Phys.* **9**, 372.
Beckers, J. M. and Schröter, E. H.: 1968a, *Solar Phys.* **7**, 22.
Beckers, J. M. and Schröter, E. H.: 1968b, *Solar Phys.* **4**, 303.
Beckers, J. M. and Schröter, E. H.: 1968c, *Solar Phys.* **4**, 142.
Beckers, J. M. and Schröter, E. H.: 1969, *Solar Phys.* **10**, 384.
Beckers, J. M. and Tallant, P. E.: 1969, *Solar Phys.* **7**, 351.
Bhatnagar, A.: 1970, Hale Observatories, preprint.
Bray, R. J. and Loughhead, R. E.: 1964, *Sunspots*, Chapman and Hall, London.
Bumba, V.: 1962, *Bull. Astron. Inst. Czech.* **13**, 42, 48.
Bumba, V.: 1967a, *Solar Phys.* **1**, 371.
Bumba, V.: 1967b, *Solar Phys.* **1**, 371.
Bumba, V. and Howard, R.: 1965, *Astrophys. J.* **141**, 1492, 1502.
Bumba, V. and Howard, R.: 1969, *Solar Phys.* **7**, 28.
Bumba, V., Howard, R., Kopecky, M., and Kuklin, G. V.: 1969, *Bull. Astron. Inst. Czech.* **20**, 18.
Deubner, F. L. and Göhring, R.: 1970, *Solar Phys.* **13**, 118.
Deubner, E. L. and Liedler, R.: 1969, *Solar Phys.* **7**, 87.
Dubov, E. E.: 1965, *Soln. Dann.* **12**, 53.
Göhring, R.: 1969, *Solar Phys.* **8**, 271.
Gopasyuk, S. I.: 1967, *Publ. Astrophys. Obs. Crimea* **37**, 29.
Grigorjev, V. M.: 1969a, *Soln. Dann.* **6**, 77.
Grigorjev, V. M.: 1969b, *Solar Phys.* **6**, 67.
Guseynov, M. J.: 1970, *Publ. Astron. Obs. Crimea* **39**, 253.
Havnes, O.: 1970, *Solar Phys.* **13**, 323.
Hénoux, J. C.: 1968, *Solar Phys.* **4**, 315.
Hénoux, J. C.: 1969, *Astron. Astrophys.* **2**, 288.
Howard, R.: 1967, *Ann. Rev. Astron. Astrophys.* **5**, 1.
Ikhsanov, R. N.: 1966, *Publ. Astron. Obs. Pulkovo* **24**, 41.
Ikhsanov, R. N.: 1967a, *Astron. Zh.* **44**, 1048.
Ikhsanov, R. N.: 1967b, *Astron. Zh.* **44**, 1211.
Ikhsanov, R. N.: 1968, *Izv. Astron. Obs. Pulkovo* **184**, 94.
Ioshpa, W. A. and Obridko, W. N.: 1965, *Soln. Dann.* **5**, 62.
Jäger, F. W.: 1964, *Proc. Meeting on Solar Magnetic Fields and High Resolution Spectroscopy*, Rome, p. 166.
Jayanthan, R.: 1970, *Solar Phys.* **12**, 104.
Jensen, E., Brahde, R., and Ofstad, P.: 1969, *Solar Phys.* **9**, 397.

Kai, K.: 1968, *Publ. Astron. Soc. Japan* **20**, 154.
Kassinskij, V. V.: 1967, *Soln. Dann.* **11**, 59.
Kiepenheuer, K. O. (ed.): 1968, 'Structure and Development of Solar Active Regions', *IAU Symp.* **35**.
Kjeldseth Moe, O. and Maltby, P.: 1969, *Solar Phys.* **8**, 275.
Kolpakov, P. E.: 1968, *Publ. Astrophys. Obs. Crimea* **38**, 59.
Kopecký, M.: 1969, *Solar Phys.* **7**, 26.
Künzel, H.: 1967, *Astron. Nachr.* **289**, 233.
Kusnezov, D. A.: 1968, *Soln. Dann.* **9**, 96.
Makita, M.: 1963, *Publ. Astron. Soc. Japan* **15**, 145.
Makorov, W. T.: 1968, *Soln. Dann.* **3**, 88.
Mallia, E. A.: 1970, *Solar Phys.* **11**, 31.
Maltby, P. and Kjeldseth Moe, O.: 1968, *Astrophys. Letters* **1**, 189.
Mamadazimov, M. M.: 1969, *Soln. Dann.* **7**, 94.
Mattig, W.: 1969a, *Solar Phys.* **6**, 413.
Mattig, W.: 1969b, *Solar Phys.* **8**, 291.
Mattig, W. and Mehltretter, J. P.: 1967, in K. O. Kiepenheuer (ed.), 'Structure and Development of Solar Active Regions', *IAU Symp.* **35**, 187.
Mehltretter, J. P.: 1969, *Solar Phys.* **9**, 387.
Mogilevsky, E. I., Demnika, L. B., Ioshpa, B. A., and Obridko, V. N.: 1967, in K. O. Kiepenheuer (ed.), 'Structure and Development of Solar Active Regions', *IAU Symp.* **35**, 215.
Livingston, W. C. and Harvey, J.: 1969, *Solar Phys.* **10**, 294.
Lüst, R. (ed.): 1965, 'Stellar and Solar Magnetic Fields', *IAU Symp.* **22**.
Ness, N. F. and Wilcox, J. M.: 1966, *Astrophys. J.* **143**, 23.
Obridko, V. N.: 1968, *Bull. Astron. Inst. Czech.* **19**, 186.
Rayrole, J. and Semel, M.: 1968, in K. O. Kiepenheuer (ed.), 'Structure and Development of Solar Active Regions', *IAU Symp.* **35**, 134.
Rayrole, J. and Semel, M.: 1970, *Astron. Astrophys.* **6**, 288.
Ringness, T. S.: 1965, Scientific report No. 6 (55) of contract No. AF 61 (052)-743.
Ringness, T. S. and Jensen, E.: 1960, *Astrophys. Norv.* **7**, 99.
Rossbach, M. and Schröter, E. H.: 1970, *Solar Phys.* **12**, 95.
Ruhm, H.: 1969, *Solar Phys.* **10**, 104.
Schleicher, H. and Schröter, E. H.: 1971, *Solar Phys.* **17**, 31.
Severny, A. B.: 1965, *Astron. Zh. U.S.S.R.* **42**, 217.
Severny, A. B. and Deubner, E. L.: 1968, in K. O. Kiepenheuer (ed.), 'Structure and Development of Solar Active Regious', *IAU Symp.* **35**, 230, 233.
Sheeley, N. R.: 1967, *Solar Phys.* **1**, 171.
Stankiewicz, A.: 1967, *Acta Astron.* **17**, 141.
Staude, J.: 1970, *Solar Phys.* **12**, 84.
Stellmacher, G. and Wiehr, E.: 1970, *Astron. Astrophys.* **7**, 432.
Stenflo, J. O.: 1968, *Acta Univer. Lundensis*, Sec. II Nr. 2.
Steshenko, N. V.: 1967, *Publ. Astrophys. Obs. Crimea* **37**, 21.
Surkov, E. P.: 1969, *Soln. Dann.* **5**, 91.
Tepliskaja, R. B. and Turchina, V. D.: 1969, *Astron. Zh.* **46**, 74.
Vitinslij, I. J.: 1969, *Solar Phys.* **7**, 210.
Wiehr, E.: 1969, Thesis, Göttingen.
Wilcox, J. M. and Howard, R. F.: 1968, *Solar Phys.* **5**, 564.
Wilson, P. R.: 1968, *Solar Phys.* **5**, 338.
Wilson, P. R.: 1969, *Solar Phys.* **10**, 404.
Wilson, P. R. and Cannon, C. J.: 1968, *Solar Phys.* **4**, 3.
Wilson, P. R. and McIntosh, P. S.: 1969, *Solar Phys.* **10**, 370.
Wittmann, A.: 1969, *Solar Phys.* **7**, 366.
Wittmann, A. and Schröter, E. H.: 1969, *Solar Phys.* **10**, 357.
Wöhl, H., Wittmann, A., and Schröter, E. H.: 1970, *Solar Phys.* **13**, 104.
Yun, Hong Sik: 1968, Thesis, Indiana.
Yun, Hong Sik: 1971, *Solar Phys.* **16**, 379, 398.
Zwaan, C.: 1965, *Rech. Astron. Obs. Utrecht* **17** (4).

Zwaan, C.: 1967, *Solar Phys.* **1**, 478.
Zwaan, C.: 1968, *Ann. Rev. Astron. Astrophys.* **6**, 135.

Discussion

Wilson: With regard to the depression of the level $\tau = 1$ in the umbra of a sunspot, I would like to point out that we find that the surface ($\tau = 0.01$ say) is depressed by ~ 500 km but that the level surface $\tau = 1.0$ is somewhat more depressed to 600 km below the corresponding level in the photosphere.

Schröter: According to Jensen *et al.* the difference of depression between the 'surface' and optical depth units (in an umbra) amounts roughly to 600 km. According to Wittmann and Schröter the depression vanishes for very small depths. Therefore – in my scheme – I left $\tau = 10^{-2}$ undepressed and $\tau = 1$ to be depressed by 600 km. The large scale height for $\tau = 1$ and the 'photospheric' scale height (~ 100 km) in the upper layers (as found by Mattig and others) favors this model.

MAGNETIC FIELD AND TURBULENCE IN SUNSPOTS

JEAN RAYROLE

Observatoire de Paris-Meudon, France

Abstract. We study the local variations of η_0 and ξ (Milne-Eddington model) which are necessary in order to bring in agreement three independent determinations of the total field H, obtained from simultaneous observations of the two lines $\lambda 5250,2$ Fe I and $\lambda 5225,5$ Fe I. The distribution of the broadening as a function of field direction is consistent with the motions induced by hydromagnetic waves.

1. Introduction

Observations of turbulent velocities in sunspots show that they are comparable with those in the undisturbed photosphere. The physical nature of these motions is an interesting problem since the magnetic field in sunspots may control the motions to a large extent. The low temperature of sunspots has been explained by Biermann (1941) as being caused by the inhibition of convection in a strong magnetic field. The possibility that travelling hydromagnetic waves are an important mode of energy transport has led to calculations on the generation of these waves in thermally un-stable layers. De Jager (1964) shows that near a critical level 'h_c' at a depth of five to ten thousand km (depending on the magnetic field) the convective motions are transformed into hydromagnetic and sound waves. The greater part of these hydro-magnetic waves is reflected downward and only sound waves can reach the surface. Danielson (1965) shows that hydromagnetic waves can be emitted from regions above 'h_c'. Musman (1967) and Savage (1969) have computed a model in which Alfvén waves generated in convectively unstable layers are permitted to propagate upward. The broadening of spectral lines by Alfvén waves has been computed by Maltby (1968).

In this work we study in a large penumbra the local variations of saturation and Doppler broadening (η_0 and ξ in Milne-Eddington model) and their correlations with the field direction.

2. The Material

For a large sunspot (October 23, 1969 – N10-E31) we have obtained simultaneous spectrograms with right and left circular polarisation for the two lines of equal excitation potential:

$\lambda 5250.22$ Fe I   Zeeman pattern (0,00) 3,00
$\lambda 5225.53$ Fe I   Zeeman pattern (1,50) 1,50 3,00

The different spectra (about 100) cover the sunspot with 1″ resolution, and their effective resolution is about 2–3″. However most of the data points discussed below belong to the penumbra of the large spot or to minor satellite umbrae, since in the largest umbra many points were underexposed.

Howard (ed.), Solar Magnetic Fields, 181–189. All Rights Reserved.

3. Data Reduction

The method used is a refinement of the one described by Rayrole (1967).
Let us introduce the following parameters.

OX direction of spectrograph slit;
OY scanning direction ($\perp$ to OX);
$\left.\begin{array}{l}F_1\\F_2\end{array}\right\}$ slits of the Lambdameter;
$\delta\lambda$ width of the slits F_1 and F_2;
$\Delta\lambda$ distance between the slits F_1 and F_2;
γ contrast factor of the photographic plate;
$RI_R\,(\lambda)$ depression of the line for right circular polarisation;
$RI_L\,(\lambda)$ depression of the line for left circular polarisation.

For each (X, Y) position on the spot and each observed line the lambdameter gives
us the $\lambda_R(X, Y)$ and $\lambda_L(X, Y)$ wavelength for which respectively:

$$\int_{F_1} (1 - RI_R(\lambda))^{-\gamma}\, d\lambda = \int_{F_2} (1 - RI_R(\lambda))^{-\gamma}\, d\lambda$$

and

$$\int_{F_1} (1 - RI_L(\lambda))^{-\gamma}\, d\lambda = \int_{F_2} (1 - RI_L(\lambda))^{-\gamma}\, d\lambda.$$

Thus we introduce in a computer the four following parameters:

$E\,251\,(X, Y)=\lambda_R(X, Y)-\lambda_L(X, Y)$ for $\lambda\ 5225,\ \delta\lambda=\ \ 40\,\text{mÅ},\ \Delta\lambda=\ \ 40\,\text{mÅ}$
$E\,253\,(X, Y)=\lambda_R(X, Y)-\lambda_L(X, Y)$ for $\lambda\ 5225,\ \delta\lambda=120\,\text{mÅ},\ \Delta\lambda=120\,\text{mÅ}$
$E\,501\,(X, Y)=\lambda_R(X, Y)-\lambda_L(X, Y)$ for $\lambda\ 5250,\ \delta\lambda=\ \ 40\,\text{mÅ},\ \Delta\lambda=\ \ 40\,\text{mÅ}$
$E\,503\,(X, Y)=\lambda_R(X, Y)-\lambda_L(X, Y)$ for $\lambda\ 5250,\ \delta\lambda=120\,\text{mÅ},\ \Delta\lambda=120\,\text{mÅ}$

and we compare it with calibration curves $F(\eta_0, \xi, \beta_0, H, \psi)$ derived from theoretical
profiles computed with Unno theory (Unno, 1956) and a Milne-Eddington model with:

$$\eta = \eta_0\, e^{-v^2} + \eta_0 a H_1\,(v) + \eta_0 a^2 H_2\,(v)$$

$$v = \frac{\lambda - \lambda_0}{\xi} \qquad a = \frac{\Gamma}{4\pi\Delta v_0} \qquad \Delta v_0 = \frac{v_0}{c}\sqrt{\frac{2KT}{M_0 A} + v_t^2} \qquad \Gamma = 4\cdot 10^9\ \text{s}^{-1}$$

and $B(\tau)=B_0\,(1+\beta_0\tau)$ for the source function variation with optical depth.

In this way $E251\,(X, Y)$ and $E253\,(X, Y)$ give us the angle with the line of sight
$\psi\,(X, Y, \eta_0, \xi)$ and the total field $H\,25\,(X, Y, \eta_0, \xi)$ while $E\,501\,(X, Y)$, $E\,503\,(X, Y)$
associated with the ψ determination give us two other independent values of the total
field $H501\,(X, Y, \eta_0, \xi)$ and $H503\,(X, Y, \eta_0, \xi)$. The local variations $\eta_0\,(X, Y)$ and
$\xi\,(X, Y)$ are selected in order to bring in agreement the three determinations of the
total field H. We take as a test of interval coherence, and therefore as indication for

the choice of an optimal model (η_0, ξ) the mean square root value $ECQM(X, Y, \eta_0, \xi)$ of these three H determinations.

$$ECQM(X, Y, \eta_0, \xi)$$
$$= \sqrt{\frac{(H25 - HM)^2 + (H501 - HM)^2 + (H503 - HM)^2}{9}}$$

where

$$HM(X, Y, \eta_0, \xi)$$
$$= \frac{H25(X, Y, \eta_0, \xi) + H501(X, Y, \eta_0, \xi) + H503(X, Y, \eta_0, \xi)}{3}.$$

In each (X, Y) position the η_0 and ξ values are chosen in order to bring the $ECQM$ function to a minimum.

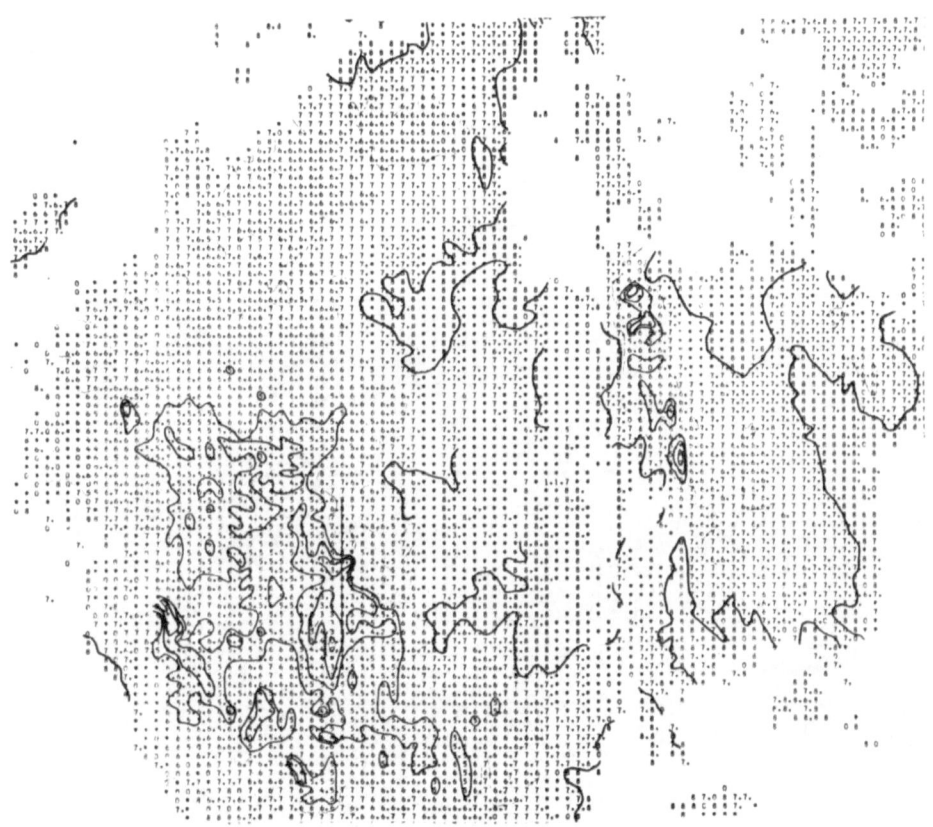

Fig. 1a.   Map of the angle Ψ between the field and the line of sight for Model 1. The distance between two points is 1″. The Ψ values are given by the following symbols:

Symbols	0	0.	1	1.	2	2.	...	
Ψ greater than (degrees)	0	5	10	15	20	25	...	and so on.

4. Observational Results

In a first approach we have taken only three models characterized by

$$\text{Model 1} \quad \eta_0 = 1 \quad \xi = 40\,\text{mÅ}$$
$$\text{Model 2} \quad \eta_0 = 20 \quad \xi = 40\,\text{mÅ}$$
$$\text{Model 3} \quad \eta_0 = 20 \quad \xi = 20\,\text{mÅ}$$

Model 1 represents the undisturbed photospheric conditions while Models 2 and 3 are in good agreement with the great saturation given by an empirical model of the umbra (Henoux, 1969), and represent respectively the conditions where the turbulent velocities are the same as those of the undisturbed photosphere or only due to thermal motions.

Figure 1 shows the values of $\psi\,(X, Y)$ for the three models and therefore the great influence of η_0 and ξ on the determination of the angle ψ with line of sight. The differences between the H determinations are shown in Figures 2, for a small region.

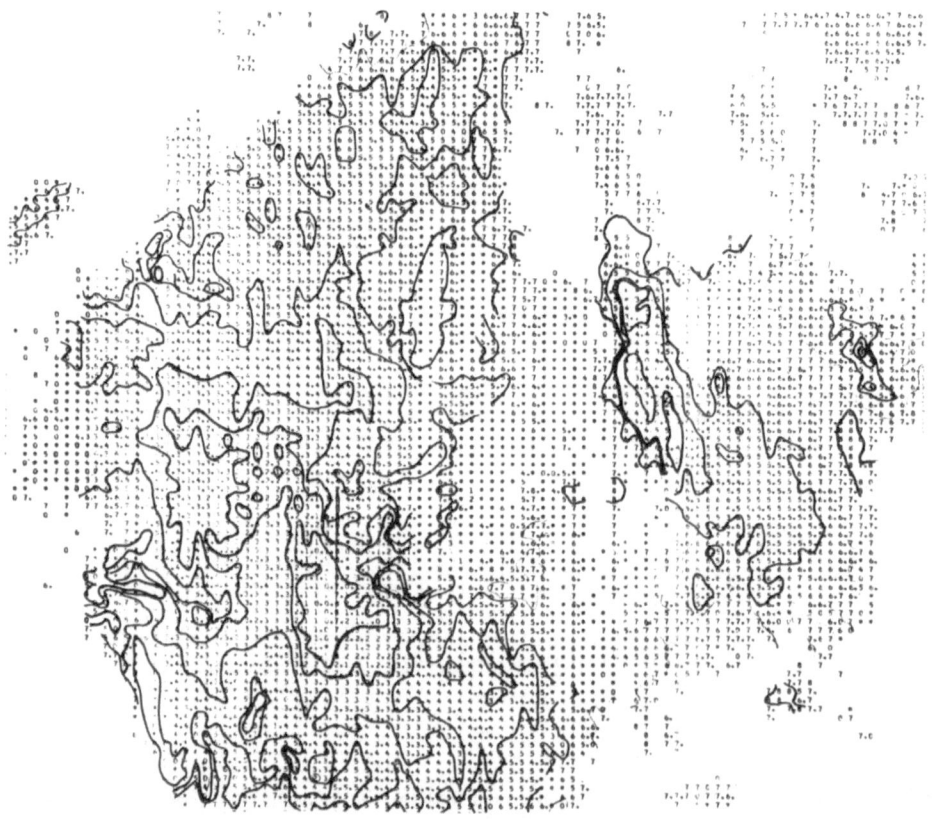

Fig. 1b.   Map of the angle Ψ between the field and the line of sight for Model 2. The symbols are the same as in Figure 1a.

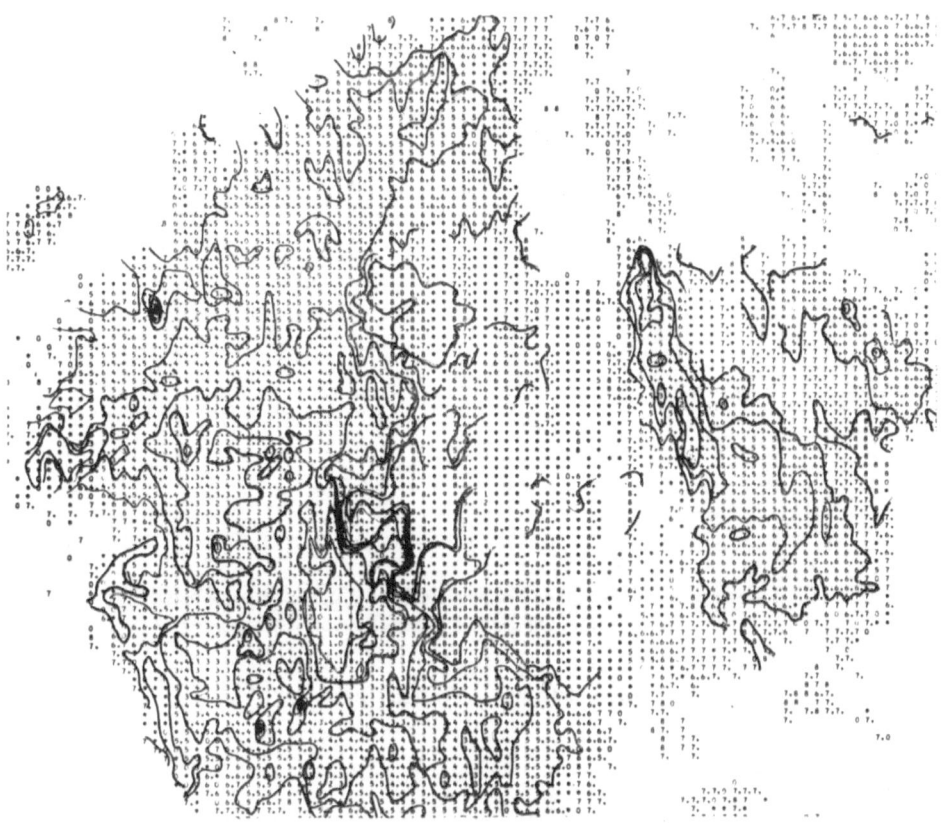

Fig. 1c.   Map of the angle Ψ between the field and the line of sight for Model 3. The symbols are the same as in Figure 1a.

We found that for 70% of the 4000 measured points the minimum of $ECQM$ is less than the noise for only one of the three models. The distribution of the most suitable model is shown in Figure 3. For all the points where $ECQM$ is less than the noise a statistical study gives

Model 1 $\eta_0 = 1$ $\xi = 40$ mÅ   21% of measured points.
Model 2 $\eta_0 = 20$ $\xi = 40$ mÅ   14% of measured points.
Model 3 $\eta_0 = 20$ $\xi = 20$ mÅ   64% of measured points.

Therefore for the majority of measured points the turbulent motions are weak and perhaps negligible, since $\xi = 20$ mÅ describes the thermal motions for $\lambda 5250$ and $T = 3700$ K.

If the only contribution to the line broadening, except the thermal motions, is due to travelling hydromagnetic waves this effect must depend on the field direction. The percentage of measured points leading to the weak broadening (Model 3) and the large broadening (Model $1+2$) is shown in Figure 4, as a function of ψ.

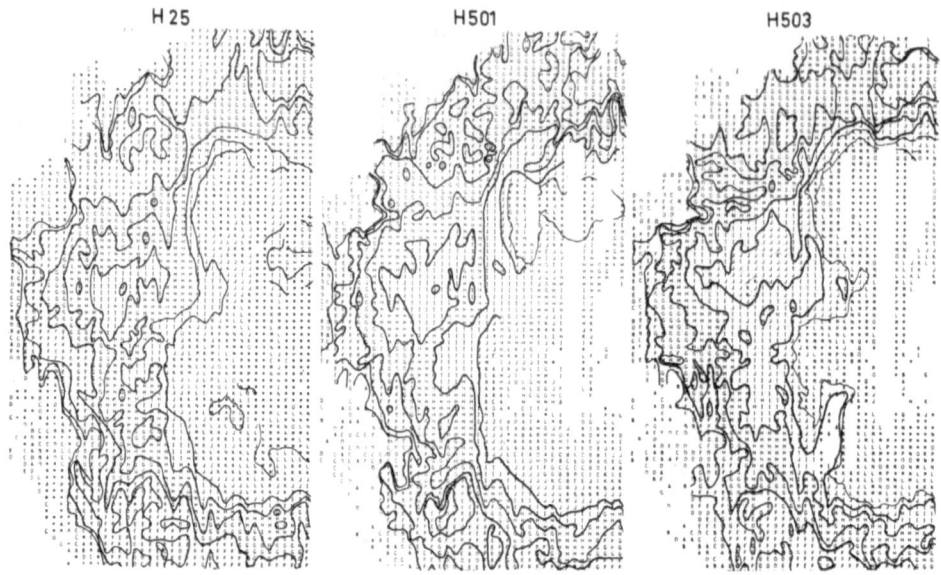

Fig. 2a.   Map of the three H values of the total field for Model 1. The distance between two points is
1″. The H values are given by the following symbols:

Symbols	A	B	C	D	E	⎫	
H greater than (gauss)	200	400	600	800	1000	⎬	and so on.

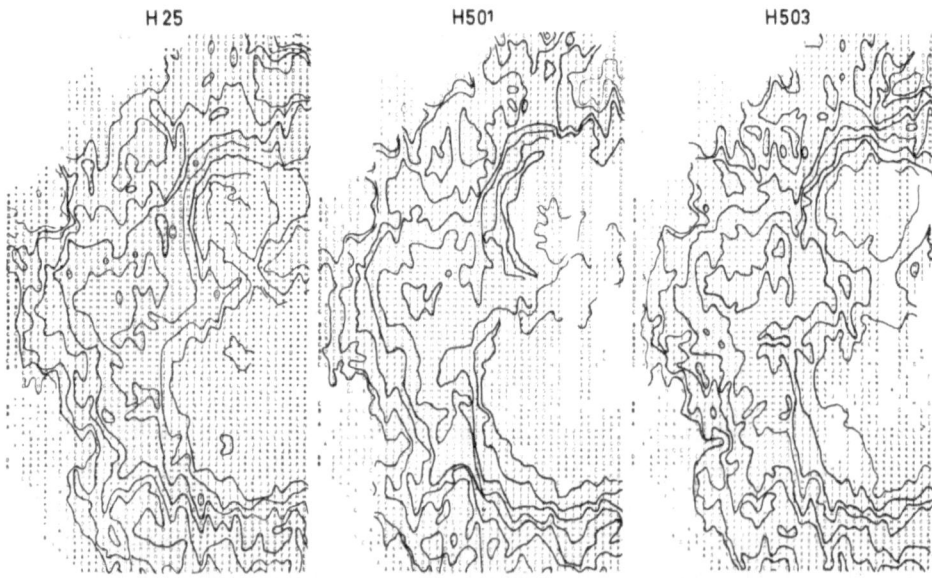

Fig. 2b.   Map of the three H values of the total field for Model 2. The symbols are the same
as in Figure 2a.

H 25   H 501   H 503

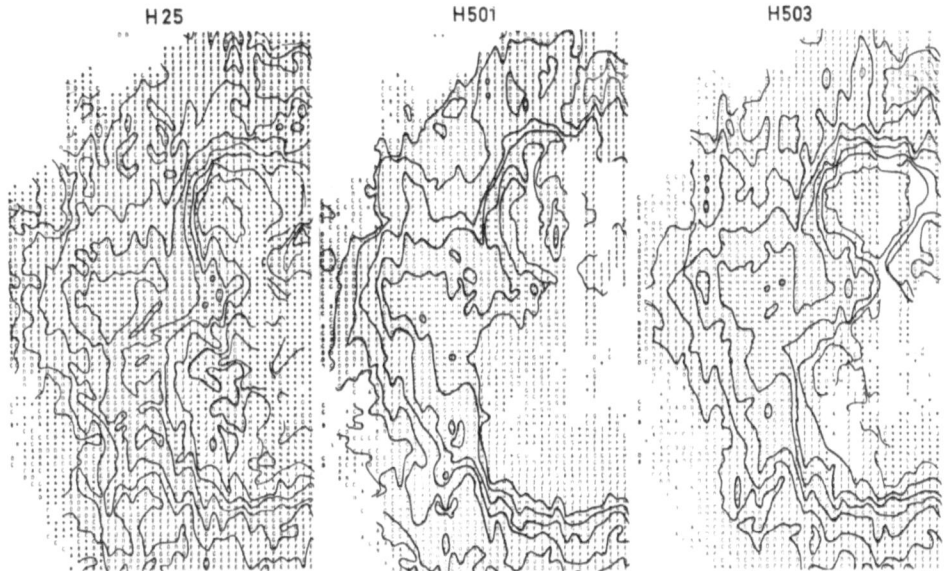

Fig. 2c. Map of the three H values of the total field for Model 3. The symbols are the same as in Figure 2a.

Clearly large turbulent motions are associated with large ψ values, while at low ψ the profiles are generally consistent with no appreciable turbulent broadening. At very small ψ ($<15°$) however it becomes impossible to select one model rather than another because the calibration functions F become insensitive to the model (η_0, ξ).

The above correlation is consistent with the motions induced by Alfvén waves, or fast magnetosonic waves since $V_a = H/(4\pi\varrho)^{1/2}$ is greater than the sound velocity in sunspots. The tendency to equipartition of the different models for great angles can be explained by the fact that:

(a) We have only taken three models to reduce the data.

(b) In an inhomogeneous atmosphere (Cowling, 1957) the vertical component of the motions induced by hydromagnetic waves is rapidly suppressed if $\delta\rho/\delta z$ is great enough. In this case the motions are horizontal and cannot produce a broadening when the field is in the vertical plane containing the line of sight.

5. Conclusion

The results obtained in this work bring us to make the following remarks.

(a) For the study of fine structures of magnetic field in sunspots it is not possible to discard the line broadening variations.

(b) This first approach points out possible observational evidence for the existence of hydromagnetic waves in sunspots. However a more accurate reduction with numerous models is needed to study the distribution of the broadening as a function of the

Fig. 3. Distribution of the three Models 1, 2, 3 as a function of the X, Y positions. The distance between two points is 1″.

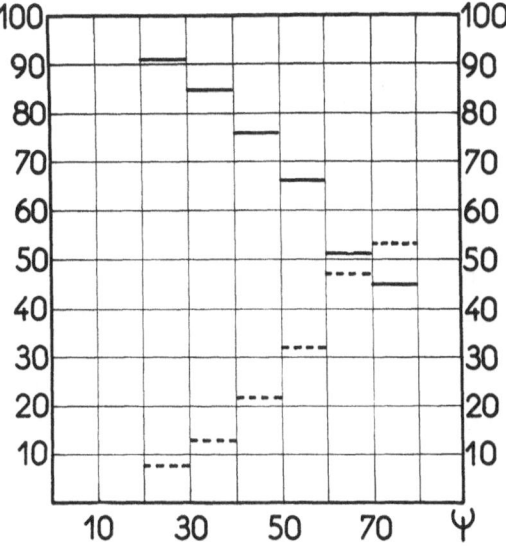

Fig. 4. Percentage of the models as a function of Ψ. —— Model 3, weak broadening; ----- Model 1 + 2, large broadening.

field direction. It will be necessary to separate more definitely the relative influences of variations in saturation and broadening.

Acknowledgements

The author wishes to thank Mrs Savinelli for help with the lambdameter reduction.

References

Biermann, L.: 1941, *Vierteljahresschr. Astron. Ges.* **76**, 194.
Cowling, T. G.: 1957, *Magnetohydrodynamics*, Interscience Publishers, New-York.
Danielson, R. E.: 1965, in R. Lüst (ed.), 'Stellar and Solar Magnetic Fields', *IAU Symp.* **22**, 314.
De Jager, C.: 1964, *Bull. Astron. Inst. Neth.* **17**, 253.
Henoux, J. C.: 1969, *Astron. Astrophys.* **2**, 288.
Maltby, P.: 1968, *Solar Phys.* **5**, 14.
Musman, S.: 1967, *Astrophys. J.* **149**, 201.
Rayrole, J.: 1967, *Ann. Astrophys.* **30**, 257.
Savage, B. D.: 1969, *Astrophys. J.* **156**, 707.

PHOTOELECTRIC MEASUREMENTS OF SUNSPOT MAGNETIC FIELDS

F.-L. DEUBNER and R. GÖHRING

Fraunhofer Institut, Freiburg i. Br., Germany

Photoelectric polarization measurements in a stable sunspot (type H) with a particularly dark umbra, where 'umbral dots' were virtually lacking, have been carried out with the Capri magnetograph (Deubner, 1969). The measurements were evaluated in terms of Unno's theory to give the value and direction of the magnetic field vector. The parameters $\eta_0 = 5$, $\beta_0 = 2.5$ and $\Delta\lambda_D = 40\,\text{mÅ}$ have been adopted for the Fe I 5250 line. Taking the configuration of the sunspot into account as well as simple conditions of steadiness of the distributions to be obtained, it is possible to derive the magnetic vector field from two-dimensional records of circular and linear polarization without ambiguities.

Since the ratio of circular and linear polarization is only little influenced by stray light, the photoelectric method gives the angular distribution with particular accuracy. A linear increase of the inclination angle with distance from the spot center up to $r \geqslant 1.2\,R_0$ results, as originally found by Hale and Nicholson (1938) with a different procedure. At the penumbral border an inclination angle of 75° is observed.

If the measured values are corrected for unpolarized stray light ($\approx 40\%$ in the umbra, cf. Kneer and Mattig, 1968), a maximum field strength of 3250 G is obtained, which agrees well with the amount of splitting of the σ-Components in the control spectra. At the penumbral border the value is 15% of the maximum field strength, still continuously decreasing outside the spot.

In several parts of the spot regions where the projected field direction deviates considerably from radial symmetry, the azimuthal component of the magnetic field nearly equals the value of the radial component. The direction of field is always in good agreement with that of the overlying chromospheric structures.

References

Deubner, F.-L. and Liedler, R.: 1969, *Solar Phys.* 7, 87.
Hale, G. E. and Nicholson, S. B.: 1938, *Magnetic Observations of Sunspots 1917–1924*, Part I, Publ. Carnegie Inst. No. 498.
Kneer, F. and Mattig, W.: 1968, *Solar Phys.* 5, 42.

Discussion

Giovanelli: (1) It is interesting to note that the magnetic field is inclined to the horizontal by some 15° at the penumbra-photosphere boundary, since Hα fibrils in such regions can also be seen inclined at rather similar angles for sunspots near the limb. This is in accordance with the general view that Hα fibrils mark out lines of magnetic force.

Howard (ed.), Solar Magnetic Fields, 190–191. All Rights Reserved.
Copyright © 1971 by the IAU.

(2) The sunspot examined showed a bright Hα marking across the middle. Does this not mean that the spot was far from being simple and structureless?

Deubner: The marking was the remainder of an Hα flare, which took place in this spot group on September, 28. The analysis carried out and presented here was based on data of October, 1, when the spot group was pretty calm and did not show any umbral structure, as also in the first figure. This does not imply, that the spot was 'simple'.

Brueckner: How did the maximum field strength in the spot change when the spot changed its position on the disk?

Deubner: The maximum field strength has only been evaluated for this particular position.

Wiehr: Have you any explanation for the fact that the discontinuity of the field strength only appears at the one but not at the other penumbra-umbra border?

Deubner: It is possible that instrumental effects, e.g. the time constant of the intensity compensator produced this asymmetric discontinuity.

OBSERVATIONS OF MAGNETIC FIELDS IN QUIESCENT PROMINENCES

EINAR TANDBERG-HANSSEN

High Altitude Observatory, National Center for Atmospheric Research,* Boulder, Colo., U.S.A.

Abstract. The longitudinal component of the magnetic field, $B_\parallel$, has been recorded in about 135 quiescent prominences observed at Climax during the period 1968–1969. The measurements were obtained with the magnetograph which records the Zeeman effect on hydrogen, helium and metal lines. The following lines were used, $H\alpha$; He I,D_3, He I, 4471 Å; Na I, D_1 and D_2, and the observed magnetic field component in these prominences was independent of the line. The overall mean value of the field $B_\parallel$ for all the prominences was 7.3 G. As a rule, the magnetic field enters the prominence on one side and exits on the other, but in traversing the prominence material, the field tends to run along the long axis of the prominence.

1. Introduction

It is of considerable interest to study prominence magnetic fields using the Zeeman effect on lines of different elements. In this study we have used the $H\alpha$ hydrogen line, neutral helium lines (D_3 at 5876 Å and the 4471 Å line) as well as several metal lines (Na I, D_1 and D_2 at 5889 Å and 5896 Å, Mg I, b_1 at 5184 Å). Since June 1968 helium and metal line observations of prominence magnetic fields have been secured on a nearly routine basis. In addition, many hundred $H\alpha$ observations also are available for comparison.

The original version of the High Altitude Observatory solar magnetograph at Climax, Colorado has been described by Lee *et al.* (1965). The instrument measures the longitudinal component of the magnetic field through the Zeeman effect on the $H\alpha$ line, by recording the difference between the oppositely circularly polarized σ-components of the line. The instrument is servoed to let the entrance slits to the magnetograph seek out the proper intensities in the two wings, regardless of line asymmetries. In 1967 the instrument was modified and put under computer control, allowing the measurements to be recorded on magnetic tape. The tapes are processed with the CDC 6600 computer of the National Center for Atmospheric Research. Also, a new entrance slit assembly was constructed that allows narrow metal lines to be used (Lee *et al.*, 1969).

The accuracy of the measurements depends both on the nature of the object under observation and on the spectral line used. If the prominence is very active, the servoing of the Doppler movements leads to a noisy signal. Typically, the accuracy is about ± 0.5 G for $H\alpha$, 1 G for He I,D_3, and 2 G or more for metal lines in quiescent prominences. All measurements discussed here pertain to emission lines in prominences seen above the solar limb.

The Zeeman effect displaces a σ-component of the line used from its normal

---

* The National Center for Atmospheric Research is sponsored by the National Science Foundation.

position λ_0, by an amount $\Delta\lambda_B$, which is related to the magnetic field B by the expression

$$\Delta\lambda_B = \frac{eB\lambda_0^2}{4\pi m_e c^2} g,$$ (1)

where g is the Landé g-factor, and e and m_e are the absolute charge and mass of an electron. For the fairly complicated atomic transitions in question we compute mean g-factors, $\bar{g}$, by weighting the individual g-factors by the relative intensities of the σ-components of the lines (see for instance White (1934) p.220). The magnetograph's data acquisition routine treats all data as if recorded with the $H\alpha$ line, and a correction factor must be applied to the output data for any other line.

One of the difficulties in calibrating the magnetic observations is to properly assess the influence of different processes on the line profile. Considerable effort has been employed to study the Hanle effect (Hanle, 1924) and its influence on measurements on prominence magnetic fields (Hyder, 1968; Stenflo, 1969; Lamb, 1970; and House, 1970a, b). For the lines affected, the mutual perturbations of the atomic levels alter the observed polarization. This, in turn, can be interpreted as caused by a magnetic field. The magnitude of the effect differs widely for different lines. In the following we have neglected the effect.

2. Observations, Hα

The data analyzed pertain to the 1968–69 calendar years. Altogether, about 1100 $H\alpha$ observations in slightly more than 400 prominences have been studied.

Of these prominences about 135 were considered quiescent and they are treated here in more detail. Figure 1 shows a histogram of the distribution of $\bar{B}_\parallel$ for this class of prominences. We find that the overall mean value is 7.3 G, and 52% of the prominences have mean values satisfying the relation

$$3 \text{ G} \leqslant \bar{B}_\parallel \leqslant 8 \text{ G}.$$ (2)

Rust (1966) found $\langle \bar{B}_\parallel \rangle \approx 5$ G for data from 1965, and Harvey (1969) found $\langle \bar{B}_\parallel \rangle = 6.6$ G. His data were obtained mainly in 1967. The differences between the mean values quoted (≈ 5 G, 6.6 G and 7.3 G) may not be significant, and may be due to selection effects. On the other hand, Harvey pointed to the possibility that the general level of prominence-supporting fields could have been greater in 1967 than closer to sunspot minimum in 1965, thereby explaining the difference between his and Rust's results. The present study supports this point of view, and it will be of interest to continue this type of observation to see if the level will decline again as we progress past the 1969 solar maximum.

When we observe the magnetic field in prominences with only one spectral line, say $H\alpha$, a series of measurements of the field from many different parts of the object is often possible. We find then as a general rule that even though the field may change some in strength across the prominence, the sense of direction ($-B$ or $+B$) is the same. Those relatively rare instances where we see both polarities may be explained in

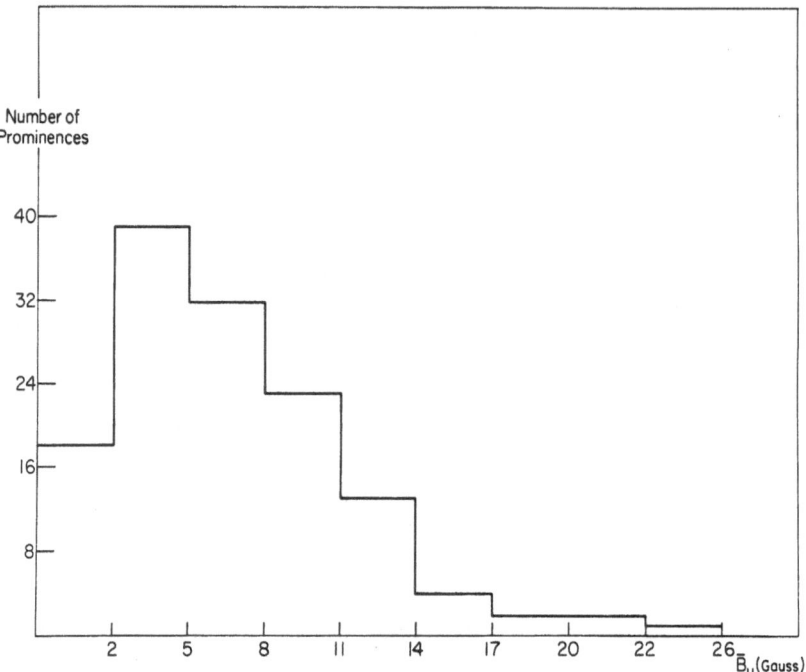

Fig. 1.   Histogram giving the distribution of mean longitudinal magnetic field $\bar{B}_{\parallel}$ for 135 quiescent prominences observed in 1968–69.

terms of geometry (the prominence sheet is bent) or localized activity in the promi-nence.

Apart from these situations, we observe the magnetic field either as it enters one side of the prominence (B negative), or as it emerges (B positive). Figure 2 shows some typical examples.

Rust (1966) and Harvey (1969) found that the observed longitudinal field tends to increase with height in the prominence. The present study confirms this general tendency, see Figure 3, but there are many prominences where this increase is masked by internal noise in the data.

3. Observations Using Several Different Spectral Lines

If we want information about the magnetic field using several different spectral lines, often there is time only to make the measurements on a few selected spots in the prominence. Table I summarizes the pertinent data for a number of quiescent pro-minences whose magnetic field has been determined by observing two or more of the hydrogen, helium or metal lines mentioned in Section 1.

The most striking result of a perusal of Table I is the impression that the magnetic field is the same whether observed with hydrogen, helium, or metal lines, i.e.

$$B_{\parallel}(\text{H}\alpha) \approx B_{\parallel}(\text{D}_3) \approx B_{\parallel}(\text{D}_1). \tag{3}$$

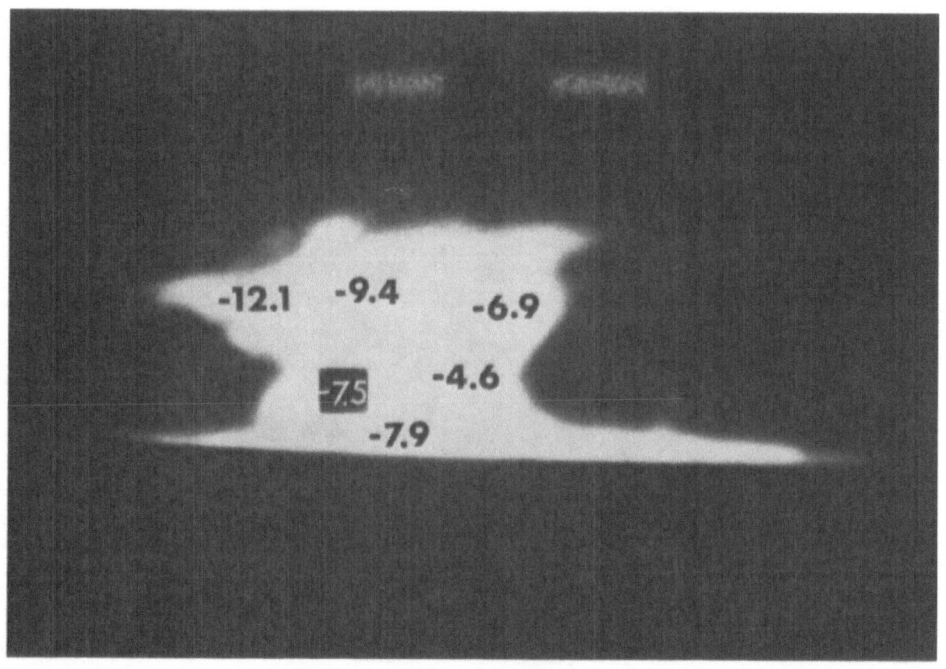

March 3, 1968

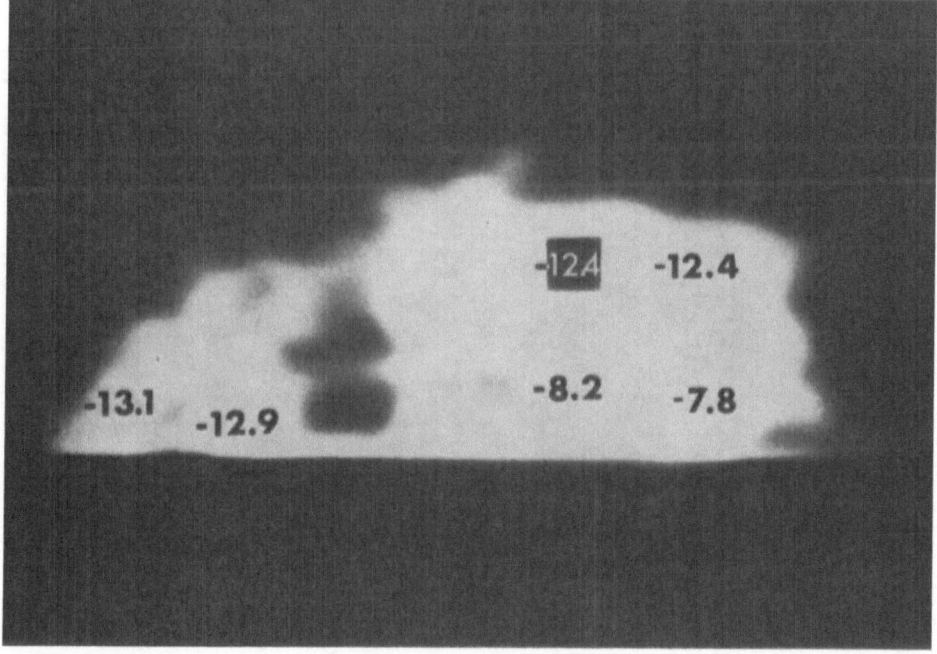

March 4, 1968

Fig. 2.   Quiescent prominences showing longitudinal magnetic fields typically of only one polarity.

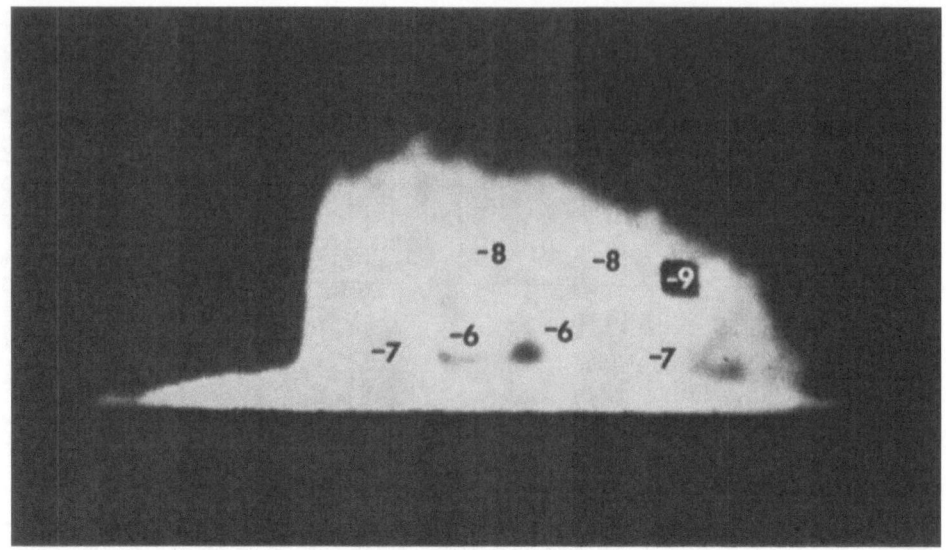

June 5, 1968

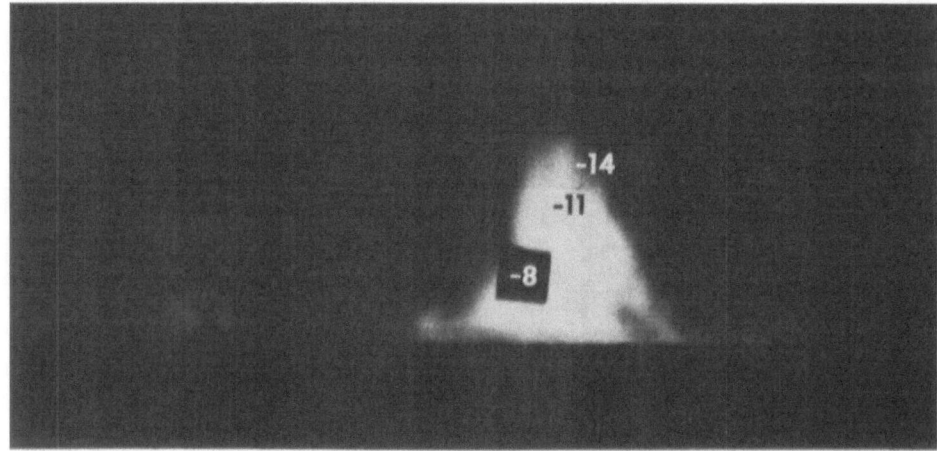

March 1, 1968

Fig. 3.   Quiescent prominences showing longitudinal magnetic fields increasing with height.

TABLE I

Observed longitudinal component of magnetic field, $B_\parallel$, in quiescent prominences

Date	Position angle	$B_\parallel$ λ	Gauss	Gauss	Object
June 26, 1968	250	D_3	12.6	± 2	Small quiescent
		D_1	10	± 4	
		$H\alpha$	9.0	± 0.7	Same prominence June 25
Sept. 8, 1968	325	$H\alpha$	-0.1	± 0.7	Small quiescent
		D_3	$+0.3$	± 0.7	
Sept. 24, 1968	315	$H\alpha$	6	± 0.5	Big quiescent
		D_3	5.5	± 0.5	
		4471	7	± 3	
Nov. 27, 1968	90	D_3	21	± 0.5	Semi-quiescent
		4471	21	± 4	
		D_1	18	± 2	
		D_2	23	± 2	
		b_1	15	± 4	
Jan. 16, 1969	59	$H\alpha$	-16.6	± 1	
		D_3	-18	± 3	
		D_3	-15	± 1	
		4471	-21	± 3	
		D_1	-20	± 2	
		D_2	-27	± 2	
Jan. 21, 1969	86	$H\alpha$	-15.7	± 0.5	Part of big quiescent
		4471	-12	± 6	
		D_1	-15	± 2	
		D_2	-14	± 3	
April 5, 1969	310	$H\alpha$	-11.5 to -14.4	± 0.5	Small quiescent
		D_3	-15.4	± 1	
April 24, 1969	297	$H\alpha$	-29.4	± 4	Semi-quiescent
		4471	-17	± 5	
		D_3	-24	± 1	
		D_1	-26	± 2	
		D_2	-31	± 2	
Aug. 16, 1969	130	$H\alpha$	5 to 8	± 1	Small quiescent
		D_3	8 to 10	± 1	
Jan. 4, 1970	75	D_3	13	± 1	Big quiescent
		D_3	18	± 2	
		D_3	14	± 1	
		D_1	20	± 4	
		D_1	15	± 4	
		D_2	10	± 8	
		D_2	13	± 6	
		b_1	12	± 2	
Jan. 6, 1970	100	$H\alpha$	-29	± 0.5	
		D_3	-28	± 1	

4. The Orientation of the Magnetic Field in Quiescent Prominences

We have seen in Section 2 that the supporting magnetic field in quiescent prominences generally enters the prominence on one side, goes through the prominence plasma (with components both along and at right angles to the long axis of the prominence),

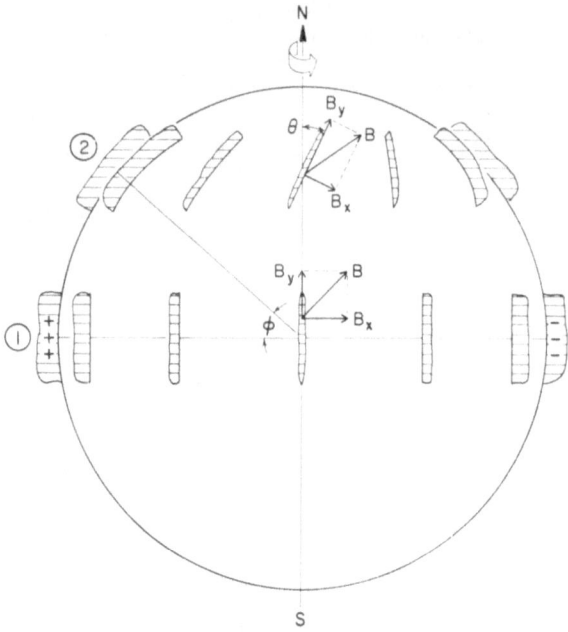

Fig. 4.   Parameters defining (a) a prominence in the plane of the sky on the solar equator (Prominence 1), and (b) a prominence at latitude ϕ and not in the plane of the sky (Prominence 2).

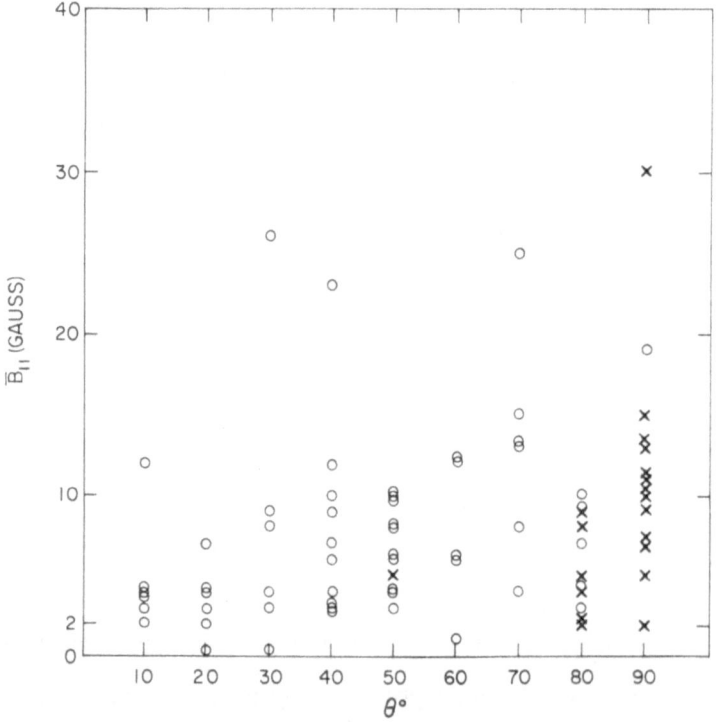

Fig. 5.   The observed mean longitudinal component of the magnetic field, $\bar{B}_\parallel$, plotted against the angle θ between the north-south direction on the Sun and the long axis of the prominence (see Figure 4).

and exits on the other side. The question arises whether we can say anything about the relative importance of the two components of the field. Let us consider an orthogonal coordinate system with x-axis horizontal and perpendicular to the (idealized) prominence sheet, y-axis horizontal and along the prominence and z-axis perpendicular to the photosphere. The observations on the limb of a prominence on the solar equator and in the plane of the sky give $B_{\parallel} = B_x$, see Figure 4, prominence 1. Often the prominence sheet does not form a plane, and even when it does, the plane may not coincide with the plane of the sky. We have studied those prominences that seem to form plane sheets and, as they move onto the disk, we have determined the angle θ that this assumed vertical plane makes with the north-south direction on the Sun, see Figure 4, prominence 2.

About 50% of the quiescent prominences observed could be approximated with the ideal cases of Figure 4. Figure 5 shows how the magnetic field $\bar{B}_{\parallel}$ relates to the angle θ. We emphasize the uncertainties involved in determining the angle θ for any prominence, but taken at face value Figure 5 indicates that the observed field $\bar{B}_{\parallel}$ increases with increasing angle θ, i.e. $\bar{B}_{\parallel}$ is greater the more the prominence is aligned parallel to the solar equator. This result indicates that even though the magnetic field enters a prominence on one side and exits on the other, inside the object the field tends to run along the long axis of the prominence.

References

Hanle, W.: 1924, *Z. Phys.* **29–30**, 93.
Harvey, J.: 1969, Thesis, Univ. of Colorado.
House, L. L.: 1970a, *J. Quant. Spectrosc. Radiative Transfer* **10**, 909.
House, L. L.: 1970b, *J. Quant. Spectrosc. Radiative Transfer* **10**, 1171.
Hyder, C. L.: 1968, *Solar Phys.* **5**, 29.
Lamb, F. K.: 1970, *Solar Phys.* **12**, 186.
Lee, R. H., Harvey, J., and Tandberg-Hanssen, E.: 1969, *Appl. Opt.* **8**, 2370.
Lee, R. H., Rust, D., and Zirin, H.: 1965, *Appl. Opt.* **4**, 1081.
Rust, D.: 1966, Thesis, Univ. of Colorado.
Stenflo, J. O.: 1969, *Solar Phys.* **8**, 260.
White, H. E.: 1934, *Introduction to Atomic Spectra*, McGraw-Hill, New York, London.

Discussion

Wiehr: How did you calibrate your magnetograph signals? I think a simple Doppler calibration is impossible since it requires a photospheric absorption line which does not exist for He D₃. Furthermore it would be impossible to use the calibration from an absorption line for the data from an emission line.

Tandberg-Hanssen: We calibrate it in the fairly conventional way as a Babcock magnetograph, and all the uncertainties regarding the interpretation of polarization in terms of magnetic fields are certainly with us.

Severny: (1) Do you have histograms showing the distribution of noise (modulator off), or what is the rms of noise at the brightness of a prominence?

(2) Due to thermal velocities the profiles of lines in prominences are very asymmetrical, and so a false signal can appear just due to the asymmetry of the line.

Tandberg-Hanssen: For integration times of around 10 min the noise in Hα is about 0.5 G.

As long as the right and left circularly polarized components have the same shape, asymmetries are eliminated.

Jordan: In the active prominences, are the high excitation lines associated with high magnetic fields or vice-versa?

Tandberg-Hanssen: Both cases are observed. I am afraid that there is, as yet, no clearcut picture to present.

Rust: Concerning the question raised by Dr. Wiehr, I think Dr. Tandberg-Hanssen tried to answer it on a more sophisticated level than it required. The Climax magnetograph is calibrated by artifically shifting the line under observation. No reference to absorption lines on the disk is necessary.

Brueckner: Is the magnetograph arrangement built in such a way that any phase shift in the instrument will not be seen by the circular analyser, so that the known high degree of linear polarization, caused by the Hanle effect and measured in prominences cannot influence the longitudinal field measurements?

Tandberg-Hanssen: The polarization analyser is located at the prime focus, hence there is no phase shift in front of the electro-optic plate.

SOME REMARKS ON THE STATICS AND DYNAMICS OF MAGNETIC FIELD STRUCTURE DEVELOPMENT IN ACTIVE REGIONS

V. BUMBA and J. SUDA

Astronomical Institute of the Czechoslovak Academy of Sciences, Ondřejov, Czechoslovakia

Abstract. Some comments are given concerning the fine structures in the umbra and penumbra of sunspots and their changes on the basis of high resolution photographs.

1. Introduction

During recent studies of the large-scale distribution of solar magnetic fields the question of the mode of magnetic field motion in the solar photosphere arose for us once again (the problem: when is it the real transport of lines of force and when the intensification of pre-existing fields). The same investigation brought us some experience about how important is the manner of illustration of the features studied. For example, sometimes it is enough to separate in synoptic charts of magnetic fields both polarities to see just different regularities.

The investigation of magnetic field motions can be made on the most concentrated fields in sunspots. To see better the details we can use good photographs of sunspots in integrated light with a resolution better than 1″ giving us a picture of magnetic field distribution with the aid of darker features on the positives of the photosphere, as is now generally accepted. This is the way we are accustomed to study the question. But looking at the negatives of sunspot pictures we may see many more details usually lost on positives between bright photospheric and penumbral formations. And so we would like to add a few remarks to the problem of the statics and dynamics of magnetic fields in sunspots on the basis of photographic material taken during the last year (1969) with our Clark refractor (20.5cm in diameter) in a relatively narrow spectral region of about 100 Å around $\lambda 5890$ Å with an exposure of the order of 0.001 s., assuming that the bright features on our negatives coincide to the smallest details with the fine-structure of the magnetic field distribution in sunspots. Because of the lack of time we shall concentrate our preliminary results on one sunspot group only: it is the large group in the southern hemisphere of the Sun, first seen on July 1 and last seen on July 13, 1969 (C.M.P. July 7).

2. Magnetic Field Fine Structure

A view of our sunspot negative at once shows the most interesting behavior of the magnetic field: never can we see a concentration of magnetic field in the form of a dark (on negative bright) dot. Always it has the form (with the exception of nuclei of very large and dark spots) of intensified intergranular space with greater concentration in

Howard (ed.), Solar Magnetic Fields, 201–211. All Rights Reserved.
Copyright © 1971 by the IAU.

the crossings of this space. This 'lacy' structure with bright granular-like grains (bright umbral dots) inside the open spaces of this delicate network is striking, and certainly it is one of the main characteristics of magnetic field finest-scale distribution. The width of individual fibrils of the network is often only a few hundred kilometers; they are just observable with our instrument (Figures 1, 2, 3, 4, 5). It is a situation we can probably not simulate in the laboratory, where one usually has the plasma held inside a magnetic vessel. Here the situation seems to be reversed; the lines of the magnetic field are pushed out from the photospheric plasma forming the 'lacy' structure, sweeping through the individual channels of the network out from the main nucleus.

Fig. 1.   Negative of the sunspot group investigated during the PFP; July 2, 1969 (7h 17m UT).

Often the smaller spot nuclei, not very dark, consist of such 'lacy' structures or have at least in a part of their peripheries such structure. Individual fibrils of the structure continue in dark fibrils (on negatives bright) of penumbra and go on to individual fibrils of intergranular space in the photosphere surrounding the spot (Bumba, 1965), (Figures 1, 2, 3, 4, 5). On negatives the magnetic bright fibrils of the penumbra always

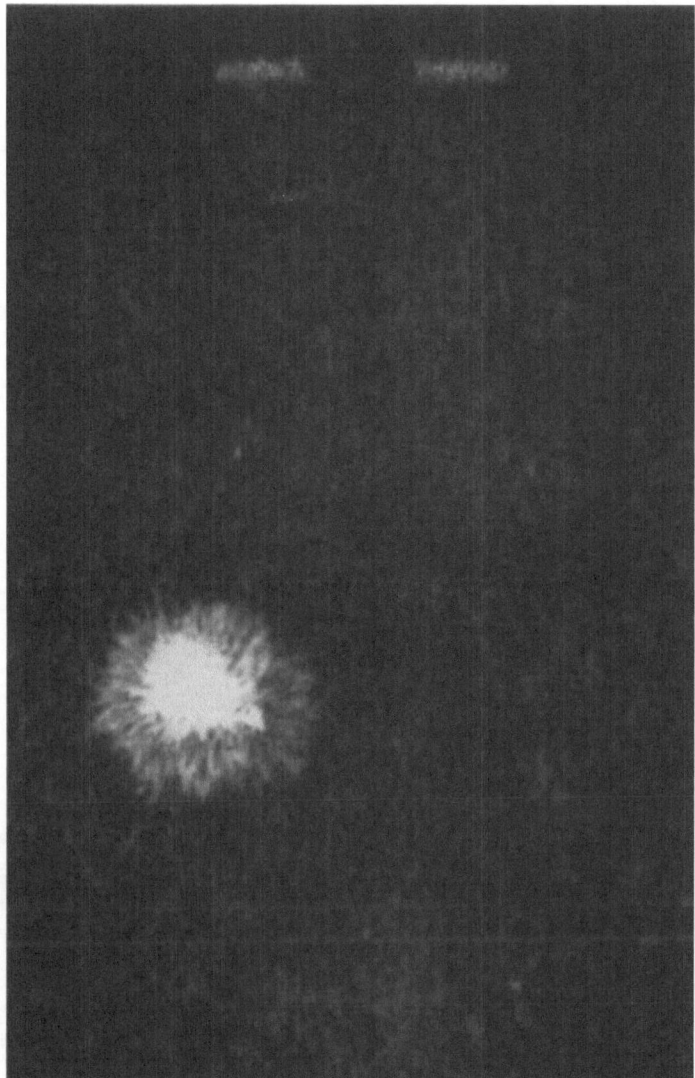

Fig. 2.   Negative of a diminishing spot; July 3, 1969 ($5^h 48^m$ UT).

seem to be predominantly pressing between themselves small elongated dark areas (on positives bright) which seem to be the remains of photospheric granules.

3. Motions

On the spot shown we can demonstrate some rules of the development we may find on the majority of large complex spots. The spot came from the invisible disk of the Sun in the form of a large drop. During its course on the visible disk it developed two

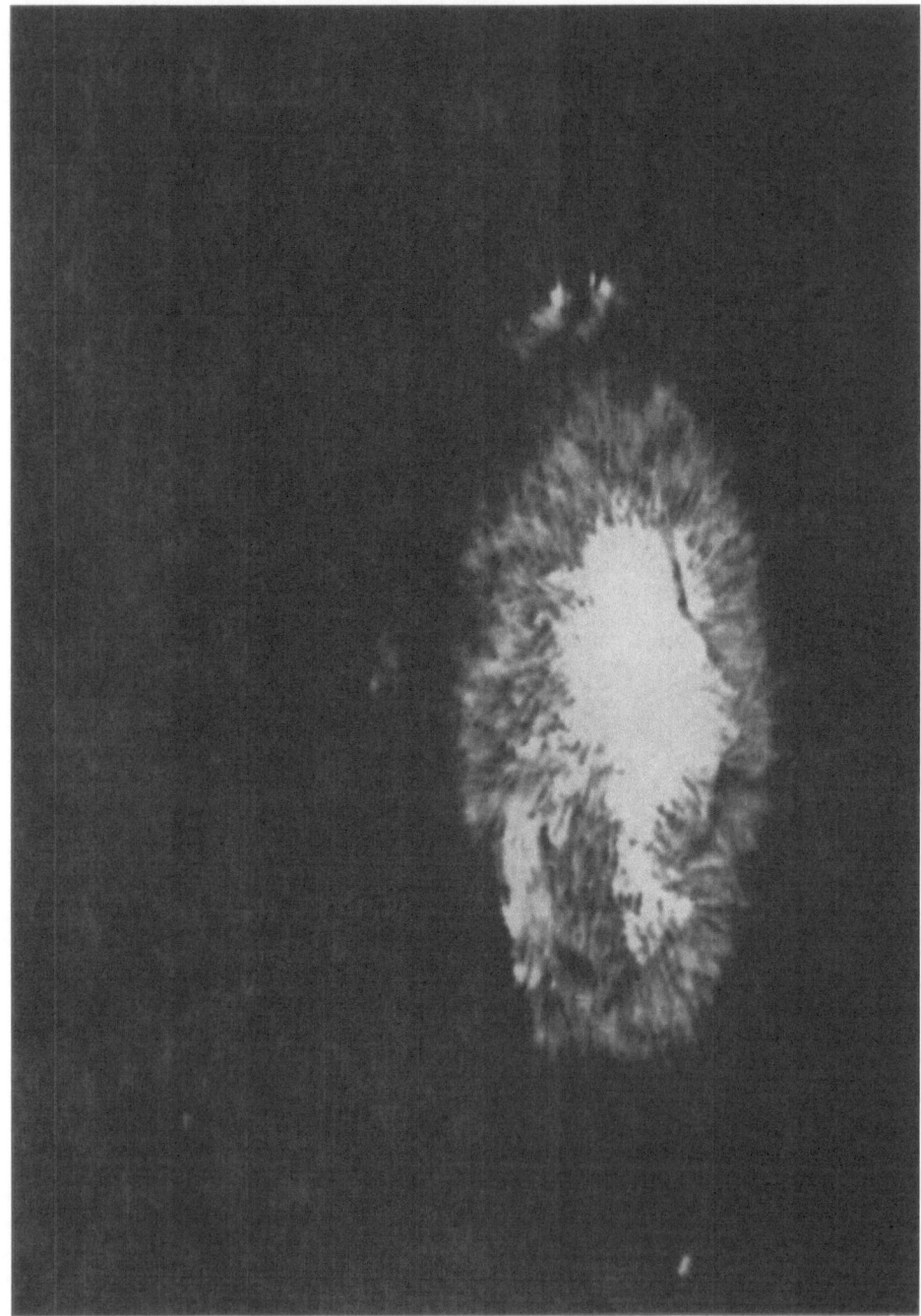

Fig. 3.   Negative of the large complex spot investigated in the research note; July 3, 1969 (5^{h}42^m UT).

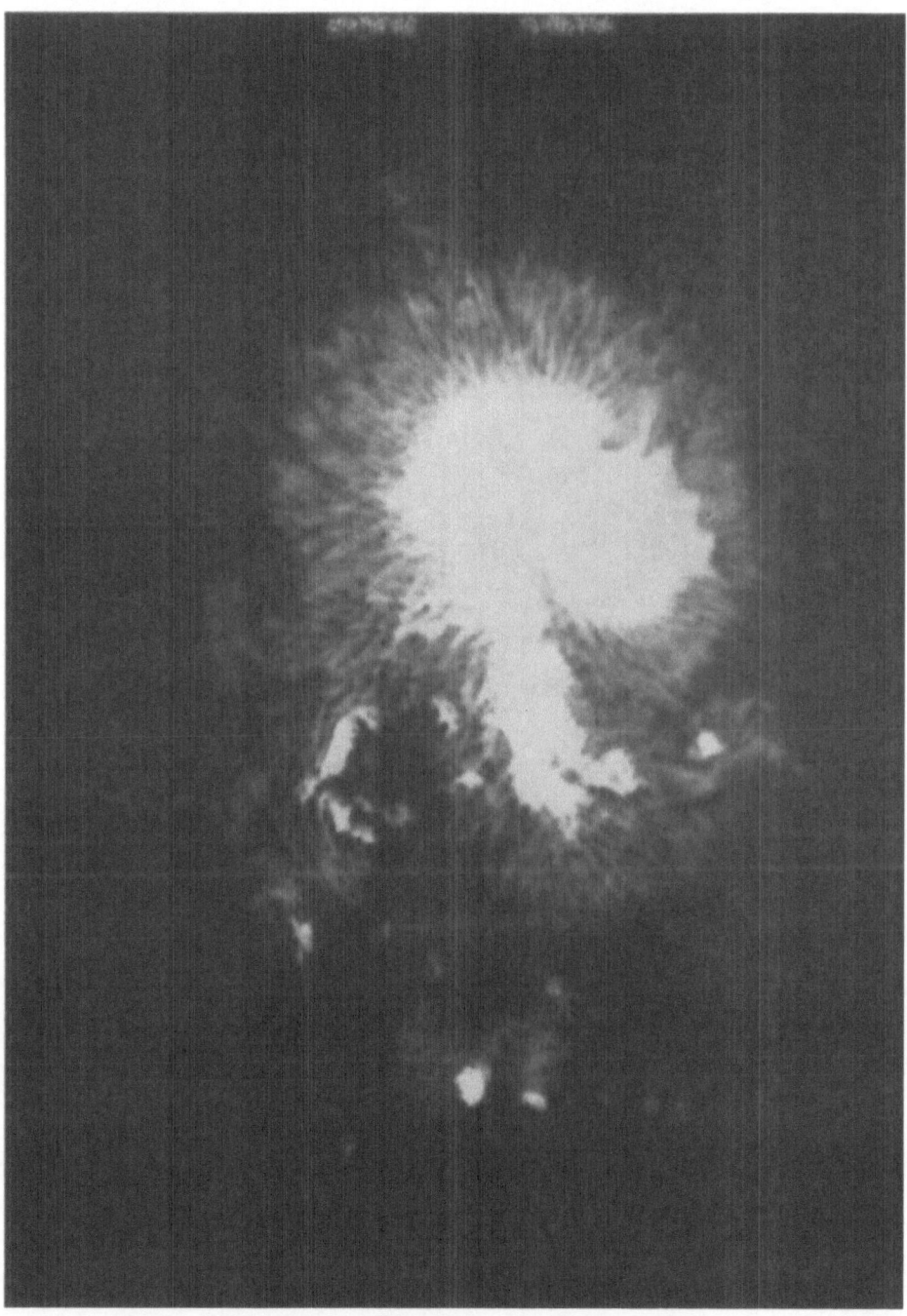

Fig. 4.   The same spot on July 5, 1969 ($8^h 44^m$ UT).

Fig. 5.   Same spot on July 7, 1969 (8ʰ07ᵐ UT).

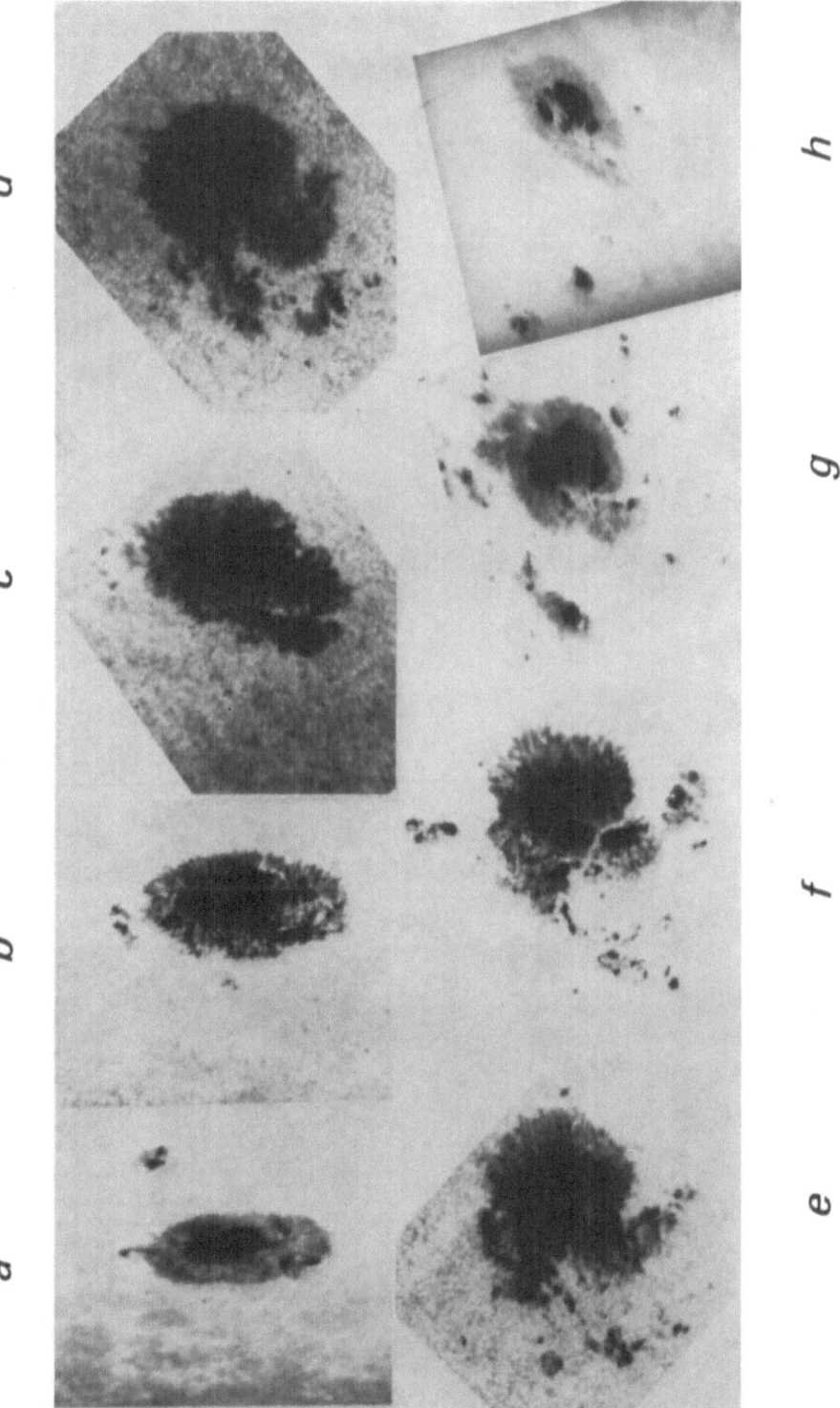

Fig. 6. Series of positives of the investigated large complex spot during its passage on the solar disc (a–h: July 2, 1969 ($7^h 25^m$ UT); July 3, ($5^h 42^m$), July 4 ($6^h 09^m$), July 5 ($8^h 44^m$), July 6 ($15^h 16^m$), July 7 ($5^h 41^m$), July 9 ($12^h 35^m$), July 11 ($11^h 31^m$)). The development of the spiral-like branches may be seen.

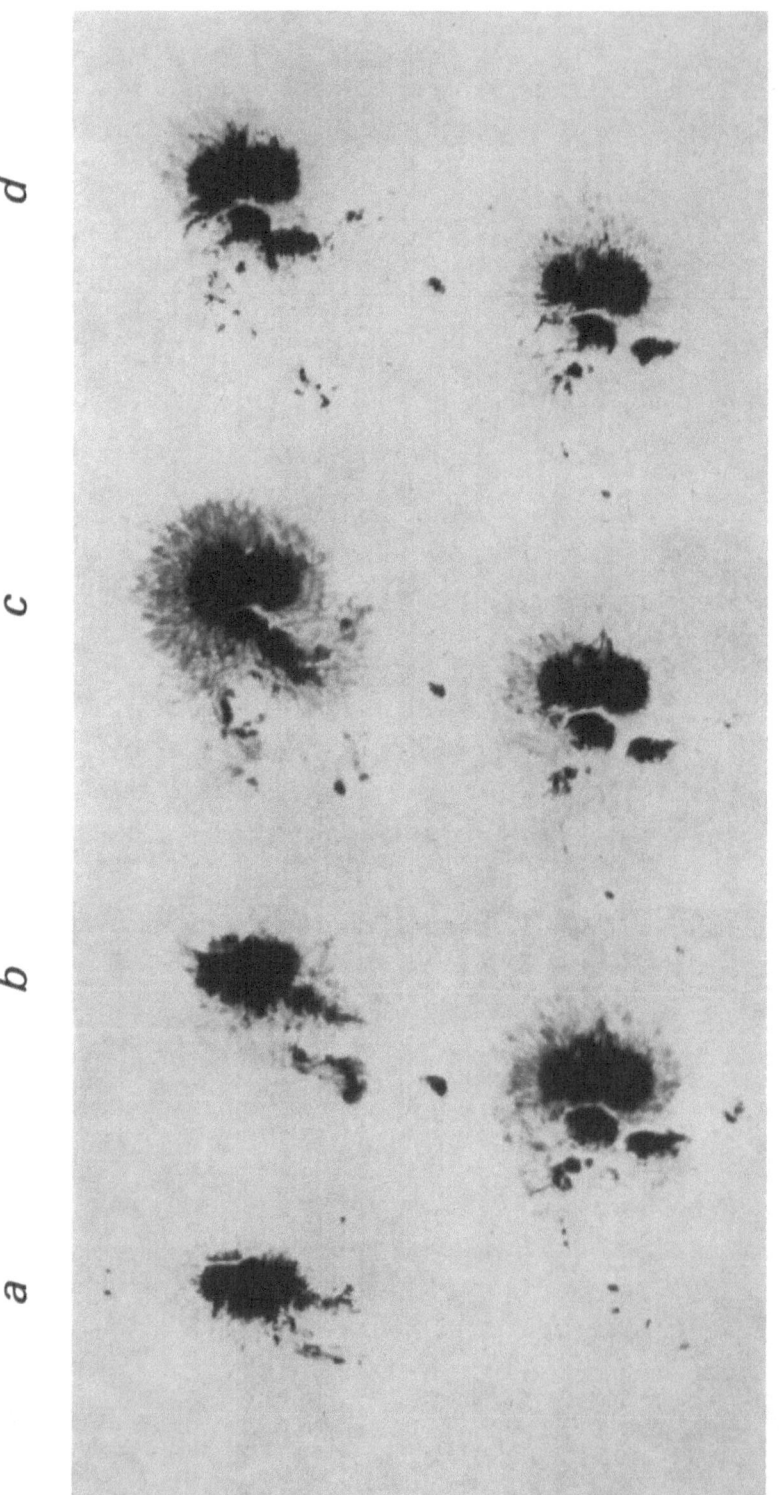

Fig. 7. Series of positives of the investigated spot demonstrating the changes in umbra and penumbra (a-g: July 3, 1969 ($5^h 42^m$ UT), July 4 ($6^h 09^m$), July 5 ($8^h 44^m$), July 6 ($15^h 16^m$), July 7 ($5^h 24^m$, $5^h 41^m$, $8^h 07^m$)).

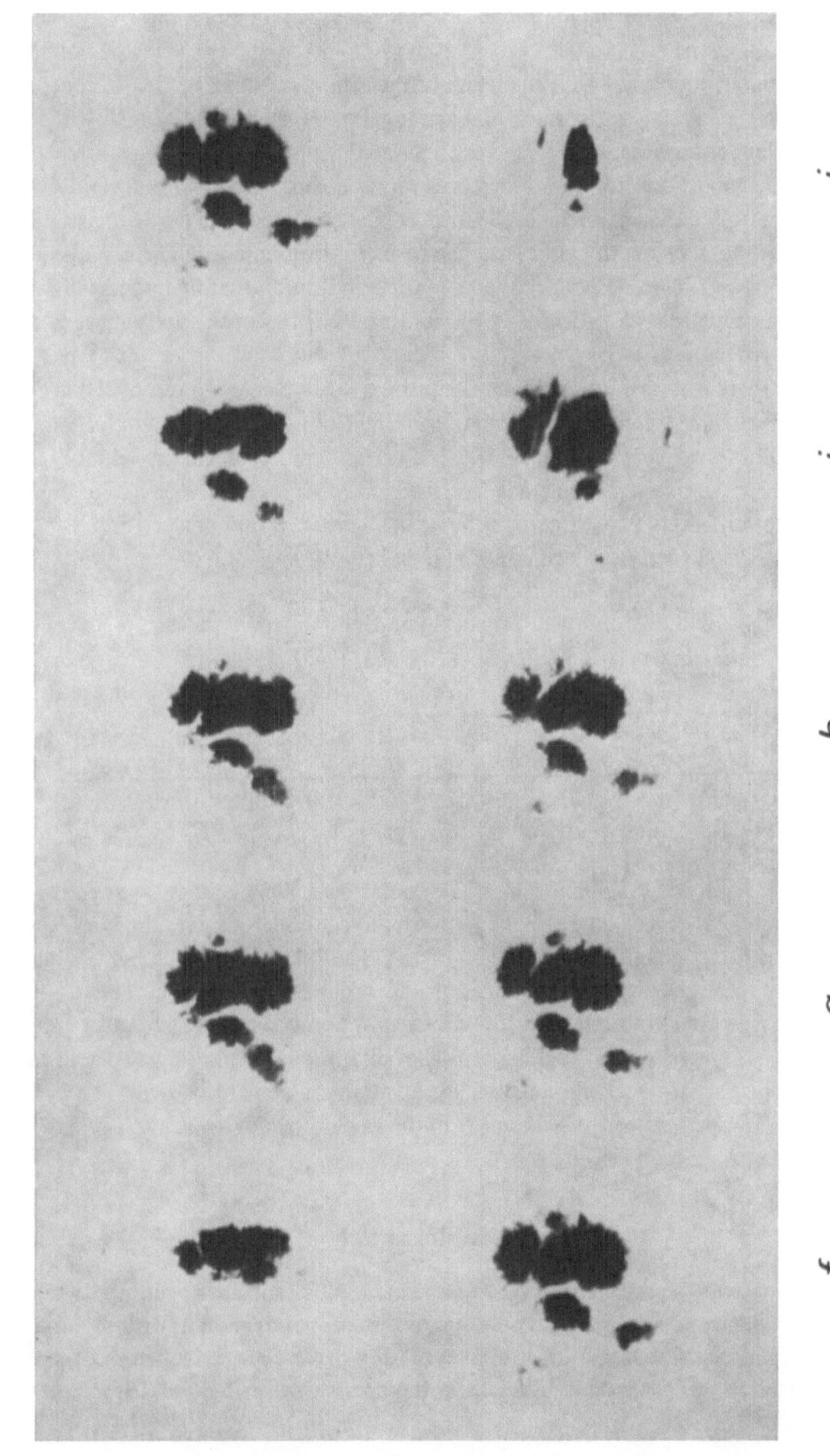

Fig. 8.   Series of positives of umbrae of the investigated spot showing the changes in light bridges (a-j): July 3, 1969 ($15^h 09^m$ UT), July 5 ($8^h 44^m$, $9^h 59^m$), July 6 ($15^h 16^m$), July 7 ($5^h 24^m$, $5^h 26^m$, $5^h 41^m$, $8^h 07^m$), July 9 ($12^h 35^m$), July 11 ($11^h 31^m$).

main spiral-like branches formed from small nuclei, and parts of the 'rudimentary' penumbra separated from the main nuclei of the spot shifting along these spiral-like branches around the spot (clockwise) and away from the spot at velocities of the order of 50 m/s (Bumba, 1964). The picture of the spot in maximal phase of branches development reminds one very much of a spiral galaxy (Figure 6).

The problem of how the small nuclei are separated from the main one or formed in its closest neighborhood we cannot yet answer. We think that the magnetic field squeezes out from the main nuclei in the form of the 'lacy' structure we demonstrated on our negatives. (During this process a decrease of central intensity of the magnetic field in the main spot was observed). We have some indication that the process of formation of such a structure proceeds in the direction out from the spot but we need more studies to be sure.

However there is a position in which the process takes place: the tail of the drop, the point in which the continuous formation of light bridges, fast changes in umbra and penumbra forms, may be observed. It is a certain kind of singularity in the spot, the place where the stability of the magnetic field of the main umbra seems to be diminished. A spectrographic investigation of such a situation would be very useful.

There are several such singularities at the periphery of the spot, connected with the light bridge formation. The main characteristic of such places seems to be their persistence. They may be observed practically during the whole passage of the spot on the disk in the same position (Figure 7) (Bumba, 1965). Many times one observes the connection of the light bridge in the singularity with formations outside the spot (Figure 6). These singularities are also places, where appear in certain stage of development very bright points often with diameter less than 1″, very short-living (of the order of seconds). There may be observed very bright penumbral structures lasting for days (Bumba, 1965). Shortly, these places in spots have to be studied more carefully.

There are several types of light bridges, as is well known. Some of them are relatively persistent, some of them change rapidly. In the observed spot are several singularities – the places with fast changing bridges (Figure 8). The velocity of some light bridge progressions is of the order of 1 km/s (of the order of Evershed motions). The problem is: is it a real motion or is it the intensification of very faint bright grains, which we may observe as a rule in the position of disappeared light bridges? There are indications that it is a real motion, but again more studies are needed. The light bridges may proceed in their development in the same direction from the penumbra as toward the penumbra (Figure 8).

4. Closing Remarks

From all we tried to say one conclusion can be made: the problem of sunspots seems to be principally connected with the solution of the sunspot dynamics, with the study of different kinds of stability of tubes of lines of force of their magnetic fields and with the investigation of plasma-field relations in the fine-scale distribution. The greatest difficulty of such research is connected with the smaller resolution of practically all direct magnetic methods inside the sunspots in comparison with short-exposure photographs of sunspots used as an indirect method.

References

Bumba, V.: 1964, *Prace Wroclawskiego tow. naukowego, Seria B,* No. 112, p.31.
Bumba, V.: 1965, in R. Lüst (ed.), 'Stellar and Solar Magnetic Fields', *IAU Symp.* **22,** 192.

ON THE STRUCTURE OF MAGNETIC FIELD AND ELECTRIC
CURRENTS OF A UNIPOLAR SUNSPOT

V. A. KOTOV

Crimean Astrophysical Observatory, Crimea, U.S.S.R.

Abstract. This talk deals with the spatial distribution of the total magnetic vector H and of electric currents in the big unipolar sunspot of October 17, 1966.

The simple classical model representing the magnetic structure of the sunspot as a 'sheaf' of the diverging field lines, as we have at the top of a solenoid, can not satisfy the recent observations. The appearance of the triplet Zeeman pattern in the umbrae of some spots, the vortex structure and the effect of rotation of the transversal field with height, originally found by Severny (1964), indicate the great complexity of the sunspot magnetic configuration. It is becoming more clear that the magnetic fields of spots have very pronounced fine structure not only along the solar surface but also in the vertical direction.

For detailed study of the field configuration at different geometrical heights in the sunspot we have measured the vector H with the Crimean magnetograph in two spectral lines Fe I $\lambda 4808$ and Ca I $\lambda 6103$ simultaneously (Rowland intensities 0 and 9 respectively). The distance between the levels of formation of these lines calculated for the photospherical model is 170 km (Buslavskij, 1969). The longitudinal field was also measured in the Hα line.

We used the empirical calibration curves found by Severny (1967) which seems to us as most reasonable at the present time.

Figure 1 shows the distribution of the transversal vector $H_\perp$ at two levels in the spot. (As the spot has south polarity, the vector directions are drawn for convenience oppositely to real ones; the same is true on Figures 3 and 4). We see essentially azimuthal asymmetry of the field structure at the lower level (where the $\lambda 4808$ line is formed) particularly in the south part of the umbra with $H_\perp \approx 4000$ G. It is also important to note the differences (up to 90°) in some parts of the spot between the azimuth of the transversal field determined at two different levels (Figure 2).

The behaviour of the inclination angle γ of the total vector at two different levels is illustrated by Figure 3 for two vertical cross-sections *N-S* and *E-W* drawn across the spot center. We see a great deal of inhomogeneity of the field directions. An asymmetry is especially pronounced at the lower level ($\lambda 4808$) for the *S-N* cross-section. Here the field is mainly horizontal with maximum field strength ≈ 3500 G. At the higher level ($\lambda 6103$) the distribution is more uniform. In the umbra there are some points where the field is almost horizontal (angle $\gamma \approx 50-70°$); this agrees with some earlier investigations (Nishi, 1962; Iošpa and Obridko, 1965).

The appearance of the azimuthal component of the field at two levels is shown in Figure 4. The clockwise twist of the field lines is stronger at the lower level, pointing

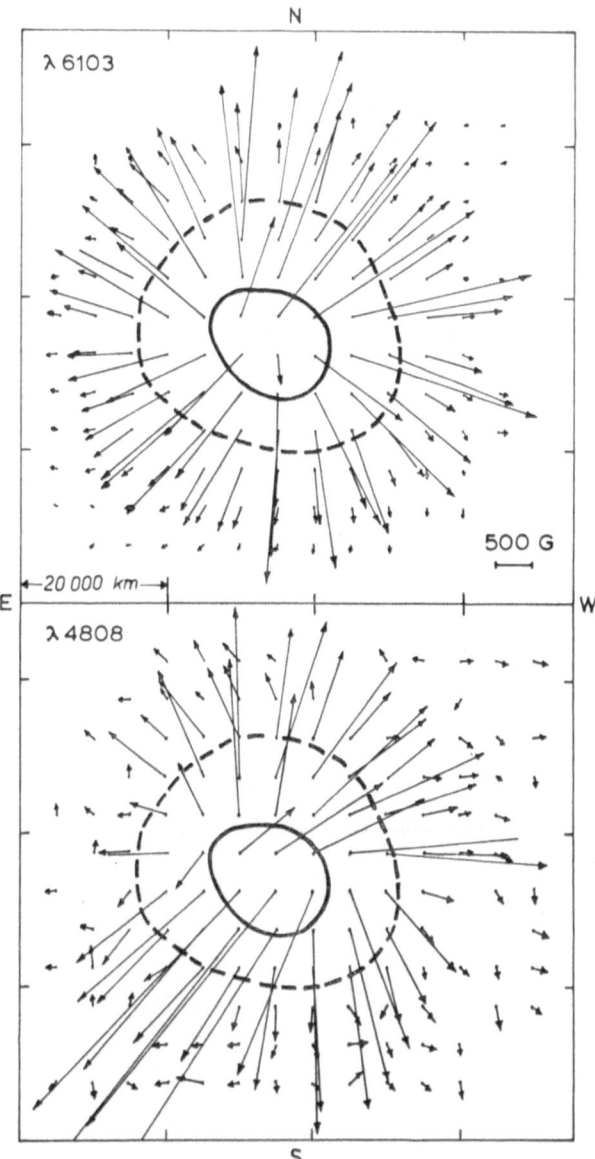

Fig. 1.   Transversal field distributions at two levels in the spot.

to the increase of twisting with depth. The same sense of spiraling is observed on the Hα-filtergrams taken at the same time.

Our data permit us also to calculate the electric currents in the spot from the relation $\mathbf{j} = c/4\pi$ rot $\mathbf{H}$. There is an appreciable difference between the distributions of the vertical currents j_z (Figure 5) obtained at two levels. At the lower level there is a large area of electric currents inside the spot flowing upwards, which is surrounded by two regions of downward currents (shaded area). But at the upper level we have only

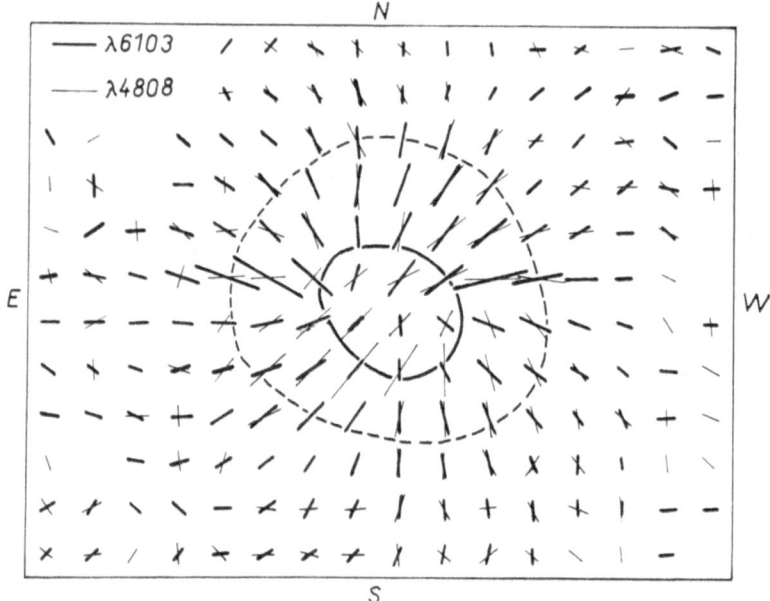

Fig. 2. The comparison of the transversal field directions.

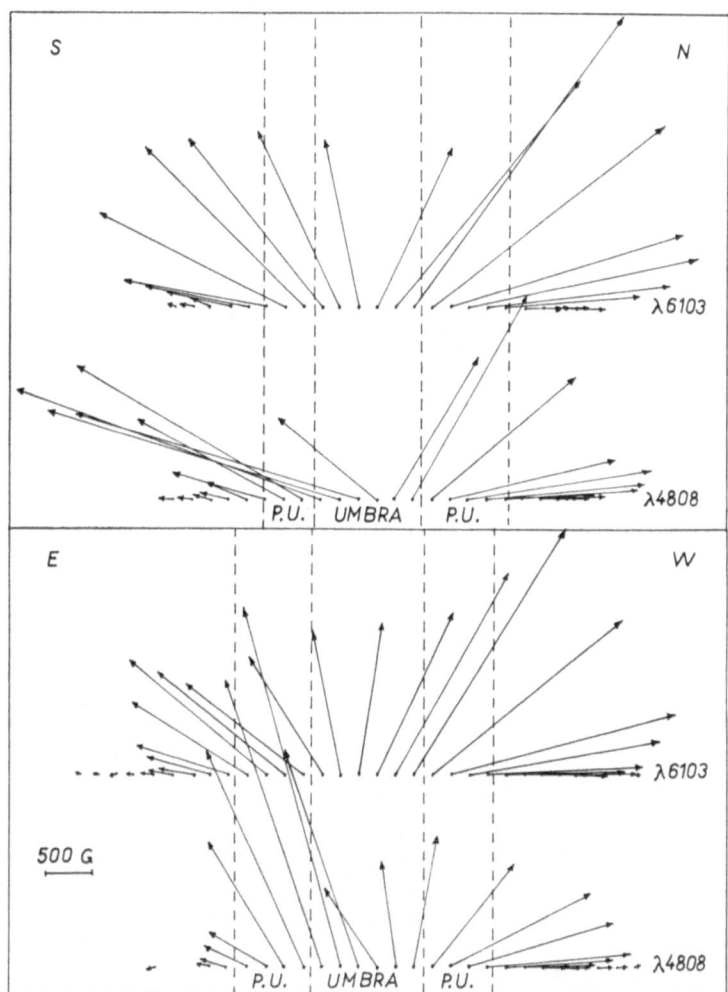

Fig. 3.   The field vectors at two levels for two cross-sections drawn across the spot center.

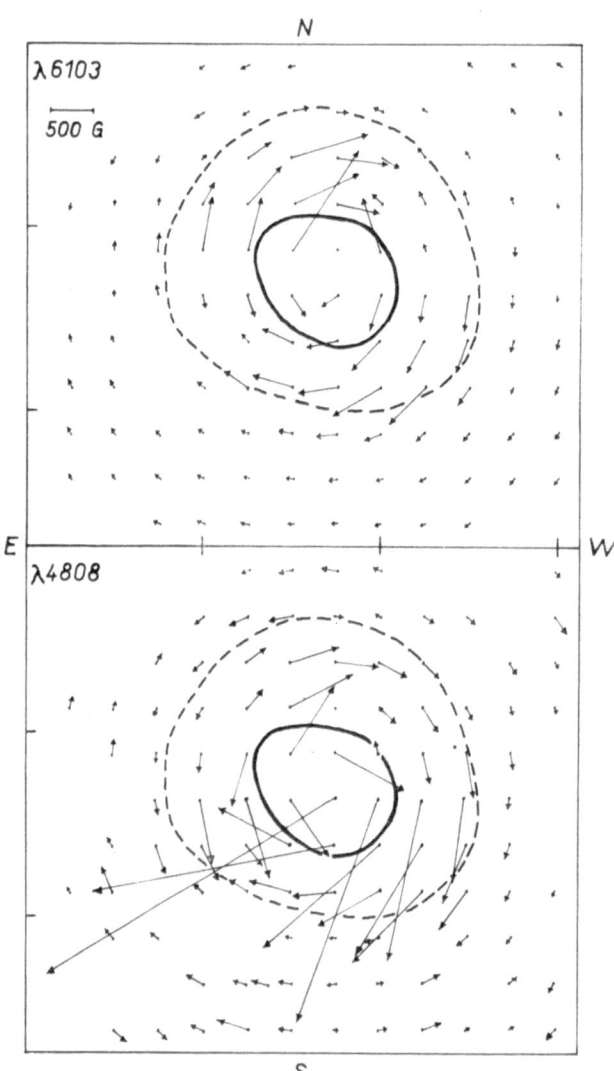

Fig. 4.   The azimuthal component of the field.

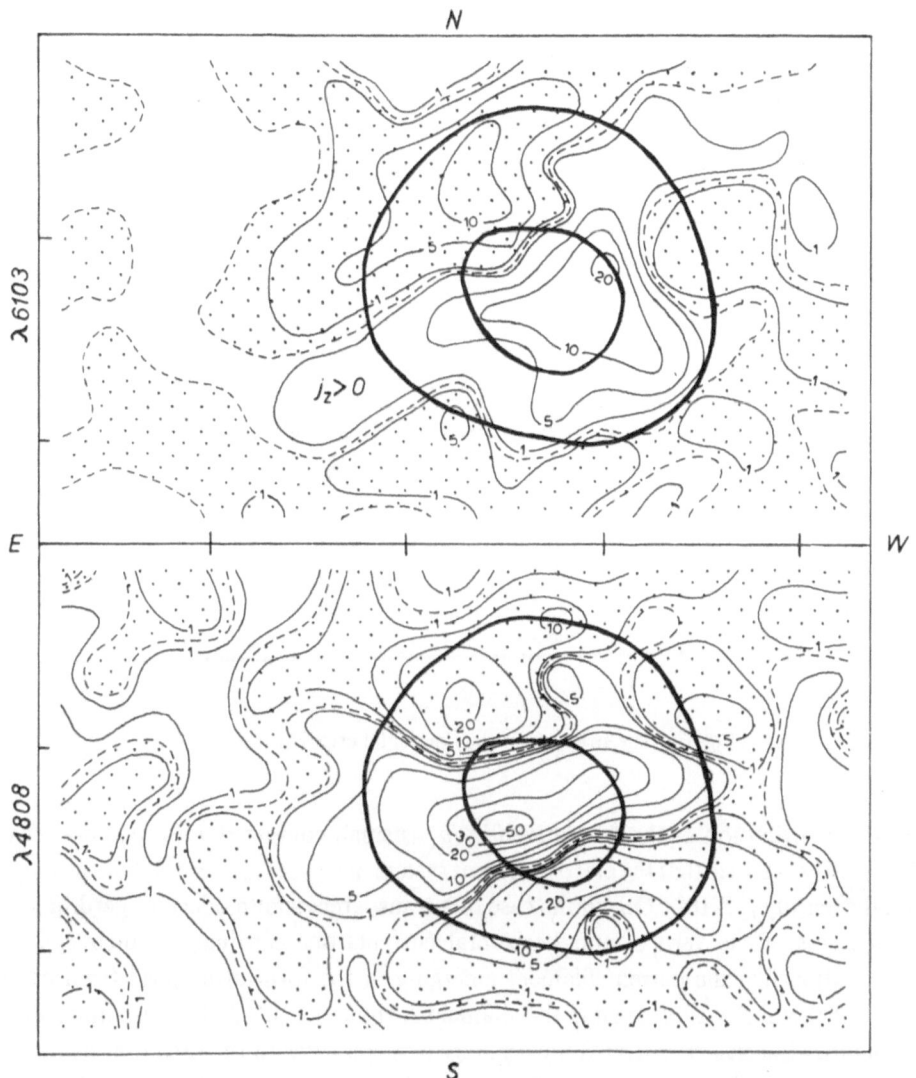

Fig. 5.   The maps of the vertical electric currents j_z (10^{-2} G/km) inside the spot.

V. A. KOTOV

two areas inside the spot with currents flowing up and down. The value of the vertical current density is $j_z \sim 10^3$–10^4 CGS.

Figure 6 shows the structure of the horizontal electric currents found from measurements of the transversal component $\mathbf{H}_\perp$ at two levels. (Note: the calculations show

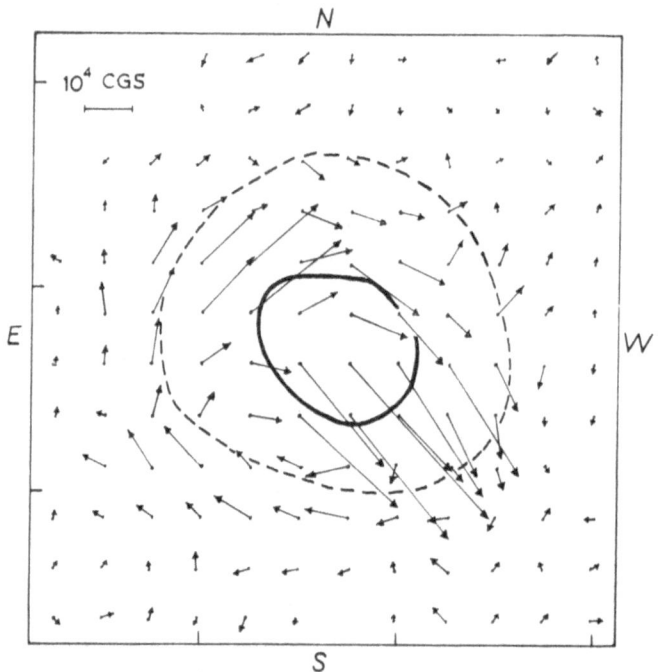

Fig. 6.   The structure of the horizontal electric currents.

that the contribution of the derivative $\partial H_\parallel / \partial_\perp$ is small compared with the derivative $\partial H_\perp / \partial_\parallel$).* We can see that the electric current system in a spot represents a kind of loop, with the density $j_\perp \sim 10^4$ CGS (in the case of a transparent spot model). Regarding the distributions of the vertical and horizontal currents one can imagine their spatial configuration like the current filament emerging in the form of a spiral from deep layers. The total electric current in this filament is about 6×10^{12} A. It is also important to note that the electric currents flow mainly perpendicular to the direction of transversal field; this is not in favor of the force-free character of the magnetic field. Using the photographic method Rayrole and Semel (1970) have come recently to the same conclusion on the field structure of a sunspot.

A calculation of magnetic force shows that it is $\sim 10^{-3}$ d/cm^3 and it is directed mainly along the solar longitude and downwards inside the spot. This is accompanied by an inflow of gases as observed in the $\lambda 6103$ line.

* See footnote on following page.

References

Buslavskij, V. G.: 1969, *Izv. Krymsk. Astrofiz. Obs.* **39**, 317.
Iošpa, B. A. and Obridko, V. N.: 1965, *Soln. Dann.* No. 3, 54.
Nishi, K.: 1962, *Publ. Astron. Soc. Japan* **14**, 325.
Rayrole, J. and Semel, M.: 1970, *Astron. Astrophys.* **6**, 288.
Severny, A. B.: 1964, *Izv. Krymsk. Astrofiz. Obs.* **31**, 126.
Severny, A. B.: 1967, *Izv. Krymsk. Astrofiz. Obs.* **36**, 22.

Discussion

Schröter: (1) The two lines used by Kotov are supposed to originate in different heights, therefore they will show a different sensitivity to temperature. As we learned here calibration errors for magnetographs are a direct function of the line's temperature sensitivity. How did Kotov assure that such errors do not occur?

(2) Have you now at your observatory enough material to decide whether the observed rotation of the field azimuthly with depth does depend on the hemisphere or on the polarity of the spot?

Severny: (1) I think it is too premature to reconsider the problem of calibration due to the effects of temperature sensitivity without thinking it over and without checking it by further observations.

(2) We have a lot of material for the unipolar spot under consideration covering all of the passage from the east to the west limb and for two rotations. This material is now being reduced and considered by Kotov.

Brueckner: What is the optical depth of the two lines used for these measurements?

Severny: I do not remember exactly, but I think that for the deep seated line $\lambda 4808$, τ is about 1.2 and for the line $\lambda 6103$ it is about 0.3.

Semel: I think you are really very lucky to obtain the magnetic field measurements in three dimensions. Could you test if your measurements yield $\nabla \cdot H = 0$?

Severny: Kotov has made this check with div $H = 0$ and there is as usual in these cases disagreement between the $\partial H/\partial z$ found from this equation and that found by direct measurements.*

---

* This fact points at the difficulties of the usual interpretation of the H-recordings in spots. We have discussed recently (Kotov, V. A.: *Publ. Crim. Astron. Obs.* **46**, in press) the considerable discrepancy between the height gradients of the vertical field $\partial H_z/\partial z$ obtained in two ways: (a) from the measured vertical field at two levels ($\sim \pm 3$ G/km) and (b) from the use of the equation div $H = 0 (\sim \mp 0.5$ G/km). This discrepancy leading to seeming violation of the equation div $H = 0$ could be mostly attributed to the great difference between very high depth- and low surface-resolutions. Here one should take into account the fine structure of the field which is less than the resolving power of the instrument.

MAGNETIC FIELD STRENGTHS DERIVED FROM VARIOUS LINES IN THE UMBRAL SPECTRUM

C. ZWAAN and J. BUURMAN

Astronomical Institute, Utrecht, The Netherlands

Abstract. Profiles of various lines in the spectra of umbrae are discussed, emphasizing the use of purely umbral lines. In the dark core of a large umbra the field strengths measured from various lines are the same, yielding 3080 ± 50 G. In regions with umbral dots the field strengths do depend on the spectral line; the strengths are lower than in the dark core. Some conclusions are discussed.

For the investigation of the magnetic field in umbrae of sunspots, purely umbral lines, i.e. Fraunhofer lines which are very weak in the photosphere and enhanced in the umbra, offer some advantages. First, the line profiles are not affected by stray light (Zwaan, 1966). Second, these lines originate mainly in the coolest columns of the umbra. Thus, even if the fine structure cannot be resolved in the spectrograms, some information on the correlation between magnetic and thermal inhomogeneities may be obtained from the comparison between measurements from purely umbral lines and deductions from lines which are also present in the photospheric spectrum.

We prepared a list of purely umbral lines with simple or otherwise interesting magnetic splitting patterns, selected on the appearance of the profiles. We used photographic spectrograms of a very large spot close to the center of the disk ($\vartheta = 13°$) recorded at Sacramento Peak Observatory on September 18, 1966. Line profiles were obtained in both opposite directions of circular polarization and, by subtracting, in Stokes parameter V. For comparison we could use density tracings from spectrograms of the same umbra recorded at Kitt Peak National Observatory. The latter spectrograms were obtained without a polarizing analyser, but the spectral resolution is higher and the noise is less.

One of the most suitable lines was found to be

$$\text{Ti \textsc{i}, } \lambda 6064,626, \text{ low E.P.} = 1.05 \text{ eV, } {}^{\text{ph}}W_{\lambda} = 7 \text{ mÅ, simple triplet, } g = 2.00.$$

In the umbral spectrum this line is relatively free from blends, it is fairly strong and therefore it can also be observed in pores.

We recommend to use this line for routine measurements of umbral field strengths and for investigations where effects from the varying amount of stray light should be excluded, e.g., into the dependence of the field strength on the umbral area and into the changes of the field with time.

Up to now we measured magnetic field strengths from the profiles of 6 purely umbral lines (equivalent widths in the photosphere ${}^{\text{ph}}W_{\lambda} \lesssim 10$ mÅ), 7 low-excitation photospheric lines (${}^{\text{ph}}W_{\lambda} > 30$ mÅ, strengthened or about equally strong in the umbra), and 1 high-excitation photospheric line (weakened in the umbra). The spectrograms refer to the *dark core* in the umbra where we have never seen any fine structure in the intensity, neither visually nor on overexposed photographs, during the full week we observed the spot. From these 14 lines we obtained

$$B = 3080 \pm 50\,\text{G},$$

without any dependence on the type of the line. The uncertainties in the field strengths determined from individual lines range from 70 to 150 G.

From this result we conclude that in the dark core there were no substantial magnetic inhomogeneities correlated with invisible inhomogeneities in temperature. Moreover, the components of the magnetic lines show widths very similar to the widths of purely umbral non-magnetic lines, which suggests little spread in the field strengths in the region of the dark core where the purely umbral lines are formed. In purely umbral lines with patterns where both the σ- and the π-components show a wide gap at the position of the unshifted line, we find, with Mehltretter (1969), no visible central component. This excludes that cool columns with much weaker fields would occupy a substantial fraction of the area.

From the discussion thus far we cannot exclude a possible existence of invisible hot elements with very weak fields. However, even if all umbral dots would reach photospheric conditions during a part of their evolution, transitional conditions are to be expected, both in space and in time. The results reported here indicate that the transitional conditions, if present, occupy a small area in the dark core.

The π-components do not seem abnormally strong in the purely umbral lines, even in the fairly strong line Ti I λ6064.6 the π-component is weak. 'Abnormal' shifts of the π-components (Severny, 1959; Hénoux, 1968; Moe and Maltby, 1968) cannot be observed in these lines. In some photospheric lines (e.g., λ6302.5 and λ6173) there may be a slight indication that the fairly strong π-components are somewhat shifted. However, we cannot find convincing evidence for an inversion in the origin of the V-profile, found by Beckers and Schröter (1968), which they tentatively interpreted as the result of inclusions of weaker fields with opposite polarity.

It should be mentioned that spectrograms from the dark core of the same spot, further from the center of the disk, do show strong π-components which are shifted. We have not yet deduced V-profiles from these spectrograms.

On the same date spectrograms were recorded from a region within the same umbra where rather faint umbral dots were visible. Provisional measurements with a comparator in the line Ti I λ6064.6 showed that the magnetic field strength is several hundreds of gauss smaller in the area with fine structure than in the dark core. This fits in with measurements in smaller spots (Zwaan, 1968). The conclusions may be summarized as follows:

(1) The strength of the umbral magnetic field depends on the fine structures such that the strongest fields are measured in dark cores, where no fine structures are visible. We suggest that this is at least one of the reasons for the intrinsic scatter in the routine measurements of field strengths in large spots.

(2) There is at least one case that in a dark core without visible fine structures the same field strength is obtained from different types of Fraunhofer lines.

(3) In regions with visible fine structures the measured field strength does depend on the type of the spectral lines, in the sense that the fields measured in purely umbral lines are stronger than the fields obtained from photospheric lines (Zwaan, 1968). Apparently this effect cannot be explained by stray light alone, so it appears that the magnetic field strength is smaller in the hotter elements than in the cooler material in between.

From the above results it follows that the inhomogeneities should be taken into account in the determination of any depth dependence of the magnetic field from line profiles. For instance, it seems pointless to determine the vertical gradient $\partial B/\partial z$ from the field strengths measured in lines of different excitation and ionization potentials on the assumption of a homogeneous model, unless it can be demonstrated that the spectrograms from the investigated umbral region can indeed be explained by a homogeneous model.

Lists of suitable lines of different excitation and ionization potentials, with brief descriptions of the line profiles in the umbral spectrum, will be published elsewhere.

Acknowledgements

During the stay of one of us (C.Z.) at Sacramento Peak Observatory as a member of the High Altitude Observatory Solar Project, the observers Mr. H. A. Mauter and Mr. L. Gilliam have helped with skill and enthusiasm to secure the spectrograms. Dr. K. O. Kiepenheuer and Dr. W. Mattig enabled the other author (J.B.) to use the digitizing microphotometer of the Fraunhofer Institut in Freiburg. Dr. A. K. Pierce has lent high-dispersion spectrograms he recorded at Kitt Peak National Observatory.

References

Beckers, J. M. and Schröter, E. H.: 1968, *Solar Phys.* **7**, 22.
Hénoux, J. C.: 1968, *Solar Phys.* **4**, 315.
Mehltretter, J. P.: 1969, *Solar Phys.* **9**, 387.
Moe, O. K. and Maltby, P.: 1968, *Astrophys. Letters* **1**, 189.
Severny, A. B.: 1959, *Astron. Zh.* **36**, 208 = *Soviet Astron.* **3**, 214.
Zwaan, C.: 1966, in *Proceedings of the Meeting on Solar Magnetic Fields* (ed. by M. Cimino), G. Barbèra, Florence, p. 169.
Zwaan, C.: 1968, *Ann. Rev. Astron. Astrophys.* **6**, 135.

Discussion

Mattig: Can you tell me something about the intensity-ratio between the dark part in the umbra and that part which is covered by umbral dots?

Zwaan: We have not evaluated the continuum intensities yet. The intensities relative to the photosphere in the green spectral region may amount to about 0.07 in the dark core and to about 0.10 in the region showing the faint dots.

THE MAGNETIC FIELDS AT DIFFERENT LEVELS IN THE ACTIVE REGIONS OF THE SOLAR ATMOSPHERE

T. T. TSAP

Crimean Astrophysical Observatory, Nauchny, Crimea, U.S.S.R.

Abstract. The strengths of the longitudinal magnetic fields recorded at different depths of active regions with a double magnetograph of the Crimean Astrophysical Observatory are compared.

The recordings of the magnetic fields were made in the lines Fe I λ5250Å, Ca I λ6103Å, Na I D_1, BII λ4554Å, Mg I λ5184Å, Hα, Hγ, Hδ.

It is shown, that there is a close correlation between the longitudinal magnetic field at different levels.

The double solar magnetograph of the Crimean Astrophysical Observatory permits one to observe magnetic fields simultaneously in any two spectral lines (Severny, 1966). A number of magnetographic records of active regions have been obtained for the period from 1965 to 1969 simultaneously in different lines. The following lines were chosen for these records Ca Iλ6103–Fe Iλ5250, Ca Iλ6103–Na I D1, Ca Iλ6103–Ba IIλ4554, Ca Iλ6103–Mg Iλ5184, Ca Iλ6103–Hδ, Ca I 6103–Hγ, λ5250–Hα. The resolution of these records is $3'' \times 9''$.

A magnetic field calibration was carried out by means of the signals of radial velocity $(E\text{-}W)$.

The comparison of the longitudinal magnetic fields of active regions, recorded in the above mentioned lines, shows very good agreement between the distributions of magnetic fields at different levels. All magnetic features seen at some level can be seen at any other level in the majority of the cases (see examples in Figures 1–4). There were observed a few cases when magnetic fields were observed only at one level, or when magnetic fields at the different levels for the same point of the solar surface showed opposite polarities. This is in accordance with the data of Severny (1966). The examples of the correlation between the strengths of the longitudinal magnetic fields in lines λ6103–λ5184, λ6103–λ4554, and λ5250–Hα are shown in Figures 5–7. A close correlation between the magnetic field strengths at different depths in the active region can be seen in Figures 5–7. The values of the correlation coefficient are 0.8–0.9. But the ratio of magnetic field strengths at different levels can vary from point to point in different ways.

The field strength at the same points of active regions on the upper level appears to be larger than on the lower (see Figures 1–4 and 5–7). Probably this can be connected with the change of the line of force direction at different depths relative to the line of sight.

Reference

Severny, A. B.: 1966, *Astron. Zh.* **43**, 465.

Howard (ed.), Solar Magnetic Fields, 223–230. All Rights Reserved.
Copyright © 1971 by the IAU.

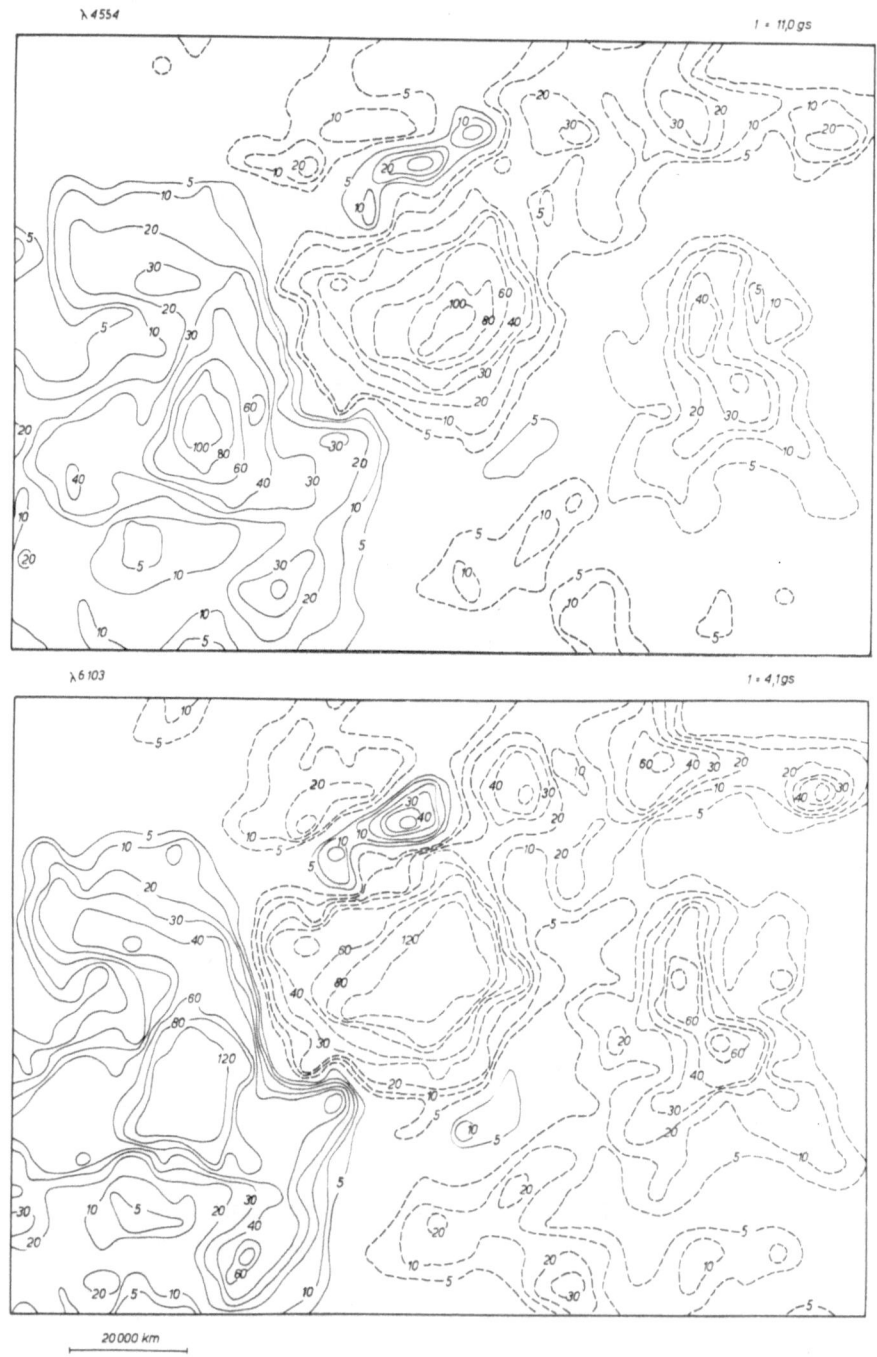

Fig. 1.   The comparison of longitudinal magnetic fields recorded in the lines Ca I λ6103–Ba II λ4554.

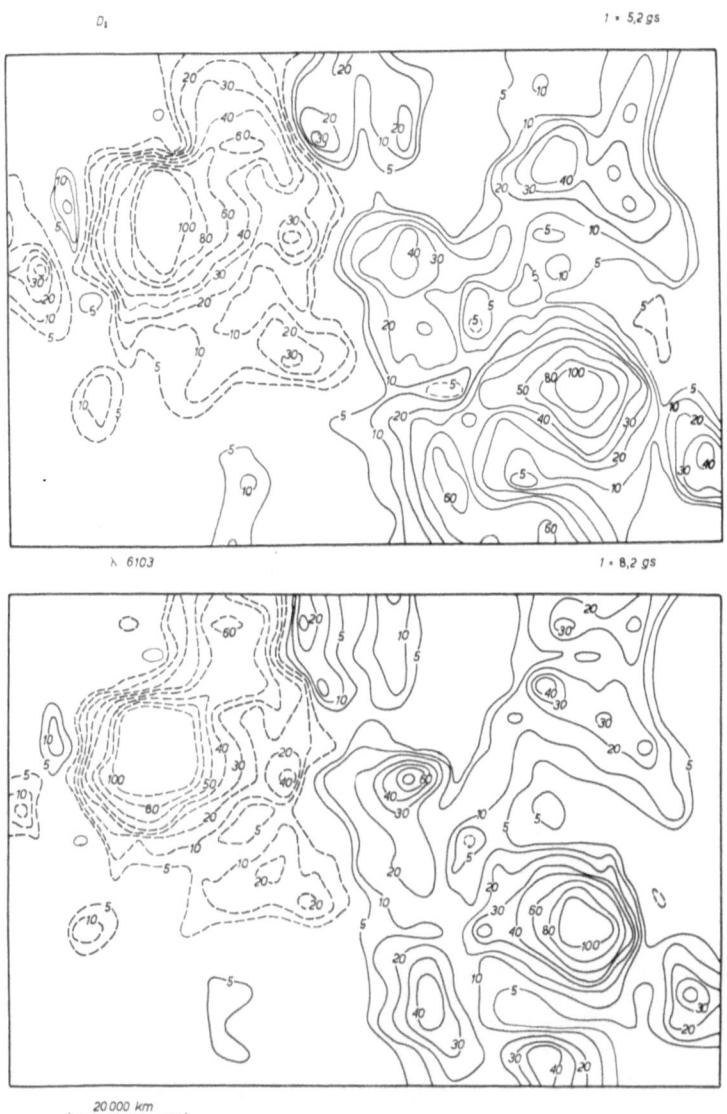

Fig. 2.   The comparison of longitudinal magnetic fields recorded in the lines Ca ɪλ6103–Naɪ D1.

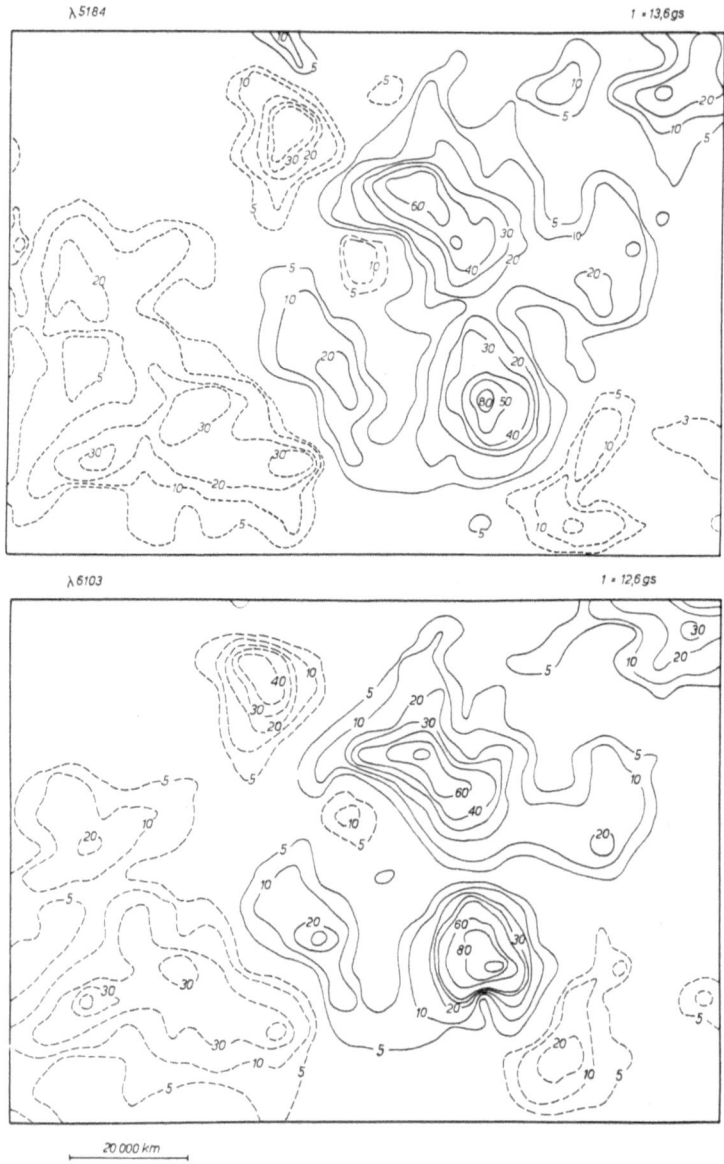

Fig. 3.   The comparison of longitudinal magnetic fields recorded in the lines Ca I λ6103–Mg I λ5184.

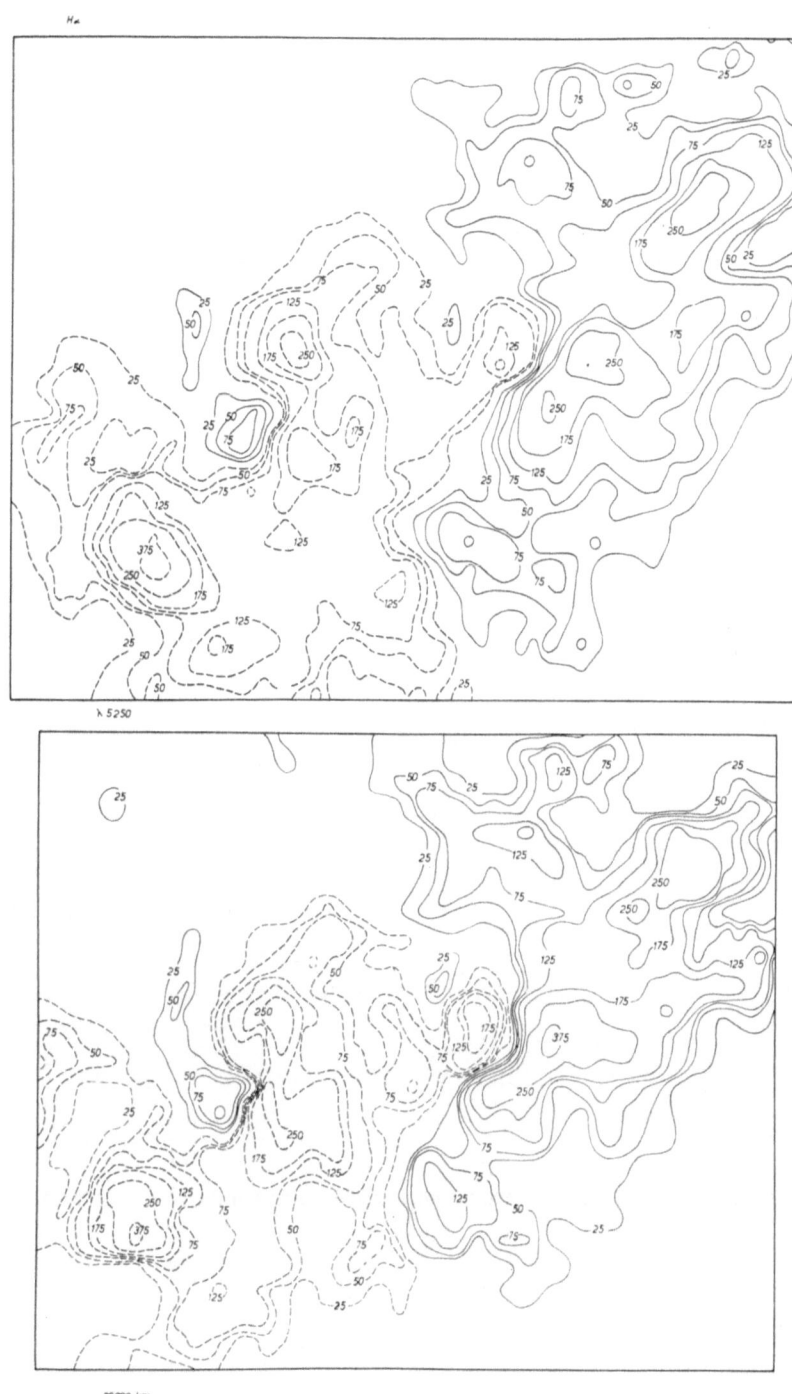

Fig. 4.   The comparison of longitudinal magnetic fields recorded in the lines Fe ɪ λ5250 and Hα.
Gauss levels are shown on the figure. Solid lines indicate south polarity.

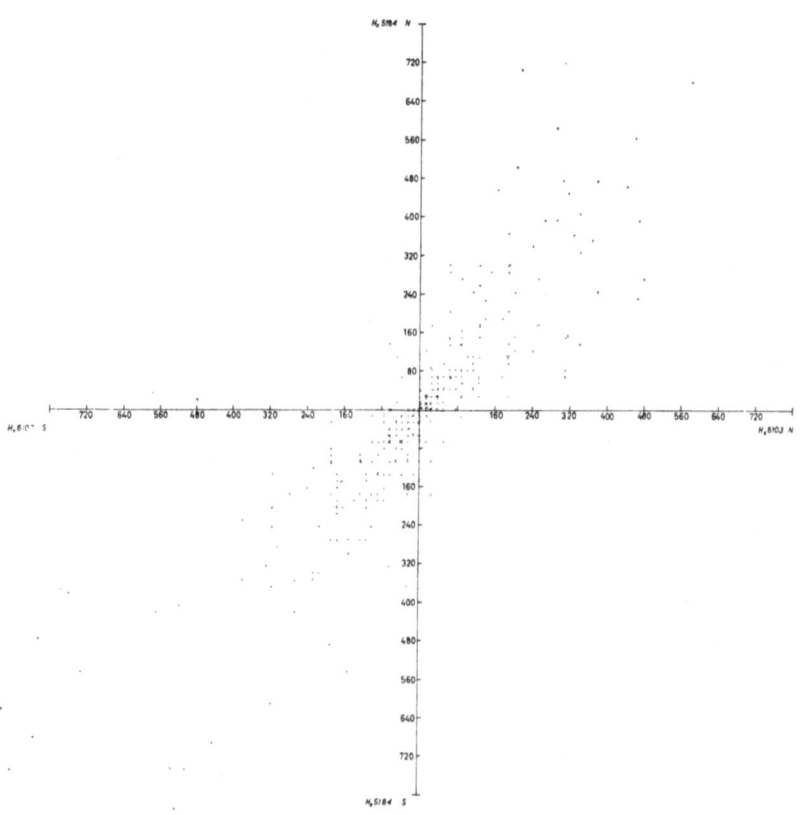

Fig. 5.   The correlation between magnetic field strengths in the lines $\lambda\lambda$6103–5184.

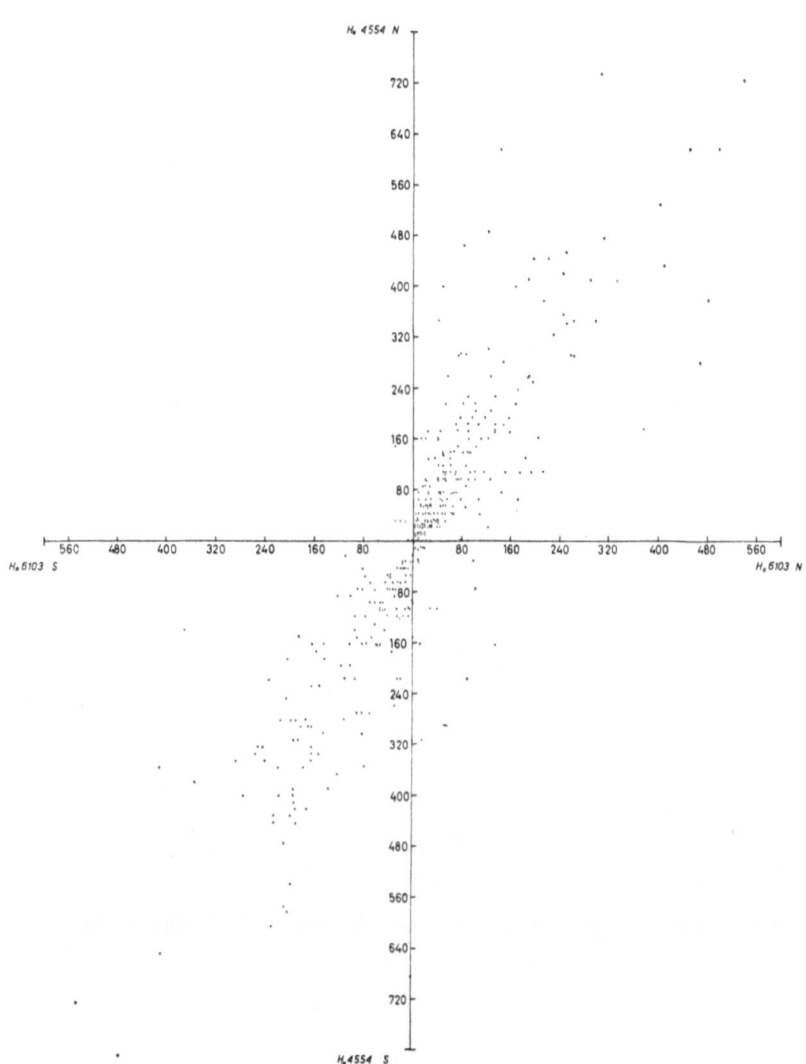

Fig. 6. The correlation between magnetic field strengths in the lines $\lambda\lambda$ 6103–4554.

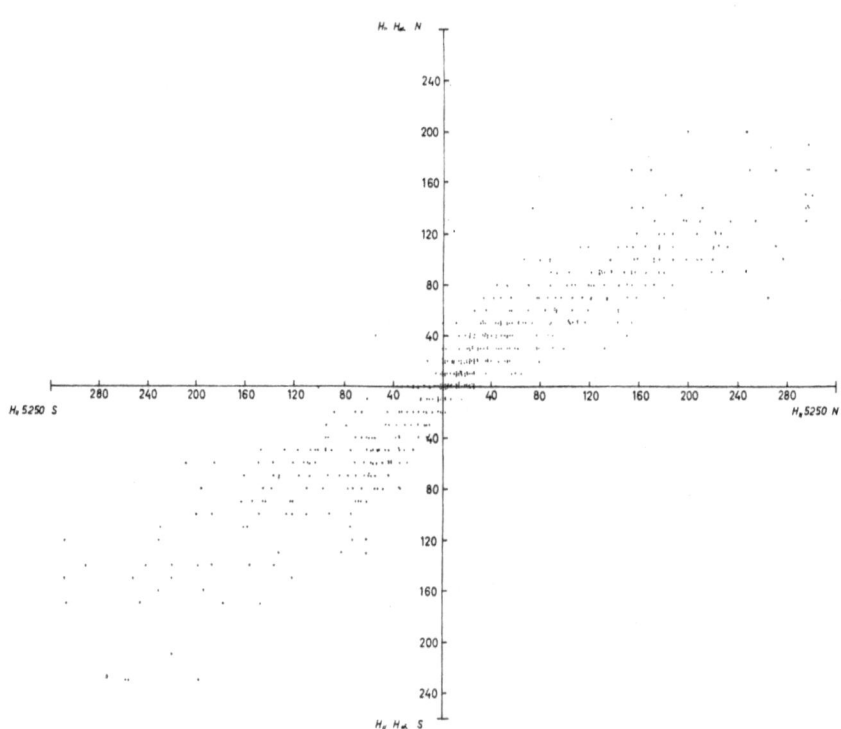

Fig. 7.   The correlation between magnetic field strengths in the lines $\lambda 5250$–Hα.

OBSERVATIONS OF THE TWO-LEVEL STRUCTURE OF SUNSPOT MAGNETIC FIELDS

H. I. ABDUSSAMATOV

Pulkovo Observatory, U.S.S.R.

Inhomogeneity of magnetic field structure, 'granulation' in sunspot umbrae and fine structure of the Evershed motions lead to the conclusion that the sunspot umbra is composed of magnetic ropes (or plaits) with dimensions near the limit of resolution. Progress in the study of these ropes is closely connected with the possibility of obtaining extensive spectroscopic information about some selected regions on the solar disc. It is extremely interesting to obtain a picture of short-time-scale variations of the magnetic field strength and the radial velocity field in connection with the transfer of energy from the photosphere to the upper layers of the solar atmosphere (chromosphere, corona).

The four-camera isothermic spectrograph of the Pulkovo Observatory is a very suitable instrument for scanning an area amounting to $180'' \times 45''$ per 53 s. The scanning is performed by continuous moving of the solar image along the diurnal parallel and simultaneous automatic moving by steps of the four photographic plates after exposure. On each plate we take a series of spectra for neighboring positions of the slit on the disc. The number of spectral sections amounts to 24.

When observing the structure of magnetic fields, a Wollaston prism is set behind the spectrograph slit and an achromatic $\lambda/4$ plate – in front of the slit (Abdussamatov, 1970). In this case the observations are made in the 4th order only in three spectral regions (Hα, $\lambda 6302$Å and $D_{1,2}$ Na) with a linear dispersion of 3.7 mm per Å at Hα.

A series of such plates characterizes the processes evolving in time and space (at least at two levels) in the solar atmosphere. The resolution is defined only by the seeing and image 'turbulence'. On June 22, 1968 the seeing conditions were good (resolution $2''$). The leading and the following spots Group 241 (numeration from the *Solnechnije Dannije*) (Figure 1) were observed. We have taken 8 spectra on each plate at a distance of $5.''8$ from one another (the exposure time is 2s).

On Figure 2 (a, b and c) we see that the magnetic field, being somewhat diluted, penetrates into the chromosphere both over the leading and the following spots ($H_{chr} \approx 0.5 H_{ph}$). If the geometrical difference of the two levels is 1500 km (White and Wilson, 1966) we have $dH/dh \approx 0.5 - 0.8$ G/km.

Investigation of the following spot shows that bright bridges can appear only as a consequence of fragmentation of magnetic fields of spots. Bright bridges are the inherent part of the structure of a spot umbra and not a formation lying over it (Abdussamatov, 1970).

Figure 2 (a, b and c) shows that the inhomogeneity of brightness in the spot is in good negative correlation with H. Dimensions of the structure elements nearly

Howard (ed.), Solar Magnetic Fields, 231–234. All Rights Reserved.
Copyright © 1971 by the IAU.

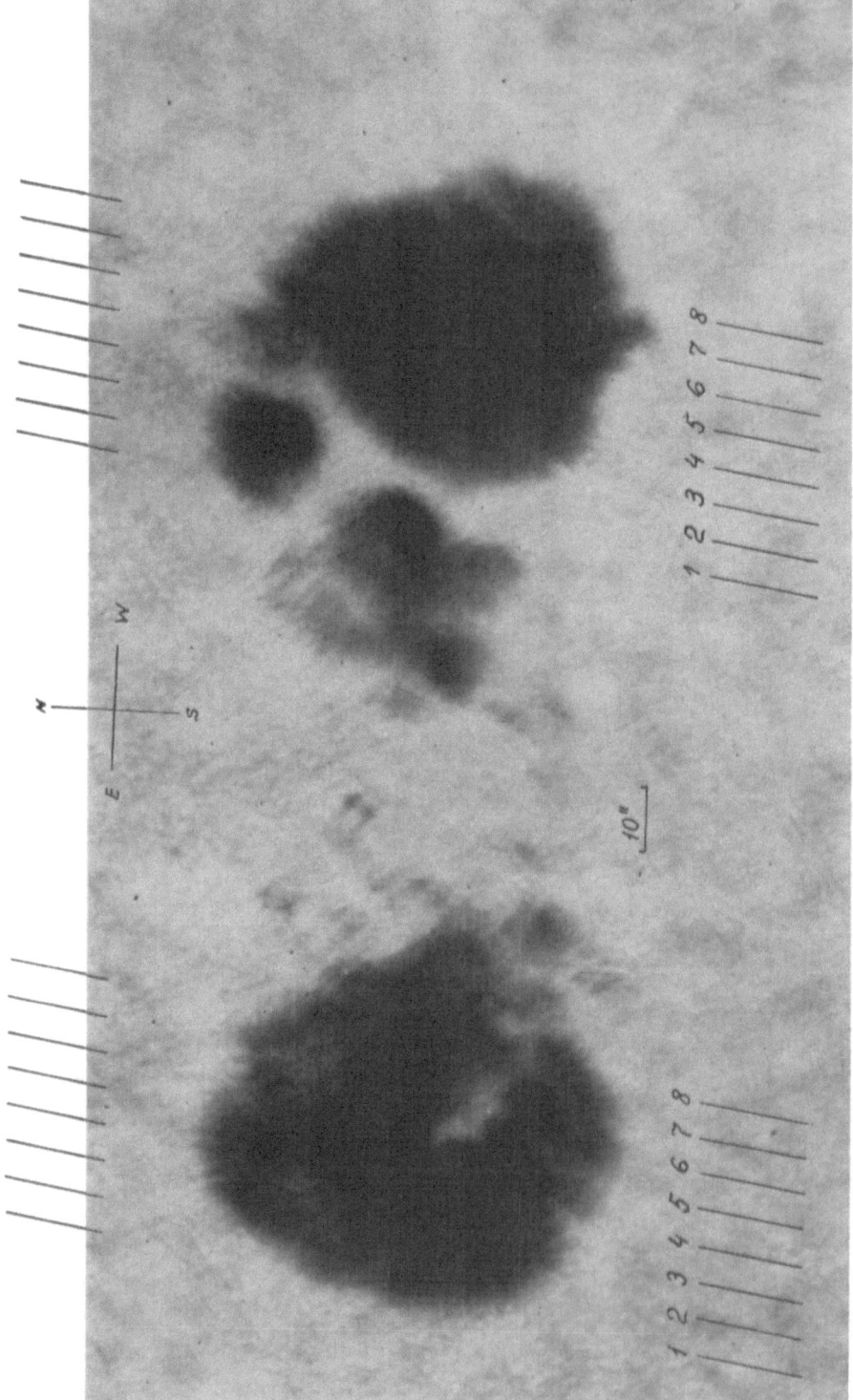

Fig. 1.   Photoheliogram of the investigated sunspot group. The slit position at different scanning steps is shown by parallel lines.

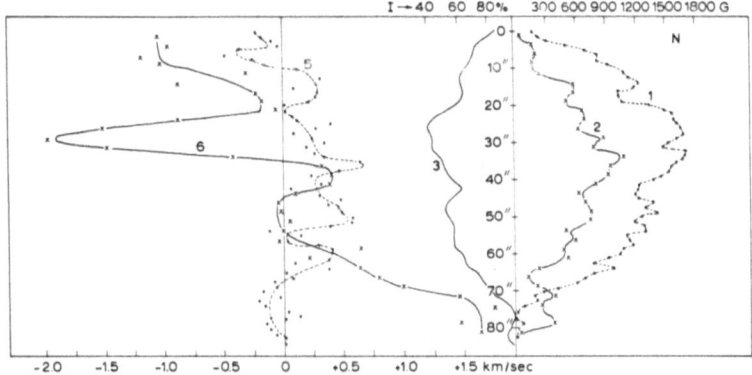

Fig. 2a.

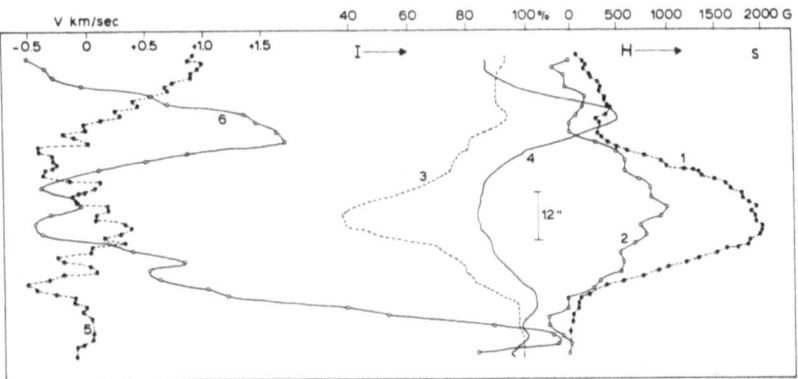

Fig. 2b.

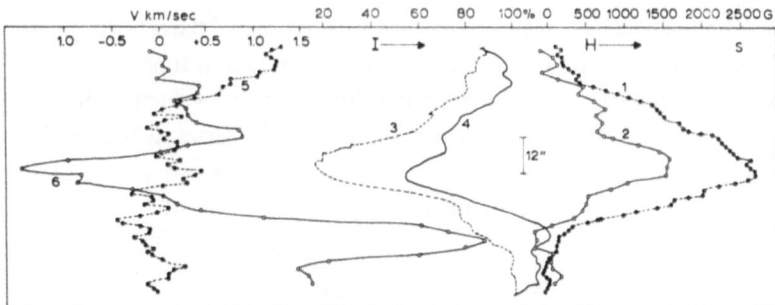

Fig. 2c.

Fig. 2a-c.   (a) Photometric section of the east spot (follower) according to slit position 5 in Figure 1.
(b) Photometric section of the west spot (leader) according to slit position 4. (c) Photometric section
of the west spot according to slit position 5. 1-2 the distribution of the magnetic field strength on the
photospheric and chromospheric levels respectively; 3-4 brightness distribution in the continuum and
the center of the Hα line; 5-6 radial velocities at the photospheric and chromospheric levels.

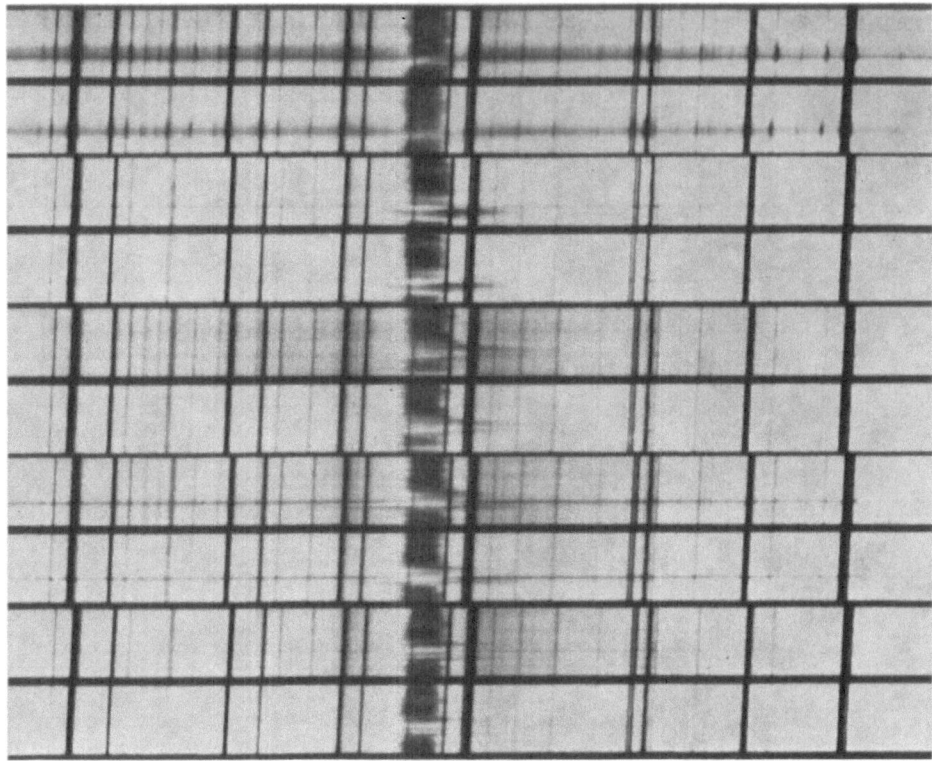

Fig. 3.   Spectrograms of a solar flare obtained by the scanning method with the application of
polarizational optics. (Hα region, the length of the slit-80″.)

correspond to that of the so-called magnetic 'knots'. It speaks in favor of the hypothesis (Abdussamatov and Krat, 1969) that the sunspot umbrae are composed of
many magnetic ropes twisted.

The method described is very useful for investigations of flares; whereas the method
of scanning permits us to obtain simultaneously (with good time resolution) not only
a picture of the fine structure of magnetic fields and radial velocities of flares but also
a series of characteristics derived from the profiles of spectral lines (Figure 3). By
repeating the scans of a given area with short time intervals we can compare physical
conditions in this region at the moment of the flare and after it.

References

Abdussamatov, H. I.: 1970, *Astron. Zh.* **47**, 82.
Abdussamatov, H. I. and Krat, V. A.: 1969, *Solar Phys.* **9**, 420.
White, O. R. and Wilson, P. R.: 1966, *Astrophys. J.* **146**, 250.

ON THE CIRCULAR POLARIZATION IN ACTIVE REGIONS

E. WIEHR

Universitäts-Sternwarte, Göttingen, Germany

Advantages of 'circular polarimetry' (determination of the Stokes parameter V) in active regions as compared to 'elliptical polarimetry' (determination of all Stokes parameters) are discussed:

(1) Complicated and doubtful reduction procedures by means of line formation theory generally are needed for 'elliptical polarimetry'. This is not the case for 'circular polarimetry'.

(2) Even if the difficulties (1) could be removed, 'elliptical polarimetry' would not help to determine the total magnetic vector outside the disc center. This is due to the fact that the azimuth φ of the linear polarization is only defined for $0° \leqslant \varphi < 180°$. When converting the measured inclination ψ to the line-of-sight into the true inclination γ to the solar surface normal for active regions outside the disc center, the ambiguity of φ always yields an ambiguity of γ. The normal field component and thus the flux can therefore not be determined without making an important assumption on the unknown field structure (see Wiehr, 1970a).

(3) For active regions near the disc center, however, where the difficulty (2) disappears, 'circular polarimetry' is sufficient for the determination of the *normal field component* and *flux*. It furthermore avoids the difficulty (1).

Some useful applications of such 'circular polarimetry' near the disc center are mentioned:

(a) Exact determinations of the magnetic flux for several active regions near the disc center show that the disbalances between the flux in preceding and following regions are within the error limits when using a Zeeman line which remains unchanged going from photosphere to spot (e.g. Fe λ 6302.5). The often reported flux disbalances (Stenflo, 1967) as observed in Fe λ 5250 or Hα, therefore, seem to be mainly due to the differences between the line absorption coefficients in the photosphere (where the calibration is carried out) and the spots or faculae (where the magnetic fields are measured).

(b) Differences between the normal field components as obtained from 'circular polarimetry' in different lines (Severny and Bumba, 1958) seem to be doubtful if the line profile changes mentioned in (a) are not taken into account.

(c) The development of active regions near the disc center can be studied very well by means of 'circular polarimetry'. An example for a young active region is shown. The development of the normal field component generally confirms the observations by Gopasyuk (1967). The development of the flux had to be extrapolated backward in time since the very beginning of this region unfortunately was not observed. (For details see Wiehr, 1970b.)

Further observations of this kind could yield important information on the first

Howard (ed.), Solar Magnetic Fields, 235–236. All Rights Reserved.
Copyright © 1971 by the IAU.

occurrence and the development of active regions as well as possible correlations with flares.

References

Gopasyuk, S. I.: 1967, *Publ. Crim. Astron. Obs.* **36**, 56.
Severny, A. B. and Bumba, V.: 1958, *Observatory* **78**, 33.
Stenflo, J. O.: 1968, in K. O. Kiepenheuer (ed.), 'Structure and Development of Solar Active Regions', *IAU Symp.* **35**, 47.
Wiehr, E.: 1970a, *Solar Phys.* **11**, 399.
Wiehr, E.: 1970b, *Solar Phys.* **15**, 148.

Discussion

Giovanelli: Is there any intrinsic difficulty in calibrating a longitudinal magnetograph by measuring the slope of the line profile at the point on the solar disk where the field is being measured?

Wiehr: In principal such a calibration is possible, but in practice it would require us to measure point by point the polarization and line profile on the solar disk. This would extend the already long observation time by at least a factor two.

APPLICATION OF THE CHROMOSPHERIC MAGNETOGRAPH
TO ACTIVE REGIONS

H. ZIRIN

Hale Observatories, Carnegie Institution of Washington,
California Institute of Technology, Pasadena, Calif., U.S.A.

Abstract. We show how to determine the magnetic field structure in active regions from the Hα morphology. We also show the role of the EFR (emerging flux region) as a bipolar region of velocity downflow. Finally, we point out that since all new magnetic flux emerges in strictly bipolar form, complex spot groups must result from surface interaction, hence most of the solar surface field may be produced on the surface.

1. Introduction

In two previous papers (Veeder and Zirin, 1970; Zirin, 1970) we have shown how chromospheric morphology may be used to deduce magnetic field structure. The bases of this program are the facts that any field strong enough to be of interest should produce tangible effects in the chromosphere, and that I am too lazy to build one more magnetograph. Unfortunately, our understanding of the Sun is not so far advanced that we can predict the relation between field and morphology, so we must find a Rosetta stone in the form of a high quality pair of magnetogram and filtergram. This problem is particularly difficult in active regions, where field patterns are particularly complex.

In the present paper I present new field-morphology correspondence drawn from 'fine scan' magnetograms made at Mt. Wilson and high resolution filtergrams made at Big Bear. From two such sets we draw the following rules, which should be added to those of our preceding papers:

(1) Every magnetic structure on the fine scan magnetogram appears as a chromospheric feature of some sort.

(2) Field transitions are marked by brightness transitions or by dark regions.

(3) The plage-antiplage discrepancy found by Veeder and Zirin is not correct; preceding plage is as bright or brighter than following plage; but *p* spots often have extensive regions of horizontal field surrounding them, which are the 'antiplage'.

(4) Emerging flux regions (EFR) are marked on magnetograms by bipolar magnetic symmetry and velocity downflow.

2. October 25, 1969

Our first exhibit is an Hα-magnetogram pair made October 25, 1969; the magnetogram (Figure 2) was made by Dr. A. Bhatnagar at Mt. Wilson. The magnetogram must be rotated somewhat to conform to the Hα picture, but we see that each fine detail appears on both pictures. I have marked salient regions on Figures 1 and 2. We see

Howard (ed.), Solar Magnetic Fields, 237–242. All Rights Reserved.
Copyright © 1971 by the IAU.

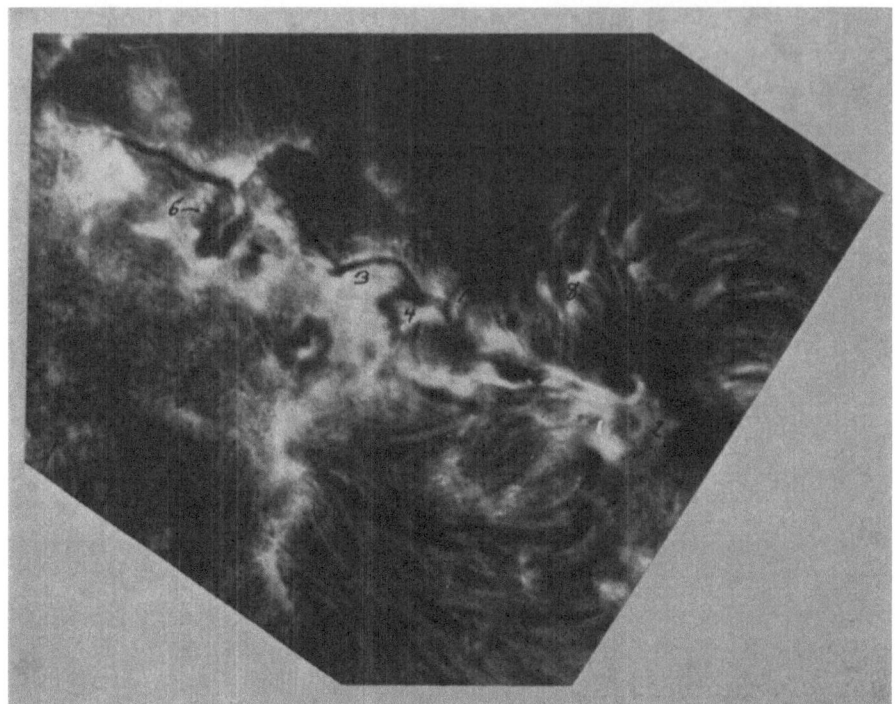

Fig. 1.   Hα picture made October 25, 1969.

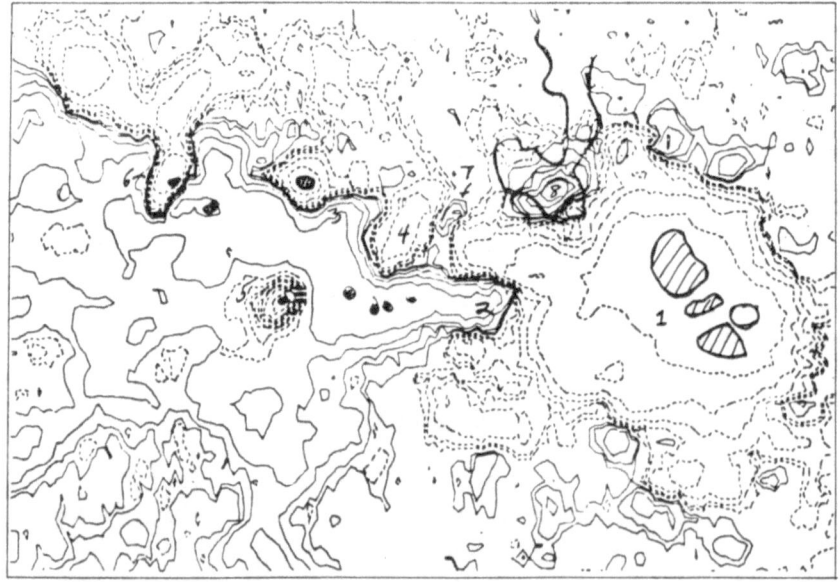

Fig. 2.   Magnetogram made by Dr. A. Bhatnagar at Mt. Wilson Observatory October 25, 1969.

a large, complex group, dominated by preceding spots (1). The boundary to following polarity is not well marked but may be distinguished at (2). From this point to the left we see a ragged division line running NE-SW, dividing the following plage from an antiplage of preceding polarity; the dividing line is clearly marked by the filaments (3). Note the inclusions of preceding polarity (4) and (5). These are seen in white light to have small spots, and are clearly defined as polarity changes by the dark radial Hα structure surrounding their bright centers. The inclusion (6) is bipolar and inverted; a small following spot leads the preceding spot in (6); considerable flare activity occurred in this region.

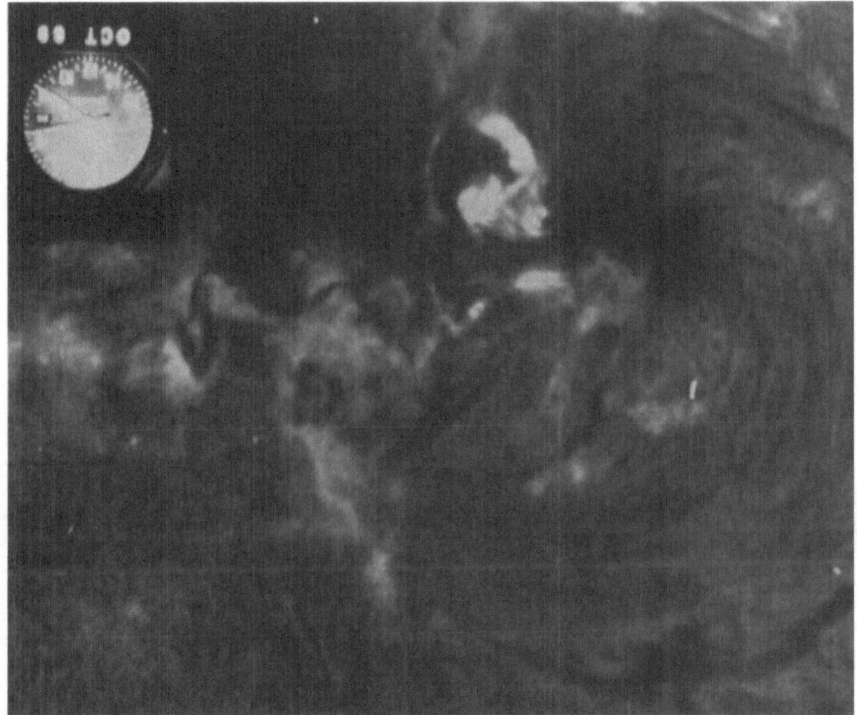

Fig. 3.   Picture of following polarity.

Regions 7 and 8 are included areas of following polarity; 7 is particularly well marked by a dark boundary. A sizeable flare occurred here on the 24th; in the picture (Figure 3) we see how closely it sticks to the region of following polarity. In summary, we see that the Hα morphology in this group corresponds strictly to the magnetic field, and given adequate optical resolution, one can normally draw a magnetogram from the Hα pictures.

3. April 10–11, 1970

The second comparison set was made April 10 and 11, 1970; fine scan magnetograms

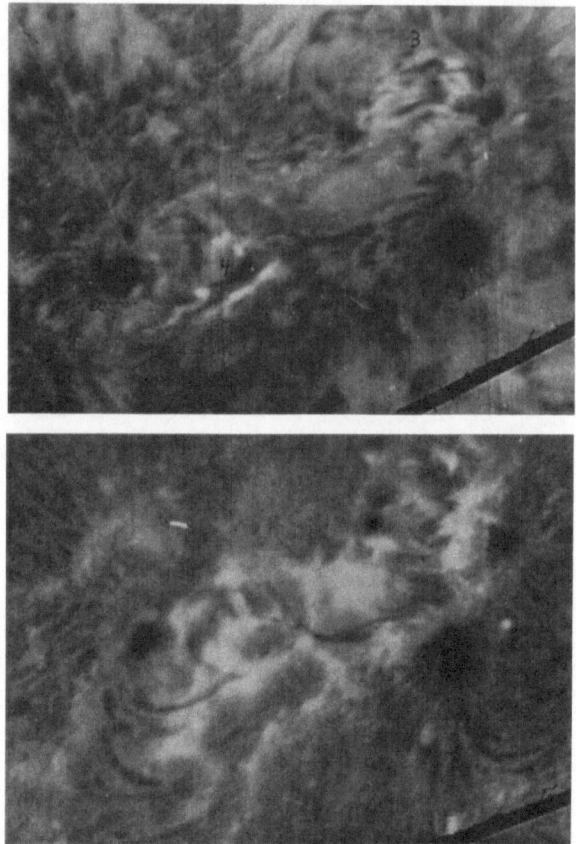

Fig. 4.   Big Bear off-band pictures showing emerging flux region. Top: April 10, 1970.
Bottom: April 11, 1970.

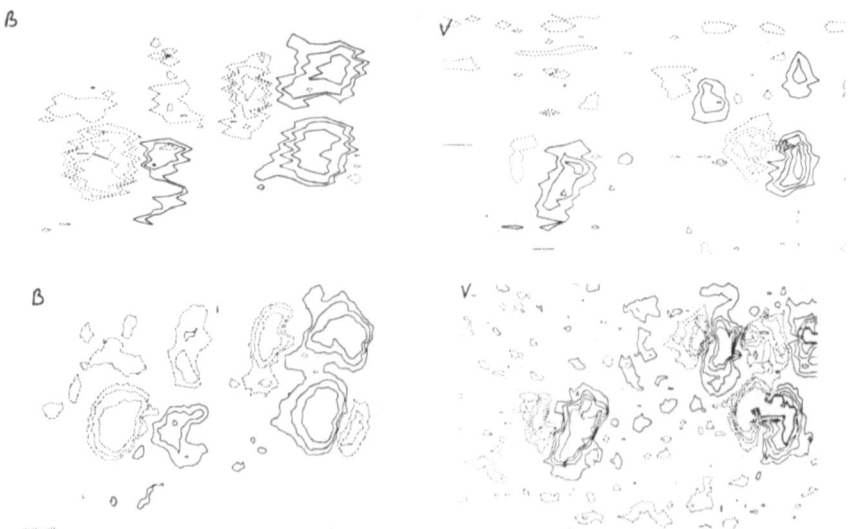

Fig. 5.   Magnetograms showing emerging flux region.

were obtained by John Adkins. The resolution of the magnetograms was only 5×25 s on the 10th, but was 5 arc sec square on the 11th.

We see in Figure 4a mature bipolar spot group (McMath plage 10669) with large preceding (1) and following (2) spots. The field distribution varies from that common in bipolar spots because this is an older group, several rotations old, which is primarily divided by the filament connecting the two spots into a following plage above and preceding polarity below. At 1622 on April 10 there was a large flare in two parts of the preceding region, the first part occurring behind the preceding spot, followed by brightening in the area (4) just ahead of the following spot. Figure 4 shows Big Bear off-band pictures; we also have smaller scale centerline Hα pictures from Pasadena. Unfortunately, the sensitivity of magnetogram (Figure 5) is low so that it does not show the weaker plage fields detectable in Hα; however, since most people are still more willing to believe the Babcock magnetogram, no matter how poor, before the chromospheric magnetogram, we make the instructive comparisons.

The most interesting feature in Figures 4 and 5 is the presence of an emerging flux region (EFR) directly N of the spot group.

These have been referred to as AFS-arched filament systems by Bruzek, (1969) and as BRL-bright regions with loops by Weart and Zirin (1968), but it is now well established that they are emerging flux regions, and I shall use that term. The EFR are typically bipolar, and, as was established by Roberts (1969), they show downflow at either end. In Figure 4b we see that the EFR matured the next day. We also see on this more sensitive velocity picture that there is an upflow in the center of the region. We also see a new region of following polarity (5) emerging ahead of the large spot. This never amounted to much.

I should like to point out an important feature of the EFR's. Flux invariably emerges in bipolar form, yet important spot groups such as that in Figure 1, almost never show this bipolar form. Very often we find an inverted polarity resulting from rapid growth of a bipolar EFR ahead of a preceding spot – the preceding part of the EFR fades and we are left with a strong polarity inversion. More often, we find an EW dividing line following big preceding spots, as in Figure 1. It would appear that the form of sunspot groups, as well as most of their flux, is the result of surface interactions which amplify and distort the new fields into large and complex spots of different orientations. It is perhaps significant that normal bipolar spot groups, which might be termed 'unevolved' are small and show little activity.

In summary, we have shown that further progress in the chromospheric magneto-graph now permits the measurement of active region fields with reasonable accuracy. All field changes correspond to visible features in the chromosphere. The interpretation of these features is somewhat harder, but it seems clear that we see dark regions in Hα wherever the field is horizontal.

Acknowledgements

I am indebted to A. Bhatnagar and J. Adkins for magnetograms, and to Peter Foukal for numerous discussions. This work was supported by NASA and by the Atmospheric Sciences section of the NSF.

References

Bruzek, A.: 1967, in K. O. Kiepenheuer (ed.), 'Structure and Development of Solar Active Regions', *IAU Symp.* **35**, 293.
Roberts, P. R.: 1969, Thesis, California Institute of Technology.
Veeder, G. and Zirin, H.: 1970, *Solar Phys.* **12**, 391.
Weart. S. and Zirin, H.: 1969, *Publ. Astron. Soc. Pacific*, **81**, 480.
Zirin, H.: 1970, Paper presented to Royal Society Special Discussion on Solar Studies.

EVOLUTION OF THE MAGNETIC FIELD
CONFIGURATION IN AN ACTIVE REGION

DENNIS L. SCHATZ*

CSIRO National Standards Laboratory, Division of Physics, Sydney, Australia

Abstract. Described and discussed is the evolution of the magnetic field configuration in an Active Region from observations made with high time resolution.

1. Introduction

At the Culgoora optical observatory a solar magnetograph of unique design using three fabry-perot interferometers in series has been developed (Giovanelli, 1967; Ramsay, 1967). The design of the instrument is such that an area of the Sun approximately 8 min in diameter can be observed through a tunable filter with a bandpass of $\frac{1}{20}$ Å. In conjunction with a polarizing beam-splitter and $\frac{1}{4}$ wave plate, simultaneous observations can be made in both circular polarized components. Using a technique similar in principle to Leighton's method of photographic subtraction (Leighton, 1959), a representation of the magnetic field can be obtained. Consequently, the Culgoora magnetograph offers a fast method of obtaining the magnetic configuration over a large area of the Sun.

2. Observations

In order to take advantage of the high time resolution allowed by this instrument, observations were made to obtain the configuration of the magnetic field in an Active Region at short time intervals and to see how the configuration evolved. The observations discussed in this paper were made on 9 and 10 February 1970. On both days observations were started at approximately 2300 UT (observations actually began 8 and 9 February UT) and ended at approximately 0700 UT. Exposures were taken approximately every two minutes with several large time gaps interspersed throughout the day. Figure 1 shows the magnetic field configuration of the active region studied near the time that the observations commenced on each day. On 9 February the sunspot was located at N19 E25 at 2200 UT (February 8) and on 10 February it was located at N19 E12 at 2200 UT (February 9).

All observations were made in the light of Ca I 6102.8 Å, and the magnetic configuration was observed to be independent of the filter position in the blue wing of the line profile over a wavelength range which varied from the wavelength corresponding to a position on the line profile $\frac{1}{8}$ of the line depth from the line center to the wavelength corresponding to $\frac{5}{8}$ of the line depth from line center. The observations were made near the center of this range.

Each exposure simultaneously produces two images in opposite circular polar-

---

* Now at the Space Sciences Laboratory, University of California, Berkeley, California.

Howard (ed.), Solar Magnetic Fields, 243–248. All Rights Reserved.
Copyright © 1971 by the IAU.

ized light on 35 mm Kodak Kodalith film. These two images must then be photographically subtracted to give the desired magnetogram. The basic method used is
the same one used by Leighton (1959). However only the first subtraction is carried
out rather than the three made by Leighton. This is quite sufficient for our purpose
since we are only concerned with the qualitative aspects of the field configuration.

As this observing program produced a large number of photographs to be subtracted,
it was necessary to develop a means of subtracting photographs on a mass production

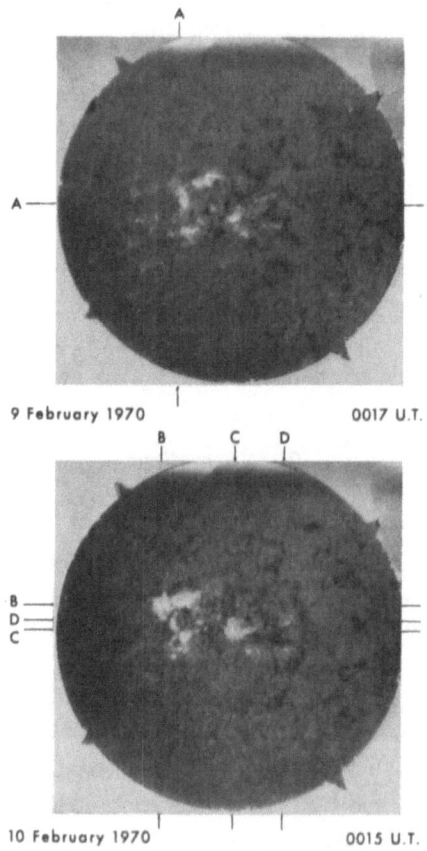

Fig. 1.   Magnetic field configuration of active region near the beginning of observing period.

basis. This was done with the help of Mr. H. Gillett in the Division's photographic
laboratory and a commercial film company (Supreme Films Pty. Ltd.) using their
optical printer. The optical printer is a pin registered system which can accurately
project the image of the negative of one of the two images in a pair on the positive of
the second image. A third film is then exposed in optical contact with the second film
giving the desired magnetogram. The final film is then developed to $\gamma = 1.0$ with an accuracy that guarantees the γ along the length of the entire film to vary by only ± 0.05.

By reregistering the magnetograms to eliminate the movement of the active region due to solar rotation, a cine film was produced in order to study the evolution of the magnetic field configuration. This was then projected and could be studied one frame at a time or accelerated to a rate of motion showing the evolution at 2500 times the original speed.

3. Results

A study of the Ca I magnetograms reveals a number of structures that should be noted. There are fields which are observed to be associated with sunspots. These can be seen at points A, B, and C in Figure 1. Likewise there are fields which are not associated with sunspots, but are associated with the white light faculae (Chapman and Sheeley, 1968; Giovanelli, 1970). This association is further supported by the fact that the conglomerate of magnetic granules away from the sunspots which are observed in the magnetogram have been shown to have approximately the same size and lifetime as the white light facular granules (Schatz, 1970). We also see one sunspot whose magnetic field appears to be insensitive to the magnetograph (point D, Figure 1). However, at the same time magnetic structures, which appear to be associated with facular fields near the spot, encroach in upon the spot and appear to cover a portion of the penumbra. Further observations with the magnetograph are necessary before the nature of this overlying material can be understood.

If we now look at the motions which are observed when we examine the cine film for each day, we again encounter a number of different phenomena that should be noted. These motions can best be divided into five types:

(1) Transverse movement of field on large scale (movement $> 30''$).

(2) Intrusion of one polarity on the opposite polarity.

(3) Disappearance of one polarity with accompanied appearance of opposite polarity.

(4) Disappearance or appearance of just one polarity.

(5) Small scale movement of individual magnetic granules or conglomerate of magnetic granules.

In Figure 2, we can easily see the transverse movement of field on a large scale (pts. A, B). This is clearly associated with expansion of sunspots underlying these points. It is interesting to note that as these spots expand they tend to clear a path through the surrounding magnetic field and push the dark polarity features to one side. This apparently causes a greater concentration of dark polarity field to be built up in between the two areas of expansion (pt. C, Figure 2). Likewise, we see the same type of large scale expansion at point A in Figure 3. Again, we see the same phenomena where the darker polarity features are swept aside. As before, this motion can be associated with the expansion of sunspots.

The intrusion of one polarity upon an area of opposite polarity is most clearly seen at pt. D in Figure 2. At the beginning of the day, there is a large dark polarity object bordered by two much smaller bright polarity objects. As the day progresses, the most western bright polarity object begins to intrude into the dark polarity object first, and then later the more eastern bright polarity object begins to intrude. This

DENNIS L.SCHATZ

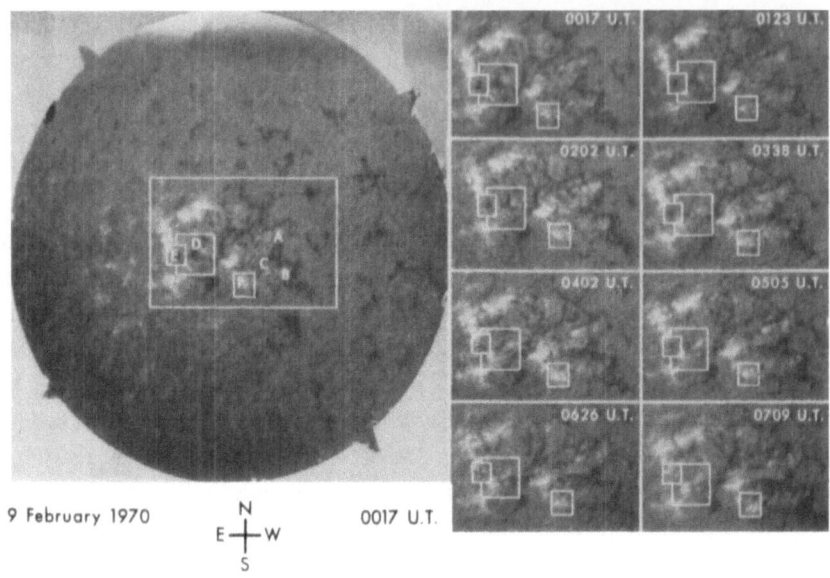

9 February 1970   N
                    E ─┼─ W   0017 U.T.
                       S

Fig. 2.   Evolution of magnetic field configuration on 9 February 1970.

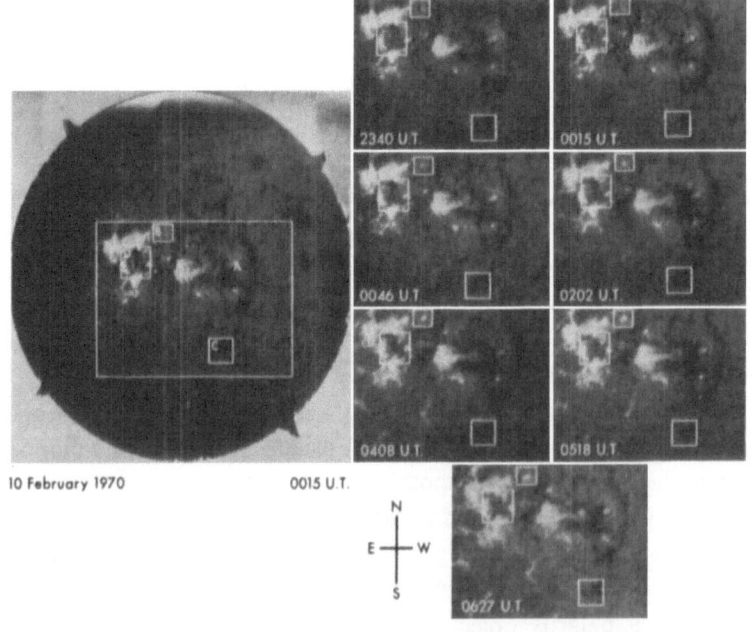

10 February 1970   0015 U.T.
                        N
                     E ─┼─ W
                        S

Fig. 3.   Evolution of magnetic field configuration on 10 February 1970.

continues until the end of the day when the bright fields have apparently connected and the dark field has almost completely disintegrated.

This phenomenon is very similar to the next type of evolution observed, the disappearance of one polarity with the simultaneous appearance of the opposite polarity nearby. This is seen at pt. E in Figure 2. The dark polarity object is very dark at the beginning of the observations, but slowly declines in intensity throughout the day. By 0625 UT, the dark field has almost disappeared and a bright object has developed in its place or just adjacent to it. By 0710 UT, the dark object has completely disappeared and only the bright object is left.

It is difficult to determine from the white light observations whether these changes are accompanied by development of small sunspots or whether they are associated with the facular regions. Observations of the magnetic fields near the limb with periodic observations made by tuning the magnetograph into the continuum would be necessary to resolve this question.

At the same time, there are occasions where the appearance or disappearance of fields is not accompanied by a clear cut change in the opposite polarity nearby. In Figure 2 (pt. F) we see the slow emergence of a bright polarity object as the day progresses without any change in a dark polarity which could be clearly associated with its emergence. In Figure 3 (pt. B) we see the development of a bright feature which appears during one of the gaps in our observations, but also has no clear opposite polarity evolution associated with it.

In addition to the large scale transverse movement stated earlier, there are clear examples of movement on a smaller scale. At pt. C in Figure 3 we can see three isolated granules that come together during the day and coalesce into two granules, and then finally into one large feature. At the same time these features as a unit have been moving closer to the larger conglomerate of magnetic field lying to the west of the three features. At pt. D, Figure 3, we see a dark polarity feature developing along a line going from N to S. When this developing feature is confronted by an object of opposite polarity, it appears to spread out along the northern perimeter of the object. This development is different from the feature at pt. F (Fig. 2) since it appears to be due to the actual emergence of magnetic field rather than just transverse movement.

4. Discussion

Sheeley (1969) in his study of the photospheric network noted the disappearance of magnetic field was a 'mild process' at the level of the photosphere at which he was observing. This is confirmed in this study since at no time is there dramatic change in the field configuration. At the same time he speculates that if the disappearance of magnetic field was accomplished by field annihilation then we should see the simultaneous disappearance of fields of opposite polarity. He found no clear examples of this in his observations, and in agreement with this (although this is only an introductory study) we found no clear observations of the simultaneous disappearance of opposite polarity fields.

On the contrary, there seems to be a number of occasions where there is the appearance or increase of one polarity and the disappearance or decrease of the opposite polarity adjacent to it or in place of it. (See Figure 3, pt. D and E.) It is interesting to note that this is the type of evolution observed by Ribes (1969) for EMF regions near the location of flares.

The description of the evolution seen in these high time resolution observations is a beginning to the possibilities that are open with observations of this sort. It would be very useful to compare, on a comparable time scale, the changes which occur in the chromosphere in Hα with the changes in the magnetic field configuration. The changes which occur during flare activity, surges or active filaments can now be more easily studied. In addition, the velocity of features in the chromosphere can be compared to motions of the magnetic field.

References

Chapman, G. A. and Sheeley, N. R.: 1969, *Solar Phys.* **9**, 347.
Giovanelli, R. G.: 1967, *Proc. Astron. Soc. Australia* **1**, 39.
Giovanelli, R. G.: 1970, *Proc. Astron. Soc. Australia* **1**, 363.
Leighton, R. B.: 1959, *Astrophys. J.* **130**, 366.
Ramsay, J. V.: 1967, *Proc. Astron. Soc. Australia* **1**, 66.
Ribes, E.: 1969, *Astron. Astrophys.* **2**, 316.
Schatz, D.: 1970, *Proc. Astron. Soc. Australia* **1**, 371.
Sheeley, N. R.: 1968, *Solar Phys.* **5**, 442.

LINE PROFILES IN SUNSPOT UMBRAE AND PENUMBRAE
BY ATOMIC BEAM SPECTROSCOPY

F. RODDIER

Faculté des Sciences de Nice, Laboratoire d'Astrophysique, Nice, France

We present profiles of the Sr resonance line obtained with an atomic beam spectrograph and the Kitt Peak main heliostat.

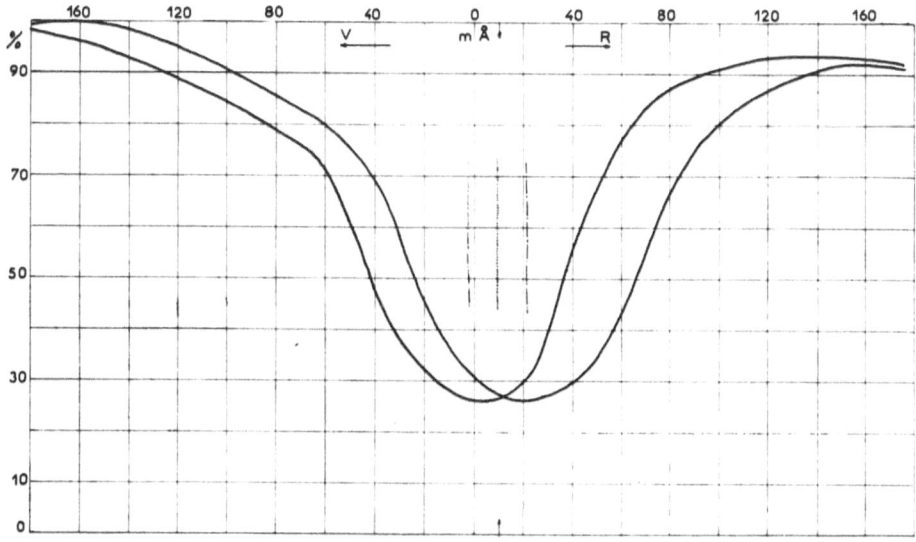

Fig. 1.

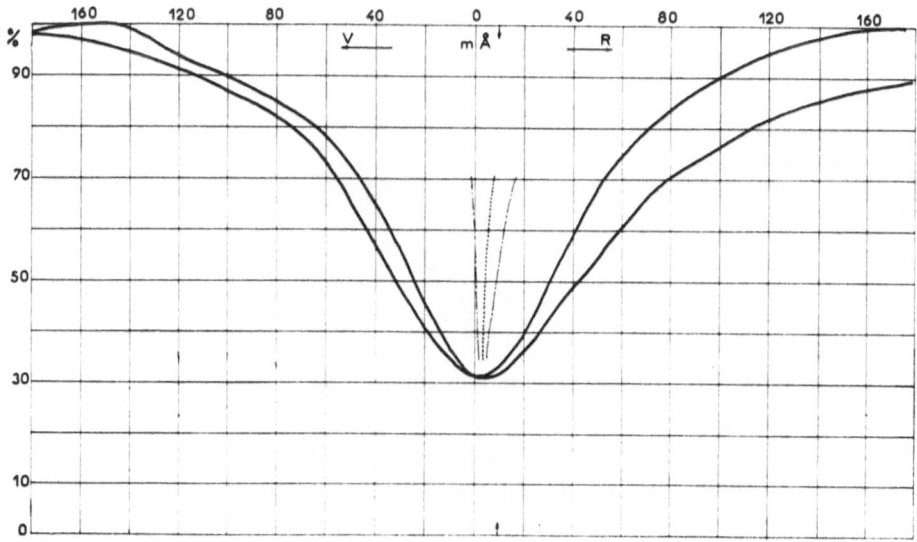

Fig. 2.

Howard (ed.), Solar Magnetic Fields, 249–251. *All Rights Reserved.*

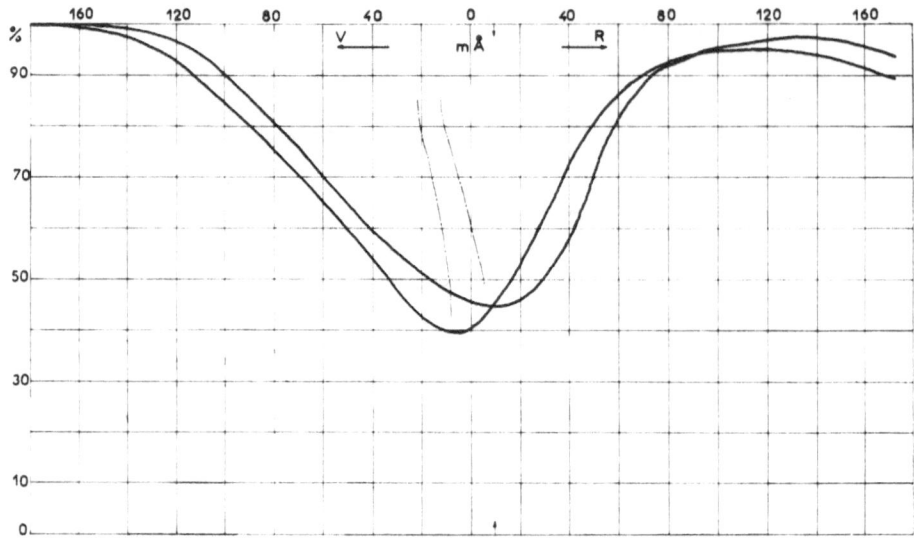

Fig. 3.

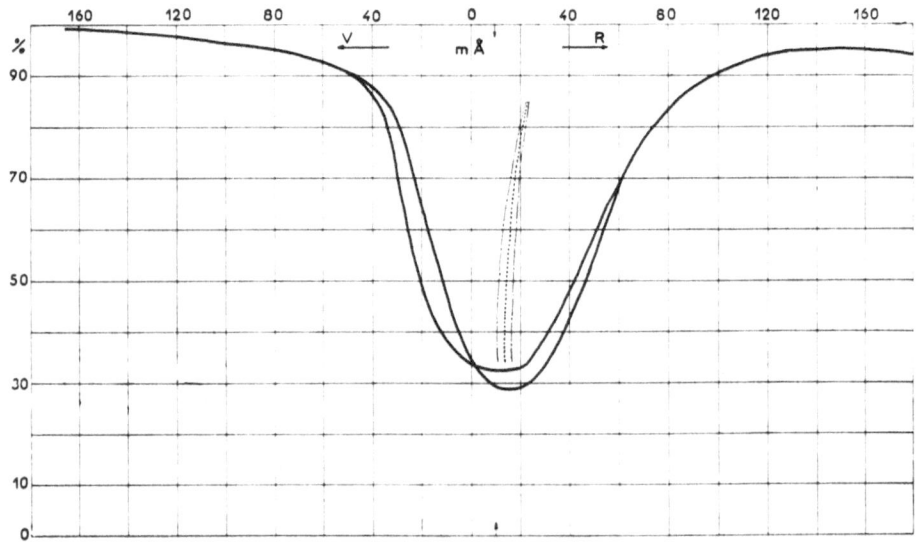

Fig. 4.

Figure 1 shows two profiles of opposite circular polarization in a sunspot umbra. The axis is vertical and split symmetrically with respect to the theoretical wavelength predicted by the Einstein theory of gravitation (arrow). This seems to show the absence of convective motion in the umbra. The splitting corresponds to a magnetic field of a little more than 1000 G.

Figure 2 shows the same profiles in a different sunspot umbra. We see a resulting violet shift decreasing toward the continuum while the splitting increases. This may show both magnetic field and convection with opposite vertical gradients, the magnetic field increasing with depth while the convection decreases.

Figure 3 shows these profiles in a sunspot penumbra between the umbra and the center of the solar disk. Both shift and assymmetry are due to the Evershed effect. Velocities are now increasing with depth. There is a systematic difference between the widths and central intensities of the two profiles which can be explained by a negative correlation between velocities and magnetic fields in the penumbral structures.

Figure 4 shows the same profiles in the penumbra between the umbra and the solar limb. We can see the same effects reversed.

Discussion

Wiehr: I want to mention that the increase of Evershed asymmetry towards the continuum does not come from an increase of the velocity with optical depth. Schröter showed in 1965 that the asymmetry comes from a satellite which is superimposed on the main line.

ON THE FINE STRUCTURE OF THE MAGNETIC FIELD
IN THE UNDISTURBED PHOTOSPHERE

V. M. GRIGORYEV and G. V. KUKLIN

Siberian Institute of Terrestrial Magnetism,
Ionosphere and Radio Propagation, Irkutsk, U.S.S.R.

1. Some authors have recently proposed a set of fine structure models (Severny, 1968; Stenflo, 1966; Grigoryev, 1969) to explain the reaction of the solar magnetograph to the structure of the magnetic fields on the Sun. We tried to make a more sophisticated model in order to obtain more realistic results. Let the magnetic fields observed in the undisturbed photosphere be described as a totality of basic elements with equal area and with different magnetic field strengths. They cover the solar surface independently and closely. Let us consider a case of three kinds of elements. The first kind of elements has a magnetic field strength H_1 and an occurrence ξ, the second kind of elements has a magnetic field strength H_2 and the occurrence η and the third kind of elements without magnetic field has the occurrence $1 - \xi - \eta$. If an ideal magnetograph has an entrance slit of area equal to the area of the η elements, then the recorded magnetic field is a mean field of these η elements and its values have a trinomial distribution. We assume that over large areas a balance of opposite polarity fluxes exists

$$\xi H_1 + \eta H_2 = 0.$$

Further let us assume that deviations from this distribution, atmospheric scintillations, instrumental optical errors, electronic noise, etc. as a whole lead to distortions in the distribution which have a distribution close to a Gaussian law. Then the distribution of observed values of the magnetic field must be a composite of the normal and of the trinomial distributions. The parameters of the proposed model are connected with the moments of the observed distribution. The system of equations of the 9th order may be solved.

Using scans in an undisturbed region of the photosphere made with different resolutions we obtained a large number of observed distributions. The determination of the model parameters leads to the best agreement of different scan treatments if the magnetic field in the basic elements is from tens up to hundreds of gauss, the size of the elements is close to 2″, and the occurrence of them does not exceed some percent. In rare cases the opposite polarity magnetic fields may differ up to 10 times. The detailed results of this study will be published soon.

2. Statistical analysis of weak magnetic field using the autocorrelation function method was done first by Howard (1962) and later this method was used to analyze the fine structure of weak magnetic fields by Vasiljeva (1964), Severny (1968), Beckers and

Howard (ed.), Solar Magnetic Fields, 252–259. All Rights Reserved.
Copyright © 1971 by the IAU.

Schröter (1968), and Livingston (1968). In these papers autocorrelation functions were used mainly to determine the characteristic scale of the magnetic field fine structure. We intend to study periodic structure of the magnetic field spatial distribution in the undisturbed photosphere in high latitude regions of the Sun.

Conformity of the calcium network and supergranular cells on the one hand and a correlation between magnetic field and brightness in K Ca II on the other hand permit us to make the indirect conclusion that the strongest magnetic fields are concentrated on the borders of supergranular cells (Simon and Leighton, 1964). Direct evidence of the existence of the magnetic field regular structures with a scale corresponding to the supergranular cells may be obtained analyzing records of the magnetic field fluctuations in undisturbed regions with the help of the autocorrelation function method.

To detect the periodic component in a signal $H(t)$ one may use the autocorrelation function

$$B(\tau) = \langle H(t) H(t + \tau) \rangle_t.$$

If we assume that the magnetic field is concentrated on the borders of the supergranulation cells and the magnetic field polarity is the same one at the opposite sides of cells then $B(\tau)$ will be similar to a curve in Figure 1 (a). The distance between maxima must be of the order of the regular component period C, and the width of the regular component peaks must influence the value of the correlation interval $\tau_{0.5}$

$$B(\tau_{0.5}) = \tfrac{1}{2} B(0).$$

But if the magnetic field polarity at the opposite sides of the cells is opposite, then the form of $H(t)$ may be of such a kind as is shown in Figure 1(b). The period of the regular component is two times more than in the first case. Unlike the first case $B(\tau)$ must have a minimum at $\tau = c, 3c, 5c, \dots$. If we used absolute values of $H(t)$, then the autocorrelation function would be of the same kind as in the first case, that means the maxima located at $\tau = c, 2c, 3c, \dots$ but the width of the maxima would be narrower.

In the general case the function $H(t)$ really may be a superposition of both kinds of curves considered, and therefore it is useful to analyze separate possibilities for $H(t)$ computing autocorrelation functions of $H(t)$ and of $|H(t)|$. Comparing them one may obtain more exact information about the magnetic field spatial structure.

We used records of magnetic field in high-latitude regions of the Sun with different resolutions. The scan length was of the order 280″–290″. About 150 autocorrelation functions for the northern hemisphere and 80 for the southern hemisphere were computed. The atmospheric oscillations, the finite aperture and the time constant of the magnetograph were not taken into consideration.

The correlation interval may be used as a size of field structural elements. In Table I the average values of the correlation intervals $\tau_{0.5}$ are presented for both the hemispheres and for the different resolutions.

At the resolution $1''.8 \times 4''.2$ the average value of $\tau_{0.5}$ is equal to $3''.3$ or 2300 km. A smaller value was obtained by Howard (1962) with the resolution $2'' \times 2''.5$. A similar

TABLE I

Resolution	N-hemisphere		S-hemisphere	
	ACF of $H(t)$	ACF of $|H(t)|$	ACF of $H(t)$	ACF of $|H(t)|$
4″	3″.4	2″.0	3″.2	2″.3
8″	5.0	3.9	4.9	–
17″	5.3	3.4	6.2	–
32″	–	–	5.3	–

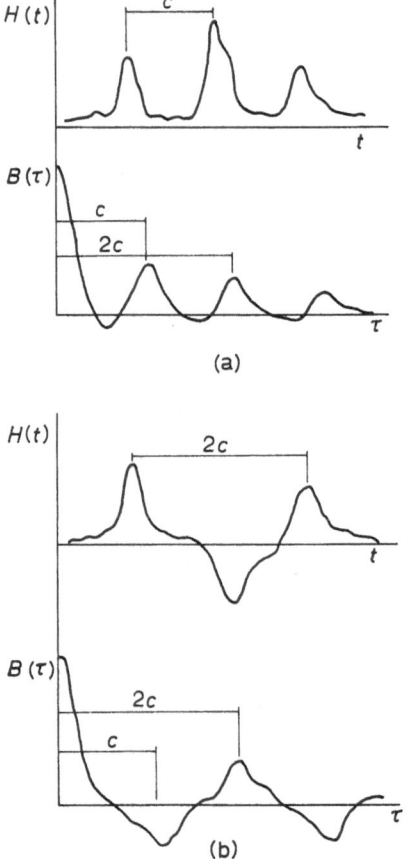

(a)

(b)

Fig. 1. (a) The form of the signal $H(t)$ with the magnetic fields of the same signs at the edges of a supergranular cell and its auto-correlation function; (b) the form of the signal $H(t)$ with magnetic fields of opposite signs at the edges of a supergranular cell and its auto-correlation function.

result was obtained by Vasiljeva (1964) with the resolution 2″.4 × 6″. A sufficiently greater correlation interval 7″.7 was obtained by Severny (1968) with the resolution 2″.5 × 9″. The data of Table I show that the mentioned discrepancies may be caused by different resolutions. However such deviations in the structural element sizes may be real and depend on the time and element location.

The other peculiarity of the autocorrelation functions is the presence of secondary maxima and minima within ranges 16−27″ and 34−47″. Looking through all the autocorrelation functions one may find that minima occur in the same range as maxima. In Table II the summary of maxima and minima locations is given. The secondary maxima near 40″ occur 2–3 times more often than the minima. Usually in scans the

TABLE II

Total number of ACF	131
Number of ACF having maximum in the range of 34 − 47″	83
Number of ACF having minimum in the range of 34 − 47″	34
Number of ACF having maximum in the range of 16 − 23″	30
Number of ACF having minimum in the range of 16 − 23″	62

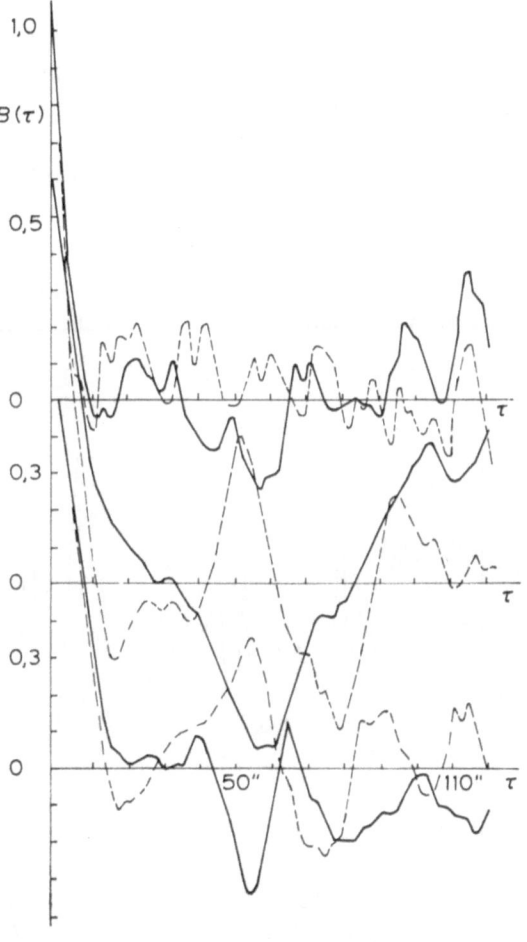

Fig. 2.   Examples of ACF of some scans. The solid line presents ACF of the original signal $H(t)$ and the dashed line presents ACF of $|H(t)|$.

minimum near 40″ in the case of $H(t)$ corresponds to the maximum in the case of $|H(t)|$ (Figure 2). This means that sometimes a regular structure of the magnetic field exists which corresponds to the supergranulation cell size and has strong magnetic fields of opposite polarity on the borders of cells.

It is useful to note that this fact may explain why the autocorrelation function computed by averaging data of different scans has no evident extrema close to 40″. Moreover the differences of autocorrelation functions of neighboring scans reflect the spatial distribution of such regular structures. It is interesting that the sharpest maxima near 40″ belong to scans located at distances of about 40″, and the autocorrelation functions of more close scans may be extremely different even without extrema.

The next peculiarity of computed autocorrelation functions is the increased fine structure in them when the resolution is better. It is evidence of the other characteristic size present in the spatial structure of the magnetic fields.

3. In order to study the fine structure of the magnetic field we tried to estimate the power spectra. Using the original scan records we computed periodograms for each scan and further averaged them. One may make an approximate estimation of the power spectra without use of the autocorrelation functions but determined within wider range of wave numbers $K = 2\pi/\lambda$.

The true power spectrum of the magnetic field spatial distribution is distorted by the atmosphere turbulence, by the instrument, and after adding the instrumental noise by the time constant. We tried to exclude these effects. After the reduction for the time constant the periodogram became as is shown in Figure 3. We assumed the instrumental noise to be of the Poisson type – an additive one. Then the smooth curve

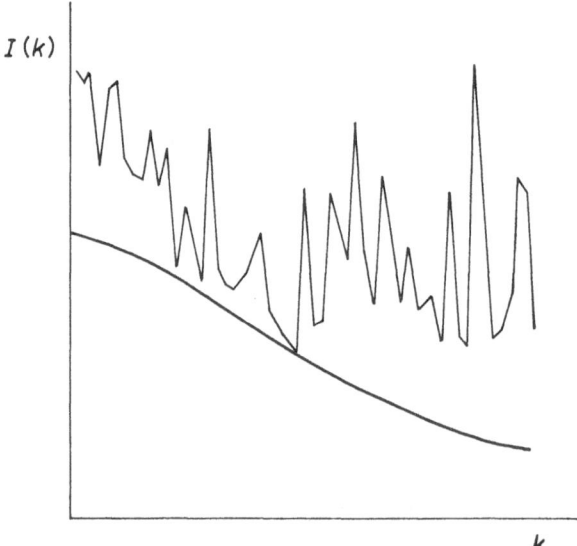

Fig. 3.   The mean power spectrum corrected for the time constant. The solid line corresponds to the estimation of the noise power spectrum.

below the periodogram fitting only the deepest minima may be adopted as an estimation of the noise power spectrum. Unfortunately it is difficult to determine exactly the
instrumental noise during the scans because the separate records of the instrumental
noise give power spectra which agree poorly with the presented periodogram. This is
the weakest point of our investigation.

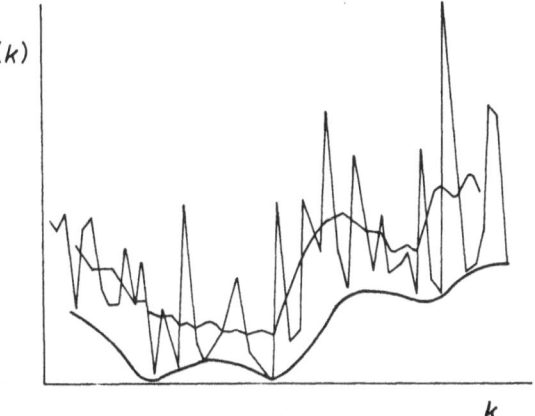

Fig. 4.   The mean power spectrum corrected for the time constant without noise. The solid curve
corresponds to a 'background' spectrum, the thin curve corresponds to a greatly smoothed spectrum.

Subtracting the noise power spectrum from the periodogram we obtain an estimation of the structure power spectrum (Figure 4). It is interesting that this spectrum is
a superposition of a sharp spectral line system and of a smoothed spectrum. One may
detect among the lines a set of equidistant maxima. Let the lower smooth curve be the
estimation of the smoothed 'quasi-continuum'. If we determine this quasi-continuum
as a background spectrum then the corresponding amplitudes of the equidistant peaks
form a set of Fourier coefficients of a certain regular structure. So we can synthesize
the estimation of the autocorrelation function of this regular structure. Using various
weighting procedures (for example Gibbs' factors etc.) we obtain similar curves
(Figures 5). As the main period of such structure is close to 50″, we presume to identify

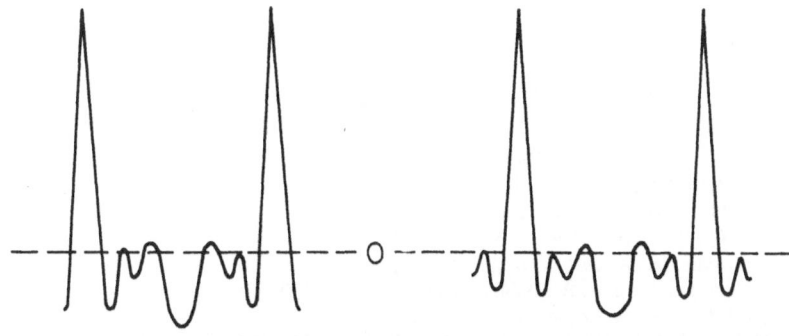

Fig. 5.   Possible ACF of the internal magnetic field distribution in the supergranular cell.

this structure with the fine structure of the magnetic field inside the supergranular cell.

One may see a strong concentration of the magnetic fields on the periphery of the cell and the presence of a magnetic field of opposite polarity in the center of the cell. Remembering the data given in Table II we may say that the higher frequency of the autocorrelation minima near 20″ is connected with this structure. It is obvious that the averaging of periodograms reduced the amplitudes of extrema (both the central minima and the periphery maxima) from 2.4 to 3 times, but the ratio of these amplitudes is only slightly influenced.

Considering the background spectrum one may see two distinct maxima within the large values of K. If we smooth very much the spectrum (a thin curve in Figure 4) then its form confirms also the presence of these maxima. These maxima may be caused by groups of basic elements of the magnetic field spatial structure and correspond to the values of λ 4 and 3 times larger than the basic element size. Therefore the basic element size is equal to 1″8 or 1300 km. This estimation is in accordance with the results by Livingston (1968).

4. Discussing the results above we may make the following conclusions. As the magnetic fields at the supergranulation cell periphery occupy an area which is about some tens of per cent of the whole cell area, so in this part of the cell according to the average occurrence of the basic elements obtained by us, the density of those elements is approximately ten times higher than in other parts of the supergranular cell. This ring consists of closely packed elements.

The most optimistic estimation of fluxes leads us to a flux of the ring 14 times larger than the flux of the central part with approximately the same density of the elements. This means that about 93% of the supergranular cell flux is linked with the magnetic fields of other cells. Such great interdependence of the supergranulation magnetic fields is remarkable. Therefore the large-scale magnetic fields are not only a result of the chance cluster of the supergranular cells but on the contrary the system of these cells is formed and controlled by the large-scale magnetic field which is a more fundamental formation. Probably this is the reason why the interplanetary magnetic fields correlate better with the background magnetic fields, because they form together the fundamental slow varying large scale supersystem of magnetic fields.

References

Beckers, J. M. and Schröter, E. H.: 1968, *Solar Phys.* **4**, 303.
Grigoryev, V. M.: 1969, *Soln. Dann. (U.S.S.R.)* No. **6**, 77.
Howard, R.: 1962, *Astrophys. J.* **136**, 211.
Livingston. W. C.: 1968, *Astrophys. J.* **153**, 929.
Severny, A. B.: 1968, *Izv. Krymsk. Astrofiz. Obs.* **38**, 3.
Simon, G. W. and Leighton, R. B.: 1964, *Astrophys. J.* **140**, 1120.
Stenflo, J. O.: 1966, *Arkiv Astron.* **4**, 2.
Vasiljeva, G. Ya.: 1964, *Izv. Glavnoj Astron. Obs. (Pulkovo)*, No. **175**, 28.

Discussion

Lamb: I would like to ask two questions concerning the magnetic field autocorrelation function which you obtained. First, for what time scales do the features in the spectrum which you have described persist? Second, how have you defined the so-called 'background' autocorrelation function used in your analysis?

Kuklin: In the case of the smallest height of the entrance slit equal to 4".2 the scanning of the whole region (more than 50 scans) requires 60–80 min, with a scanning speed equal to 2" per s.

We may consider this point as one of the weak points of our study because it was made in some sense arbitrarily. We select the 'background' power spectrum as a bottom of the linear spectrum minima (a curve tangent to those minima).

SUPERGRANULATION AT THE CENTER OF THE DISK

EDWARD N. FRAZIER

The Aerospace Corporation, El Segundo, Calif. U.S.A.

Abstract. The Kitt Peak multi-channel magnetograph was used to make raster scans of the super-granulation at the center of the disk. The scan pattern was arranged to largely cancel out the effects of the 5 min oscillations and the granulation and to enhance visible effects of the supergranulation.

The results show vertical supergranular motions. These motions consist of relatively isolated 'patches' of downward flowing material, or 'downdrafts', with an average speed of 0.1 km/s. These downdrafts coincide with patches of magnetic field, and the speed of the downdraft is linearly proportional to the magnetic field strength (50 to 100 G). These downdrafts are also regions of increased brightness in the chromosphere and photosphere.

1. The Observations

The multi-channel magnetograph at the Kitt Peak Solar telescope was used to observe the magnetic, velocity and brightness properties of the supergranulation at the center of the disk. This has been a difficult observation not only because the magnetic, doppler and brightness signals are very weak but also because the doppler and brightness signals are masked by the stronger effects of the granulation and the 5 min oscillations. The limited success of previous attempts to view the supergranulation vertically have left a gap in our observational knowledge of this phenomenon.

Sufficient sensitivity can be achieved by using a Babcock type magnetograph. The unwanted effects of the granulation and the oscillations can be effectively filtered out on the basis of their size and their time behavior by modifying the scanning procedure as follows:

(1) A 2.4 arc s entrance aperture was used, which 'smeared out' much of the granulation.

(2) Every point in the raster was observed twice, with a time interval of 150 s. The two values were later averaged, cancelling out the effects of the 5 min oscillations.

(3) The area was scanned repeatedly for four hours and the results again averaged, further reducing all short-lived effects and enhancing the long lived effects.

Since the data is in digital form, it can be fully processed by a computer and displayed in the form of digital 'photographs'. Figure 1 shows the photospheric magnetic field. The field is clearly broken up into isolated knots and attains surprisingly high field strengths (50 to 100 G). Figure 2 shows the Ca II K chromospheric network. This quiet region shows the same very high correlation between calcium emission and magnetic field that the active regions do. Figure 3 is the photospheric brightness network (as seen in the core of the Fe λ 5250.2 line). The photospheric network is sharper and even more highly correlated with magnetic field. The vertical velocities can be seen on Figure 4. Quite prominent are the downdrafts which coincide with the magnetic regions. With the possible exception of the Calcium emission there is almost no visible indication of a cellular structure.

Howard (ed.), Solar Magnetic Fields, 260–267. All Rights Reserved.
Copyright © 1971 by the IAU.

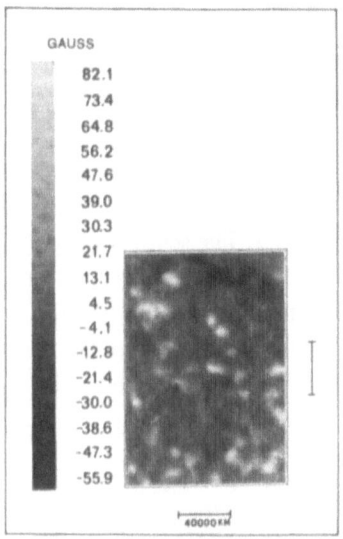

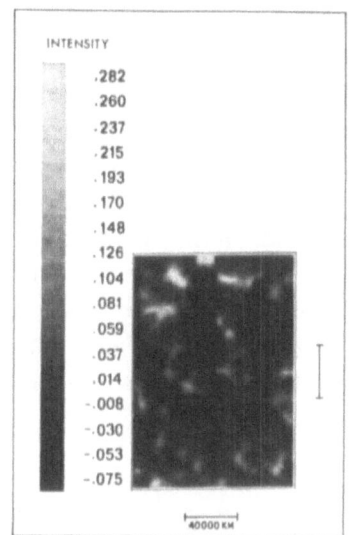

Fig. 1.   Fig. 2.

Fig. 1.   Photographic representation of a magnetogram. North is at the top, west at the right. Magnetic field as measured in the wings of Fe λ 5233.0. Magnetic fields greater than + 82.1 G or − 55.9 G are present but are displayed as totally white or totally black, respectively, to allow a more effective scaling. This procedure is followed on all such magnetograms.

Fig. 2.   The brightness field in the core of the Ca II K line. Intensity is expressed as $(I-I)/I$.

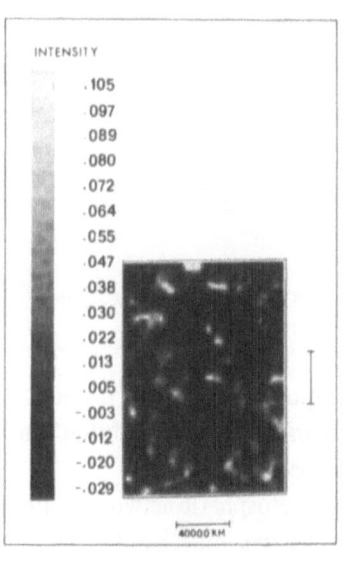

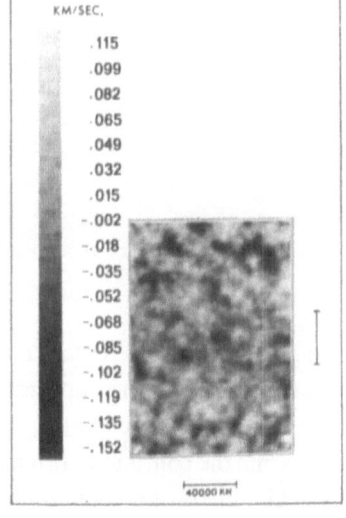

Fig. 3.   Fig. 4.

Fig. 3.   The brightness field in the core of the Fe λ 5250.2 line. The bright patches are slightly sharper than those of Figure 2.

Fig. 4.   Vertical velocity measured in the wing of Fe λ 5233.0. Positive velocities are rising.

2. The Cross Correlations

The correlations between the various arrays can be quickly investigated in more detail through the use of scatter plots. Figure 5 shows that the vertical velocity is approximately a linear function of the magnetic field strength. The large amount of scatter is most likely caused by remnants of the granular and oscillatory velocities. Figure 6 shows the relation between the photospheric network (core of Fe λ 5250.2) and the magnetic field. The relation is again approximately linear. The scatter, although much smaller than in Figure 5, is almost certainly not caused by either the oscillations or the granulation. There is one very tempting explanation for the source of the scatter: There are a few points in Figure 6 that deviate widely from the average. They represent very bright network points with almost zero longitudinal magnetic field. These points

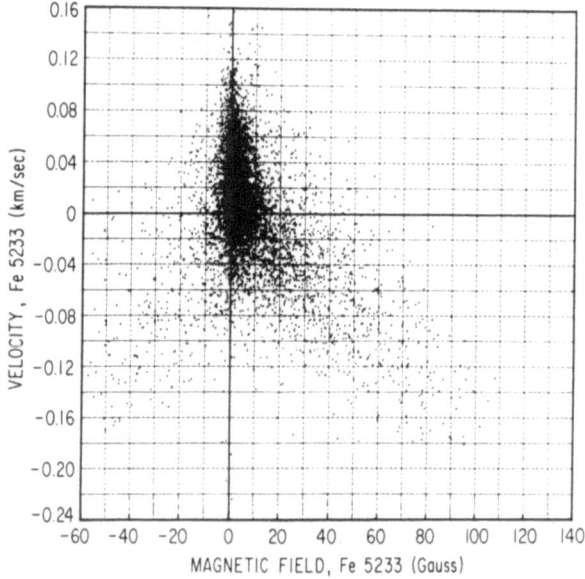

Fig. 5.   Scatter plot of the vertical velocity against the vertical magnetic field.

can be traced back to the centers of very small bipolar regions. So it could well be that the scatter is due to the unobserved transverse magnetic field and that the photospheric network is a very accurate indicator of total field strength.

By comparing the continuum brightness with the photospheric network an important result is achieved. If the supergranulation is convective, one would expect to see the continuum show a temperature excess at the center of the cells and a decrease at the edges (i.e. at the network points). Figure 7 shows that the observed effect is exactly the opposite of the expected: The supergranule boundaries are hotter (roughly 10 K using $E = \sigma T^4$) than average. This should not however be construed to mean that the convective nature of supergranules is disproved. Indeed if one applies a mixing length argument to a convective supergranular cell, one finds that, by virtue of the cell's great size and slow speed, a temperature excess of only 0.03 K is needed to drive it. Thus the convective fluctuation is unobservable, and we are instead observing a secondary heating mechanism.

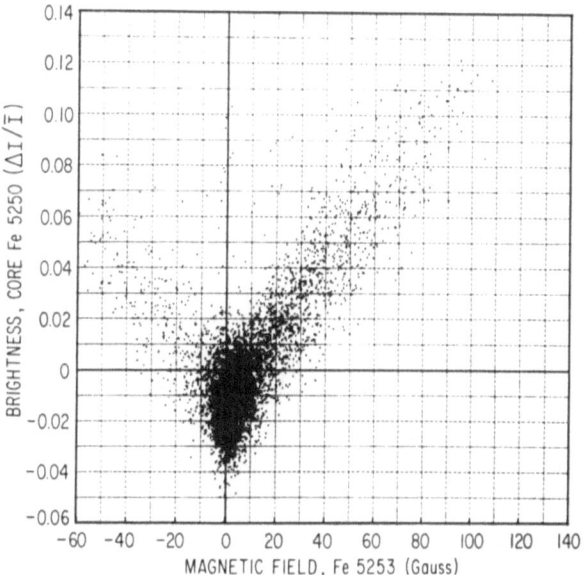

Fig. 6.   The brightness-magnetic field relation.

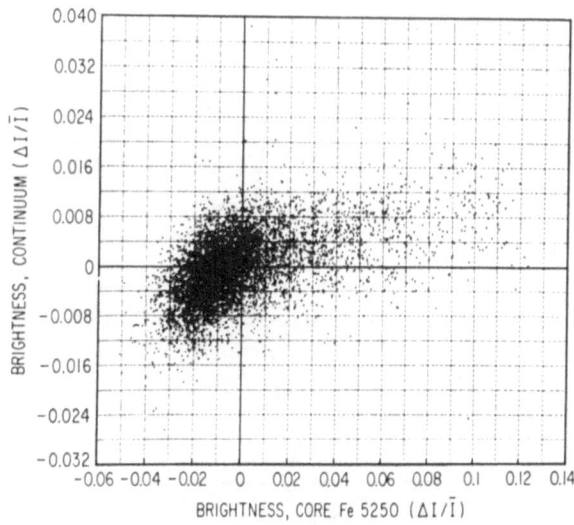

Fig. 7.   The continuum brightness compared with the photospheric network. Bright points in the network are also bright in the continuum.

3. Morphology

We now examine the shape and structure of the 'cells', 'patches', 'network', 'knots' and 'downdrafts'. First, let us look quickly at those isolated individual features that may be called patches, clumps, knots, or downdrafts. 39 of them were identified in the field of view. The average profile of them is shown in Figure 8. The full width at half maximum is about 7000 km.

The much discussed supergranular 'cells' are best visible in the Calcium emission

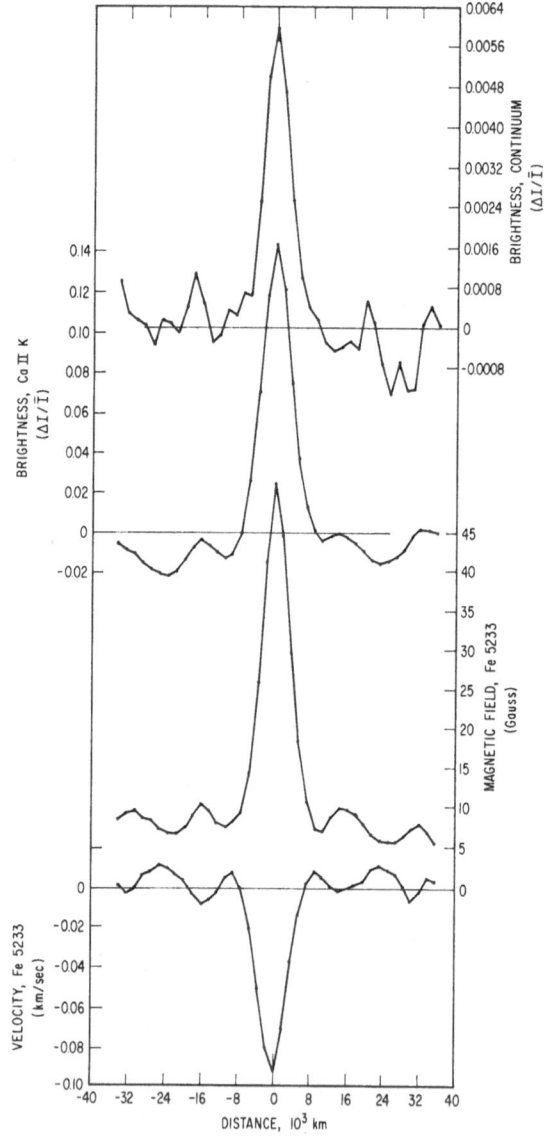

Fig. 8. Average downdraft profiles. Each profile is the average of scans taken in both the X and Y directions across the same 39 points in the raster.

(Figure 2). They are roughly 30 000 km in diameter and are partially (sometimes completely) bounded by visible 'walls'. The bright patches are far more visible than the walls and generally lie at the vertices of three or more cells. One can easily identify the more prominent cells; the less prominent ones require much more judgement to define. I have defined 20 cells and computed average 'cell profiles'. This profile extends from one vertex across a cell to the most distant vertex (Figure 9). It is interesting that the velocity profile (the lowest curve) exhibits a strong difference with the classic case of stationary convection, the Bénard cell. In Bénard cells the

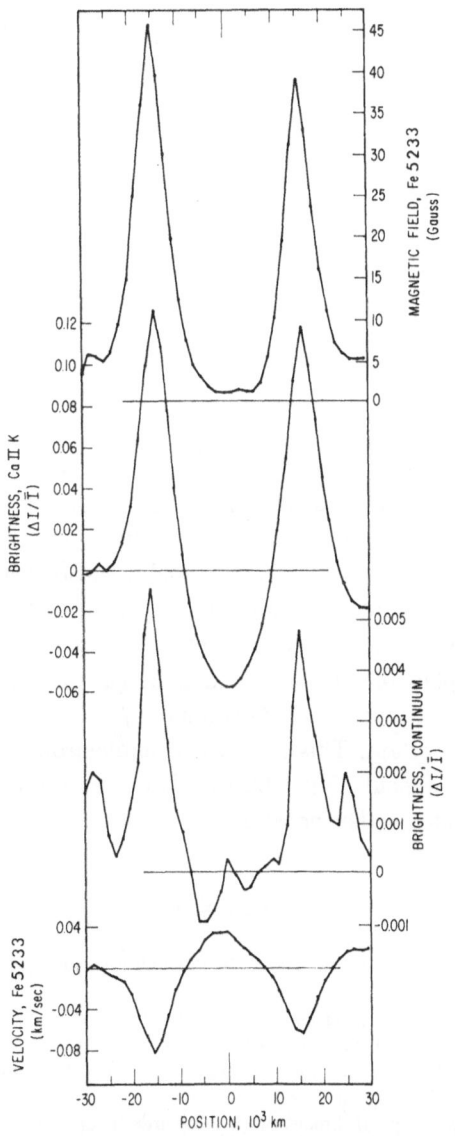

Fig. 9.   Average supergranule profiles. Each profile is the average across the same 20 supergranular cells. The diameter of each cell was first normalized to be equal to the average diameter.

upflow at the center attains twice the amplitude as the downflow at the vertices, whereas almost the opposite is observed here. Evidently the magnetic field and/or stratification effects are exerting a strong influence on the flow pattern.

Note also that the magnetic profile has a very low and flat minimum in the center. If magnetic fields are being swept to the boundaries by the supergranular motions, then the process must be very efficient. However, the minimum is not at 0 G. In fact, a very careful inspection of the magnetic field array reveals a background component that is totally uncorrelated with the supergranulation. This background field has a strength of about ± 5 G, with opposite polarities being quite common.

4. Conclusion

These results fill in most of the observational gaps in our knowledge of the super-granulation. The flow pattern of supergranulation is now completely known. The brightness network is now observed at all levels of the solar atmosphere, and the presence of magnetic fields and their high correlation with the flow pattern and the network is clearly established. However, some questions remain unanswered and at least one new question has been raised.

Is the supergranulation indeed a convective phenomenon? These observations could not answer this question. They are consistent with a convective hypothesis, but they indicate that the behavior is much more complicated than a simple stationary flow pattern.

What is the source of the secondary heating effect indicated by the presence of the bright network? Is it even a heating effect, or perhaps a density condensation? Heating could be caused by the dissipation of Alfvén waves via collisions between ions and neutral atoms. The downflowing material will also have an effect.

Perhaps the most important of all the results is the surprisingly high magnetic field strength at the vertices. Previously the magnetic field strength was thought to be only a few gauss. This implies that $\beta \gg 1$ and that the gas flow dominates the magnetic field. Now, with field strengths of 100 G common, $\beta \ll 1$ and the role of the magnetic field becomes very important. This makes magnetohydrodynamic models of super-granulation extremely difficult, but I believe that if we are to fully understand this phenomenon, such models will be necessary.

Discussion

Kuperus: Did you find both polarities concentrated in small areas along the boundary of the super-granulation network?

Frazier: There was a reversal of polarity (more accurately a small bipolar region) at only 3 out of 39 supergranular vertices. I suspect that this low frequency is just a result of the fact that I happened to observe a region that was primarily unipolar. There could well be large quiet regions where the po-larities are much more mixed.

Kuperus: Could you give a typical dimension of these areas over which you find the magnetic field reversal?

Frazier: About 8000 to 10000 km.

Simon, G. W.: Two comments: First, it is not surprising to find field strengths of 50 G at supergranule boundaries; in fact, since supergranules are formed in the convection zone where equipartition energies are much higher than in the photosphere where the observations are made, I would expect to find fields of several hundred gauss at the boundaries. Second, theoretical incompressible fluid convection models show down drafts at boundary vertices with larger velocities than the central up drafts, so your observations of the velocity distributions are in good agreement with what one would expect, if the results of the incompressible fluid theory can be applied to the case of the solar atmosphere.

Frazier: Answer to comment #1. If one makes the assumption that has usually been made, that is that the magnetic volume force is negligible, and that $B^2/8\pi = \frac{1}{2}\varrho v^2$, then one must apply this equality at all levels. This means that although the magnetic field might be very strong in the convection zone, it would still decrease rapidly with height and be very small at the photospheric levels. One must still investigate the magnetic volume force to see if it can maintain a field strength greater than equipartition of energy would imply.

Answer to comment #2. I was using a very crude model of convection, namely Bénard cells. as my reference. If a more sophisticated model exists which reproduces the observed assymmetry, then I would be very happy to see it.

Deubner: Do you have a feeling for a time scale of the weak background fields, whether these could possibly be as long as e.g. the supergranules, or only for intermediate periods between the granular and supergranular lifetimes?

Frazier: I observed one raster scan per hour for four hours, so neither my time resolution nor my time span was adequate to answer this question well. Most of the background fields did not change at all throughout the four hours. A few showed detectable changes within a few hours and in two cases, very weak bipolar regions evolved abruptly in less than one hour (one appeared and the other disappeared). This question certainly deserves further observations.

Weiss: I would like to comment on Dr. Frazier's remark that fields concentrated by subphotospheric convection should decrease rapidly with height. If the field is concentrated at about 2000 km depth into regions with a similar scale, then the photospheric field intensity need not be too drastically reduced and surface fields of several hundred gauss might be expected.

Frazier: Again, I would say that this cannot be done without the magnetic volume force. If it indeed can be done with the use of this term, then I would feel very gratified.

MAGNETOGRAPHIC AND SPECTROGRAPHIC OBSERVATIONS
OF WEAKLY ACTIVE REGIONS

C. J. DURRANT

Cambridge Observatories, Cambridge, England

Abstract. The value of high resolution photo-electric observations of plage fields and their photospheric effects is demonstrated, with especial regard to obtaining an empirical model of a weakly active region.

1. Introduction

Observations of the photosphere and low chromosphere allow us to probe the structure of active regions with the minimum of difficulties arising from complex geometrical structure and non-LTE effects.

This is a preliminary report of Cambridge work on photospheric line central intensities in plage regions.

2. Observations

The lines observed (Table I) are all in the region of 5200 Å and are mostly iron lines. They were chosen to include lines of widely different atomic characteristics. I shall

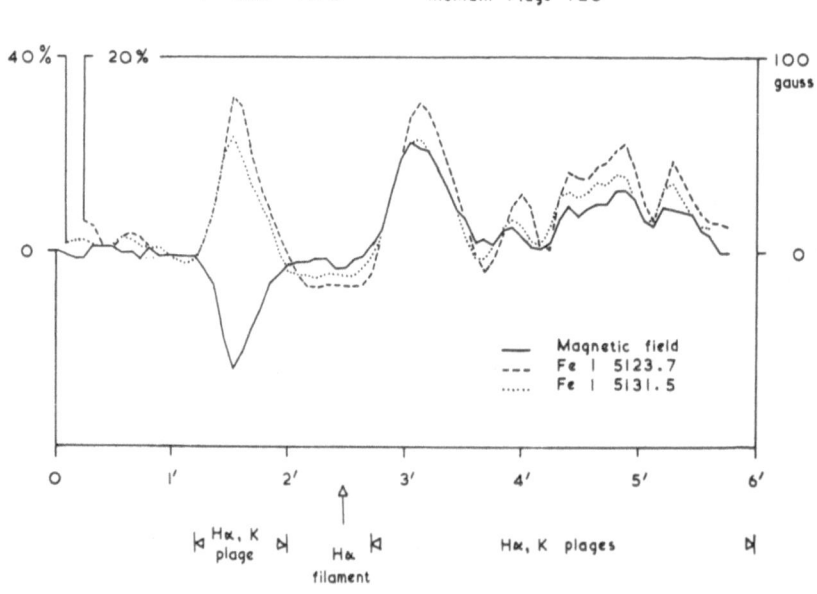

Fig. 1. Sample scan of 6 arc min of the solar disc showing longitudinal magnetic field strengths and the percentage variation of the residual intensities of the Fe I lines at each point. Also indicated are the extents of visual features of Hα and K spectroheliograms.

TABLE I

Lines observed

Line	5123.7	5131.5	5247.1	5250.2	5250.6	5455.5	5455.6	5247.6	5234.6	5242.5
Order of grating	V	V	V	V	V	V	V	V	III	III
Projected slit width, mÅ	12	12	11	11	11	10	10	11	30	30
Ion	FeI	FeI	FeI	FeI	FeI	FeI	FeI	CrI	FeII	FeI
Residual intensity, %	23	29.5	40	39	27	23	17	35	47	45.5
Excitation potential, eV	1.01	2.22	0.09	0.12	2.20	4.32	1.01	0.96	3.22	3.63
Mean Landé	0	2.5	1.75	3	1.67	1.33	0.75			1
Number of traces	7	7	4	4	31	5	5	4	4	13
Mean linear correlation coefficient	78	89	76	82	80.5	84	78	82	66	73
Percentage variation in residual intensity, per G	0.27	0.38	0.51	0.58	0.31	0.33	0.46	0.60	−0.03	0.14

discuss mainly the observations in the V order where the slit is equal to a magnetic splitting of about 500 G or a doppler shift of 1 km s^{-1}.

The magnetograph has an entrance slit of $5'' \times 4''$ and the spectrograph slit is $5'' \times 0.4''$. The line observations are smoothed so both sets of observations refer to the same effective area on the solar disc. The magnetograph employs the Fe I 5250.2 line. The magnetic and intensity records are obtained within about 3 min of one another. Such a set of records is shown in Figure 1 which shows the field record and simulta-

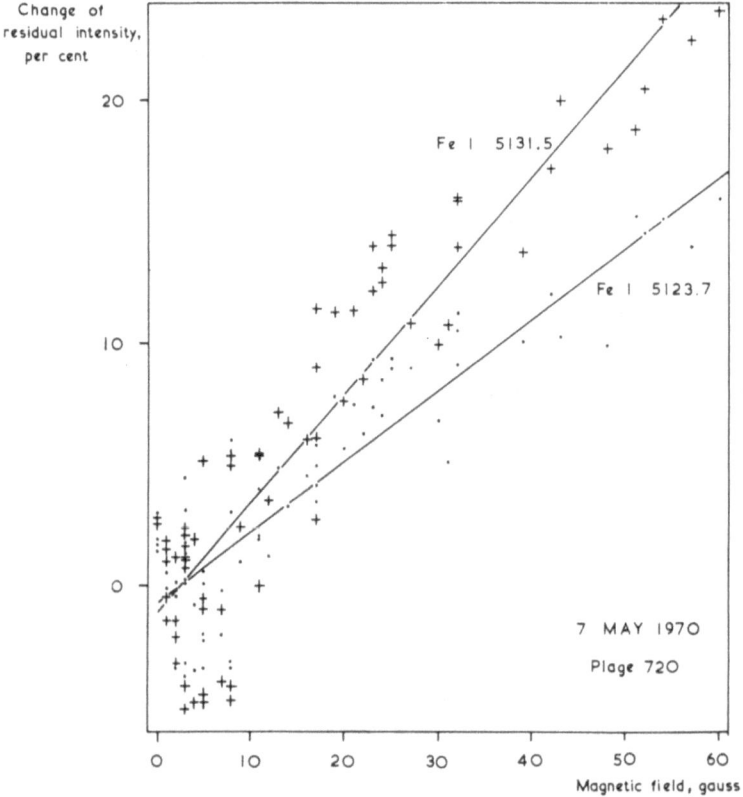

Fig. 2.   The data from Figure 1 plotted to display the dependence of the percentage change of the residual intensities of the Fe I lines on the longitudinal field strength. The two lines were obtained by least squares fitting.

neous records of the Fe I lines 5123.7 (a magnetically null line) and 5131.5 (a Zeeman simple triplet). Typically the residual intensities of neutral ions change by 6% of continuum intensity for typical plage fields strengths which von Klüber has been careful always to describe as effective fields. It can be seen that the line variations follow the *longitudinal* field extremely closely for all field strengths.

Figure 2 shows a plot of percentage change of residual intensity against magnetic field strength and reveals a roughly linear relation, which is the same for fields of both

polarities. The slope of the relation can vary markedly from plage to plage and sometimes within the same plage, by a factor of two. This must be borne in mind when
averaging linear correlation coefficients and linear regression coefficients so that the
'average' behavior of the various lines in plage regions may be compared.

Table I gives the mean slope (percentage variation in line centre intensity per gauss)
weighted according to the linear correlation coefficient found for each trace for each
line studied. Each trace represents at least 60 independent samples. More accurate
slopes will be given when more traces have been measured and prepared for computer
reduction.

3. Interpretation

Whilst many factors can affect line centre intensities, the observations show that
temperature must be the dominant factor. The Fe I 5242.5 and Fe II 5234.6 lines
observed in the III order are formed in similar regions of the atmosphere and would
show the effects of velocity fields, both micro- and macro-, with similar magnitude
and sense. However this is not observed. Whereas the Fe I line weakens in plage
regions the Fe II line in 3 out of 4 statistically significant cases strengthened by a small

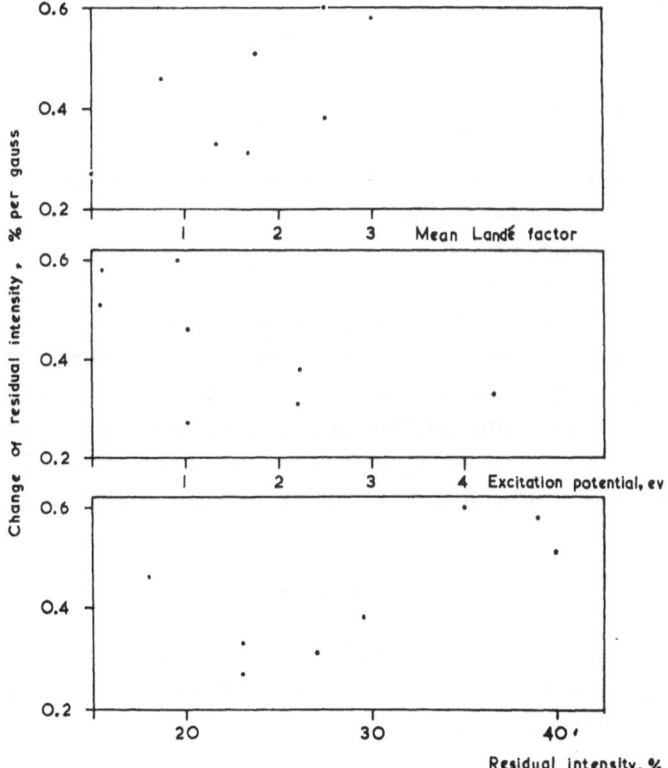

Fig. 3.   Plots showing the dependence of the percentage change of residual intensity, per gauss, of
the lines observed in the Vth order (Table I) on mean Landé factor, excitation potential and undisturbed residual intensity.

amount. Further, the magnetically null Fe I line, 5123.7, shows the same intensity variations as the other Fe I lines.

The Figure 3 shows the dependence of the slope on various line parameters. There is no strict dependence on mean magnetic splitting, but there is a suggestion of a dependence on excitation potential and on central intensity, i.e. height of formation in the atmosphere.

We are computing models in hydrostatic equilibrium using a temperature structure which is perturbed from the quiet structure. Preliminary trials suggest that the observations cannot be fitted by a temperature perturbation which monotonically decreases with depth. Crude estimates, neglecting saturation and magnetic effects, suggest temperature increases of 100 K at $\log \tau_{5000} \sim -0.7$ and less than 50° at $\log \tau_{5000} < -1.0$.

We cannot insist on this conclusion until the fuller calculations are complete.

4. Theory

The solar atmosphere is seen to be heated in active regions by 10–100 K at optical depths of 10^{-1} to 10^{-2}. The source of this heating is strictly correlated with effective field strength and the accepted source is dissipative processes within shock fronts. Chromospheric effects arise from magnetohydrodynamic shocks, but MHD effects will be present at optical depths of 10^{-1} to 10^{-2} only if the ratio of magnetic to gas pressure is greater than unity. This requires fields of 800 – 1200 G. If such fields were present, they would reduce the dissipation which is inversely proportional to the phase velocity of the wave mode carrying most of the energy. Then a vastly increased energy flux in the form of waves would be required to counter this effect and to produce the observed temperature rise.

It seems more plausible that the heating in the photospheric regions is due to ordinary hydrodynamic shocks as is that in quiet regions. Clearly the flux must be increased to account for the extra heating which occurs through earlier development of the shock front and by increased dissipation within the shock. Calculations by Michalitsanos confirm that temperature increases of the right order of magnitude can be obtained by increasing the flux by a factor of 10 in active regions with fields of the order of 100–200 G. This is in good agreement with the MHD flux required in the chromosphere.

Acknowledgements

This work was carried out jointly with Dr H. von Klüber, Mr D. W. Beggs, Miss H. A. Couper, Miss J. V. Field and Mr A. G. Michalitsanos, and was supported by the Science Research Council.

Discussion

Wiehr: Did you take care for the variations of the line depression due to the splitting? I think, realistic variations of central line depressions are only given for the non split line Fe 5123.7. Furthermore, in my opinion there was a correlation in your last slide between the Landé factor and variation of the line depression.

Durrant: I emphasized that the temperatures I mentioned were preliminary order-of-magnitude estimates. This question can only be answered by complete computations now in progress.

Sheeley: At Kitt Peak Dr Gary Chapman and I worked on this problem. We found that the relative effect of Zeeman sensitivity and temperature on line intensity depends on the line used. As a rule-of-thumb the effect is 50–50 for 5250.

Pasachoff: With regard to your graph of the correlation between line intensity and magnetic field, could you please assess the errors in each of those quantities. Also, in what direction did you fit the regression line, i.e. from which quantity onto the other.

Durrant: The intensity variations were fitted to the magnetographic observations as the errors in intensity are larger.

Pasachoff: In a case like this, when both quantities are subject to errors of measurement, the least-squares method of computing a regression line is not valid.

Durrant: I agree, but fitting a regression line by eye yields relations which are not significantly different.

THE MAGNETIC AND VELOCITY FIELDS AND BRIGHTNESS
IN THE SOLAR ATMOSPHERE

S. I. GOPASYUK and T. T. TSAP

Crimean Astrophysical Observatory, Nauchny, Crimea, U.S.S.R.

Abstract. Simultaneous observations of the magnetic fields, the line-of-sight velocities and brightness were made in active and quiet regions with the Crimean double-magnetograph in the following lines: Hα, K₃ Caɪɪ, Hβ, Hγ, Hδ, Mgɪ λ 5184 Å, Caɪ λ 4227 Å, D₁ Naɪ, Baɪɪ λ 4554 Å, Caɪ λ 6103 Å, Feɪ λ 5250 Å.

It is shown, that in the active regions the horizontal velocity is larger than the vertical one.

The mean velocities in the quiet solar photosphere have an isotropic distribution (Gopasyuk and Kalman, 1971).

The mean vertical velocities increase exponentially with height in active and quiet regions.

The correlation between velocities at different levels in active and quiet regions decreases with the distance between the levels of the formation of spectral lines, and it disappears for the velocities recorded in λ 6103 Å and Hβ, for λ 5184 and Hα lines in active regions and for the velocities recorded in λ 5250 Å and Hα lines in quiet regions.

The position of the maximal field strength within a magnetic hill coincides statistically with the zero line of the line-of-sight velocities for active as well as for quiet regions.

Simultaneous observations of the magnetic fields, the line-of-sight velocities and brightness were made with the Crimean double-magnetograph (Severny, 1966). This magnetograph permits one to record magnetic fields, velocities and brightness in any two spectral lines simultaneously.

We used the following lines: Hα, K3 Caɪɪ, Hβ, Hγ, Hδ, Mgɪ λ 5184, Caɪ λ 4227, D₁ Naɪ, Baɪɪ λ 4554, Caɪ λ 5103, Feɪ λ 5250.

We have obtained many records in active and quiet regions at different distances from the center of the disk.

The entrance aperture was 2.8 × 9.0 and sometimes 2.7 × 4.5 s of arc.

From maps of velocities we find the mean velocity for each region of outward and downward motions.

A comparison of these velocities for active regions at the center of the disk and near the limb shows that, in general, near the limb they are higher than at the center, which means that the horizontal motions have higher velocities than vertical ones. For example for the Hα-line the horizontal component exceeds the vertical one by about a factor of two. At the same time, as was shown by Gopasyuk and Kalman (1971), the mean velocities in the quiet solar photosphere have an isotropic distribution (Figure 1).

The comparison of mean vertical velocities recorded in different lines at the disk center shows an increase of this velocity outwards with height for active regions as well as for quiet regions. Assuming that the spectral lines in active regions are formed at the same levels as in quiet regions, we find an exponential increase with the height.

$$v(h) = V_{5250} \exp(h/h_0) \tag{1}$$

where $h_0 = 3350$ km and the level of the line formation $\lambda 5250$ is taken as zero.

Howard (ed.), Solar Magnetic Fields, 274–278. All Rights Reserved.
Copyright © 1971 by the IAU.

We also found that the vertical component of the mean velocity at the disk center (averaged over many active regions) is nearly equal to that obtained for the quiet Sun.

All said above is illustrated in Figure 2. The open circles show the downward motions and the crosses show the upward motions in active regions. The mean velocities in quiet regions are shown by solid dots. The solid line is the exponential curve which approximates the observations.

The correlation coefficients between velocities at different levels were computed for active and quiet regions at the disk center and we found that the correlation decreases with the distance between the levels of the formation of spectral lines. It practically disappears for the velocities recorded in Ca I λ 6103 and Hβ or for Mg I λ 5184 and

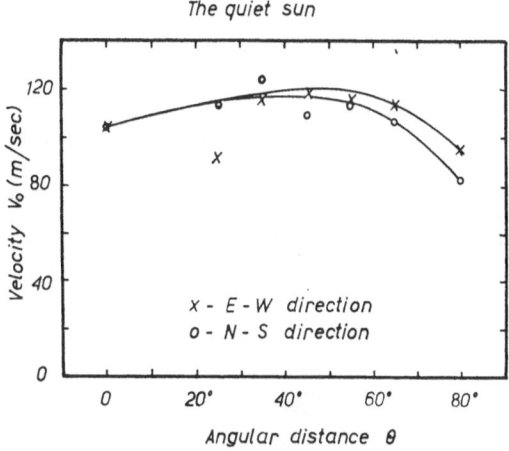

Fig. 1. The variation from center to limb of the mean line-of-sight velocity in the quiet solar photosphere. The records were made in the λ 5250 line.

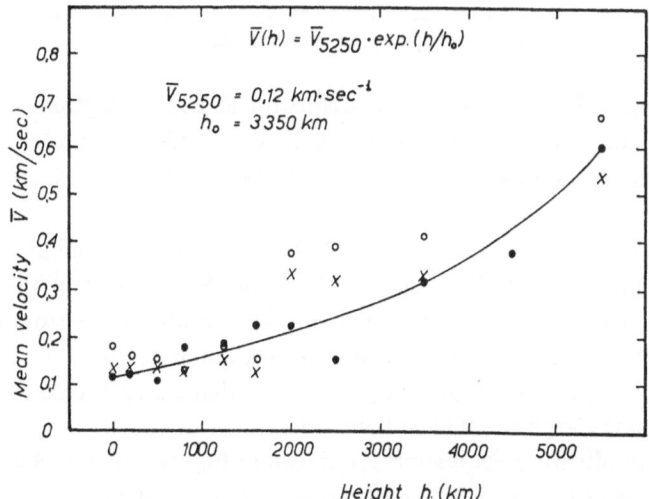

Fig. 2. The variation of the mean vertical velocities with height in active and quiet regions.

$H\alpha$ lines in the active regions and for the velocities recorded in Fe I λ 5250 and the $H\alpha$ line in quiet regions. In both cases of active and quiet regions the decrease of the correlation coefficients with distance between the levels can be described by the relation:

$$R(h) = 0.7 - \left(\frac{h}{2h_0}\right)^2. \tag{2}$$

Here $h_0 = 2750$ km for an active regions and $h_0 = 3350$ km for a quiet region.

The variation of the correlation with distance in active and quiet regions is shown in Figure 3. The solid line corresponds to active regions and the dotted line to quiet regions.

Now, further we considered the correlation between the line-of-sight velocities and the longitudinal magnetic fields for active and quiet regions everywhere over the disk

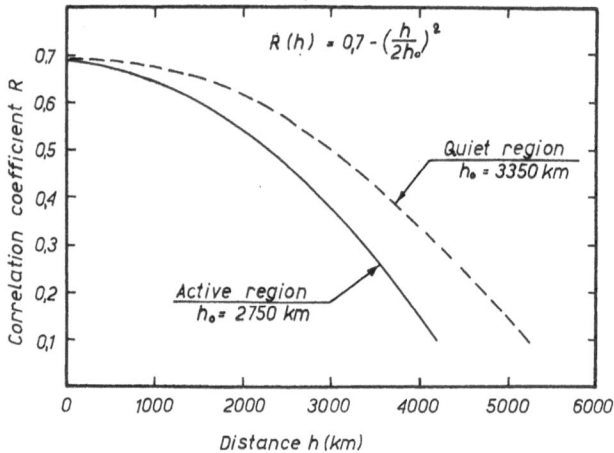

Fig. 3.   The variation of the correlation with the distance between the velocities
in active and quiet regions.

for different lines. The correlation coefficients for all lines except one corresponding to K3 Ca II in active regions do not exceed 0.10–0.15. This leads probably to the conclusion that there is no correlation between line-of-sight velocities and longitudinal magnetic fields.

We have also found that the position of the maximum longitudinal field strength within a magnetic hill coincides statistically with the zero-line of the line-of-sight velocity for active as well as for quiet regions. This result follows from the measures of the shortest distances between the point of maximum field and the corresponding adjacent zero-line of line-of-sight velocity. Besides that we determine the distribution of the velocities in the points of maximum fields.

Both these results for quiet regions are shown in Figure 4. Figure 4 corresponds to the observations in the Fe I λ 5250 line in quiet regions at the center of the disk. The upper curve shows the distribution of the shortest distances mentioned above. This is essentially a Gaussian distribution as can be seen from comparison of open circles with the solid line.

The curve below shows the distribution of the line-of-sight velocities at the points of the maximum field. The solid line is the exponential curve which approximates the observations.

In Figure 5 we have the same distributions found from the observations in the CaI λ 6103 line in active regions near the center of the disk ($\theta \leqslant 30°$).

The dispersion for the various distributions is 2600 km in the case of quiet regions and 3000 km for active regions.

We wish to emphasize that this kind of connection between the motions and magnetic

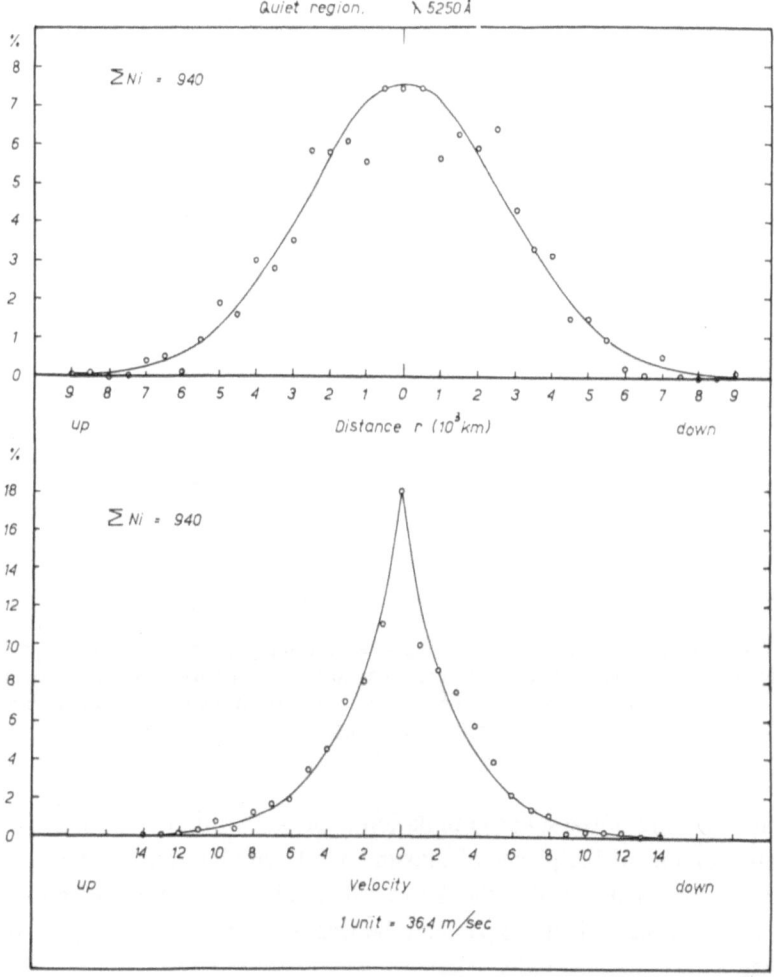

Fig. 4. The distribution of the shortest distances between the point of the maximum longitudinal field strength within a magnetic hill and the corresponding adjacent zero-line of line-of-sight velocity (the upper curve). The distribution of the line-of-sight velocities at the points of the maximum field (the lower curve) in the quiet regions at the center of the disk.

fields remains the same at different levels in the solar atmosphere and at different distances from the center of the disk.

We should also point out that the maximum fields are usually positioned at the points of maximum brightness.

Finally we can conclude that (a) our observations are not in good agreement with

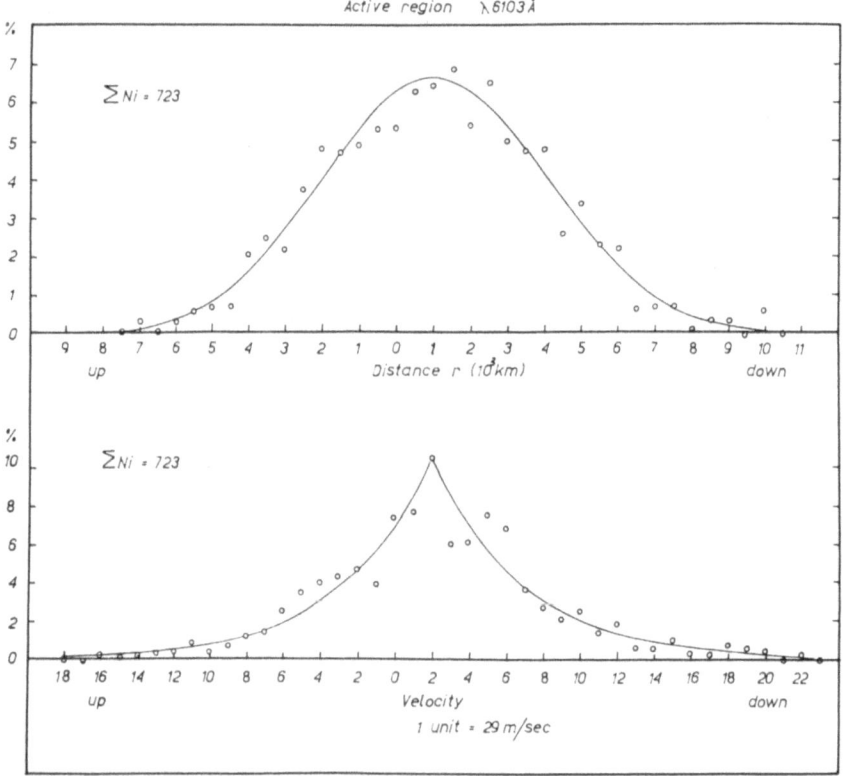

Fig. 5. The distribution of the shortest distances between the point of the maximum longitudinal field strength within a magnetic hill and the corresponding adjacent zero-line of line-of-sight velocity (the upper curve). The distribution of the line-of-sight velocities at the points of the maximum field (the lower curve) in active regions near the center of the disk.

the common picture of the super-granulation because they do not show predominant downward motions at the places of maximum field strength and brightness (b) our observations show that probably shock waves do not heat the chromosphere.

It seems to us that Ohmic losses may be important for the heating of the solar atmosphere.

References

Severny, A. B.: 1966, *Astron. Zh.* **43**, 465.
Gopasyuk, S. I. and Kalman, B.: 1971, *Izv. Krymsk. Astrofiz. Obs.* **44**, in press.

MAGNETIC FIELDS MEASURED WITH THE 10830 Å HeI LINE

J. HARVEY and D. HALL

Kitt Peak National Observatory*, Tucson, Arizona, U.S.A.

Abstract. Several advantages of near infrared spectral lines for magnetic field measurements are listed. In particular, the 10830 Å multiplet of HeI is well suited for observations of chromospheric magnetic fields.

New photoelectric spectroheliograms made with the 10830 Å line reveal a large amount of filamentary fine structure in active regions. This fine structure has important consequences on the interpretation of 10830 Å magnetograms. Except for an association of 10830 Å disk filaments with polarity reversals there is little correlation between absorption features and the 10830 Å longitudinal field. Comparisons of chromospheric and photospheric observations show that the chromospheric field is spatially more diffuse and weaker than the photospheric field.

1. Introduction

Many attempts have been made to observe magnetic fields in the chromosphere. These observations are difficult to interpret for various reasons:

(1) The inhomogeneous structure of the chromosphere manifest in chromospheric spectral lines has a profound effect on magnetic measurements made with such lines. This structure cannot be ignored in interpreting chromospheric magnetograms.

(2) The wings of some 'chromospheric' lines have large contributions from photospheric heights.

(3) Blending of chromospheric lines with photospheric lines reduces the relative contribution of the chromosphere to the final magnetic measurement (Zhulin *et al.*, 1968).

(4) Atomic level interference can produce polarization in chromospheric lines which affects some magnetic measurements (Lamb, 1970).

At Kitt Peak, magnetograms have been made with K_3, $H\beta$ and $H\alpha$ for several years but, as representative measurements of the magnetic field in the chromosphere, our observations are unsatisfactory because of the problems listed above. In an effort to minimize some of these problems, we examined the chromospheric lines in the near infrared spectrum. We have selected the lines of the HeI multiplet (λ 10829.08, λ 10830.25 and λ 10830.34 Å) as the most suitable for chromospheric magnetic field measurements in the accessible spectrum. Since these lines are formed only in the chromosphere problem 2 is avoided and, by chance, problem 3 is also not important. Problems 1 and 4 are still important. Several SiI lines close to 10830 Å are suitable for photospheric field measurements.

The location of the lines in the infrared might seem to be a disadvantage but there are several advantages to using infrared lines compared with visible lines for magnetic measurements. Among these are:

---

* Operated by the Association of Universities for Research in Astronomy, under contract with the National Science Foundation.

Howard (ed.), Solar Magnetic Fields, 279–288. All Rights Reserved.

(1) Better seeing (Turon and Lena, 1970).

(2) Less scattered light (Staveland, 1970).

(3) Less atmospheric extinction.

(4) Better reflectivity by aluminized mirrors and, as a result, less instrumental polarization.

(5) Better magnetic sensitivity since the ratio of Zeeman splitting to line width varies roughly linearly with wavelength.

(6) Less noise produced by image motion while measuring sunspots because of the smaller contrast between umbral and photospheric intensities.

(7) Less precise optical surfaces are required.

(8) Fewer blending problems because there are fewer lines.

Some of the disadvantages of infrared lines for magnetic measurements are:

(1) Solar energy flux is smaller (by a factor of about 3 between 5000 and 10000 Å).

(2) Unless special precautions are taken, noise during a magnetic measurement with existing photodetectors is due to dark current rather than shot noise.

(3) Spectral lines tend to be weak.

Considering the good and bad features, we believe that the near infrared is a very promising spectral region for magnetic field measurements.

2. Observations

To observe at wavelengths around one micron we replaced the photomultipliers in the Babcock-type magnetograph of the McMath Solar Telescope (Livingston, 1968; Livingston and Harvey, 1970) with silicon photodiodes (Electro-Nuclear Laboratories, Type PDS-050L). Low noise preamplifiers were constructed so that the photodiodes looked electrically like photomultipliers to the rest of the magnetograph electronics.

We made spectroheliograms in addition to magnetograms by placing one exit slit of the magnetograph in the core of the line and the other exit slit in the continuum and recording the signals separately while making a raster scan.

The 10830 Å multiplet consists of three lines, two of which overlap to produce a line as much as eight times stronger than the remaining line. We used the strong line for our observations and adopted an effective Landé g-value of 1.47 which should not be in error by more than ± 0.03 due to optical depth variations. For photospheric measurements we used the 10827 Å line of Si I which has a simple triplet Zeeman pattern with a Landé g-value of 1.50.

During magnetic measurements, an effort to compensate for large scale variations of the strength of the 10830 Å line was made. We used the Doppler servo to produce a small sinusoidal wavelength oscillation of the line with respect to the exit slits at a frequency different from that used to analyze the magnetic field. By doing this we produced an artificial longitudinal Zeeman effect of known, constant strength. Any variation in the strength of the signal derived from the artificial Zeeman effect is due to a change in the observed line profile which affects the actual longitudinal field observation in essentially the same way. By dividing the actual signal by the artificial signal at each observed point, we obtain measurements of the average longitudinal Zeeman effect produced in the 10830 Å absorption features. Essentially the same technique was used on the original Climax magnetograph (Lee *et al.*, 1965).

We calibrated our observations in the usual manner by placing a circular polarizer in front of the analyzer and displacing the spectrum line on the exit slits a distance corresponding to a known magnetic field strength. Some of the observational parameters are listed in Table I.

TABLE I

Observational parameters

Parameter	Magnetograms	Spectroheliograms
Entrance aperture (arc s)	4.8 × 4.8	1.4 × 1.4
Exit slit (Å)	0.46	0.46
Integration time (s)	0.3	0.03
Modulation frequency (Hz)	10^3 (Zeeman) 45 (reference)	800
Area scanned (arc s)	163 × 288	160 × 186

3. Results

Previous spectroheliograms made with the 10830 Å line (D'Azambuja and D'Azambuja, 1938; Zirin and Howard, 1966) as well as spectra (Mohler and Goldberg, 1956) have shown the distribution of 10830 Å absorption on the disk to be very irregular. (The line strength fluctuates in a pattern similar to other chromospheric lines, i.e. strong in plages and filaments). However, the spatial resolution of previous observations of 10830 Å has not been adequate to reveal the fine structure (Giovanelli, 1967).

Figure 1 is one of our spectroheliograms taken in good seeing in which we begin to resolve fine structure. Filamentary details dominate this picture. In the lower portion of the illustration which shows the ratio of 10830 Å to continuum brightness we see that the sunspots nearly vanish. The 10830 Å absorption is still present in the locations of the spots although it is very weak in the darkest parts of the umbrae.

Figure 2 shows an older active region in which the 10830 Å absorption is weaker than in the region shown in the first figure. Except for one dark disk filament, filamentary structure is not prominent. Instead, there are extended, dark, mottled areas.

Figure 3 shows a region near the limb. The continuum picture shows photospheric faculae faintly and very little limb darkening, especially compared with the limb darkening in the 10830 Å line. Both pictures are reproduced with equal contrast. There is good, but not perfect, correspondence between the positions of 10830 Å absorption and continuum faculae. The ratio picture reveals the chromosphere as a bright band roughly 9000 km high. The sunspot does not vanish in the ratio picture, unlike the spots closer to the disk center shown in Figures 1 and 2. This behavior can be explained as a difference in the height of the spot seen in the continuum and in the

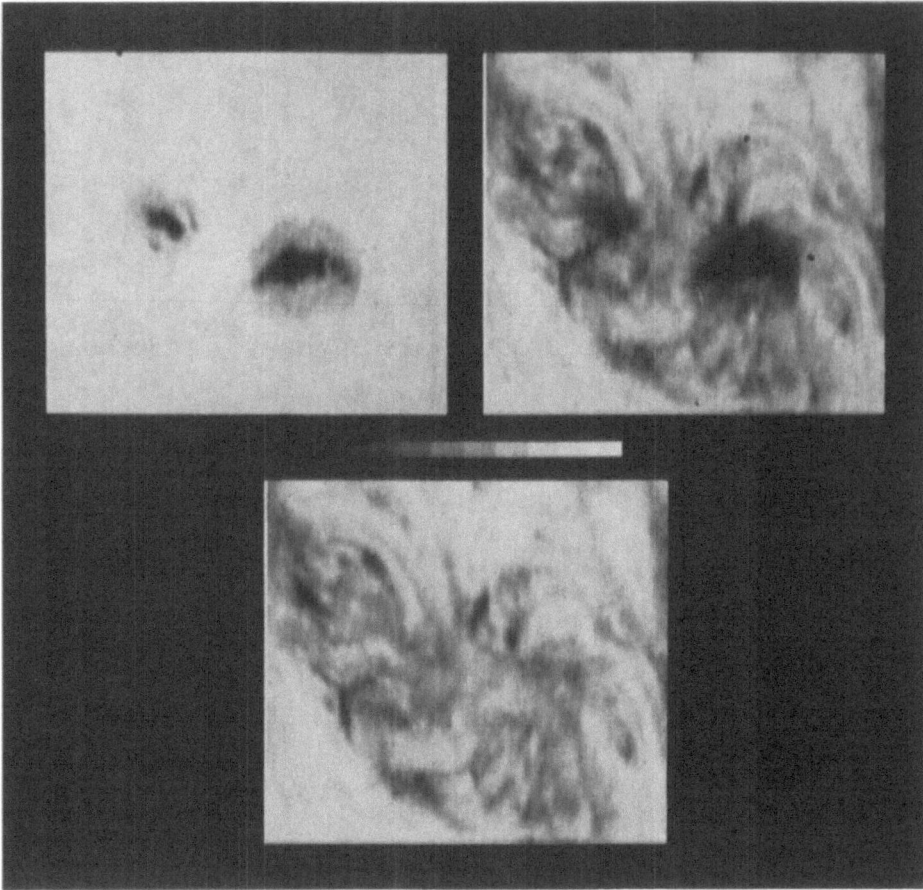

Fig. 1.   Simultaneous photoelectric spectroheliograms taken on June 19, 1970 with the core of the 10830 Å line (right) and the nearby continuum (left). The lower picture shows the ratio of the core to the continuum measurements. The pictures have been generated with a computer simulating an effective photographic gamma of 4.

10830 Å line. Adopting this explanation, we find an apparent height difference of roughly 1000 km. Thus our 10830 Å observations refer to features which probably lie roughly between 1000 and 9000 km in height on the average.

Our spectroheliograms show filamentary structures (disk filaments and fibrils) as the most prominent features of active regions. While our spatial resolution is not yet adequate to clearly resolve the other mottled areas, it is probably safe to assume that better spatial resolution will resolve these areas into tiny fibrils and fine mottles similar to those visible on high-quality Hα filtergrams.

A somewhat confusing situation results from comparisons of 10830 Å magnetograms and spectroheliograms such as shown in Figures 4 and 5. We see that the strongest longitudinal fields are associated with sunspots. The darkest, filamentary features in 10830 Å, which we presume to be disk filaments, are generally well

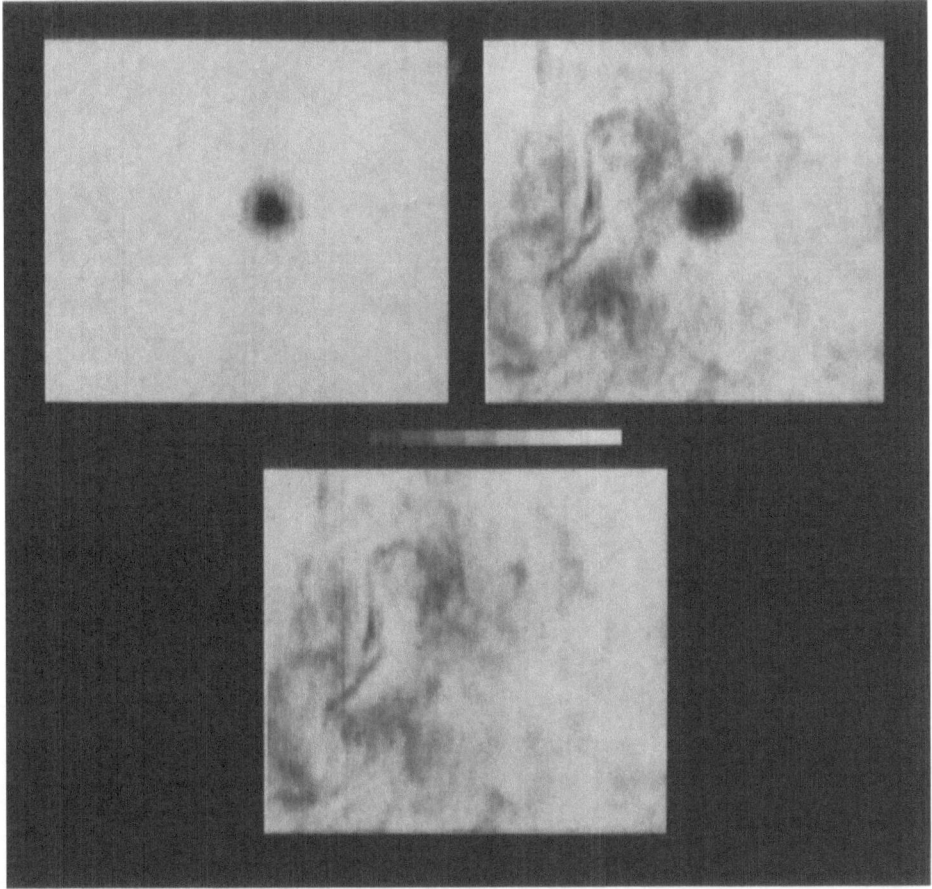

Fig. 2.   Same as Figure 1 except an older active region near the disk center.

associated with reversals of the polarity of the 10830 Å longitudinal field. Beyond these two expected associations, there is very little distinct relationship between 10830 Å features and the 10830 Å magnetic field. We can see that wherever there is a significant field in 10830 Å there is also some 10830 Å absorption. But in many cases, such as the lower left center of Figure 4, the absorption is very weak even though the field is strong. In other cases, away from polarity reversals, the field is fairly weak but the absorption is relatively strong. The older active region shown in Figure 5 exhibits the same poor correlation between field strength and absorption so that the state of development of the active region does not seem to be a significant factor. Other regions we have studied show the same behavior.

Comparisons of 10830 Å and photospheric (10827 Å) magnetograms such as shown in Figures 6 and 7 give consistent results. The most obvious result is that the general pattern of fields is nearly identical. The next result, which is well shown in Figure 6, is that the average longitudinal field in the chromosphere is spatially more

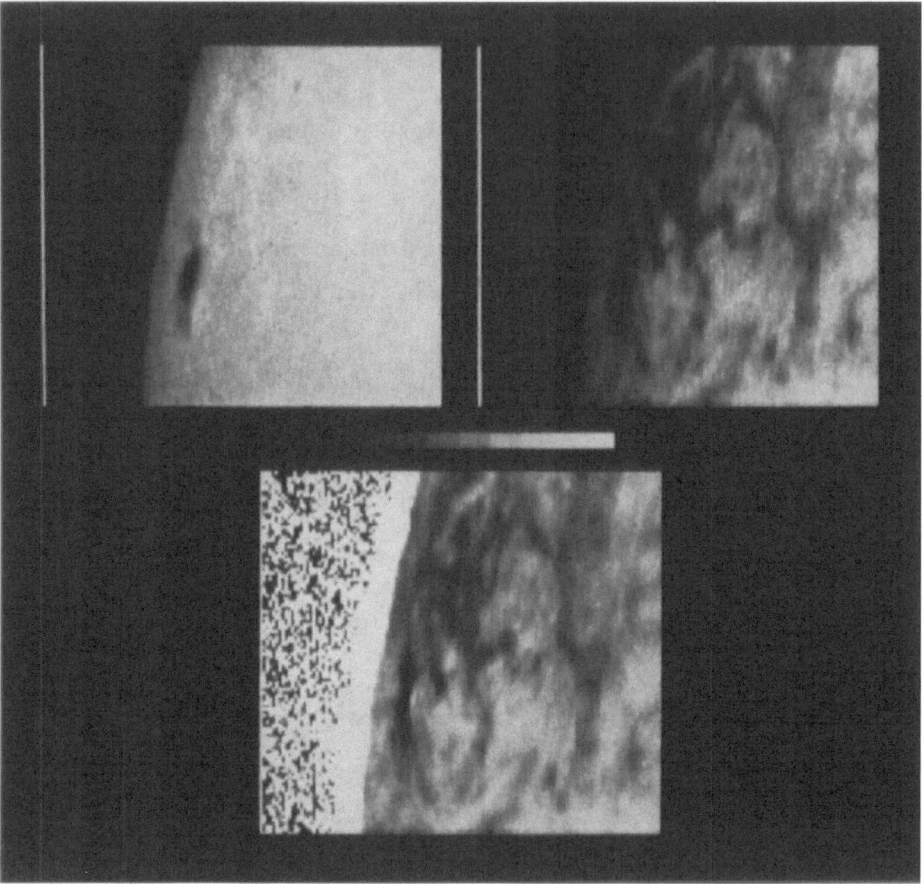

Fig. 3.   Same as Figure 1 except a region near the west limb. In the ratio picture, measurements in the chromosphere above the limb show as a bright band with a height of about 9000 km. Above that height only noise is visible as checkered pattern of light and dark elements.

diffuse than the photospheric field. This is particularly clear in the contour maps. The final obvious result, which is well shown in Figure 7, is that the average longitudinal field strength we measure with 10830 Å is roughly $\frac{1}{2}$ to $\frac{3}{4}$ of the value we measure at the same point with the 10827 Å line.

4. Discussion

It is premature to attempt to construct a model of the chromospheric magnetic field from these preliminary observations. Our most serious problem is the effect of un-resolved fine structures on our magnetic observations. The compensation technique we use to correct for line profile variations cannot remove the effects of unresolved

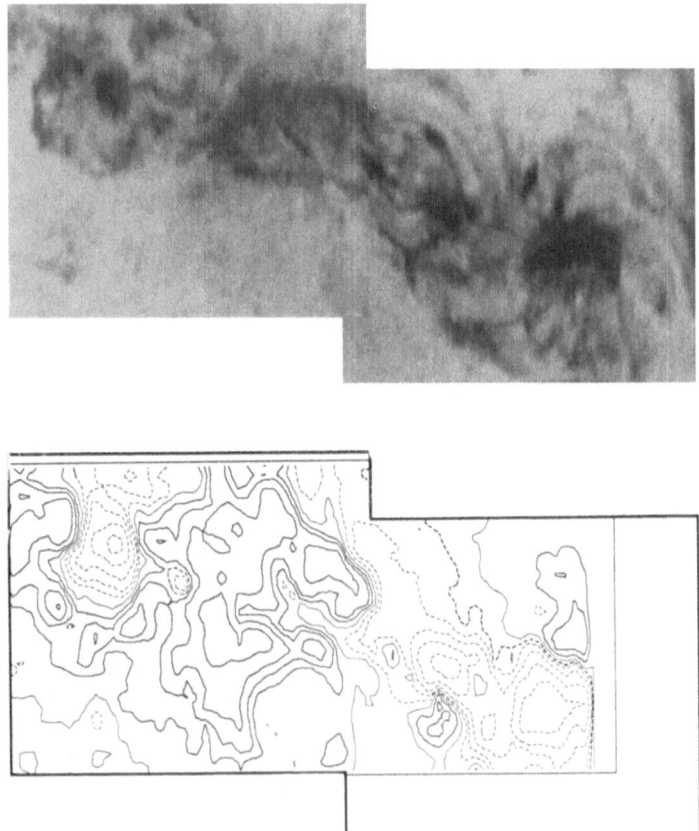

Fig. 4.   Comparison of a nonsimultaneous 10830 Å spectroheliogram (top) and magnetogram (bottom). North is up and east to the left. The contours of the magnetogram are drawn for average longitudinal field strengths of 25, 50, 100, 200 and 400 G.

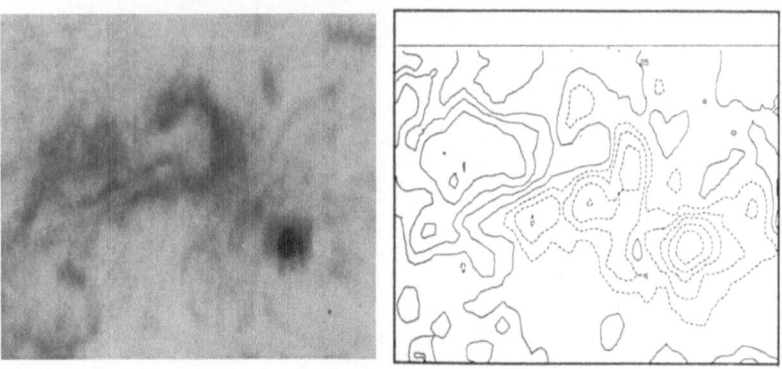

Fig. 5.   Same as Figure 4 except an older active region.

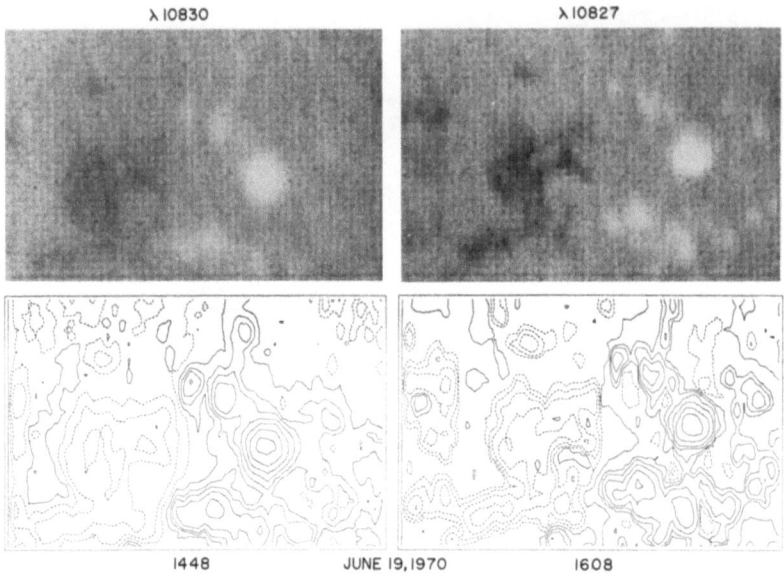

Fig. 6.   Magnetograms taken with the 10830 Å line (left) and the 10827 Å line (right). In the top photographs the departure from gray is proportional to the average longitudinal field strength and polarity. The contours are drawn at values of 25, 50, 100, 200, 400, 800 and 1600 G. The field strength is underestimated in the sunspot at the right because of excessive Zeeman splitting. Note the diffuseness of the field in 10830 Å compared with 10827 Å.

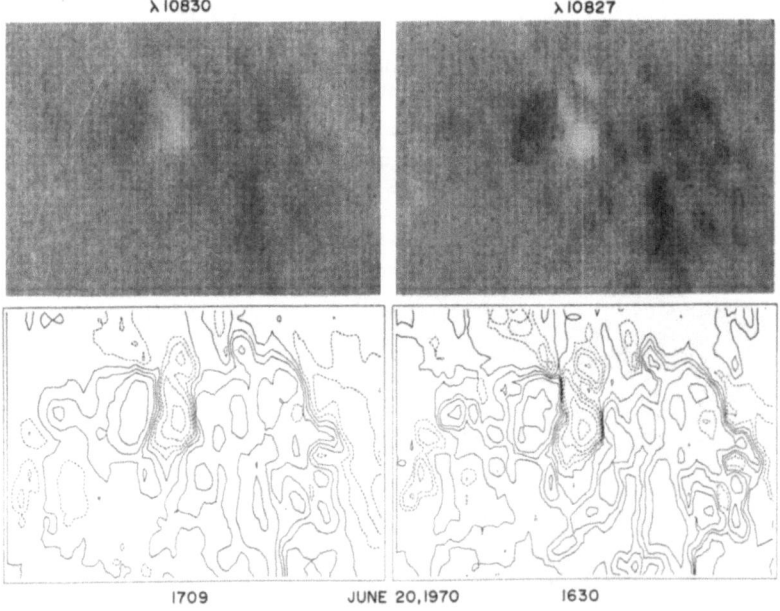

Fig. 7.   Same as Figure 6. Note the weakness of the field in 10830 Å compared with 10827 Å.

fine structures and the effect on our magnetic measurements is different for the two lines we used. To illustrate, suppose we have a simple two component model of the solar atmosphere of the sort proposed by Namba (1963) and de Jager *et al.* (1966). Imagine that the magnetograph aperture is partially filled by each type of component and that one component has a longitudinal magnetic field and that the other does not. In the case of the photospheric line, suppose that the line profile does not differ significantly in the two components. Therefore in our magnetic measurement we get an equal contribution from the magnetic and nonmagnetic components and we measure a magnetic field strength which is the average of the field including both magnetic and nonmagnetic components. That is, we measure the average longitudinal magnetic field integrated over the aperture of the magnetograph. If the region is at the center of the disk this quantity would simply be the magnetic flux passing through the photosphere.

In the case of the chromospheric line let us suppose that the line is formed only in the component with a magnetic field. Then our magnetic measurement includes no contribution from the nonmagnetic components and we measure the average magnetic field *excluding* the nonmagnetic elements. Similarly if the line is formed only in the nonmagnetic elements, only they would contribute to our measurement (which would be zero in this case).

Of course, the actual situation in the solar atmosphere is not so simple as this and extreme care is necessary in interpreting our observations. We have no evidence of significant variations of the 10827 Å line profile outside sunspots so we believe that our measurements with this line actually give the longitudinal magnetic field strength averaged over our 5 arc s aperture. Since we know that the photospheric field tends to be clumped into small elements which do not completely fill our large aperture, the field strengths in these elements are always greater than the average field strengths we measure.

Our spectroheliograms in 10830 Å give abundant evidence of line profile variations so with our observing technique the 10830 Å magnetic measurements are weighted averages of the longitudinal magnetic field in those 10830 Å absorption features falling within the aperture. The weighting function is simply the relative contribution of each feature to the total observed line profile. One can now imagine many models of the 10830 Å absorption features and their magnetic fields which can explain the observed decrease in average field strength in the chromosphere. For example, if we assume that all the magnetic flux we observe in the photosphere passes through the level we observe in the chromosphere then the observed decrease in field strengths might be due to a channeling of the flux somewhat away from the 10830 Å features so that the field strength in those features is small.

Another explanation which we feel is more likely to be correct is simply that the magnetic field is more horizontal in the 10830 Å absorption features. This would reduce the observed line-of-sight field component and would be consistent with a spreading of the field with increasing height implied by the diffuseness of the 10830 Å fields.

References

D'Azambuja, L. and D'Azambuja, M.: 1938, *Bull. Astron.* **11**, 349.
De Jager, C., Namba, O., and Neven, L.: 1966, *Bull. Astron. Inst. Neth.* **18**, 128.
Giovanelli, R. G.: 1967, in *Solar Physics* (ed. by J. N. Xanthakis), Interscience, London, p. 353.
Lamb, F. K.: 1970, *Solar Phys.* **12**, 186.
Lee, R. H., Rust, D. M., and Zirin, H.: 1965 *Appl. Opt.* **4**, 1081.
Livingston, W. C.: 1968, *Astrophys. J.* **153**, 929.
Livingston, W. C. and Harvey, J.: 1971, in preparation.
Mohler, O. C. and Goldberg, L.: 1956, *Astrophys. J.* **124**, 13.
Namba, O.: 1963, *Bull. Astron. Inst. Neth.* **17**, 93.
Staveland, L.: 1970, *Solar Phys.* **12**, 328.
Turon, P. J. and Lena, P. J.: 1970, *Solar Phys.* **14**, 112.
Zhulin, I. A., Ioshpa, B. A., Mogilevskiy, E. I., and Obridko, V. N.: 1968, *Solar Activity*, No. 3, Nauka Press, Moscow, p. 34.
Zirin, H. and Howard, R.: 1966, *Astrophys. J.* **146**, 367.

Discussion

Tandberg-Hanssen: Did you have Hα filtergrams of the same region where you saw dark features in 10830; and if so, what did the 10830 dark features correspond to?

Harvey: We did not have good Hα filtergrams corresponding to our 10830 observations. The comparisons we did indicate that the darkest features in 10830 are disk filaments in Hα. We intend to make simultaneous 10830 and Hα observations in the near future.

OBSERVATIONS AND INTERPRETATION OF SUPERGRANULE VELOCITY AND MAGNETIC FIELDS

STEVEN MUSMAN

*Sacramento Peak Observatory, Air Force Cambridge Research Laboratories,
Sunspot, New Mex. 88349, U.S.A.*

Abstract. I have studied the observed concentrations of vertical velocity and vertical magnetic field in the corners of the coarse network. Using a horizontal velocity inferred from the vertical velocity, I have computed the possible rate of concentration of the field. The rate turns out to be much higher than observed. I conclude that the observed motions in supergranules are not concentrating the observed field at the corners of the network. I have suggested four possible alternate situations consistent with the observations.

I am going to discuss an interpretation of observations of quiet region magnetic fields. I will start with observations and try to see what physical situation the observations compel one to believe in.

Magnetograph observations give the longitudinal components of both the velocity and the magnetic field. These furnish only a partial description of a magnetohydrodynamic flow, but even this partial description can provide some useful information.

The most questionable assumption which I make is that the magnetograph measurement may be interpreted as a velocity and magnetic field at some height in the solar atmosphere. The measurements can at best be only weighted averages of these quantities.

When my observational program at Sac. Peak was held up due to instrumental difficulties Dr. Edward Frazier graciously provided me with the results of his observations. These are a portion of the work which he has described earlier in this session. I used the results for the neutral iron line at 5233, which is formed in the upper photosphere. The area of each of Frazier's scans contains about 20 super granules, and each supergranule contains about 300 measured points. Each measured point is the average of two pairs of successive observations taken $2\frac{1}{2}$ min apart in order to eliminate the effect of the five minute oscillations.

The most conspicuous feature of the observations is the correspondence between strong features in the velocity and magnetic fields. I have taken the product of the velocity and magnetic field and chosen the ten largest local extrema. In all cases these correspond to conspicuous features in both fields separately. I have superimposed these features, averaged them, and also averaged over various azimuths.

Figure 1 shows the results of this process for the average vertical component of the magnetic field. The center of the coordinate system corresponds to a vertex of the coarse network. The signs of two of the ten magnetic fields used were negative. I have reversed the signs of these fields before averaging.

Figure 2 shows the average vertical velocity. This is downward in all cases regardless of the sign of the magnetic field.

Howard (ed.), Solar Magnetic Fields, 289–292. All Rights Reserved.
Copyright © 1971 by the IAU.

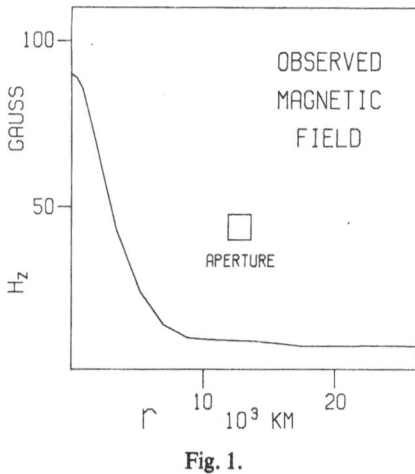

Fig. 1.

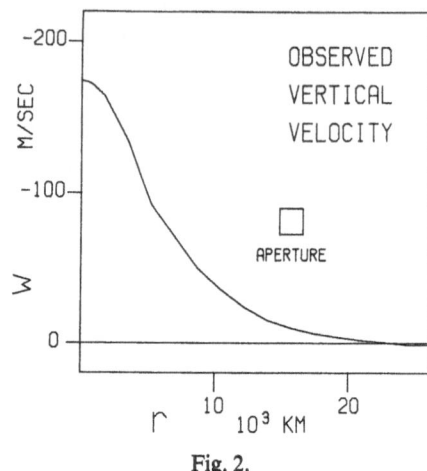

Fig. 2.

I will now describe a method for deriving the horizontal component of the velocity from the vertical one. Consider the continuity equation:

$$\frac{\partial \varrho}{\partial t} + \nabla \cdot \varrho \mathbf{v} = 0, \tag{1}$$

where ϱ is density, t is time, and $\mathbf{v}$ is velocity. A scale analysis for large slow flows reveals that the first term is much smaller than the second. Also, in a cylindrical coordinate system with r and u the radial coordinate and velocity and z and w the vertical coordinate and vertical velocity, this can be written as:

$$\frac{1}{r}\frac{\partial}{\partial r}(r\varrho u) + \frac{\partial}{\partial z}(\varrho w) = 0 \tag{2}$$

for the case of axial symmetry. If one assumes that the density ϱ does not vary with r and that

$$\frac{\partial}{\partial z}(\varrho w) = -\alpha \varrho w, \tag{3}$$

then the horizontal velocity can be found by

$$u(r, z_0) = \frac{\alpha(z_0)}{r} \int_0^r r' w(r', z_0)\, dr' \tag{4}$$

where z_0 is the observed level and α is assumed to be independent of r. This is equivalent to assuming a separable solution to a partial differential equation. Or, expressed in a different way, that the form of the components of velocity is the same at all heights.

The horizontal velocity derived from Equation (4) is shown in Figure 3. The value of α has been chosen to fix the maximum at 500 m/s in agreement with independent measurements. The horizontal velocity shown is one integral of the continuity equa-

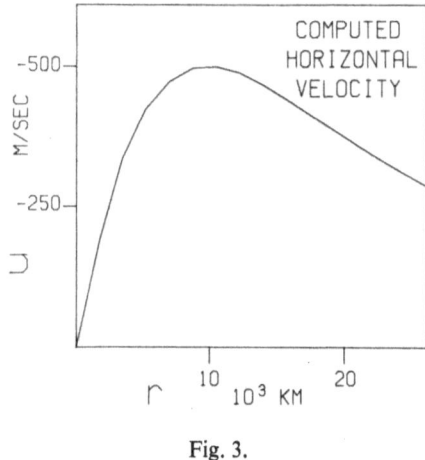

Fig. 3.

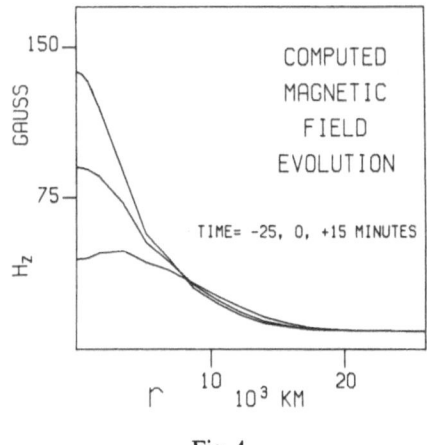

Fig. 4.

tion. It is not *the* unique integral as I have had to assume some information about the vertical structure of the velocity field.

I will now consider the effect of this velocity on the magnetic field. The magnetic diffusion time for this problem is on the order of months so that the field can be considered as 'frozen in'. The induction equation can be written as:

$$\frac{\partial \mathbf{H}}{\partial t} = \nabla \times (\mathbf{v} \times \mathbf{H}).$$ (5)

The vertical component of this equation in the case of axial symmetry is

$$\frac{\partial H_z}{\partial t} = \frac{1}{r} \frac{\partial}{\partial r} \left[r \left(w H_r - u H_z \right) \right],$$ (6)

where the subscripts r and z denote the radial and vertical components of the magnetic field. We know that $|w| < |u|$, and there is good evidence for $|H_r| < |H_z|$, so that if we assume that $|w H_r| \ll |u H_z|$ Equation (6) can be integrated for $\partial u / \partial t = 0$. The result is shown in Figure 4. The consequences of concentration are very marked, observable changes in the field strength should occur in less than an hour. No one has reported changes of this magnitude taking place in so short a time. Thus it appears that the observed velocity field is not in the process of concentrating the observed magnetic field.

I will outline briefly four possible alternate physical situations which are more consistent with the observations. This list is not meant to be exhaustive or the items on it even mutually exclusive.

Situation Number 1. The electrical conductivity is several orders of magnitude less than believed so that the field is no longer 'frozen in'.

Situation Number 2. The velocity cannot be described by a simple separable solution of the continuity equation. That is, although there may be a horizontal velocity of

500 m/s it is not where I have predicted. This can be checked by observation away from the center of the solar disk. I am planning such observations.

Situation Number 3. The flow is along the field lines. In most cases this would also imply Situation Number 2.

Situation Number 4. The magnetic field is concentrated in knots, and the material flows around the knots.

Discussion

Leighton: Is the vertical velocity field seen in the photosphere simply the downward extension of the chromospheric infall of matter seen in Hα?

Musman: This would be a consistent interpretation of the photospheric observations. It would also be included under Situation Number 3 which I have suggested.

Cowling: I suggest that you do not observe the concentration of fields simply because the filaments of strong field are relatively permanent, and the further concentration is unnecessary.

Musman: I agree. If concentration is not currently going on it must have occurred at some other time or place.

VERTICAL VELOCITIES ASSOCIATED WITH PLAGE
REGION MAGNETIC FIELDS

R. G. GIOVANELLI and J. V. RAMSAY

CSIRO Division of Physics, National Standards Laboratory, Sydney, Australia 2008

The Culgoora magnetograph (Ramsay *et al.*, 1970) produces simultaneous filtergrams in opposite circular polarizations at a wavelength selected by a filter bandwidth 0.005 nm (Ramsay *et al.*, 1970). In the blue wing of the 610.27 nm line of CaI, regions of magnetic fields in strong or weak plages are very obvious in one or other polarization, depending on polarity, even before subtraction; in one polarization they are bright, but almost invisible in the other. They are more difficult to discern at equal intervals from the line centre in the other wing (Figure 1). When subtractions are carried out to yield magnetograms of the same sense, the two magnetograms from opposite wings give results which appear to be much the same. An example is shown in Figure 2. Similar results are obtained over a wide range of positions in the wings of the 610.27 nm line.

Our observations have involved many differing states of adjustment of the filter for which the instrumental line profile must have varied over wide limits without any substantial change in the relative appearances of the red- and blue-wing filtergrams. Rotation of the $\lambda/4$ plate through 90° has no significant effect beyond a reversal of the polarized filtergrams. Hence we conclude that the asymmetry in the appearances of the filtergrams is not due to the instrument but is of solar origin. We interpret it as due to a more-or-less uniform inwards velocity in magnetic regions.

The 610.27 nm absorption line is not as deep in magnetic as in non-magnetic regions, a result first found by Sheeley (1967) and Chapman and Sheeley (1968). Figure 3 shows a schematic line profile in the absence of magnetic field together with the combined effects of a longitudinal Zeeman splitting and a uniform Doppler shift to the red in a magnetic region. Filtergrams obtained in the blue wing will show magnetic regions much brighter than the background in one polarization, and of reduced contrast in the other. In the red wing, the maximum intensity difference from background is less than in the blue wing. However the intensity differences between the two polarizations are identical for the two wings.

A non-uniform vertical velocity would have the effect of making one wing of the line steeper than the other. This would cause magnetograms in one wing to show magnetic fields with greater contrast than in the other. Whilst magnetograms in opposite wings are not always identical, there does not appear to be any systematic difference, nor have we found any differences that could not be explained by seeing or the state of instrumental adjustment. We therefore conclude that the vertical velocity is more-or-less uniform.

There is another way of demonstrating and measuring downwards velocities in magnetic regions. The filter, consisting of a series of interferometers, transmits a pass

Howard (ed.), Solar Magnetic Fields, 293–297. All Rights Reserved.
Copyright © 1971 by the IAU.

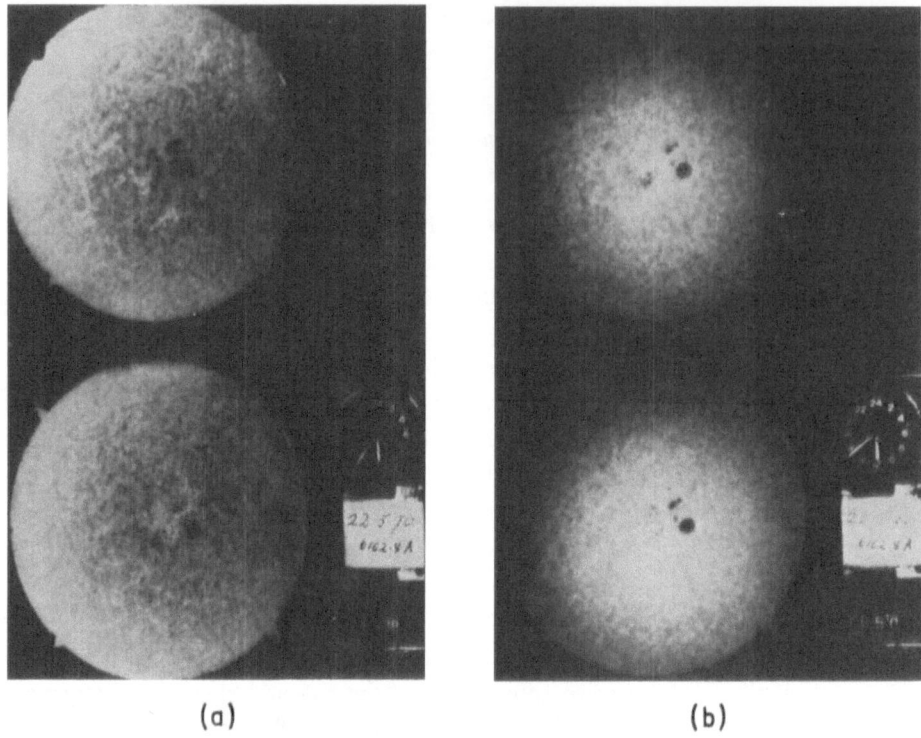

Fig. 1.   Filtergrams in the Ca I 610.27 nm absorption line, blue wing (a) and red wing (b). The two upper photographs are in one circular polarization, the two lower in the other.

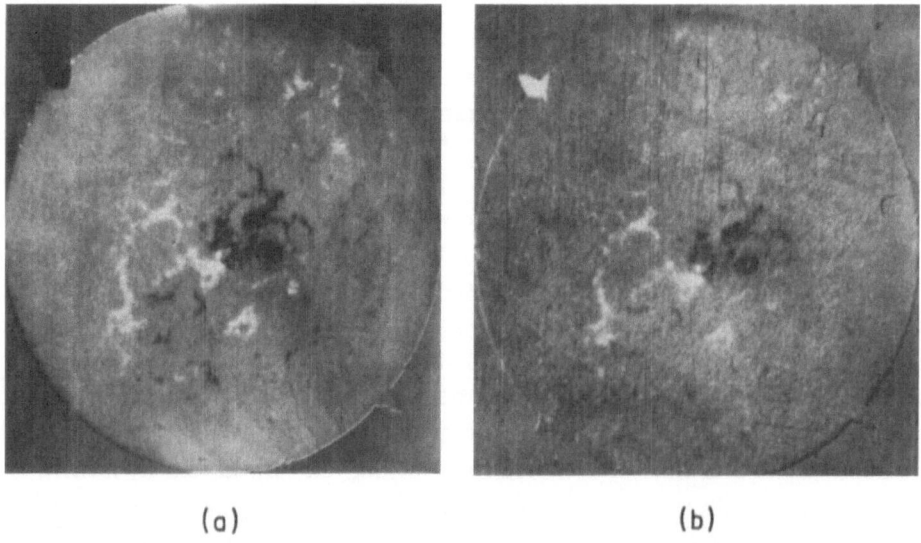

Fig. 2.   Magnetograms obtained from filtergrams in Figure 1. (a) is from the blue wing, (b) from the red wing.

band whose peak intensity varies in wavelength towards the blue by 0.005 nm from the centre of the field to the edge, the wavelength shift being proportional to the square of the distance from the centre. Thus a filtergram obtained with the centre of the field 0.0025 nm to the red of line centre has the edge of the field at 0.0025 nm to the blue, while at 0.707 of the radius from centre to edge there is a minimum intensity corresponding to line centre (Figure 4c). If two such filtergrams obtained in opposite circular polarizations are subtracted, magnetic fields will be seen near the

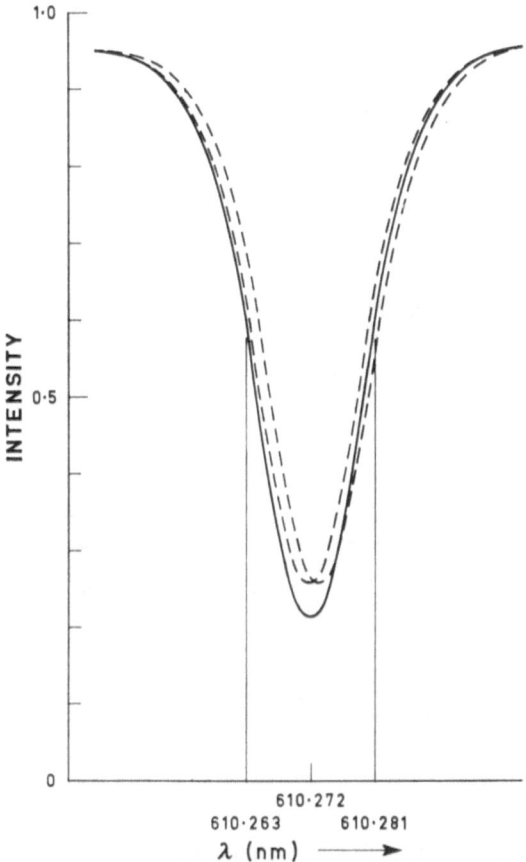

Fig. 3.   Schematic line profile in a non-magnetic region (deepest core), and Doppler-shifted Zeeman components in a magnetic region.

centre of the picture in the same sense as in the red wing. Near the edge, they appear in the same sense as in the blue wing, while at the line centre of the *magnetic region* they are invisible. If the magnetic region has a Doppler shift, the reversal in sense does not occur exactly at the zone corresponding to the normal line centre. Figure 4 shows the effect, in which the reversal occurs at 0.0010 nm to the red of line centre (i.e., closer to the centre of the field), corresponding to a downwards velocity of 0.5 km s^{-1}. This result is not affected systematically by seeing.

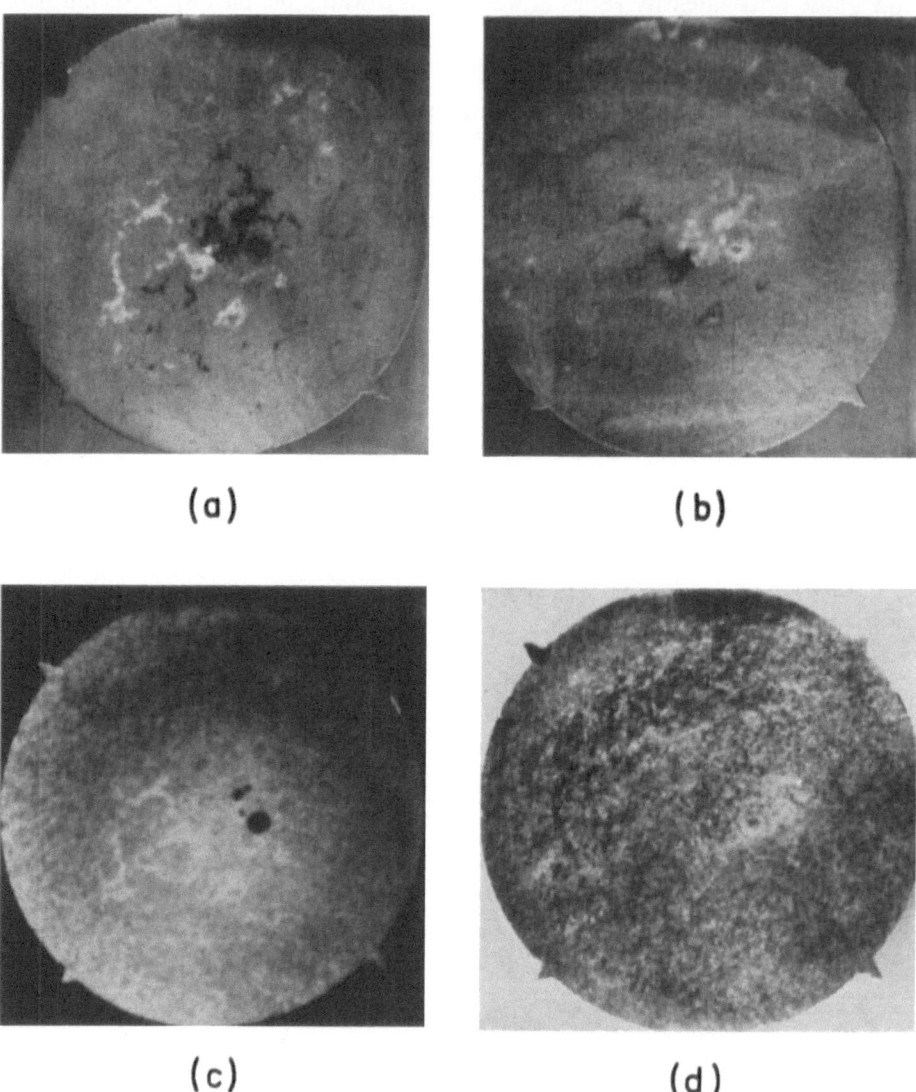

(a)

(b)

(c)

(d)

Fig. 4.   (a) Magnetogram in CaI 610.27 nm absorption line, blue wing; (b) magnetogram near line centre; (c) one of the filtergrams from which (b) was derived. Note the dark ring in (c) corresponding to line centre. The corresponding velocity field with dark denoting a downwards velocity is shown in (d).

Both methods yield velocities in the same sense (inwards) and of the same magnitude.

When we compare these results with velocities observed in non-magnetic regions, an important difference emerges. As Leighton (1962) has shown for many lines including 610.27 nm, there is a strong correlation between intensity at a given interval from the line centre and velocity as measured at that part of the line profile, such that the brighter (and hence hotter) regions are moving upwards. But where there are magnetic fields, the monochromatic intensity is greater and material is moving downwards!

Velocity photographs obtained by appropriate subtraction of filtergrams from opposite wings do not show an obvious pattern corresponding to the magnetic field, though comparisons with the magnetograms in a blink comparator show, in conformity with the results described above, that in magnetic regions the velocities are downward (Figure 4d). Thus the magnetic field does not appear to be the *cause* of the associated downward velocities. Beckers and Schröter (1968) have previously stated that magnetic regions are almost always associated with downward velocities, the rms value being about the same as in the undisturbed photosphere. Our results are in agreement with theirs.

References

Beckers, J. M. and Schröter, E. H.: 1968, *Solar Phys.* **4**, 142.
Chapman G. A. and Sheeley, N. R.: 1968, *Solar Phys.* **5**, 442.
Leighton, R. B.: 1962, *Astrophys. J.* **135**, 474.
Ramsay, J. V., Kobler, H., and Mugridge, E. G. V.: 1970, *Solar Phys.* **12**, 492.
Ramsay, J. V., Giovanelli, R. G., and Gillett, H. R.: 1971, this volume, p. 24.
Sheeley, N. R.: 1967, *Solar Phys.* **1**, 171.

SPECTRA-SPECTROHELIOGRAPH OBSERVATIONS

ALAN M. TITLE and JOHN P. ANDELIN, JR.

Harvard College Observatory, Cambridge, Mass. 02138, U.S.A.

Abstract. This paper describes a photographic technique for recording solar spectra. The technique, which we call spectra-spectroheliography, requires the same amount of observing time as conventional spectroheliography but yields much more information. Data reduction techniques have been developed to use this information for the construction of spectroheliograms and other contour maps. Parameters investigated by this method include the vector magnetic field, line-of-sight velocity, continuum intensity, and line strength. Results indicate (1) good correlation between bright regions in spectroheliograms and small vertical magnetic fields; (2) poorer correlation between larger magnetic fields and bright regions; (3) a feature in the continuum resembling a pore but having no vertical magnetic field; and (4) a gradient in the vertical magnetic field of 12 kG/arc s.

1. Introduction

We would like to discuss a technique for recording and analyzing solar spectra that is similar to spectroheliography (SHG) but yields much more information. This technique, which we call spectra-spectroheliography (S^2HG), appreciably extends the number of physical parameters that can be studied from a single observation. It also eliminates many of the difficulties encountered in interpreting conventional spectroheliograms, and achieves these results without introducing any additional complexity during observation. After outlining the principles underlying this technique, we will present some representative spectroheliograms and contour maps of magnetic fields and continua reconstructed from S^2HG spectra obtained at the Kitt Peak Solar Tower in 1969.

2. System Description

S^2HG is an extension of SHG and, as such, is closely related to it. In both instances, the Sun's image is moved with respect to the entrance slit of a spectrograph and recorded on film that is simultaneously being moved with respect to the exit slit. The most important difference between S^2HG and SHG is that S^2HG records a wide spectral region and SHG records only a narrow wavelength band.

Ordinarily, spectroheliograms are made by fixing both the film and the Sun's image and moving the spectrograph. Exposure is made through a narrow exit slit. The result is a photograph of the Sun with variations in density representing variations in the intensity of solar radiation (integrated over a small fixed wavelength band). It is difficult to determine from a single spectroheliogram whether these changes in density are due to Doppler shifts, Zeeman splittings, or variations in continuum intensity or line strength. Various techniques using combinations of spectroheliograms have been developed in an attempt to distinguish these different physical effects.

The S^2HG spectrograph (Figure 1) has a wide exit aperture and a specially constructed movie camera capable of rapidly advancing the film. In recording spectra,

Howard (ed.), Solar Magnetic Fields, 298–309. All Rights Reserved.
Copyright © 1971 by the IAU.

the spectrograph remains stationary and the Sun's image is allowed to drift past the entrance slit. A single frame of the film is exposed while the Sun moves a distance Δ with respect to the entrance slit. The film is then rapidly advanced and the next frame is exposed. Except for the time required to advance the film, exposure times are the same as in making conventional spectroheliograms. (Our film advance times were typically 2 to 5% of our exposure times.) Thus, S^2HG spectra are frames on the film, each frame containing the entire spectral profile along a line on the Sun's disk (Figure 2, a and b). Since successive frames contain spectral profiles for adjacent regions on the Sun, the film contains the spectral line profile for each position on a two-dimensional grid on the solar disk. The grid spacing is Δ perpendicular to the slit

SPECTRASPECTROHELIOGRAPH

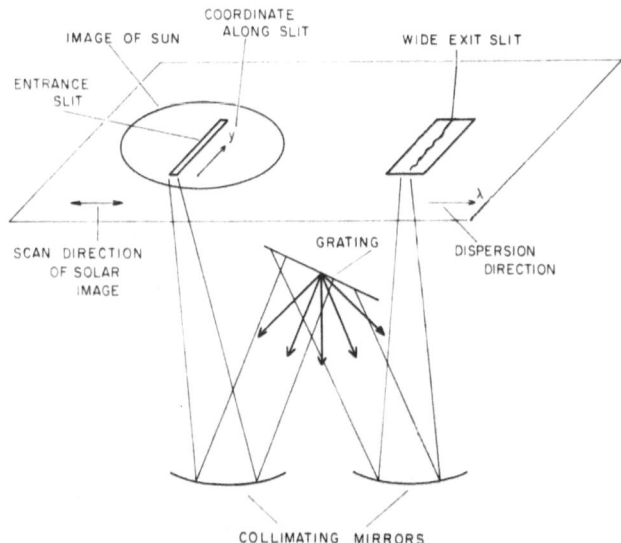

Fig. 1.   Schematic of S^2HG optics.

and the resolution of the spectrograph along the slit. By inserting quarter- or half-wave plates and a polarization-dependent beam splitter into the system, we obtain spectral profiles in polarization pairs, as shown in Figure 2c.

Figure 3 illustrates how conventional spectroheliograms can be constructed from S^2HG film strips. The wavy line represents a particular spectral line as it appears in each frame. In principle, to produce a spectroheliogram at a given wavelength, we need only remove a thin strip of film from each frame at the appropriate position for that wavelength. The desired spectroheliogram can then be built up merely by placing adjacent strips side by side. In practice, of course, we do not cut up our original film. The only information we require is the relative intensity of the original solar radiation at the given wavelength for each position on the Sun. To obtain this, we microdensitometer the film and digitally record the output on tape. We then construct the spectroheliogram from the data on the tape.

Since the tape contains far more information than we need for the spectroheliogram, we can use the remaining information to construct contour maps of other interesting physical parameters. For example, by selecting different positions in each frame, we can vary wavelength and produce a spectroheliogram of the center of a line even in the presence of large Doppler shifts. Moreover, we can make contour maps

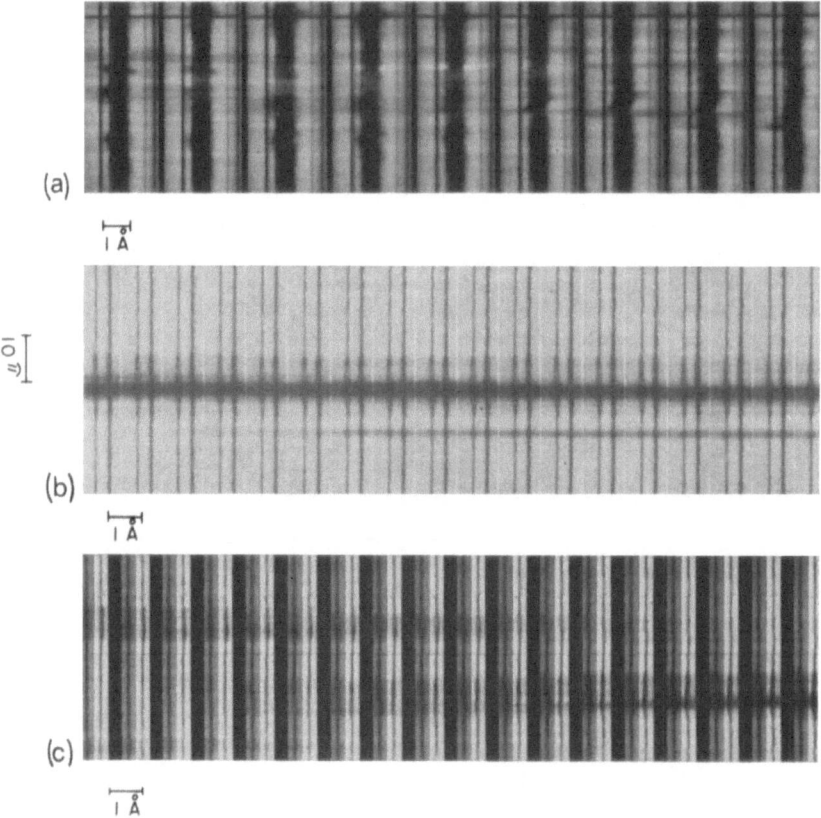

Fig. 2.   Representative S²HG spectra. The film strips are positives. (a) Hα spectra. Adjacent frames are separated by one arc second on the solar disk. (b) Fe I (λ 5250.2) spectra (left line). Adjacent frames are separated by one-half arc second on the solar disk. (c) Fe I (λ 5250.2) spectra in right and left circular polarization. Adjacent frames are separated by one-half arc second on the solar disk.

of the Doppler shift itself as well as of the continuum intensity and the total absorption of the line. If a line displays a Zeeman effect and we take spectra in right and left circular polarizations, then we can construct longitudinal magnetic field maps. If Zeeman sensitive lines are taken in several pairs of linear polarizations, then transverse magnetic field maps can be constructed.

Whenever the magnetic field is small, our computer program is essentially a digital analog of an Evans magnetograph. For fields above a kilogauss, however, the program directly measures the splitting of the Zeeman components. It adjusts the separation of the peak-detecting slits to detect peculiar line profiles, thus alerting us to those interesting regions of the Sun where the profile of an observed line turns out to be different from the theoretical prediction. Such discrepancies can only be detected by examining an entire line profile. An important advantage of S^2HG spectrograms is that if subsequent theoretical work should suggest other spectral measurements, we could imme-

SPECTRA (S^2HG)

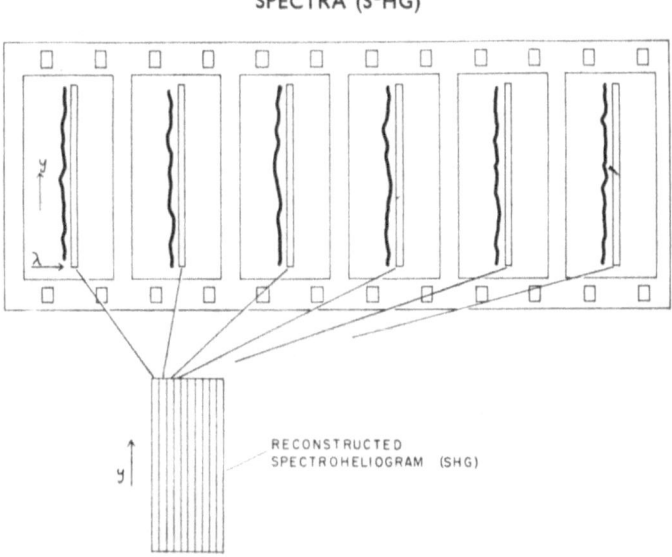

Fig. 3.   Principle of constructing spectroheliograms from S^2HG spectra.

diately investigate these by rewriting the S^2HG computer program and analyzing existing data. We would not need additional observing time for new data.

An important practical feature of S^2HG is that it does not require the precise mechanical motion nor delicate operation of a spectroheliograph. Thus, the increased rate of data acquisition and variety of physical parameters investigated are achieved concurrently with simplification of the process of solar observation.

3. Observations

In August 1969 we used the S^2HG system to make observations in $\lambda 5250.2$ (Fe I) using the Kitt Peak Solar Tower. Spectra were photographed on 70 mm Linagraph Shellburst film approximately once a second, during which time the Sun moved one-half arc second relative to the entrance slit. Seeing was often better than one-half arc second, so we were able to record the spectrum of $\lambda 5250.2$ at every point on a one-half arc second grid. Since we photographed the Sun's surface at a rate of one-half square arc minutes per minute, we were able to record a complex spot group three or four times in 15 min, alternately analyzing for linear and circular polarizations.

4. Reproducibility of Results

Most of the data reduced so far are from gap and pore regions rather than from well-developed spots. To make certain that our data reduction procedures were free from ambiguities, the films of these regions were subjected to two tests: (1) Different individuals independently analyzed the same strip of film, from setting up the micro-densitometer to generating the digital data for construction of contour maps. In each case the contours generated agreed to better than one-half second of arc. The magnitudes of the longitudinal magnetic field, for example, agreed to a few gauss in quiet regions and to a few percent in fields of 500 to 1000 G. This small difference could easily be due to differences in positioning of the data grids with respect to the solar features. (2) Contour maps constructed from two different films of the same pore taken one-half hour apart were compared. The central longitudinal magnetic field differed by 5%; the contours agreed to better than one-half arc second. We do not know how much of this 5% difference is due to positioning of the data grids, to a change in the seeing conditions or to an actual change in the field of the pore.

5. Experimental Results

A. CONTOUR MAPS

We would like to present some spectroheliograms and contour maps of magnetic fields of corresponding regions. The spectra are from gaps and pores up to two arc seconds in diameter. A 9×12 arc s region around one such gap is shown in Figure 4. In contrast to a map of the continuum (not presented here), which had no particularly distinguishing features, the spectroheliogram (Figure 4a) clearly shows a very bright region (A) and several smaller bright regions nearby. A comparison of this spectroheliogram and the corresponding vertical magnetic field map (Figure 4b) indicates a very high correlation. The very bright region (A) is associated with a vertical magnetic field of nearly 500 G and most of the smaller bright regions, only 1 or 2 s in diameter, have vertical magnetic fields of 100 to 300 G, (C) and (D) for example.

Figure 5 is another set of contour maps showing (a) the continuum, (b) the spectroheliogram at line center, and (c) the vertical magnetic field of a small pore having a central field exceeding one kilogauss. The narrow lane of low vertical field in the top-center of the magnetic field map is apparently real – it is also present in a vertical magnetic field map (not shown) constructed from film exposed $\frac{1}{2}$ h earlier.

In contrast to the very high correlations between vertical magnetic fields and bright regions in Figure 4, we see that the location cospatial with the highest field [(E) in Figure 5b] occurs 2 to 3 arc s from the bright patches in the spectroheliogram. The correlation does not improve when other positions in the spectral line are examined. This, then, is a pore that is clearly visible in the continuum but in a spectroheliogram does not show up as a bright region over its entire area.

The last set of contour maps (Figure 6) is of an interesting region containing some small pores. The continuum map (a) shows pores at F and G. The line center spectroheliogram (b) indicates that one of the pores correlates with a bright region and the other with a dark one. It also shows an additional bright feature (H) to the left and below the two pores. The vertical magnetic field map (c) is qualitatively correlated with the spectroheliogram in that the vertical magnetic fields and bright regions in

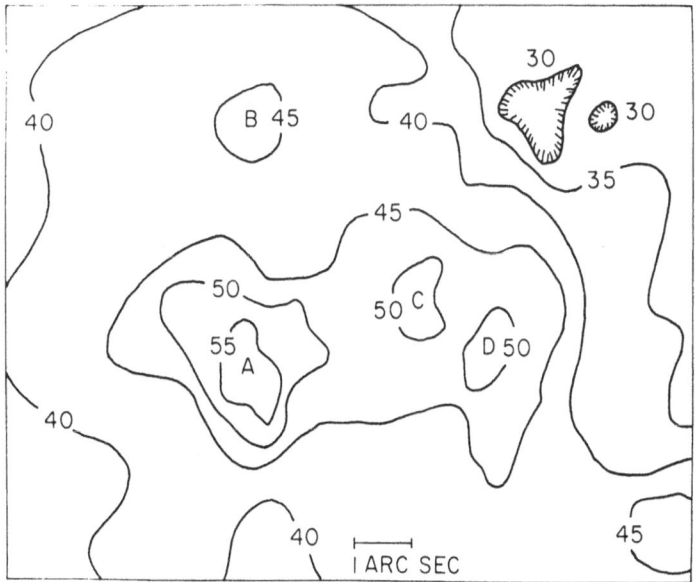

Fig. 4a.

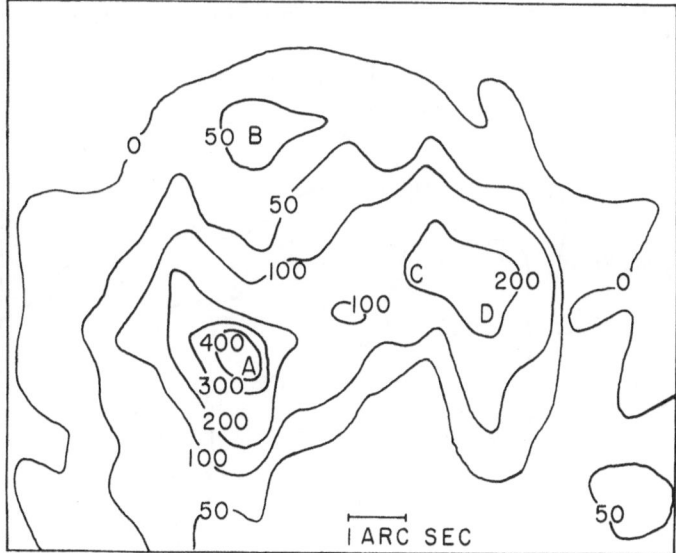

Fig. 4b.

Figs. 4a–b.   Contour maps of a gap region. (a) Spectroheliogram at line center FeI (λ5250.2). (b) Longitudinal magnetic field.

the spectroheliogram are roughly cospatial. However the vertical magnetic field map is distinctly different from the continuum map; that is, the pore (F), clearly visible in the continuum, has no vertical magnetic field. The linearly polarized spectra from this pore differ from spectra in neighboring regions. Thus, even though the data are not completely reduced, we suspect the presence of a significant horizontal field.

B. SPECTRA

Shown in Figure 7 are 4 traces of the $\lambda5250.2$ spectrum in a pair of orthogonal linear

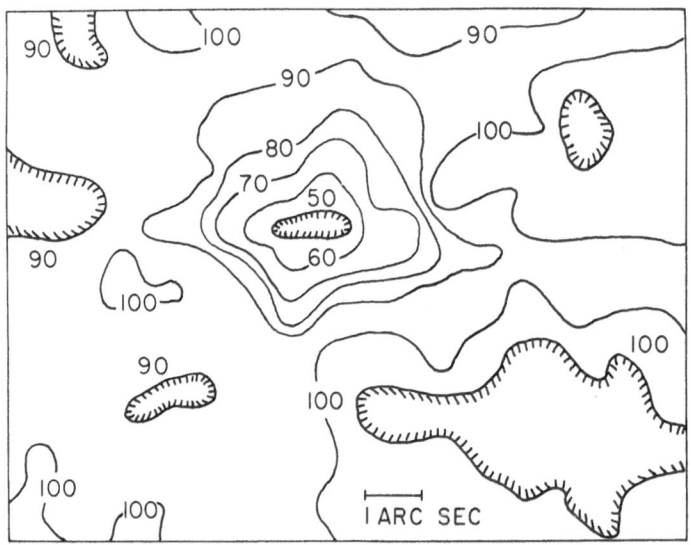

Fig. 5a.

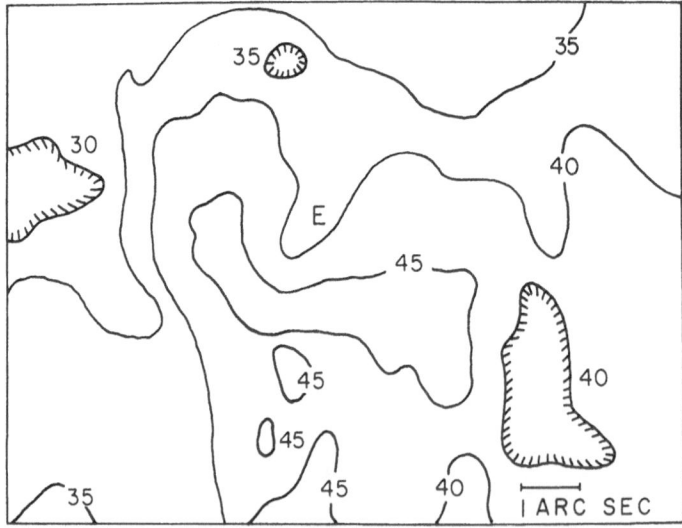

Fig. 5b.

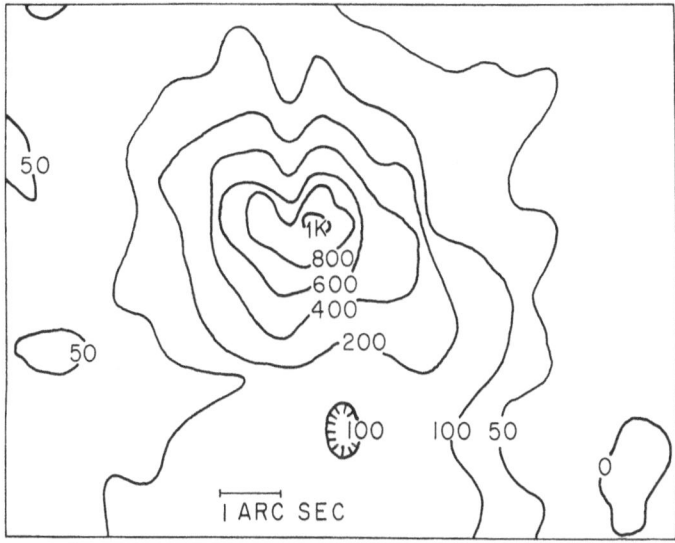

Fig. 5c.

Figs. 5a–c.   Contour maps of pore region I. (a) Continuum. (b) Spectroheliogram at line center.
(c) Longitudinal magnetic field.

polarizations. The spectra are from points on the Sun spaced one arc second apart along a line that starts just inside the umbra of an ordinary round sunspot and ends just outside, in the penumbra. From these tracings we infer that the transverse component of the magnetic field rotates about 90° in 2 arc s, that the field strength is on the order of 2000 G, and that the change of direction occurs precisely at the umbra-penumbra boundary. Inside the umbra the transverse component of the field is in the east-west direction. Inspection of the circularly polarized spectra indicates that the vertical component of the field does not do anything unusual in this region.

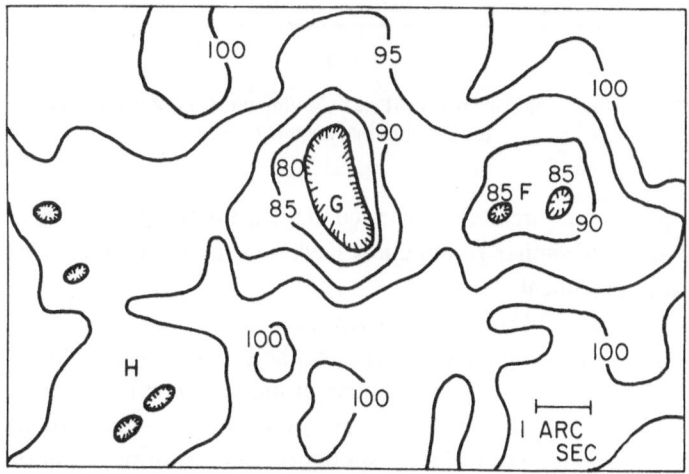

Fig. 6a.

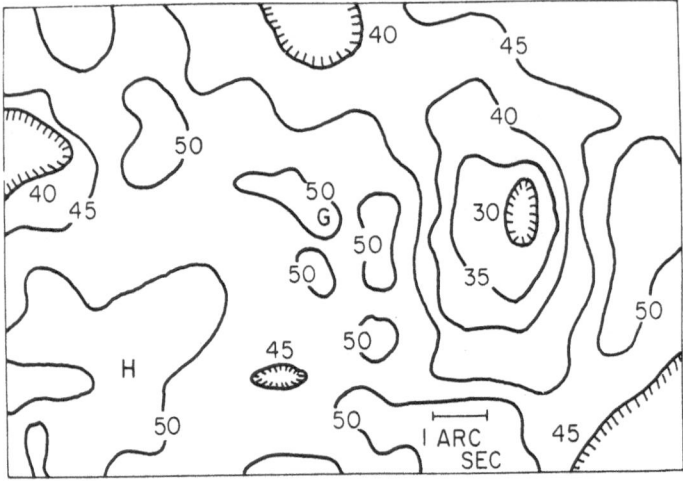

Fig. 6b.

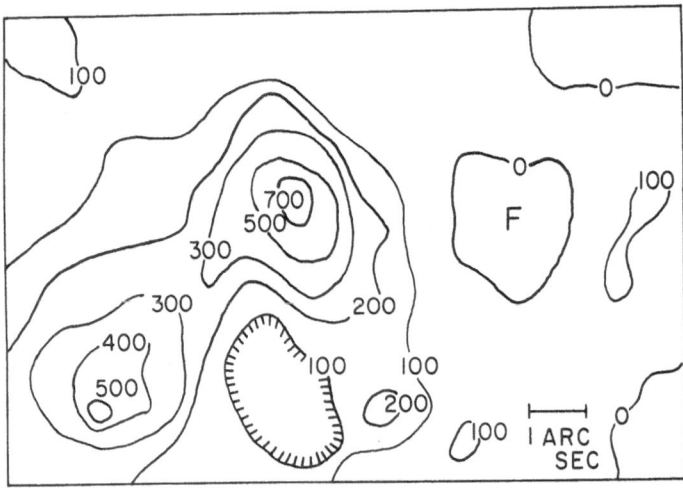

Fig. 6c.

Figs. 6a–c.   Contour maps of pore region II. (a) Continuum. (b) Spectroheliogram at line center.
(c) Longitudinal magnetic field.

In fact, comparison of circularly polarized spectra for this high transverse gradient sunspot and those of similar round spots having no unusual transverse fields shows them to be almost identical.

We cannot now state whether the sharp gradient in the transverse field develops with time or if the spot originates with this characteristic. However, the feature is not uncommon. We have observed it in a number of similar ordinary round sunspots.

High gradients in the transverse field are by no means limited to regular round sunspots. We mentioned the ordinary spots first only to draw attention to the fact that high gradients and high currents can and do exist in what might be considered

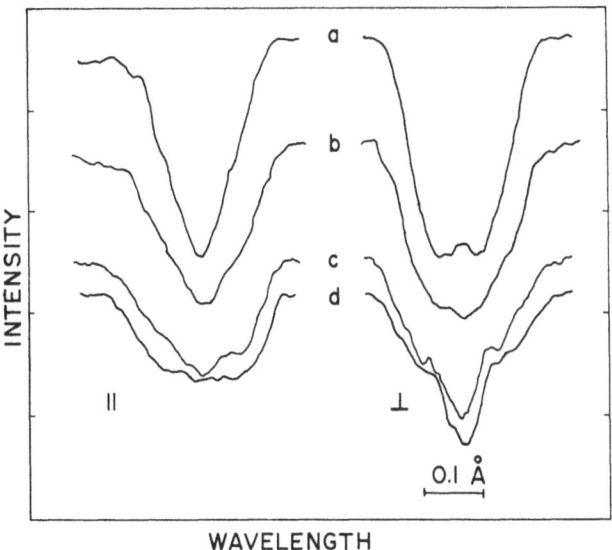

WAVELENGTH

Fig. 7.   The traces are of Fe I λ 5250.2 in orthogonal polarizations. Points *a*, *b*, *c* and *d* lie along a radius of the sunspot. Traces for *a* are from a point 1 arc s inside the penumbra; for *b*, from a point on the umbra-penumbra boundary; *c* and *d*, from points 1 and 2 arc s, repectively, inside the umbra.

uncomplicated field regions. Shown in Figure 8 is a series of spectra in pairs of orthogonal linear polarizations taken of a complex sunspot having several umbra. Note that a transverse field of 2500 G rotates by 90° in an arc second or less.

Figure 9 shows spectrograms of the same complex sunspot in circular polarization. From these spectra we infer that a 2900 G field reverses direction in one-half arc second or less; that is, there exists a gradient in the magnetic field on the order of 12 kG per arc second. This high gradient region occurs along a line 8 arc s long. The field is vertical and upward on one side of the line and vertical and downward on the other. Note also that the magnitude of the field increases toward the line of discontinuity; that is, the field is 2900 G at the reversal but 2500 G only a few arc seconds away. Further examination of the transverse field shows it to rotate by about 45° cospatially with the reversal. This discontinuity persisted for at least a week during which time the length of the line increased. No flares were observed in this high vertical gradient region.

6. Summary of Results

A. CONTOUR MAPS

(1) The gradient of the vertical magnetic field in regions of pores and gaps is often at least as great as 500 G/arc s for pores and 250 G/arc s for gaps.

(2) Regions with moderate magnetic fields (less than 700 G) that are not in the immediate neighborhood of large magnetic fields are brighter than average in the center of λ 5250. The magnetic field regions in the magnetogram and the bright regions in the spectroheliogram are coincident within our one-half arc second resolution.

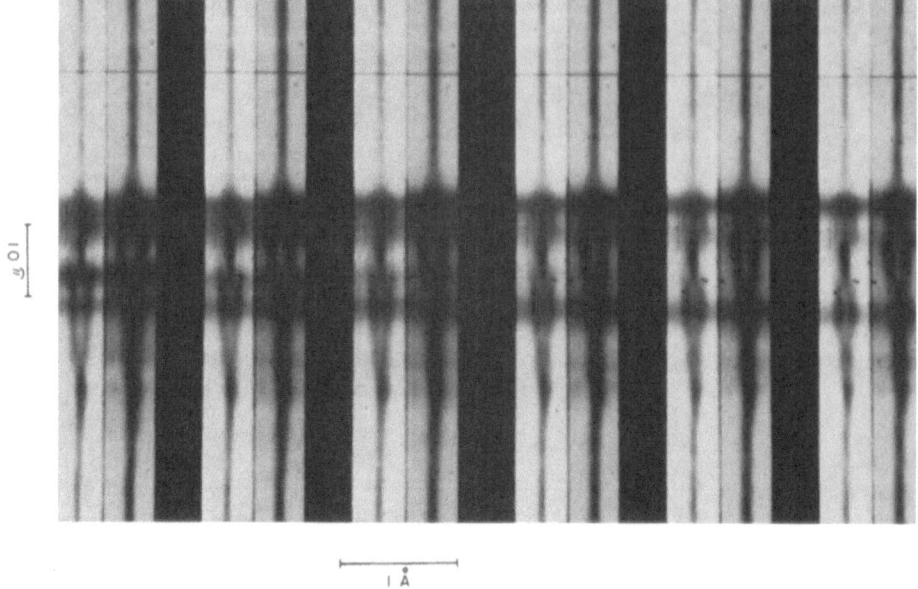

Fig. 8.   Six pairs of spectra from the region around λ5250.2. One member of each pair was taken in right circularly polarized light, the other in left circularly polarized light. Each pair of spectra is displaced from the next by a distance corresponding to one-half arc second on the Sun. The displacements are in the R.A. direction.

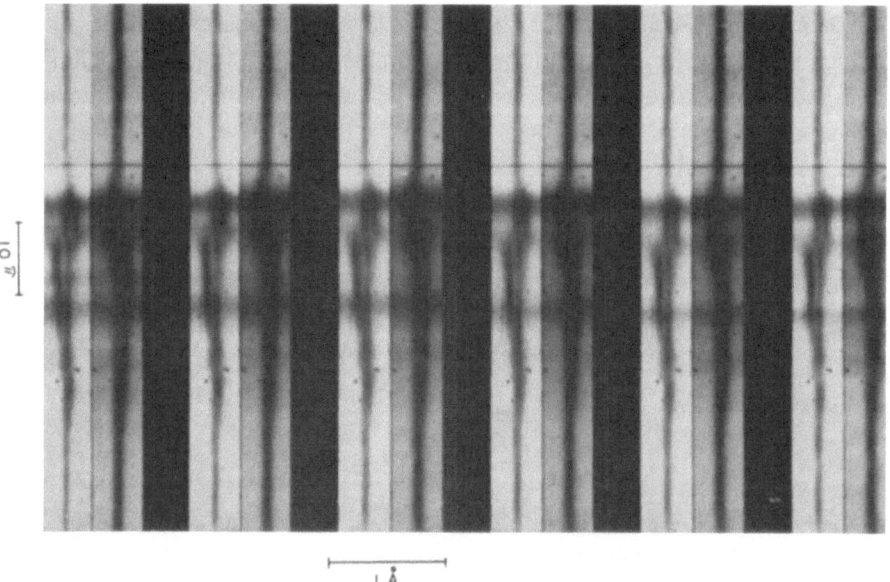

Fig. 9.   Six pairs of spectra from the region around λ5250.2. One member of each pair was taken in a linear polarization parallel to the slit, the other in the orthogonal direction. Each pair of spectra is displaced from the next by a distance corresponding to one-half arc second on the solar disk. The displacements are in the R.A. direction. The variation in intensity between members of a pair is due to different spectrograph efficiencies for the two polarizations.

(3) Regions with magnetic fields greater than 700 G are darker than average in the continuum. [The converse is not necessarily true, as shown in (4) below.] The high magnetic field region in the magnetogram and the dark region in the continuum map are coincident within our one-half arc second resolution.

(4) Regions that are darker than average in the continuum and of an appropriate size to be a pore sometimes have no vertical magnetic field.

B. SPECTRA

(1) The transverse field in apparently normal round sunspots may change direction by approximately 90° in 1 to 2 arc s, with the reversal occurring at the umbra-penumbra boundary.

(2) The transverse field in complex spots may change direction by 90° in one half arc second or less.

(3) In complex spots large vertical magnetic fields may change direction by 180° in one half arc second or less. This implies a gradient of 12 kG/arc s.

Acknowledgements

The work described here was based on data taken with the McMath Solar Telescope of the Kitt Peak National Observatory. We are extremely grateful for the help and encouragement given to us by the staff of the Solar Division there. Without the considerable aid of Keith Pierce we could not have obtained data of such quality.

Funding for the rapid-film-advance camera and the early years of S^2HG operation was provided by Sacramento Peak Observatory, Air Force Cambridge Research Laboratory under Air Force Contract No. F19628-67-C-0734. Additional support has come from NASA Contract No. NAS 5-3949.

Discussion

Jakimiec: I would like to emphasize that the observed large gradient of the magnetic field strength does not necessarily imply the presence of a strong current at the place. Large gradients of the field strength may exist also in the regions where the magnetic field is already nearly potential.

Meyer: It would be interesting to compare the magnetic field gradients that you observe with the distances that one would expect theoretically if one considers the diffusive penetration by finite conductivity of opposing fields. For a quasistationary diffusive equilibrium at the border of regions of opposite field direction one has to require that the plasma being swept up by the diffusive cancellation of opposite fields has time to settle vertically. For electrical conductivities suggested by Schröter and Kopecký for sunspots, one determines typical widths for the transition region between opposite strong fields of less than 100 km.

Michard: Since you told us that this line of abrupt inversion of $H_{\parallel}$ was stable in time, it is not surprising that no flares were associated with it. Changes in the magnetic patterns on an hour-to-hour scale are necessary for flare productivity.

THE TIME DEPENDENCE OF MAGNETIC, VELOCITY, AND INTENSITY FIELDS IN THE SOLAR ATMOSPHERE

NEIL R. SHEELEY, JR.

Kitt Peak National Observatory*, Tucson, Ariz. U.S.A.

Abstract. Spectroheliogram movies of magnetic, velocity, and intensity fields have been obtained in a wide variety of spectral lines. The time resolution was typically 10–30 s. Some of the lines used were: CN λ 3883; CaI λ 6103; FeI λ 4071; FeI λ 5434; FeII λ 4924; CaII λ 3934; CaII λ 8542.

During the summer of 1970, spectroheliogram movies were made daily using the 38 cm solar image provided by the east auxiliary telescope at the Kitt Peak National Observatory. The use of a very fast film (Kodak 2485) permitted rapid scanning of the spectroheliograph resulting in a time resolution ranging from 10 s to 30 s depending

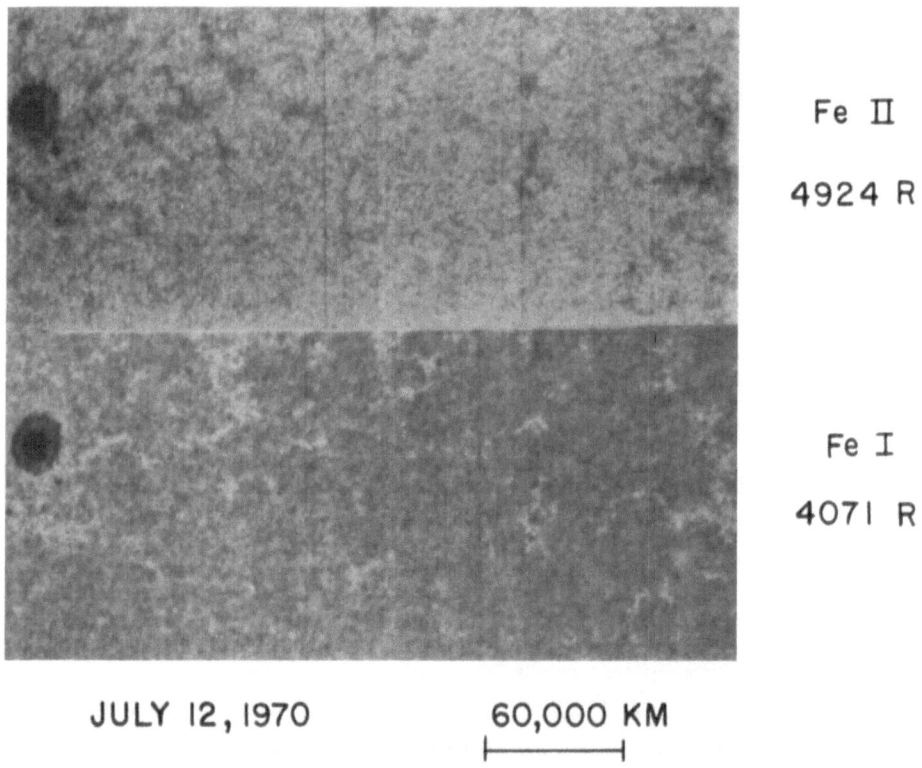

Fe II

4924 R

Fe I

4071 R

JULY 12, 1970   60,000 KM

Fig. 1.   Comparison between the dark network in FeII λ 4924 and the bright network in FeI λ 4071.

* Operated by the Association of Universities for Research in Astronomy Inc., under contract with the National Science Foundation.

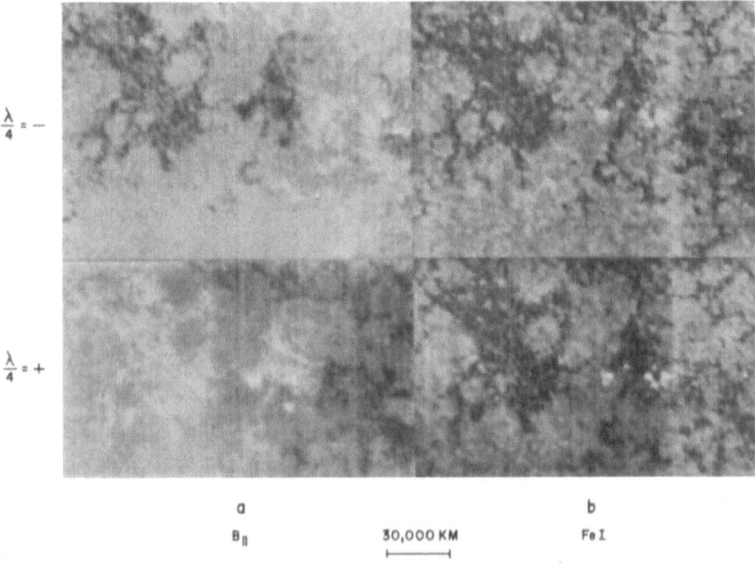

Fig. 2.   Comparison between the bright network in Fe I λ 4071 and the magnetic field distribution obtained using Ca I λ 6103.

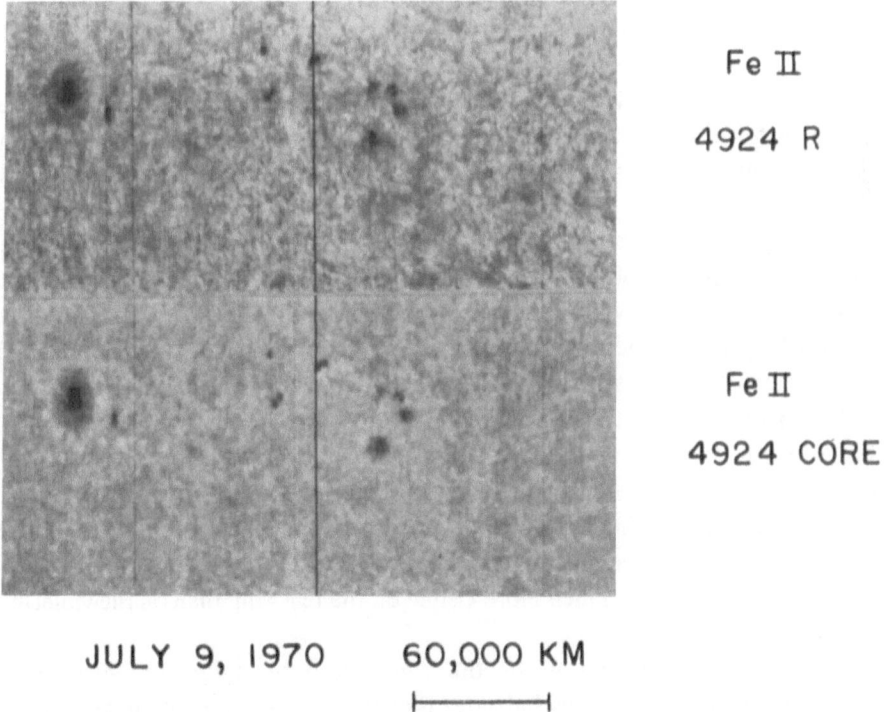

Fig. 3.   Comparison of the appearance of spectroheliograms taken on the λ 4924 red wing and in the λ 4924 core respectively.

on the line studied. A spatial resolution of 1 s was achieved for periods of 30 min to 1 hr on about 25% of the mornings observed. Although the study of these movies is only beginning, a few of the most striking features will be listed in this paper. Here we shall describe some preliminary observations made using the 4924 Å line of Fe II.

Figure 1 compares simultaneous spectroheliograms made in the red wings of Fe II λ4924 and Fe I λ4071. Dark plages and network appear on the λ4924 spectrohelio- gram and correspond roughly to the bright plages and network on the λ4071 spectro- heliogram. It is already known (see Figure 2 taken from Sheeley and Engvold, 1970) that this bright network in λ4071 is cospatial with photospheric magnetic fields.

Figure 3 compares spectroheliograms made in the λ4924 red wing and core res- pectively. Although these spectroheliograms were not taken simultaneously, they were taken within 30 min and the comparison is typically what one observes with simultaneous pictures. The core spectroheliogram shows active region brightenings characteristic of the middle chromosphere and a somewhat confusing diffuse brighter- than-average network similar to that seen in the cores of so many medium-strong lines such as Mg I λ5183 (Chapman and Sheeley, 1968). However, in Figure 3 the dark network in Fe II λ4924 is not as striking as that of Figure 1. In particular, the network seems to be visible, not because it is much darker than average, but because the velocity granulation characteristic of nonmagnetic regions does not occur here. An examina- tion of the data to determine the reason for this difference has not yet begun.

Figure 4 emphasizes the relation between the dark network in λ4924, the magnetic field pattern, and the bright network of λ4071. In this figure, the λ4071 spectrohelio- gram was taken simultaneously with the λ4924 red-wing spectroheliogram first in right-circularly-polarized light and 40 s later in left-circularly-polarized light. It is seen that the magnetic signal alternately reinforces and cancels the dark plages in the λ4924 red wing depending on the magnetic polarity and the 'handedness' of the circu- lar polarization.

Figure 5 shows how the λ4924 line can be used to make a Zeeman photograph. Simultaneous red-wing λ4924 spectroheliograms in right and left circularly-polarized light are shown together with the Zeeman photograph obtained by cancelling them. This figure represents one frame of a sequence taken every 30 s for one hour. This magnetic movie looked essentially the same as those seen in Ca I λ6103 (Vrabec, 1971) and the brightness field in CN λ3883 (Sheeley, 1969).

Figure 6 summarizes combinations of spectroheliograms taken on red and violet wings of λ4924. Simultaneous violet-wing and red-wing λ4924 spectroheliograms are shown together with the intensity field obtained by adding them and the velocity field obtained by subtracting them. Several features are clear:

(1) The dark network is much more visible on the red-wing than on the violet-wing spectroheliogram.

(2) The dark network appears on the 'sum' spectroheliogram in which doppler shifts have been cancelled. This indicates that the network is really darker than its surroundings and is not *solely* the result of doppler shifts.

(3) The region of the dark network appears on the velocity spectroheliogram and

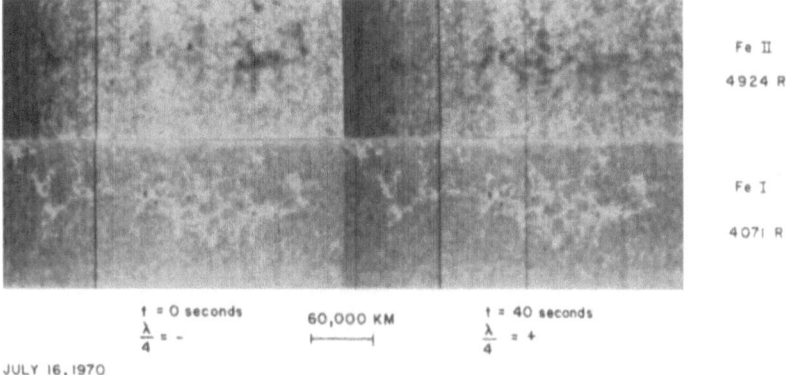

Fe II
4924 R

Fe I
4071 R

t = 0 seconds   60,000 KM   t = 40 seconds
$\frac{\lambda}{4}$ = - $\frac{\lambda}{4}$ = +

JULY 16, 1970

Fig. 4.   Comparison between the $\lambda 4071$ bright network and the magnetically polarized $\lambda 4924$ dark network.

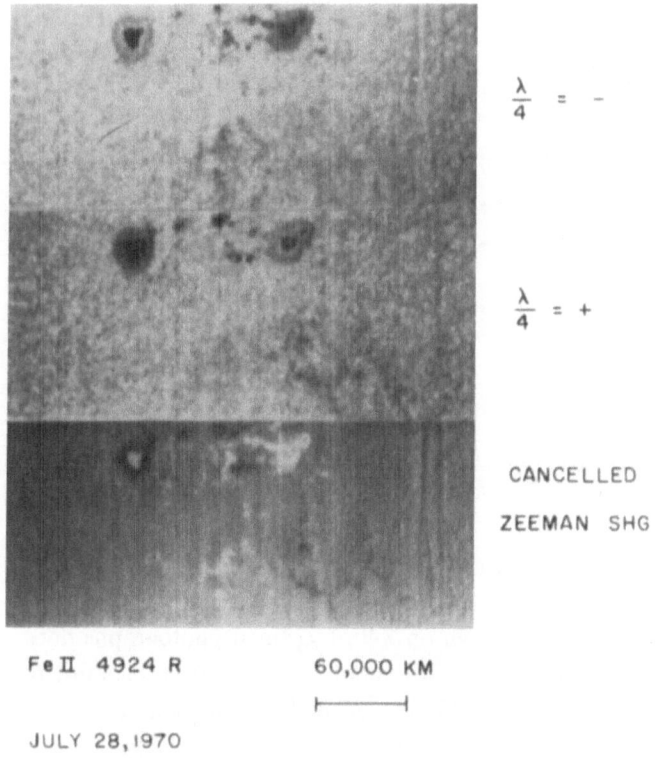

$\frac{\lambda}{4}$ = -

$\frac{\lambda}{4}$ = +

CANCELLED

ZEEMAN SHG

Fe II 4924 R   60,000 KM

JULY 28, 1970

Fig. 5.   Comparison between uncancelled and cancelled $\lambda 4924$ Zeeman spectroheliograms showing the appearance of the magnetic field as determined with this Fe II line.

in this region the small-scale velocity is severely inhibited. Furthermore there is a definite downward direction to the velocity in the dark plage region.

A number of properties of the $\lambda 4924$ dark plage regions have been learned by viewing the movies:

(1) The visibility of the red-wing dark plages oscillates with a definite 5 min period.

(2) The visibility of the plages on the 'sum' spectroheliogram (obtained by adding the violet and red-wing spectroheliograms) does not oscillate. In particular, when frames in which a dark plage is no longer visible (due to its being 180° out of phase

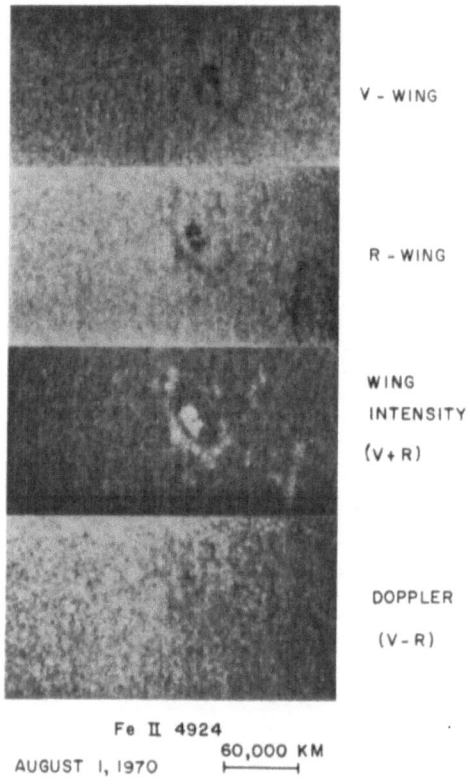

Fe II 4924

AUGUST 1, 1970   60,000 KM

Fig. 6. Comparison between (a) violet wing; (b) red wing; (c) average wing intensity; (d) velocity field, as determined with the Fe II $\lambda 4934$ line.

with its time of maximum visibility) are added, the dark plage appears clearly on the sum picture.

(3) The magnetic field shown on $\lambda 4924$ Zeeman photographs does not share this 5 min oscillation. Rather it shuffles about on the solar surface just the way the CN bright structures and the Ca I $\lambda 6103$ magnetic fields are known to move (see references above).

Further study of these movies as well as movies obtained in other lines listed above is presently underway.

In summary, although the study of these movies is only beginning, a few of the most striking features are:

(1) On a short time scale, the 5 min oscillations of velocity and intensity are clearly visible.

(2) On a long time scale, the 15 min shuffling of brightness and magnetic fields are clearly visible. Surrounding sunspots, 'spheres of influence' have been observed in which the surface motions of the fields are predominantly outward from the spots.

(3) In magnetic regions the velocity field behaves differently than in non-magnetic regions. In field regions, the velocities are generally downward and the amplitude of the 5 min oscillation appears considerably reduced from the non-field amplitude.

(4) Dark plages, visible on the red wing of Fe II λ 4924, correspond spatially to the magnetic field pattern and show a striking oscillation of contrast with a 5 min period. Neither the magnetic fields nor the average wing intensity (obtained by adding V-wing and R-wing spectroheliograms) share in this oscillation.

Acknowledgements

The help and collaborations of A. Bhatnagar, C. C. Curtis, B. Gillespie, S.-Y. Liu, and R. A. Shine is gratefully acknowledged.

References

Chapman, G. A. and Sheeley, Jr., N. R.: 1968, *Solar Phys.* **5**, 446.

Sheeley, Jr., N. R.: 1969, *Solar Phys.* **9**, 347.

Sheeley, Jr., N. R. and Engvold, O.: 1969, 'Spectroheliograms in Fe II λ 4924', paper presented at AAS Solar Meeting, Pasadena, Calif., Febr. 18–21, 1969.

Sheeley, Jr., N. R. and Engvold, O.: 1970, *Solar Phys.* **12**, 75.

Vrabec, D.: 1971, this volume, p. 329.

ON THE REALITY OF MAGNETIC FINE STRUCTURE

CONSTANCE SAWYER

Space Disturbances Laboratory, ESSA Research Laboratories, Boulder, Colo. 80302, U.S.A.

On filter magnetograms of the Sun made at Lockheed Solar Observatory, small mottles create a salt-and-pepper appearance. Outside plages, the surface seems to be sprinkled with little magnetic elements, with opposite polarities intermingled. The many steps of the photographic subtraction process required to make these elements visible tend to cast doubt on their reality. Independent and stronger evidence for the quantization of magnetic field, recently presented by Livingston and Harvey (1969), stimulates an effort to define more carefully the characteristics of the elements seen on filter magnetograms. The purpose of this contribution is to show that these characteristics are compatible with those of the elements observed at Kitt Peak.

If one accepts the reality of the elements on the basis of the Kitt Peak observations, the Lockheed data offer a complementary view that may help us to use this new phenomenon to understand the development and decay of active regions.

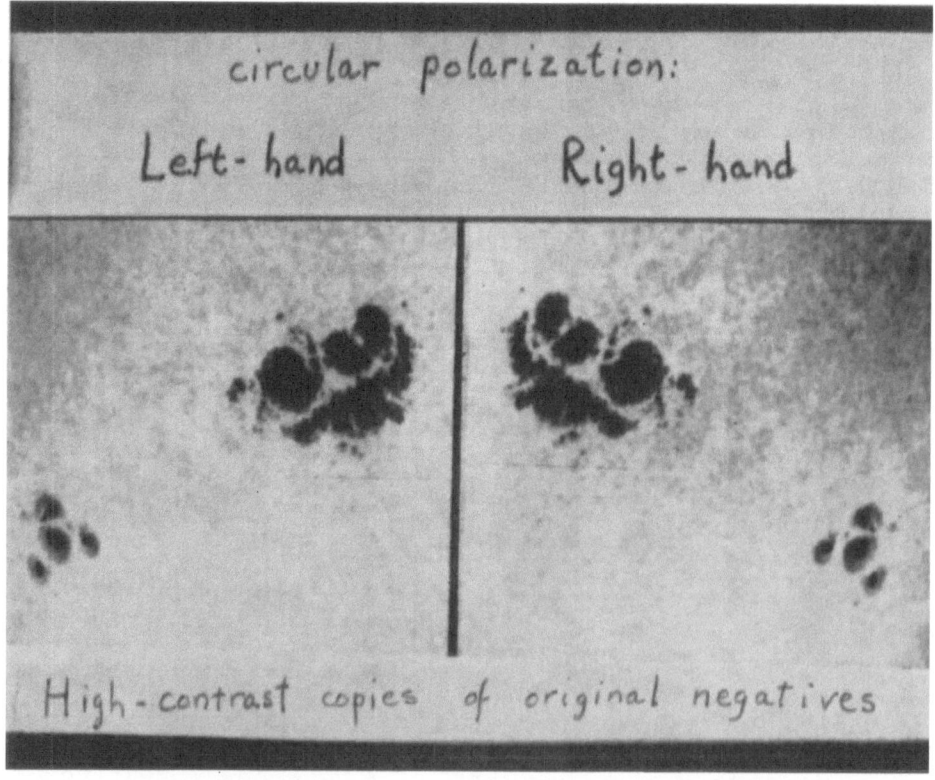

Fig. 1.   High contrast copies of original negatives with opposite circular polarization.

Howard (ed.), Solar Magnetic Fields, 316–322. All Rights Reserved.
Copyright © 1971 by the IAU.

The Lockheed filter transmits a 0.15 Å band in the blue wing of the half-Ångstrom broad, Zeeman-sensitive Fe I line at $\lambda 5324$. A quarterwave plate is rotated between successive exposures of the solar image, so that right-hand circularly polarized light is recorded on one frame, and left-hand on the next. The resulting photographs (Figure 1) show sunspots, facular lanes, and fine structure, with rather subtle differences resulting from the polarization switch between neighboring frames. These differences are enhanced, and intensity variations that do not depend on polarization are suppressed, by photographic subtraction, which combines a positive image in each

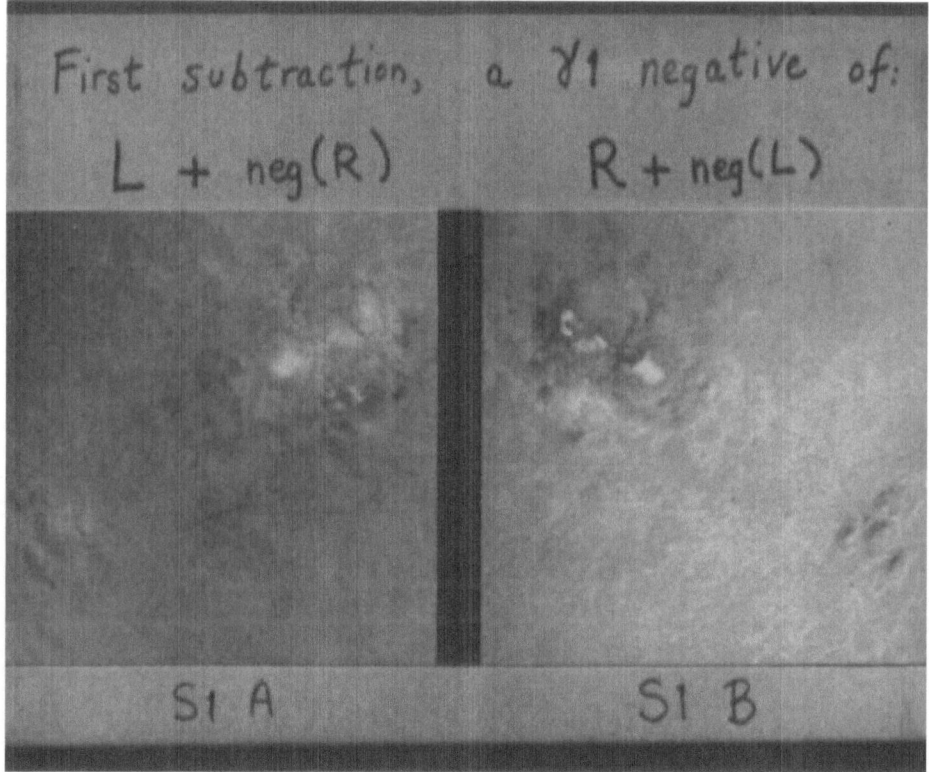

First subtraction, a γ1 negative of:

L + neg (R)   R + neg (L)

St A   St B

Fig. 2.   First subtraction: the positive of one circular polarization is combined with the negative of the opposite polarization.

polarization with a negative image in the opposite polarization. The material studied here was prepared at Lockheed Observatory by carrying out two subtractions for each polarity, (Figures 2 and 3) making a half-tone print of each, and copying these on transparencies of different color. The part of the solar intensity field that is enhanced by a positive longitudinal magnetic field appears as high density on a red transparency, and the part enhanced by negative field as high density on a blue transparency. When these are overlayed, the strong fields in an active region stand out as intensely colored, coarsely structured regions, while outside the active region, the pattern consists of a mixture of red and blue dots that tend to form narrow mean-dering strings. The lanes that stand out most clearly look like network cell boundaries and are the same size.

The observations were made April 1, 1969, and show McMath plage 10014, with Mt. Wilson sunspot groups 17207 and 17211; lat $\sim$N20, CMP about April 3.

The red and blue dots are complementary, in that dense dots on the red transpa-rency overlie blank spots on the blue transparency, and vice versa. This gives a pro-nounced moiré effect when the transparencies are rotated slightly from exact super-

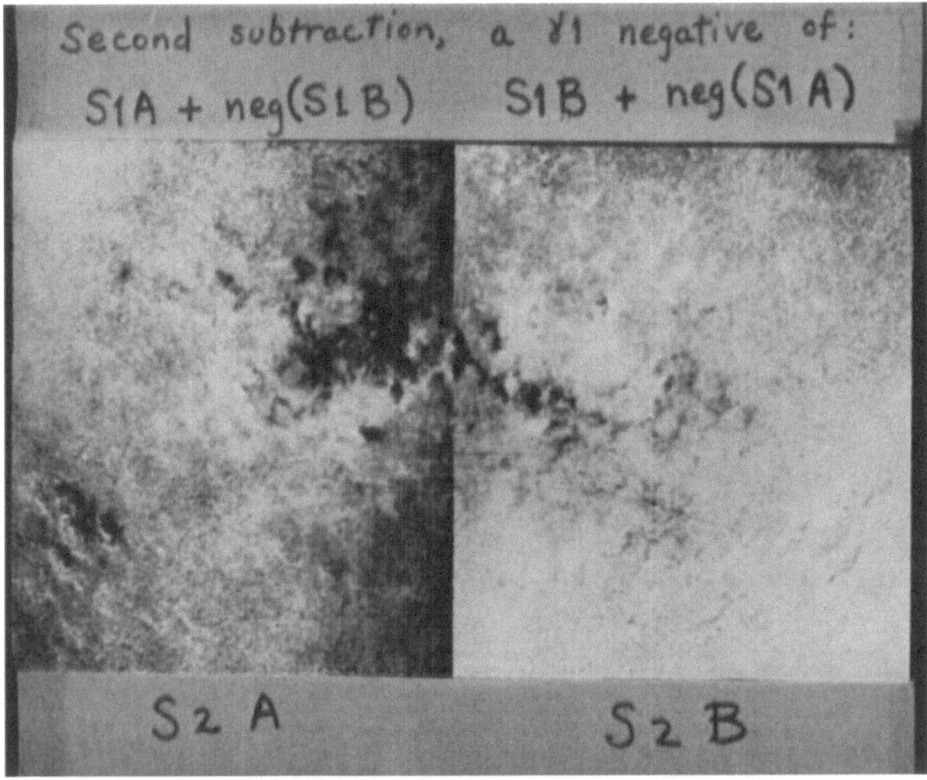

Fig. 3. Second subtraction: one first subtraction (Figure 2) is combined with the negative of the other first subtraction.

position. This effect is distinct from the moiré pattern produced from the half-tone grid. The grids on the two transparencies are at about 45° to each other, so that the half-tone moiré pattern appears when the transparencies are rotated through an angle of 45° from superposition. The 'magnetic dots' are much larger than the half-tone dots, by a factor of 8 or 10, and also considerably larger than the grain of the photo-graphic emulsion. They are smaller than the mottles that can be seen before subtrac-tion. In the various steps of the photographic process, the magnetic dots first become prominent (compared to the emulsion grain, for example) after the first subtraction.

The smallest dot that can be recognized on these prints includes four half-tone dots, an area of 1.25 (arc s)^2. Only one out of 6 dots are larger than 3 (arc s)^2, and many of these large dots are irregularly shaped and could well be clusters of individual dots. The mean area for 153 dots is 2.05 (arc s)^2, corresponding to a diameter of 1.8 (an upper limit, because of the clustering and limited spatial resolution). The frequency distribution of dot area can be represented by an exponential relation:

$$-\ln \frac{N}{No} = \frac{\text{element area in (arc s)}^2}{1.2} = \frac{\text{area in } 10^6 \text{ km}^2}{0.6}.$$

Red dots and blue dots were counted in areas of about the same size as the Mount Wilson magnetograph aperture (275 (arc s)^2), and the corresponding magnetic field measure, ranging from less than 5 G up to 80 G, was found from the published Mount Wilson magnetogram (*Solar-Geophysical Data*). The signed difference, (number of red dots – number of blue dots), is correlated with the magnetic field, with correlation coefficient 0.90 (39 pairs of values) and regression relation close to:

one uncompensated dot/aperture = one gauss.

The *total* number of dots, red + blue, in each area is also related to the field, with

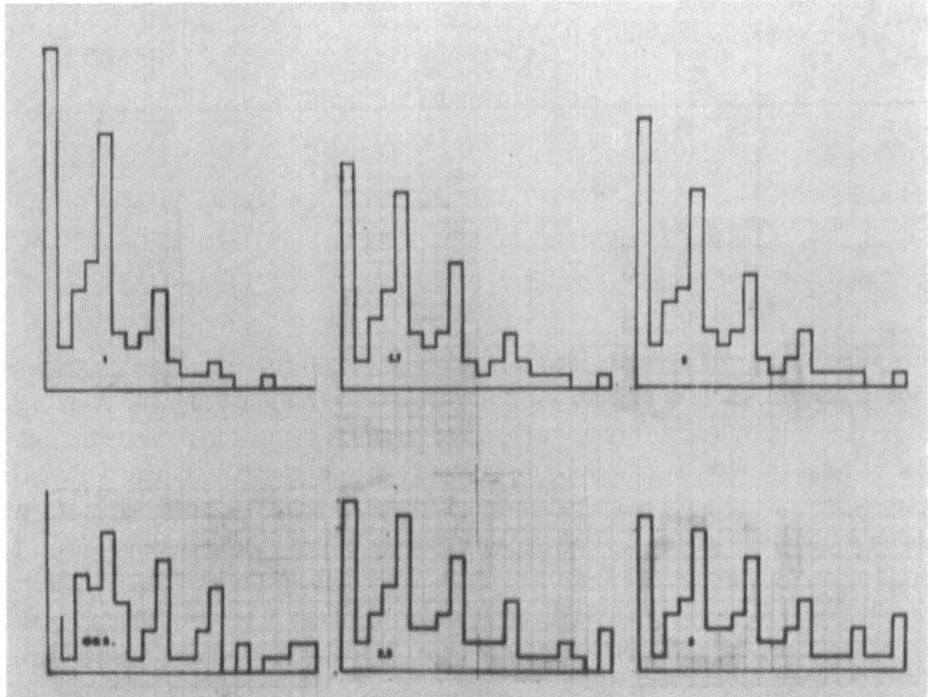

Fig. 4. Number of cases with magnetograph signal in a given range. Top row: computed for $x = np/q = 1, 1.7, 2$. Bottom row: the distribution observed by Livingston and Harvey (1969), computed for $x = 2.5$, $x = 3$.

correlation coefficient 0.69. This reflects the higher density of dots in plage regions.

The relation: 1 dot/aperture $= 1$ G, with the Mount Wilson aperture area of 275 (arc s)$^2 = 1.45 \times 10^{18}$ cm^2, leads to a value for the flux per dot of 1.4×10^{18} Mx. Temperature sensitivity of the line $\lambda 5250$ used for the Mount Wilson measurements leads to underestimation of the flux by a factor of 2 or 3 (Harvey and Livingston, 1969), so a more realistic estimate of the flux per dot would be $\sim 3 \times 10^{18}$ Mx. This may be compared with flux of 2.8×10^{18} Mx found at Kitt Peak.

An estimate of the dot density in regions of weak field can be made directly from counts: in 17 count areas where the Mount Wilson field measurement is $\leqslant 10$ G, the mean sum (the number of red dots + the number of blue dots) is 32 in 275 (arc s)^2 or 2.9 in 25 (arc s)^2. The mean difference, or number of uncompensated dots in these low-field regions, is 10 in 275 (arc s)^2 or 0.9 in 25 (arc s)^2.

Livingston and Harvey scanned near the center of the quiet solar disk using a Babcock-type magnetograph with aperture $5'' \times 5''$. When an isolated magnetic feature was detected, it was centered on the aperture, and the magnetic signal recorded. These signals tend to cluster at values that are multiples of 21 G, as can be seen in the observed histogram reproduced in Figure 4. The distribution within each signal range is skewed toward smaller values as would be expected, Livingston and Harvey point out, whenever inclination of the field direction or scattering beyond the aperture causes some loss of signal. Therefore, we consider all the cases with signal in the range 10 to 30 G to represent apertures containing a single element, those in the range 30 to 50 to contain 2 elements, etc. (Table I).

TABLE I

Livingston and Harvey's frequency distribution

Signal in gauss	10 to 30	30 to 50	50 to 70	70 to 90	90
Number of uncompensated magnetic elements	1	2	3	4	5
Number of cases observed	29	13	10	4	4

As the number of elements increases, the number of cases decreases only rather slowly, suggesting that more than one element must often fall into a single aperture area.

To determine the density of elements more precisely, we need to estimate the number of cases when the signal was zero, because either no elements, or equal numbers of positive and negative elements appeared in the aperture. As a trial estimate, we assume a binomial distribution for single elements. Suppose the elements fall into m surface areas, each $5'' \times 5''$. The probability that a certain area receive a given element is $p = 1/m \ll 1$. Let $1 - p = q \sim 1$. If there are n elements of each polarity, the number of areas containing 0, 1, 2, ..., j elements of $+$ polarity will be proportional to

$$P_0 = q^n, \quad P_1 = nq^{n-1}p, \quad P_2 = \left(n(n-1)/2\right) q^{n-2}p^2,$$

$$..., \quad P_j = \frac{n!}{(n-j)! \, j!} \, q^{n-j}p^j, \, ...,$$

and the same is true for − polarity. Then the number of cases with difference of 0, 1, 2, ..., k between the number of + elements and the number of − elements will be proportional to the products of the probabilities, summed over all combinations that result in this difference.

$$D_0 = P_0 P_0 + P_1 P_1 + \cdots + P_n P_n$$

$$D_1 = 2[P_0 P_1 + P_2 P_3 + \cdots + P_j P_{j+1} \cdots + P_{n-1} P_n]$$

(both positive and negative differences are counted)

$$D_k = 2 \sum_0^{n-k} P_j P_{j+k}.$$

The computation is simplified if we note that $n!/(n-j)! \sim n^j$, ignoring terms of order $1/n$ and smaller. Then

$$P_j \sim \frac{q^n}{j!} \left(\frac{np}{q}\right)^j = P_0 \frac{x^j}{j!},$$

where $x = np/q \sim n/m$ is approximately equal to the average number of elements of each polarity in one aperture area. The histograms in Figure 4 were constructed to fit these constraints: (1) the number of cases within each 20-G signal range is proportional to D_k defined above; (2) within each 20-G interval, the cases are distributed in the pattern 0.18, 0.23, 0.47, 0.12, the average pattern in Livingston and Harvey's distribution; (3) the number of cases with signal ≥ 10 G, $\sum_{k=1}^n D_k$, is 60, as in the observed distribution. Some characteristics of each of the computed distributions are given in Table II.

TABLE II

Characteristics of the computed frequency distributions

$x = np/q$	1	1.7	2	2.5	3
Total number of cases for $N(\geq 1) = 60$	87	78	76	74	72
Total number of elements of each polarity	86	131	150	183	213
Density ($N^+ + N^-$)	2.0	3.4	4.0	5.0	5.9
Average excess, $N^+ - N^-$	1.0	1.4	1.5	1.7	1.9
Excess/sum	0.5	0.4	0.4	0.3	0.3

From the histograms and Table II we can estimate that a random distribution of elements that fits the observed distribution has x lying between 1.5 and 3.0, density between 3 and 6, average number of uncompensated elements between 1.3 and 1.9, and ratio of difference to sum between 0.3 and 0.4.

Table III summarizes the values derived from the two aspects of magnetic elements.

The values in Table III show that both sets of observations could fit into a single model of magnetic elements that are distributed almost randomly in quiet regions that are almost field-free on a large scale.

There are facts that make us doubt the reality of magnetic dots: (1) the possibility of

TABLE III

Comparison of characteristics derived from independent sources

	Kitt Peak	Lockheed
Element area, (arc s)^2	1	2
Density (number in $5'' \times 5''$)	3 to 6	3
Average excess	1 to 2	0.9
Average flux per element, maxwells	2.8×10^{18}	3×10^{18}

spurious effects arising in the photographic subtraction process; (2) the dots have not appeared in magnetograms with high spatial resolution in which both polarizations are recorded simultaneously, made at Aerospace Corporation's San Fernando Observatory; (3) the dots do not show up in Lockheed subtractions of frames exposed several minutes apart, although the lifetime of the dots would be expected to be much longer. Against these we can set the positive evidence: (1) the continuity of the appearance of the structure from thinly scattered dots through weak chains to strong chains that have the same appearance as the chromospheric network; (2) the correlation of dot excess with magnetic field measured with a scanning magnetometer; (3) sharing of characteristics such as density, size, and magnetic flux with elements observed at Kitt Peak.

References

Harvey, J. and Livingston, W.: 1969, *Solar Phys.* **10**, 283.
Livingston, W. and Harvey, J.: 1969, *Solar Phys.* **10**, 294.

Discussion

Pasachoff: After this week of discussing bright mottles and dark mottles, I would like to thank you for introducing us to red mottles and blue mottles at this penultimate moment.

You obviously have a lot of experience in making the color overlays, and I wonder if you could please comment on what happens when you move things slightly out of registration.

Sawyer: In fact, all the photographic work was done at Lockheed Observatory. However, I have seen misregistered material, and it has a rather disturbing appearance.

Leighton: In view of the fact that one cannot distinguish perfect cancellation within some very small offset, which could introduce a spurious 'grain' with subsequent cancellations, do you prefer to think of the effects as real or as spurious?

Sawyer: I believe that the effect is not simply due to offset, because that produces dark and bright elements in a fixed geometrical relationship, like bright peaks and shadows, at least in the same neighborhood; and the dots don't look like that. I tend to think of them as real, but rather on the basis of the agreement with the Kitt Peak observations.

Leighton: At one time I noticed that one often gets a spurious, grainy, second cancellation if the two singly-cancelled pictures are copied in too sharp focus. A truer result is obtained if these second-generation pictures are printed slightly out of focus.

Frazier: Did you check the noise of the whole system by cancelling a pair of filtergrams taken with the *same* polarity.

Smith, Sara: This has been done with other filter magnetograms, and a smooth background was obtained.

Title: I believe that it is possible to subtract pairs to $\pm 0.0005''$, if every process is carefully controlled. I have produced subtracted test movies on which displacements of $0.001''$ are detectable. However, one can produce any number if any part of the process is not under control.

Hα STRUCTURES AND
SMALL-SCALE MAGNETIC FIELD CONFIGURATIONS

SARA F. SMITH

Lockheed Solar Observatory, Burbank, Calif., U.S.A.

Abstract. Magnetic field observations made utilizing a $\frac{1}{7}$ Å birefringent filter for the FeI line at 5324 Å resolve magnetic field features comparable in scale to the Hα fine structure. Without exception, low lying filaments in the center of the disk coincide with boundaries of polarity change in the line of sight component of the magnetic field. Small filaments can be identified on a scale comparable to the Hα fibril structures. The arch filament systems are associated with areas of mixed polarity. 'Satellite spots' may sometimes be identified as small localized enhancements of magnetic field of both polarities around the perimeter of a sunspot.

In November 1968, a program of observing solar magnetic fields was initiated at the Lockheed Solar Observatory. The magnetic fields are detected by means of a filter rebuilt by Ramsey (1969) at Lockheed for observing in the FeI line at 5324 Å. The technique employed is essentially the same as that developed by Leighton (1959) except that in place of the spectroheliograph we employ the $\frac{1}{7}$ Å bandpass FeI filter in a 17.5 cm aperture refracting telescope. Direct observations in the FeI (5324 Å) line as in the upper left of Figure 1 show the bright network which is cospatial with the magnetic field network as shown by Chapman and Sheeley (1968) for other photospheric lines. Comparison of the direct FeI photographs with Hα photographs of comparable resolution, as in Figure 2, shows that the magnetic field network is also precisely cospatial with Hα plage within the resolution of the photographs. The filter magnetograms have not yet been successfully calibrated to yield accurate measures of field strength. However, comparison with Kitt Peak magnetograms of lower resolution ($\sim$ 2.4 arc s) shows that even weak plage represents field strengths of several hundred gauss, consistent with Chapman and Sheeley's measurements of the field strengths in the photospheric network.

In these high resolution magnetograms it is also confirmed that low altitude filaments relate very accurately to polarity boundaries when observing the line-of-sight fields in the center of the disk. Deviations are generally within the width of the filament. On the basis of this confirming evidence, it seems reasonable to continue defining filaments as absorption features corresponding to lines or zones of 0 line of sight field. Using this definition, it is clear that filaments have a complete spectrum of sizes from those that exceed the dimensions of an active region to those having the same dimensions as fibrils. Examples of the very small filaments are seen in the middle of the active region in the lower right picture of Figure 1. The precise way in which the small filaments lie between network of opposite polarity is better illustrated when the subtracted magnetic field photographs are reproduced first as half-tone transparencies, secondly as color transparencies, and then shown superposed with an Hα transparency as is illustrated in Figure 2. The fine pattern of dots is a result of the half-tone copy.

Howard (ed.), Solar Magnetic Fields, 323–328. All Rights Reserved.
Copyright © 1971 by the IAU.

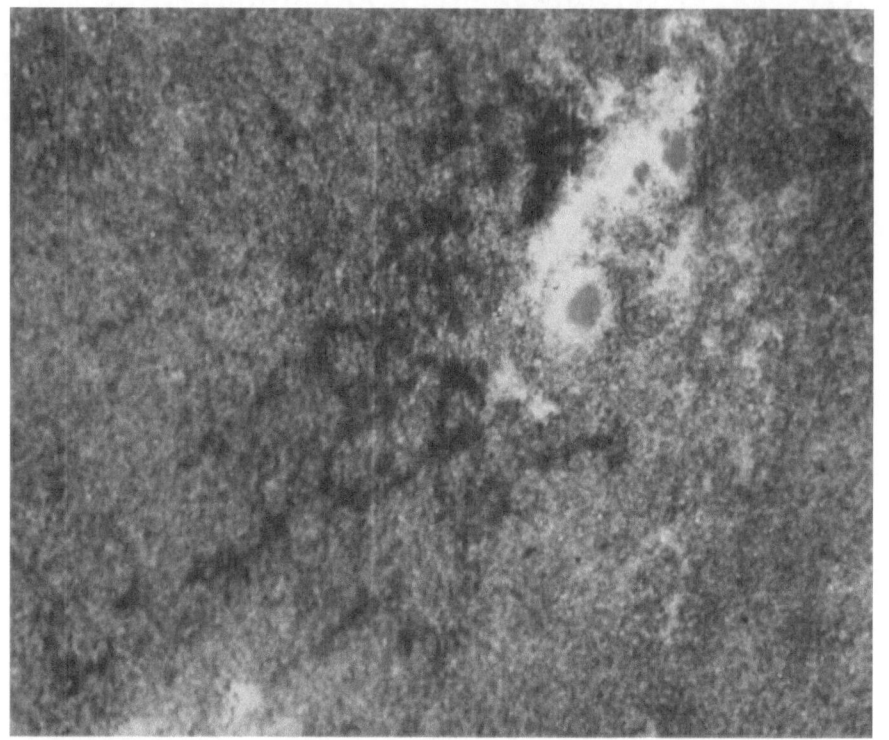

Fig. 1b.

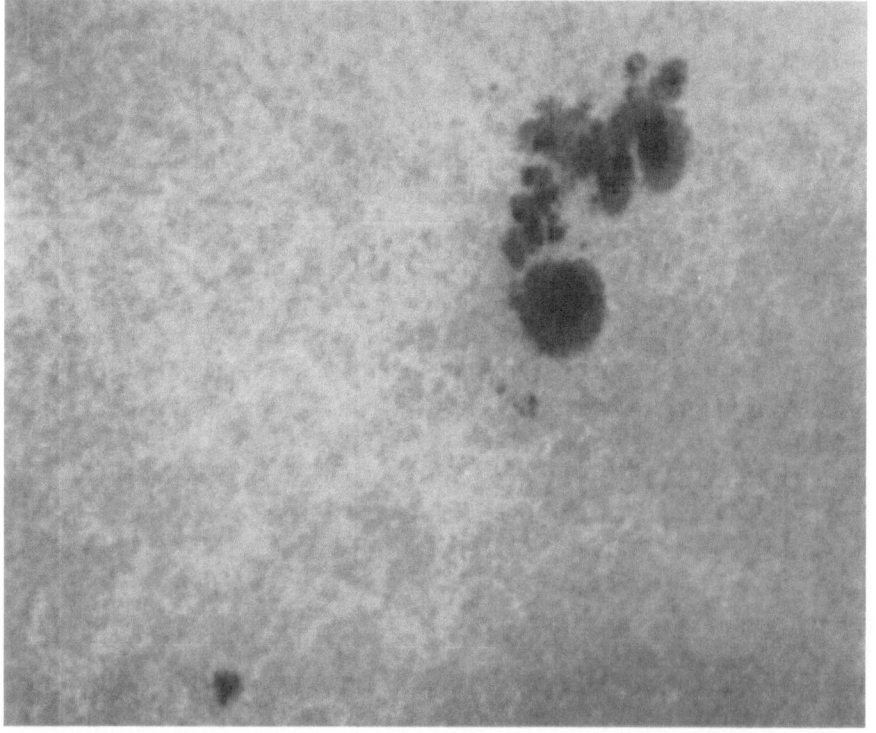

Fig. 1a.

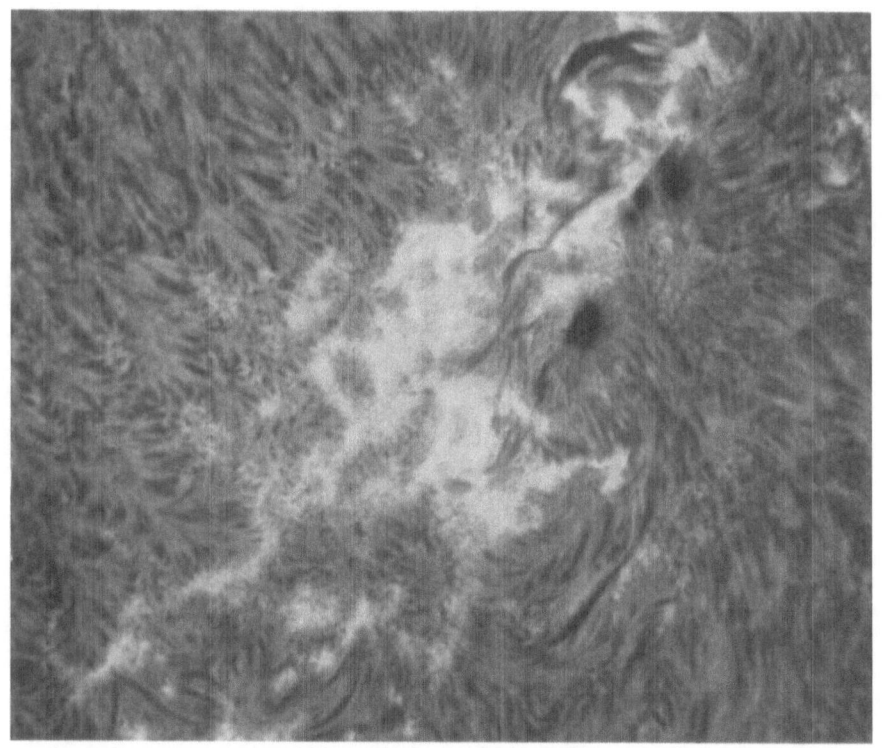

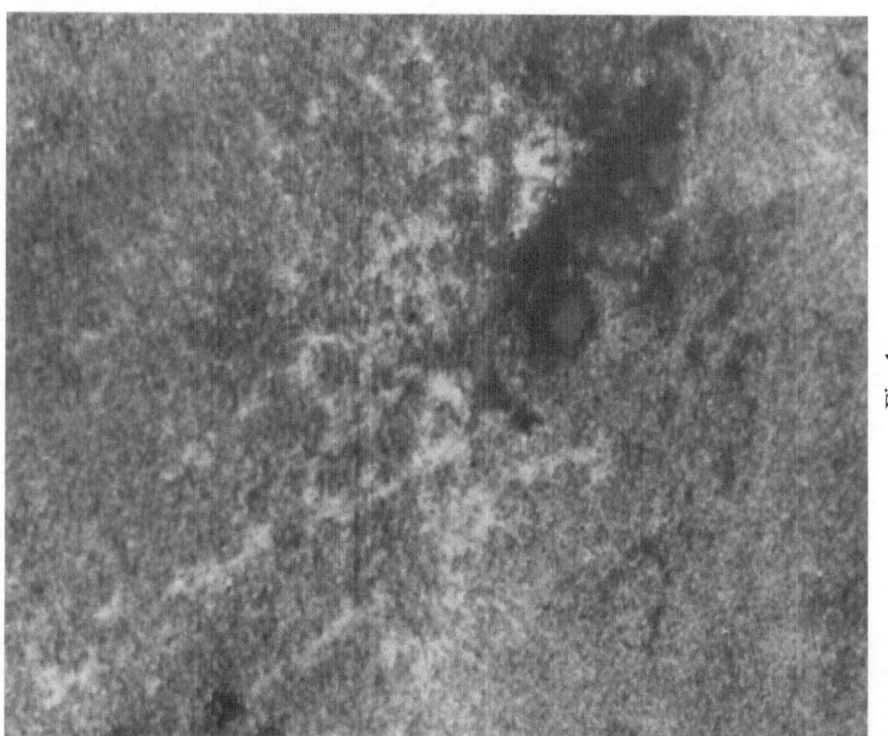

Fig. 1c.

Fig. 1d.

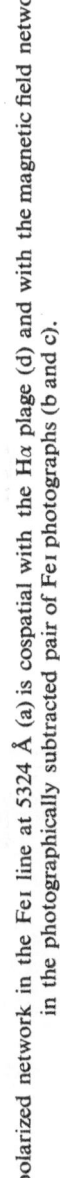

Figs. 1a-d.   The bright polarized network in the Fe I line at 5324 Å (a) is cospatial with the Hα plage (d) and with the magnetic field network shown in the photographically subtracted pair of Fe I photographs (b and c).

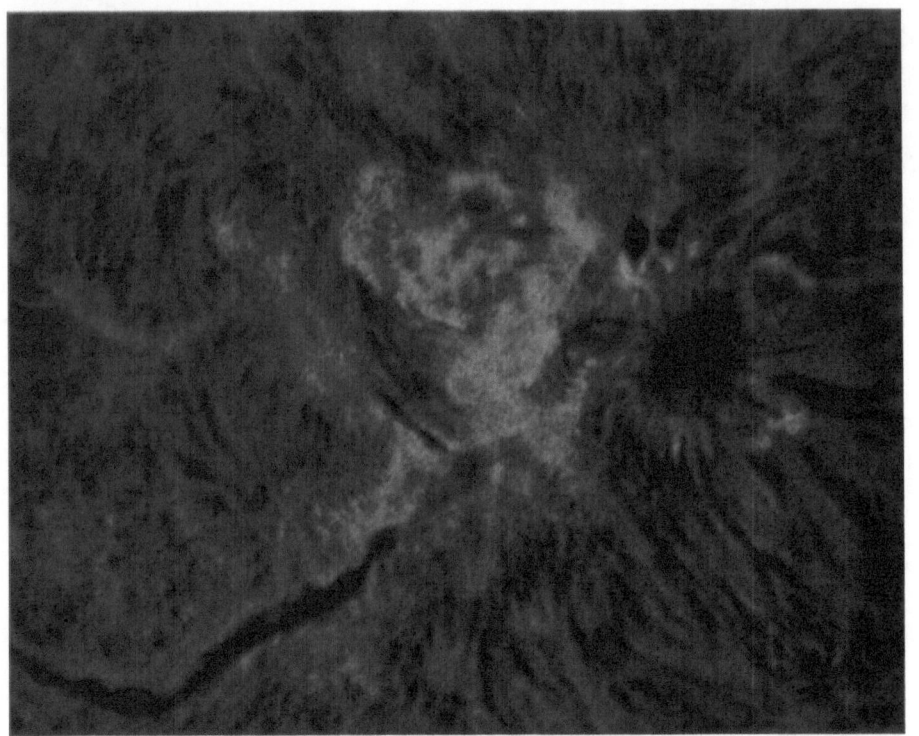

Fig. 3. This composite photograph shows a new active region (mostly orange and green) beginning to develop in an old active region (mostly red and blue). Surrounding the sunspot are small areas of enhanced field of both polarities. Small surges and dark absorption features are related to these 'satellite' poles.

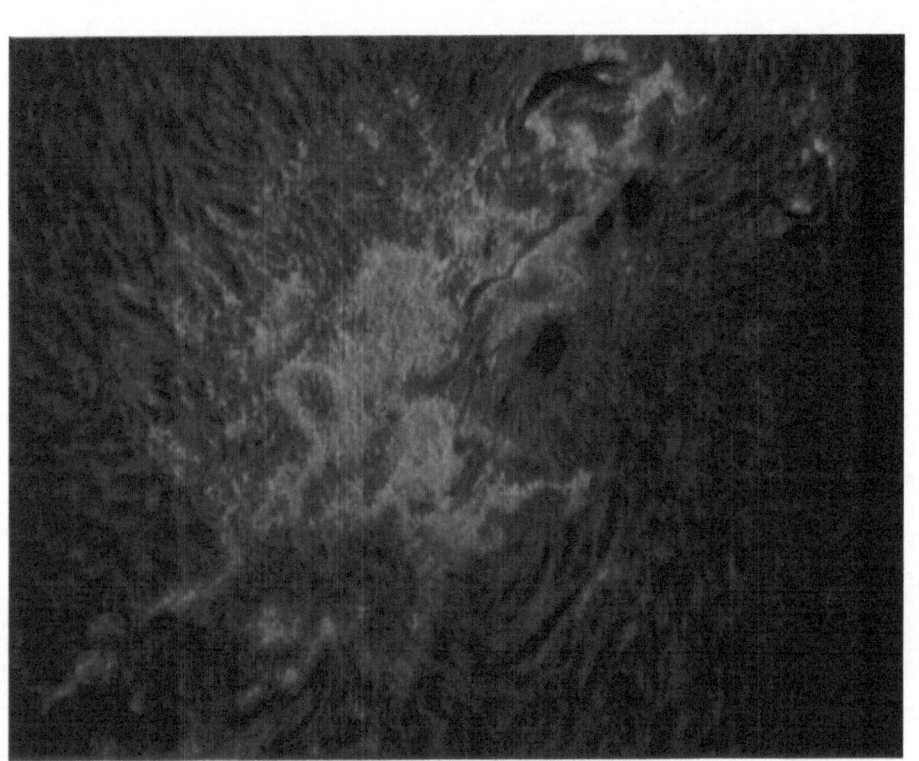

Fig. 2. This illustration is a composite of the magnetic field and Hα photographs in Figure 1. Blue represents positive polarity. Red represents negative polarity. Green represents positive polarity (blue), coincident with bright Hα plage (yellow). Orange represents negative polarity (red) also coincident with bright Hα plage. A system of small filament and fibrils marks the boundary between opposite polarities in Hα.

In most respects, the more recent high resolution magnetograms confirm earlier results and provide additional detail about established relationships between the magnetic field configuration and Hα structure. However, earlier results in which arch filament systems are consistently seen as bridges connecting areas of opposite polarity *are not satisfactorily* confirmed in the high resolution magnetograms. Indeed, *sometimes* the fibril patterns, not necessarily identified as 'arch filament systems', do appear as bridges between areas of opposing polarity. However, the relationship of the 'arch filament systems' to the magnetic field configuration as described by Bruzek (1968) seems now to be more complex. There are at least 3 field configurations which can be associated with fibrils patterns or arch filament systems as seen in Hα centerline photographs. First is the classical situation in the upper left corner of Figure 2 in which there is an apparent bridging of opposite polarities. It should be noted however, that frequently, as in this case, these observations differ in some respects from earlier lower resolution observations. The fibril system is only present in the gap of 0 or nearly 0 field between the regions of opposite polarity. The ends of the arches approach but do not overlap the strong magnetic field. The fibrils are not exceptionally dark and therefore not identified as an 'arch filament system'. A second case is near the right border of Figure 2. The dark arches which appear to overlap areas of strong field are separated by weak field. The ends may appear to approach or overlap fields of either the same polarity or opposite polarity. However, considering possible height differences between the observed field and Hα structures, it is not possible to say with any confidence, that the end of a given structure definitively ends in an area of given polarity.

The third case, shown in Figure 3, is a typical arch filament system as described by Bruzek (1968). These are often seen in new active regions; in this case a new active region growing in an old active region. Here, the arches are very prominent due to Doppler shifts (Bruzek, 1968). Instead of a simple bipolar configuration, the fields underlying the arch filament system are a complex mixture of both polarities near both ends of the arch filament system. The scale of the polarity reversals and the arches are comparable making it difficult or impossible to identify an end of an individual fibril with an area of given polarity. In fact, from the improved resolution magnetic field filtergrams, it is questionable as to whether a distinction should be made or can adequately be made between small filaments, various fibrils patterns and arch filaments systems.

Another feature to point out in Figure 3 are the satellite spots described by Rust (1967) which frequently border large sunspots. In the filter magnetograms these so-called satellites may be seen as small compact areas where there may be an enhancement of both polarities instead of just the polarity opposite to the main spot. As noted by Rust small surges may be observed in close proximity to these 'satellites poles'. The base of the surges are thus close to or coincident with the polarity boundary between the two small poles, raising the question of whether or not surges are like filaments in following channels between areas of opposite polarity but in the dimension of the magnetic field we are unable to observe with the filter.

It is evident that many more examples need to be studied before definitive relationships can be established between Hα structure and the fine-scale magnetic field patterns.
It is of interest to note, however, that there is a common denominator throughout all
of the observations described. The ends of filaments, fibrils, arch filament systems and
surges are coincident with or lie within a few seconds of arc of lines or zones of zero
line of sight fields.

References

Bruzek, A.: 1968, in K. O. Kiepenheuer (ed.), 'Structure and Development of Solar Active Regions',
    IAU Symp. **35**, 293.
Chapman, G. A. and Sheeley, N. R., Jr.: 1968, in K. O. Kiepenheuer (ed.), 'Structure and Develop-
    ment of Solar Active Regions', *IAU Symp.* **35**, 161.
Leighton, R. B.: 1959, *Astrophys. J.* **130**, 366.
Ramsey, H. E.: 1969, *Sky Telesc.* **37**, 363.
Rust, D. M.: 1968, in K. O. Kiepenheuer (ed.), 'Structure and Development of Solar Active Regions',
    IAU Symp. **35**, 77.

MAGNETIC FIELD SPECTROHELIOGRAMS FROM THE SAN FERNANDO OBSERVATORY

DALE VRABEC

San Fernando Observatory, The Aerospace Corporation, El Segundo, Calif., U.S.A.

Abstract. Zeeman spectroheliograms of photospheric magnetic fields (longitudinal component) in the Ca I 6102.7 Å line are being obtained with the new 61-cm vacuum solar telescope and spectrohelio-graph, using the Leighton technique. The structure of the magnetic field network appears identical to the bright photospheric network visible in the cores of many Fraunhofer lines and in CN spectro-heliograms, with the exception that polarities are distinguished. This supports the evolving concept that solar magnetic fields outside of sunspots exist in small concentrations of essentially vertically oriented field, roughly clumped to form a network imbedded in the otherwise field-free photosphere. A time-lapse spectroheliogram movie sequence spanning 6 hr revealed changes in the magnetic fields, including a systematic outward streaming of small magnetic knots of both polarities within annular areas surrounding several sunspots. The photospheric magnetic fields and a series of filtergrams taken at various wavelengths in the Hα profile starting in the far wing are intercompared in an effort to demonstrate that the dark strands of arch filament systems (AFS) and fibrils map magnetic field lines in the chromosphere. An example of an active region in which the magnetic fields assume a distinct spiral structure is presented.

1. Introduction

It appears that an intimate relationship exists between magnetic fields and most of the structure so far observed on the Sun. The more these magnetic fields are studied, the more evident it is becoming that magnetic features of increasingly finer detail play a fundamental role. With this in mind, The Aerospace Corporation has recently placed into operation at its San Fernando Observatory a 61-cm aperture vacuum solar telescope and spectroheliograph (Vrabec and Rogers, 1968; Mayfield *et al.*, 1969) designed to produce Zeeman spectroheliograms of magnetic fields with relatively high spatial and temporal resolution, preliminary examples of which are presented here.

2. Method

Figure 1 illustrates the basic method employed, which is the one devised by Leighton (1959). Magnetic fields are visible in the upper two spectroheliograms where they appear as light patches that are different in the two strips. The spectroheliograph is equipped with a polarizing beamsplitter which separates the dispersed light into two complementary optical channels. A bandwidth of approximately 0.1 Å set 0.07 Å in the blue wing of the Zeeman-sensitive 6102.7 Å line of neutral calcium is used, and the two spectroheliograms are exposed simultaneously so that seeing variations are identical in both channels. The two channels discriminate between the two oppositely circu-larly polarized Zeeman components produced by the longitudinal magnetic field, each transmitting only the component which is blocked in the other channel. Wherever there is a magnetic field, a differential exposure results between the two channels, the

Howard (ed.), Solar Magnetic Fields, 329–339. All Rights Reserved.
Copyright © 1971 by the IAU.

sign of which depends upon the polarity of the field. This is the effect visible in the two upper strips. Sunspots and background mottling due to photospheric granulation and velocity fields appear in both spectroheliograms. To obtain a record in which only the magnetic fields appear, the upper spectroheliogram is subtracted from the middle one by means of photographic subtraction, thereby cancelling common-mode features and enhancing the magnetic fields. Polarities are distinguished by whether the photo-

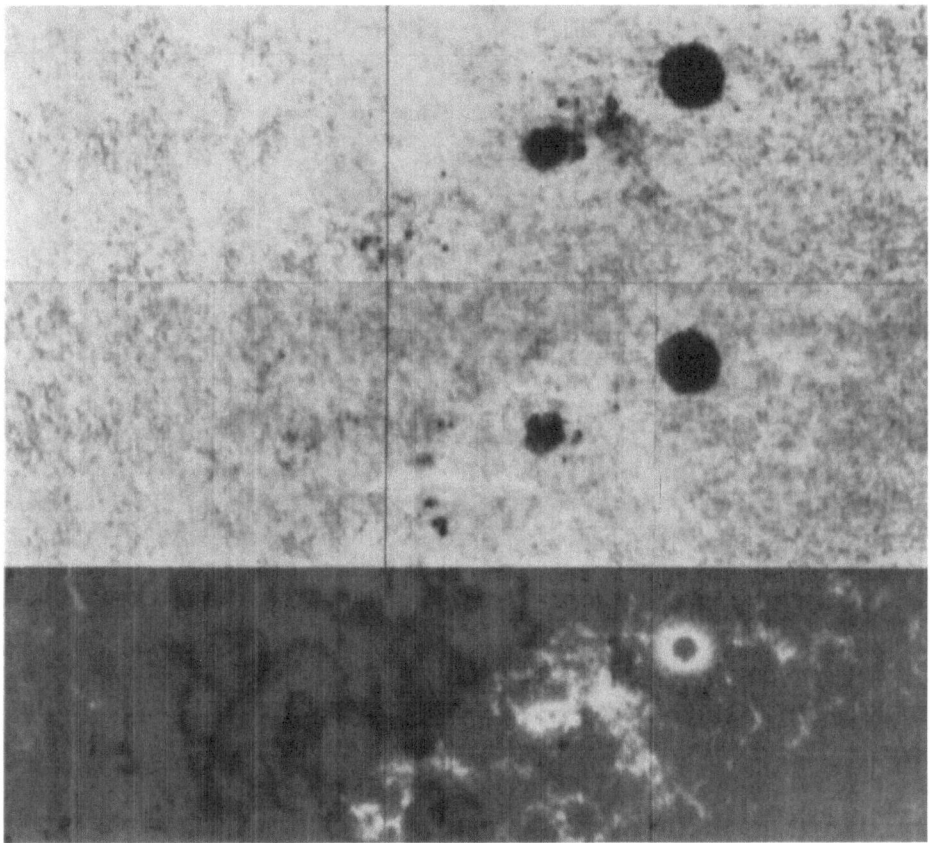

Fig. 1.   Leighton's technique to obtain Zeeman spectroheliograms of magnetic fields is illustrated by reference to the three strips. The upper two are the original spectroheliograms simultaneously exposed 0.07 Å in the blue wing of the Ca I 6102.7 Å line by means of a polarizing beamsplitter. The lower strip, obtained by photographically subtracting the upper from the middle strip, shows only the photospheric longitudinal magnetic field component.

graphic density is greater or less than the density of the field-free background. The original spectroheliograms are exposed on two separate rolls of 70-mm cine film, and a cine optical printer is utilized for the photographic subtraction. Care is necessary to maintain precise registration between the images and the perforations of the two film rolls in the spectroheliograph camera, and between the films sandwiched in the optical printer.

3. Description of a Typical Zeeman
Spectroheliogram

Figure 2 depicts a typical bipolar magnetic region (BMR) photographed on 21 February, 1970. On a large scale, opposite magnetic polarities are separated into two distinct regions which conform to Hale's law. Localized intrusions of one polarity into a region dominated by the opposite polarity occur, particularly where the two main regions interface. The fields are resolved into a partially fragmented, mesh-like, cellular network that is highly localized, but the resolution (~ 2.5 arc s, or $\sim 1\,800$ km) is insufficient to show the network resolved into individual points. We can anticipate that it should exhibit such fine structure on the basis of the superb CN spectrohelio-

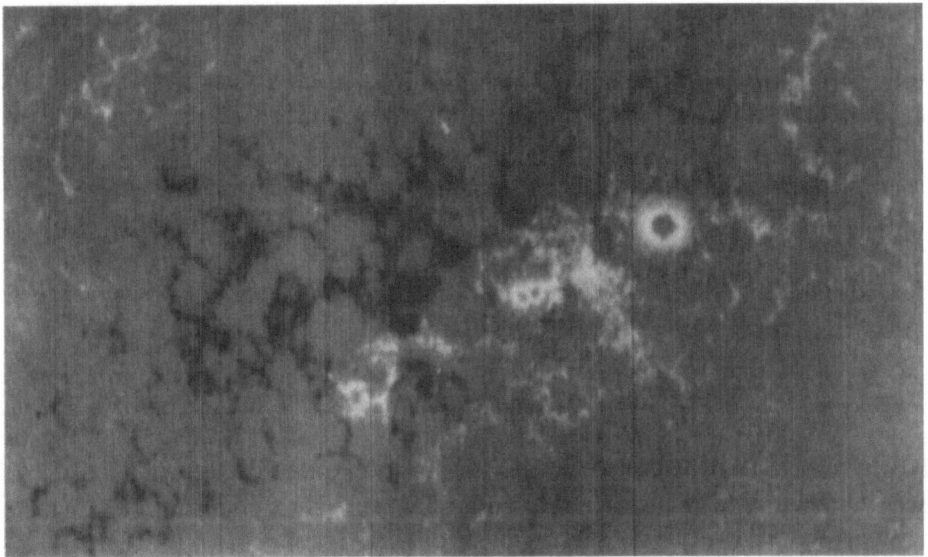

Fig. 2.   Zeeman spectroheliogram of a typical bipolar magnetic region (BMR) near the center of the disk, taken 21 February, 1970 at 1915 UT.

grams taken by Sheeley (1969). These reveal a photospheric network formed from clumps of individual bright points, where the photospheric temperature is higher than its average value. Chapman and Sheeley (1968) have shown that this bright network is co-spatial with photospheric magnetic fields. Our Zeeman spectroheliograms, which map the longitudinal magnetic field component, complement the CN spectrohelio-grams of Sheeley, and are especially useful because they reveal the *polarities* of individual features.

Several types of isolated magnetic features are visible in Figure 2. The most prominent of these (besides sunspots) are white-light pores, such as the dark magnetic feature just below the sunspot with the double umbra. Others appear to simply be network fragments. These may be associated with the network structure of the BMR, itself, or may be fragments of adjacent, intermingling magnetic regions, remnants of decaying regions which preceded the BMR, or newly emerging field. The smallest of the isolated features, and the most interesting to us, are the concentrations of both polarities which cluster around the principal sunspots in Figures 2 and 3. The majority of these do not have any corresponding continuum features that can be identified on filtergrams of approximately 1.5 arc s resolution, so the fields must be relatively strong in order to be visible on the (presently uncalibrated) spectroheliograms. Whether or not white-light features corresponding to these field concentrations exist, can only be answered when concurrently exposed, higher resolution photographs become available. It is very likely that the concentrations we observe correspond to those 'magnetic knots' observed by Beckers and Schröter (1968) that were located near sunspots, and to the 'satellite sunspots' reported by Rust (1968), though the latter were confined to features of opposite polarity to the associated sunspot. A distinctive feature of these particular magnetic concentrations is that they display a characteristic of streaming radially outward from their associated sunspots. Outflow of bright points from sunspots was first noted by Sheeley (1969) in the CN spectroheliograms referred to previously, and in the following section we will describe our own observations.

Perhaps the most important point to stress in this section is that our data support the evolving concept that outside of sunspots, magnetic fields on the Sun appear to exist in small concentrations of essentially vertically oriented field that exhibit a tendency to become clumped together, forming a localized magnetic network imbedded in the otherwise (vertical) field-free photosphere.

4. Short-Term Time-Dependent Behavior of Magnetic Fields

To study the time-dependent characteristics of the magnetic fields exhibited in our Zeeman spectroheliograms, we prepared a time-lapse movie consisting of 47 individual magnetic field spectroheliograms of a complex region crossing the central meridian on 26 January, 1970. The data span 6 hr, beginning at 1735 UT. Figure 3 shows the first spectroheliogram of this series, together with concurrent Hα filtergrams of the same region taken at selected wavelengths in the profile of this line. The fields exhibit considerable complexity. The 'stranded' appearance of the fields between the principal sunspots is seen to be associated with chains of pores visible in the far-wing filtergram. In the movie the most conspicuous changes in the fields occur along these same strands, which is also where a conspicuous flare commenced during the period of observation. The area immediately surrounding the principal sunspots is seen to be relatively free of organized magnetic network. The network nearest the sunspots appears to either assume the form of a fragmented garland encircling the sunspots, or to present a curved boundary roughly concentric with the sunspot. Not infrequently, this network is comprised of segments of both polarities.

The following types of time-dependent behavior of magnetic fields were noted in this 6 hr movie.

(1) Associated with the principal sunspots there is a persistent, approximately

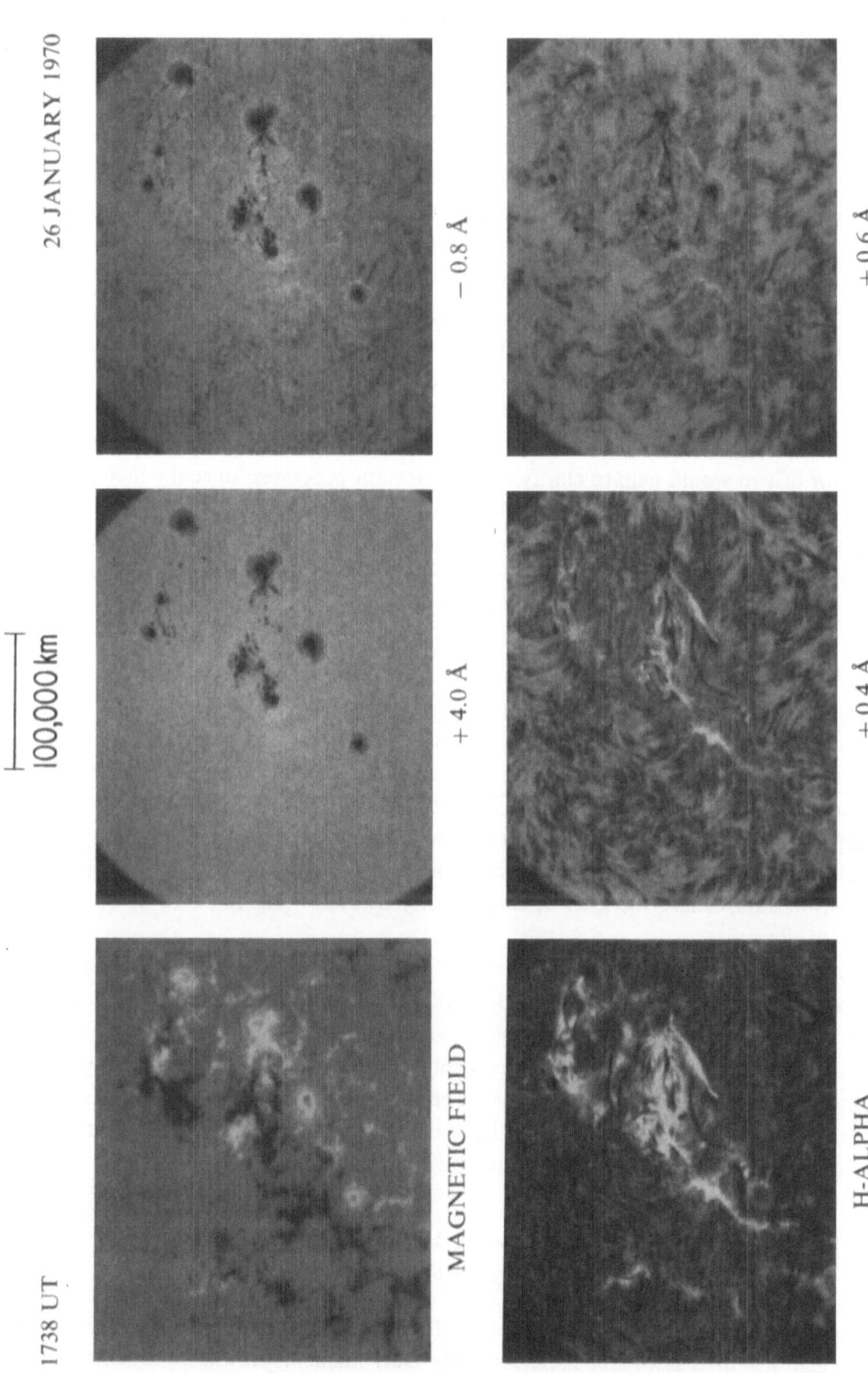

Fig. 3. The intimate relationship existing between photospheric magnetic fields and Hα chromospheric structure is seen by intercomparing the sequence of filtergrams exposed at different wavelengths in the Hα profile with the Zeeman spectroheliogram at the upper left.

radial outflow of small concentrations of magnetic field of both polarities throughout the annular region extending from the outer boundary of the sunspot penumbra, out to the radius where the network first appears. Stationary features, other than pores, occur infrequently in this annular zone, and the existence of horizontal streaming throughout this zone suggests that conditions there are unfavorable for the maintenance of a stable network structure, which therefore only begins at the periphery of the zone. What the source of this moving magnetic flux is, and to what extent this flow contributes to the flux in the surrounding network, are not presently clear. Although some of these 'streaming flux knots' first appear at the outer edge of the sunspot penumbra on our spectroheliograms, where unfortunately the penumbra is photographically saturated, many first become visible at a greater radius, especially those of opposite polarity to the sunspot. Some streaming flux knots disappear before reaching the surrounding network; others appear to merge with the network when they reach it, especially when their polarity is the same. Improvement in resolution by a factor of two would help to clarify these important processes, since the majority of these features are only marginally visible in these preliminary data.

(2) Outside of these regions we observe examples where a number of separate network fragments of the same polarity converge, become concentrated into an irregular unresolved patch which sometimes then fades. The reverse also occurs where a compact patch of network expands and separates into distinct fragments of the same polarity, some of which disappear.

(3) Movement of field concentrations along curved arcs occurs.

(4) Field concentrations underlying the extreme opposite ends, or 'feet', of arch filament systems (AFS) frequently seen in Hα, exhibit a tendency to separate with time. When this occurs near sunspots, the field concentrations may move towards these sunspots. Several AFS can be seen in the two filtergrams taken nearest line center.

(5) In several cases, apparent interaction between small concentrations of opposite polarity occurs, where the area of one polarity dwindles and the area of opposite polarity increases.

(6) The most conspicuous movement of magnetic fields occurs in the 'strands' referred to earlier. Here the magnetic features of each polarity are roughly organized into separate strands, each of one polarity. The features comprising one of these strands move coherently towards the associated sunspot to which the strand appears connected, which is of the same polarity. The free end of this strand 'slides' past the free end of the strand of opposite polarity and displaces it laterally. The movement of the strands gives the impression of causing them to straighten out.

5. Hα Structure and Photospheric Magnetic Fields

Hα photographs reveal a rich progression of complex structure suggestive of an intimate relationship to associated magnetic fields. Dark filaments are known to demarcate the boundary between regions of opposite magnetic polarity (Babcock and Babcock,

1955), and bright Hα plage to only occur over the strongest magnetic field concentrations in the network (Howard and Harvey, 1964). It seems apparent that fibrils must map magnetic field lines, yet the details remain obscure. A technique we have been using in our investigations is to obtain concurrently with the Zeeman spectroheliograms, Hα filtergrams using a Zeiss filter that is sequentially tuned through the line profile from the far red wing into the far blue wing. These filtergrams range from pure photospheric to pure chromospheric, and record the majority of Doppler-shifted features. The filtergrams and Zeeman spectroheliograms are then intercompared (Figure 3). An important result of this sequential wavelength scanning technique is that the complete locus of features which have a velocity structure can be traced, provided the Doppler shifts are within the range of tuning. This has proved to be essential in locating the 'feet' of AFS which can only be seen in the red wing of Hα, when observed near the center of the disk (Bruzek, 1969). In this manner, in all the AFS that appeared in the region shown in Figure 3 during the period of observation, the ends of the dark strands have been found to be imbedded in magnetic fields of opposite polarity, the dark arch structure in all likelihood mapping the field lines as they extend through the chromosphere.

With reference to Figure 3, in the far wings of Hα essentially only sunspots, pores, and photospheric granulation appear, except near the limb, where the photospheric network is also visible (faculae). Away from the limb, the bright photospheric network only becomes visible when the continuum begins to be significantly depressed by line absorption, and attains maximum visibility at approximately 1 Å from line center. A one-to-one correspondence between the bright elements of the photospheric network and magnetic concentrations comprising the network of photospheric magnetic fields, independent of polarity, is evident.

Elongated, dark, jet-like features bordering the bright photospheric network are the first traces of the dark Hα chromospheric network. It is seen that these dark features generally point away from magnetic field concentrations, one end beginning at the concentration, or very near it. Their upward flaring geometry when viewed obliquely towards the limb is consistent with the conjecture that these features map magnetic field lines in the chromosphere that extend out of the photosphere. As the filter is tuned towards line center many of these features develop into fibrils by lengthening one end in a curvilinear fashion, generally towards the interiors of the network cells, while the opposite end, imbedded in the photospheric fields, remains essentially fixed in position. At the same time the dark chromospheric network thickens and the number of fibrils increases. Viewed obliquely, fibrils appear either to sharply curve over to assume a nearly horizontal orientation immediately outside of the magnetic field network, or to connect with dark features which are inclined substantially out of the vertical. Outside of the magnetic field network, the absence of observable vertical components of magnetic fields implies the presence of predominantly horizontally oriented fields. Although in some areas of Figure 3 it appears possible to trace fibril structure from the fixed end almost continuously until it meets a similar structure extending from a region of opposite polarity, the majority of fibrils fade from visibility, or otherwise lose their identity before such a connection can be traced. Further support for the conjecture that fibrils map field lines comes from the observation that fibrils that are rooted in magnetic fields of the same polarity avoid connecting with each other. Because a basic characteristic of the network is that extended portions of it are of a single polarity, fibrils extending from field concentrations of the same polarity quite frequently come into proximity, in which case they interact by channeling into parallel fibril strands, rather than connecting, as in the case of AFS. Many examples of this are evident in Figure 3, but again, there is a clear need for a substantial improvement in spatial resolution. Differences between filtergrams taken at two wavelengths symmetrically displaced to the red and blue of the center of Hα are evident, so velocities are going to play an important role in any definitive interpretation of fibril structure.

Within a few tenths Ångstrom from line center the bright mottles comprising the bright Hα plage first appear and attain a maximum contrast at the line center, where they overlie most of the network of photospheric magnetic fields visible in the Zeeman spectroheliograms. Bray (1969) has shown that these bright mottles lie substantially lower in the chromosphere than the dark features we have been discussing. The inference is that the stronger fields directly over the network play a role in increasing the transparency of the chromosphere. The dark chromospheric network, when viewed obliquely towards the limb, also gives the impression of being depressed directly over the magnetic field network, as if the mapping was not as fully developed there.

In this example, as well as others for which we have similar data, both regions of leading and following magnetic polarity have associated bright Hα plage, and thus do not conform to the distinction between plage and antiplage made by Veeder and Zirin (1970); i. e., that only following magnetic fields show bright plage in the center of Hα. This does not preclude the possibility that an asymmetry in brightness between leading and following plage may exist on a statistical basis, or possibly at different stages of development of an active region.

6. Example of a Magnetic Field Region Exhibiting a Spiral Structure

Figure 4 is a Zeeman spectroheliogram taken on 29 September, 1969 at 2000 UT. The area is approximately one-quarter the solar disk, centered on a point located roughly on the solar equator, west of the meridian. Two southern hemisphere bipolar sunspot groups, each in a typical BMR, are seen at the bottom left and right. For comparison, at the upper left is a northern hemisphere bipolar group to which Hale's law applies. In the interesting active region at the upper right the distinction between polarities has made visible a pronounced spiral structure. The development of this structure is shown in Figure 5 which also shows the configuration this region had on three previous days. Referring to the magnetic fields near the center of this active region, the areas of opposite polarity lay north and south of each other on September 25, and ended up east and west by September 29; i.e., the line of magnetic polarity reversal

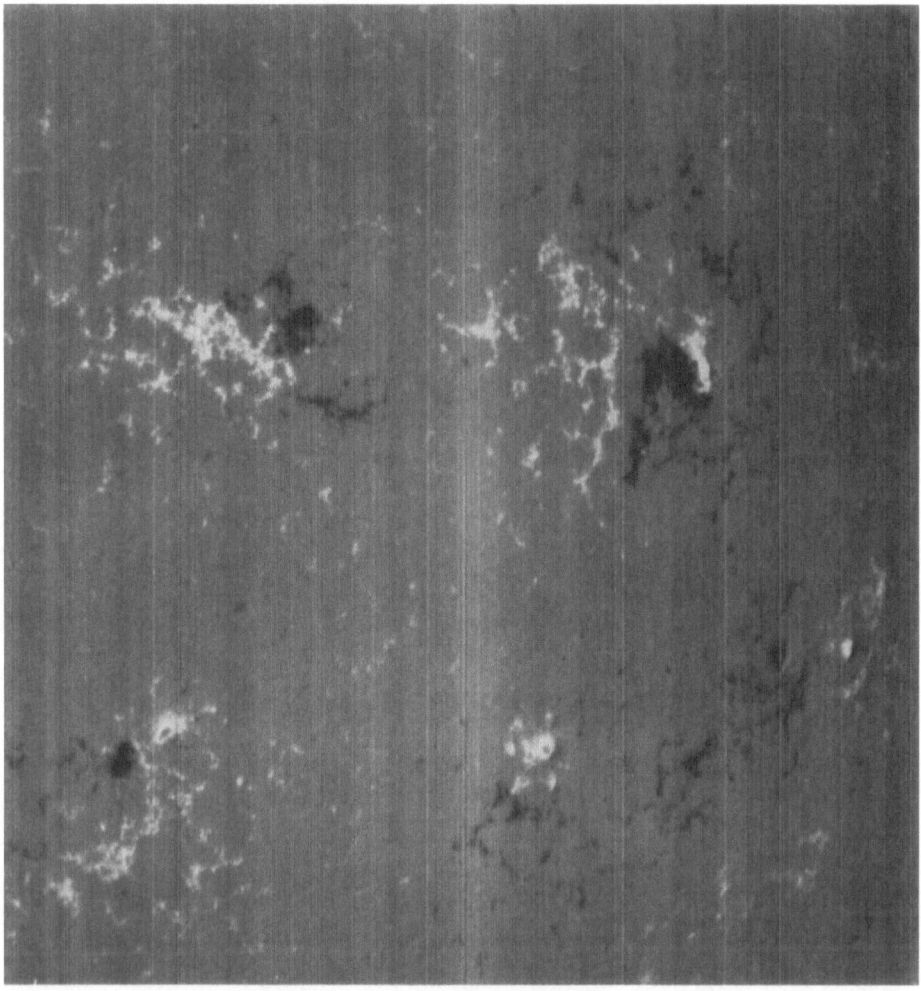

Fig. 4.   The network structure of photospheric magnetic fields in the large is seen in this Zeeman spectroheliogram taken on 29 September, 1969 at 2000 UT. Note the distinct spiral structure of the active region at the upper right. In the central region the normal order of polarities is inverted, and there is a high vertical field gradient across the line of polarity reversal.

rotated clockwise as viewed from the Earth, resulting in an inversion of the normal order of polarities. The spiral structure did not simply result from a clockwise rotation of the central area. Its development involved the formation, growth, decline, and disappearance of sunspots of both polarities. On 27 September, 1969, commencing at 0416 UT., a class 3B flare occurred in this region of inverted polarities. This type of complex evolution of sunspot groups involving rotation has been found to be associated with an increased likelihood of occurrence of energetic flares (Sakurai, 1967; McIntosh, 1970a; Sawyer and Smith, 1970). It has been estimated that only a few percent of all active regions develop a spiral geometry of this type (McIntosh, 1970b).

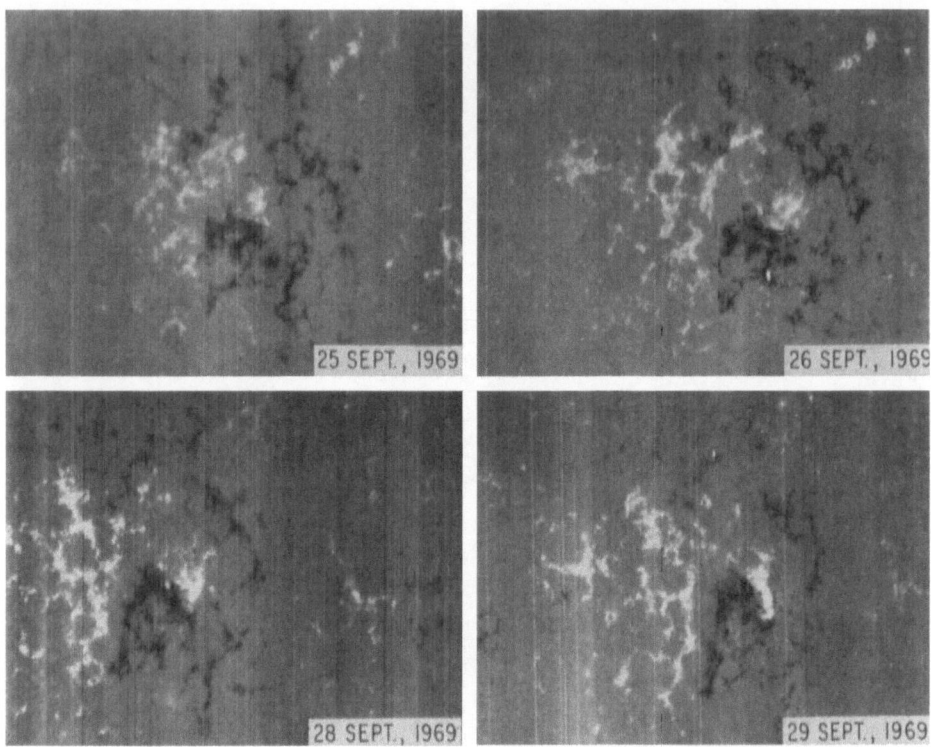

Fig. 5.   The development of spiral structure in the magnetic fields can be followed in this sequence which spans 4 days. Changes in the network between successive days are evident. (Magnetic field data for September 27th is missing.)

Acknowledgements

This paper reports the first data to be obtained with the major instrument of the Aerospace solar program. Many individuals have contributed to the accomplishment of this program milestone to obtain time-lapse Zeeman spectroheliograms. The existence of the San Fernando Observatory is largely due to the support of I. A. Getting, M. T. Weiss, and R. A. Becker. E. B. Mayfield has been in charge of this program since its inception. D. Grey and E. H. Rogers made major contributions to the design of this instrumentation, and the latter to the development of techniques for photographic subtraction of cine film. C. Denny and C. Wheeler of the Boller and Chivens Division of The Perkin-Elmer Corporation were responsible for manufacture of the 61-cm solar telescope and spectroheliograph. W. Mott and S. Wiemokly have been responsible for mechanical and electronic systems, respectively, and W. Medawar, R. Maulfair, and J. Paul for all photographic work. Finally, G. A. Paulikas has been a continuing source of support and encouragement to this program.

References

Babcock, H. W. and Babcock, H. D.: 1955, *Astrophys. J.* **121**, 349.

Beckers, J. M. and Schröter, E. H.: 1968, *Solar Phys.* **4**, 142.
Bray, R. J.: 1969, *Solar Phys.* **10**, 63.
Bruzek, A.: 1969, *Solar Phys.* **8**, 29.
Chapman, G. A. and Sheeley, N. R.: 1968, *Solar Phys.* **5**, 442.
Howard, R. and Harvey, J. W.: 1964, *Astrophys. J.* **139**, 1328.
Leighton, R. B.: 1959, *Astrophys. J.* **130**, 366.
Mayfield, E. B., Vrabec, D., Rogers, E. H., Janssens, T. J., and Becker, R. A.: 1969, *Sky Telesc.* **37**, 208.
McIntosh, P. S.: 1970a, World Data Center A – Upper Atmosphere Geophysics, Report UAG-8, Part 1, ESSA, Boulder, Colorado, 22.
McIntosh, P. S.: 1970b, Private communication.
Rust, D. M.: 1968, in K. O. Kiepenheuer (ed.), 'Structure and Development of Solar Active Regions', *IAU Symp.* **35**, 77.
Sakurai, K.: 1967, *Rep. Ionosphere Space Res. Japan* **21**, 113.
Sawyer, C. and Smith, S. F.: 1970, World Data Center A – Upper Atmosphere Geophysics, Report UAG-9, ESSA, Boulder, Colorado, 9.
Sheeley, N. R.: 1969, *Solar Phys.* **9**, 347.
Veeder, G. J. and Zirin, H.: 1970, *Solar Phys.* **12**, 391.
Vrabec, D. and Rogers, E. H.: 1968, *Astron. J.* **73**, S81.

Discussion

Sheeley: Have you attempted to measure the velocities of the magnetic points that flow out from sunspots?

Vrabec: There appears to be some spread in the horizontal velocities. A rough average is 1 km/s. But in the case of some knots, the velocity appears to be twice as large, and for others, one-half as large. These are approximate values.

Giovanelli: I disagree with you in that I believe that it is the *bright* fibrils that are related to the magnetic fields.

Vrabec: The distinction between the physical nature of bright and dark Hα chromospheric features still appears to be unresolved. This is an important area of investigation.

Leighton: I find the fact that *dark* magnetic fragments may move out of a *light* (opposite polarity) sunspot to be most striking. What flow pattern would the theorists suggest to do this?

Vrabec: Since this question is addressed to the theorists, I had better reserve comment, though we have considered the possibility of arched flux tubes originating in the sunspot whose re-intersections with the photosphere move outward.

Simon, G. W.: You could imagine magnetic flux concentrations 'shooting' up out of the sunspot and reversing their orientation as they re-intersect the photosphere.

ON THE TIME FLUCTUATIONS OF MAGNETIC FIELDS

A. SEVERNY

Crimean Astrophysical Observatory, Nauchny, Crimea, U.S.S.R.

Abstract. The simultaneous measurements of longitudinal magnetic fields in lines $\lambda 5250$ and $\lambda 6103$, the line of sight velocities and intensities in these lines at some fixed point (area $2''.3 \times 4''.5$) within the moderately strong (~ 100 G) areas of solar magnetic fields were examined to find the possible fluctuations of m.f. with time. No fluctuations were found which cannot be correlated with the simultaneously measured seeing conditions such as contrast, image excursions and scintillations for characteristic periods less than 5^m. The possibility of $\sim 5^m$ oscillations of magnetic fields correlated with the well known sight-line velocity oscillations is pointed out.

Simultaneous strip-chart records of longitudinal magnetic field, sight-line velocity and intensity in the lines $\lambda 5250$ and $\lambda 6103$ at some fixed point on the Sun within an area with a moderately strong (~ 100 G) magnetic field were made in the summer of 1968 to investigate the possible fluctuations of magnetic field with time. The image of the Sun was held fixed on the slit with the aid of two sensors positioned at the western and southern borders of the Sun (with an accuracy of $\pm 0''.2.$). This system of pointing did not permit us to follow fluctuations for periods longer than 5–10 min, because of the rapid change in the vertical diameter of the Sun ($\simeq 1''/10^m$) due to refraction early in the morning when the seeing is usually the best. The resolution was $2''.3 \times 4''.5$ and the velocity of the records, 1 cm $=8$ s at a time constant $=1$ s (cf. Severny, 1967, 1968).

Simultaneously image excursions (dancing) were also recorded, along with the contrast of the granulation to exclude the possible influence of seeing on the behavior of magnetic field records. The image excursions were measured by photoresistors (Si) in two modes: (1) behind a slit $2''$ – wide and $10''$ in length put radially at the E – border of the image, (2) behind a hole with size $1''$ receiving light (by reflection from a transparent plate) from the same point on the disk which is being recorded. The photocurrent alternating, during the image excursions, passed through an RC-filter to an amplifier with a pass-band 0–20 Hz, and the signal was registered on the strip chart. For measurement of the contrast we scanned the granulation on the image very near to the point on the disk under investigation. A very small rod with a photoresistor behind a hole with a size of $1''$ performed oscillations with amplitude $\pm 90''$ and frequency 25 Hz. The amplifier with a wide pass-band 500 Hz–3500 Hz produced a signal which was a measure of contrast. These devices are, except for excursions at the border of the Sun, essentially of the same kind as described by Deubner (1968). From more than 30 of the best early morning records we have considered only 5 showing more or less clearly the fluctuations of magnetic field which at a direct inspection did not show close correspondence between magnetic field fluctuations and fluctuations of seeing.

Figure 1 is one of these 5 selected records showing directly good correlation between fluctuations of $\delta_{\parallel}$ (5250) and those of contrast (second from the top). It also plots sight-line velocity $v_{\parallel}$ (5250) and image-excursions $\partial I/\partial t$ showing the appearance

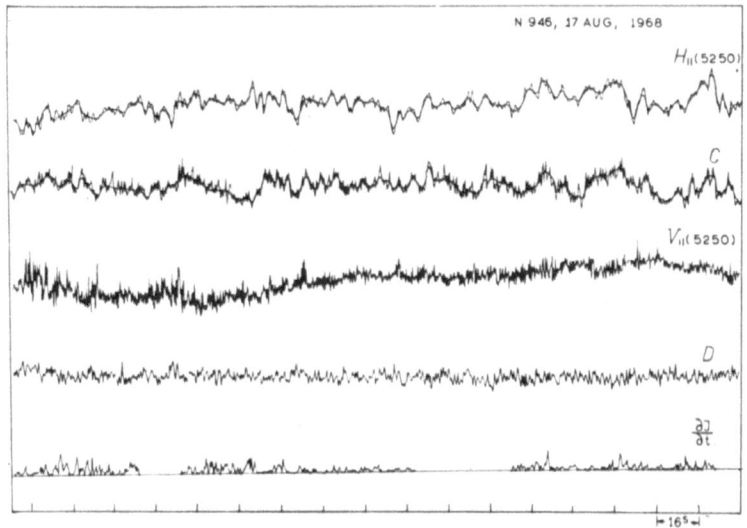

Fig. 1.   The example of original records of magnetic field $H_{\parallel}$, contrast C, sight-line velocity $V_{\parallel}$, dancing of images D, measured at the limb of the solar image, and of the rate of change of intensity of granules due to image excursions $\partial I/\partial t$.

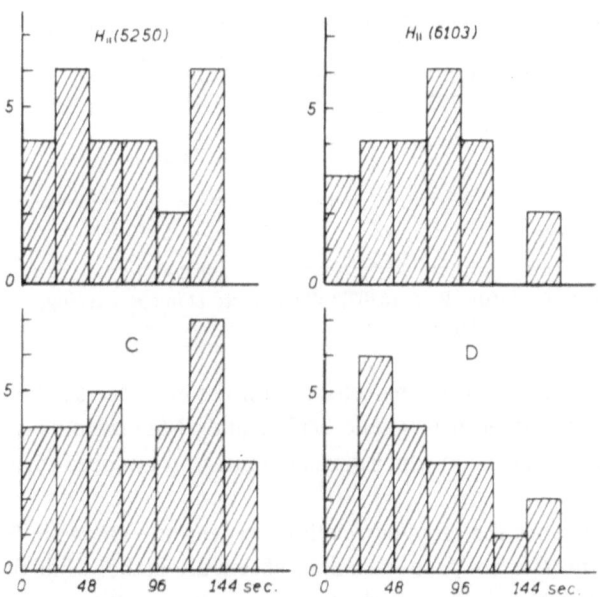

Fig. 2.   The distribution of the periods of oscillations found from autocorrelation curves for simultaneous records of magnetic fields in two lines $\lambda 5250$ and $\lambda 6103$, of contrast C and image dancing D.

of SPO (Howard, 1967) at $v_{\parallel}$ – records simultaneously with the appearance of peaks on image dancing records. There is nothing new here, and this agrees with conclusions obtained earlier by Deubner (1968) that most of magnetic field and sight-line velocity SPO fluctuations correspond to fluctuations of contrast and image-motions.

However, we should be cautious with such 'at first sight' conclusions and to try to evaluate quantitatively the correlations in question. For this purpose the auto-correlation curves for all such fluctuations as well as cross-correlations between different fluctuations were computed with a lag $\Delta m = 1$ which corresponds to 2 mm on the records or to 1.6 s of time.

In all cases the autocorrelation curves show some periodicity of fluctuations and the graph on Figure 2 shows that the distribution of periods in 7 cases considered is more or less 'grey', although there is, possibly a tendency for fluctuations in $H_{\parallel}$ (5250) to show a similarity in distribution with that of the contrast. The most suggestive are, however, the cross-correlation curves (see Table I). This table shows that in most cases the magnetic field fluctuations are well correlated with image dancing and fluc-tuations of contrast, (some time lag in cross correlation with D and C can be due to the

TABLE I

Cross correlations between fluctuations of different records

C.C.	938(I) 8 Aug. 58	938(II) 8 Aug. 68	939 9 Aug. 68	943 13 Aug. 68	946(A) 17 Aug. 68	946(B) 25 Aug. 68
AB	+0.15(0)	−0.25(3ˢ.2)	0.34(0)	+0.01(0) −0.63(−43ˢ)	−	+0.40(−6ˢ.0)
AC	0.00(0)	+0.23(0)	−0.06(0)	−0.10(0)	+0.45(0)	−0.27(+5ˢ.0)
AD	+0.59(0)	+0.50(0)	+0.28(0)	−0.02(0)	0.23(0)*	−0.40(−5ˢ.0)
AI	−	−	−	+0.56(−2ˢ.5)	−	−
BC	+0.02(0)	+0.13(0)	−0.11(0)	+0.14(0)	−	+0.42(0)
BD	+0.30(0)	−0.26(−3ˢ.2)	−0.28(0)	−0.28(0)	−	−0.10(0)
AV	−	−	−	−0.09(0) +0.32(11ˢ.0)	−	−
Source	D	D	D	I	C	C, D

Key to notations: A-$H_{\parallel}$ (5250); B-$H_{\parallel}$ (6103); C-contrast; D-image dancing, $V_{\parallel}$-sight-line velocity $\lambda 5250$; * measured in the second made.

difference in the position of seeing sensors and the entrance slit hole). The absence of cross correlation between magnetic field A and B fluctuations can frequently be due to atmospheric dispersion which is usually most pronounced early in the morning at good seeing.

However, sometimes the cross-correlation between the magnetic field fluctuations and those of C and D is not pronounced, or absent altogether while the magnetic field fluctuations clearly show the periodicity. In this respect the case of record 943 of August 13, 1968 deserves special attention. Figure 3 shows part of the original records with curves drawn to smooth out noise. We see clearly pronounced a 30ˢ

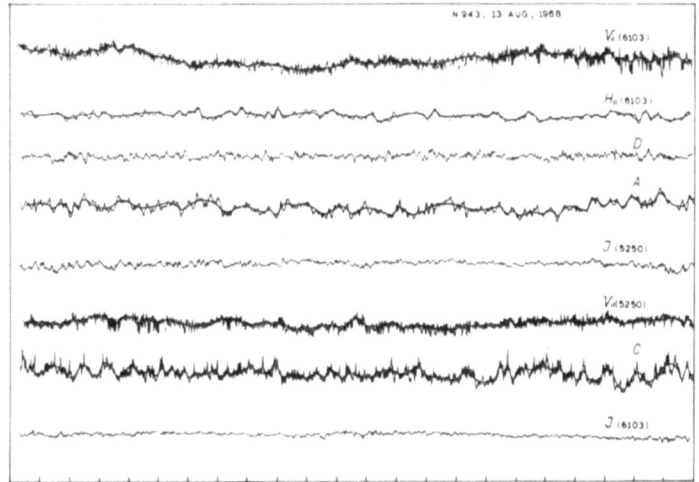

Fig. 3. The part of original simultaneous records of magnetic field in $\lambda 5250$ (A) and in $\lambda 6103$ (B) of sight line velocities $V_{\parallel}$ in these lines, contrast C, image dancing D and of intensities in the cores of both lines I.

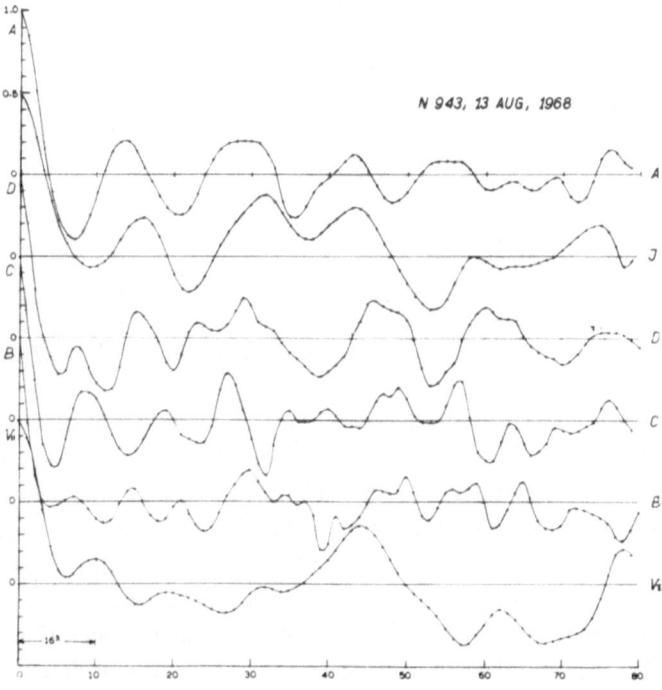

Fig. 4. Autocorrelation curves for the records presented on Figure 3.

periodicity in the magnetic field – A fluctuations ($H_{\parallel}$ (5250)) which is not reflected in magnetic field B – records ($H_{\parallel}$ (6103)). There is also no good correspondence between the green and red line magnetic field records and the records of image dancing and contrast. On the other hand we observe very good correspondence in records of magnetic-field fluctuations in $\lambda5250$ and the records of intensity in the same line. We wish to emphasize that this correspondence is one of the most typical and remarkable features of all our records – even the smallest peaks of the magnetic field records are usually reproduced in the records of intensity, as we can see from the example in (Figure 3), as well as in other cases. (This is why we do not even

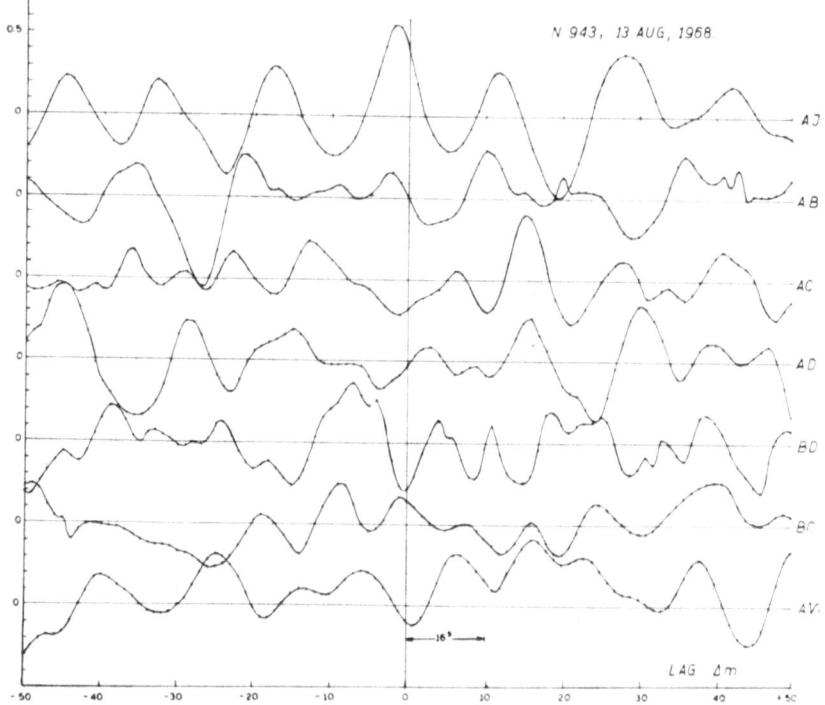

Fig. 5.   Cross correlation curves between the different fluctuations presented on Figure 3.

calculate the cross correlation in this case.) The autocorrelation curves shown in Figure 4 confirm all said above: while almost all the maxima of B – magnetic-field fluctuations correspond to those of D and C, in the case of A magnetic-field fluctuations ($\lambda5250$) there is no such agreement, and all secondary maxima are in the best correspondence with those of intensity fluctuations I. I wish to draw your attention to the fact that the characteristic periods form a sequence of successive modes, m, 15^s, 30^s, 45^s, 60^s as if we had a standing wave oscillation. The consideration of the cross-correlations brings also additional evidence of very good correlation between the magnetic field and intensity in $\lambda5250$ as can be seen in Figure 5 (according to the data in Table I, the cross-correlation coefficient is $+0.56$ at a lag $\Delta m = -2\overset{s}{.}5$).

So far only the influence of image excursions and of the contrast on magnetic-field fluctuations have been considered. But, as is well known from stellar observations, there exists one more very important factor of seeing – scintillations or twinkling. According to measurements made by Ellison and Wilson (1951), Ellison and Seddon (1952), and by Nettelblad (1953) even in the case of planets we have appreciable scintillations. For instance in the case of Mars (4″.4) the scintillation amplitude $(Z=57°)$ is $\pm 14\%$ (while for Sirius $\pm 23\%$, Ellison and Seddon (1952)), for Saturn the mean amplitude is $\Delta m = 0^{m}.10$ (at $D = 20$ cm), or $\sim 9\%$. If we use for solar observations a small entrance aperture (4″.5 × 2″.3) we should expect the amplitude of scintillations of the same order*, because the signal of the magnetic field,

$$\delta i_{\parallel} \sim \frac{\partial I_\lambda}{\partial \lambda} \Delta \lambda_H = I_\lambda \frac{\partial r_\lambda}{\partial \lambda} \Delta \lambda_H \tag{1}$$

is proportional to the intensity in the wing of the spectral line, I_λ. The amplitude of the fluctuations of I_λ actually measured is usually 3–5% in good seeing. Being cor-

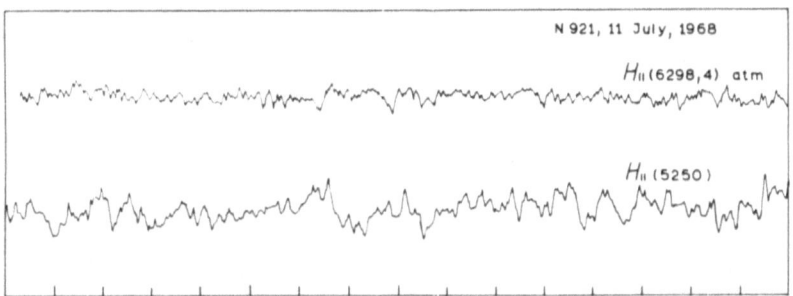

Fig. 6. Showing the absence of the influence of turbulence inside of the spectrograph on the recorded fluctuations of magnetic field $H_{\parallel}$.

rected for the change of intensity I_0 (usually measured for a large portion of the spectrum far outside the line) the signal $\delta i_{\parallel} \sim I_\lambda/I_0$ can have an amplitude twice as large because I and I_0 fluctuate non-coherently (see Zhukova, 1959) even if the distance between two spectral regions is only 500 Å. (I wish to remind you that a contrast $\sim 1\%$ corresponds to a signal ~ 10 G). The accidental excursions of the line across the exit slit uncorrected by the line-shifter can contribute still more, although a special check with a telluric line showed that the magnetic field fluctuations in this line are small, and completely independent of magnetic field fluctuations in $\lambda 5250$ (Figure 6). The effect of scintillations does not influence so strongly the magnetic field fluctuations recorded in the scanning mode, because we pass, while scanning, completely incoherent states of scintillations. We should also keep in mind that twinkling can produce immediately a signal which is equivalent to a magnetic field signal, and comparable with it if the frequency of modulation in the magnetograph is

---

* The image excursions and loss of contrast can reduce this amplitude due to additional incoherency brought from the parts of image outside the entrance aperture.

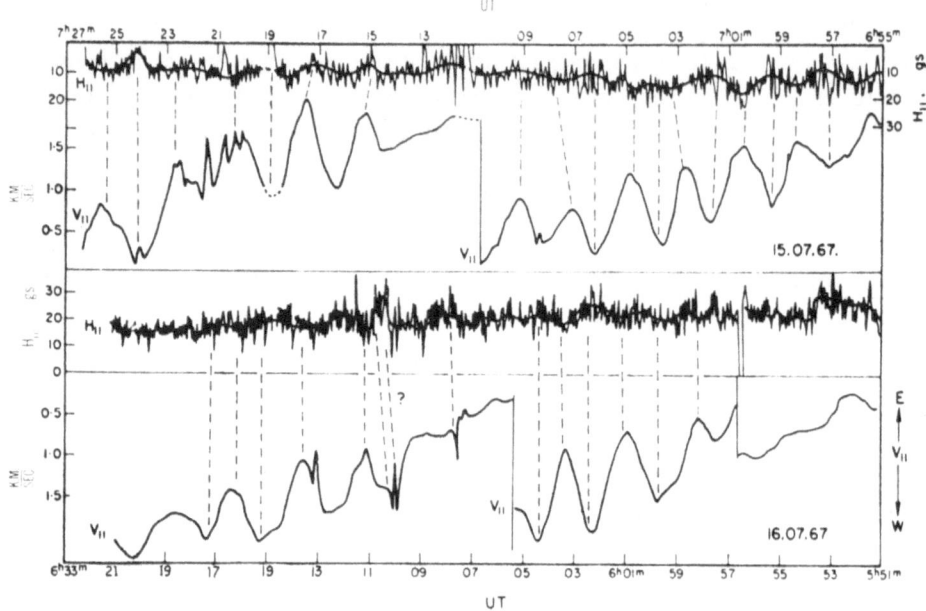

Fig. 7.   Illustrating some synchronism in fluctuations of magnetic fields $H_\parallel$ and velocities $V_\parallel$ with
the period $\sim 5^m$.

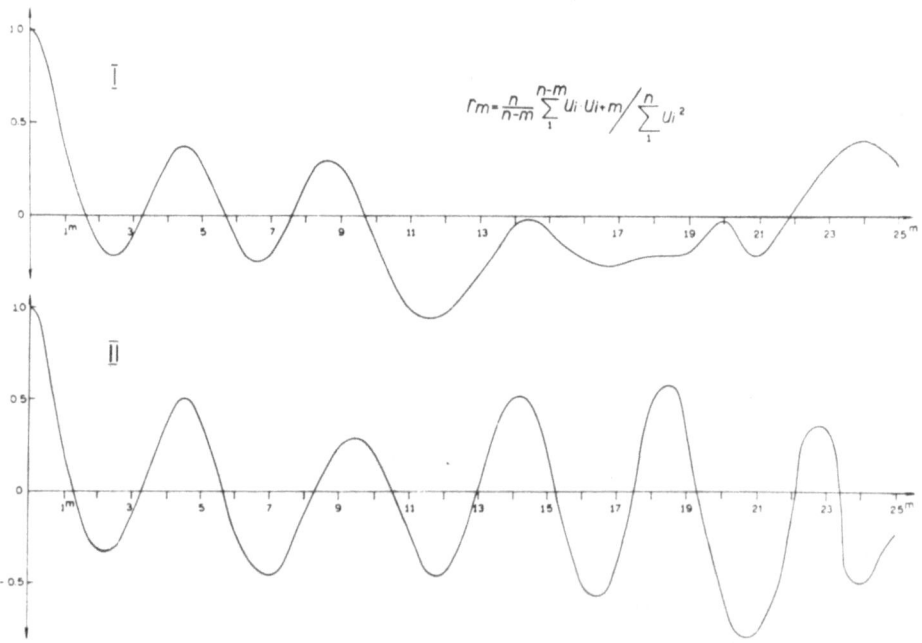

$$r_m = \frac{n}{n-m} \sum_1^{n-m} U_i \, U_{i+m} \Big/ \sum_1^n U_i^2$$

Fig. 8.   Autocorrelation curves for the data shown in Figure 7.

within the range of frequencies of scintillations, which is very wide and can reach even 100 Hz and more (Vinogradova, 1959).

So we are coming to the rather pessimistic conclusion that most of the short-period time fluctuations of magnetic field signal and velocity signal observed so far are probably due to fluctuations of seeing.

At the same time we observe sometimes good coherency between the well-known 5^m – fluctuations of sight-line velocity and fluctuations in magnetic field, taking 20–30^s averages, as can be seen from Figure 7. A similar correspondence was found by Howard (1967). The autocorrelation curve for the fluctuations shown above is in Figure 8, showing clearly expressed periods which are a little less than 5^m. There are no corresponding fluctuations in intensity for these fluctuations and they are probably too slow to be ascribed to any variations in seeing.

References

Deubner, F.: 1968, *Solar Phys.* **3**, 439.

Ellison, M. and Wilson, R.: 1951, *Observatory* **71**, 26.

Ellison, M. and Seddon, H.: 1952, *Monthly Notices Roy. Astron. Soc.* **112**, 73.

Howard, R.: 1967, *Solar Phys.* **2**, 3.

Nettelblad, F.: 1953, *Mitt. Lunds Astron. Obs.*, Ser. II, N. 130.

Severny, A.: 1967, *Astron. Zh.* **44**, 481.

Severny, A.: 1968, in K. O. Kiepenheuer (ed.), 'Structure and Development of Solar Active Regions', *IAU Symp.* **35**, 233.

Vinogradova, R.: 1959, in *Publ. Conf. on Scintill. of Stars*, Moscow 18-20 June, 1958, Akad. Nauk, U.S.S.R., p. 135.

Zhukova, L.: 1959, in *Publ. Conf. ou Scintill. of Stars*, Moscow 18-20 June, 1958, Akad. Nauk, U.S.S.R., p. 116.

FIVE-MINUTE OSCILLATIONS IN THE SOLAR MAGNETIC FIELD

ANDREW S. TANENBAUM and JOHN M. WILCOX

Space Sciences Laboratory, University of California, Berkeley, Calif. 94720, U.S.A.

and

ROBERT HOWARD

Hale Observatories, Carnegie Institution of Washington, California Institute of Technology, Pasadena, Calif. 91101, U.S.A.

Abstract. Evidence for the existence of 5 min oscillations in the photospheric and low chromospheric magnetic fields is presented, their properties discussed, and a possible production mechanism suggested.

For about a decade it has been known that there are velocity oscillations in the photosphere and low chromosphere with periods of about 5 min (Leighton *et al.*, 1962). These oscillations have also been detected in the brightness, both in the wings of lines, and in the continuum.

A logical question which follows from this is whether or not these oscillations are present in the magnetic field as well. In this paper we report the answer as – yes, they are there, but are often well hidden by the noise.

In 1967, Severny reported seeing oscillations in the magnetic field, but these were of periods seven to nine minutes rather than five minutes, and somewhat irregular.

The simplest way to look for oscillations in the magnetic field is to use the solar magnetograph to observe a single point on the Sun for several hours, scanning westward very slowly to compensate for solar rotation. The raw data for this type of observation consists of the velocity as a function of time, the magnetic field as a function of time, and the brightness as a function of time. In Figure 1 the velocity and magnetic field are shown for one 3-hr observation of this type. All the observations discussed in this paper are of this 'non-scanning' type.

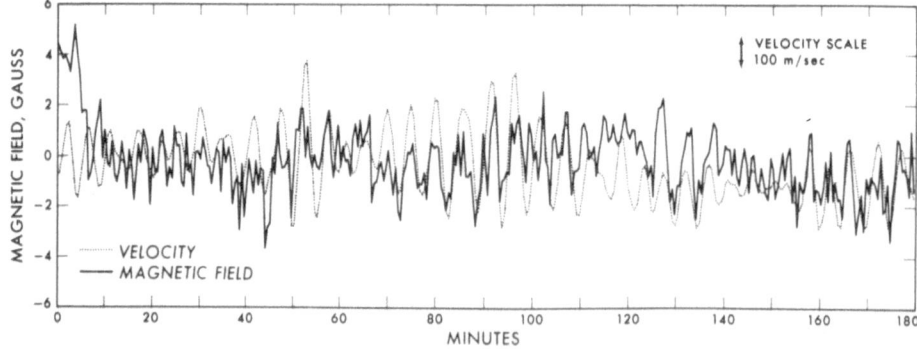

Fig. 1. Plot of velocity and magnetic field as a function of time for an observation in which the aperture was fixed on a single point on the Sun, moving slowly westward across the disk to compensate for solar rotation. The oscillations in the magnetic field can be seen here.

The velocity curve of Figure 1 shows the 5 min oscillations are sometimes present and sometimes not present, as usual. Although the magnetic signal is very noisy, if one looks carefully one can see numerous cycles where the maximum in velocity coincides with the maximum in field, and where the minimum in velocity coincides with the minimum in field.

Figure 2 is an autocorrelation of the magnetic field for four separate observations of the type just described. The four runs were scattered over a period of two years. Each of the observations was for an interval of about three hours. The top one is for the data

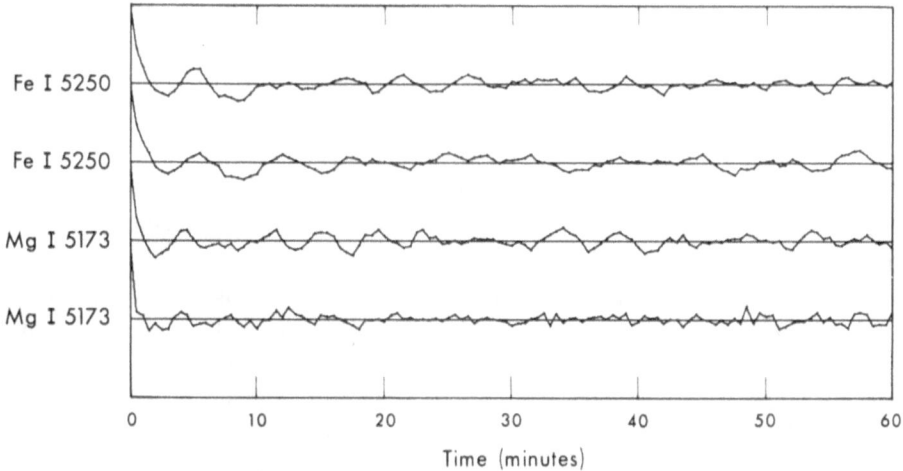

Fig. 2.   Autocorrelation of magnetic field for four separate observations. The horizontal line passing through each curve represents a correlation of 0. The lines above and below represent +1.0 and − 1.0 respectively.

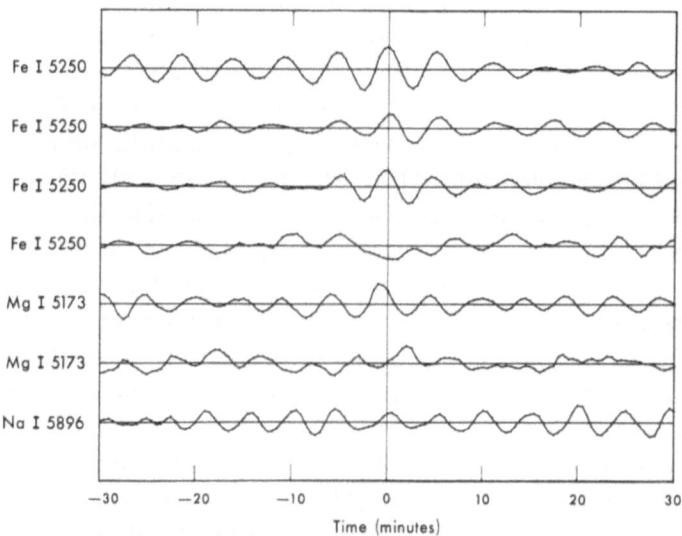

Fig. 3.   Cross correlation of velocity and magnetic field. Same scale as Figure 2.

of Figure 1. Two are for the photospheric line Fe I 5250 and two for the low chromospheric line Mg I 5173. The maximum for the two iron runs occurs at just over 5 min,
while the maximum for the magnesium runs is about forty seconds less. This is consistent with the well known fact that the period decreases somewhat with height
(Noyes and Leighton, 1963).

Figure 3 shows a cross correlation between the magnetic field and velocity for
several spectral lines. Note that all the curves are oscillatory in character for several
cycles. This means that *both* the velocity and magnetic field have the oscillations because the cross correlation of a sine wave (velocity) with random noise (magnetic
field) is essentially constant at zero. If one curve is random noise, it does not matter
how much you shift it with respect to the other one, the correlation is the same.

Note that the fourth curve has an *anti*correlation at zero lag. This is just due to the
fact that if the sign of the field is reversed, the sign of the correlation coefficient will
reverse, but of course the physical interpretation is the same. In other words, in a
region of positive field the correlation coefficient is positive when the velocity and
field are in phase, but in a negative field region there is an anticorrelation if they are
in phase.

Another approach to digging the magnetic field oscillations out of the noise is
to use a superposed epoch analysis. First the velocity and magnetic field are plotted as
functions of time. Then the starting time of the individual velocity oscillations are
determined. Then, conceptually at least, the individual velocity oscillations are cut
out and pasted vertically below one another. The same thing is done for the magnetic
field (or brightness) but with the starting times determined from the *velocity* observations. In Figure 4 and 5 the first column is velocity, the middle column is magnetic
field, and the last column is brightness. One thing that is immediately obvious is
that the magnetic signal is very noisy.

The top magnetic curve corresponds to the same time interval as the top velocity
curve. The second magnetic curve corresponds to the same time interval as the second
velocity curve, etc. At the bottom of the first column is the average velocity oscillation.
Next to it is the average magnetic field during a velocity oscillation. Next to that is the
average brightness during a velocity oscillation. The magnetic signal is still rather
noisy, not surprising considering what went in to it, so a 30 s running mean was
computed and is displayed at the bottom of each column. Here the 5 min oscillation
in the magnetic field shows up quite clearly. Remember that the time intervals were
chosen so as to line up the velocity curves in phase; the fact that the oscillation comes
through in the field means that there is a definite phase relation between the two.

Figure 6 is a summary of four superposed epoch analyses. The example in the lower
left hand corner looks like it is out of phase with the others, but that is just because it
is in a region whose field polarity is opposite to those of the other three examples, so
it is really not shifted by 180°. The graph in the lower right hand corner seems to have
the magnetic phase advanced somewhat. This is for the magnesium line. The brightness oscillations in the chromospheric lines are advanced with respect to the photospheric lines, so this may be related in some way.

The observational results may be summarized as follows:

(1) Using three techniques (namely, looking at the raw data, correlation analysis, and superposed epoch analysis) the existence of periodic oscillations in the Sun's magnetic field has been demonstrated.

(2) The periods are the same order as the velocity periods, about 5 min.

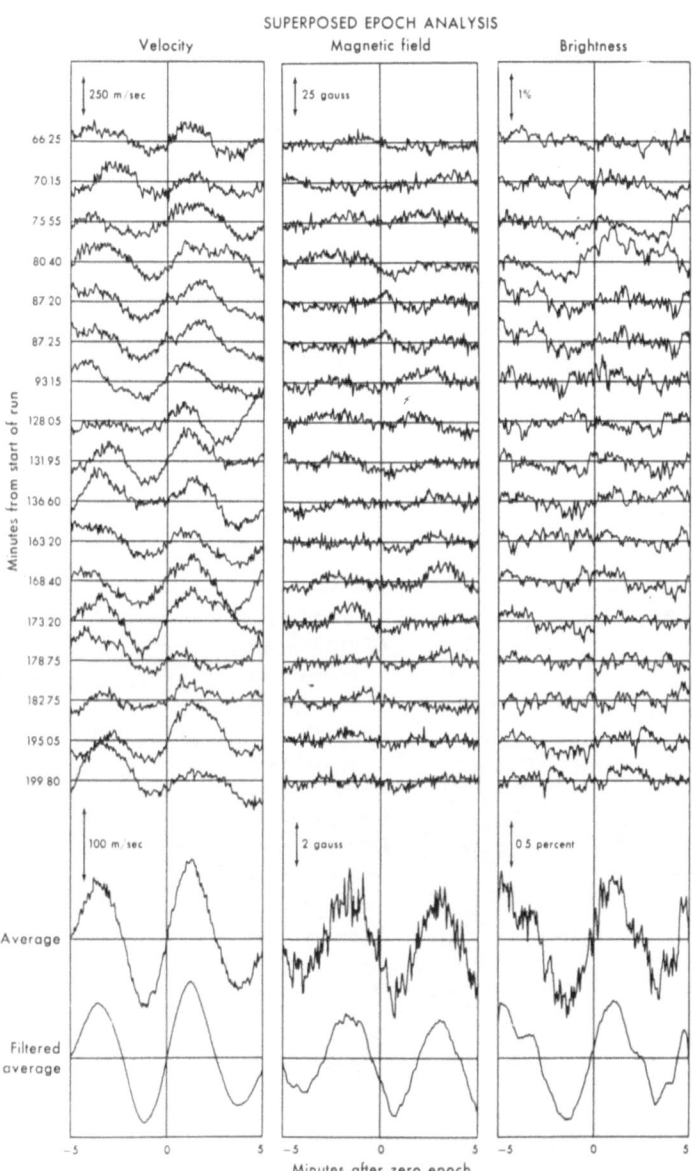

Fig. 4. Superposed epoch analysis of an observation of the type shown in Figure 1. The curves labeled 'average' represent the average of all the curves above them, although at a different scale. The filtered average is a 30 s running mean of the average to remove high frequency noise.

(3) The magnetic oscillations have a definite phase relation with the velocity oscillations.
(4) They exist at least over the range in height covered by the lines Fe I 5250
to Mg I 5173.
(5) The amplitude is around one or two gauss; consequently they can be easily
masked by noise.

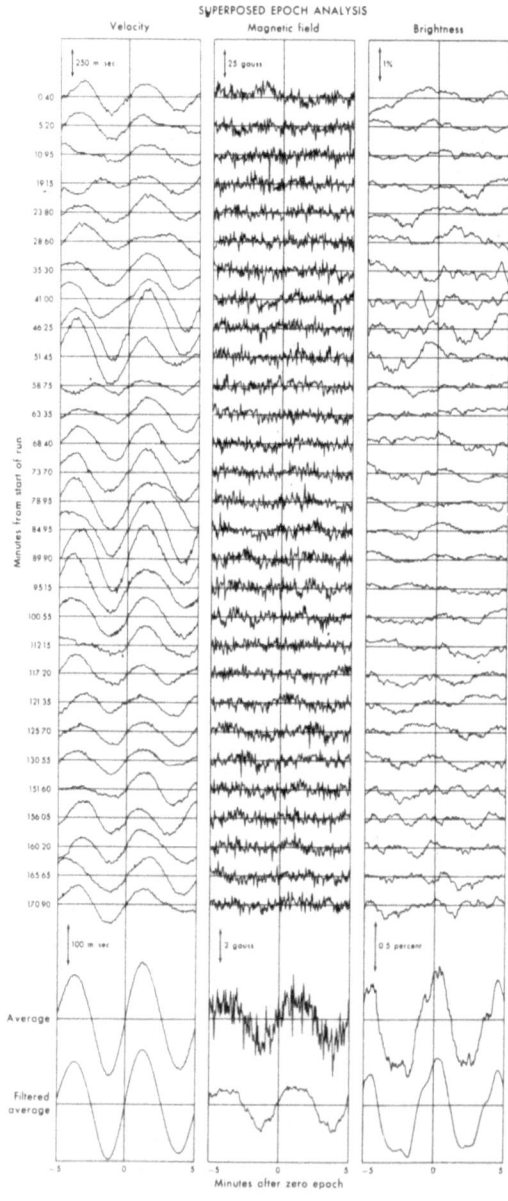

Fig. 5.   Same as Figure 4, but for different observations.

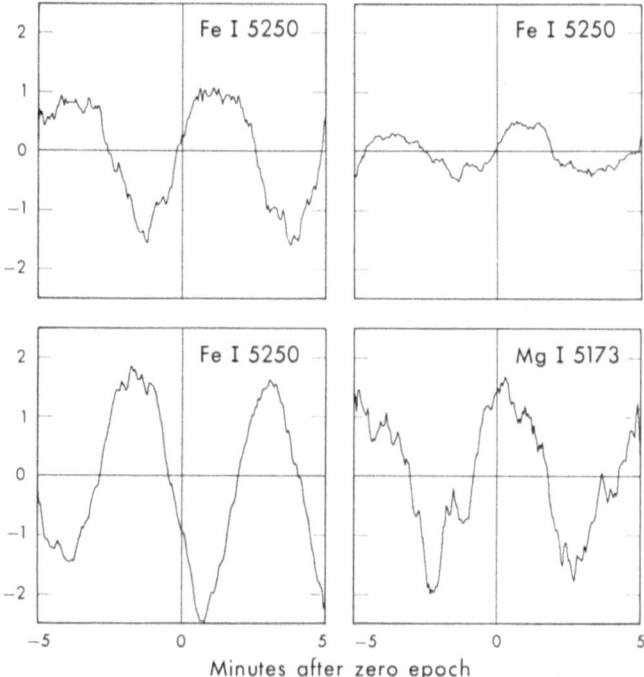

Fig. 6. Four examples of mean magnetic field oscillation profiles during a velocity oscillation as deduced from superposed epoch analysis.

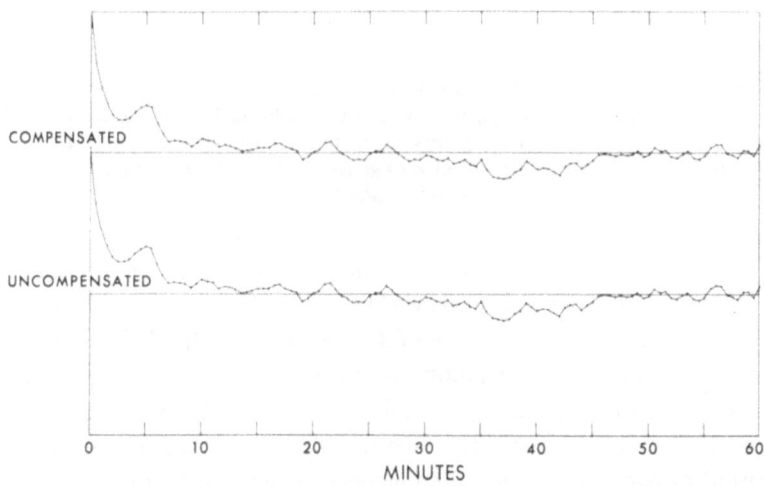

Fig. 7. Upper curve is autocorrelation of magnetic field as normally computed (Zeeman signal divided by brightness signal). Lower curve is for the same data but without compensation for brightness variations, i.e. just the Zeeman signal. The similarity of the two curves shows that the 5 min oscillations in the magnetic field is not a feed through of the brightness oscillation.

(6) They occur both in regions of weak field (<5 G) and in regions of strong field (≈ 80 G).

Now that it has been established that the 5 min oscillations exist in the magnetic field, the question of what causes them arises. One possibility is some instrumental artifact. Inasmuch as the oscillations are present in both the velocity and the brightness they might be feeding through into the magnetic field observations.

First consider the brightness. The brightness oscillations in the photosphere consist of a 1% variation in the average light level, so it is a very small effect. The magnetograph does not actually record the magnetic signal, but rather the Zeeman signal,

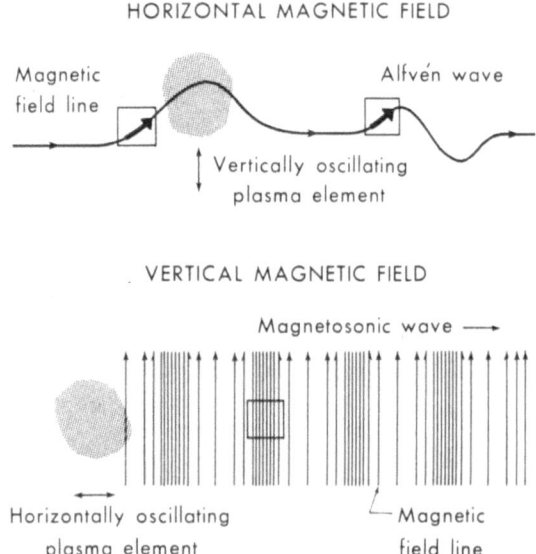

Fig. 8. Schematic representation of how oscillations in the solar plasma could be transferred to the line-of-sight magnetic field. If the magnetograph aperture coincided with any of the small squares, the magnetic field would show an oscillatory character. In the top example, the magnetic field oscillations would have an average of zero gauss; in the bottom example, the average would be non-zero.

which is proportional to the product of the magnetic field and the brightness. To get the magnetic field, a computer program divides each Zeeman reading by the instantaneous value of the brightness, so the oscillations in the brightness cancel out exactly. As a check, an autocorrelation was made both with and without this brightness compensation. Figure 7 shows the 5 min peak present in both cases.

Now consider the velocity. It is true that the spectral line is moving back and forth with a period of five minutes, but the magnetic signal is accoupled to the velocity; it depends only upon changes in the line profile occurring at the KDP frequency, which is 18000 times higher than the frequency of the velocity oscillations. In addition, it is hard to explain how an instrumental effect would appear both in the weak field case where the Zeeman signal changes sign, and in the strong field case where it does not. Furthermore, if it really were an instrumental effect, the phase shift between velocity and field ought to be constant, whereas in Figure 6 one of the examples has an appreciably different phase shift from the others.

If it is not an instrumental effect, it must be of solar origin. Because the solar magnetic field is frozen into the oscillating plasma, there are several ways in which the oscillations in the plasma (which are observed as velocity oscillations) could be transferred to the magnetic field. Figure 8 shows schematically how vertical waves could cause oscillations in a horizontal magnetic field, and how horizontal waves could cause oscillations in a vertical magnetic field.

Acknowledgements

This work was supported in part by the Office of Naval Research under contract N00014-69-A-0200-1016, by the National Aeronautics and Space Administration under grant NGL 05-003-230, and by the National Science Foundation under grant GA-1319.

References

Leighton, R. B., Noyes, R. W., and Simon, G. W.: 1962, *Astrophys. J.* **135**, 474.
Noyes, R. W. and Leighton, R. B.: 1963, *Astrophys. J.* **138**, 631.
Severny, A. B.: 1967, *Soviet Astron.* **11**, 383.

Discussion

Musman: You don't need MHD waves to produce oscillations in the magnetic field due to frozen-in field.

Tanenbaum: If a horizontal field is uniformly moved up and down there will be no oscillations in the longitudinal field, but if there are spatial variations there may be. That is what I intended to show in the last slide.

PART IV

OBSERVATIONS OF
MAGNETIC FIELDS ASSOCIATED WITH
FLARES AND OTHER TRANSITORY PHENOMENA

SOLAR MAGNETIC FIELDS IN ASSOCIATION WITH FLARES

R. MICHARD

Observatoire de Paris-Meudon, France

Abstract. The association of Hα flares with photospheric magnetic field patterns is reviewed on the basis of empirical evidence. The insertion of flares in the $H_{\parallel}$ and $H_{\perp}$ patterns is described. Present data on the evolution of magnetic structures in connection with flares suggest that flares result from special trends in this evolution leading to stressed and/or 'misconnected' fields. However no sudden changes with a flare-like time scale have been definitely observed in photospheric fields.

1. The Insertion of Flares in Magnetic Patterns

1.1. OCCURRENCE OF FLARES AGAINST 'MAGNETIC TYPE' OF A.R.

As is well known, flares (and flare-like transient events) are associated with Active Regions of we define as such an area of the Sun permeated by significant magnetic fields. This is even the case for the faint flare-like brightenings which follow the blow-off of quiescent filaments, since these filaments are separating extended regions of opposite magnetic polarities, resulting from the decay of the strong fields of Active Centers.

However, not all magnetic field patterns are likely to become the seat of a flare event. For a flare to occur the magnetic field must *deviate somehow from the simple bipolar structure*, characterized by two magnetic poles at the photosphere aligned roughly along the *E-W* directon, with the line of separation of polarities, that is the zero line of H_V (for vertical) not much inclined to the solar meridian.

Magnetic patterns departing to various degrees from the simple bipolar one are characterized in such empirical classifications as
– the classical Mt Wilson magnetic types of sunspots groups.
– the Meudon classification of $H_{\parallel}$ patterns in A.R. (Martres *et al.*, 1966a).
– the similar Mount Wilson classification (Smith and Howard, 1968).

These last two systems take into account the fields outside spots and lead to very good correlations with flare productivity.

1.2. LOCATION OF FLARES RELATIVE TO THE INVERSIONS OF POLARITIES

In complex or anomalous patterns of the magnetic field, the flares *cannot occur at any place*. We must qualify this statement, by saying that the *initial brightenings in Hα at the start of a flare* have privileged positions. As early as 1958, Severny (1958) showed these to occur only at or very close to a line of polarity inversion, provided that the *horizontal gradient* of the longitudinal field across the inversion line was strong enough. He gave (Severny, 1960) the condition

$$\Delta H_{\parallel}/\Delta x \gtrsim 0.1 \text{ G/km.}$$

Shortly after Michard *et al.* (1961) pointed out that this condition means in practice the presence of a strong horizontal field at the inversion line, ruling out any special association of flares with neutral lines or neutral points, at least at the photospheric level. The Crimean observers confirmed this in detail when they started observing transverse fields (Severny, 1961). On the other hand they showed a quantitative correlation between the strength of the local gradient of $H_\parallel$ and the importance of the associated flare (Gopasyuk *et al.*, 1963).

The condition about the strength of the horizontal gradient of $H_\parallel$ or equivalently about the strength of the transverse field at the photospheric level is by no means sufficient for the occurrence of flares (nor perhaps very strictly necessary).

Other suggestive properties of flare localization in connection with inversion lines of $H_\parallel$ have been described by Martres *et al.*, 1966 and independently or subsequently by many observers (Moreton and Severny, 1966, 1968; Smith and Ramsey, 1967).

(1) In a perturbated bipolar structure having a main 'normal' inversion line separating the *E* and *W* polarities and some *'abnormal' inversion line*, this latter one is the preferred seat of the flares. 'Abnormal' inversion lines may encircle an island or peninsula of following polarity inside the preceding one, or the reverse. Their essential property is to have the wrong orientation to the meridian, near 90° rather than near 0°.

(2) The initial Hα brightenings tend to occur close to the inversion line rather than directly on it. In the case of flares showing more than one bright knot, that is the most frequent case, these knots are located on *two different polarities*, on both side of the inversion line. When bright elongated features develop, they do it along the inversion line on both sides of it.

The large proton flares with two bright ribbons overlying row spots of opposite polarities are a particular case of this general phenomenon.

It is interesting to note that the first observation of the X-ray emission of a flare, obtained at good spatial resolution by Vaiana *et al.*, 1968, reveal a close correlation with the Hα emission. A two-ribbon structure is observed, with a bright cloud connecting the two ribbons above the inversion line.

1.3. ASSOCIATION OF FLARES WITH HORIZONTAL (TRANSVERSE) FIELDS

Besides the already mentioned result of the localization of flares at places of strong horizontal field in the photosphere, I should recall two interesting findings of the Crimean observers.

Severny (1963, 1964) studied the place of flares on maps showing the azimuth φ of the transverse $H_\perp$ component projected on the photosphere. He found that, as a rule, flares occur in regions of 'bifurcation' of $H_\perp$: this means regions where the azimuth φ changes abruptly from point to point indicating the proximity or perhaps the crossing of flux tubes with strongly different orientations. He also presented many cases where the horizontal field in a flare productive area runs nearly parallel to the inversion line, instead of crossing it frankly in the direction of the gradient of $H_\parallel$. In such cases the flare ribbons may be said to extend parallel to the inversion line or to the horizontal field indifferently.

Besides direct polarimetric measurements, further indications on the orientations of transverse fields may be derived from the examination of elongated Hα structures, fibrils and small filaments, on high resolution pictures. The available descriptions are often consistent with the idea that the tubes of force in flare productive area, cross the inversion line at small angles and show erratic angular changes or 'bifurcations'.

From complete observations of the 3 components of the magnetic field, it is possible to compute the vertical component of the electric current

$$j_z = \frac{1}{4\pi} (\nabla \times \mathbf{H})_z.$$

Severny found the location of a flare to correspond with the regions of maximum j_z. I shall not develop this point since you will hear a communication from Severny on these topics.

1.4. CONCLUSIVE REMARKS

To sum up, the empirical evidence suggests that the locus of flares are systems of tubes of force which, instead of connecting most directly the available magnetic poles, are badly elongated and stressed. Following the expression of Gold (1968), there are probably 'misconnections' in the magnetic field. Such a situation is obviously forced by the subphotospheric distribution of magnetic poles and, as a result, the field high above the photosphere may contain much more energy than a potential field.

2. Evolution of Flare Productive Magnetic Patterns

Changes of field structure in connection with flares have been a controversial subject and still are in some respects. In this section we shall talk about the less controversial part of the subject. Everybody agrees that flare productive Active Regions are progressively but fastly evolving; significant changes in spot morphology and magnetic structure may be observed in time intervals of the order of one hour. Various workers associated the productivity of flares with specific periods in the life of A.R. when significant evolution takes place (Bumba *et al.*, 1968).

Obviously the building up of flaring magnetic structures is the result of an evolution in which take part the following processes: – Arrival of new poles and magnetic ropes from subphotospheric layers, often resulting in the birth of new spots. – Random surface motion of magnetic poles probably resulting from large scale convection. – Differential rotation.

If we try to describe this complex evolution by some global parameters calculated over the whole A.R. such as spot area (Sivarenam, 1969; Sawyer, 1967), total flux of following and preceeding polarity F_f and F_p, their sum or differences (Martres *et al.*, 1967), and then to associate the changes of such parameters to the occurrence of flares, the results are often quite unclear as might be expected. However in connection with very large flares, spectacular changes of $F_f + F_p$ and $F_f - F_p$ have been recorded by Severny.

In general one should look in more detail to trace the changes possibly connected

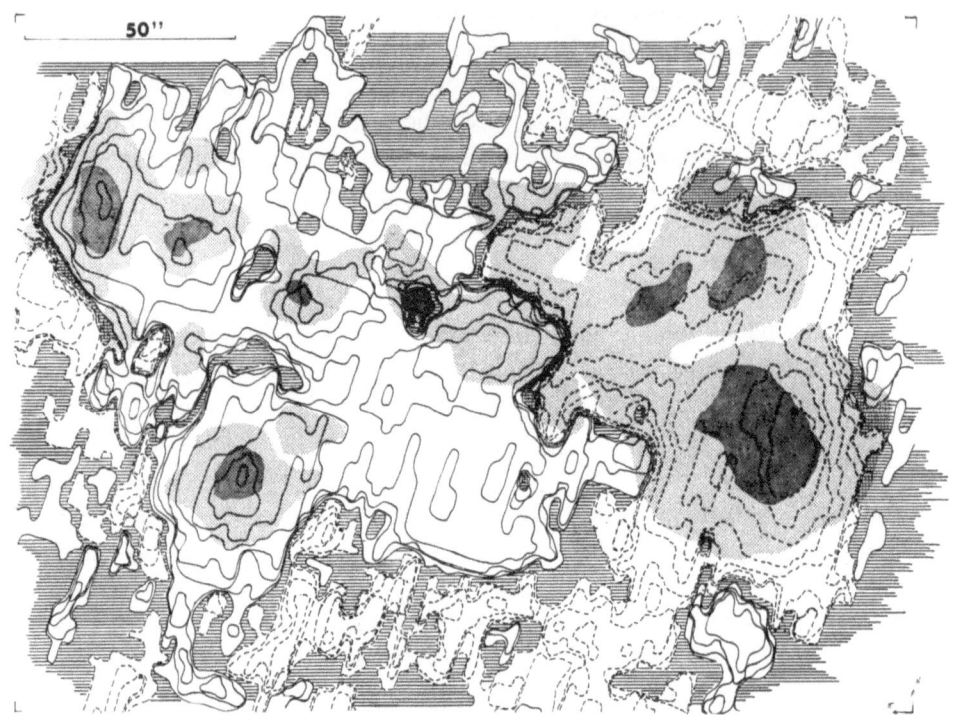

Fig. 1a.

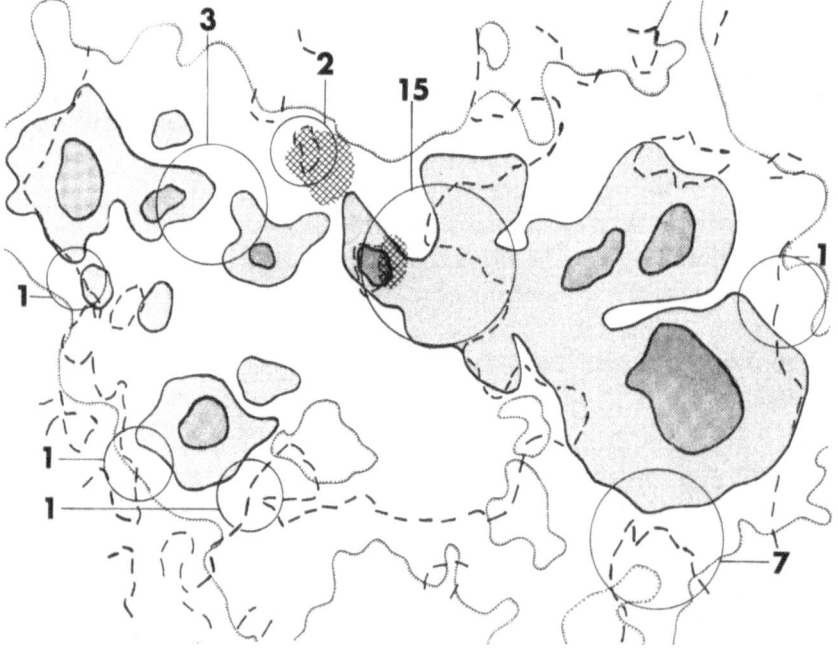

Fig. 1b.

with flares. This was done first by Severny who compared magnetic maps taken *a few hours apart* in the best cases. He observed after flares a decrease of the gradient $\Delta H_{\parallel}/\Delta x$ across the inversion line which has been the seat of the flare. Sometimes a magnetic hill involved in the flare has moved away from the inversion line, or has decreased or even disappeared (Severny, 1963).

Other observers have collected results which are more or less consistent with Severny's (Michard *et al.*, 1961; Rust, 1968). *However the word of caution of Michard et al.*, 1961 is still in order: the time distribution of observations is such that it is impossible to tell if the observed changes are *progressive, or sudden* and coincident with the flare. Their *internal accuracy also should be questioned*, particularly for the measurement of horizontal gradients which are very sensitive to image quality.

In the case of very big flares, particularly proton flares, the Crimean observers (Gopasyuk *et al.*, 1963; Severny, 1963; Howard and Severny, 1963) found very large changes in spot morphology and magnetic fluxes when comparing data taken a few hours before and after the event. For instance in the case of the cosmic ray flare of July 16, 1959 a decrease of all large umbral fields in the active center by a factor of 3 was found. However at Mount Wilson (Howard and Babcock, 1960) no flare associated change was noted in the smaller range of fields recorded (saturation at about 100 G). This different behavior is surprising.

In contradiction also with Severny, Malville and Tandberg-Hanssen (1969), found only minor changes when studying the evolution of the longitudinal field (measured in Hα) in connection with major flares of May 21–23, 1967, including a proton flare*.

In analysing data from the CSSAR, 1965, Martres *et al.* (1968a, b), found an interesting law of evolution of magnetic patterns involved in flares. These authors divide each complex pattern of longitudinal field into a number of 'Evolving Magnetic Features' or 'Structures Magnétiques Evolutives'. These features are local peaks of $H_{\parallel}$, spotted or unspotted, of characteristic dimensions 10^4–10^5 km, each of a single polarity. The evolution of the whole pattern of course results from the individual stories of each EMF.

In accordance with previous results, each flare involves at least two adjacent EMF of opposite polarities. But it is found that these two EMF have opposite senses of evolution in the period of flare occurrence, one being increasing, the other decreasing.

* This paper also contains the first observation of magnetic fields in the flare plasma emitting Hα.

Fig. 1.   The location of a series of minor flares in a longitudinal magnetic field pattern. (a) Longitudinal field of an active center 24°N and 17°E on July 27th, 1967, observed at Meudon, time 08.50 UT. North is towards the top; east towards the left. Isogauss are for 15, 30, 50, 100, 250, 500, 1000, 1500 G; horizontally hatched areas have fields smaller than 15 G. Major sunspots are also drawn and shaded from a K_1 spectroheliogram. The active center shows a number of 'inclusions' or 'parasitic polarities'. (b) On the same drawing of spots are shown the inversion lines and the regions of flare production during the same day from Quarterly Bulletin on Solar Activity: the observed number of flares and subflares at each location is noted. Hatched: starting points of a flare of importance 1n, time 08.15–09.16, from an Hα film obtained in Meudon. This flare is one of several homologous flares at exactly the same location. Note that a few flares occurred on a zero of $H_{\parallel}$ without observed inversion of polarity.

In the first work the evolution of EMF was characterized by the changes in area of the included spots. More recently, Ribes (1969) using a good series of maps of $H_\parallel$ of a single region taken at hourly intervals for three days (night excluded) could characterize each feature A by its contribution to the flux integral $\Delta F = \int_A H \, ds$ and was able to fully confirm the above law. In both work it was proven that if two adjacent EMF of opposite polarities have the same sense of variation, no flare can connect them.

It should be remarked that this law of variation of the magnetic features involved in flares, was established by the observers at Meudon on a sample of active regions producing *only minor flares*. This law is not in contradiction with the type of changes described by Severny and co-workers for active centers producing major flares, such as the decrease or disappearance of parts of the magnetic poles. However there is perhaps a difference in interpretation because we consider these changes *as progressive*. At least we do not believe that the accuracy of the available observations allows one to pin point sudden changes in direct coincidence with flares.

One may try to summarize as follows a typical evolution of flare productive magnetic patterns: these patterns contain several magnetic poles, of which two at least, being of opposite polarities, are evolving in opposite senses: the flux of pole I through the photosphere increases, while the flux of pole D decreases probably as a result of large scale convection. As a consequence part of the tube of forces connecting I and D are pulled and distorted: in the photosphere the lines of force are dragged with the high density plasma; in the chromosphere-corona however the re-arrangement of the tubes of force proceeds by sudden 'disconnections', i.e. flares.

However stresses of the magnetic ropes leading to preflare condition may also very well occur by horizontal motion of the poles (Gopasyuk *et al.*, 1963) or by rotation (Gopasyuk, 1965; Stenflo, 1969).

3. The Case for Sudden Changes of Magnetic Fields in the Photosphere in Correlation with Flares

I have presented the empirical data as being consistent with *progressive changes* of the magnetic patterns in the photosphere, inducing 'catastrophic' changes in the chromosphere-corona. We must recall here that we are dealing with a complex situation where:

– in the upper solar atmosphere $\qquad B^2/8\pi \gg \varrho v^2$

– in the deep photosphere $\qquad\qquad B^2/8\pi \ll \varrho v^2$

the magnetic observations being available only for the 'boundary' between these two regions.

If we accept that a substantial part of the flare energy is derived from the magnetic energy in the upper layer, large changes of the coronal magnetic structure are necessary which should perhaps be reflected by significant changes at the photosphere, in close time association. For instance according to many authors, the energy of a large flare is $\gtrsim 10^{32}$ ergs; on the other hand the total magnetic energy of a large active center (taken on a thickness of 1000 km) is of this same order of magnitude or lower. This, by the way, suggests that we should be less generous in our estimates of flare energy output, or give up the idea to pump it out of the magnetic field.

To come back to our subject we ask what is the evidence for abrupt changes of the photospheric field in close time association with flares? In my opinion the evidence is not conclusive from the magnetic observations alone, because they were not obtained with a sufficient resolution in time (only one map per hour in the best cases) and the effect of changes in seeing could not be checked.

Observations of the changes in photospheric structures which are closely correlated with magnetic fields, such as sunspots, did not disclose significant changes. On the other hand, chromospheric and coronal structures (including fine structure) are well known to present spectacular variations at the time of flares (Bruzek, 1968a, b; Bruzek and DeMastus, 1970; Wild, 1969).

Thus for the time being it is best to conclude that *sudden* changes of photospheric fields have not yet been definitely observed, although they probably exist since abrupt variations of the coronal fields should have some counterpart in the photosphere.

In conclusion this graph indicates the probable causal relations connecting the different classes of phenomena under discussion.

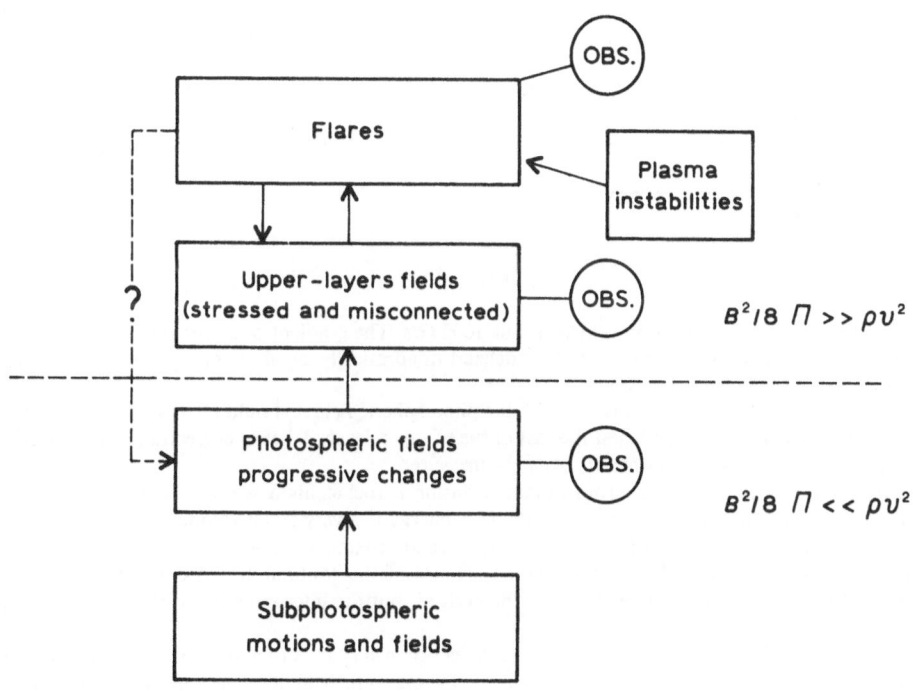

References

Bruzek, A. and DeMastus, H. W.: 1970, *Solar Phys.* **12**, 447.
Bumba, V., Krivsky, L., Martres, M.-J., and Soru-Iscovici, I.: 1968, in K. O. Kiepenheuer (ed.), 'Structure and Development of Solar Active Regions', *IAU Symp.* **35**, 311.

Gold, T.: 1968, in Y. Öhman (ed.), 'Mass Motion in Solar Flares and Related Phenomena', *Nobel Symp.* **9**, 205.

Gopasyuk, S. I., Ogir, M. B., Severny, A. B., and Shaposhnikova, E. F.: 1963, *Izv. Krymsk. Astrofiz. Obs.* **29**, 15.

Gopasyuk, S. I.: 1965, *Izv. Krymsk. Astrofiz. Obs.* **33**, 100.

Howard, R. and Babcock, H. W.: 1960, *Astrophys. J.* **132**, 218.

Howard, R. and Severny, A. B.: 1963, *Astrophys. J.* **137**, 1242.

Martres, M.-J., Michard, R., and Soru-Iscovici, I.: 1966a, *Ann. Astrophys.* **29**, 245.

Martres, M.-J., Michard, R., and Soru-Iscovici, I.: 1966b, *Ann. Astrophys.* **29**, 249.

Martres, M.-J., Michard, R., Soru-Iscovici, I. and Tsap, T.: 1968a, in K. O. Kiepenheuer (ed.), 'Structure and Development of Solar Active Regions', *IAU Symp.* **35**, 318.

Martres, M.-J., Michard, R., Soru-Iscovici, I. and Tsap, T.: 1968b, *Solar Phys.* **5**, 187.

McKim, Malville, J. and Tandberg-Hanssen, E.: 1969, *Solar Phys.* **6**, 278.

Michard, R., Mouradian, Z., and Semel, M.: 1961, *Ann. Astrophys.* **24**, 54.

Moreton, G. and Severny, A. B.: 1966, *Astron. J.* **71**, 172.

Moreton, G. and Severny, A. B.: 1968, *Solar Phys.* **3**, 282.

Ribes, E.: 1969, *Astron. Astrophys.* **2**, 316.

Rust, D. M.: 1968, in K. O. Kiepenheuer (ed.), 'Structure and Development od Solar Active Regions', *IAU Symp.* **35**, 77.

Severny, A. B.: 1958, *Izv. Krymsk. Astrofiz. Obs.* **20**, 22.

Severny, A. B.: 1960, *Izv. Krymsk. Astrofiz. Obs.* **22**, 12.

Severny, A. B.: 1963, in R. Lüst (ed.), 'Stellar and Solar Magnetic Fields', *IAU Symp.* **22**, 238.

Severny, A. B.: 1964, *Izv. Krymsk. Astrofiz. Obs.* **31**, 159.

Severny, A. B.: 1969, in Z. Švestka (ed.), 'The Proton Flare Project', *Ann. IQSY* **3**, MIT Press, Cambridge, Mass., p. 11.

Sivaraman, K. R.: 1969, *Solar Phys.* **6**, 152.

Smith, S. F. and Howard, R.: 1968, in K. O. Kiepenheuer (ed.), 'Structure and Development of Solar Active Regions', *IAU Symp.* **35**, 33.

Smith, S. F. and Ramsey, H.: 1967, *Solar Phys.* **2**, 158.

Stenflo, J. O.: 1969, *Solar Phys.* **8**, 115.

Vaiana, G. S., Reidy, W. P., Zehnpfenning, I., Van Speybroeck, L., and Giacconi, R.: 1968, *Science* **161**, 564.

Wild, J. P.: 1969, *Solar Phys.* **9**, 260.

Discussion

Title: The gradient of a vertical field is about 1000 G/s. The gradient of a horizontal field is of the same order. Since measurements of energy depend quadratically on B, errors of seeing can cause energy errors of orders of magnitudes.

Sweet: In correlating the $H\alpha$ emission with the photospheric field it should be remembered that the $H\alpha$ emission may not be occurring at the seat of the energy release. A better correlation may be forthcoming with the coronal fields when these can be measured.

Brueckner: From a XUV spectrum obtained during a rocket flight we, at the Naval Research Laboratory, think that the main source of the flare energy is a very small volume, in this case given by the resolution of the spectroheliograph, which was approximately 5 arc s. From the Stark effect splitting in the HeII line 304 and from the occurrence of the continuum, extremely high electron densities in this central part of the flare can be derived, but the interpretation of the spectra is still uncertain.

Dodson-Prince: Although it is true that most flares occur in centers of activity with well developed spots, strong magnetic fields, and steep gradients in the magnetic field, it seems appropriate to keep in mind the fact that approximately 7% of flares of importance ≥ 2 have taken place in regions with either no spots or only very small spots with area of the order of 100 millionths of the hemisphere. The magnetic fields associated with these flares appear to be primarily the bipolar fields of old plages. These rare flares in regions without large spots indicate that at least occasionally very simple magnetic circumstances may accompany the development of major $H\alpha$ flares.

ON CORONAL INSTABILITY AND MOVING RADIO FEATURES
ASSOCIATED WITH A FLARE SPRAY

G. DAIGNE

Meudon-Nançay Observatory, France

Abstract. A solar radio outburst with its different parts well seen on metric wavelengths is described. A stable source shows successively two kinds of emission before the flash-phase of the event, and a moving type IV burst ($V \simeq 1000$ km/s), initiated $\frac{1}{4}$ solar radius high, is directly associated with a spray-type prominence. Parameters of this radiating source are discussed.

A type II burst seen on lower frequencies with a spectrograph ($V \simeq 3200$ km/s) seems to start after the flash phase. Two possible cases are considered for the generation of the shock-wave responsible for type II burst.

1. Introduction

A solar radio outburst occurring on 1968 December 11th, on the East limb of the Sun, is described with its different parts: one part preceding the flash-phase, a moving type IV burst, and a type II burst. The first part of this event may be similar to the coronal instability before the onset of a flare reported by Wild (1968).

Observations of radio outbursts induced by flares or prominence eruptions (Wild *et al.*, 1968) have allowed to associate their moving part with arch expanding structures (Wild, 1969; Kai, 1970). The association of the moving type IV burst, reported here, with a spray-type prominence is of special interest because density and maybe magnetic fields are then quite different than in conditions preceding flares.

The only optical observation of this event was by an amateur solar observer. Bruzek (private communication) reported this observation: at 11^h46^m UT "a giant flame with two branches" had a height of approximately 0.25 $R_\odot$, it was disrupted at 11^h53^m UT having reached a height of 0.7 $R_\odot$, and at 12^h06^m UT its upper parts. The phenomenon was interpreted as a huge eruptive spray-type prominence; it was produced or accompanied by a flare which may have started at about 11^h40^m UT. According to Warwick (1957) flare-sprays are characterized by extraordinarily high speed, by clumpiness, and by virtually straight line trajectories outward from a small center of origin in the flare (Orrall and Smith, 1961).

2. Radio Observations

Total flux measurements were made by the Nançay station at three frequencies: 9400 MHz, 408 MHz, 169 MHz. Other data were communicated by different observatories: 900 MHz (Bordeaux), 408 MHz (San Miguel), 260 MHz (Ondřejov), 111 MHz (Potsdam) and spectral observations in the frequency range 950–32 MHz (Weissenau). Position measurements were made by the Nançay's East-West interferometer at 408 MHz and East-West radioheliograph at 169 MHz (Vinokur, 1968).

The radio event begins at 11^h41 UT and lasts until 12^h00 UT. The total flux

Howard (ed.), Solar Magnetic Fields, 367–375. All Rights Reserved.

observed at different frequencies are given in Figure 1 (9400, 900, 408, 169 and 111 MHz). The most important time for such a complex event is the occurrence of the flash phase (Wild *et al.*, 1963). Even if we have no idea on the optical flare we can take $11^h47^m30^s$ as beginning of the flash-phase. It corresponds to the start of the fast increase in flux at 9400 MHz as well as on decimetric wavelengths, and the occurrence

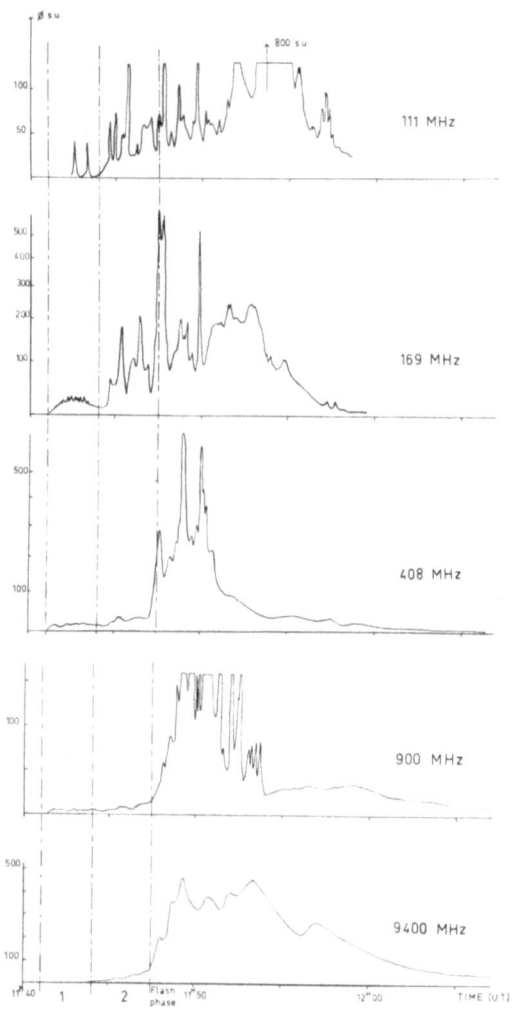

Fig. 1.   Observed flux at different frequencies for the radio event (1968, December 11).

of strong type III bursts on metric wavelengths (between 11^h48^m and 11^h50^m UT). Before this flash-phase, we can distinguish two kinds of emission occurring in the corona: first a small noise storm, and then the beginning of what may be the Flare Continuum as recently defined by Wild (1970); these two parts are respectively labelled 1 and 2 on the different figures.

A. BEFORE THE FLASH-PHASE

As can be seen from the total flux on Figure 1, the difference between the two parts 1 and 2 is mostly important at low frequencies. The part 2 may be characterized by (a) an increase of the continuum emission (169 and 111 MHz), (b) strong flux variations (169 MHz) slower than for typical type III bursts at this frequency. In the same time a classical gradual rise occurs on high frequency (9400 MHz); this precursor may have started earlier, at about 11^h41^m UT, in the same time as for the other frequencies. In such a case the delay would be due to the poor sensitivity of our total flux measurement at 9400 MHz (about 10 s.u.).

Positions measured at 408 MHz and 169 MHz are plotted on Figure 2; they remain stable during the two parts defined earlier. Two centers can be seen at 408 MHz, their distance being 2′5; only one center is present at 169 MHz, but our resolving

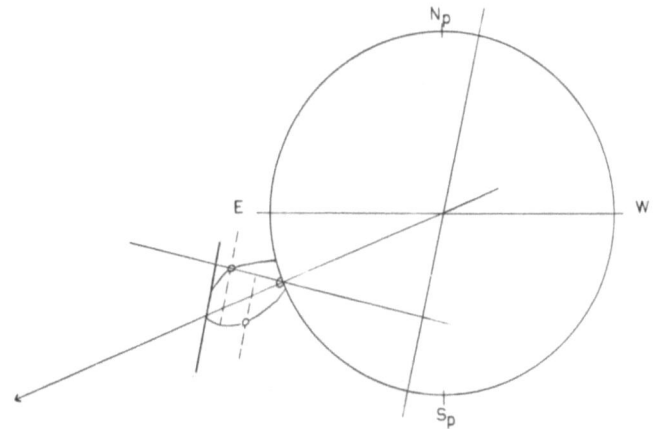

Fig. 2.   East-West positions of the radio emission at two frequencies (– – – – –:408 MHz; _____:169 MHz). A loop has been drawn over the active center (20°S).

power is only 3′4 at this frequency instead of 1′7 at 408 MHz. A hypothetic loop has been drawn over the optical center which may be an explanation for the double center at 408 MHz; this emission would come approximately from the same height in the two feet of the loop.

The altitude of the radio sources will be now measured assuming they occur radially on the active center; this is certainly not true at 408 MHz, but it enables us to compare different bursts. With this radial hypothesis, the deduced height is 0.5 $R_\odot$ at 169 MHz; it seems to show a very high density in the corona before the flash-phase if we assume that plasma radiation is the main mechanism.

On Figure 3 (a) the flux densities at different frequencies are summarized; the frequency scale is deduced from a density model (twice Newkirk's density model above active regions (Wild *et al.*, 1963), that is four times Newkirk's density model in the quiet corona in period of maximum activity $Ne = 4.2 \times 10^4 \times 10^{4.32/R}$). We assume that each frequency is emitted at the plasma level, except for the high frequencies (408 MHz to 9400 MHz) for which the model is no longer valid. This figure is only a rough picture of the event to give a more general view of the event, without physical significance for the moving type IV burst.

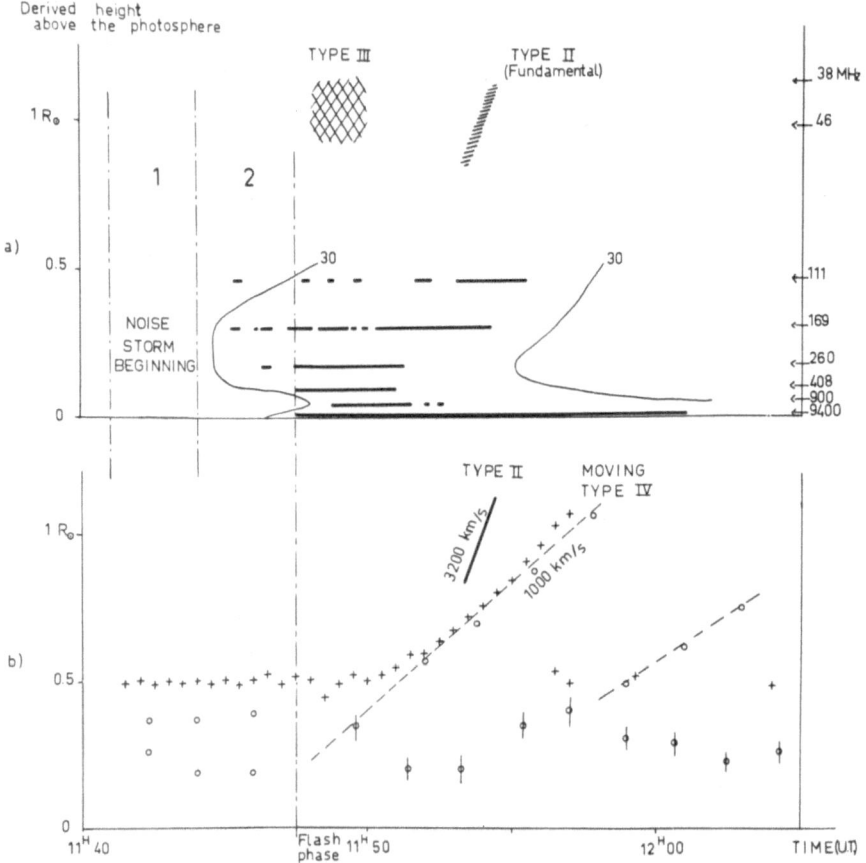

Fig. 3.   (a) Summarized picture of the radio event. Horizontal bars are for radio flux greater than 100 s.u. at the corresponding frequencies (right side on the picture). The thin line indicates the region where radio flux is greater than 30 s.u. (b) Derived height above the photosphere with a radial hypothesis from measurements at two frequencies (○ 408 MHz; + 169 MHz); vertical bars at 408 MHz are for large sources (2′ or 3′ in diameter). Altitude of the type II burst is deduced from a density model (twice Newkirk's model above active regions).

B.   THE FLASH-PHASE

Its beginning, between $11^h47^m30^s$ and 11^h48^m, is seen at all frequencies:
– start of the impulsive part of the type IVμ.
– strong burst on decimetric wavelengths, lasting about half a minute (408 and 169 MHz).
– beginning of a type III bursts cluster on low frequencies (about 40 MHz).

– start of a continuum enhancement between 300 and 600 MHz (from spectral observations).

C. AFTER THE FLASH-PHASE

A moving type IV burst is visible both at 408 MHz and 169 MHz, with the same
position at the two frequencies; these positions are plotted on Figure 3(b). According
to the radial hypothesis the moving type IV burst rises with a velocity of 1000 km/s.
At 408 MHz position measurements are separated by a lapse of time of 2 mn, due to
the fringes of the interferometer; one of this position cannot have been measured
because of the too high emission level at that time (about 11^h48^m UT). Nevertheless
we can say that the moving type IV burst is seen *first* at 408 MHz, and visible at
169 MHz only when the disturbance reaches the height of the preceding source at
this frequency.

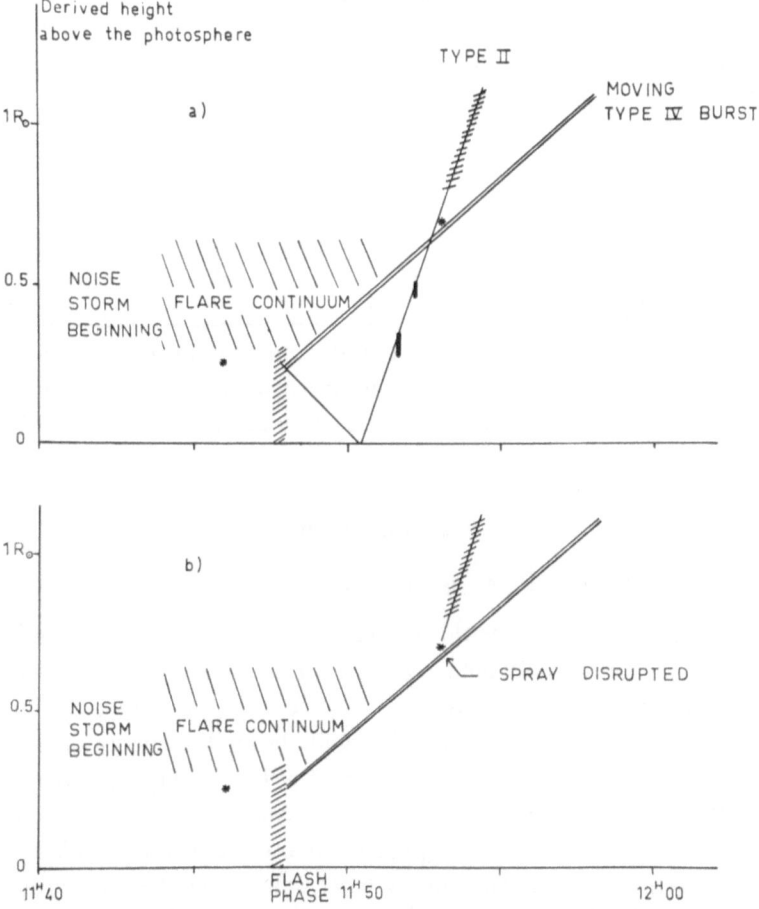

Fig. 4.   Two possible paths for the shock-wave responsible for type II burst, and positions of spray
(*). (a) Reflection on chromospheric levels (vertical bars are for peculiarities at 169 MHz and 111 MHz).
(b) Directly from the spray.

As positions at different frequencies are the same, the moving type IV burst has been interpreted as a synchrotron source (Boischot *et al.*, 1968); but in our case the source at 169 MHz nearly reaches its maximum in flux during the beginning of its movement, so that plasma radiation cannot be neglected in the emission process of some moving type IV burst, as pointed out by Wild *et al.* (1968).

The remaining source at 408 MHz is large and seems to fluctuate between the two previous positions, but this may be due to a second moving feature produced 8 mn after the first, with a velocity of about 800 km/s. It is fainter than the first moving type IV burst and seen only at 408 MHz. After that the radio sources at 169 MHz and 408 MHz are still there, with flux and position similar to what they had during the noise storm beginning (part 1); their flux decrease slowly and sources disappear 30 mn later.

D. TYPE II BURST

Spectral observations show a type II burst with harmonic; its fundamental band is the frequency range 60–38 MHz; it occurs between $11^h53^m20^s$ and $11^h54^m20^s$. The altitude of the fundamental emission, deduced from a density model (still twice Newkirk's model) is plotted on Figure 3(a) and (b). From this model, the deduced speed of the shock wave responsible for the type II burst is found to be 3200 ± 400 km/s.

Two peculiarities observed on the flux at 169 MHz and 111 MHz seem to be produced by the passage of the shock wave through the corresponding plasma levels. When it occurs at 169 MHz ($11^h51^m40^s$ UT), the diameter of the source is strongly enlarged from about 4 or 5' arc to 7' arc. These peculiarities are plotted on Figure 4(a) in the same density model. The departure time from chromospheric level, assuming a constant speed, is not greatly influenced by the density model. When changing the model, we change also the speed of the disturbance deduced from the frequency drift of the Type II burst.

3. Discussion

A. CHARACTERISTICS OF THE RADIATING ELECTRONS

The three features of the decimetric and metric radio emission are: noise storm beginning (Part 1), Flare Continuum (at 169 MHz *it lasts until the passage of the rising disturbance* – type IV burst – *through this radio source*), moving type IV burst. They may be associated with three orders of magnitude for the energy of the radiating electrons: some keV during the noise storm, tens of keV during the Flare Continuum (in fact little is known on the radiation mechanism of this part, but it cannot be synchroton radiation, the source being much too low in the corona). We can estimate the electron energy of the moving type IV burst, supposed to be a synchroton source, with the method developed by Ramaty *et al.* (1968). We obtain a mean energy of about 1 MeV in a magnetic field of 4.5 G at $0.6 R_\odot$; the density has been taken twice Newkirk's density model ($Ne = 8.3 \times 10^7$ cm^{-3} at $0.6 R_\odot$).

The Flare Continuum *starting before* the flash-phase may be due to a pure coronal acceleration mechanism by turbulance in a closed magnetic structure as proposed by Pneuman (1967) for an explanation of flares.

B. DEVELOPMENT OF THE FLARE PHENOMENA IN THE CORONA

Even if we know little on the associated optical phenomena, its nature and the position of the spray at two different times is important enough to try to associate optical and radio observations. Chromospheric material must have been expelled about 11^h40^m UT to produce the spray-time prominence. This optical feature has never been observed as the very first phenomenon of a flare (Bruzek, 1968).

The spray is directly associated with the moving type IV burst, optical and radio positions being the same at 11^h53^m UT (Figure 4), and velocities being very similar. The radio emission of the moving type IV burst starts only when the spray has reached an altitude of about $0.3\ R_\odot$, at the time of the flash-phase; may it be a critical height for the acceleration of particles in this event's conditions?

Density generally adopted for spray type prominences is about 10^{10} cm^{-3}; with a speed of 1000 kms^{-1}, its kinetic energy is balanced by magnetic energy for a field strength of 45 G, which is an order of magnitude greater than the value usually found at $0.6\ R_\odot$ above active regions (Newkirk, 1967). Whatever the magnetic structure over the ascending spray, it must be disrupted or carried out. This may explain the disappearance of the Flare Continuum at 169 MHz after the passage of the disturbance.

C. TYPE II – MOVING TYPE IV BURST

These two slow drift radio features have been attributed to a common flare-induced shock wave by Kai (1970), Stewart *et al.* (1970). In the observation reported here, they have quite different speeds, they cannot be attributed to a common shock wave. Moreover the moving type IV burst is not necessarily produced by a magnetic shock, its velocity being nearly the Alfvén velocity in the surrounding plasma (outside the spray). In the same medium the shock wave responsible for the type II burst would have a Mach number greater than 3, but according to Sagdeev (1966) a perpendicular collisionless shock breaks before its Mach number reaches 2. So that it seems necessary to invoke a different shock mechanism than the perpendicular case proposed by Zaitsev (1969) for the generation of type II burst.

The problem is now to study the travelling path of this shock wave through the inner corona; two possible cases are represented on Figure 4(a) and 4(b).

(a) The shock wave would be emitted at the time of the flash-phase, in the explosive center which produces the moving type IV burst ($0.25\ R_\odot$ high). Then it would be reflected on high density levels at 11^h50^m to escape outwards with a constant velocity. When passing through the corresponding plasma level at 169 MHz and 111 MHz it would have produced the two peculiarities previously mentioned ($11^h51^m40^s$ UT at 169 MHz).

(b) As pointed out by the amateur solar observer, the spray-type prominence disrupted at 11^h53^m UT. A straight line from the type II burst intersects the moving type IV burst trajectory at that time, so that the shock wave responsible for type II burst may be produced by the spray when it encounters a convenient magnetic structure. This interpretation is plotted on Figure 4(b). In fact the spray disruption may also be produced by the shock wave.

We need further observations to decide between the two possible generations of the shock wave. A very similar radio outburst occurring on the disk has been observed on 1970 March 25; type II and moving type IV burst have still different velocities (ratio about 3), they will be studied in more detail.

Riddle (1970) has recently reported a radio event in which the moving type IV burst is not directly associated with a flare spray, but time delayed. A type II burst occurred in the same time as the optical feature, but its deduced speed is greater than the ascending spray velocity, as in our case. If it is confirmed that type II and moving type IV bursts are associated with expelled chromospheric materials, such as flare sprays, prominence eruptions, or more generally plasma clouds, their radiating parameters could be quite different than in stable coronal conditions.

Acknowledgements

I am grateful to Dr Bruzek who has made available the optical data, to Dr H. Urbarz who sent me his radio spectral observations, and to Drs A. Krüger, M. Greco and Poumeyrol for total flux measurements at 111, 408 and 930 MHz respectively. I thank Dr. M. Pick for helpful discussions.

References

Boischot, A. and Clavelier, B.: 1968, *Ann. Astrophys.* **31**, 445.
Bruzek, A.: 1968, *COSPAR Symp. Tokyo*, 61.
Kai, K.: 1970 *Solar Phys.* **11**, 310.
Newkirk, G. A.; 1967, *Ann. Rev. Astron. Astrophys.* **5**, 213.
Orrall, F. Q. and Smith, H. J.: 1961, *Sky Telesc.* **22**, 330.
Pneuman, G. W.: 1967, *Solar Phys.* **2**, 462.
Ramaty, R. and Lingenfelter, R. E.: 1968, *Solar Phys.* **5**, 531.
Riddle, A. C.: 1970, *Solar Phys.* **13**, 448.
Sagdeev, R. Z.: 1966, *Rev. Plasma Phys.* **4**, 23.
Stewart, R. T. and Sheridan, K. V.: 1970, *Solar Phys.* **12**, 229.
Vinokur, M.: 1968, *Ann. Astrophys.* **31**, 457.
Warwick, J. W.: 1957, *Astrophys. J.* **125**, 811.
Wild, J. P.: 1968, *Proc. ASA* **1** (4), 137.
Wild, J. P.: 1969, *Solar Phys.* **9**, 260.
Wild, J. P.: 1970, *Proc. ASA* (to be published)
Wild, J. P., Smerd, S. F. and Weiss, A. A.: 1963, *Ann. Rev. Astron. Astrophys.* **1**, 291.
Wild, J. P., Sheridan, K. V. and Kai, K.: 1968, *Nature* **218**, 536.
Zaitsev, V. V.: 1969, *Soviet Astron.* **12**, 610.

Discussion

Sweet: What electron energies are required during the pre-flash phase?

Daigne: Some keV for the noise storm, and may be tens of keV for continuum enhancement 3 mn before the flash phase.

Erratum

The optical observation reported in the Introduction of this paper has to be corrected in the following way (Bruzek, second private communication):

"At 11^h46^m UT it (the spray-type prominence) had the shape of a slightly distorded Y or of a fork (with a height of 0.5 $R_\odot$) and was very bright; at 11^h53^m UT the branches of the Y were disrupted into at least 3 knots in such a way that the center of gravity does not appear to be lifted much; the brightness was reduced considerably. The stem of the Y kept its place and existed at a reduced height and brightness probably at least until 12^h55^m UT. Unfortunately, there was no observation before 11^h46^m and none between 11^h46^m and 11^h53^m UT! Therefore we know virtually nothing about the rate of growth. The only thing we may be sure of is that the upper part of the prominence was disrupted between 11^h46^m and 11^h53^m UT."

I apologize for having built the discussion only on two optical positions; part B of the discussion 'Development of the Flare Phenomena in the Corona' is quite erroneous, also the position of the spray-prominence reported on Figure 4 at two different times.

The main problem would be to know at what time, between 11^h46^m and 11^h53^m, occurs the disruption of the Y shape? One possibility is that it occurs at the beginning of the flash phase, seen on the radio emission, between $11^h47^m30^s$ and 11^h48^m UT. In such a case the moving type IV burst, emitted at that time, and from a height of about 0.25 $R_\odot$, would be associated with the spray-type prominence disruption. It is to be noted that the height of 0.25 $R_\odot$ corresponds approximately to the upper part of the stem of the Y shape, and this stem remains stable.

There is no evidence to associate the type II burst with the optical phenomena, so that I retain only the first interpretation: the shock wave responsible for type II burst would be seen after reflection on high density levels.

I am grateful to S. F. Smerd, doubtful on the direct association of radio and optical observations; his criticisms enable us to determine precisely the optical phenomena and I thank A. Bruzek very much for it.

MAGNETIC FIELDS ASSOCIATED WITH SOLAR FLARES

EARLE B. MAYFIELD

San Fernando Observatory, The Aerospace Corporation, El Segundo, Calif., 90045, U.S.A.

Abstract. An investigation of strong magnetic fields associated with flares was made with the 61 cm vacuum telescope and spectroheliograph at the San Fernando Observatory. Magnetograms of the longitudinal component of the field were made daily during the period of September 25–29, 1969. Observations were also made during this same period of the enhanced radio emission at 3.3 mm wavelength. On September 27 an importance class 3 or 4 flare occurred in the region studied. Total magnetic flux was determined for September 25, 26, 28 and 29 for the region which included the flare. In an area of about 190 arc s by 250 arc s the flux values for these dates were respectively 2.4, 2.5, 2.2 and 2.8×10^{22} Mx. Following the flare of September 27 the flux decreased significantly. Magnetic energy change in the region of the flare can be determined if an appropriate height is known. Following Howard and Severny (*Astrophys. J.* **137**, 1242, 1963) a height of 10^9 cm was used. This yields a value of 5×10^{31} ergs for the decrease of magnetic energy in the longitudinal component of the field.

1. Introduction

A number of recent theories of solar flares have proposed that magnetohydrodynamic processes, primarily in or near the region separating magnetic fields of opposite polarity, are responsible for flares. These include Severny (1958, 1961, 1963), Jaggi (1963), Petschek (1964), Syrovat-skii (1966), Sturrock and Coppi (1966), Alfvén and Carlqvist (1967) and Sturrock (1968). Although these theories differ in specific details of pre-flare field configuration, on-set and energy transfer, there is agreement that significant field changes should be observed in the region associated with a flare. Several models have been used to explain the observed plasma instability and the release of flare energy in times of 10^2 to 10^3 s. Although these models postulate initial conditions in the fields prior to a flare, e.g. large magnetic gradient at the neutral line, complex structure with interpolated regions of opposite polarity, etc., and predict certain observable characteristics, the experimental verification of any theory is difficult or equivocal.

Previous investigations of magnetic fields associated with flares have measured magnetic flux changes during times of major flares. These measurements of fields have included both longitudinal and transverse components of the field. The results have shown significant flux reductions following major flares and have been interpreted in terms of energy loss by the fields to flare energy. Reductions of magnetic energy of order 10^{31} to 10^{32} ergs have been estimated.

This investigation of magnetic fields was directed to determining field configurations and flux changes associated with active regions which were expected to flare. Such a region studied was associated with McMath plage region 10333 near Mount Wilson sunspot 17503 which had a central meridian passage on September 27. 3, 1969. An importance class 3 or 4 flare occurred in this region at NO7, EO2 on September 27, 1969 beginning at about 0345 UT. High resolution magnetograms of the longitudinal component of the fields were obtained daily during the period of September 25

Howard (ed.), Solar Magnetic Fields, 376–389. All Rights Reserved.
Copyright © 1971 by the IAU.

through 29. These were made with the 61 cm vacuum telescope and spectroheliograph at the San Fernando Observatory and have a seeing limited resolution of about 1 arc s. Simultaneous radio maps at 3.3 mm wavelength with a resolution of about 2.8 arc min were made. Regions of millimeter emission which have been previously reported by Mayfield *et al.* (1970) have been observed to be associated with the neutral line of primarily bi-polar magnetic fields and to have enhancements prior to major flares. For the flare of September 27 an enhancement of more than 10% in the millimeter emission over an undisturbed region of the disk was observed about 36 hr prior to the flare. This region was near the neutral line of the magnetic field associated with the flare.

2. Experimental

The instrument used for this investigation was the vacuum telescope and spectroheliograph of the Aerospace Corporation San Fernando Observatory. Since the instrument has not been described elsewhere, a brief description is given.

The solar telescope is a Coudé system consisting of a primary 24 in. (61 cm) diam and a secondary 12 in. (30 cm) diam cassegrain which feeds a gregorian mirror. This three mirror configuration produces both a flat field and a stigmatic image with two image sizes. The primary image diameter is 115 mm and the secondary is 50 mm. All three of the surfaces of the telescope are aspheric. The cassegrain figures are ellipsoidal for the first surface and hyperboloidal for the second surface. The gregorian is an ellipsoid. All of the mirrors are made of Cer-Vit to reduce thermal problems and coated with silver to enhance the reflectivity. A dielectric coating is applied over the silver to reduce oxidation. Four plane mirrors are used to direct the beam through the axles to the spectroheliograph. The entrance window is a plane surface and the exit window is a field lens which images the entrance pupil of the telescope on the grating of the spectroheliograph.

The spectroheliograph uses a Littrow mounting with a three element lens and a 6×10 in. (15×25 cm) grating of 1200 grooves/mm. Dispersion with the grating used in second order is about 1 mm/Å. Slit lengths are 57 mm for 70 mm film format and there are two exit slits to obtain both polarized spectroheliograms simultaneously for magnetic field data. A spectrum line servo eliminates thermal drifts or displacement of the spectrum during scanning. An automatic exposure control is incorporated in the scan drive. Scanning is accomplished by rotation of the spectroheliograph about an axis located in the base normal to the optic axis. To accommodate for rotation of the solar image and to permit scanning in any direction the system can be rotated around the optic axis. Both the telescope and spectroheliograph have a focal ratio of $f/20$.

For this investigation, the Ca I 6102. 7Å line was used. This line was selected because it is well suited for moderate field strength up to about 1500 G. It was also the one used by Leighton (1959) in his development of the method for magnetic field measurements and subsequent studies (Simon and Leighton, 1964). The exit slit of the spectroheliograph was positioned in the blue wing, 0.07 Å from the line center. The blue wing was selected to reduce the contrast in the spectroheliogram pairs. Both entrance and exit slits were set to $60 \mu m$ which, with the grating used, gave a scan time of typically 140 s.

The spectroheliograms were recorded on S0392 film and processed to an average density of about 1.0 and a gamma of 4.0. Careful process control was maintained to insure uniform development and contrast of the two original films. Photometric density scales were recorded on each film for subsequent calibration. The solar spectrum in the region of $\lambda 6102$ was also photographed on each film to be used for reduction of the magnetic field film. One of these films was then contact printed to produce a transparent positive print. This print was processed to the same density as the original and to a gamma of 1.0. To produce the magnetic field film, this positive reversal film is placed in registered contact with the other original negative and a photographic copy made of the two. This photographic subtraction of the positive and negative transparencies cancels any density variations between the two oppositely polarized spectroheliograms which are not caused by magnetic fields.

The reduction of the data to obtain field strength is based on the photometric calibration and the line profile obtained from the spectrum. The film is read with a computer controlled precision cathode ray tube densitometer which scans the film and records the density for a matrix of points separated by $100 \mu m$ and computes the field strength. At present, uncertainty in the method yields an error of about ± 85 G in the determination of field strengths.

3. Results

Magnetograms were obtained daily during the period of September 25/29, 1969. For the flare of September 27, magnetograms were obtained about 8 hr prior to and about 16 hr after onset. However, the data for September 27 were not suitable for reduction because of instrumental errors. Figure 1 is a magnetogram for September 25 taken at 2200 UT. Region 10333 is located near the center of the figure, north is at the top and west is on the right. Positive fields out of the surface are lighter and negative fields into the surface are darker than the average, zero fields. Typical field strengths range from a minimum of about 85 G to a maximum of about 1450 G. Although much stronger fields exist in the sunspot umbras, these regions are very underexposed in the original spectroheliograms and hence do not yield magnetic field data. Magnetograms for September 26, 28, and 29 are shown in Figures 2 through 4. Region 10333 can be seen in each of these. Its structure is primarily bi-polar but both polarities have a spiral structure which changes with each day. This region can be readily identified, however, after 4 days. Maximum field strengths measured in this region were as high as about 1450 G but typically a few hundred gauss.

Radio maps at 3.3 mm wavelength were also obtained during the same period. These were made with the 15 ft (4.57 m) diameter radio telescope at Aerospace in El Segundo. The telescope has a half-power beam width of 2.8 arc min and is computer controlled to map the Sun in a flexible manner. For the data used in this study, the antenna mapped a matrix of 21×21 points centered on the disk. The emission of

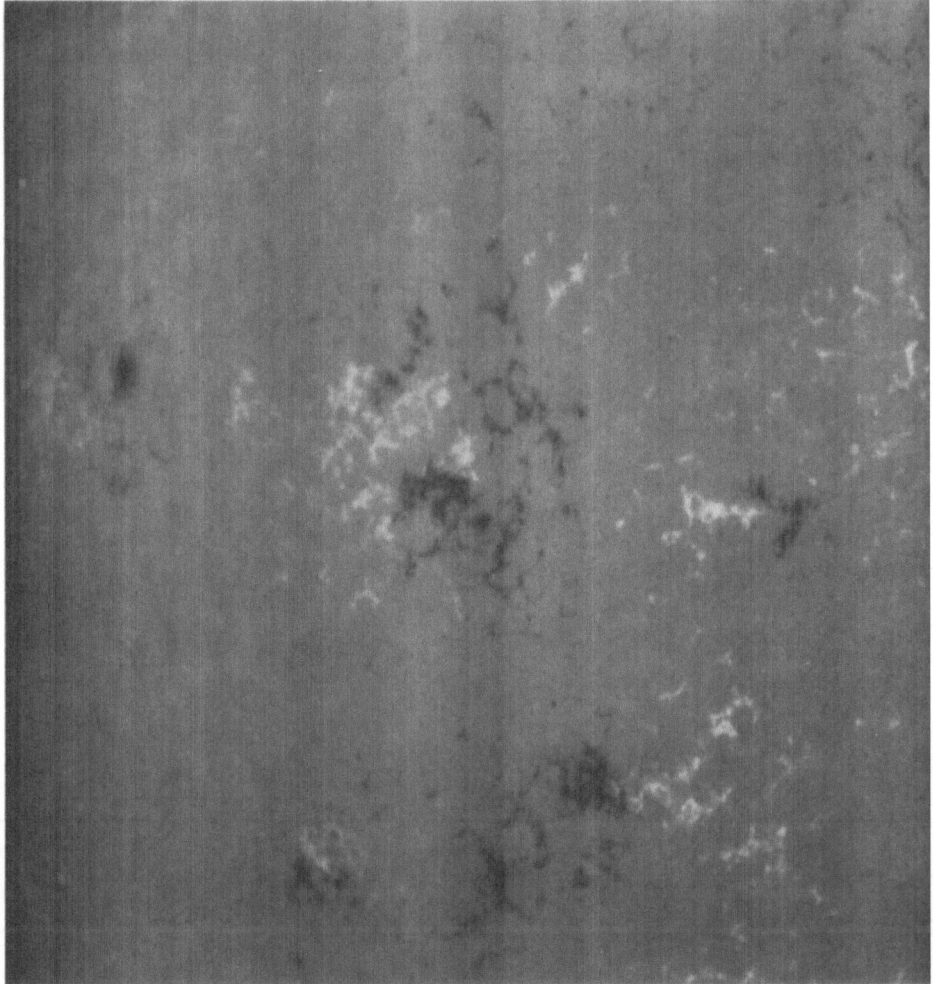

Fig. 1.   Magnetogram of the longitudinal component of the fields for September 25, 1969. North
is at the top, west is at the right.

each of these points was normalized to an undisturbed region near the center of the
disk and the percent of enhancement of these points plotted. Isotherms were drawn on
the map to identify enhanced regions of mm emission. Figures 5a and 5b show mm
maps taken on September 25 and 26 prior to the flare.

4. Data Analysis

The magnetograms were reduced with the cathode ray tube densitometer. For the
pictures of September 25/29, an area of about 190 arc s × 250 arc s was used. A total
of 1.56×10^5 density values were read. To increase the signal to noise and to reduce
the number of values for computation, each four adjacent points were averaged.

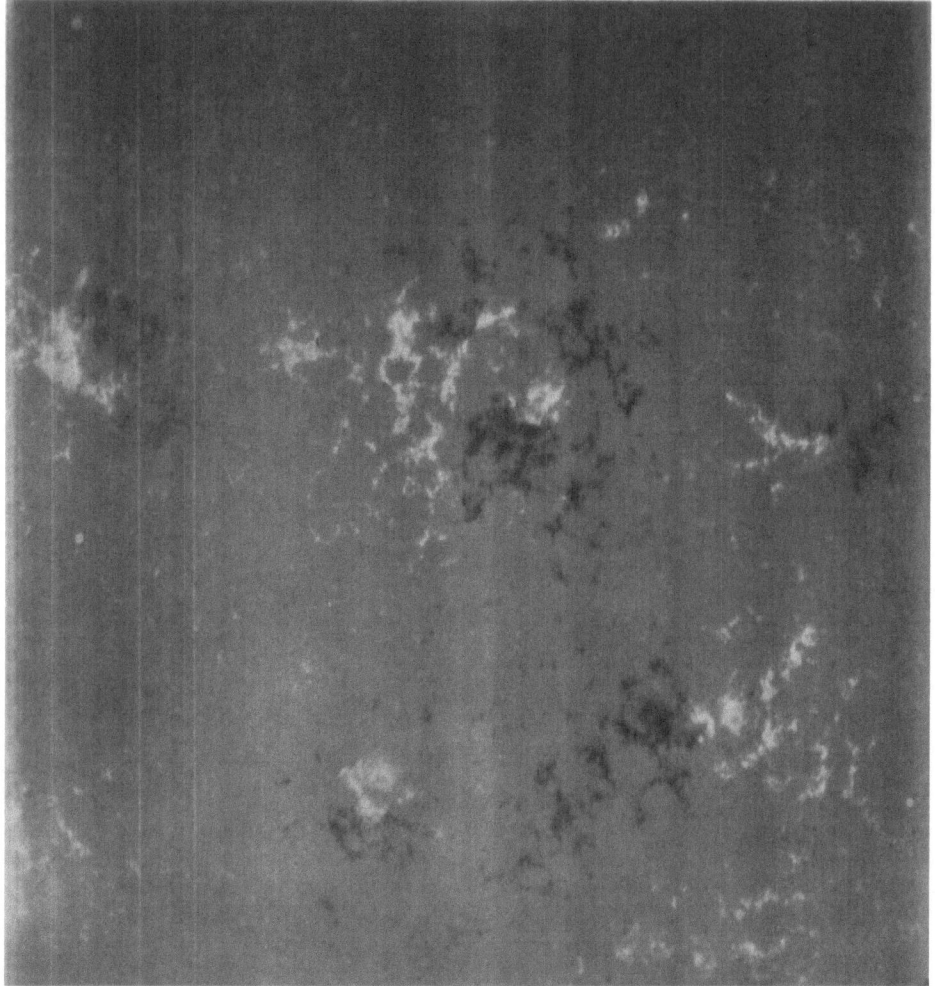

Fig. 2.   Magnetogram of the longitudinal component of the fields for September 26, 1969.

This gave 40000 final values. The mean of these was determined and used as the average density corresponding to zero magnetic field.

Plots of the density data were made in frequency distribution histograms. Since the distribution is skew because of saturation in the densitometer only the low density half was plotted. This is shown in Figures 6a, b, c, and for d September 25, 26, 28 and 29. These show the frequency of occurrence of magnetogram density from the mean, zero field, to the minimum density corresponding to the maximum fields measured. It is apparent that significant changes in the distribution histograms occurred associated with the flare of September 27. The marked increase in frequency at the origin indicates a reduction in stronger fields. These distributions also show an increase in occurrence for densities associated with strong fields. This is caused by saturation of the line and

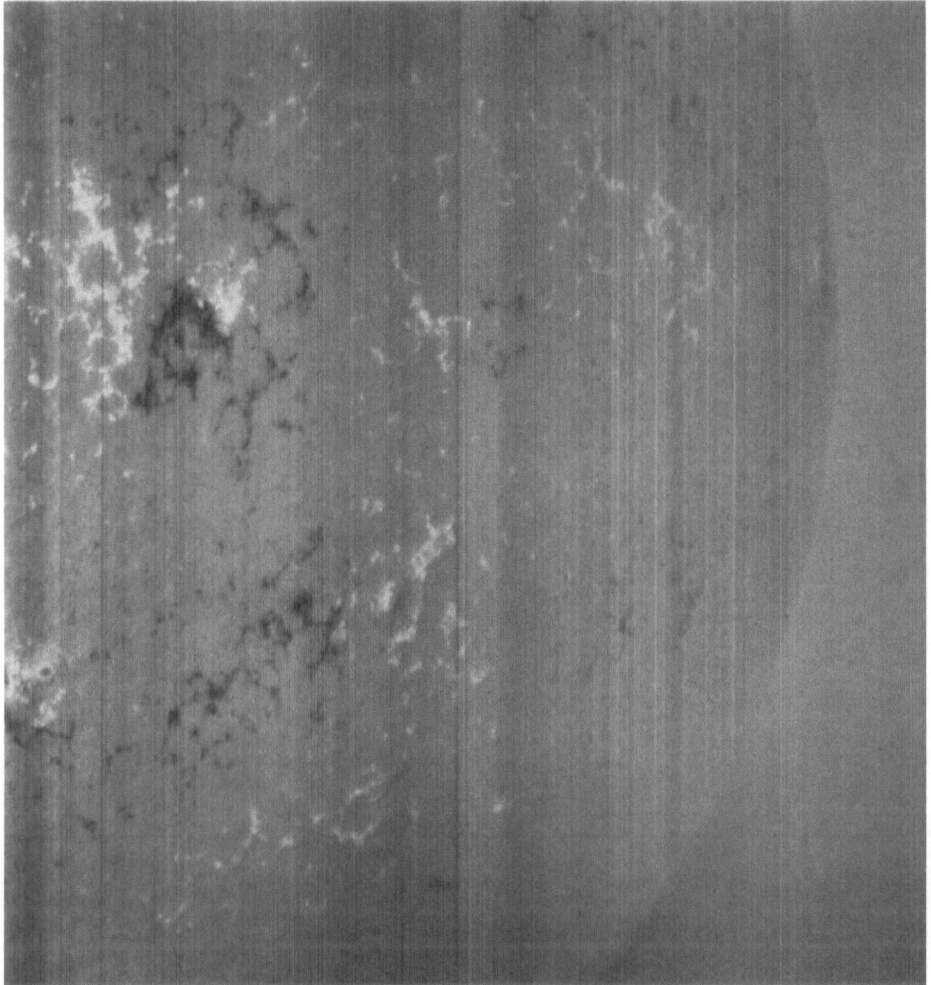

Fig. 3. Magnetogram of the longitudinal component of the fields for September 28, 1969.

by drifting of the line position and setting at the exit slit of the spectroheliograph. The calibration spectrum used to obtain $dD/d\lambda$ for field strength determination shows that fields of ± 1450 G could be measured without saturation. This agrees well with Leighton's (1959) estimate of ± 1500 G.

Total magnetic flux was computed from the histograms and the area data determined from the known size of the densitometer spot on the original film. The flux for September 25, 26, 28 and 29 are shown in Table I.

These values show a decrease on September 28 following the flare of September 27. The general increase in magnetic flux of this region during the period was in agreement with other data which show that the region was growing and increasing in complexity.

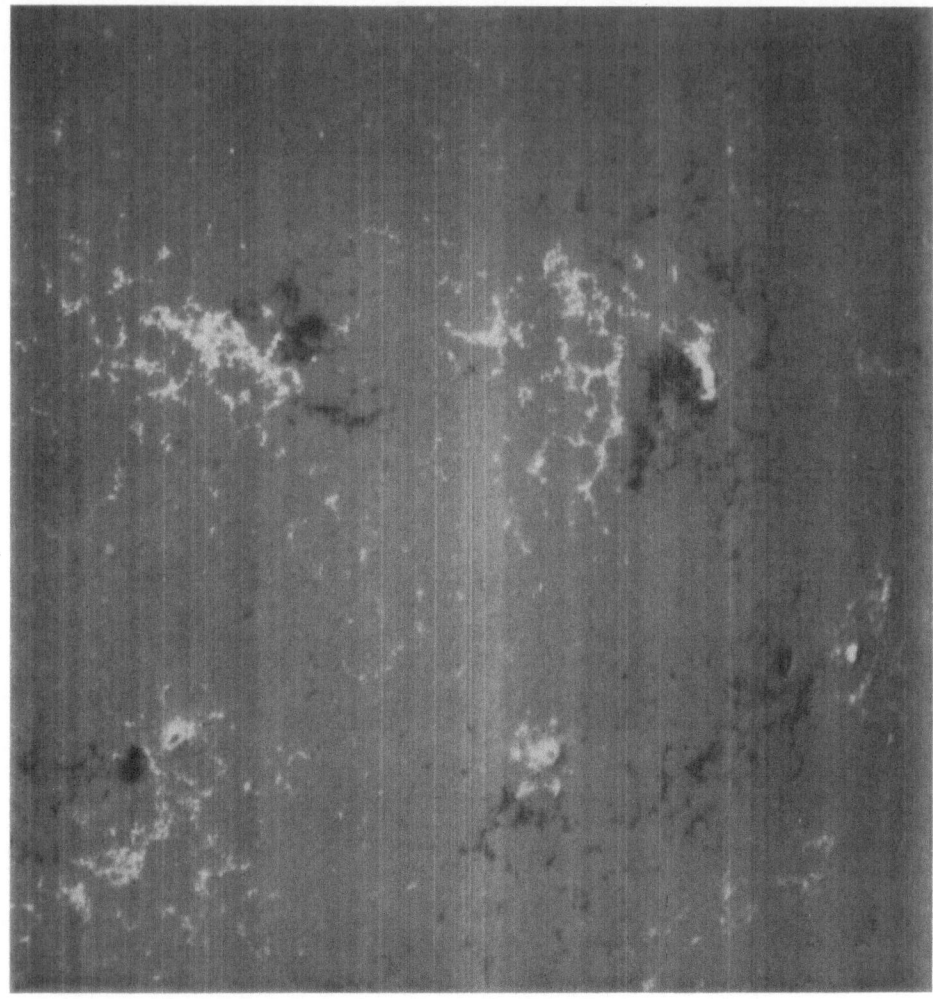

Fig. 4.   Magnetogram of the longitudinal component of the fields for September 29, 1969.

TABLE I

Magnetic flux observed in an area of
190×250 arc s for September 25/29, 1969

Date	Magnetic flux (Maxwells)
9/25	2.4×10^{22}
9/26	2.5×10^{22}
9/28	2.2×10^{22}
9/29	2.8×10^{22}

5. Discussion

The flare of September 27 was a major event of geophysical significance. The *ESSA Solar Geophysical Data* (1970) gave an importance classification number of 3N or 4B. It produced *X*-rays in the 0–20 Å interval and a large increase in charged particles

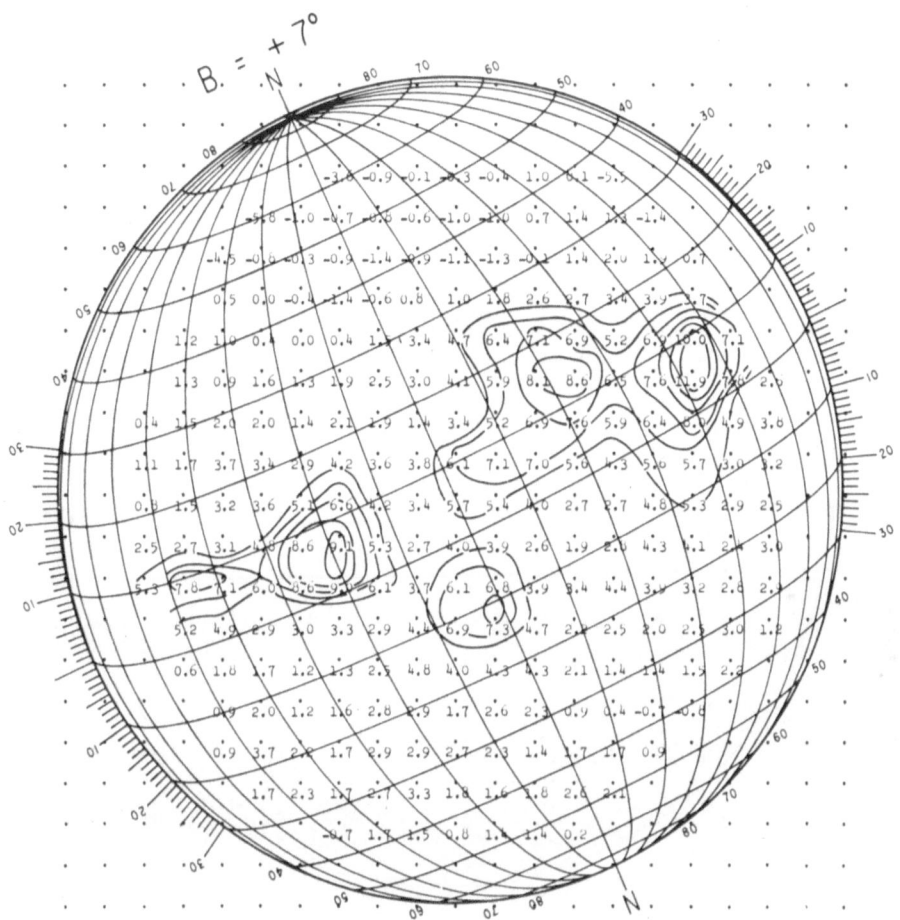

Fig. 5a. Map of 3.3 mm emission for September 25, 1969. Numbers are percent of enhancement over an undisturbed region, isotherms are for enhancements of 5% and greater for 1% increments.

for *E* > 10 MeV at satellite altitudes. It also caused a large geomagnetic storm on September 29, 30 although the effects of the flare of September 25 may have contributed. The peak radio flux was reported as 2000 flux units (1 unit = 10^{-22} w/m² Hz) and types II and IV bursts were recorded in m and dm wavelengths followed by a noise storm in meter wavelengths.

The decrease in magnetic flux from September 26–28 appears to be associated with the flare on the 27th. Previous determinations of magnetic energy changes associated with large flares particularly those with charged particle emission have been reported. Evans (1959) observed magnetic field changes during an importance 1+flare on April 30, 1959. Assuming a height of 5×10^8 cm, an energy change of 4×10^{31} ergs was

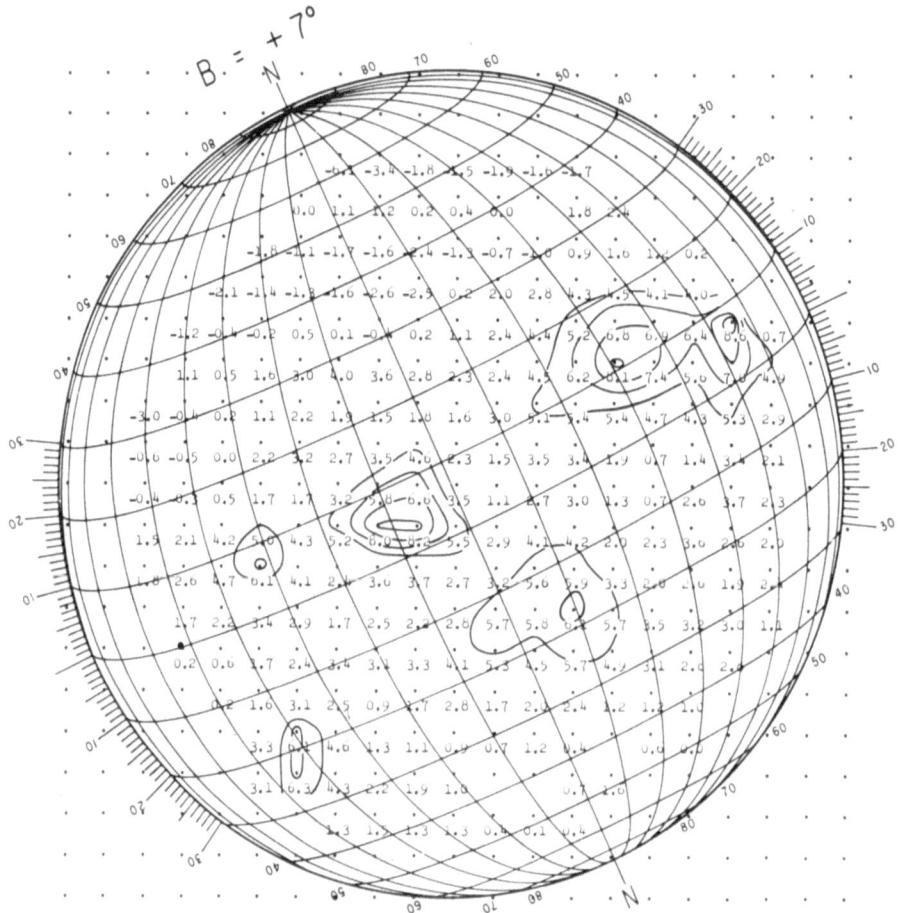

Fig. 5b. Map of 3.3 mm emission for September 26, 1969. Numbers are percent of enhancement over an undisturbed region, isotherms are for enhancements of 5% and greater for 1% increments.

determined. Gopasyak (1961) obtained a value of 10^{32} ergs for several importance class 2+ or 3+ flares. Howard and Severny (1963) have measured an energy of 4×10^{32} ergs for the flare of July 16, 1959. This flare produced geophysical effects and charged particle radiation. Malville and Tanberg-Hanssen (1969) have speculated that energy of approximately 10^{31} ergs could have been dissipated by the magnetic field for the flare of May 21, 1967. Severny (1969) has measured energy changes in magnetic fields of about 5×10^{32} ergs for the flare of July 7, 1966. These experimentally measured values of flare energy were obtained from observations of the longitudinal component of the magnetic field.

Determination of energy in the magnetic field requires an assumption of the effective height of the field. Following Howard and Severny (1963) a value of 10^9 cm was used. This yields 5×10^{31} ergs for the change in energy in the longitudinal component of the magnetic field for the flare of September 27. This change which occurred in

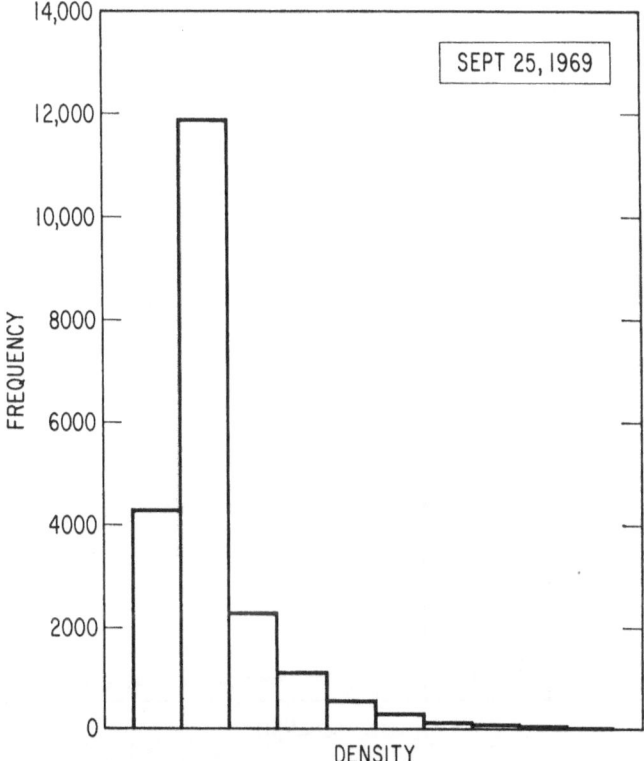

Fig. 6a.   Frequency distribution histogram of film density for the positive field component of the magnetograms for September 25, 1969.

about 48 hr was based on data taken 8 hr before and 40 hr after the flare. In the day preceding the flare the magnetic energy increased by about 10^{32} ergs. This is similar to the observations of Severny (1969) who noted that the energy typically increased before a flare. During the second day after the flare an increase in field energy was observed.

Estimates of energy change based on measurements of the longitudinal component of photospheric magnetic fields are subject to considerable uncertainty. Chromospheric currents significantly alter the assumed scalar potential as Mogilevsky and

Shelting (1966) have shown. Solar rotation also leads to errors both in foreshortening of the area and change in the longitudinal field strength with aspect. Corrections can be made for foreshortening. However, corrections for field changes with aspect are questionable. The flare of September 27 occurred near central meridian and the magnetic field observations were made within 25° of the center of the disk. This significantly reduced the problem of correcting for observations made near the limb.

As has been previously reported by Mayfield *et al.* (1970) there is enhanced radio emission at millimeter wavelengths about one day prior to importance class 2 or 3,

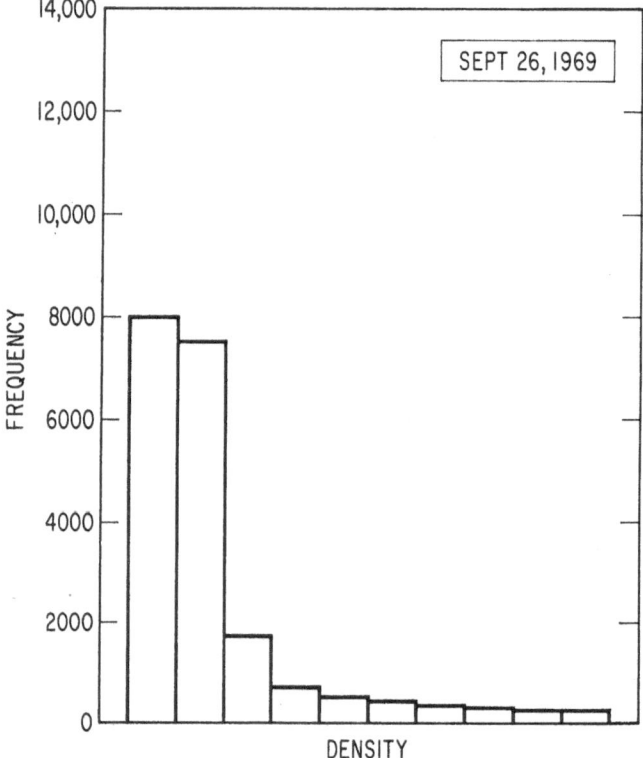

Fig. 6b. Frequency distribution histogram of film density for the positive field component of the magnetogram for September 26, 1969.

flares. This enhancement occurs near the neutral line of primarily bipolar magnetic fields and is generally confined to small regions of less than 3 arc min in extent. Typically the enhancement is greater than 8% and the temperature gradient is greater than 0.5%/deg. prior to these large flares. For the days of September 25 and 26 the enhancements were about 10% and the gradients greater than 0.5%/deg. The maximum coincided with the central region of the spiral field where the two polarities are contiguous. For September 26 the 8% isotherm conforms closely with the neutral line both in length and direction. However, lack of antenna resolution prevents identifying the shape with certainty.

These observations of magnetic energy changes associated with a large flare accompanied by charged particle radiation are consistent with other measurements and with theoretical predictions. The occurrence of enhanced millimeter radio emission consistently observed to precede large flares is not explicitly predicted. Although the models proposed by Severny and by Jaggi can explain this observed heating at the neutral line, their analyses were made to explain instabilities associated with the flash phase of the flare which yield time constants of 10^2 to 10^3 sec. However, Furth et al. (1961) have considered instabilities in sheet pinches in laboratory plasmas and

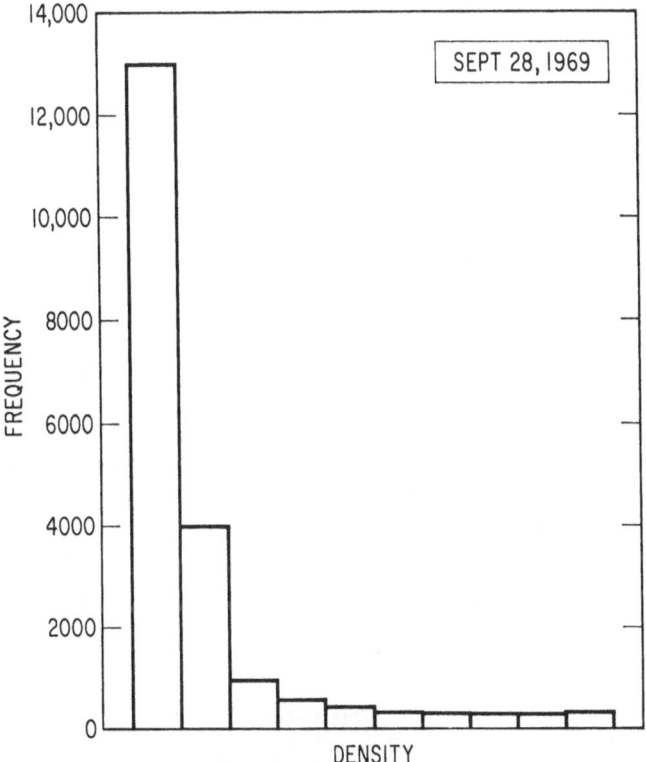

Fig. 6c. Frequency distribution histogram of film density for the positive field component of the magnetogram for September 28, 1969.

have shown the existence of several modes with characteristic time constants for local and non-local instabilities. The results of this investigation and previous results by Mayfield et al. (1970) suggest that similar processes may occur in flares.

Acknowledgements

The author thanks Dr F. I. Shimabukuro who kindly provided the 3.3 mm radio observations and Mr Dale Vrabec and Mr Neal Baker who collaborated in the

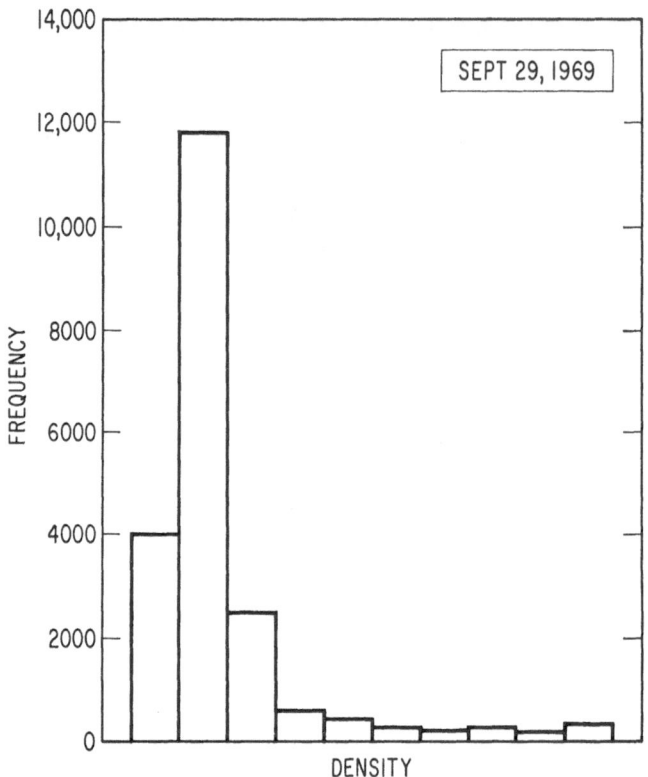

Fig. 6d.   Frequency distribution histogram of film density for the positive field component of the
magnetogram for September 29, 1969.

magnetic field measurements. Mr Kenneth Reed assisted with the densitometer
reduction and numerical computations.

References

Alfvén, H. and Carlqvist, P.: 1967, *Solar Phys.* **1**, 220.
ESSA Solar Geophysical Data: 1969, ESSA Research Laboratories, U.S. Dept. of Commerce, No.
    307, Part II.
Evans, J. W.: 1959, *Astron. J.* **64**, 330.
Furth, H. P., Killeen, J., and Rosenbluth, M. N.: 1963, *Phys. Fluids* **6**, 459.
Gopasyak, S. I.: 1961, *Soviet Astron.* **5**, 158.
Howard, R. and Severny, A. B.: 1963, *Astrophys. J.* **137**, 1242.
Jaggi, R. K.: 1963, *J. Geophys. Res.* **68**, 4429.
Leighton, R. B.: 1959, *Astrophys. J.* **130**, 366.
Malville, J. M. and Tandberg-Hanssen, E.: 1969, *Solar Phys.* **6**, 278.
Mayfield, E. B., Higman, J., and Samson, C.: 1970, *Solar Phys.* **13**, 372.
Mogilevsky, E. I. and Shelting, B. D.: 1966, *Atti del Convegno Sui Campi Magnetici Solari*, (ed. by
    M. Cimino), G. Barbera, Firenze.
Petschek, H. E.: 1964, AAS/NASA Symposium on the Physics of Solar Flares, Washington, D.C.
Severny, A. B.: 1958, *Soviet Astron.* **2**, 310.
Severny, A. B.: 1961, *Soviet Astron.* **5**, 299.

Severny, A. B.: 1963, *Soviet Astron.* **6**, 747.

Severny, A. B.: 1969, in *Annals of the IQSY*, MIT Press, Cambridge, Mass., **3**, 11.

Simon, G. W. and Leighton, R. B.: 1964, *Astrophys. J.* **140**, 1120.

Sturrock, P. A.: 1968, in K. O. Kiepenheuer (ed.), 'Structure and Development of Solar Active Regions', *IAU Symp.* **35**, 471.

Sturrock, P. A. and Coppi, B.: 1966, *Astrophys. J.* **143**, 3.

Syrovat-skii, S. J.: 1966, *Soviet Phys. JETP* **23**, 754.

Discussion

Deubner: How large was the area actually included in the analysis shown in the histograms?

Mayfield: The area analyzed was 190×250 arc s.

Wiehr: Your third and fourth observations were carried out far from the disk center, where the Leighton technique does not lead to the normal field component. A correction of this effect would require the knowledge of the transverse field component. If you did not consider this for your flux determinations I would suggest that the flux variations are not realistic.

Mayfield: Corrections for foreshortening were made. Changes in flux may be due to growth or decay of the magnetic region, changes between the longitudinal and transverse components and to loss to the flare. The observed change in flux is assumed to be associated with the flare but this cannot be proved.

Cowling: Decreases in magnetic flux should occur only through magnetic arches being drawn below the photosphere. Is there any sign of magnetic regions of opposite polarity flowing together?

Mayfield: Yes. We have observed such flows in time-lapse magnetograms where in complex fields small regions of opposite polarity merge. These will be shown by Vrabec later.

SUNSPOT MAGNETIC FIELDS AND HIGH ENERGY ELECTRONS
IN FLARES

TATSUO TAKAKURA

Tokyo Astronomical Observatory, University of Tokyo, Mitaka, Tokyo, Japan

Abstract. A balloon observation of an impulsive hard X-ray burst on September 27, 1969 showed the size of the source to be one arc minute or less. It was remarkably smaller than the associated Hα flare with a size of 3 arc min.

The efficient acceleration of electrons and the trigger of the flares are suggested to be attributed to a large scale electric potential field caused by a gas motion near the photosphere. The primary cause of the onset of flares would be the acceleration of electrons. The electrons excite plasma waves which make the conductivity lower by several orders, so that the electromagnetic energy I^2L stored before the onset of the flare would be suddenly converted into the heat due to the ohmic loss.

At the early phase of flares, electrons are accelerated up to about 1 MeV in the sunspot magnetic field in the short time of 10 to 100 s. The total number of the electrons accelerated above 50 keV is estimated to be 10^{35}–10^{36}. The behavior of the energetic electrons in a flare region has been investigated by using hard X-ray bursts and microwave radio bursts.

The present paper will briefly show the results of a recent balloon observation of the location and size of the source of an impulsive hard X-ray burst, and propose a model to generate an electric potential field, which accelerates electrons to emit hard X-ray and microwave impulsive bursts and also to be a trigger for the flare.

1. Balloon Observation of the Position and Size of a Hard X-Ray Burst

On September 27, 1969 a balloon observation of the position and size of a hard X-ray burst was made successfully in cooperation with the Institute of Space and Aeronautical Science (Takakura *et al.*, 1971). A one dimensional modulation collimator with a half-power width of 1.3 arc min was used (Figure 1). The observed energy range was effectively 30–60 keV.

A flare of importance 3 occurred accompanied by a comparatively small hard X-ray impulsive burst and a microwave burst.

The center of the X-ray source was on the line passing through the center of a big Hα flare region of 3 arc min diameter (Figure 2). The size of the X-ray source was 1 arc min or less which is remarkably smaller than the Hα flare (Figure 3).

On the other hand, the position and size of the source of the associated microwave burst were observed at Toyokawa with 1-dimensional interferometers (Tanaka and Énomé, 1971). The radio position is also on a line passing through the center of the flare region although the scanning direction was different by 27° from that of the X-ray collimator as shown in Figure 2. The measurement of the radio source size was not accurate enough (less than 1.8') in this event to judge whether the sources of the X-rays and the microwaves are coincident or not.

The observed size of 1 arc min or less of the hard X-ray source is comparable with or smaller than the source size of normal microwave impulsive bursts so far observed at Toyokawa. On the other hand, there is an excess of a factor 10^2 in the total number of electrons emitting the hard X-ray bursts compared with the electrons emitting the concurrent microwave bursts (Takakura, 1969; Takakura and Scalise, 1970). This implies that the acceleration of a majority of the electrons is made in a small region in which the gyro-synchrotron emission of radio waves is negligible but hard X-rays

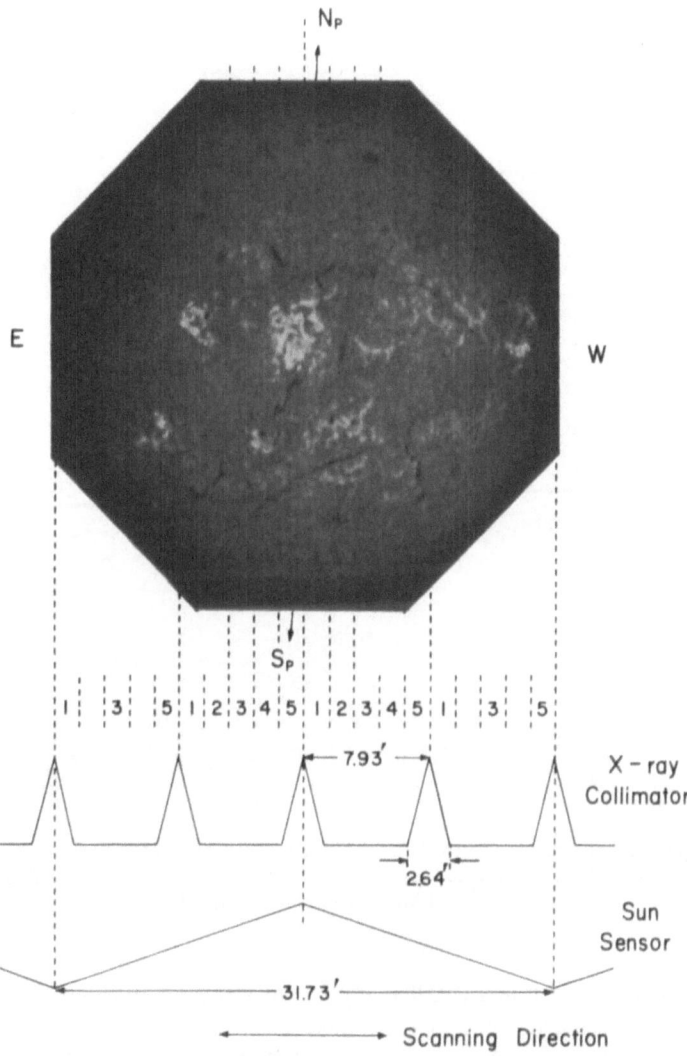

Fig. 1. One-dimensional transmission responses of the X-ray modulation collimator and the Sunsenor (lower curves). The Hα photograph of the hard X-ray burst, 03ʰ57ᵐ06ˢ UT on September 27, 1969 is available by courtesy of the Carnarvon Tracking Station and the World Data Center A, ESSA, Boulder.

are predominant. This condition is satisfied in a plasma with a density of $10^{10}\,\mathrm{cm}^{-3}$, if the magnetic field is less than 30 G, or if the magnetic field is less than 100 G and the energies of the electrons are less than 1 MeV. In these cases low frequencies are suppressed by the Razin effect (Ramaty, 1969), while the emissivity at the higher frequencies is low due to the weak magnetic field. The microwave impulsive burst

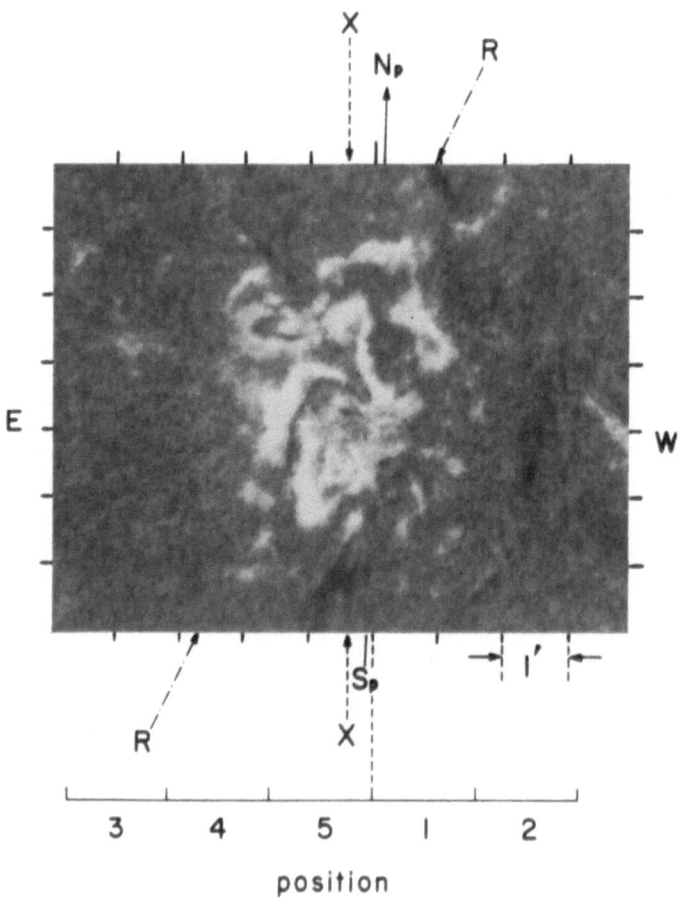

Fig. 2.   The Hα flare region (cf. Figure 1) and the location of the source of the hard X-ray burst. The center of the X-ray sources was located on the line X. The center of the source of associated radio burst at 3750 MHz was located on the line R.

may be emitted from a much smaller number of electrons diffused into the regions with higher magnetic fields or accelerated there.

In order to accelerate 10^{36} electrons above 50 keV up to 1 MeV in 10 to 100 s in a region of the order of 10^4 km, the efficiency of acceleration must be very high. Therefore, I would suggest the following process for the generation of large scale electric potential fields to accelerate electrons and to trigger the flares.

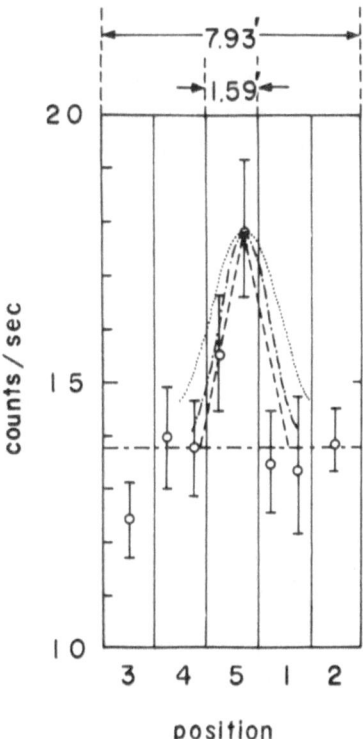

Fig. 3.   One-dimensional intensity distribution of the hard X-rays during the burst from 0356 to 0358.5 UT. Position numbers correspond to those given in Figures 1 and 2. Dashed curve: expected profile for a point source. Dot-dashed curve: expected profile for a gaussian source with a half power width of 1.1 arc min. Dotted curve: gaussian source of 2.2 arc min.

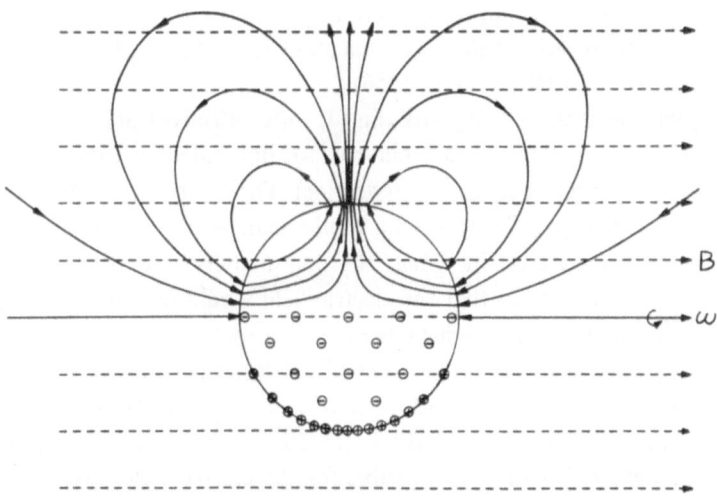

Fig. 4.   The electric lines of force for E_0 and E_t (effective). They also show the paths of the steady current. The steady charge distribution is shown schematically in the lower half.

2. Generation of Potential Electric Fields by a Rotating Gas Cloud in a Magnetic Field

Suppose that a gas cloud rotates in a uniform magnetic field near the photosphere, where the gas pressure is greater than the magnetic pressure, and the conductivity is nearly *isotropic* due to the fact that the collision frequency is higher than the gyro-frequency.

In the rotating cloud, a Lorentz force $q(\mathbf{v} \times \mathbf{B})$ produces a charge separation. The charge distributions in the cloud and at its surface produce potential electric fields not only in the cloud but also outside the cloud. Since the outside of the cloud is also a conducting gas, a current flows from the outside through the rotating cloud.

In the steady state, the following quadrupole potential field and a steady current can be obtained. It is assumed that the conductivity σ is isotropic and uniform, and the rotation axis is parallel to a uniform magnetic field.

In spherical coordinates (r, θ, φ) and in MKS units, we have the following potential field,

$$r \geqslant a: \quad V_0 = \frac{B\omega a^2}{15} \left(\frac{a}{r}\right)^3 (1 - 3\cos^2 \theta),$$

$$\mathbf{E}_0 = -\operatorname{grad} V_0$$

$$= \frac{B\omega a}{5} \left(\frac{a}{r}\right)^4 \{(1 - 3\cos^2 \theta)\, \mathbf{e}_r - 2(\sin\theta \cos\theta)\, \mathbf{e}_\theta\},$$

$$r \leqslant a: \quad V_i = B\omega a^2 \left\{ -\frac{1}{3} + \frac{1}{5}\left(\frac{r}{a}\right)^2 (2 - \cos^2 \theta) \right\},$$

$$\mathbf{E}_i(\text{effective}) = \boldsymbol{\omega} \times \mathbf{r} \times \mathbf{B} - \operatorname{grad} V_i$$

$$= \frac{B\omega a}{5} \left(\frac{r}{a}\right) \{(1 - 3\cos^2 \theta)\, \mathbf{e}_r + 3(\sin\theta \cos\theta)\, \mathbf{e}_\theta\},$$

where ω is the angular frequency of the rotating gas-sphere with a radius a. (Incidentally, if the outside of the rotating sphere is a vacuum, $\mathbf{E}_0$ is $\frac{5}{2}$ times the above value and E_i (eff.) $= 0$, e.g., Davis, 1947; Backus, 1956).

The steady current is given by $\mathbf{J}_0 = \sigma\mathbf{E}_0$ and $\mathbf{J}_i = \sigma\mathbf{E}_i$(effective), and $\operatorname{div}\mathbf{J} = -\partial\varrho/\partial t = 0$ is satisfied. The paths of the steady current are illustrated in Figure 4. They also indicate the lines of force of $\mathbf{E}_0$ and $\mathbf{E}_i$ (effective). The external electronic field $\mathbf{E}_0$ has generally a component parallel to the magnetic field. A part of the electric field pe-netrates through the corona, in which the conductivity is *anisotropic* due to the smaller collision frequency, so that the electric field would be modified, and also the current flows almost along the magnetic field.

The electric field strength is of the order of $0.3\, B\omega a$ near the surface $(r = a)$ of the rotating gas, and the maximum potential difference is $0.2\, B\omega a^2$. If we take $\omega a = 4$ ms^{-1} and $B = 0.3$ Wbm^{-2} (3000 G), we have $E_{\max} \simeq 10^{2 \cdot 5}$ Vkm^{-1} and $V_{\max} \simeq 10^{6 \cdot 5}$ V for $a = 10^4$ km. Therefore, it would be possible for the electric field along the magnetic field in the lower corona to be 1–10 Vkm^{-1} or more.

The time scale to come to a steady state would be long. However, due to a skin effect, any rapid variation in the electric field would be followed by a current, although the current is confined in thin layers or thin pipes or threads. The thickness is given by the skin depth $\delta \simeq \sqrt{D\tau}$, where D is the diffusion coefficient and τ is the time scale for the variation of the electric field ($D = 1/\sigma\mu_0 \simeq 10^6$ m^2/s at the photosphere). It may be noted that the frozen condition is not necessarily satisfied if the skin depth is reasonably taken as the characteristic length.

Before the onset of the flare, a current may be flowing in the lower corona due to an electric field of a few V/km. The gas motion which deforms the sunspot configuration or differential rotation would generate the electric field. If the current is I, and an equivalent inductance of the circuit is L, the energy $I^2 L$ is stored. The L is equivalent to the magnetic field created by I. Both the current and the inductance would not be in a steady state as Alfvén and Carlqvist (1967, 1969) have suggested but would increase with time according to an increase in the skin depth. The L and I are estimated to be 10^{-3} Henry and 10^{14} A, respectively, for a time scale of one day and for an

TABLE I

Summary of the flare model

	Energy storage	Trigger
Gas motion near the photosphere	Gradual $\tau \simeq$ day	Impulsive $\tau \simeq 10\text{--}100$ s
↓ Charge distribution in the moving gas ↓		
Electric field outside and inside	$E \simeq 5$ V/km in the lower corona	$E \geqslant 15$ V/km in the lower corona
	↓ I and L (both are functions of τ)	↓ Acceleration of electrons ($K \leqslant 1$ MeV)
		↓ Electron plasma waves
	↓ Stored energy $I^2 L$	↓ Decrease in conductivity
	↓ Energy release due to ohmic loss ($\tau \simeq 1\text{--}10$ s) ‖ Flare	↓ Hard X-ray burst Microwave impulsive burst Type III bursts
	Sporadic coronal condensation Gas ejection M.H.D. waves ↓ Acceleration of protons and electrons to higher energies	

electric field strength of 5 V/km in the lower corona. In this case the stored energy is 10^{25} joules which is required for a big flare.

Now, if the electric field exceeds about 10 V/km in the lower corona due to an additional faster impulsive gas motion near the photosphere, almost all of the thermal electrons are accelerated to higher energies, since the acceleration rate exceeds the collisional loss rate. *This would be a critical condition for the trigger of flares.* In this electric field one second is enough to accelerate electrons up to relativistic energies. However, in order to accelerate 10^{36} electrons above 50 keV, 10^2 s may be required in order to increase the skin depth.

The accelerated electrons excite electron plasma waves due to the two beam instability, and the waves spread over the regions in which I is flowing. The plasma waves decrease the conductivity by 5 orders in the corona (Buneman, 1959; Hamberger and Friedman, 1968). Accordingly, the stored energy $I^2 L$ would be converted suddenly into the thermal energy due to the ohmic loss. This time scale L/R is estimated to be 1–10 s.

The accelerated electrons also emit hard X-rays, and a part of the electrons diffuse into the stronger field region to emit a microwave impulsive burst.

More details will be published elsewhere.

3. Summary

A summary of the flare model is shown in Table I.

References

Alfvén, H. and Carlqvist, P.: 1967, *Solar Phys.* **1**, 220.
Backus, G.: 1956, *Astrophys. J.* **124**, 508.
Buneman, O.: 1959, *Phys. Rev.* **115**, 503.
Carlqvist, P.: 1969, *Solar Phys.* **7**, 377.
Davis, L., Jr.: 1947, *Phys. Rev.* **72**, 632.
Hamberger, S. M. and Friedman, M.: 1968, *Phys. Rev. Letters* **21**, 674.
Ramaty, R.: 1969, *Astrophys. J.* **158**, 753.
Takakura, T.: 1969, *Solar Phys.* **6**, 133.
Takakura, T. and Scalise, E., Jr.: 1970, *Solar Phys.* **11**, 434.
Takakura, T., Ohki, K., Shibuya, N., Fujii, M., Matsuoka, M., Miyamoto, S., Nichimura, J., Oda, M., Ogawara, Y., and Ota, S.: 1971, *Solar Phys.* **16**, 454.
Tanaka, H. and Énomé, S.: 1971, *Solar Phys.* **17**, 408.

Discussion

Cowling: Could the problem of flare energies have been solved along these lines, it would have been solved 30 yr ago. Takakura has ignored induction effects corresponding to changes $\partial B/\partial t$. I conservatively estimate that these may reduce the efficiency of his mechanism in the ratio 10^{-10}.

Takakura: I have obtained the solution in the steady state at which $\partial B/\partial t = 0$. The time to be steady would be long. However, as I have suggested, if we take the 'skin effect' into account any rapid variation of the electric field is followed by the current, though the current would be concentrated in thin filaments or thin pipes. This effect has been taken into account in my numerical estimates for the flare energy and for the trigger. Furthermore, if we take the skin depth as a characteristic length, the frozen condition is not necessarily satisfied near the photosphere.

THE X-RAY CORONA AND THE PHOTOSPHERIC
MAGNETIC FIELD

A. S. KRIEGER, G. S. VAIANA and L. P. VAN SPEYBROECK

American Science and Engineering, Cambridge, Mass., U.S.A.

Abstract. Soft X-ray (3–60 Å) photographs of the solar corona have been obtained on four flights of a rocket-borne grazing incidence telescope having a resolution of a few arc seconds. The configuration of the X-ray emitting structures in the corona has been compared to the magnetic field distribution measured by photospheric longitudinal magnetograms. The X-ray structures trace the three dimensional configuration of the magnetic field through the lower corona.

Active regions in the corona take the form of tubular structures connecting regions of opposite magnetic polarity within the same or adjacent chromospheric active regions.

Higher, larger structures link widely separated active regions into complexes of activity covering substantial fractions of the disk. The complexes are separated by areas of low average field in the photosphere. Interconnections across the solar equator appear to originate over areas of preceding polarity.

Enhanced X-ray emission is observed, outside the active belt, over areas of enhanced general magnetic field. Bright point-like X-ray features are observed above bipolar areas in the general coronal field.

Two, probably related, classes of X-ray features are associated with points of high field gradient along the longitudinal magnetic, 'neutral line'. Both in the non-flaring active region and in flares the brightest emission is observed from one or more small line sources (at our present resolution). In flares this can account for a substantial fraction of the total soft X-ray emission of the flare.

1. Introduction

In order to derive meaningful models of the physical processes taking place in the solar corona, one must know the structure of the corona and the configuration of the magnetic field through the corona. For example, the direction and strength of the magnetic field in the lower corona will determine the type of waves that may propagate there and the mechanism by which they dissipate their energy. Also, the problems of energy storage in non-flaring active regions and energy release in flares require the knowledge of the configuration of the magnetic field of the active region in the lower corona.

Mathematical extension of the photospheric fields into the corona is a difficult exercise. Observationally, all three components of the field must be measured, and theoretically, one must solve for a stable, force-free field configuration. Direct observation of the coronal structures are needed to acquire confidence in the calculations. In visible light, coronal structures can be observed only at the limb, and, with sufficient resolution, only during total solar eclipses, and occasionally, under good seeing conditions, in a few selected lines.

Soft X-rays provide an advantageous waveband for examination of the lower corona. Soft X-rays are thermally emitted by plasmas with temperatures in the range of 10^6 K or more, hence they are the characteristic radiation of the coronal plasma. Moreover, there is no photospheric or chromospheric background radiation at these

Howard (ed.), Solar Magnetic Fields, 397–412. All Rights Reserved.
Copyright © 1971 by the IAU.

wavelengths, therefore, coronal structures can be seen in projection on the solar disk. In addition, X-ray optics with angular resolution approaching that attainable in visible light are now available.

Over the past several years our group has conducted four successful sounding rocket flights of a grazing incidence X-ray telescope with a resolution of a few arc seconds. A full description of the instrumentation has been reported in the literature (Giacconi *et al.*, 1969; Vaiana *et al.*, 1968). We have obtained photographs in the 3 to 60 Å wavelength range of the X-ray emitting structures of the solar corona as projected on the disk. Two of the flights were devoted to the study of solar flares, and the other two were directed primarily toward investigations of the general corona. On all four flights non-flaring active regions were observed.

An examination of an X-ray photograph obtained during our most recent flight together with a map of the photospheric longitudinal magnetic field (Figure 1) shows the highly ordered structures observed in the corona. They are clearly associated with the underlying photospheric fields. It is natural to suppose that the X-ray structures reflect the configuration of the coronal magnetic fields. To make a plausibility argument, one need only say that in order to produce a detectable photographic image with the X-ray telescope currently in use, a coronal region must have a temperature greater than 10^6 K and a density greater than a few times 10^8 cm^{-3}. Thus the highly ordered structures observed in X-rays can be contained by magnetic fields of only a few gauss in strength. Accordingly, we discuss here the relationship between the observed coronal X-ray structures and the underlying photospheric magnetic field in order to understand the configuration of the coronal magnetic field. Other aspects of our X-ray photographs have been discussed elsewhere (Van Speybroeck *et al.*, 1970; Vaiana and Giacconi, 1969).

Although the discussion in this paper is still entirely qualitative the results display the general characteristics of the coronal magnetic field. We have observed X-ray structures associated with a variety of solar phenomena. Active regions appear in the corona to be composed of tubular structures connecting regions of opposite magnetic polarity on the photospheric longitudinal field map. Widely separated active regions are interconnected over significant fractions of the solar disk by X-ray emitting structures. Bright, hot, narrow features have been observed bridging the 'neutral line' of the longitudinal magnetic field in active regions and flares. Relatively diffuse X-ray emitting features overlie areas of enhanced field on the quiescent disk. These features can be identified as the bases of white light coronal structures. Small, bright, X-ray features in the quiescent corona are associated with bipolar areas in the general solar field.

2. Active Regions

When active regions are observed in X-rays at the limb of the Sun (Figure 2) they are seen to be complex tubular arches and/or loops of enhanced density and temperature rising to heights of more than 10^5 km above the photosphere. Thus, X-ray observations of active regions on the disk are two-dimensional projections of three-dimensional

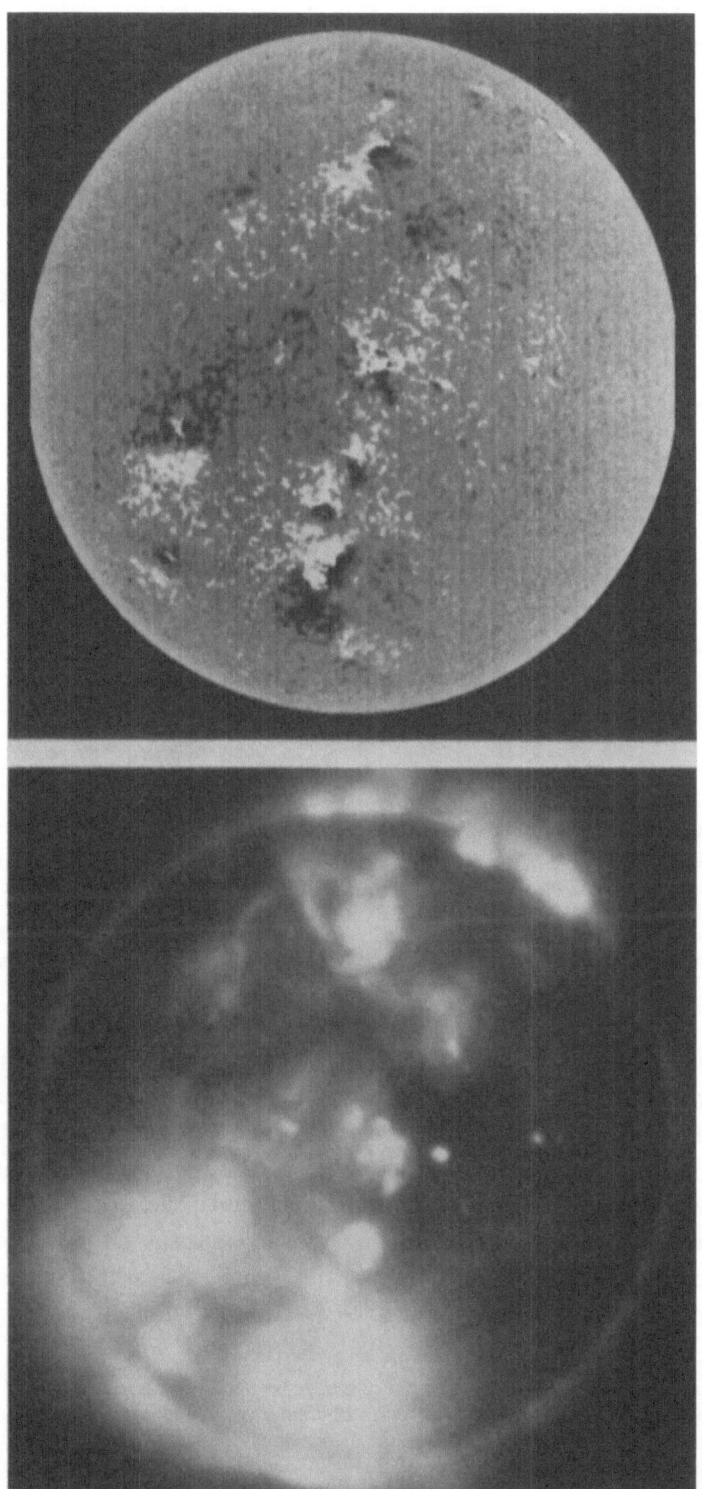

Fig. 1. A comparison between the appearance of the Sun in soft X-rays and the photospheric magnetic field. Left: An X-ray exposure in the wavebands 3–16 Å, 44–47 Å taken March 7, 1970 shortly after fourth contact of the solar eclipse. The limb of the Moon is visible to the southeast of the limb of the Sun. Right: A map of the longitudinal component of the photospheric magnetic field on March 7, 1970 as observed by Livingston et al. (1970).

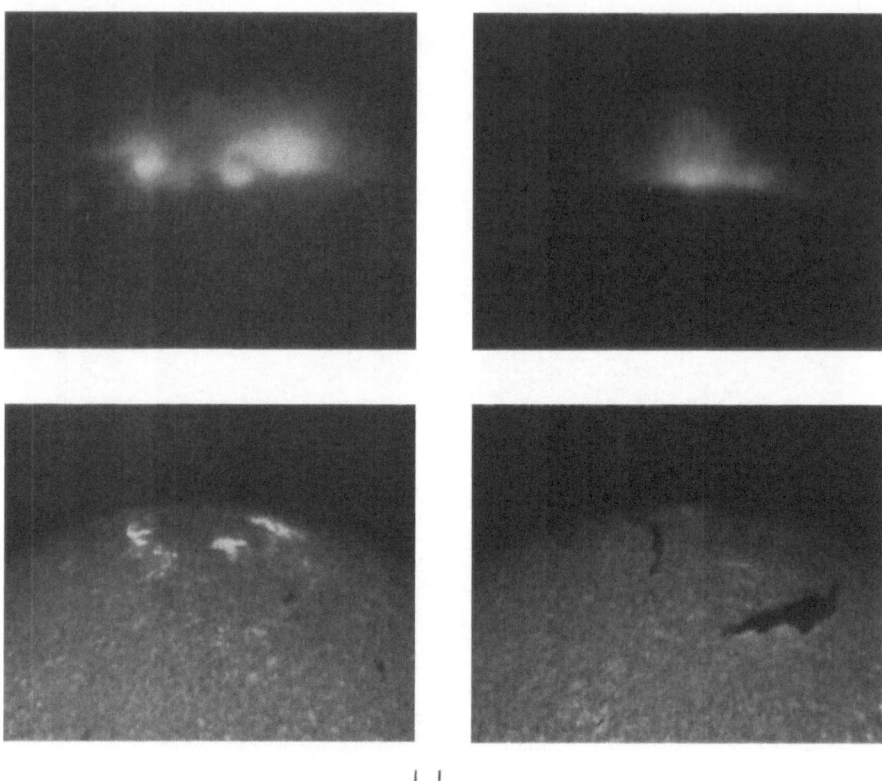

Fig. 2.   The appearance of active region structures in X-rays at the limb. Top: X-ray photographs in the 3–17 Å passband of active region associated coronal features. Left: A group of active regions near the limb. At least three arches connecting different portions of the group can be distinguished. Right: A loop structure associated with an active region very close to the limb. The coronal loop extends to an altitude of at least 150000 km. Bottom: Hα photographs of the corresponding portions of the disk taken two hours before the rocket flight (June 8, 1968) by ESSA Boulder Observatory. The length of the bracket below the photographs is one arc-minute.

structures. The appearance in projection on the disk of the coronal structure of three active regions is shown in Figure 3 together with Hα, and CaK spectroheliograms of the regions, and their longitudinal magnetic field distribution. A close examination of the photographs shows that the structures observed in X-rays interconnect regions of opposite magnetic polarity, as displayed by the longitudinal field map. The interconnections proceed between portions of the same or adjacent active regions. In general, the intensity of the observed X-ray emission is highest near the 'neutral line' of the longitudinal magnetic field.* Often, where a large field gradient is observed, a

---

* Almost all of our comparisons of the X-ray structures have been made with regard to the longitudinal field maps because of the difficulties involved in obtaining vector magnetic field information. For our purposes here this is satisfactory. However, the reader should remain aware of the ambiguities involved. For example, the location of the longitudinal field 'neutral line' may not be neutral at all, but might possess transverse fields.

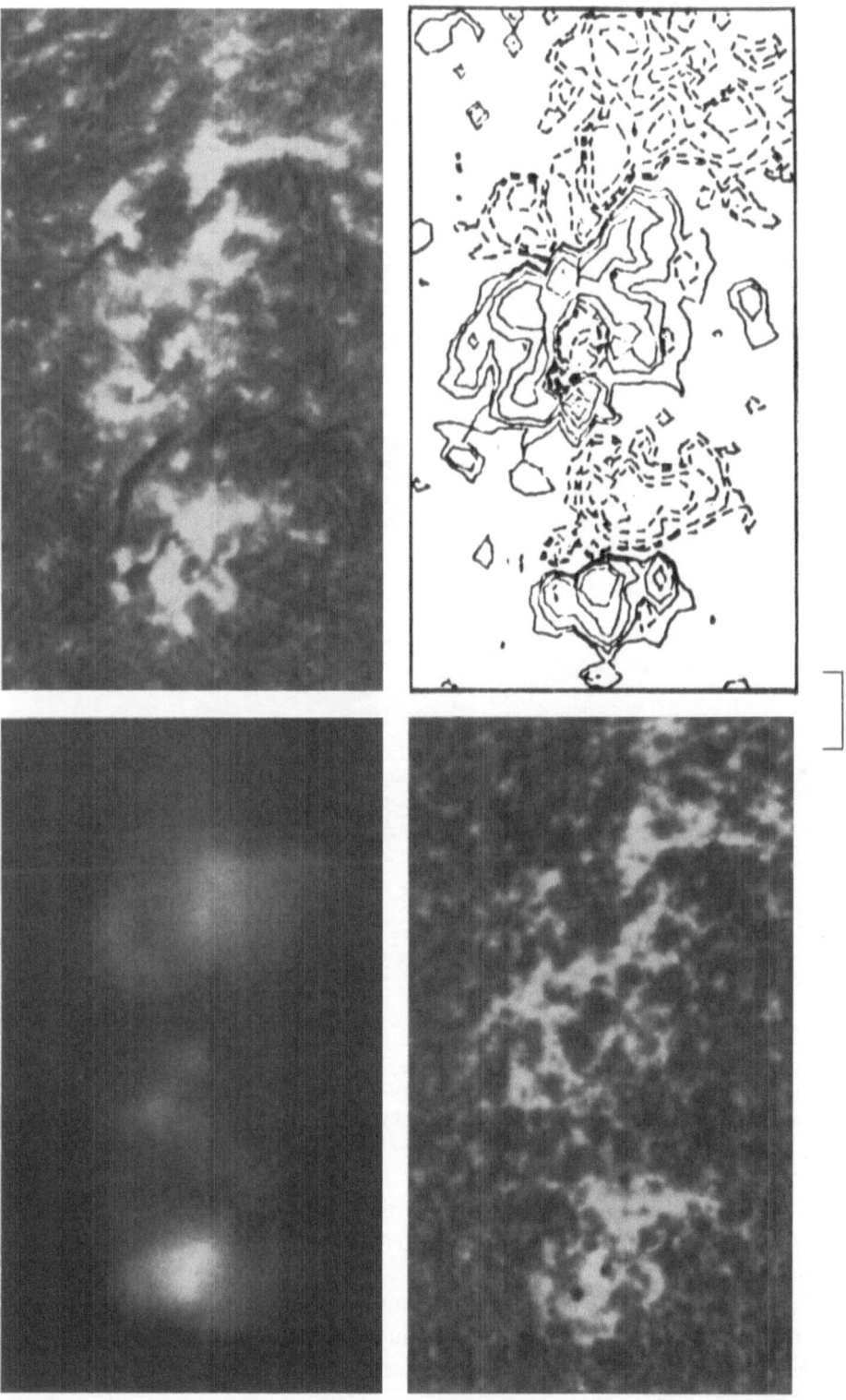

Fig. 3. The appearance of active region structures in projection on the disk. Top left: A group of active regions observed November 4, 1969 in the 3–23 Å, 44–56 Å soft X-ray wavebands. Top right: The same regions observed in Hα (courtesy of Sacramento Peak Observatory). Bottom left: CaK (courtesy of Sacramento Peak Observatory). Bottom right: Longitudinal component of the photospheric magnetic field (courtesy of Mt. Wilson Observatory). The length of the bracket below the photographs is one arc minute.

bright core, whose width is unresolvable in our telescope, connects the regions of preceding and following polarity across the neutral line. The ends of the bright X-ray core overlie the areas of maximum Hα and CaK intensity. The core structure passes directly over the region of reduced intensity in Hα and CaK. The X-ray spectrum of the core is harder than that of the rest of the active region when observed through two different passband filters. For example, in the case shown in Figure 3, we have measured the ratio of surface brightnesses of the core, at our resolution, through

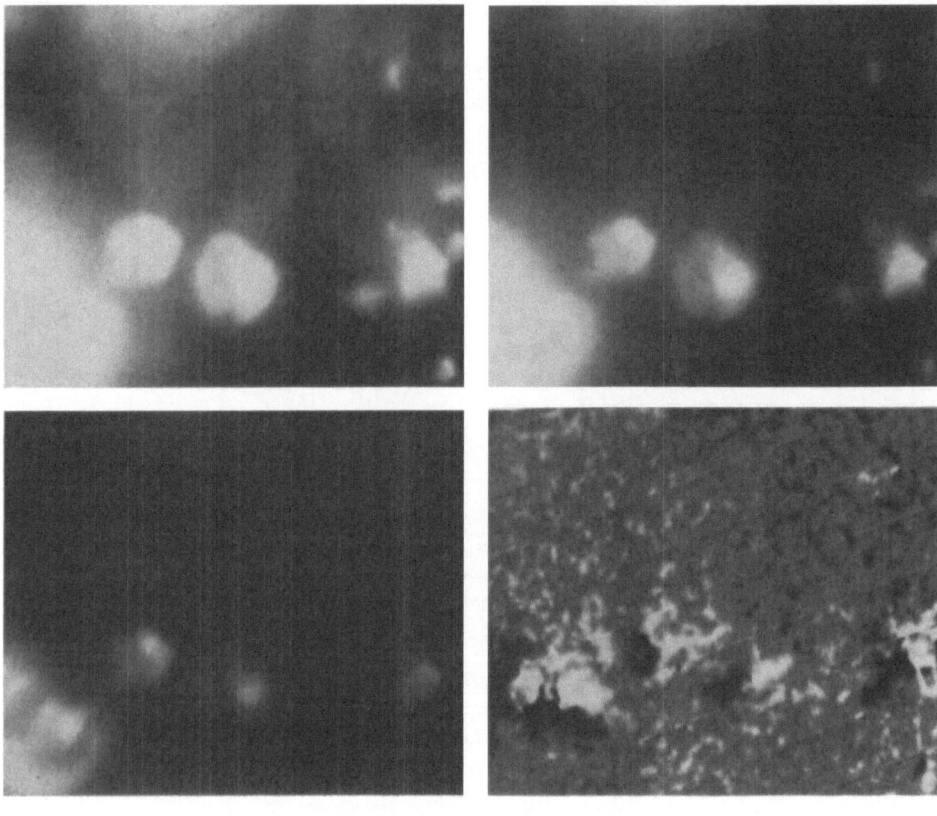

Fig. 4. Comparison of the X-ray emitting structure of active regions as observed in three passbands. Top left: 3–36 Å, 44–64 Å. Top right: 3–16 Å, 44–47 Å. Bottom left: 3–13 Å, 18–24 Å. Bottom right: The corresponding photospheric magnetic fields measured by Livingston *et al.* (1970). The bracket below the photographs is one arc-minute in length.

3–18 Å and 3–23 Å, 44–64 Å waveband filters. It is at least twice that of the remainder of the active region. If one assumes that the X-ray spectrum is of thermal origin, then this implies a higher temperature for the core.

Another group of active regions (Figure 4) is shown in three X-ray passbands (to bring out different aspects of the coronal structures) along with the longitudinal magnetic field. In the softest wavelengths we observe interconnections between the active regions as well as the interconnections between portions of the same active region observed in the shorter wavelength exposures. At the shortest wavelengths we note several of the bright core structures. The configurations of the X-ray structures indicate the complexity of the magnetic field pattern. We point out the presence of high field gradients induced by the incursion of following polarity into areas of preceding polarity or vice versa at the sites of some of the bright X-ray cores. Also note that the tubes of coronal plasma interconnecting preceding and following polarities are bent out of the vertical plane. This effect is especially noticeable in the lower left corner of Figure 4, (see also Figure 1). One result of the bending of the X-ray structures is a projection effect which, in the upper left photograph of Figure 4, is noticed as a non-uniform brightness across the active regions. Examination of the shorter wavelength exposures (upper right, and lower left) clarifies the situation.

3. Groups of Active Regions

High sensitivity, long wavelength X-ray exposures reveal interconnections between active regions covering extensive areas of the solar disk. One such set of interconnections, on the eastern portion of the disk in Figure 1, covers about 60° of solar longitude and about 90° of latitude. The structures between the active regions are 5 to 10% as bright as the brightest of the active regions they interconnect.

Examination of the underlying photospheric fields indicates that, just as within individual active regions, the interconnections link opposite longitudinal magnetic polarities. Moreover, connections between active regions on opposite sides of the solar equator, link the preceding polarities of the two hemispheres, not the following

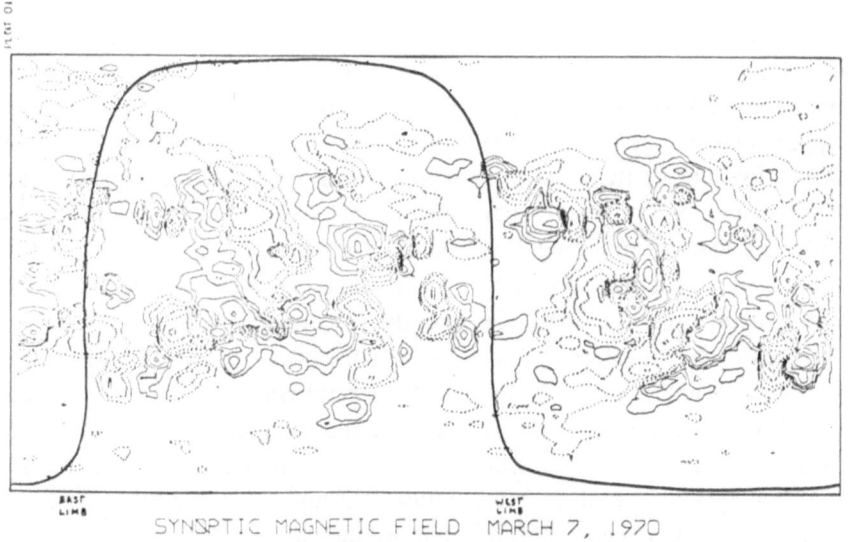

SYNOPTIC MAGNETIC FIELD MARCH 7, 1970

Fig. 5. A synoptic chart of the photospheric magnetic field on March 7, 1970 (from Livingston *et al.*, 1970).

polarities. Figure 4, a detail of Figure 1, shows bridges crossing the equator which commence over the preceding portions of three southern hemisphere active regions. We interpret these interconnections as indications of the severing and reconnection of the magnetic fields of the active regions as postulated by Babcock (1961).

The active regions are joined into 'complexes of activity' as observed in the magnetograms of Bumba and Howard (1965). Figure 5 is a synoptic chart of the solar magnetic field on March 7, 1970, the date of the rocket flight, as observed by Livingston *et al.* (1970). It shows that the coronal interconnections which we observe delineate the areas of related photospheric fields. In the areas underlying coronal regions where no interconnections are observed, the average photospheric field appears to be reduced. In Figure 1, two complexes of X-ray interconnection are sharply delimited by a region from which no X-ray emission was observed somewhat to the west of the central meridian. A large dark filament underlies this boundary in the chromospheric spectroheliograms, and the average longitudinal photospheric magnetic fields are significantly weaker there.

The eastern of the two complexes observed on the disk may be identified with a zone of activity. The active regions comprising this complex can be traced back in the CaK reports published in *Solar Geophysical Data* (1970) for at least 10 rotations. The members of the western group of regions, which also show interconnections across the solar equator, are only two or three rotations old. These new active regions have formed in the magnetic remnants of older regions. The coronal interconnections that we observe in X-rays may be a manifestation of older features than the active regions presently forming the complex.

4. X-Ray Emission from the Quiescent Corona

Faint patches of diffuse X-ray emission are observed on quiescent portions of the solar disk in long exposures with filters having long wavelength X-ray passbands. An example of these features can be seen in Figure 1. In the 44–50 Å waveband these patches are approximately $\frac{1}{2}$ as bright as the peak limb brightening. Within the errors in our determination, these patches have the same general X-ray spectral distribution as the limb brightening. We conjecture that in fact the patches of soft X-ray emission seen on the disk are regions of enhanced density in the quiet corona.

Figure 6 shows such an area of X-ray emission along with the underlying chromospheric structure seen in CaK, Hα, and the longitudinal component of the solar magnetic field. It is evident that the X-ray emission is associated with an area of increased photospheric magnetic field in the general corona. The X-ray emission of Figure 6 overlies a unipolar magnetic region and its 'ghost' (Bumba and Howard, 1965). The patches of X-ray emission are associated with enhanced areas of the CaK network but they appear to be more diffuse than the corresponding chromospheric features. This is the phenomenon that one would expect if the magnetic field in the corona were to spread from the supergranule boundaries to fill in the supergranule. At the line where a filament is observed in Hα and CaK separating the unipolar

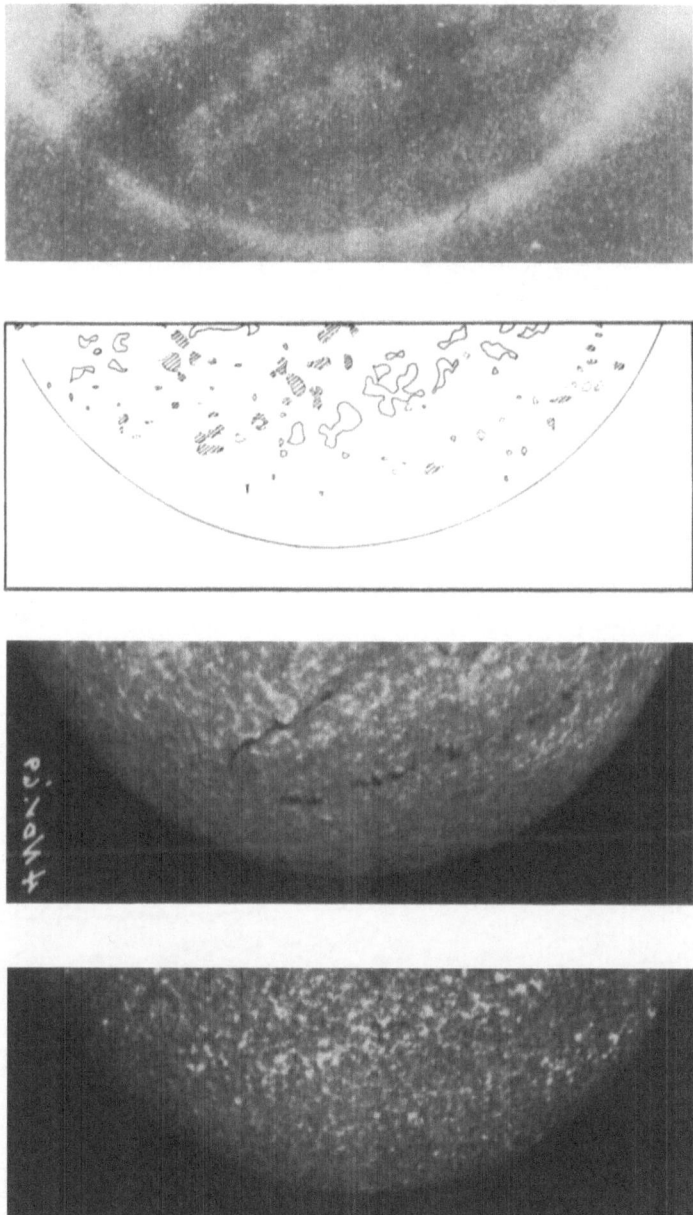

Fig. 6. The appearance of a unipolar magnetic region in (from top to bottom): 3–23 Å, 44–64 Å X-rays; photospheric magnetic field (courtesy of Mt. Wilson Observatory); Hα and CaK. The Hα and CaK photographs were provided by the Sacramento Peak Observatory.

magnetic region from its ghost, no X-ray emission is apparent. We take these facts to indicate that over the unipolar region and ghost region the coronal density is enhanced, whereas it is reduced over the filament region. Perhaps, it is the coronal material which should have filled this region which has formed the filament. In any case, the UMR delineates a region of enhanced density in the lower corona. When areas which appear on the disk as diffuse coronal emission reach the limb, they are observed as enhancements of the limb brightening. An X-ray photograph obtained in conjunction with the March 7, 1970 solar eclipse was compared with a radial density

Fig. 7. A composite print of the solar corona on March 7, 1970 composed of an X-ray exposure in the 3–36 Å, 44–64 Å wavebands and the radial density gradient filter white light exposure of Newkirk and Lacey (1970).

gradient photograph of the white light corona (Newkirk and Lacey, 1970). There was a one to one correspondence between enhanced X-ray limb brightening and the bases of white light coronal rays and helmet streamers (Van Speybroeck *et al.*, 1970). This fact is evident in a composite print of the X-ray and white light exposures (Figure 7).

Interspersed in the diffuse patches of X-ray emission on the disk are relatively intense spots of brighter emission. These are most noticeable in Figure 8. They appear to be somewhat higher in temperature than the rest of the general corona on the basis of comparisons of exposures with different filters. These point-like regions are associated with all small bipolar areas in the general magnetic field (Figure 1 and Figure 9) and the Ca network is enhanced at their sites. Both from their relative X-ray brightness, and from their bipolar nature, it seems reasonable to assume that these points represent low lying closed structures in the general coronal field. It is possible

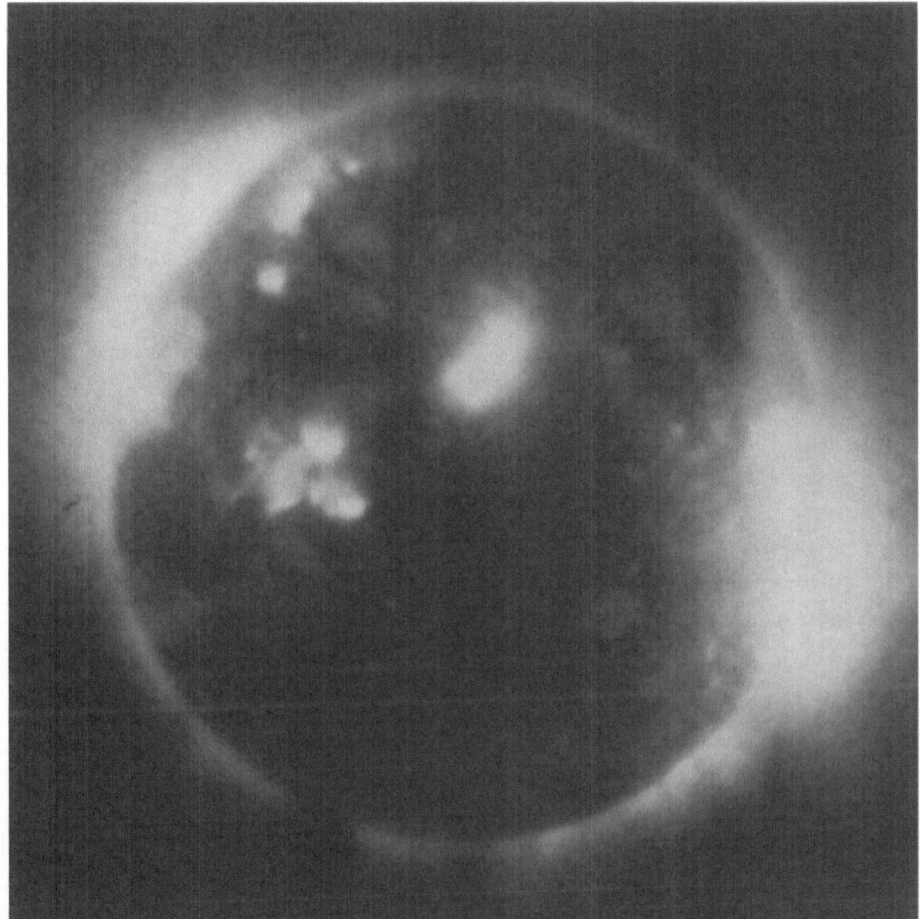

Fig. 8.   A photograph of the Sun in the 3–18 Å, 44–46 Å wavebands obtained on April 8, 1969.

that they are regions of particularly intense coronal heating or that they are miniature active regions. Further investigation of these features is required.

5. Flares

On two of our four rocket flights (June 8, 1968 and November 4, 1969) we have observed importance 1 solar flares in progress. Figure 10 shows a comparison of the

appearance of the flare observed June 8, 1968 in 3 to 18 Å X-rays, Hα, and CaK along
with a magnetogram of the region in which the flare took place. As we have noted
elsewhere (Vaiana and Giacconi, 1969), the X-ray emission follows the form of the Hα
emission in general with the important exception of a bright spot bridging the magnetic
neutral line of the flare. The bright spot emits more than 50% of the total energy
observed from the flare in the 3–18 Å passband.

The X-ray appearance of the flare observed on November 4, 1969, seems very
different from its appearance in on-band Hα, however, as is shown in Figure 11. The
X-ray emission, or at least 98% of the total energy in the 3–18 Å band, is produced in
two regions narrower than the resolution of the telescope (∼4 arc sec) and no more
than 10 or 15 arc sec long. The two spots of X-ray emission correspond to the region
of brightest emission in the wings of Hα. We therefore interpret these X-ray spots as

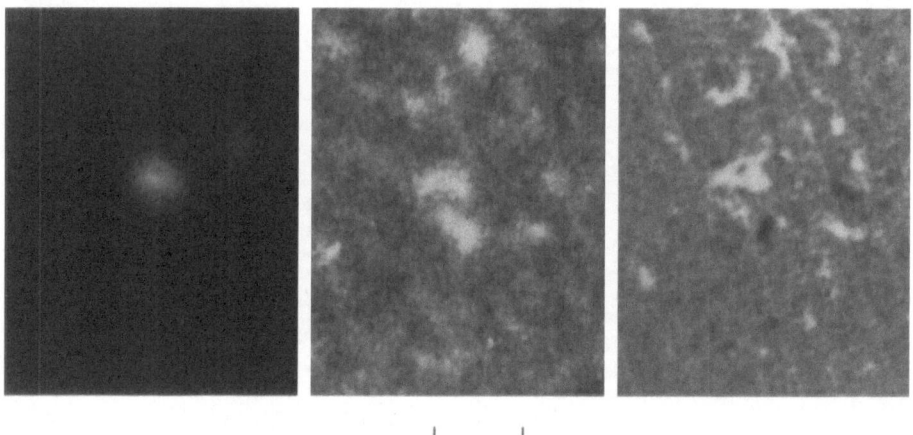

Fig. 9.   An X-ray 'bright point' in the quiescent corona and the underlying structures in CaK.
(courtesy of Sacramento Peak Observatory) and the photospheric magnetic field (Livingston *et al.*,
1970). The bracket below the photographs is one arc minute in length.

being asociated with the hottest points in the Hα flare. Unfortunately, magnetic field
information was not available for this region for times close to the flare, because the
flare was near the limb. However, we note that in the 3–12 Å waveband the June 8, 1968
flare appears strikingly similar to the November 4, 1969 flare (Figure 12) in that the
majority of the emission is from a restricted region. It would appear that the two
bright spots observed on November 4, 1969 are analogous to the bright core of the
June 8, 1968 flare although flare X-ray emission from the rest of the X-ray plage at
longer wavelengths was substantially weaker.

The bright cores which we have recently observed in non-flaring active regions bear
a strong resemblance to the situation observed during the June 8, 1968 flare despite the
difference of a factor of 50 to 100 in the surface brightness of flares and active regions.
Both the cores of active regions and the spot of most intense emission in the flare are
narrow linear structures bridging the neutral line of the longitudinal field at a point of

high gradient. With regard to the November 4, 1969 flare the situation is less clear because of the lack of magnetic field data and the position of the flare near the limb. It might be that the two spots of flare X-ray emission bridge the neutral line at two nearby points as we have occasionally observed in active regions (Figure 4). Alternatively, it might be that these spots which are associated with the region of maximum Hα emission are analogous to the foot points of the active region core. It may be

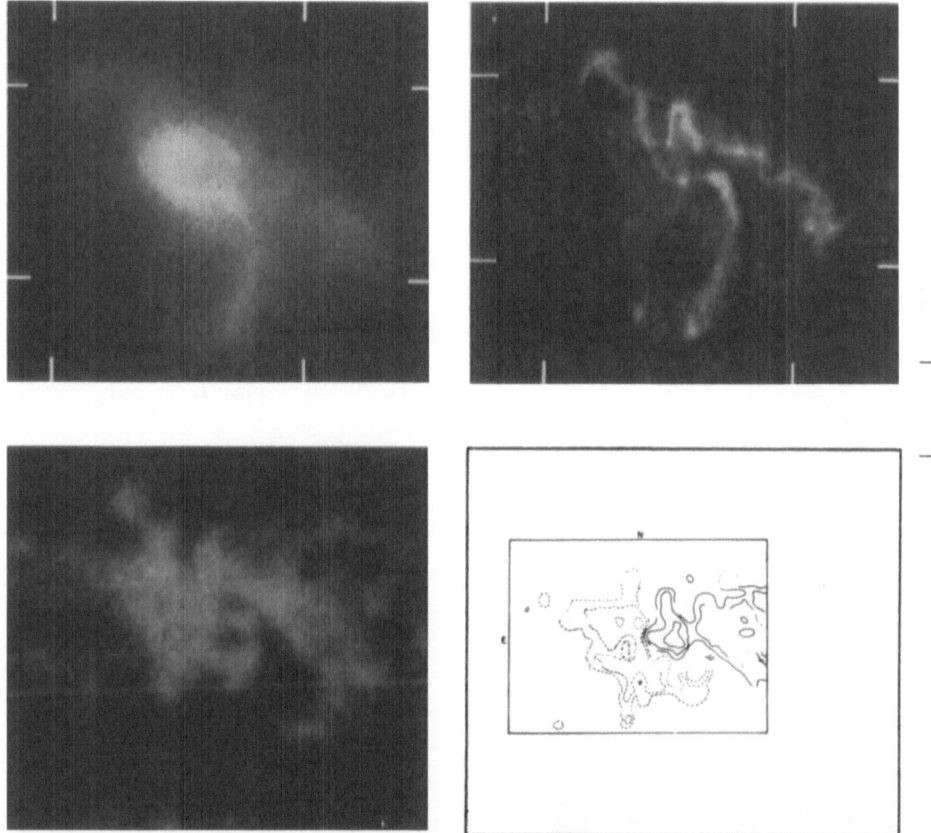

Fig. 10.   Appearance of the solar flare of June 8, 1968 (1742 UT) in X-rays, Hα and CaK and the magnetic field configuration at the time of the flare. Top left: 3–18 Å X-rays. Top right: Hα (courtesy of ESSA Boulder Observatory). Bottom left: CaK (courtesy of McMath-Hulbert Observatory). Bottom right: Photospheric longitudinal field (courtesy of J. Harvey, Kitt Peak National Observatory). The bracket beside the photographs is one arc-minute in length.

highly significant that the June 8 observations took place at the peak of the flare while a non-thermal radio burst was in progress. On the other hand, the November 4 observations were made during the decay phase of the flare. In any case, our observations suggest that the bright, X-ray core of the active region becomes the site of the most intense X-ray emission in flares. The bright X-ray core may play an important role in the flare process.

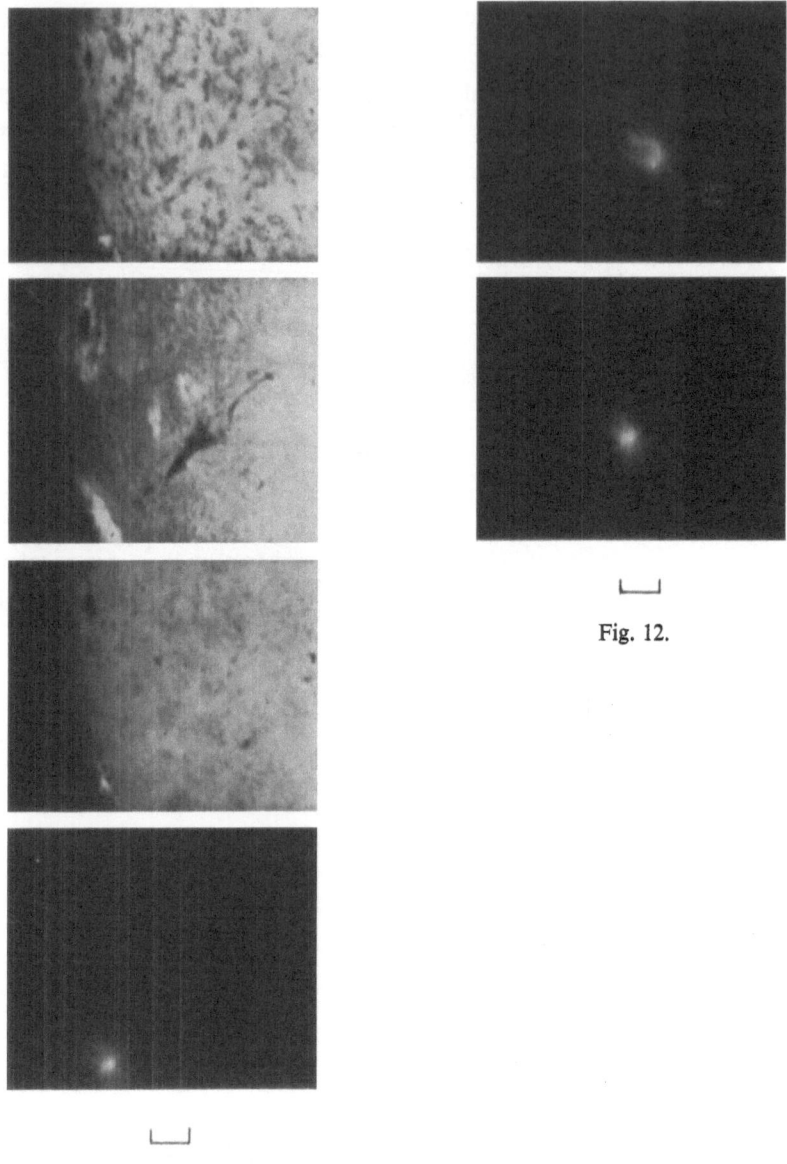

Fig. 12.

Fig. 11.

Fig. 11. The appearance of the solar flare of November 4, 1969 in Hα and X-rays. Top to bottom: Red wing Hα; On-band Hα; Blue wing Hα; 3–12 Å X-rays. The photographs were taken at about 2030 UT. The Hα photographs were provided by H. Zirin, California Institute of Technology. The bracket below the photographs is one arc minute in length.

Fig. 12. A comparison of the appearance of the June 8, 1968 flare (top) and the November 4, 1969 flare (bottom) in the 3–12 Å waveband. The bracket below the photographs is one arc minute in length.

6. Conclusions

Direct comparisons of the soft X-ray images of the corona with the photospheric magnetic fields indicate that the X-ray structures do indeed trace the configuration of the magnetic field through the lower corona. It appears that given sufficient sensitivity and resolution X-ray emitting coronal structures can be observed wherever there is magnetic field penetrating the corona. A range of phenomena have been observed showing field lines which close at a variety of heights in the solar atmosphere.

Active regions appear to be composed of tubular structures often taking the form of loops or arches, which have their feet in regions of opposite magnetic polarity. These loops usually interconnect portions of the same or adjacent active regions.

Higher, larger structures link widely separated active regions into complexes of activity covering substantial fractions of the disk. The complexes are separated by areas of low average longitudinal field in the photosphere and by quiescent filaments in the chromosphere. Interconnections across the solar equator appear to originate over areas of preceding polarity.

Enhanced X-ray emission is observed, outside the active belt, over unipolar magnetic regions. The X-ray features appear more diffuse than the underlying chromospheric network. Reconnection of these field lines takes place high in the corona where the structures can only be observed in white light. Bright point-like X-ray features are observed above bipolar areas in the general coronal field. Presumably these indicate structures which reconnect low in the corona.

Two, probably related, classes of X-ray features are associated with points of high field gradient along the longitudinal magnetic, 'neutral line'. Both in the non-flaring active region and in flares the brightest emission is observed from one or more line sources (at our present resolution) associated with the 'neutral line' at the point of maximum gradient. In flares this can account for a substantial fraction of the total soft X-ray emission of the flare.

To date, only four rocket flights of the high resolution grazing incidence X-ray telescope have been conducted. The total observation time involved is about 15 min. Despite the limited number of observations, X-ray imaging has obtained new information on the configuration of the lower corona under conditions ranging from flares to unipolar magnetic regions. Each of the rocket flights has revealed at least one phenomenon that had not been observed before and that had not been anticipated.

Acknowledgements

The group of investigators who have participated in the solar X-ray rocket program includes, in addition to the authors, R. Giacconi, W. Reidy, T. Zehnpfennig, J. Davis and M. Zombeck. This work was supported by NASA under contracts NASW-2027 and NAS5-9041.

References

Babcock, H. W.: 1961, *Astrophys. J.* **133**, 572.

Bumba, V. and Howard, R.: 1965, *Astrophys. J.* **141**, 1502.

Giacconi, R., Reidy, W., Vaiana, G., VanSpeybroeck, L., and Zehnpfennig, T.: 1969, *Space Sci. Rev.* **9**, 3.

Livingston, W., Harvey, J., and Slaughter, C.: 1970, *Nature* **226**, 1146.

Newkirk, G. and Lacey, L.: 1970, *Nature* **226**, 1098.

Solar-Geophysical Data Bulletin: 1970, Environmental Science Services Administration, U.S. Dept. of Commerce, Boulder, Colo., IER-FB-309.

Vaiana, G., Reidy, W., Zehnpfennig, T., VanSpeybroeck, L., and Giacconi, R.: 1968, *Science* **161**, 564.

Vaiana, G. and Giacconi, R.: 1969, in *Plasma Instabilities in Astrophysics*, (ed. by D. A. Tidman and D. G. Wentzel), Gordon and Breach, N.Y., p. 91.

Van Speybroeck, L. P., Krieger, A. S., and Vaiana, G. S.: 1970, *Nature* **227**, 818.

MAGNETIC FIELDS IN THE LOWER CORONA ASSOCIATED WITH THE EXPANDING LIMB BURST ON MARCH 30TH 1969 INFERRED FROM THE MICROWAVE HIGH-RESOLUTION OBSERVATIONS

SHINZO ÉNOMÉ and HARUO TANAKA

The Research Institute of Atmospherics, Nagoya University, Japan

Abstract. An expansion of the source of a great solar microwave burst was observed a little beyond the west limb on March 30, 1969. This expansion is interpreted in terms of diffusion of energetic electrons in a turbulent magnetic field in the flare region. The height of the source is estimated to have been 10^4 km.

A remarkable expansion of the radio source was observed by the quick-scan radio interferometer at 3.75 GHz (Tanaka *et al.*, 1969) during the solar radio burst on March 30, 1969. This burst occurred 10° behind the west limb of the Sun. A simple calculation of the geometry revealed that the height of the source of the burst was 10^4 km and that of the parent S-component was not higher than 10^4 km at 3.75 and 9.4 GHz. The extension of the expanded source was as high as $\sim 0.3\,R_\odot$, which indicates a stretching-out of the magnetic field of, say, tens or a hundred of Oersteds so far, according to a theory of radiation of solar microwave bursts (Takakura and Scalise, 1970).

The expansion started at 0250 UT just after the time of maximum flux and continued about 10 min as shown in Figure 1. The upper curve shows the intensity-time profile

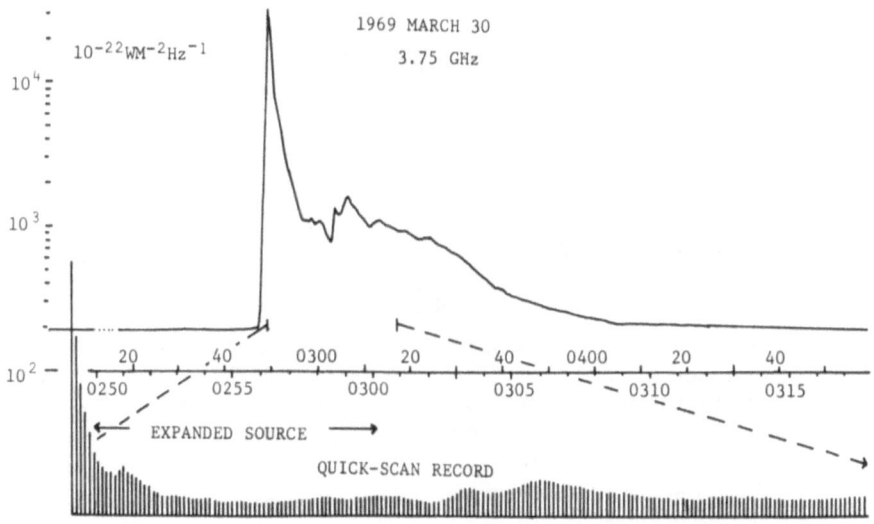

Fig. 1. The curve in the upper part shows the intensity-time profile of the burst at 3.75 GHz with the upper time scale. The row of bars in the lower part shows a schematic drawing of the quick-scan record at 3.75 GHz, representing a temporal change of the brightness of the source in an arbitrary unit in every 10 s with the lower time scale.

Howard (ed.), Solar Magnetic Fields, 413–416. All Rights Reserved.
Copyright © 1971 by the IAU.

of the burst with the upper time scale. The quick-scan record of every 10 s of the burst is illustrated schematically in the lower picture with the lower time scale.

As for the interpretation of the phenomenon, physical expansion of the active region itself may be rejected, since it is generally believed that the flare energy is fed by the magnetic energy of the active region, it is not possible to blow up the active region as a whole by the flare energy.

Figure 2 shows the quick-scan record of the Sun. East is to the right and west is to the left. The way of expansion suggests diffusion of high energy electrons, probably tens of keV and more, through the magnetized solar plasma.

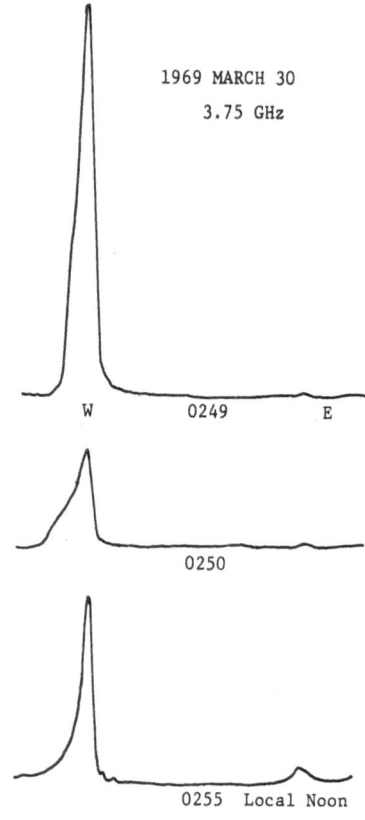

1969 MARCH 30

3.75 GHz

W   0249   E

0250

0255   Local Noon

Fig. 2.   Quick-scan records at 3.75 GHz showing strip scan of the solar disk at 0249, 0250, and 0255 UT. The expansion started at 0249 UT and was remarkable at 0250 UT. The interference pattern, seen at the right-hand side of the source at 0255 UT, means the obscuration of the source by the solar disk.

Following the diffusion model, which has been often applied to the solar cosmic ray propagation through the interplanetary space (Parker, 1963), and with an assumption of spherical symmetry, the diffusion equation is

$$\frac{\partial U}{\partial t} = \frac{D}{\varrho^2} \frac{\partial}{\partial \varrho} \left(\varrho^2 \frac{\partial U}{\partial \varrho} \right) - KU,$$

where U, D, and K stand for the particle density, the diffusion coefficient, and the loss rate of particles respectively.

A solution of this equation is

$$U(\varrho, t) = \frac{N}{4(\pi Dt)^{3/2}} \exp\left(-\frac{\varrho^2}{4Dt} - Kt\right),$$

where N denotes the particle number injected initially.

Taking the x-axis along the scanning direction of the interferometer, we integrate the density $U(\varrho, t)$ over two dimensions perpendicular to the x-axis. And we obtain the one dimensional particle density $U(x, t)$ given by the following relation:

$$U(x, t) = \frac{N}{\sqrt{(\pi Dt)}} \exp\left(-\frac{x^2}{4Dt} - Kt\right). \tag{1}$$

Here we further assume that the radio source was optically thin at 3.75 GHz, which may be assured in this case. Then the particle density $U(x, t)$ is proportional to the strip brightness distribution, which is the observable quantity. Thus, from the Equation (1) we obtain the following equation for the half-power source size $W(t)$:

$$W(t) = (\ln 2 \cdot 4Dt)^{1/2},$$

namely, $W(t)$ is proportional to the square root of time t. This relation is seen in Figure 3 to hold well. From this figure we derived that

$$D = 6.5 \times 10^{17} \text{ cm}^2 \text{ s}^{-1}.$$

Since the solar atmosphere is regarded as collision-free, we have to consider the collision between energetic electrons and magnetic field irregularities to explain the diffusion process.

Motion of a charged particle in a random magnetic field is discussed by Jokipii (1966). Assuming a superposition of a constant magnetic field and homogeneously

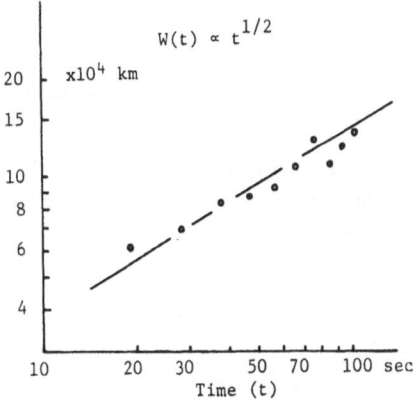

Fig. 3.   The half-power width of the source is plotted against time. The solid line shows a relation that the width is proportional to the square root of time.

turbulent fields of small quantities, he has noted that a particle undergoes a random walk in a plane perpendicular to the constant field.

Here, however, we shall follow a fundamental procedure discussed by Schatzman (1967). Let us consider a magnetic field consisting of a constant field B_0 and a fluctuating part δB, and put the characteristic correlation length of the irregular field L and the transverse velocity of a charged particle V. And if we denote the transverse displacement of the guiding center of a particle over L as δL, it may be given by the following equation (Alfvén and Fälthammar, 1963);

$$\delta L = (\delta B/B_0)\, L.$$

Thus the transverse motion of a charged particle is described as a random walk in a plane perpendicular to the constant field, with a step δL and a correlation time $\tau = L/V$, in a case of isotropic pitch angle distribution.

Then the transverse diffusion coefficient D is given by the following relation (Chandrasekhar, 1943):

$$D = (\delta L)^2/2\tau = LV\,(\delta B/B_0)^2/2.$$

The observed value of D is 6.5×10^{17} cm^2 s^{-1}, and V may be 10^{10} cm s^{-1}, that is 30 keV. Then

$$L\,(\delta B/B_0)^2 = 1.3 \times 10^8 \text{ cm}.$$

Since the correlation length may be much smaller than the scale of the active region:

$$\delta B/B_0 = 0.36 \quad \text{for} \quad L = 10^9 \text{ cm}.$$

The substance of δB may be inferred as finite amplitude MHD waves, which may play an important role in heating the flare region and in acceleration of solar cosmic rays.

References

Alfvén, H. and Fälthammar, C.-G.: 1963, *Cosmical Electrodynamics*, Oxford University Press, London.
Chandrasekhar, S.: 1943, *Rev. Mod. Phys.* **15**, 1.
Jokipii, J. R.: 1966, *Astrophys. J.* **146**, 480.
Parker, E. N.: 1963, *Interplanetary Dynamical Processes*, Interscience Publishers, New York.
Schatzman, E.: 1967, Lecture Notes of City College Lectures given at City College of City University of New York.
Takakura, T. and Scalise, E.: 1970, *Solar Phys.* **11**, 434.
Tanaka, H., Kakinuma, T., Énomé, S., Torii, C., Tsukiji, Y., and Kobayashi, S.: 1969, *Proc. Res. Inst. Atmospherics, Nagoya Univ.* **16**, 113.

ELECTRIC CURRENTS CONNECTED WITH THE PROTON
FLARES OF 7 JULY AND 2 SEPTEMBER, 1966

A. B. SEVERNY

Crimean Astrophysical Observatory, Nauchny, Crimea, U.S.S.R.

Abstract. It is observed that the change of the net magnetic flux associated with flares can exceed
10^{17} Mx/s, which corresponds according to Maxwell's equation to the e.m.f. $\sim 10^9$ V which is specific
for the high energy protons generated in flares. It is shown that this value of e.m.f. can hardly be
compensated by e.m.f. of inductance which should appear due to the actually measured motions in a
flare generating active region. The values of electric field strength thus found, together with measured
values of electric current density (from rot H), leads to an electric conductivity which is 10^3 times
smaller than usually adopted.

Figure 1 shows the behavior of the net magnetic flux

$$\Phi = F_s - F_n$$

connected with the proton flares of 7 July and 2 September, 1966. Values of Φ were
obtained by planimetry of isogauss contour charts, an example of which is seen in
Figure 2, for July 6 (Zvereva and Severny, 1970). The value Φ is measured for one and

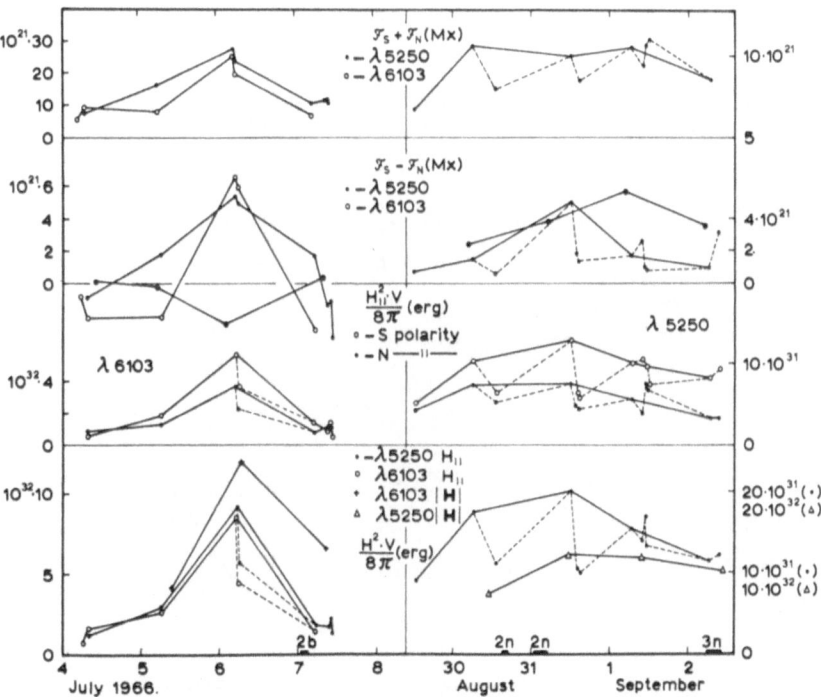

Fig. 1.   The progressive change of the total flux $F_S + F_N$, net flux $F_S - F_N$ and of the energies of
longitudinal H (separately for N-solid dots, and S-open circles polarities) and total $H_\parallel$-fields associated
with the flare of July 7, 1966 and of August 30 and September 2, 1966. Dashed lines in the later case
are actual measurements, solid lines are drawn through the most reliable data (best seeing).

Howard (ed.), Solar Magnetic Fields, 417–421. All Rights Reserved.
Copyright © 1971 by the IAU.

the same fixed geometrical region. The boundary of this region is fixed to some extent arbitrarily just to include the main magnetic features. However an inspection of the whole series of maps shows that this is not important, because the main changes in Φ are connected with changes in position of isogauss lines *inside* this fixed region in

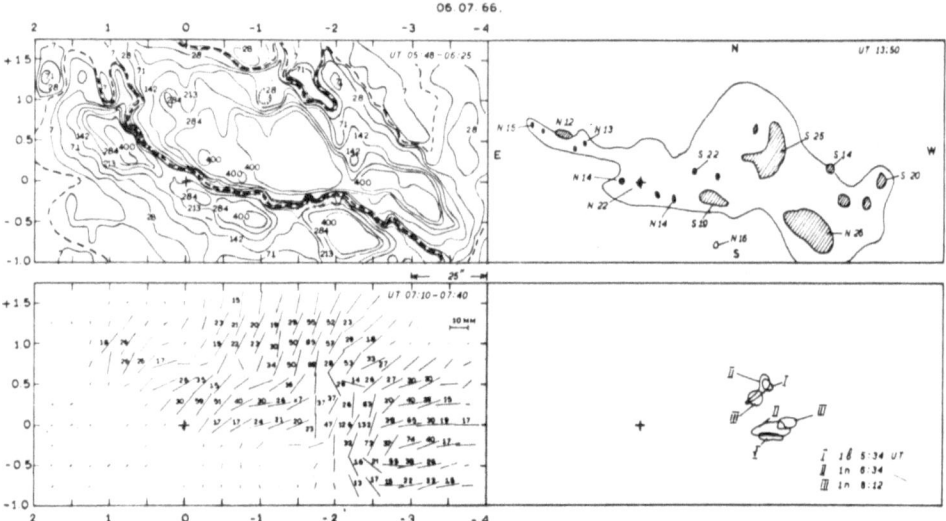

Fig. 2.   An example of maps of longitudinal and transverse fields before the flare of July 7, 1966 (left), and the drawing of sunspots and flares (right) for the same area.

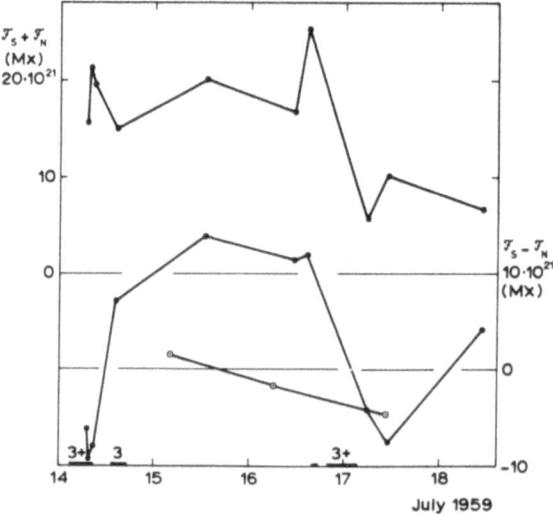

Fig. 3.   The same as in Figure 1 but for the flare of July 16, 1959 (solid dots). Open circles are the values of the net flux from routine measurements of maximum fields in the umbrae of sunspots and their areas; the main contribution to the flux over the area is due to fields surrounding visible sunspots.

the range of *strong* fields ($\geqslant 300$–500 G), while the contribution from the periphery is negligible, not exceeding 10–15%.

In both cases we have a bias of flux of one polarity before the flare (S-polarity) which is characteristic of many other flares. A similar behavior of Φ shows also in the well known flare of 16 July 1959 as may be seen in Figure 3. A similar phenomenon for about 10 less important flares was found by Martres *et al.* (1968). For the flares under consideration we have Table I.

TABLE I

No.	Flares	Import	Φ(before)	Φ(after)	$\dfrac{\Delta\Phi}{\Delta t}$	Δt
1	17 July 59	3+	$+12 \times 10^{21}$ Mx	$-8\ \times 10^{21}$ Mx	21.75×10^{17} Mx/s	$0.^{d}8$
2	7 July 66	2+	-2.0×10^{21} Mx	$+7.0 \times 10^{21}$ Mx	$1.1\ \times 10^{17}$ Mx/s	$1.^{d}0$
3	2 Sept. 66	2+	$+0.4 \times 10^{21}$ Mx	$+4.3 \times 10^{21}$ Mx	0.33×10^{17} Mx/s	$1.^{d}25$
4	22 July 62	2	-1.0×10^{21} Mx	-0.2×10^{21} Mx	$1.1\ \times 10^{17}$ Mx/s	$1.^{h}7$

Here we have added also the flares of 22 June 1962 and 17 July 1959 for which a set of maps before, during and after the flares were available (Howard and Severny, 1963; Severny, 1963). The values $\Delta\Phi/\Delta t$ in this table are in general agreement (a little higher) with (Martres *et al.*, 1968). There is also a tendency for $\Delta\Phi/\Delta t$ to increase with the increase of importance of the flare.

It seems reasonable to adopt

$$\frac{\Delta\Phi}{\Delta t} = \begin{cases} \leqslant 0.1 - 0.3 \times 10^{17} \text{ Mx/s} & \text{for small flares} \\ \sim 1.0 \times 10^{17} & \text{Mx/s} \quad \text{for important (proton) flares} \end{cases} \tag{1a}$$

According to Maxwell's equation we have for the e.m.f.:

$$\oint \mathbf{E}_{ds} = -\frac{1}{c}\frac{\Delta\Phi}{\Delta t} = \begin{cases} (1\text{–}3) \times 10^8 \text{ V} & \text{small flares} \\ 1.0 \times 10^9 \text{ V} & \text{important flares} \end{cases} \tag{1b}$$

The following values for the electric field strength result (at $L=10^{10}$ cm for the length of a contour)

$$E = \begin{cases} 3.3 \times 10^{-4} & \text{small f.} \\ 3.3 \times 10^{-5} & \text{large f.} \end{cases} \text{(CGSE)} \tag{2}$$

This value corresponds closely to the energy of protons generated by flares, but this refers to a fixed contour, and the question arises whether this e.m.f. can really appear. It can, if the e.m.f. of inductance

$$\frac{1}{c}\mathbf{V} \times \mathbf{H}$$

can be disregarded as compared with the value $\mathbf{E}$ in the expression

$$\mathbf{E}' = \mathbf{E} + \frac{1}{c}\mathbf{V} \times \mathbf{H}$$

determining the e.m.f. through the moving contour, or if motions are too small. To evaluate the e.m.f., $1/c(\mathbf{V} \times \mathbf{H})$, we have measured the motions in three different ways: (a) from the contraction of the region of S-polarity on isogauss-contours on the charts, (b) by the increase of the sizes of penumbra and umbra, (c) from the distances between the sunspots inside the group considered, (d) from instantaneous simultaneous measurements of line-of-sight velocities with the magnetograph when a group is not far from the limb.

The results are summarized in Table II.

TABLE II

Maximum velocities (cm/s)

Method	Region N	Region S
a	0.85×10^4	1.05×10^4
b S penumbra	0.95	1.30
S umbra	1.20	0.80
c	0.50	...
d	$\leqslant 1.50$	$\leqslant 1.50$

So we have $v \leqslant 100-150$ m/s and

$$\frac{1}{c} \mathbf{V} \times \mathbf{H} \leqslant 10^{-4} \left(= \frac{10^4 \times 3 \times 10^2}{3 \times 10^{10}} \right)$$

taking the field at the periphery of the region in question $= 300$ G (actually it is a little less). The ratio:

$$\frac{1}{c} vH/E = \begin{cases} < 1.0 & \text{for small flares} \\ < 0.1 & \text{for large flares} \end{cases}. \tag{3}$$

If we take the isogauss line 10^3 as the periphery, we should decrease L by 5–10 times; that means a 5–10 fold increase in E, and the ratio (3) becomes even smaller.

By the way, from the direct measurement of the field it follows that

$$|\text{rot } \mathbf{H}| \leqslant 1 \text{ G/km} = 10^{-5} \text{ G/cm}$$

and the current density is $\mathbf{j} = c/4\pi \, |\text{rot } \mathbf{H}| \leqslant (3 \times 10^{10}/1.26 \times 10) \, 10^{-5} = 2.4 \times 10^4 \text{ CGSE}$.
Hence from the equation

$$\mathbf{j} = \sigma \left(\mathbf{E} + \frac{1}{c} \mathbf{V} \times \mathbf{H} \right)$$

taking $E = 3 \times 10^{-5}$ we obtain

$$\sigma = \frac{\gamma}{E} = \frac{2.4 \times 10^4}{3 \times 10^{-5}} \simeq 10^9 \text{ CGSE}$$

which is 10^3 times smaller than usually adopted, (10^{12}) (cf. Kopecky and Kuklin, 1966). The tendency for bright plages and flares to appear on the line $V_{\parallel} = 0$ (Severny, 1960;

Gopasyuk and Tsap, 1969) irrespective of whether they occur near the center or the edge of the disk, is also in favor of the considerations we have presented here. If further observations support our results we may consider the flare phenomenon as connected with strong electric currents and the corresponding mechanism can be the one presented by Alfvén and Carlquist in their theory of flares (Alfvén and Carlquist, 1967).

References

Alfvén, H. and Carlqvist, P.: 1967, *Solar Phys.* **1**, 220.
Gopasyuk, S. and Tsap, T.: 1969, *Astron. Zh.* **46**, 923.
Howard, R. and Severny, A.: 1963, *Astrophys. J.* **137**, 1242.
Kopecky, M. and Kuklin, G.: 1966, *Bull. Astron. Inst. Czech.* **17**, 45.
Martres, M., Michard, R., Soru-Iscovici, I., and Tsap, T.: 1968, *Solar Phys.* **5**, 187.
Severny, A.: 1960, *Izv. Krymsk. Astrofiz. Obs.* **24**, 281.
Severny, A. 1963, *Izv. Krymsk. Astrofiz. Obs.* **30**, 161.
Zvereva, A. and Severny, A.: 1970, *Izv. Krymsk. Astrofiz. Obs.*, in press.

Discussion

Sweet: In deriving the induction field it is necessary to observe the actual fluid velocity to an accuracy sufficient to compare it conclusively with the magnetic flux changes. It is not clear that you have attained this accuracy.

Severny: We estimated velocities, as I said in my talk, by four different methods – three of them are related essentially to the horizontal motions and one to the line of sight motions. All these bring us to the consistent result and values about 100 m/s. The same value comes out of measurements of line of sight velocities when the active area is near the border of the disk and when the sight-line velocity is essentially the velocity of horizontal motion. We think this estimate 100 m/s is sufficiently reliable.

Maltby: Regarding the time scale involved in completing one magnetogram have you considered measuring a smaller area in order to look for short time scale variations?

Severny: The characteristic time-scale was determined by the interval between successive magnetograms which is 40^m. So we were concerned with the mean values for approximately this time-interval.

Harvey, J.: In order to compute the electric current from vector magnetograms, it is necessary to resolve the 180° ambiguity inherent in transverse magnetic field measurements. What technique do you use to resolve this ambiguity?

Severny: This ambiguity can be avoided from the careful consideration how lines of force can go and how they cannot at a given position, and the strength of the 'sources' of lines of force – sunspots and magnetic hills. In most cases the discrimination is unambiguous.

Wiehr: As I pointed out yesterday, the flux disbalances are due to variations of the line profile from photosphere to spot, and they will disappear when using an unchanged line (e.g. Fe 6303 instead of Fe 5250). I wonder, therefore, whether the net flux and moreover its changes have real physical meaning.

Severny: (1) Usually the main contribution to the flux comes not from the spots themselves but from the much bigger area around; moreover the contribution from umbrae can and must usually be disregarded due to their much smaller area and due to saturation effects inside sunspots. You should just do yourself the planimetry of isogauss magnetograph maps to realize this and compute the contribution of sunspots from the usual routine data.

OBSERVATIONS OF MAGNETIC FIELD CHANGES
IN ACTIVE REGIONS

K. L. HARVEY

Lockheed Solar Observatory, Saugus, Calif., U S.A.

and

W. C. LIVINGSTON, J. W. HARVEY, and C. D. SLAUGHTER

Kitt Peak National Observatory*, Tucson, Arizona, U.S.A.

Abstract. A time sequence of longitudinal magnetograms of two active regions, McMath Regions 9281 and 9760, have indicated magnetic field changes occurring in localized areas with time scales of the order of hours. We believe the observed field changes are evolutionary in nature, rather than related to the occurrence of small flares. Three examples of evolutionary magnetic changes are discussed.

During the past two years, Kitt Peak National Observatory and the Lockheed Solar Observatory have participated in a cooperative program to observe simultaneously the magnetic fields and Hα activity in four active regions. The main purpose of this program was to attempt to observe magnetic field changes using high-time resolution magnetograms. This paper is a brief report of the analysis of the observations of two regions, McMath Regions 9281 and 9760.

At Kitt Peak, a time sequence of longitudinal magnetograms was made using a Babcock-type magnetograph (Livingston, 1968) at a rate of one magnetogram per 80 and 120 s for McMath Regions 9281 and 9760, respectively. Spatial resolution of the magnetic observations was no better than 2.5 arc s (size of the entrance aperture). During the study of these two regions, approximately 800 magnetograms were made over a total of 20 hr observing time. For most of the duration of the magnetic observations at Kitt Peak, simultaneous center Hα or off-band Hα filtergrams were obtained by the Lockheed Solar Observatory.

The longitudinal magnetograms of McMath Regions 9281 and 9760 have shown field changes occurring in localized areas having a time scale of the order of hours; Ribes (1969) has found similar time scales for field changes in active regions. We could infer no relation between the field changes and the 55 flares observed in Region 9760, though no flare of importance greater than 1 was observed; this confirms the results of an earlier study by Godovnikov *et al.* (1964). Most of the observed magnetic field changes occurred outside the areas which flared. Although no flares were observed in Mc Math Region 9281, a filament eruption occurred during the course of an underlying field change.

Our observations of magnetic field changes suggest that the field changes we

---

* Operated by the Association of Universities for Research in Astronomy, Inc., under contract with the National Science Foundation.

Howard (ed.), Solar Magnetic Fields, 422–427. All Rights Reserved.
Copyright © 1971 by the IAU.

observed are evolutionary in nature, rather than being associated with Hα flare activity. Three examples of this behavior are as follows:

(1) Observations of McMath Region 9281 began on 21 March 1968 approximately 13 hr after the birth of the region and continued through 24 March. As shown in Figure 1, the magnetic polarities were initially aligned north-south; by 22 March the region had developed to a more nearly east-west orientation, which it retained. Weart and Zirin (1969) and Weart (1970) have also noted some new regions emerging with high inclinations and developing toward an east-west orientation.

The change in orientation of Region 9281 appears to be due to the emergence of new flux within the boundaries of the region, rather than a horizontal movement of

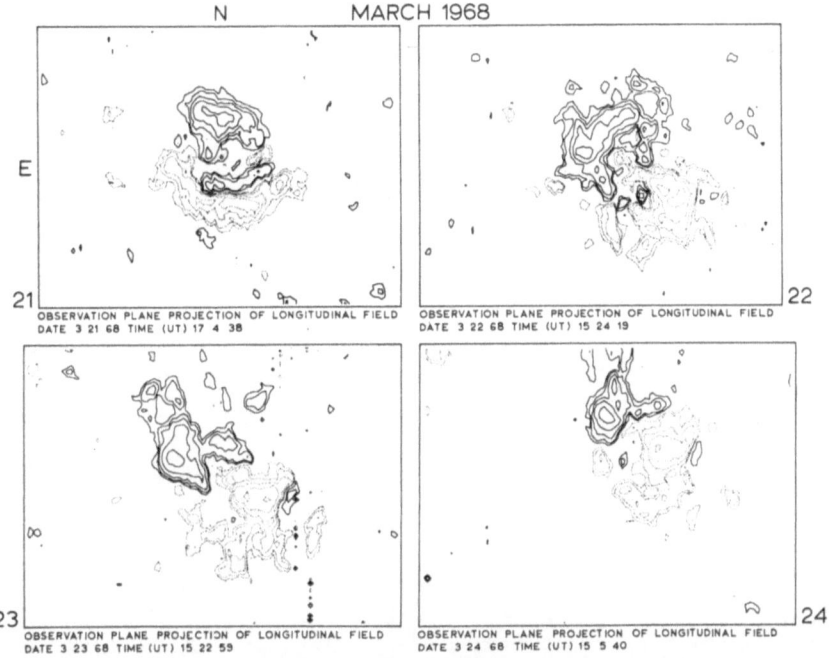

Fig. 1.   Isogauss contours of the longitudinal magnetic field of McMath Region 9281 for 21 through 24 March 1968. The contour levels are 100, 200, 400, 800, and 1600 G.

the established flux. On 21 March 1968, as shown in Figure 2, two significant field changes, occurring over a period of several hours, were observed: (1) a strengthening of the positive field in an area of negative polarity along the eastern border of the region, and (2) a strengthening of negative fields along the western border of the region. In both cases, the field changes preceded and accompanied the growth of sunspots, which developed into the principal spots of the region on 22 March 1968, and thus established the more east-west orientation of the region. The birth of sunspots in areas of increasing field strength has been observed by Gopasyuk (1967) and Ogir and Shaposhnikova (1965).

(2) Our observations suggest that McMath Region 9281 emerged with a definite non-potential field configuration and evolved to a more potential-like configuration. In Figure 3, we have compared the orientation of the Hα fine structure with the computed transverse component of a potential field configuration calculated using the observed longitudinal field distribution. The Hα fine structure is assumed to be aligned along field lines as indicated by the work of Bruzek (1969) and Tsap (1965).

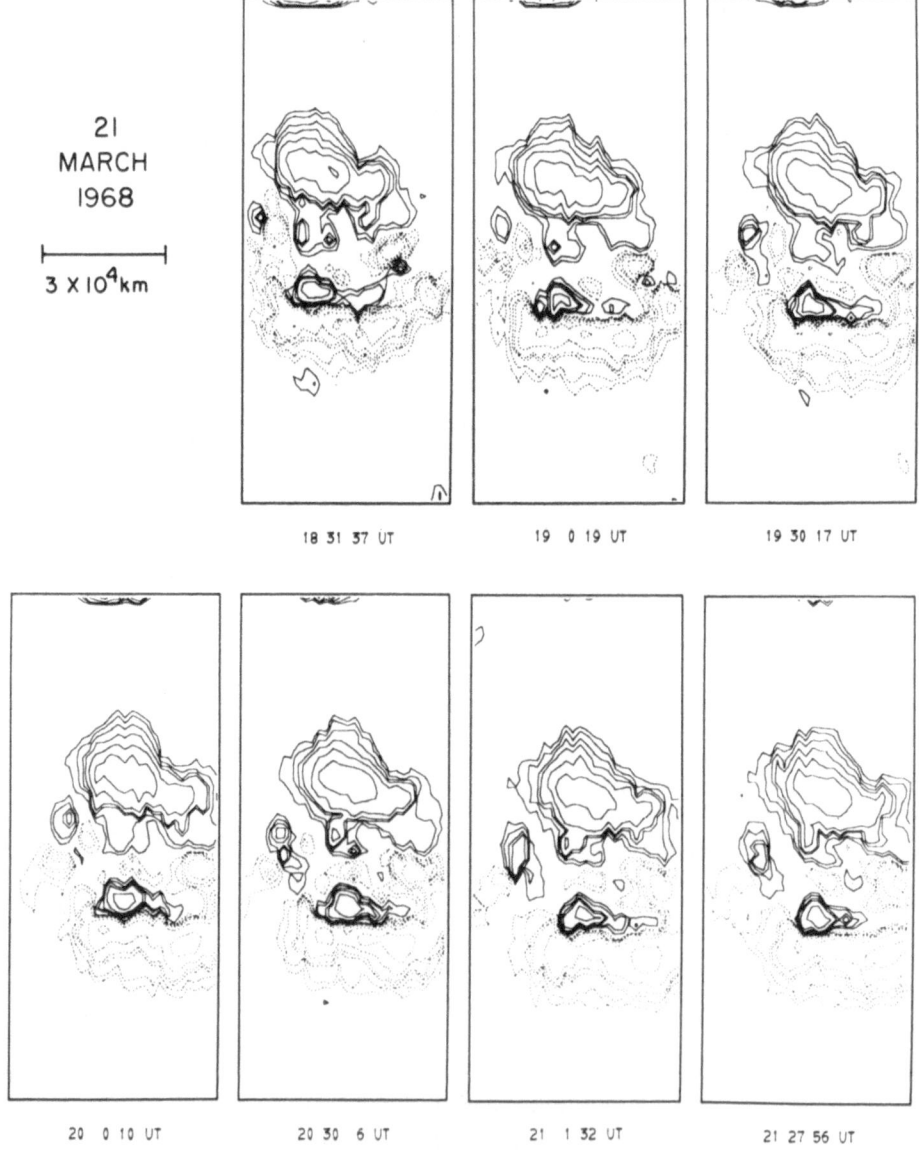

Fig. 2. Isogauss contours of the longitudinal magnetic field shown at roughly one-half hour intervals on 21 March 1968 of Mc Math Region 9281.

On 21 March, the Hα inter-region fine structure showed poor correspondence with the calculated transverse field. The alignment improves by 22 March and is quite good on 23 March 1968 (not shown in Figure 3) when the region has reached its maximum development.

(3) Our observations of McMath Region 9760 indicate a change in the direction of

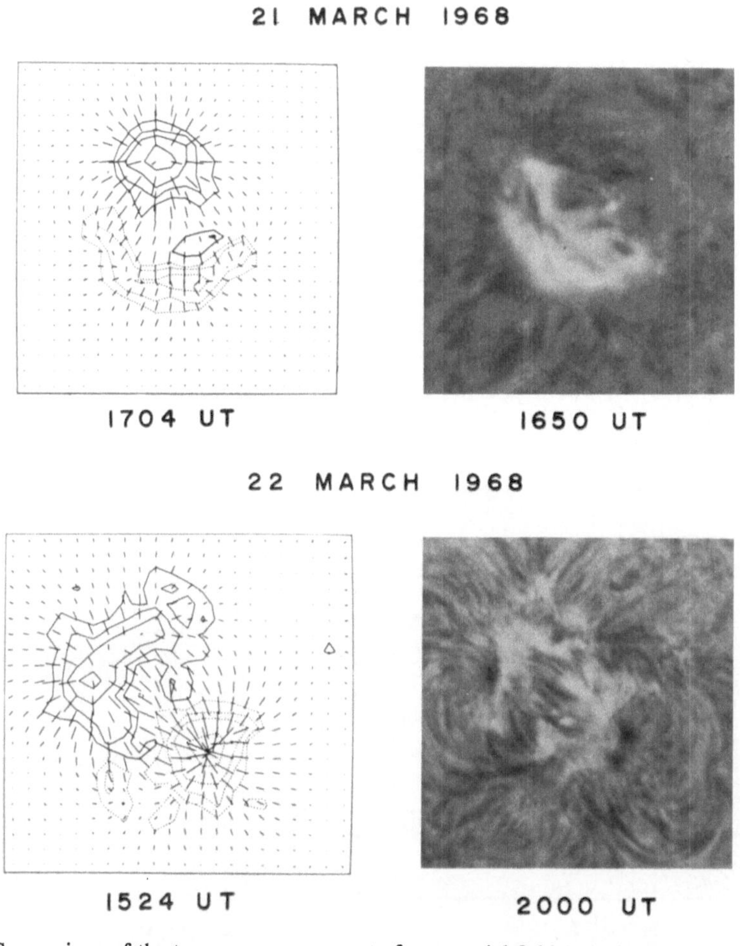

21   MARCH   1968

1704 UT   1650 UT

22   MARCH   1968

1524 UT   2000 UT

Fig. 3.   Comparison of the transverse component of a potential field (calculated from the observed longitudinal field distribution) with Hα fibril structures in Region 9281.

the transverse field from 8 to 11 November 1968 in the vicinity of the longitudinal neutral line, the line of polarity reversal. The observed transverse field is shown in Figure 4 for 8, 9, 10 and 11 November. The orientation of the lines indicates the direction of the transverse field and the dark line represents the position of the polarity reversal line with respect the transverse field. On November 8, the direction of the transverse field is inclined to the neutral line. On successive days, the inclination decreased and by November 11, the transverse field is parallel with the neutral line. Similar alignments of the transverse field with the $H_{\parallel}=0$ line have been noted by Moreton and Severny (1968) and Rayrole and Semel (1970). The Hα fibril structures in the area of the neutral line also showed a similar behavior.

Because of the many problems in observing and interpreting measures of the transverse field, it is difficult to be certain that the observed change in the transverse field direction is entirely real.

In addition to the longitudinal magnetic field observations discussed above, a time

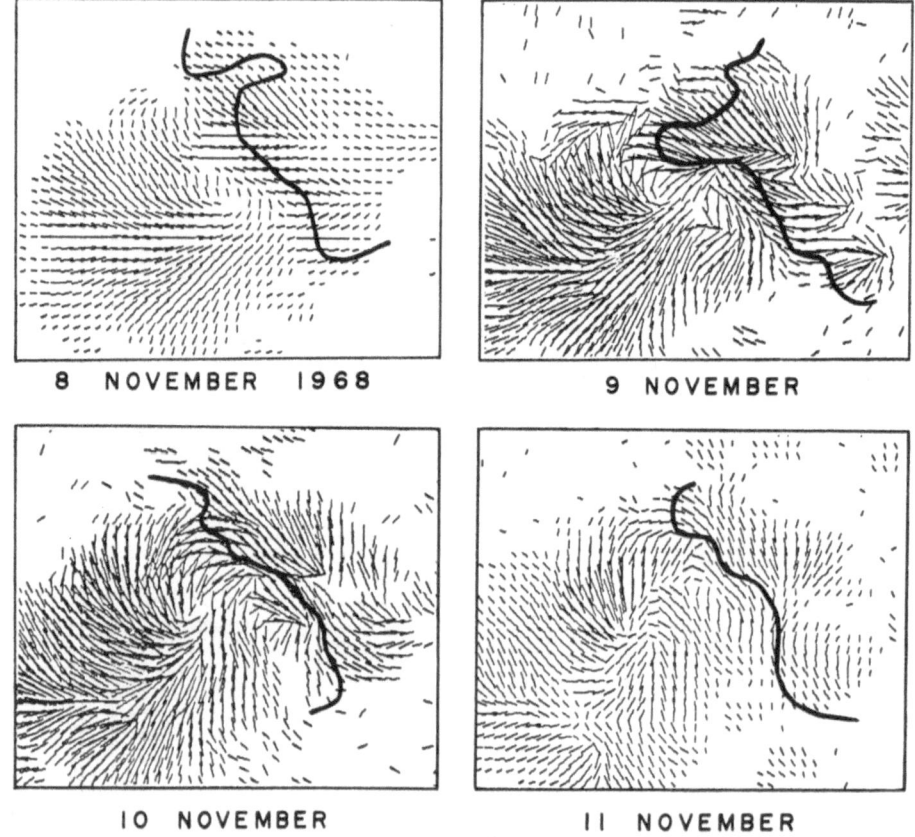

Fig. 4.   Position of the polarity reversal line (dark line) relative to the observed transverse field.

sequence (12 magnetograms per hour) of (1) the transverse and longitudinal magnetic fields of McMath Region 10148 and (2) the longitudinal magnetic fields measured simultaneously in Hα and Fe I 6569 Å of McMath Region 10385 have been made.

The results of these observations and, separately, a more detailed report of the longitudinal magnetic field observations will be published in the future.

References

Bruzek, A.: 1969, *Solar Phys.* **8**, 29.
Godovnikov, N. V., Ogir, M. B., and Shaposhnikova, E. F.: 1964, *Izv. Krymsk. Astrofiz. Obs.* **31**, 216.
Gopasyuk, S. I.: 1967, *Izv. Krymsk. Astrofiz. Obs.* **36**, 56.
Livingston, W. C.: 1968, *Appl. Opt.* **7**, 425.
Moreton, G. E. and Severny, A.: 1968, *Solar Phys.* **3**, 282.
Ogir, M. B. and Shaposhnikova, E. F.: 1965, *Izv. Krymsk. Astrofiz. Obs.* **33**, 92.
Rayrole, J. and Semel, M.: 1970, *Astron. Astrophys.* **6**, 288.
Ribes, E.: 1969, *Astron. Astrophys.* **2**, 316.
Severny, A.: 1969, in *Annals of the ISQY*, MIT Press, Cambridge, Mass., **3**, 11.
Tsap, T. T.: 1965, *Izv. Krymsk. Astrofiz. Obs.* **33**, 92.
Weart, S. R.: 1970, *Astrophys. J.* **162**, 987.
Weart, S. R. and Zirin, H.: 1969, *Publ. Astron. Soc. Pacific* **81**, 270.

Discussion

Michard: The kind of changes of magnetic patterns that you showed, with one pole increasing and another (of opposite polarity) decreasing, are consistent with the description by the Meudon observers (Martres *et al.*). But do you agree with their finding that Hα brightenings in flares are located on these features evolving in opposing directions?

Harvey, K.: No, not for the regions we observed.

Meyer: You mentioned that the evolution of the active region considered in your first slides shows a tendency to approach a potential field configuration. Does this also hold for the increasing alignment of the tangential field component with the neutral line of the vertical flux distribution shown in your last slide?

Harvey, K.: No.

THE MAGNETIC FIELDS AND THE POLARIZATION OF RADIO
EMISSION IN THE ACTIVE CENTER OF OCTOBER 1968

N. ERUSHEV, A. B. SEVERNY, and T. TSAP

Crimean Astrophysical Observatory, Nauchny, Crimea, U.S.S.R.

During the period of strong flare activity of the active center McMath No. 9740, from October 25 to November 2, 1968, 32 magnetograph records of longitudinal $H_\parallel$ and 20 of transverse fields, $H_\perp$, were obtained for this center with the Crimean magnetograph. Sometimes 5–7 daily records were made. Hα and K spectroheliograms, Hα cine film and routine determinations of sunspot magnetic fields were also obtained frequently. However the most important point is that all these optical observations were accompanied by simultaneous and continuous measurements of intensity and polarization of radio emission from this active center at a wavelength of 3 cm, recorded with the aid of the big 22-m radiotelescope of the Crimean Observatory (the resolution is 8′ for this wavelength), (Erushev and Zvetkov, 1970).

The present communication contains some preliminary results about the behavior of the magnetic fields and polarization of radio emission associated with flares. Figure 1 is an example of magnetic charts before and after the flare of importance 2n, October 27, 1968 (14:30–16:30, Moscow time) showing very strong changes with time in the configuration of the magnetic fields. We may see in Figure 1 the disappearance of an S-polarity 'neck' (near A) connecting two magnetic hills of the same (S) polarity B and C, so that the hill, A, of S-polarity disappeared completely. Also very strong changes are connected with the flare of importance 2b, 1 November 1968, (11:01–12:03, Moscow time) as can be seen in Figure 2. We see here an enormous simplification of the magnetic structure and a decrease of gradients after the flare.

In both cases flares appear very close to or on the neutral line $H_\parallel = 0$ in the regions of the strongest gradients of the longitudinal field $H_\parallel$, and also near the top of the big magnetic hill of N-polarity which underwent the most rapid changes in time (cf. the similar cases in Severny and Zvereva, 1970).

From planimetry of magnetic charts of the kind we have just shown we found the total magnetic flux $F_N + F_S$ and the net flux $F_N - F_S$, correcting the corresponding area for the effects of foreshortening. Figure 3 shows the variations with time of the total flux (crosses) as compared with the plot of polarization of radio emission (dots) at 3 cm

$$p = \frac{I_\circlearrowleft - I_\circlearrowright}{I_\circlearrowleft + I_\circlearrowright}$$

(at the bottom). We also plot the net flux $F_N - F_S$ in comparison with the polarization $I_\circlearrowleft - I_\circlearrowright$ (arbitrary units, at the top). On the horizontal axis flares of importance > 1 are plotted.

Howard (ed.), Solar Magnetic Fields, 428–431. All Rights Reserved.
Copyright © 1971 by the IAU.

We observe a kind of synchronism in the behavior of magnetic fluxes and polarization of radio emission. Especially clearly pronounced is the correspondence in the decrease of fluxes $F_N + F_S$ and the decrease of polarization associated with each flare, (27, 29 October and 1 November flares). Sometimes, for instance in the cases of 28, 29 October, even small separate fluctuations in the polarization of radio emission can be traced in corresponding changes of magnetic fluxes when magnetic measurements

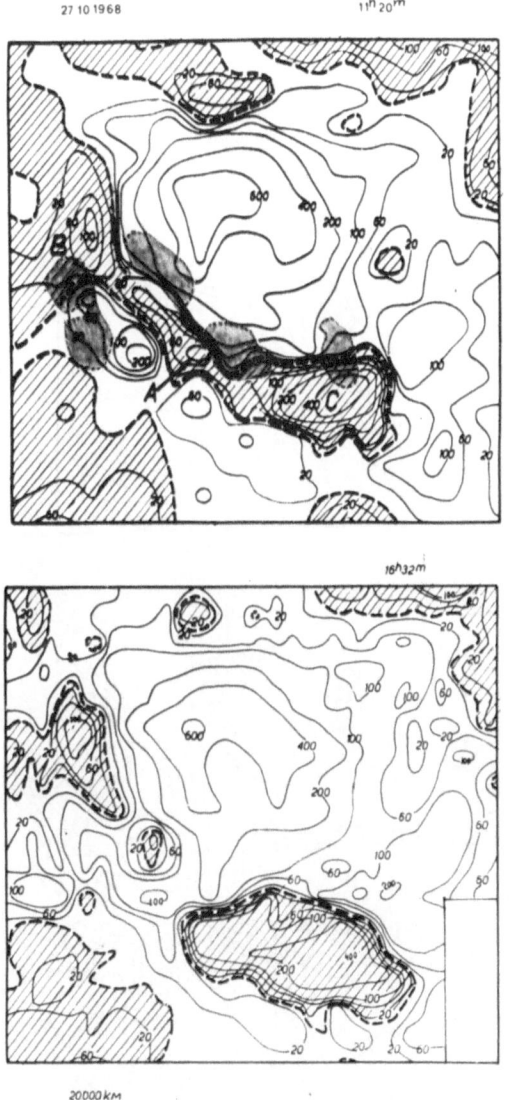

Fig. 1. Showing the position of the flare 27 Oct. 1968 on magnetic map before (at the top) the flare, and the change in the pattern of magnetic field when comparing it before and after (at the bottom) the flare.

are frequent (the resolution in time of magnetic data is 40^m in the best case). Besides this, we observe also that the general day-to-day variations are roughly the same, namely gradual increase of radio and magnetic parameters to the 29th of October and then decrease. Although the magnetic data are rather scarce, the correspondence of both magnetic and radio data can hardly be accidental.

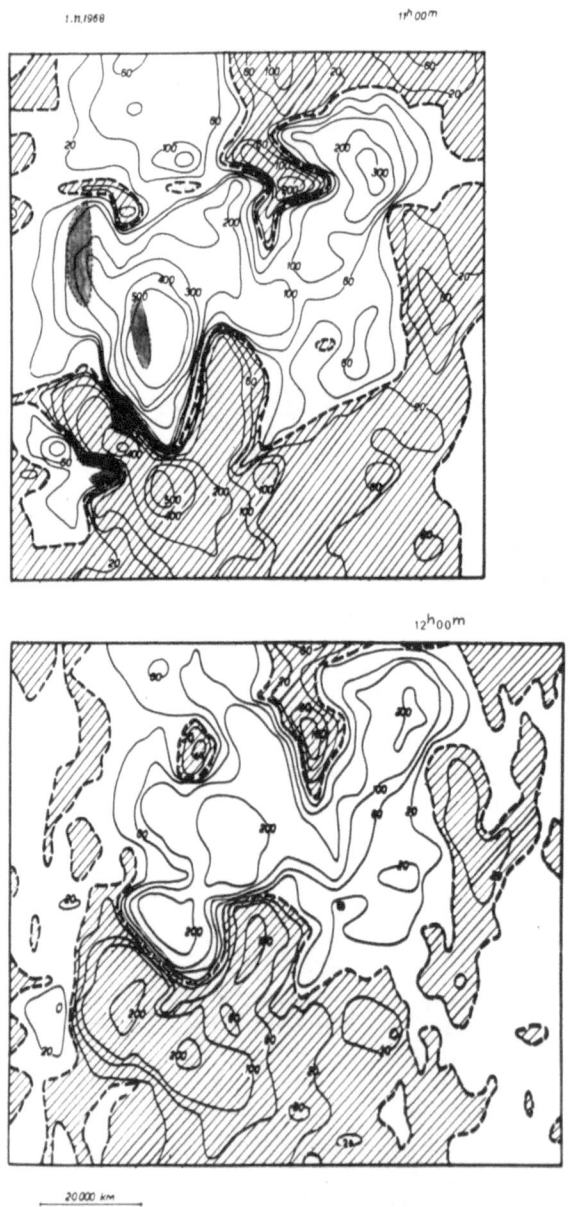

Fig. 2.   The same as on Figure 1 for the flare 1 November 1968.

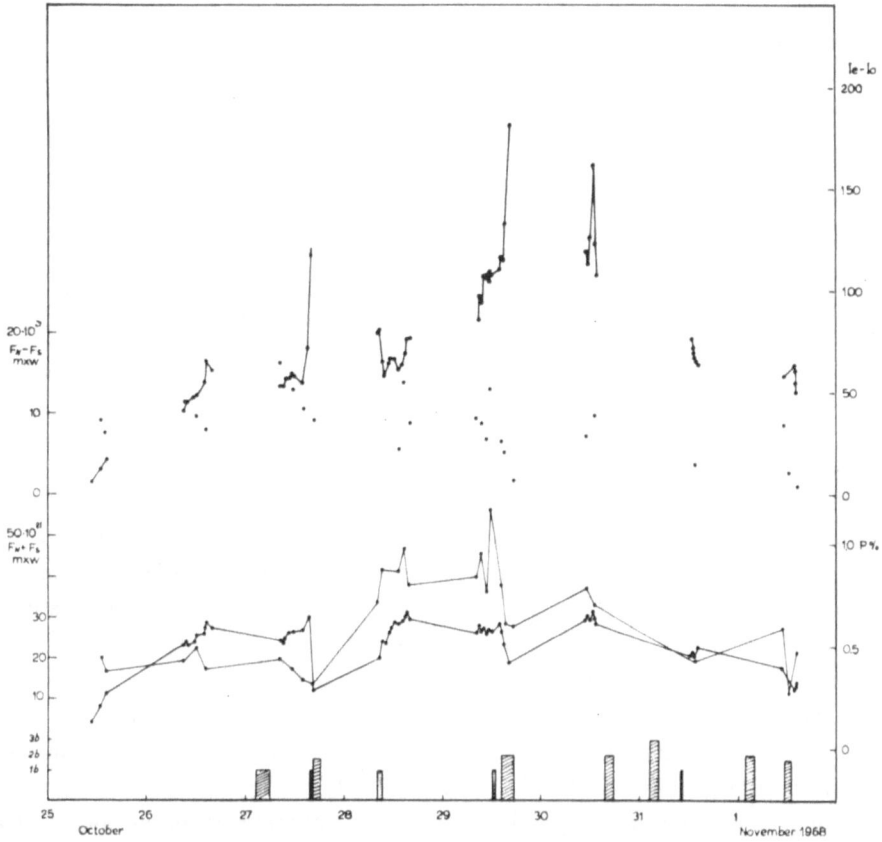

Fig. 3. The change of the percentage of polarization P of radio emission at $\lambda = 3$ cm (solid dots) compared with the total magnetic flux $F_S + F_N$ (open circles, below). At the top the comparison of the polarization $I_e - I_o$ (arbitrary units) of radio emission with the net magnetic flux $F_N - F_S$.

Similarly, as in an earlier paper by Martres *et al.* (1968), the main change in magnetic fields is the decrease of one polarity flux. In the case considered here the other polarity flux remains practically the same, and here this was the flux of N-polarity of the leading spot that showed pronounced variations with time (the group was in the Southern hemisphere).

References

Erushev, N. and Zvetkov, L.: 1970, *Izv. Krymsk. Astrofiz. Obs.*, in press.
Martres, M., Michard, R., Soru-Iscovoco, I., and Tsap, T.: *Solar Phys.* **5**, 187.
Severny, A. and Zvereva, A.: 1970, *Izv. Krymsk. Astrofiz. Obs.* **41–42**, in press.

THE POSITION REGULARITIES OF FLARES RELATED TO THE FIELD MAXIMUM IN SUNSPOT GROUPS

V. V. KASINSKY

Siberian Institute of Terrestrial Magnetism, Ionosphere and Radio Propagation, Irkutsk, U.S.S.R.

Abstract. In this paper common regularities in flare positions have been studied in relation to spots as the extreme points of the magnetic field. The angular distribution diagrams of flares reveal that equatorial-western shifts of flares occurred during the I.G.Y. period (1957-58) in both hemispheres. This shift with some peculiarities is confirmed in the material of the 19th solar cycle, using the *Solar Data* bulletins, U.S.S.R., 1955-66. One result is that the western component of the shift does not change its sign and increases steadily during the cycle, while the equatorial component reverses sometimes and decreases steadily in amplitude to the end of the cycle.

A flare is associated with neutral points (or lines) of magnetic field. On the other hand an active region is a region of enhanced magnetic field. Sunspots are regions where a magnetic field has its maximum strength. The position of a spot unlike a neutral line is always fixed strictly, therefore in this paper common regularities in flare positions referring to spots as the extreme points of the field – center of the spots – have been studied.

Mean distance spot – flare is quite short (Waldmeier, 1938) therefore a principal piece of information about flare positions is the angular distribution of them related to the magnetic poles. The material used for this study was the synoptic maps of solar activity containing sketches of flares of importance 1+ and higher at moments close to the maximum brightness in Hα (*I.G.Y. Maps*, 1961).

The angular distribution diagrams are drawn by counting the occurrence of flares in a discrete number of sectors arranged around the spot and then a summation over all spots. Flares of the northern and southern hemispheres are analyzed separately. Analysis of the diagrams leads to the conclusion that flares in the northern hemisphere seem to show a concentration southward while the southern ones are concentrated northward in relation to their centers (spots). It may be said that there is an equatorial shift of flares compared to spots in both hemispheres. It seems that this asymmetry is conserved for the strongest flares of importance 2+ and 3 as well as for the leader and the follower spots in bipolar groups. (See Figure 1 and Figure 2.)

Analysis of the diagrams reveals that a western shift of flares is also present as it is noticed for some diagrams. To summarize over all spots, one can say that there is an equatorial – western shifting tendency of the position of flares during 1957-58.

The measure of positional asymmetry can be estimated by the formulae:

$$P = \frac{n_N - n_S}{\sqrt{2(n_N + n_S)}} \qquad Q = \frac{n_E - n_W}{\sqrt{2(n_E + n_W)}},$$

where $n_i (i = N, S, E, W)$ in the number of flares moving in each direction, the quantities $\sqrt{2(n_N + n_S)}$ and $\sqrt{2(n_E + n_W)}$ give the random excess in one direction when the

Howard (ed.), Solar Magnetic Fields, 432–434. *All Rights Reserved.*

normal law distribution for excesses is assumed. If $P>1$, $Q>1$ the excess may be qualified as nonrandom and this is fulfilled for most cases.

These results are verified with the material of solar cycle 19 with about 12000 flares using the *Solar Data* bulletin (1954-1966, Nauka, U.S.S.R.). The relative positions are found by means of coordinates for flares and sunspots with an accuracy of $\pm 2°$.

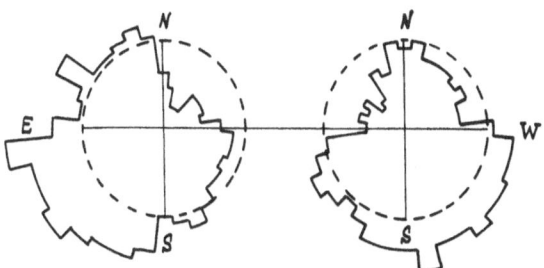

Fig. 1.   The angular distribution of flares relative to the leader and follower spots in the northern hemisphere. Broken circle represents equable distribution.

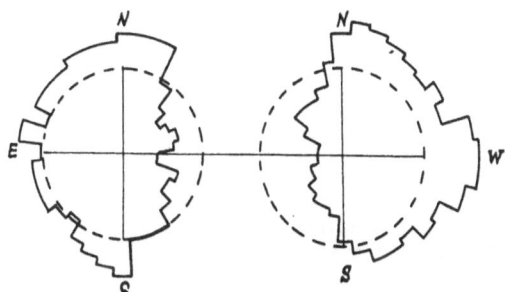

Fig. 2.   The angular distribution of flares relative to the leader and follower spots in southern hemisphere.

The values of the asymmetries $P(t)$ and $Q(t)$ averaged over each month are plotted as a function of the time. $P(t)$ and $Q(t)$ show in such a run that on the average during the cycle the equatorial-western shifting tendency of flare positions seems to be confirmed, although sometimes an equatorial asymmetry changes its sign to the opposite (to the pole). This change occurred in the beginning of the 19th cycle in the southern hemisphere and in the beginning of the 20th cycle in the northern one.

It is significant to note that the equatorial shift reverses sometimes. Plots of functions P and Q also revealed the 'impulsive' character of the asymmetries especially noticeable when the numerical differentiation of $P(t)$, $Q(t)$ with respect to time is taken. A half-width of impulses is of the order of 0.5–1 year. On the other hand, by numerical integration of the asymmetry values one can show that the cumulative quantities $\sum P(t)$ and $\sum Q(t)$ grow significantly toward the end of the cycle. Cumulative asymmetries are reproduced in Table I where in the second column each hemisphere probability, assuming a normal law of distribution for excesses of flares, is given. It can be seen from the table that probabilities are negligible, that is, the asymmetry of flare positions is nonrandom in character.

TABLE I

ΣP	-12.8	10^{-66}	$+2.1$	$3 \cdot 10^{-2}$
ΣQ	-8.9	10^{-34}	-8.0	10^{-27}

Northern hemisphere	Southern hemisphere

Another interesting peculiarity of the flare positions follows from vector-diagrams if the vector shift of the center distribution of flares is plotted against time. For this mean year vector diagram it can be seen that the equatorial shift component increases to the maximum and decreases to the end of the cycle, while the western shift component increases steadily from the beginning to the end of the cycle.

The peculiarities in flare positions are difficult to interpret in terms of the well-known magnetic field behavior. Perhaps it is connected with the geometry and development of large scale magnetic fields. But independent of the hypothetical reasons, the flare positions compared to field maximum points yield no less information than their positions relative to the zero points of the magnetic field.

References

I.G.Y. *Solar Activity Maps:* 1961, DI, vxx, Ann., I.G.Y. Pergamon Press.
Solar Data bulletin, Nauka, U.S.S.R. 1954–1966.
Waldmeier, M.: 1938, *Z. Astrophys.* **10**, 276.

Discussion

Meyer: If the occurrence of flares is spatially correlated with the 'neutral-line' in active regions, one also expects an asymmetry of the kind you discussed, due to the systematic difference in latitudinal position and concentration between leading and following polarities in sunspot groups. Does the observed systematic distribution of flares that you find correspond to such an effect or is it at variance with it?

Kuklin: If the flares are concentrated near the zero-line between leading and following sunspots then the distribution asymmetry will be opposite for these two centers. However Figures 1 and 2 show the similarity of the distribution asymmetries as for the leading sunspot so for the following one in both hemispheres. Therefore the mentioned agreement is absent.

AN ATTEMPT TO ASSOCIATE OBSERVED
PHOTOSPHERIC MOTIONS WITH THE MAGNETIC FIELD
STRUCTURE AND FLARE OCCURRENCE IN
AN ACTIVE REGION

M. J. MARTRES, I. SORU-ESCAUT, and J. RAYROLE

Observatoire de Paris-Meudon, France

Abstract. We have tried to find empirical evidence for the role of photospheric motions in the building up of the flare productive magnetic patterns in Active Regions.

The bright Hα faculae are associated with $V_{\parallel}$ structures different from a classical Evershed flow and particularly 'anomalous' in the regions and periods of high flare occurrence. The flares observed occurred at 'crossings' of the lines $V_{\parallel}=0 (V \neq 0)$ and $H_{\parallel}=0$ and at places where $V_{\parallel}=0$ showed abrupt changes of direction. It is suggested that these anomalous $V_{\parallel}$ structures are evidence of vortex motions.

1. Introduction

The location and occurrence conditions of Hα flares in an eruptive Active Region (A.R.) have been roughly defined in relation with the magnetic field pattern:

– existence of an inversion line $(H_{\parallel}=0)$ with strong gradient of $H_{\parallel}$ and often anomalous orientation to the meridian (Severny and Moreton, 1966; Martres *et al.*, 1966; Smith and Ramsey, 1967). The flares start on both sides of this line, whatever the position of the A.R. on the disk, implying that the inversion line of $H_{\parallel}$ nearly coincide with the line H_V (for vertical) $=0$.

– the two 'evolving magnetic features' (E.M.F.) of opposite polarities separated by this inversion line evolve in opposite senses (Martres *et al.*, 1968; Ribes, 1969).

The equation for the field changes

$$\frac{\partial H}{\partial t} = \text{rot}\,(\mathbf{V} \times \mathbf{H}) + \eta \nabla^2 \mathbf{H}$$

shows that the term rot $\mathbf{V} \times \mathbf{H}$ dominates the field variation (unless a fast dissipation of the field in unresolved fine structures occurs). Therefore the structure of the $\mathbf{V}$ field in relation to $\mathbf{H}$ should be essential for an understanding of the field changes associated with flares (see also the review by R. Michard in the same Proceedings).

We have tried to get some empirical evidence of these processes by a study of chromospheric structures, longitudinal magnetic fields, line of sight velocities patterns $(V_{\parallel})$ and flare localization in the A.R. No. 1512-16 (numeration according to *Cartes Synoptiques*) observed in Meudon from September 17th to 23rd, 1966 (60 °E to 20 °W). This A.R. at 20 °N was then young and increasing in dimensions, of mild magnetic complexity (class Cp in Meudon classification) and produced about 50 little flares. $H_{\parallel}$ and $V_{\parallel}$ were observed in the line 6173 FeI.

Line of sight velocities patterns are difficult to interpret in terms of actual motions. Short period changes (5 min oscillations for instance) are superimposed on more or

Howard (ed.), Solar Magnetic Fields, 435–442. *All Rights Reserved.*

less stationary and systematic motions: since the solar photosphere should remain spherical and conserve its matter, stationary $V_{\parallel}$ may result either from horizontal systematic motions, or from vertical motions involving an unresolved fine structure whose ascending and descending elements have different 'weights' in the resulting line profile. Of course intermediate or mixed cases are also possible!

2. Classification of Line of Sight Velocity Features in the A.R.

As a help to sort out different possible classes of $V_{\parallel}$ features, we look in Figures 1 and 2 at the $V_{\parallel}$ pattern in connection with the $H_{\parallel}$ pattern (Figure 1) and $H\alpha$ structure (Figures 1 and 2).

It is possible to recognize the following classes of $V_{\parallel}$ features corresponding perhaps to different physical motions.

1.   Around some of the sunspots, especially the extreme leading and following spots, we note the usual Evershed flow, which can be recognized from the orientation of the line $V_{\parallel}=0$ (normal to the direction of the center of the disk) and from the sign of velocities on both sides. At the poor resolution of our $H\alpha$ pictures there are no particular relations between the regions of normal Evershed flow and $H\alpha$ features.

2.   Around the central spot, which is of following magnetic polarity but surrounded by 'parasitic' poles of leading polarity, we have (Figure 1) *an irregular $V_{\parallel}$ pattern* (different from the Evershed flow). In this same area we have very bright $H\alpha$ features, better delineated at $H\alpha \pm 0.5$ Å (Figure 1). This is also the region where flares occurred on the 18th and following days. We suggest this 'irregular' pattern to be due to horizontal motions, similar to a strongly distorted Evershed flow.

3.   There are islands of 'negative' $V_{\parallel}$ (away from the observer or downwards) overlying regions of $H\alpha$ brightness above the average. At $H\alpha \pm 0.5$ Å (Figure 1) such an area is seen just above the following spot (left of the picture). On September 22, when the Active Region was relatively close to the center of the disk and flare activity had disappeared, these features represent a large part of the $V_{\parallel}$ pattern.

4.   At the periphery of the A.R. we have $V_{\parallel}$ features of doubtful association with $H\alpha$ features. We are looking at the transition between the A.R. and the surrounding photosphere and probably the complex superposition of supergranulation and oscillations dominates the V field. In the 'quiet' photosphere the relations between motions and the $H\alpha$ pattern are rather obscure as yet: In our observation of September 18 there is an indication of an association of $V_{\parallel}>0$ (upwards) velocities, with dark $H\alpha$ features at ± 0.5 Å bordering the magnetic poles east of the A.R.

3. Comparison of $H_{\parallel}$ and $V_{\parallel}$ Patterns in Relation to Flare Occurrence

In the following we shall concentrate on the comparison of the observed $H_{\parallel}$ and $V_{\parallel}$

patterns, which were observed simultaneously; 64 maps of these two quantities were available for the A.R. and period under consideration.

The relative dispositions of $H_{\parallel}$ features termed E.M.F. (for Evolving Magnetic Features) and $V_{\parallel}$ features are very diverse.

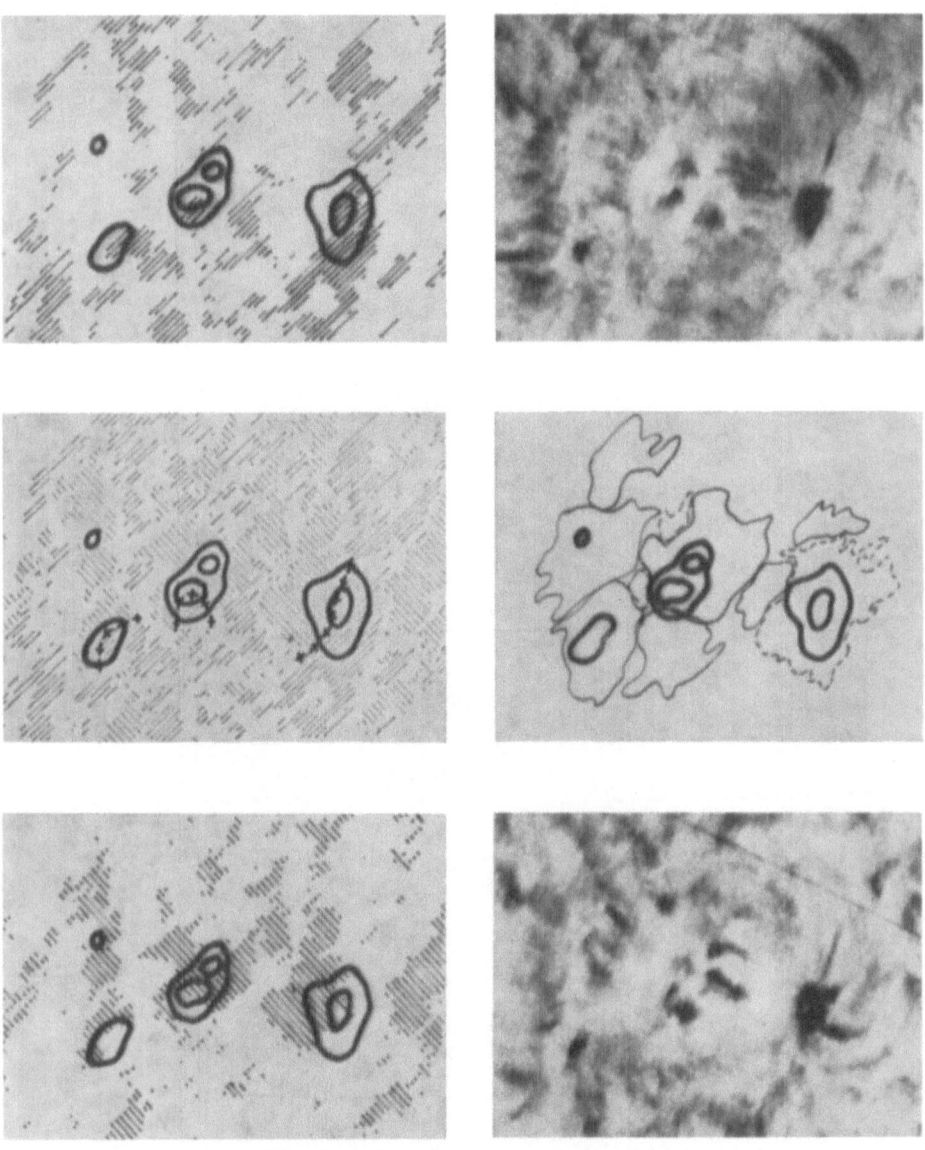

Fig. 1. 18–09–1966: Active Center at 42°E C.M. Left: radial velocity field 8h10 UT, from top to bottom: (1) component towards the observer (>0); (2) the two components; (3) component away from the observer (<0). Right: (1) spectroheliogram $=$ Hα—0.5 Å, 9h07; (2) principal magnetic features of the active center according to the longitudinal magnetic map at 8h10; (3) spectroheliogram $=$ Hα $+ 0.5$ Å, 9h10.

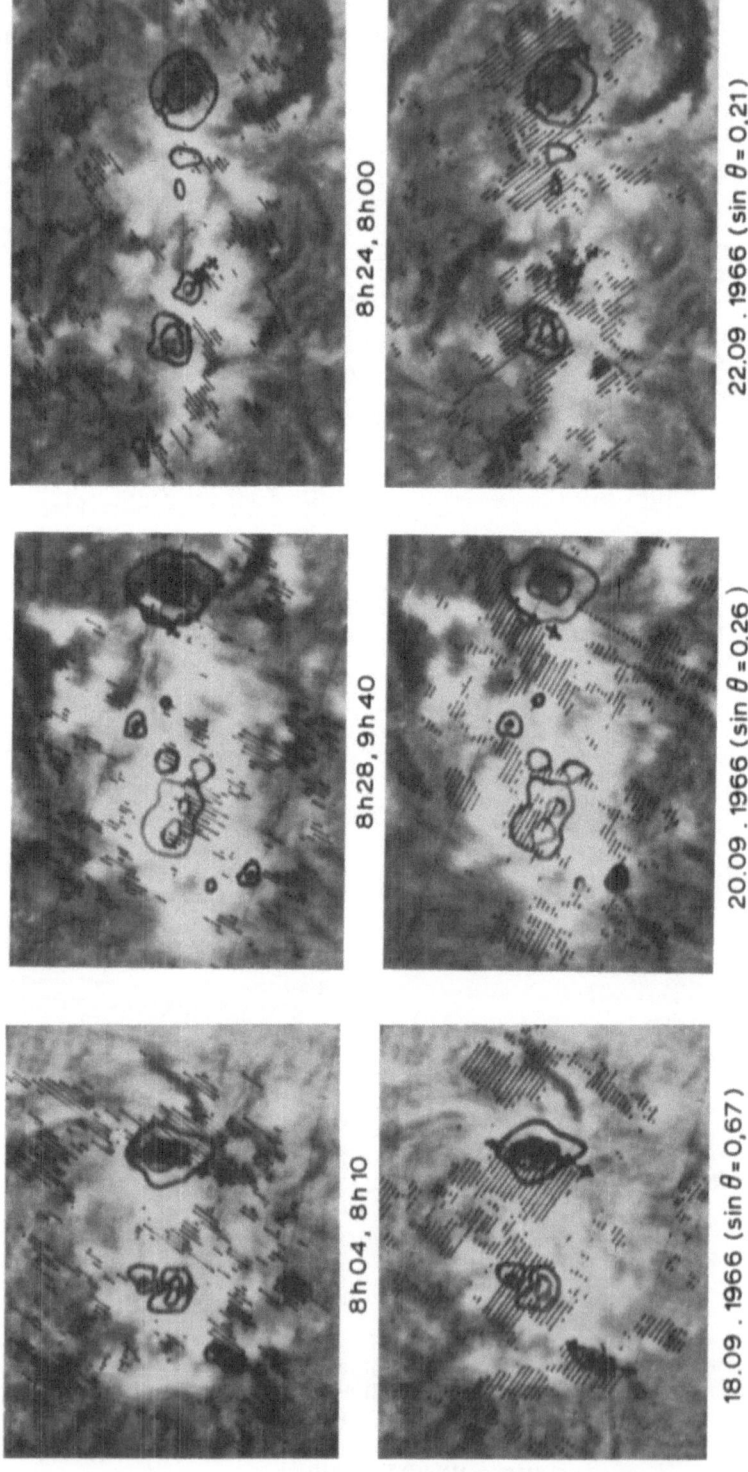

Fig. 2.   Comparison between the radial velocity distribution and bright Hα faculae. The first hour indicated is that of the observation of the radial velocities.

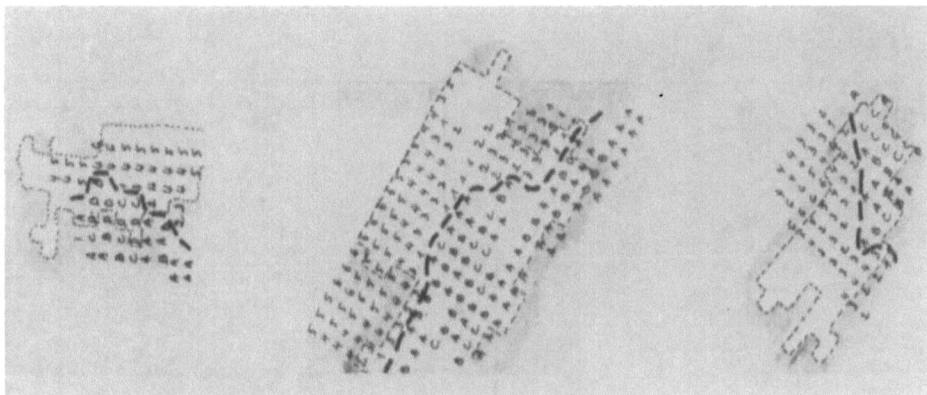

Fig. 3.   Sample of islands of velocities crossed by $H_\parallel = 0$ curve. 18 September 1966, 8h10; 19 September 1966, 7h50; 20 September 1966, 14h20. Numbers and letters correspond to $H_\parallel$ values in the magnetic map scale. The thick dashed line is the magnetic inversion line. The fine dashed line and the dotted line demark the regions where the velocity 'islands' are respectively positive and negative. The shadowed areas are regions without measurable velocities.

3.1. *There are inversion lines* $H_\parallel = 0$ *(with opposite magnetic polarities on both sides) crossing area where* $V_\parallel$ *keeps the same sign (Figure 3)*

Such situations have been discussed already by Semel (1967) and Stepanov (1965), who proved that they cannot be due to errors in $V_\parallel$.

We have found that such E.M.F. pairs, having the same line of sight velocities, do not change much or evolve in the same sense. Although they may present large gradients of $H_\parallel$ at the inversion line, *they do not take much part in the flare activity*.

3.2. *There are lines with* $V_\parallel = 0$, *with change of the sign of* $V_\parallel$, *crossing E.M.F. of strong* $H_\parallel$ *without magnetic inversion (Figure 4)*

Such situations were also pointed out by Semel (1967). They are fairly classical because they occur for instance in connection with the normal Evershed flow of an isolated sunspot.

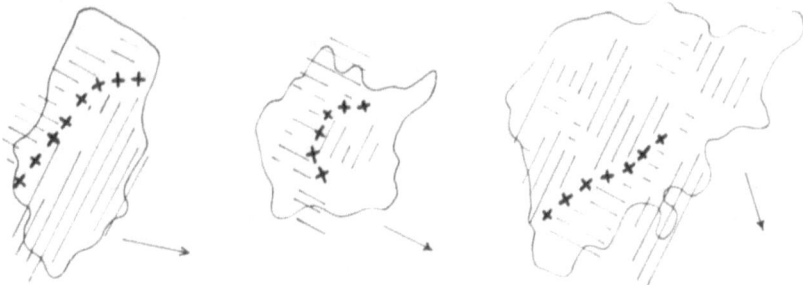

Fig. 4.   Sample of magnetic structure with strong magnetic fields crossed by curves $V_\parallel = 0$ which delimit islands of velocities of opposite sense. 17 September 1966; 20 September 1966; 21 September 1966. /// regions of positive velocities; \\\ regions of negative velocities. The crossed line is the $V_\parallel = 0$ line. The arrows show the direction of the center of the solar disk. When these dispositions are observed, the regions are not flaring.

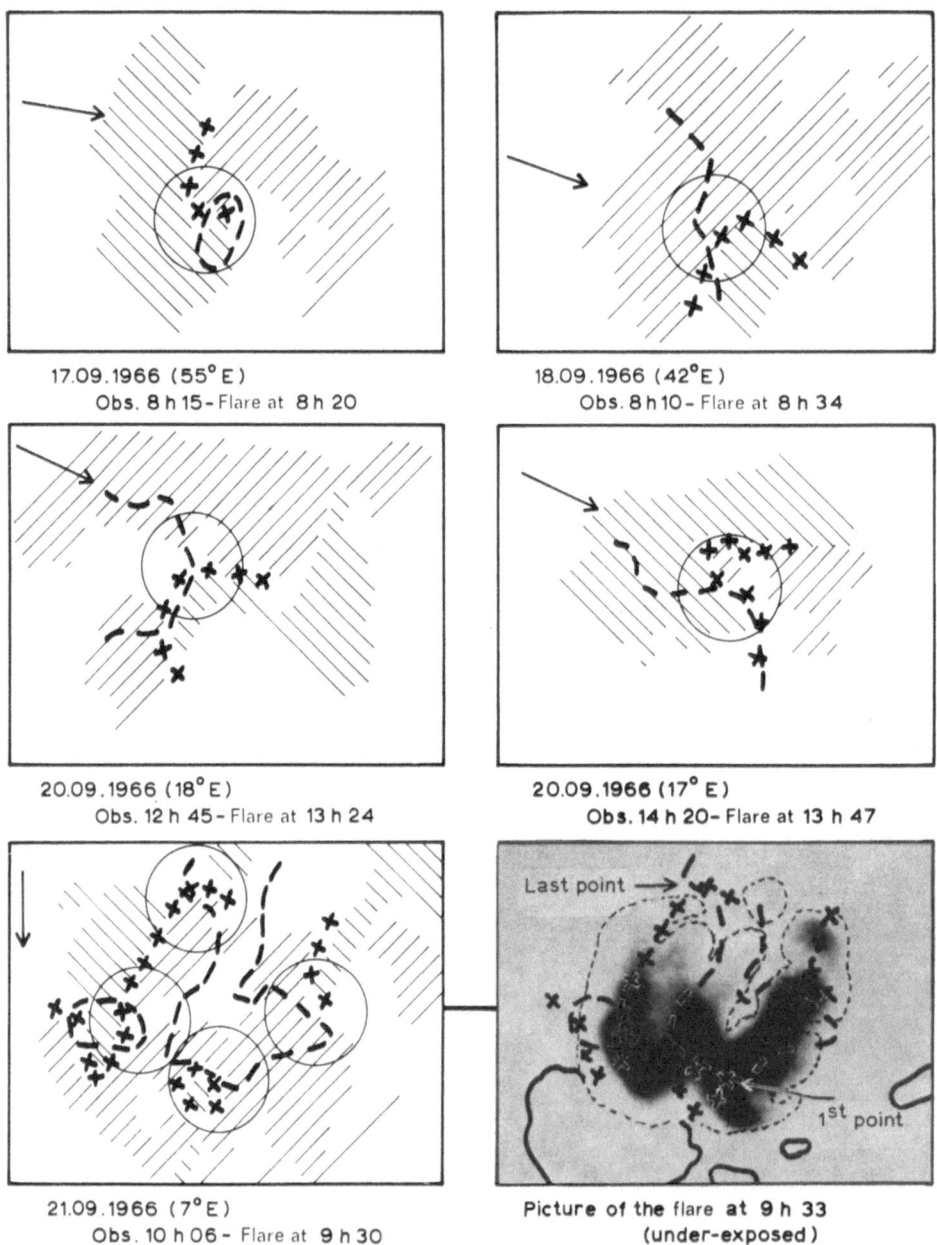

17.09.1966 (55°E)
Obs. 8 h 15 - Flare at 8 h 20

18.09.1966 (42°E)
Obs. 8 h 10 - Flare at 8 h 34

20.09.1966 (18°E)
Obs. 12 h 45 - Flare at 13 h 24

20.09.1966 (17°E)
Obs. 14 h 20 - Flare at 13 h 47

21.09.1966 (7°E)
Obs. 10 h 06 - Flare at 9 h 30

Picture of the flare at 9 h 33
(under-exposed)

Fig. 5.   Sample of flare localization (the notation is the same as before). The circles indicate the region of flare occurrence where the lines $H_\parallel = 0$ and $V_\parallel = 0$ are crossing. At the bottom, on the right, Meudon photographic observation of the flare at 9h33 (low exposure).

In the case of $V \neq 0$ the position of $V_{\parallel} = 0$ is fairly well determined and it should be, for a horizontal flow, the locus of the points where V is perpendicular to the direction of the center of the disk.

However if $V = 0$ at the line $V_{\parallel} = 0$ the position of this line may be very uncertain due to the imprecision of the zero of line of sight velocities in the observations.

In part of the cases where $V_{\parallel} = 0$ crosses an E.M.F. we clearly recognize from the orientation of this line, the sign and size of $V_{\parallel}$ on both sides of it, the *classical Evershed flow* with a distribution of V having approximate axial symmetry. This occurs primarily on the leading and extreme following spots, which take not much part in the flare activity. In other cases such a simple situation cannot be recognized.

3.3. *There are regions where the lines* $H_{\parallel} = 0$ *and* $V_{\parallel} = 0$ *cross each other (the measurements of both quantities being significant).*

The E.M.F. involved in such configurations are bright in Hα and are generally the support of flare activity. The position of the line $V_{\parallel} = 0$ is not related to the contours of the E.M.F. The geometrical relations between the lines $V_{\parallel} = 0$ and $H_{\parallel} = 0$ in such regions do not form clear cut patterns.

Among the recorded flares we were able to examine the location of 15 events, occurring between the 17th and 22d of September at less than one hour of available $H_{\parallel}$ and $V_{\parallel}$ records.

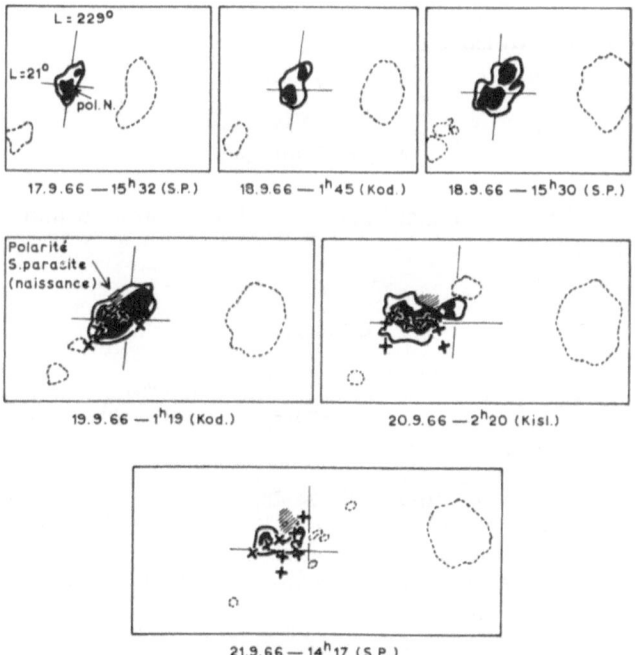

Fig. 6. Sample of horizontal displacements of spots and magnetic poles. The dashed line shows the localization of the other main spots.

12 of these started in regions of crossing of the lines $H_\parallel = 0$ and $V_\parallel = 0$ (examples in Figure 5) independently of the A.R. position on the disk.

Two started at places where the two lines were nearly crossing (less than 5″ distance).

One near a $V_\parallel = 0$ line without an associated $H_\parallel = 0$ line: however a parasitic magnetic polarity appeared a few hours after, leading to a situation similar to the above cases.

This result – the crossing of lines $H_\parallel = 0$ and $V_\parallel = 0$ in the flaring area – makes more precise Severny's result (Severny, 1960). Indeed he indicated that: "the neutral point is practically always found on the neutral lines of velocity maps", and he considered that the neutral point denotes the place of the most frequent appearance of solar flares.

In most cases we observe in the flare productive area an abrupt change of the direction of the line $V_\parallel = 0$, one of its branches being perpendicular to the direction of the center of the disk, the other parallel. This suggests a kind of *vortex motion*.

Some support for this idea may be found in Figure 6 showing the evolution of the central most eruptive part of the A.R. where 43 minor flares where located. A vortex motion of the spot and of the parasitic polarity born in the night of September 18–19 is apparent.

4. Conclusion

These observations suggest that certain systems of photospheric motions may be the cause of the changes of surface magnetic fields, which influence the distribution of Hα structure, and eventually lead to the build up of flare productive situations.

Further observations may help to clarify the nature of the interactions between **V** and **H**; however the knowledge of only one component of these vectors (particularly in the case of **V**) will make the problem very difficult.

Acknowledgement

We are grateful to Prof. R. Michard and we wish to express to him our thanks for many helpful discussions.

References

Martres, M. J., Michard, R., and Soru-Iscovici, I.: 1966, *Ann. Astrophys.* **29**, 245.
Martres, M. J., Michard, R., Soru-Iscovici, I., and Tsap, T.: 1968, *Solar Phys.* **5**, 187.
Ribes, E.: 1969, *Astron. Astrophys.* **2**, 316.
Semel, M.: 1967, *Ann. Astrophys.* **30**, 513.
Severny, A. B.: 1960, *Izv. Krymsk. Astrofiz. Obs.* **24**, 281.
Severny, A. B. and Moreton, G. E.: 1966, *Astron. J.* **71**, 172.
Smith, S. and Ramsey, H. E.: 1967, *Solar Phys.* **2**, 158.
Stepanov, V. E.: 1965, in R. Lüst(ed.), 'Stellar and Solar Magnetic Fields', *IAU Symp.* **22**, p. 267.

VOLUME CHARACTERISTICS OF MAGNETIC-CHANNEL FLARES

L. KŘIVSKÝ

Astronomical Institute, Czechoslovak Academy of Sciences, Ondřejov, Czechoslovakia

Abstract. On the basis of Hα pictures of two proton flares of medium size, the volume of the separate parts of magnetic-channel flares, are given. At the time of the main release of energy the volume of the whole system is determined at $\sim 2.5 \times 10^{27}$ cm^3 and the size of areas III after Petschek model at $\sim 7.0 \times 10^{20}$ cm^3.

1. Areas of the Origin of Accelerated Particles in Flares

Vaiana *et al.* (1968) found from pictures of the flare of 8/6/1968, importance 1N in the X-ray emission 3.5–14 Å, that the volume in which this two-ribbon flare, connected at the narrowest point by a loop coupling, radiated, amounted to 10^{28} cm^3; this was roughly at the time of maximum brightness of the flare in Hα. The determination of the volume is based on the permissible assumption that the ribbons of the flare have a circular cross-section diameter equal to the width of the ribbons measured on the photographs. It is necessary to point out that the flare was in an active region, where a proton flare with the usual typical development for this type of flare, occurred a day later, on 9/6/1968: two parting emission ribbons, a weaker diffusion ribbon in the emission in the center between them being generated, and an absorption filament (Křivský and Švestka, 1969) coming from the mouth of this flare channel. The flare of 8/6/1968, recorded in the X-ray emission by Vaiana *et al.*, therefore, was only one of the precursors of the later typical proton flare, and it was of a kind which does not differ much from the mighty, fully developed proton flares (Křivský, 1963, 1970). Provided considerations are made along the lines of Vaiana *et al.* (1968), i.e. that the emission areas indicate the space where the particles are accelerated (which, it seems, are identical with this type of flare to volumes 'enveloped' by layers radiating in Hα) the volumes one obtains at the beginning of the development of proton flares, when the main emission of energy occurs in the form of accelerated particles, will be sufficiently representative for determining the volumes where the principal acceleration of the particles occurs.

If one accepts the fact that in the case of the flare observed in the X-ray emission by Vaiana *et al.* (1968), the processes in the three coordinate components, inherent to mighty proton flares (Křivský, 1968, 1970), had already developed at least to a minor extent, this is proof that the accelerating processes are basically of two types and that they are located in two volumes, differing in shape: (1) in two flare quasi-parallel ribbons, located horizontally with respect to the surface of the Sun, and (2) in an arched space, vertically developed between both flare ribbons. These two types of volumes can easily be distinguished at first glance on photographs of the flare in the X-ray emission of 8/6/1968 (Vaiana *et al.*, 1968, Figure 2A; Neupert, 1969, Figure 11; Vaiana, 1969). The determination of these two fundamental volumes for accelerating

Howard (ed.), Solar Magnetic Fields, 443–449. All Rights Reserved.
Copyright © 1971 by the IAU.

the particles, along the neutral lines (in the flare ribbons) on the one hand, and in the magnetically closed, generating structure of the vertically developing loop tunnel with the peak emission ribbon, formed the contents of the model of proton flares published earlier. This model is founded on the interaction of an emerging magnetic channel with a magnetic system of spot groups (Křivský, 1968, 1970).

It will not be an error, at least as far as the order is concerned, if the characteristic volumes for the initial rapid particle acceleration of an impulse nature (at the time of the Y-phase), as well as for the later development of the acceleration system in the volumes of the higher ascending parts of the flare channel, will be derived from the photographs of proton flare in Hα (Křivský, 1969); this is substantiated by the conclusions in the paper of Vaiana et al. (1968) on the identity in shape and position of the flare ribbons of the X- and Hα-emissions.

The fundamental volume characteristics of the individual phases of proton flares of an intermediate size, as well as of their physically differing volumes, have been determined from two flares, i.e. the disc flare of 12/7/1961 and the limb flare of 18/11/1968 (Křivský, 1970). Also the critical volume for the narrow regions of the microturbulent layers, important for accelerating the particles according to the Petschek wave model, (denoted in Petschek's model as regions III, 1964), was determined.

2. Volume Characteristics of Proton Flares

The determination of volumes of flares with an emission of fast particles has been the subject of a number of papers. For example, De Jager (1967) for all types of hard X-emission gives the volume $10^{25}-10^{28}$ cm^3; in (1969) he proved theoretically that the active flare plasma comes from a volume of 10^{29} cm^3. For 'optical' flares he gives the volume of just under 10^{26} cm^3, and high-energy particles are considered to come from a volume of 10^{27} cm^3. Vaiana et al. (1968), using the photographs of the X-ray emission flare of 8/6/1968, give a volume of 10^{28} cm^3. Friedman and Hamberger (1969) determined theoretically the volume of slow stationary shock fronts of regions III (according to Petschek) in which the magnetic energy transforms to plasma energy accompanied by a microwave emission as $10^{19}-10^{20}$ cm^3. In (Křivský, 1969) the volumes of the proton flare of 26/9/1963 were determined by measuring the photographs in Hα at various stages of development: at the beginning during the Y-phase as $10^{26}-10^{27}$ cm^3, and less for the hard X-ray emission; the initial volume of the ribbon as the peak of the loops about 10^{26} cm^3, later 10^{28} cm^3; the volume of the whole channel of a perfectly developed flare as $10^{29}-10^{30}$ cm^3. Takakura (1969) has fixed the volume as 10^{29} cm^3, in which at the explosive phase, a hot coronal condensation (10^7-10^8 K) with hard X-ray bursts, originated. Švestka (1970) determined the volume of the region of origin of a hard X-ray burst as 5×10^{27} cm^3 and of a hot condensation at the peak of development as about 10^{29} cm^3. Important results are also in Zirin (1964) and in Acton (1968).

With a view to the known phase of development of proton flares, which are called magnetic-channel flares (Křivský, 1963, 1968, 1969a, b, 1970) and which have also been described in a later phase of their development by Bruzek and DeMastus (1970). Also for a number of reasons following from the time connections of the phases of development with the γ-emission, with the X-ray emission, with the radio emission, and with white-light radiation, it is necessary to divide the measured and computed volume charateristics of this type of flare into three stages:

Stage I – Y-phase (beginning of the splitting of the flare ribbon into two), at the time of the occurrence of the hard and very hard impulsive X-ray emission (Křivský in Švestka, 1966; Valníček, 1967; Křivský and Švestka, 1969), at the time of the first radio burst (mm, cm) and at the time of the integral flare-emission (Švestka, 1970); this phase at the very beginning of the flare is considered to be a process in which the main transformation of the magnetic energy in the fast particle flux takes place (cosmic and subcosmic radiation).

Stage II – approximately at the time of the maximum flare brightness. This stage was originally considered by a number of authors to be the phase in which the very fast particles are emitted (e.g., Ellison *et al.*, 1961). It seems that this stage is characteristic for the generation and emission of a cloud of plasma, or for the escape of particles, accelerated to a lesser extent in the peak regions on the loops (Křivský, 1970), which takes place a little later.

Stage III – the largest spatial expansion of the flare channel; owing to the weak radiation in Hα, one speaks of the late, decay phase of the flare. The peak 'ribbon' on the loops still radiates intensely even in the coronal lines (Křivský, 1969a). This stage is probably responsible for the late phase of radio type IV, and the escape of particles and plasma will drop off.

For each of the three stages mentioned, the volumes of the following regions, according to the diagram in Figure 1, will be given: V_A = volume of ribbon A; V_B = volume of ribbon B; $V_A + V_B$; V_C = volume of 'ribbon' C at the peaks of the loops; V_{LFA} = volume of the foot of the loops above ribbon A; V_{LFB} = volume of the foot of the loops above ribbon B; $V_C + V_{LFA} + V_{LFB}$ = volume of loops; $V_{ribbons+loops} = (V_A + V_B) + (V_C + V_{LFA} + V_{LFB})$, i.e. the whole formation of the flare channel; and further

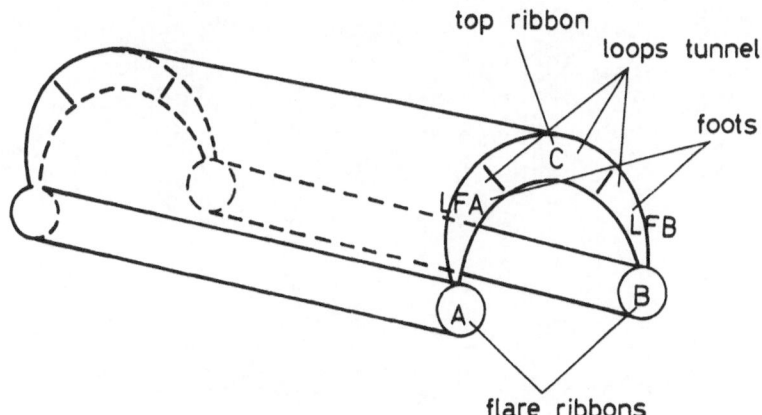

Fig. 1.   Sizes of the separate parts of magnetic-channel flares.

according to the diagram of the Petschek model (1964) the theoretical volumes of regions III (see Figure 2), applied to the model of the magnetic channel (Křivský, 1968, 1970), the dimensions of which have been derived from the flares measured (only for the layer III theoretical width 10^2 cm is applied, Friedman and Hamberger, 1969): $V_{P\ III\ A}$ = system around the first X-type zero point (ribbon A); $V_{P\ III\ B}$ = system around the second X-type zero point (ribbon B); $V_{P\ III\ A} + V_{P\ III\ B}$ = volumes of both systems around zero lines.

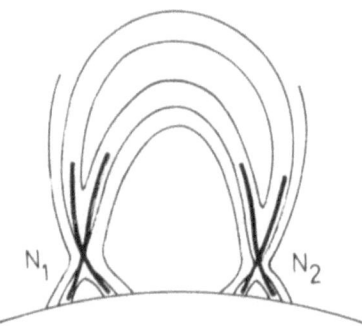

Fig. 2.   Vertical section of the flare channel, N_1 and N_2 are intersections of the zero lines through the plane of the vertical section. Weak lines – magnetic field lines, full lines – regions III after Petschek (1964).

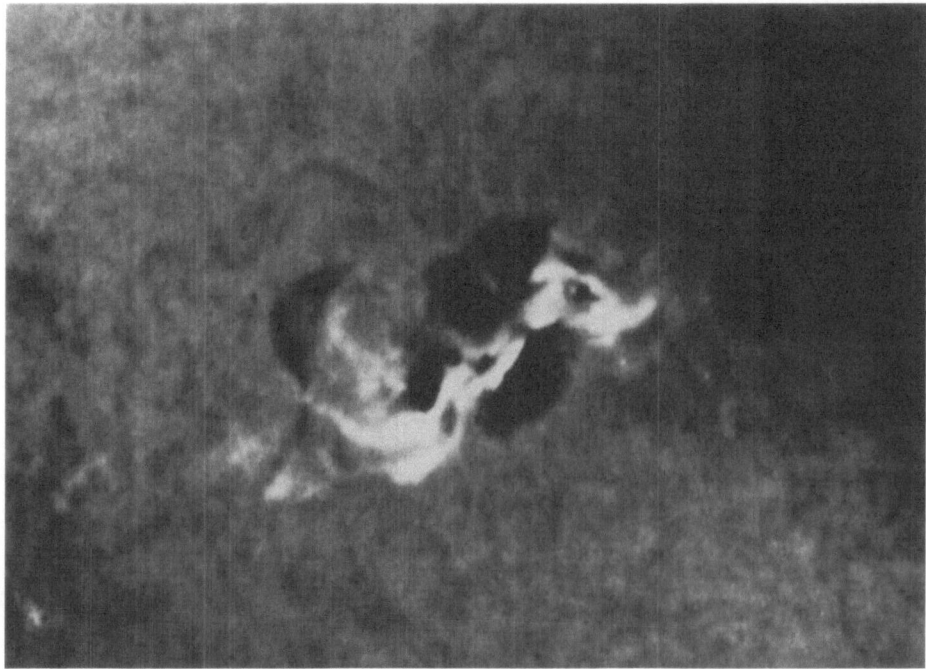

Fig. 3.   Proton flare Hα, 12 July, 1961, 10 23 UT (Černošice near Prague).

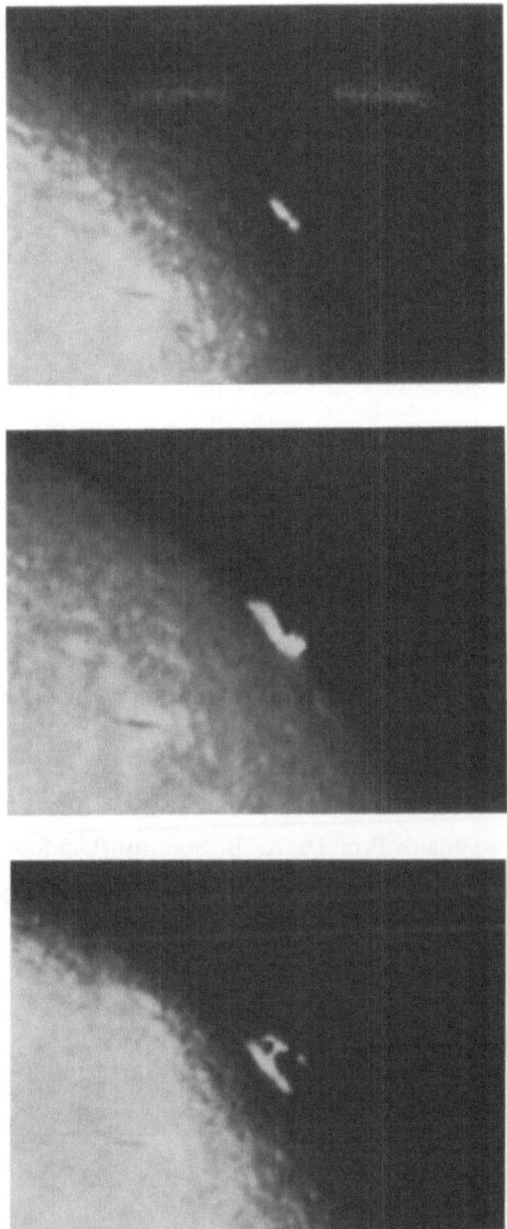

Fig. 4.   Proton flare Hα, 18 Nov., 1968, (a) 10 30; (b) 10 40; (c) 11 10 UT (Catania).

The volumes of the individual parts of the system of the magnetic channel of the flare according to the stages mentioned, are contained in Table I. The volumes were determined from the measurements on the photographs of two proton flares, which may be considered as being of intermediate size: (1) flare of 12/7/1961, beginning 10:00, phase Y 10:20–10:22, maximum 10:25–10:31, end (13:03), position 7°S,

TABLE I

Volume characteristics of proton flares of middle size

Developing stages volumes	I cm³	II cm³	III cm³
V_A	$\sim 9.6 \times 10^{26}$	$\sim 4.5 \times 10^{27}$	$\sim 4.8 \times 10^{27}$
V_B	$\sim 6.3 \times 10^{26}$	$\sim 1.8 \times 10^{27}$	$\sim 1.9 \times 10^{27}$
ribbons: $V_A + V_B$	$\sim 1.6 \times 10^{27}$	$\sim 6.3 \times 10^{27}$	$\sim 6.7 \times 10^{27}$
V_C	$\sim 7.0 \times 10^{26}$	$\sim 1.6 \times 10^{27}$	$\sim 3.0 \times 10^{27}$
V_{LFA}	$\sim 1.2 \times 10^{26}$	$\sim 4.8 \times 10^{26}$	$\sim 2.4 \times 10^{27}$
V_{LFB}	$\sim 1.2 \times 10^{26}$	$\sim 4.8 \times 10^{26}$	$\sim 2.4 \times 10^{27}$
loops: $V_C + V_{LFA} + V_{LFB}$	$\sim 9.4 \times 10^{26}$	$\sim 2.6 \times 10^{27}$	$\sim 7.8 \times 10^{27}$
$V_{\text{ribbons+loops}}$	$\sim 2.5 \times 10^{27}$	$\sim 8.9 \times 10^{27}$	$\sim 1.4 \times 10^{28}$
$V_{P\,III\,A}$	$\sim 3.4 \times 10^{20}$	$\sim 9.0 \times 10^{20}$	$\sim 1.2 \times 10^{21}$
$V_{P\,III\,B}$	$\sim 3.4 \times 10^{20}$	$\sim 9.0 \times 10^{20}$	$\sim 1.2 \times 10^{21}$
$V_{P\,III\,A} + V_{P\,III\,B}$	$\sim 6.8 \times 10^{20}$	$\sim 1.8 \times 10^{21}$	$\sim 2.4 \times 10^{21}$

22 °E, importance 3 + ; Hα photographs were obtained from the Crimean Astrophysical Observatory (U.S.S.R.), from Dr Otavsky's station near Prague and from the Astronomical Institute Ondřejov (Figure 3); (2) flare of 18/11/1968, beginning 10:26, phase Y 10:26–10:35, maximum (10:35), end 11:57, position 22 °N, 87 °W, importance 2N. Hα photographs made at the Catania Astrophysical Observatory (Italy) were used (Figure 4).

Acknowledgements

The author wishes to thank Prof Dr A. B. Severny (U.S.S.R., Crimea) and Prof. Dr G. Godoli (Italy, Catania) who so kindly furnished Hα-pictures of flares.

References

Acton, L. W.: 1968, *Astrophys. J.* **152**, 305.
Bruzek, A. and DeMastus, H. L.: 1970, *Solar Phys.* **12**, 447.
De Jager, C.: 1967, in *Electromagnetic Radiation in Space*, (ed. by J. G. Emming), D. Reidel Publ. Co., Dordrecht-Holland, p. 101.
De Jager, C.: 1969, in Y. Öhman (ed.), 'Mass Motions in Solar Flares and Related Phenomena', *Nobel Symp.* **9**, 171.
Ellison, M., McKenna, S. M. P., and Reid, J. H.: 1961, *Dunsink Observ. Publ.* **1**, 53.
Friedman, M. and Hamberger, S. M.: 1969, *Solar Phys.* **8**, 104.
Křivský, L.: 1963, *Bull. Astron. Inst. Czech.* **14**, 77.
Křivský, L.: 1968, in K. O. Kiepenheuer (ed.), 'Structure and Development of Solar Active Regions', *IAU Symp.* **35**, 465.
Křivský, L.: 1969a, *Bull. Astron. Inst. Czech.* **20**, 139.
Křivský, L.: 1969b, *Bull. Astron. Inst. Czech.* **20**, 163.
Křivský, L.: 1970, *Bull. Astron. Inst. Czech.* **21**, 67.
Křivský, L. and Švestka, Z.: 1970, *Space Research X*, COSPAR XII, Prague 1969 (ed. Donahue *et al.*), North-Holland, Amsterdam, p. 817.
Neupert, W. M.: 1969, *Ann. Rev. Astron. Astrophys.* **7**, 121.
Petschek, H. E.: 1964, in *AAS-NASA Symp. Physics of Solar Flares 1963*, Washington, p. 425.
Švestka, Z.: 1966, *Space Sci. Rev.* **5**, 388.

Švestka, Z.: 1970, *Solar Phys.* **13**, 471.

Takakura, T.: 1969, *Solar Phys.* **6**, 133.

Vaiana, G. S., Reidy, W. P., Zehnpfennig, T., Van Speybroeck, L., and Giacconi, R.: 1968, *Science* **161**, 564.

Vaiana, G. S. and Giacconi, R.: 1969, in *Plasma Instabilities in Astrophysics*, (ed. by Wentzel and Tidman), Gordon and Breach, New York, p. 91.

Valníček, B.: 1967, *Bull. Astron. Inst. Czech.* **18**, 249.

Zirin, H.: 1964, *Astrophys. J.* **140**, 1216.

THE RELATION BETWEEN DASHES AND FLARES
(PHYSICAL NATURE OF THE DASH PHENOMENA)

D. A. KUZNETSOV

Siberian Institute of Terrestrial Magnetism,
Ionosphere and Radiowave Propagation, Academy of Sciences, Irkutsk, U.S.S.R.

and

A. A. SHPITALNAYA

The Pulkovo Observatory, Leningrad, U.S.S.R.

Abstract. The phenomenon of dashes in the eruptive loop prominences is found to represent the part of the solar flare process which includes all the levels of the Sun's atmosphere. For 75% of observations a coincidence of dashes and flares in space and time was discovered. The rest of the dash observations may be interpreted as the excitation of dashes by the flares arising outside the limb.

Following the solar flare theory of Alfvén and Carlquist, the possibility to interpret the dashes as the eruptive instability of the pinch-effect is discussed. The computation of flare-current magnetic field gives the value $H \sim 3 \cdot 10^4$ oersted, its order being in agreement with the observed value.

1. Disturbed short-lived ($1–25^m$) small features, the so-called 'dashes', are observed at a height range of 10000 km to 70000 km above the solar limb (Young, 1895; Ellison, 1959; Billings, 1957; Prokofjeva, 1957; Shpitalnaya, 1964; Krat, 1968). The dashes arise in the legs of arches of loop eruptive prominences (Ellison, 1959; Billings, 1957). One of the present authors has recently discovered that dashes possess strong distinctly localized magnetic fields (Shpitalnaya and Vjalshin, 1970).

Magnetic field strengths as high as 10000 oersted were measured on spectrograms of D_3 and $H\beta$ lines taken with a Wollaston prism and a $\frac{1}{4} \lambda$ plate attached to the Pulkova coronal spectrograph.

An asymmetric line-of-sight velocity structure characteristic of dashes was observed in spectra obtained in the region of the strong $H\alpha$, $H\beta$, He D_3, Na D_1 D_2, etc. lines (Young, 1895; Ellison, 1959; Billings, 1957; Prokofjeva, 1957; Shpitalnaya, 1964; Krat, 1968). The velocities of the dashes are within the limits of 25–500 km/s. Plasma outflow from the dashes activates the surrounding corona. The density of the latter is increased (Kleczek and Hansen, 1962; Shpitalnaya, 1964).

The spectrograms of D_3 taken at Pulkovo on August 19, 1964 are shown in Figures 1a and 1b. During a period of less than one minute the plasma outflow velocity increased from 107 km/s to 200 km/s, i.e. the plasma was ejected with an acceleration of about 2 km/s², which is higher than the acceleration of gravity.

2. It is usually supposed that flares activate prominences if the angular distance between them does not exceed 20° and if dashes originate during the existence of the flare. A detailed comparison was made between the location and the time of development of these non-steady phenomena. The flare data and coordinates are given in *Solar*

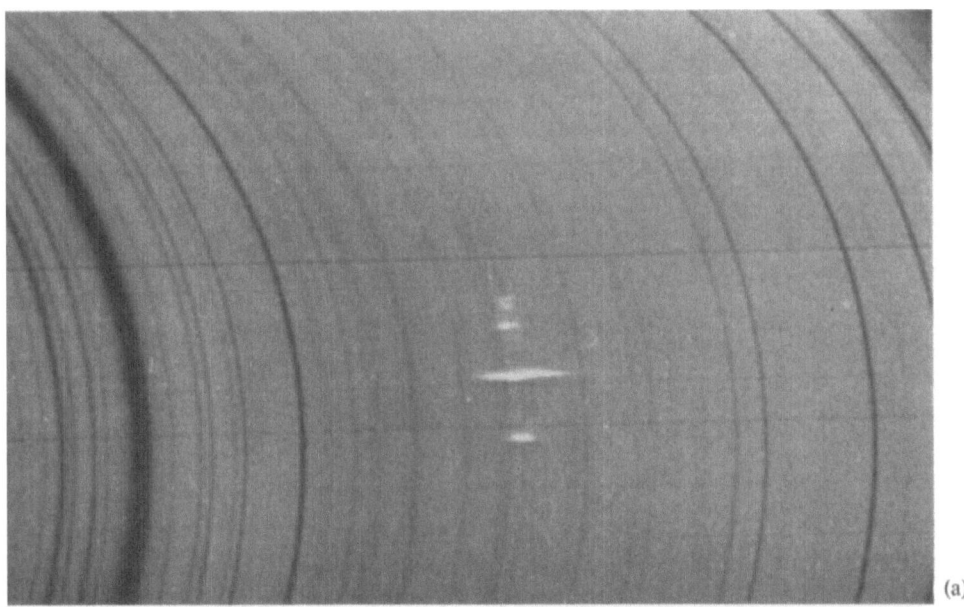

(a)

(b)

Fig. 1.   He–D₃ limb spectra obtained at Pulkovo observatory on August 19, 1964; the height above
the solar limb was about 12″. (a) 13^h10^m UT the maximum line-of-sight velocity of the dash (V_{max})
is about 100 km/s; (b) 13^h11^m UT, $V_{max} = 200$ km/s. The acceleration of plasma ejection
was about 2 km/s².

TABLE I

Illustration of the connection between dashes and flare activity

Date	Number of dash events	Number of coincident flare events	Percentage of coincidence
1960	15	14	93%
1961	3	3	100%
1962	11	8	73%
1963	7	5	71%
1964	4	2	50%
1965	4	2	50%
1966	16	11	69%
	60	45 mean	75%

Data, Zurich Quarterly Bulletins, Fraunhofer Institute Maps, and observations obtained at Pulkovo. The results are summarized in Table I.

It follows that some of the dashes may be associated with flares on the invisible solar hemisphere (Westin and Liszka, 1970).

3. This fact does not raise doubts that dashes are part of a complex flare process. According to the Alfvén and Carlquist mechanism (1967) the origin of a flare is due to an electric circuit breakdown in a vertical current, $I \sim 10^{11}$–10^{12} A, passing through an area of the photosphere with a diameter of $\sim 10^9$ cm, and is accompanied by the generation of an oscillatory voltage of about 10^9 V. It is proposed that dashes are formed as a consequence of the so-called 'eruptive instability' of a pinch (Arzimovich, 1961). Plasmoids are ejected from the surface of the pinch with velocities close to those of plasma compression. The latter can be as high as several hundred kilometers per second in laboratory experiments (Arzimovich, 1961). The total energy applied to the plasma during one discharge cycle, $(t_2 - t_1)$ must be higher than all the forces acting in the plasma and the radiative loss. As $m_i \gg m_e$ the power applied to a pinch of unit length may be expressed as

$$\int_{t_1}^{t_2} W \, dt \gtrsim \int_{t_1}^{t_2} kN_i(t) \, T_i(t) \, dt + \tfrac{1}{2}MV^2 + \frac{1}{8\pi} \int_{r_2}^{r_1} (H^2 + H_c^2) \, r \, dr, \qquad (1)$$

where W is the power, r is the radius of the pinch, N is the concentration, T is the temperature, k is the Boltzmann constant, H is the magnetic field of the current, H_c is the 'catching' magnetic field, M is the pinch mass, and V is the return velocity. From this relation the magnetic field strengths compressing the plasma may be obtained. It is assumed that the time for magnetic field penetration into the plasma is much greater than the period of the oscillation of the flare instability (Alfvén and Carlquist, 1967). It is also assumed that the velocity of the plasmoids ejected from the pinch is equal to that of the compression of the pinch.

Omitting intermediate calculations we find for the mean magnetic field strength $\bar{H}$:

$$\bar{H} \approx \left[\frac{16\pi b I U_0}{\bar{V}(r_1 + r_2)} \right]^{1/2} \tag{2}$$

where U_0 is the initial voltage value, $\bar{V}$ is the mean velocity of eruptive motions, b is the factor of proportionality.

The total flare current branches in a large number of channels in the corona. Each of these channels is a current circuit from prominence knots.

We consider that there are at least 10^3 separate channels representing circuits from 10 dashes. This is in accordance with the observed fine structure of prominences. If the initial radius of a pinch is $r_1 \sim 10^9$ cm, $\bar{V} \sim 10^7$ cm/s, $I \sim 10^{11}$ A, and the voltage is 10^9 V (Severny, 1965; Alfvén and Carlquist, 1967) then

$$H \sim 3 \cdot 10^4 \text{ Oe}.$$

This estimate of the magnetic field strength coincides with the data within an order of magnitude (Shpitalnaya and Vjalshin, 1970).

This discrepancy may be due to the fact that effects distorting the measured line profiles and masked Zeeman line splitting are present in these observations as this investigation indicates a lower limit of $\bar{H}$ in the dashes.

We note that fields with strength 10^4 Oe are necessary to explain the observed metal line splitting in flares (Alikaeva, 1969).

References

Alfvén, H. and Carlquist, P.: 1967, *Solar Phys.* **1**, 220.
Alikaeva, K. V.: 1969, *Astron. Astrophys.* **8**, 92.
Arzimovich, L. A.: 1961, *Ruled Thermonuclear Reactions*, Moscow.
Billings, D. D.: 1957, *Publ. Astron. Soc. Pacific* **69**, 407.
Ellison, M. A.: 1959, *The Sun and Its Influence*, London.
Kleczek, J. and Hansen, R. T.: 1962, *Publ. Astron. Soc. Pacific* **74**, 441.
Krat, V.: 1968, in Y. Öhman (ed.), 'Mass Motions in Solar Flares and Related Phenomena', *Nobel Symp.* **9**, 93.
Prokofijeva, I. A.: 1957, *Solar-Geophysical Data* **8**, ESSA Boulder.
Severny, A. B.: 1965, *Astron. J.* **42**, 217.
Shpitalnaya, A. A.: 1964, *Publ. Pulkovo Obs.* **177**, 60.
Shpitalnaya, A. A. and Vjalshin, G. F.: 1970, *Solar-Geophysical Data*, **4**, ESSA Boulder.
Westin, H. and Liszka, L.: 1970, *Solar Phys.* **2**, 3.
Young, H. A.: 1895, *The Sun*, London.

Discussion

Kuperus: Could you give a reference of the experimental results mentioned in your paper on the excitation of plasmoids by the interruption of a current?

Soboleva: Arzimovich, L. A.: 1961, *Controlled Thermonuclear Reactions*, Moscow.

PART V

THEORIES OF SMALL SCALE MAGNETIC FIELDS

THEORIES OF SMALL-SCALE MAGNETIC FIELDS

P. A. SWEET

Dept. of Astronomy, The University, Glasgow, U.K.

Abstract. This review is concerned with the origin of the fine structure of the fields and their relationship to the heating of the solar chromosphere and corona, the structure of prominences and the production of energetic particles in solar flares. The dynamics of sunspot formation, and the large-scale structure of individual sunspots have not been dealt with, although the evolution of AR fields has been considered insofar as it affects the flare problem.

1. The Fine-Structure at Photospheric Level

Starting with the mean fields measured within and in the neighbourhood of spots, and in the quiet photosphere, the theoretical problem is to account for the observed fine structure and to deduce the non-radiative energy fluxes implied.

1.1. QUIET PHOTOSPHERE

At photospheric level there is no sharp distinction between the quiet Sun and an AR. Veeder and Zirin (1970), however, have pointed out that the fibril structure in the overlying chromospheric Hα-network sets in fairly sharply at the 5 G contour of the mean longitudinal field, and that there is no mean transverse field outside this contour. The quiet photosphere is therefore chracterized by a mean magnetic field normal to the atmosphere and of strength varying up to 5 G. The localization of this flux within relatively small areas at the junctions of three or more supergranules was first observed by Leighton *et al.* (1962). These concentrations, which are of the same polarity, have diameters of the order 5000–10000 km. Within the cells a weak fine structure, of mixed polarity, of the order of 2000 km in diameter is present, q.v. Livingston (1968), as shown in Figure 1. There is no evidence, however, that this is correlated with the granules. The discovery by Howard (1967) of rapidly oscillating velocity fields with periods of the order of a few seconds raises the question whether yet a finer magnetic field structure may be present beyond the limit of spatial resolution.

The main intersupergranular concentrations, with peak field strengths up to 100 G, are generally attributed to convection by the horizontal motions within the super-granules as first suggested by Pikel'ner (1963) and considered quantitatively by Parker (1963). Parker's two-dimensional treatment was extended by Clark and Johnson (1967) to a three-dimensional hexagonal geometry, as shown in Figure 2. Although diffusion due to finite electrical conductivity allows a steady state to be set up in time-scales comparable with the lifetime of the supergranules, the field amplification, of the order of $R_m = \sigma v_{\text{hor}} L/c^2 \sim 5 \times 10^4$ at photospheric level, involve field strengths far too high to be achieved by the conceivable mechanical forces available. In the above expression L is the diameter of a supergranule. The concentrations must there-fore be determined by the dynamics of the supergranules. A recent dynamical numer-

Howard (ed.), Solar Magnetic Fields, 457–474. All Rights Reserved.
Copyright © 1971 by the IAU.

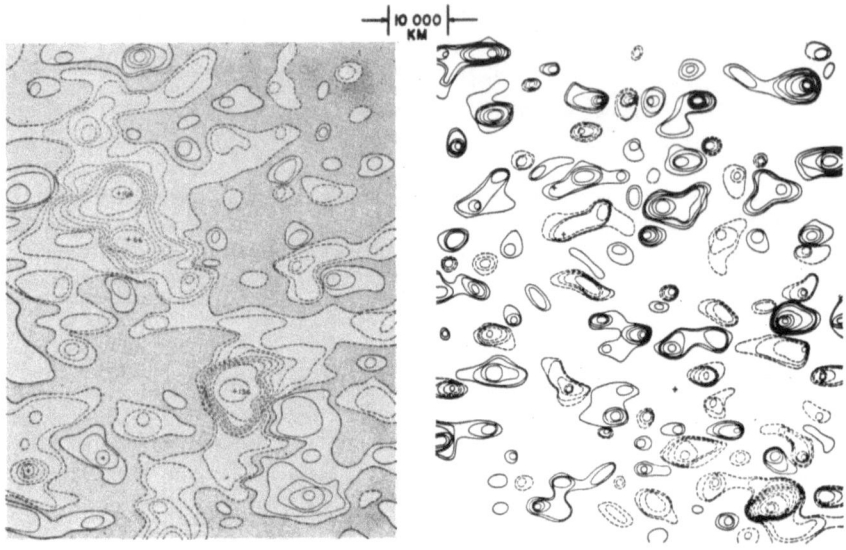

Fig. 1.   *Left*, magnetic map of quiet photosphere. Isogauss lines are at 2, 4, 8, 16, 20, 24, 32, 64 G, with solid being (+), or outgoing flux, dashed (−) or ingoing flux. *Right*, velocity map made simultaneous to the magnetic map showing distribution of oscillatory elements. Isovelocity lines are 0.2, 0.3, ..., 1.2 km/s dashed is up, solid is down (Livingston, 1968).

ical model of a supergranule by Weiss (1970) has resulted in a field amplification of the order of 10, in agreement with observation.

A considerable amount of analytical work has been done on the possible fine structure which may be produced by smaller scale turbulent motions as distinct from the large scale persistent supergranular motions. Again, dynamical effects are often neglect-

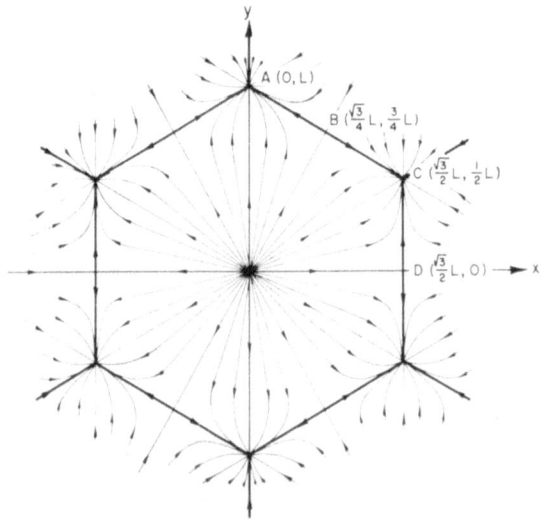

Fig. 2.   Schematic of the streamlines of the horizontal motions at the top of a hexagonal supergranule (Clark and Johnson, 1967).

ed, perhaps more justifiably in this case since in practice a statistically steady state is more likely to be determined by field diffusion than by equipartition of energy. There are, however, conflicting opinions on this point. Parker (1969) shows that if the prescribed motions are such that the velocity autocorrelation time at all wavenumbers is short compared with the time taken to travel one wavelength at the corresponding Fourier velocity amplitude then the magnetic field is amplified indefinitely at all wavenumbers. The system will therefore attain a steady state determined by the dynamics, i.e. by some form of equipartition of energy. An extension of the statistical approach has been made recently by Nagarajan (1970) to include both the dynamics and finite time ratios. The results show that, although equipartition is attained for large wavenumbers at the outset, the maximum power later shifts indefinitely towards lower wavenumbers. On the other hand a computational approach with prescribed random motions by Moss (1970) has shown that a statistically steady state is in fact reached. The rms amplification of the mean field is approximately $\frac{1}{2}(R_m)^{0.35}$, where $R_m = \tau \sigma v_{turb}^2/c^2$ and τ is the mean lifetime of a turbulent element. The amplification is almost independent of the ratio of the mean lifetime to the time taken to travel one wavelength, at least for the small values of this ratio adopted in the computation. Values $v_{turb} = 0.4$ km/s, $\sigma = 10^{11}$ e.s.u. and $\tau = 300$ s, typical of the photospheric granulation, lead to $R_m \sim 50$. The resulting amplification by granular motion would be of the order of 2 only, and not necessarily correlated with the instantaneous velocity field. A typical magnetic configuration after a statistically steady state has been reached is shown in Figure 3. This is consistent with observations. It is also to be noted that oscillations of the same amplitude, with periods of the order of 10 s, treated as turbulence, would yield $R_m \sim 1$, i.e. would produce no amplification and hence no finer structure.

The reason for this apparent contradiction with the analytical approaches is not clear. It may be because Moss' model is only two-dimensional, or because the evolution has not been continued long enough in time.

Fig. 3.   Field distortion by random velocity field with $R_m = 80$. (Ordinate is vertical direction; uniform magnetic flux is prescribed at upper and lower boundaries) (Moss, 1970).

1.2. Active Region Photosphere

Within an AR the simultaneous longitudinal and transverse field measurements of Severny (1965) confirmed that the mean field is roughly what would be expected from flux tubes emerging through the photosphere, in the spot umbrae, into a tenuous force-free atmosphere. The field rapidly becomes transverse outside the umbral regions and areas of opposite polarity are connected. Studies of the motion of gas in the arch filaments connecting spots, by Bruzek (1969), confirm that spots grow by the emergence of subphotospheric flux tubes. The vortical structure hinted at by the chromospheric network was also confirmed, in particular by Steshenko (1969), and is consistent with the idea that the subphotospheric flux tube giving rise to the AR must have been appreciably twisted. It has been emphasized by Veeder and Zirin (1970) that the transverse field usually extends as far as the quiet photosphere. It is tempting, therefore, to think of the AR field as embedded in the general field, the magnetic surface which separates the two fields intersecting the photosphere at the boundary of the transverse field. The helmeted structures in the corona, q.v., Newkirk (1967), support this picture.

As in the quiet photosphere the mean field of an AR is shredded into flux tubes with diameters ranging from the order of 5000 km down to at least the limit of resolution, of the order of 1000 km, with hints of sizes down to 300 km. q.v., Abdusamatov and Krat (1969). The presence of transverse fields within spot umbrae themselves, as shown by Severny (1965) is striking evidence of this, as are the magnetic knots and pores with longitudinal field strengths of the order of 1000 G outside the spots, studied by Sheeley (1967) and Steshenko (1967). These flux tubes must protrude through the tangle of flux tubes of the transverse field. The apparent motion of gas across lines of force is taken by Severny (1965) to indicate that this is occurring in the relatively field-free space between flux tubes. Regarding the connectivity of these flux tubes, the work of Beckers and Schröter (1968a) shows that part of the flux from a spot can return into the surrounding photosphere in the form of the magnetic knots. In one AR containing a unipolar spot some 3000 knots were measured, of which 2000 were of opposite polarity to the spot, 1000 were of the same polarity, and the net flux of the knots probably balanced the flux from the spot. In this case the entire flux from the spot would appear to have re-entered the photosphere as flux tubes, while the remainder of the knots returned as loops into the photosphere in the form of knots of opposite polarity. An evolutionary study of the photospheric network within an AR by Sheeley (1969) revealed a number of bright points in the photospheric emission, moving away from the spot, with velocities of the order of 1 km/s, and collecting in the surrounding supergranular network. If these are identified with flux tubes this strengthens the picture of flux tubes being convected away from the spot by the supergranular motion, and this could be the mechanism by which the spot decays. It is not obvious, however, that the disappearance of the bright points in groups indicates the mutual annihilation of flux tubes as Sheeley suggests. It could equally well be due to changes in the magnetic forces on coalescence. Regarding the areas of network which arise in situ as diverging bright points, it seems likely that these are produced by flux tubes emerging from below the photosphere in much the same way as the whole AR develops.

No satisfactory theory of the formation of these filaments is available. While it is true that an area of magnetic flux can be concentrated by twisting, as in the theories of Gold and Hoyle (1960) and Alfvén (1968), the result obtained by Anzer (1968), that a flux tube with a free cylindrical boundary cannot be twisted at all without destroying its cylindrical geometry, indicates that the flux tubes in previous theories are likely to be unstable. A new approach to the effect of turbulence in respect of the shredding of a magnetic field has been made by Mogilevski (1970) in terms of the statistical properties of a system of granules with isolated magnetic fields. Under solar conditions granule diameters down to 100 km are estimated, and the shredding could be due to collective effects of the granules. A gravitationally governed Rayleigh-Taylor instability would seem to be another possibility, but a model has not yet been worked out.

As was noted by Steshenko (1967) magnetic knots only cause a darkening of the photosphere if the field strengths exceed 1200 G. Some progress has been made towards a theoretical explanation of this by Simon and Weiss (1970) who have shown that the critical flux for penumbral formation is 10^{20} Mx, with a radius of 1500 km, corresponding to a mean field strength of 1400 G.

There is no considerable difference in the granular motion within an AR as compared with the quiet photosphere. These motions are therefore likely to be decoupled, or almost decoupled, from the magnetic field, as in the quiet photosphere. Unlike the quiet photosphere, however, the bulk of the magnetic flux, of the order of 10^{22} Mx is in the form of flux tubes some 1000 km in diameter and with field strengths of the order of 1000 G, moving outwards from the spot with transverse velocities of the order of 1 km/s. But like the large flux concentrations in the photospheric network, ohmic diffusion is unlikely to be important.

1.3. Spot umbrae

In the umbra the mean field, of the order of 3000 G, contains an uncertain number of independent flux tubes inclined at considerable angles to the vertical, as Severny's (1965) work shows. In addition to these there appear to be gaps in the field which correspond to the umbral dots studied, notably by Beckers and Schröter (1968b). The dots are probably some 200 km in diameter and are as bright as the quiet photosphere. Their lifetimes are about 1500 s. A spot umbra normally shows 20 or so dots situated largely in the outer part of the umbra. Their origin poses a difficult theoretical problem; it has been suggested by Wilson (1969) that they cannot be very much larger vertically than horizontally. If this is true they do not represent the tops of deep-seated convection columns and therefore could not have the observed lifetimes without a source of energy. Joule heating would appear to be insufficient even allowing for the reduction of the effective conductivity due to the Piddington drift of the neutral gas through the plasma ions. New calculations of this conductivity by Kopecký and Kuklin (1969) and by Oster (1968) have shown values of the order of 10^9

462 P. A. SWEET

e.s.u. However, the dots do not appear to have large motions, and it is unlikely that they contribute to any non-radiative flux from the umbra. Spot umbrae as a whole do not, in fact, exhibit direct Doppler velocities comparable with the granular velocities of the surrounding photosphere. The turbulent velocities of the order of 2 km/s inferred from line widths and curves of growth, q.v., Zwaan (1968) must represent a finer-scale turbulence than in the granulation.

2. Chromospheric and Coronal Heating

2.1. THE QUIET ATMOSPHERE

Observational models show that about 2×10^6 erg/s/cm^2 of non-radiative energy input is required, of which about 10% is needed for the upper chromosphere and corona. Ulmschneider (1970) shows that the lower chromosphere can be heated by non-magnetic shock-waves as Osterbrock (1961) originally suggested, although the short time interval of 10 s required between shocks does not agree with the granulation periods of 300 s. The sensitivity of the heating to the interval between shocks is indicated by the curves in Figure 4. Moreover, Frazier's (1968a, b) analysis of the granulation velocities shows that the granulation flux is well below the total required. Flux with

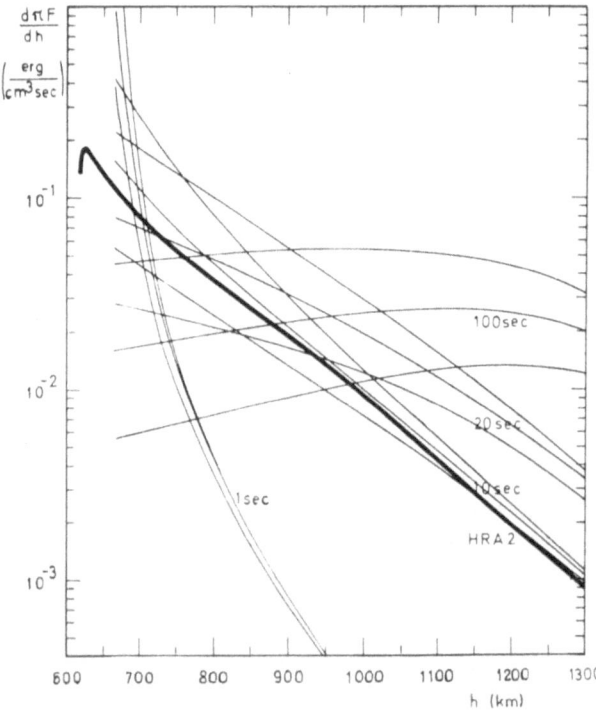

Fig. 4. Radiative energy loss (heavily drawn) compared with shock heating in the HRA2 model. There are 4 triplets of curves differing by the indicated wave periods P. The initial flux at $h = 600$ km (above $\tau = 1$) in each triplet is, from bottom to top: 1.0×10^6, 2.0×10^6, 4.0×10^6 erg/s/cm^2 (Ulmschneider, 1970).

wavelengths below the present resolving power would be necessary on this theory. Certainly the 10 s period is closer to the theoretical period of maximum acoustic power output from the convection zone calculated by Stein (1968), and it may be that the short period oscillations observed by Howard (1967) contain the flux required.

If studied alone, and without reference to AR's, it is not entirely clear whether the quiet corona is due to magnetic effects or not. It may be significant that non-magnetic shock-wave models, notably by Kuperus (1965) and Ulmschneider (1967), cannot be extended downwards into the lower chromosphere. This may be due to a lack of understanding of the sharpness of the chromosphere – corona boundary. It was first pounted out by Kuperus and Athay (1967) that the sudden change in the temperature

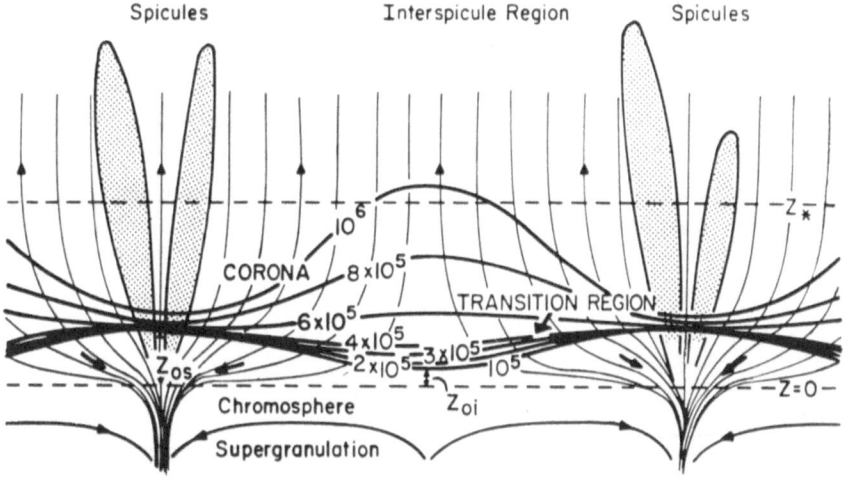

Fig. 5.   A schematic cross-section of the upper chromosphere – lower corona showing magnetic field
lines, isotherms and directions of the observed motions (Kopp and Kuperus, 1968).

gradient at the base of the transition region implies an input of thermal conduction energy in excess of the observed radiative loss. Any input of shock energy only aggravates this problem. Kopp and Kuperus (1968) have suggested that the channeling of the magnetic field into the intersupergranular regions concentrates the thermal conduction flux into these regions of the chromosphere, thus giving a possible origin for the spicules, as indicated in Figure 5. This mechanism depends on whether the channeling of the photospheric flux is continued up to the transition region. Clark (1968) shows that this depends sensitively on any vertical motions present, but is not inconsistent with observed supergranular motions.

A recent review of coronal heating has been given by Kuperus (1969). If coronal heating is magnetic in origin it would seem likely that the energy is propagated along the intersupergranular flux tubes. The fields in these would have the triple effect of increasing the acoustic generation rate in the convection zone, lowering the level in the chromosphere at which the magnetic stress becomes comparable with the gas pressure, and raising the level at which magnetohydrodynamic waves become shocks. The propagation of torsional waves along such flux tubes has been considered by Howe (1969), using an extension of the Hill spherical vortex, torsional motions being less affected by gravity. It is evident from these calculations that dissipation by the Piddington mechanism has to be allowed for before it can be certain that the wave will penetrate into the corona.

2.2. ACTIVE REGIONS

Although the coronal emission can increase by factors of 10–100 relative to the quiet Sun it is not known to what extent the net radiative loss from the low chromosphere is changed. Since this latter is so much the greater part of the total for the atmosphere the enhancement of the total non-radiative flux from the photosphere is uncertain. It is possible that the corona is heated independently of the low chromosphere by flux propagated along the magnetic knots. In this connection the ohmic dissipation process at the boundary of intertwined flux tubes, first suggested by Hoyle and Wickramasinghe (1961), might be worth looking into again. A sketch of the random walk of flux tubes due to supergranulation motions, which would effect such an inter-twining is given in Figure 6.

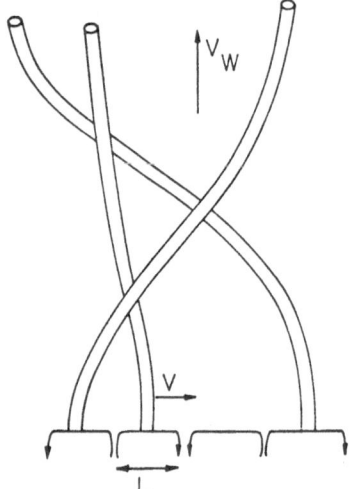

Fig. 6.   Intertwining of flux tubes due to random walk produced by supergranular motions (Jokipii and Parker, 1969).

2.3. SPOT UMBRAE

If the turbulent velocities of 2 km/s deduced from line widths were real and represented travelling waves, a wave flux of the order of 10^{10} erg/s/cm^2 would be implied, i.e., an appreciable fraction of the photospheric radiative flux, and more than sufficient for the entire coronal AR. It has been difficult to account for a wave flux of this order theoretically, q.v., Musman (1967). Syrovatskii and Zhugzdha (1968), however, have shown that if the Boussinesq approximation is dropped, a new mode of instability arises which would permit oscillatory convection. In this mode the gas oscillates in columns entirely along the field and interacts only radiatively with adjacent columns. Thus the criterion for instability differs only slightly from the Schwarzschild criterion, e.g., for a polytropic layer with $n (= d \log \varrho / d \log T) = 2$, the criterion for the second harmonic with a large field is $\gamma > 1 \cdot 32$ as compared with the Schwarzschild criterion $\gamma > 1 \cdot 5$ with no field. The usual form of the local Schwarzschild criterion with a field, q.v., Gough and Taylor (1966) is

$$B_v^2 / (B_v^2 + \gamma P) > \frac{1}{\gamma} - \frac{n}{n+1},$$

where B_v is the vertical field strength, which of course would predict stability for a sufficiently large field. A similar study, but without an explicit application to the sunspot problem, has been made by van der Borght (1969).

3. Prominence Fields and Stability

3.1. FIELD TOPOLOGY

Studies of the field configuration in the neighbourhood of prominences, deduced partly by calculating the potential field corresponding to the observed photospheric longitudinal fluxes, and by making use of the observed longitudinal and transverse components at photospheric and chromospheric levels, have been made by Rust (1967, 1970), Iospha (1968) and Malville (1968). Figure 7 shows a typical configuration. These have confirmed the Kippenhahn and Schlüter (1957) and Dungey (1958) theoretical models of the gravitational support of a gas filament or sheet by sagging lines of force of the AR field, showing particularly that there is a component of the field along the

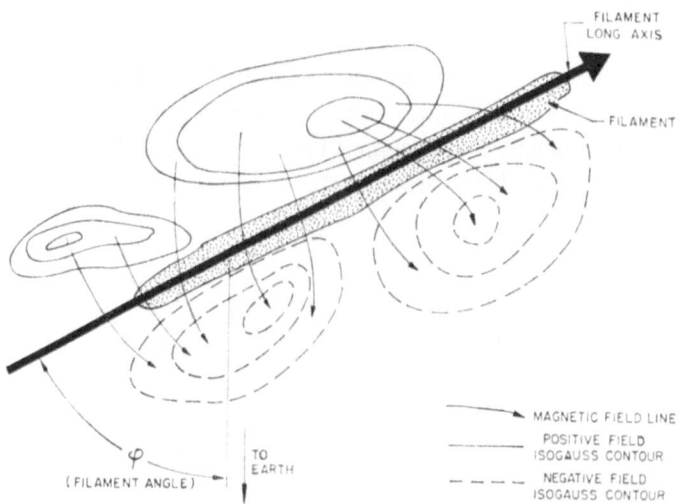

Fig. 7. Magnetic configuration in the neighbourhood of a filament inferred from photospheric magnetic fluxes (Rust, 1967).

filament. If the field lines of the component at right angles to the filament are closed up round it the filament field is topologically a twisted flux tube embedded in the general AR field. A recent theoretical model on these lines has been by Anzer and Tandberg-Hanssen (1970). It differs from Dungey's in that, although the filament as a whole is in hydrostatic equilibrium with gravity, there is no detailed equilibrium within the filament itself.

3.2. OVERALL STABILITY

The important question of the dynamical stability of prominences, treated as current sheets, rather than filaments, has been dealt with rigorously by Anzer (1969). The necessary and sufficient conditions for stability are

$$d|B_x|/dz \geqslant 0 \quad \text{and} \quad d(\log|B_x|)/dz \geqslant d\log\sigma/dz,$$

where z is the height coordinate, B_x is the horizontal field crossing the sheet normally and σ is the mass density within the sheet. It is pointed out that the first condition is often observed to be satisfied, although it is difficult to assess the second criterion, involving the density.

3.3. INTERNAL STABILITY

The work of Lerche and Parker (1968) in the context of the interstellar medium and the parallel work of Nakagawa (1970) applied to the prominence problem, shows that Rayleigh-Taylor instabilities put constraints on the geometry of a quiescent prominence. Figure 8 shows how an element of gas can be detached from its surroundings by a local interchange of flux tubes. This work shows that, in addition to the well-known thermal instability of an optically thin plasma whose emission rate decreases with increasing temperature, a plane plasma supported against gravity by a magnetic field in a constant direction is always unstable to non-thermal modes. The effect of changing the direction of the field with height, the so-called onion-skin configuration, can be tested in a preliminary way by considering the instability at an interface where the field direction changes discontinuously. This situation may also arise at the interface between the prominence and its supporting vacuum AR field.

Such an interface is unstable for wavenumbers along and perpendicular, respectively,

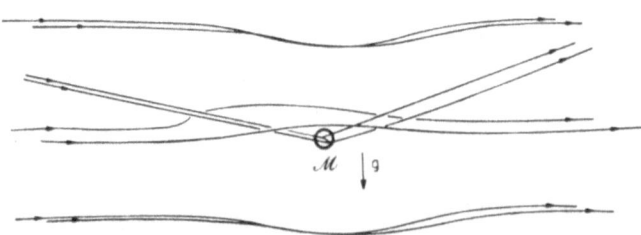

Fig. 8.   Sketch of the displaced lines of force in the neighbourhood of a small element of gas $\mathcal{M}$ suspended in a magnetic field (Lerche and Parker, 1968).

to the prominence field given by

$$k_x < (V_A^v/V_A)^2 \, g \cos^2 \theta/c^2$$
$$k_y < g/(V_A^v)^2 \sin^2 \theta$$

where V_A^v is the Alfvén speed corresponding to the supporting field, but with the prominence density, V_A is the Alfvén speed within the prominence, c is the thermal velocity of sound, g is the acceleration of gravity and θ is the inclination of the supporting field to the prominence field. Thus, as expected, the prominence is most stable where the two fields are at right angles. The second criterion shows that instability might be suppressed if the width of the prominence were less than something of the order of $(V_A^v)^2/g$. This latter value would seem amply large enough to allow the observed widths of prominences.

The possibility of a microscopic plasma instability, namely the two-stream instability between electrons and ions, is raised in the current interruption theory of filament discharges by Alfvén and Carlquist (1967), who point out that in a tenuous plasma the currents induced by twisting a flux tube can involve electron-ion drifts exceeding the local velocity of sound.

3.4. METASTABILITY

Finally, the usual warning should be heeded, namely, that the stability criteria as usually derived are marginal stability criteria depending on the sign of the energy taken to the second order, and are not sufficient to show whether a dynamical instability can actually arise in a given situation. For this purpose it is necessary to show that the system can be made to pass continuously from a sequence of stable configurations to a configuration in which, on continuing the sequence, no adjacent equilibrium is attainable, i.e., is in a metastable state. This requires that in marginal stability the next non-vanishing energy perturbation of higher order is negative.

4. Flares in Relation to AR Fields

It has long been realised that flares are associated with complex AR's during phases of rapid development, q.v. recent survey and analysis of the development of AR fields by Bappu et al. (1968) and the association with flares by Martres et al. (1968). Figure 9 shows a typical flare-producing magnetic configuration. The relationship at present is only a statistical one in the sense that within a given AR a number of flares can occur at random during a given phase, and it has been impossible to predict a flare from the instantaneous magnetic configuration. This may be partly because the only flare effect measurable with sufficient spatial resolution, i.e., the Hα-emission does not occur at the seat of the energy release. Neither has it been possible to measure the field sufficiently soon before and after a flare to be able to say how much of the change observed is due to the secular evolution of the AR and how much to the flare itself. In any case it is possible for the emerging and re-entering flux to change connectivity without changing the observed longitudinal flux distribution. It is necessary to examine the transverse field in order to establish this. A study by Bappu *et al.* (1968) of the field of an AR developing on the border of a pre-existing AR shows evidence of some reconnection, as shown in the example in Figure 10.

Flares are more strongly correlated with the motions of gas in the surges and loops and with the evolution of coronal filaments, q.v., survey by Öhman (1968). Thus the sudden disappearance of filaments on the disk, and eruptive prominences at the limb,

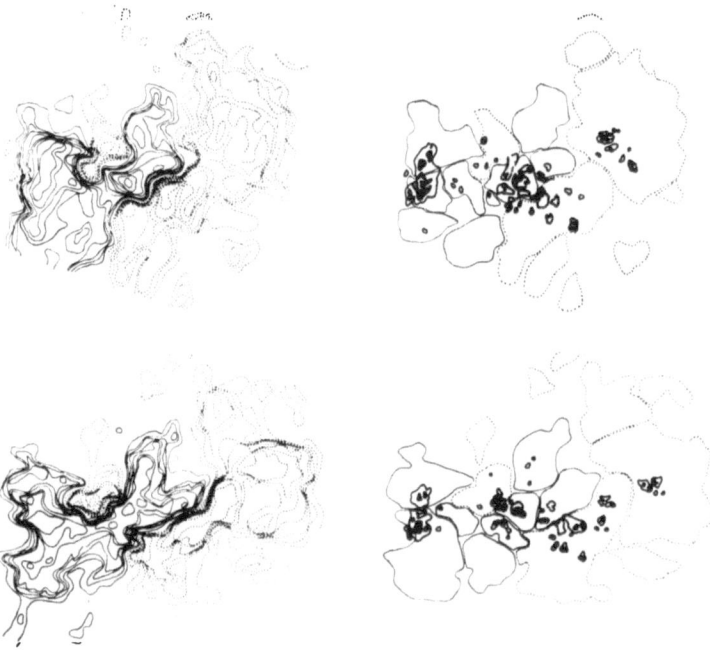

Fig. 9.   An example of the development of a complex flare – producing AR, CSSAR 9. *Upper*, 30/09/65; *lower*, 1/10/65. *Left*, isogauss of longitudinal field; *right*, spots and flare positions (hatched) (Martres *et al.*, 1968).

formerly regarded merely as the consequences of flares may now indicate that the filaments are the primary seat of the energy release and the Hα-emission is due to high energy particles arriving in the photosphere or chromosphere along the field lines. The tendency of the Hα-flare to be located in a series of localized regions on either side of filaments supports this view. Flare models on these lines have been put forward by Carmichael (1964), Sturrock (1968) and, in somewhat more detail in respect of the topology of the field, by Křivský (1968) and Rust (1970). q.v., also the author Sweet (1969) for a recent review of possible mechanisms of flares. In these models it is implied that the triggering of the filament would be effected by the changes in the large-scale field of the AR, and would be governed by criteria concerning the interaction of the entire filament with the AR field, along the lines given by Anzer (1968). The current interruption theory of Alfvén and Carlquist (1967) involves the internal stability of the filament.

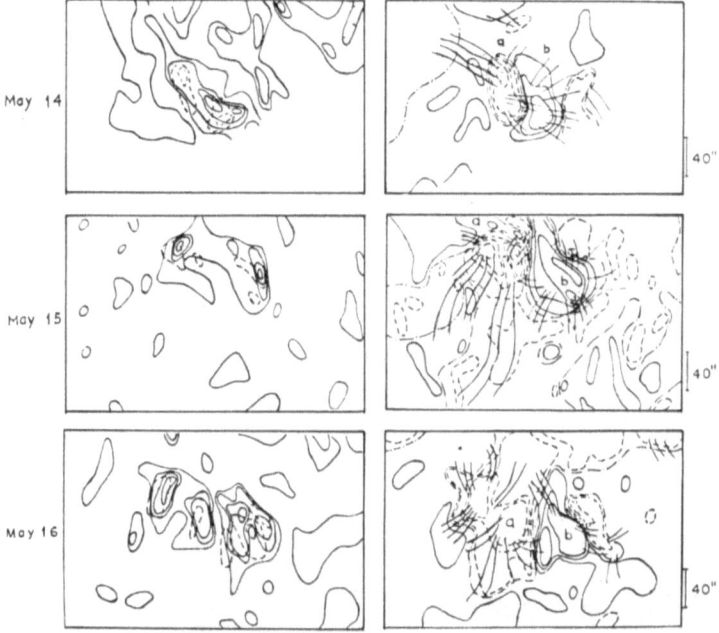

Fig. 10.   Changes in the transverse field during a period of rapid evolution of an AR. *Left*, full lines
are isogauss of the transverse field at 50 G intervals, commencing at 100 G; dotted lines are longitudinal
field concentrations. *Right*, isogauss of the longitudinal component and sketches of the lines of force
drojected on to tangent plane. Full lines indicate *S* polarity, dotted lines *N* polarity (Bappu *et al.*, 1968).

5. Particle Acceleration

5.1. NUMBERS INFERRED FROM OBSERVATIONS

The onset of a flare is characterized by radio and hard X-ray bursts indicating the
production of electrons in the energy range 20 keV − 3 MeV. The upper end of the
range is inferred, somewhat uncertainly, from the radio bursts. The lower end is uncertain since the soft end of the X-ray spectrum is of longer duration than the hard end
and is of a form that could be due to thermal electrons at temperatures of the order
of 10^8 K. The spectrum in the range 20 keV − 100 keV can be represented by $N(E) \propto$
$\propto E^{-n}$ where $N(E)$ is the density of electrons with energies greater than E, and n is
in the range 2 to 4, assuming that the hard X-rays are due to bremsstrahlung, q.v.,
Holt and Ramaty (1969). The spectrum falls off more rapidly at energies exceeding
100 keV. Assessments of the total number of electrons involved in the hard X-ray
bursts show that it is not impossible that the total energy emitted in the flare in the
form of plasma ejection, EUV and Hα, subsequent to the initial burst, is derived from
the energy of these high energy electrons by absorption and transformation into K.E.
of mass motion, q.v., Kane and Anderson (1970). In a 2B flare the total number of
high energy electrons is of the order of 10^{38}.

    In a proton flare protons up to energies of 30 GeV are observed, the total energy involved being much less than that of the electrons. Owing to uncertainties of propagation in interplanetary space it is not possible to tell whether these are produced simultaneously with the electrons.

5.2. TURBULENT ACCELERATION

In the theories of Pneuman (1968) and Elliot (1969) high energy particles would be produced by an enhancement, within the AR, of the shockwave turbulence normally responsible for the heating of the chromosphere and corona. While there is considerable literature on the rate of acceleration of particles for given turbulence spectra, little is known about the nature of the turbulence. If, for example, the corona could be regarded as an assembly of magnetohydrodynamic shocks with a mean interval τ at any point, at a mean Mach number M, then statistical Fermi acceleration would apply. In this case the time taken for an acceleration from a thermal energy to a relativistic energy E is given by

$$t \simeq \frac{c\tau}{2MV_A} \left(1 + \tfrac{1}{2} \log_e (E/E_0)\right)$$

where E_0 is the rest mass energy of the particle and V_A is the Alfvén speed. With plausible values in the corona, $\tau = 10$ s, $M = 10$ and $V_A = 1000$ km/s, the time scale is 10^2 s. This is of the right order of magnitude for flares. On the other hand if the magnetohydrodynamic waves are broken down into solitons then, according to Gintsberg (1966) the maximum energy attainable by an electron is $\tfrac{1}{2} m_p M^2 V_A^2$, where m_p is the proton mass. Under the same conditions this latter amounts to approximately 10 keV, which is insufficient.

Again, if sufficient energy is present in the form of electron plasma waves with a uniform spectrum then, according to Pikel'ner and Tsytovich (1969), an electron with energy E is accelerated according to the relation

$$\frac{d(E/E_0)}{dt} = \alpha (E/E_0)^{-1/2}$$

where $\alpha \sim (v_{Te}/c)^2 R \omega_{pe}$, and v_{Te} is the electron thermal velocity, R is the ratio of Langmuir to thermal energy and ω_{pe} is the electron plasma angular frequency. With values of v_{Te} and ω_{pe} typical of the corona, the above mechanism is potentially capable of very rapid acceleration. The difficulty is in assessing the magnitude of the Langmuir energy.

5.3. DIRECT ELECTRIC-FIELD ACCELERATION

Only the component of the electric intensity in the direction of the magnetic field can be effective; the component at right angles to it merely causes the particle to participate in the mass fluid motion. The current-sheet mechanisms of Petschek (1964) and Syrovatskii (1966), and the current interruption mechanism of Alfvén and Carlquist (1967) during the flash phase, essentially provide such component fields. All three mechanisms are undoubtedly successful in producing electric currents intense enough to cause the electrons to drift relative to the ions with the velocity of sound. The subsequent evolution of the high current region is governed by the turbulence caused by the resulting electron-ion two-stream instability.

Since the exterior magnetic fields driving the currents are the same in both the Petschek and Syrovatskii mechanisms it would seem that they must develop in exactly the same way. The situation has been treated in some detail by Friedman and Hamberger (1969). The electric current is determined not by Coulomb collisions but by collisions with the small-scale fluctuations of electric intensity produced by the turbulence. Under these circumstances the plasma behaves as if it had the effective conductivity $0 \cdot 1 \, (4\pi n_e e^2/m_e)^{1/2}$ esu first established by Buneman (1959). The ambient plasma is heated on entering the high current region, the magnetic energy carried in dividing roughly equally between thermal and kinetic energy. Taking an ambient magnetic field of the order of 500 G a temperature of 10^8 K, as indicated by the soft X-rays, is therefore attainable at electron densities of the order of $4 \times 10^{11} \, \mathrm{cm}^{-3}$ i.e., at chromospheric level. As in the case of ohmic currents, particles of a few times the mean thermal energy, i.e., from a threshold energy $\simeq 50$ keV are able to run away down the mean electric field. The full voltage drop available in the Petschek mechanism is of the order of 30 GeV. The authors are also able to calculate the particle spectrum.

It is tempting to believe that much the same physics might govern the Alfvén and Carlquist mechanism. There are two differences, however, which must be borne in mind. In the first place the heating of the plasma in the Petschek mechanism is derived from the magnetic energy of plasma entering the current region from outside as distinct from the former mechanism where it is heated in situ. Secondly the Alfvén and Carlquist mechanism appears to be the result of a self-triggering instability, whereas the current-sheet mechanism is not. It would therefore seem that the former

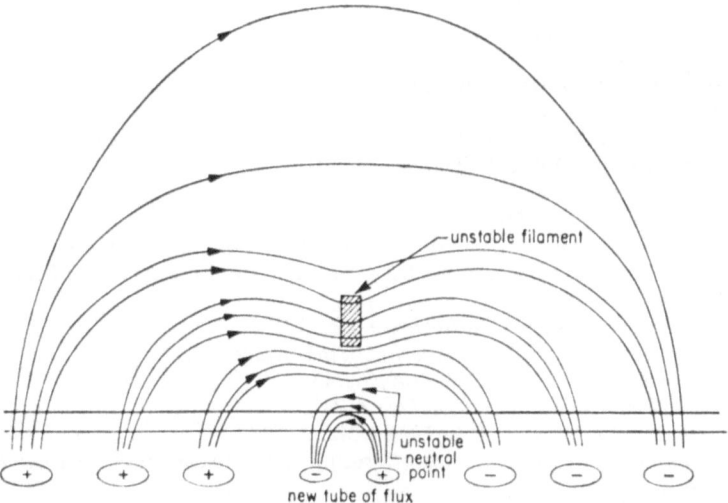

Fig. 11.   Schematic representation of Kippenhahn-Schlüter-Dungey model of a prominence. The intrusion of a new flux tube from below the photosphere as shown could make the prominence unstable and also give rise to a field topology suitable for the Petschek mechanism (Rust, 1970).

mechanism must pass the test of metastability within itself, while the latter needs an external metastability. In this connection it must be noted that the untwisting of the flux tube, which releases the energy in the former mechanism, would tend to remove the two-stream instability.

In conclusion, the Petschek mechanism, operating at chromospheric level, possibly below a chromospheric filament as in the topology of Rust (1970), shown in Figure 11, is fast enough and would appear capable of producing high energy particles in the required energy range. The total energy available in the Rust filament model, however, seems a little low for a large flare, and should be looked into more carefully. Again, the theory is incomplete in that the ability of the filament overall to become metastable has not been established.

References

Abdusamatov, H. I. and Krat, V. A.: 1969, *Solar Phys.* **9**, 420.

Alfvén, H.: 1968, *Ann. Geophys.* **24**, 341.

Alfvén, H. and Carlquist, P.: 1967, *Solar Phys.* **1**, 220.

Anzer, U.: 1968, *Solar Phys.* **3**, 298.

Anzer, U.: 1969, *Solar Phys.* **8**, 37.

Anzer, U. and Tandberg-Hanssen, E.: 1970, *Solar Phys.* **11**, 61.

Bappu, M. K. V., Grigorjev, V. M., and Stepanov, V. E.: 1968, *Solar Phys.* **4**, 409.

Beckers, J. M. and Schröter, E. H.: 1968a, *Solar Phys.* **4**, 142.

Beckers, J. M. and Schröter, E. H.: 1968b, *Solar Phys.* **4**, 303.

Buneman, O.: 1959, *Phys. Rev.* **115**, 503.

Bruzek, A.: 1969, *Solar Phys.* **8**, 29.

Carmichael, H.: 1964, in *AAS-NASA Symp. Phys. Solar Flares*, (ed. by Hess, W. N.) U.S. Govt. Printing Office, Washington, p. 451.

Clark, A.: 1968, *Solar Phys.* **4**, 386.

Clark, A. and Johnson, A. C.: 1967, *Solar Phys.* **2**, 433.

Dungey, J. W.: 1958, *Cosmical Electrodynamics*, p. 54, C.U.P.

Elliot, H.: 1969, in *Proc. XI COSPAR Symp. Tokyo*, p. 356.

Frazier, E. N.: 1968a, *Astrophys. J.* **152**, 557.

Frazier, E. N.: 1968b, *Z. Astrophys.* **68**, 345.

Friedman, M. and Hamberger, S. M.: 1969, *Solar Phys.* **8**, 104.

Gintsberg, M. A.: 1966, *Soviet Astron.* **10**, 434.

Gold, T. and Hoyle, F.: 1960, *Monthly Notices Roy. Astron. Soc.* **120**, 89.

Gough, D. O. and Taylor, R. J.: 1966, *Monthly Notices Roy. Astron. Soc.* **133**, 85.

Holt, S. S. and Ramaty, R.: 1969, *Solar Phys.* **8**, 119.

Howard, R.: 1967, *Solar Phys.* **2**, 3.

Howe, M. S.: 1969, *Astrophys. J.* **156**, 27.

Hoyle, F. and Wickramasinghe, N. C.: 1961, *Monthly Notices Roy. Astron. Soc.* **123**, 51.

Iospha, B. A.: 1968, in K. O. Kiepenheuer (ed.), 'Structure and Development of Solar Active Regions', *IAU Symp.* **35**, 261.

Jokipii, J. R. and Parker, E. N.: 1969, *Astrophys. J.* **155**, 777.

Kane, S. R. and Anderson, K. A.: 1970, *Astrophys. J.* **162**, in press.

Kippenhahn, R. and Schluter, A.: 1957, *Z. Astrophys.* **43**, 36.

Kopecký, M. and Kuklin, G. V.: 1969, *Solar Phys.* **6**, 241.

Kopp, R. A. and Kuperus, M.: 1968, *Solar Phys.* **4**, 212.

Křivský, L.: 1968, in K. O. Kiepenheuer (ed.), 'Structure and Development of Solar Active Regions', *IAU Symp.* **35**, 465.

Kuperus, M.: 1965, *Rech. Astron. Obs. Utrecht* **17**, 1.

Kuperus, M.: 1969, *Space Sci Rev.* **9**, 713.

Kuperus, M. and Athay, R. G.: 1967, *Solar Phys.* **1**, 361.

Leighton, R. B., Noyes, R. W., and Simon, G.: 1962, *Astrophys. J.* **135**, 474.

Lerche, I. and Parker, E. N.: 1968, *Astrophys. J.* **154**, 515.

Livingston, W. C.: 1968, *Astrophys. J.* **153**, 929.

Malville, J. McKim: 1968, *Solar Phys.* **5**, 236.

Martres, M., Michard, R., Soru-Iscovici, I., and Tsap, T.: 1968, *Solar Phys.* **5**, 187.

Mogilevski, E. I.: 1971, this volume, p. 480.

Moss, D. L.: 1970, *Monthly Notices Roy. Astron. Soc.* **148**, 173.

Musman, S.: 1967, *Astrophys. J.* **149**, 201.

Nagarajan, S.: 1971, this volume, p. 487.

Nakagawa, Y.: 1970, *Solar Phys.* **12**, 419.

Newkirk, G.: 1967, *Ann. Rev. Astron. Astrophys.* **5**, 213.

Öhman, Y. (ed.): 1968, in 'Mass Motions in Solar Flares and Related Phenomena', *Nobel Symp.* **9**, 13.

Oster, L.: 1968, *Solar Phys.* **3**, 543.

Osterbrock, D. E.: 1961, *Astrophys. J.* **134**, 347.

Parker, E. N.: 1963, *Astrophys. J.* **138**, 552.

Parker, E. N.: 1969, *Astrophys. J.* **157**, 1119.

Petschek, H. E.: 1964, in *AAS-NASA Symp. Phys. Solar Flares* (ed. by Hess, W. N.) U.S. Govt. Printing Office, Washington p. 426.

Pikel'ner, S. B.: 1963, *Soviet Astron.* **6**, 757.

Pikel'ner, S. B. and Tsytovich, V. N.: *Soviet Astron.* **13**, 5.

Pneuman, G. W.: 1968, *Solar Phys.* **2**, 462.

Rust, D. M.: 1967, *Astrophys. J.* **150**, 313.

Rust, D. M.: 1970, *Astrophys. J.* **160**, 315.

Severny, A. B.: 1965, *Soviet Astron.* **9**, 171.

Sheeley, N. R.: 1967, *Solar Phys.* **1**, 171.

Sheeley, N. R.: 1969, *Solar Phys.* **9**, 347.

Simon, G. and Weiss, N. O.: 1970, *Solar Phys.* **13**, 1.

Stein, R. F.: 1968, *Astrophys. J.* **154**, 297.

Steshenko, N. V.: 1967, *Publ. Krymsk. Astrofiz. Obs.* **37**, 21.

Steshenko, N. V.: 1969, *Publ. Krymsk. Astrofiz. Obs.* **39**, 245.

Sturrock, P. A.: 1968, in K. O. Kiepenheuer (ed.), 'Structure and Development of Solar Active Regions', *IAU Symp.* **35**, 471.

Sweet, P. A.: 1969, *Ann. Rev. Astron. Astrophys.* **7**, 149.

Syrovatskii, S. I.: 1966, *Soviet Astron.* **10**, 270.

Syrovatskii, S. I. and Zhugzhda, Yu. D.: 1968, *Soviet Astron.* **11**, 945.

Ulmschneider, P. H.: 1967, *Z. Astrophys.* **67**, 193.

Ulmschneider, P. H.: 1970, *Solar Phys.* **12**, 403.

Van der Borght, R.: *Astron. Astrophys.* **2**, 96.

Veeder, G. J. and Zirin, H.: 1970, *Solar Phys* **12**, 391.

Weiss, N. O.: 1970, private communication.

Wilson, P. R.: 1969, *Solar Phys.* **10**, 404.

Zwaan, C.: 1968, *Ann. Rev. Astron. Astrophys.* **6**, 135.

Discussion

Wilson: Would not the oscillatory convection suggested by Syrovatskii and Zhugzdha transport too much energy into the visible region of the spot so that it would be brighter than is actually observed?

Sweet: It would not be possible to assess the flux without incorporating the mechanism into a model of the overall structure of the spot. There have certainly been difficulties in the past in achieving an energy balance in spot models.

Athay: The question of magnetic field line orientation above the supergranules is of fundamental importance to problems related to the chromosphere-corona transition region. Could you comment further on what is known, or predicted by theory, concerning the field lines over the supergranules? To my knowledge we have no direct observational evidence as to the true character of the field lines in this region.

Sweet: The height to which the flux concentrations can persist has been considered theoretically by Clark. It does seem possible from his work that the concentrations can extend to the chromosphere-corona transition layer. However, his work was based on a linear theory which may be difficult to apply to the large flux variations involved.

Wild: You mentioned that in both the Petschek and the Alfvén flare mechanisms there was a problem in obtaining sufficient energy release to explain the flare. Can you indicate how one estimates an upper limit to the energy available by these mechanisms?

Sweet: In both mechanisms the energy depends on the amount which can be stored by deforming the potential field into a force-free field by motions at photospheric level. It is impossible to derive this from direct observations of the field, but from a theoretical view-point it is difficult at present to see how appreciable energy can be stored without making the field unstable. It certainly seems less than the 10^{32} ergs required for a large flare.

SUNSPOT MAGNETIC FIELDS AND UMBRAL DOTS

P. R. WILSON

Dept. of Applied Mathematics, University of Sydney, Sydney, Australia

Abstract. The fine structure features of the umbral magnetic fields (i.e. large field gradients and changes of polarity in local regions) are considered as evidence of strong non-linear interactions between magnetic and thermal or mechanical forces in the umbra. It is suggested that the umbral dots observed in white light are the optical manifestation of these interactions. A three-dimensional radiative transfer analysis of possible models for these bright features is discussed and this enables one to place limits on the geometry of these features and on the non-radiative energy requirements of the models. Of the several models considered those which were compatible with convection as the source of this energy were found to be quite inconsistent with the available data. The most likely model was found to have a diameter of 200 km, a height of 50 km and an average emission of non-radiative energy of 4×10^3 erg s^{-1}/cm^{-3} throughout this region. It is shown that this is two orders of magnitude greater than the energy available from Joule heating by locally twisted magnetic fields. However, if the energy flux transported through the umbra by Alfvén waves is partially dissipated in regions of locally twisted fields it is shown that the emission into a volume of the above dimensions is of the right order of magnitude.

1. Introduction

One of the problems of doing theoretical work in this field is the difficulty of obtaining reliable observations. If one takes a set of observations at any given time and attempts to develop a theory or model to account for them, it may be that by the time the theory is presented, no one believes the observations any more. The observational data which form the basis of the present study are the following:

(i) Umbral dots are of approximately photospheric brightness and have diameters of order 200 km. This has been deduced by Beckers and Schröter (1968) from colour intensity ratios. It is generally agreed that they have lifetimes of between 30 and 60 min.

(ii) In some local regions of the umbra the magnetic field may have a large horizontal gradient, and even change polarity within distances of order 500 km. Mogilevsky (1967) and others suggest that these variations are associated with the umbral dots. Zwaan and Buurman (1971) has studied one sunspot umbra in which a particularly dark region showed no evidence of umbral dots or field inhomogeneities, while outside this region, umbral dots were observed together with indications of small scale polarity reversals.

(iii) An umbral dot is an isolated structure; i.e. the distance between neighboring dots is large compared with their diameters (Danielson, 1964; Beckers and Schröter, 1969).

Although there are several observers who would reject some of these 'observations', there is sufficient evidence for them to warrant an investigation of some of their theoretical consequences and these turn out to be rather interesting. However, it should be emphasized that this is essentially a model atmosphere investigation and that the results must be treated with some caution. They, do however, suggest a new mechanism for small scale atmospheric heating which appears to be worth further investigation.

Howard (ed.), Solar Magnetic Fields, 475–479. All Rights Reserved.
Copyright © 1971 by the IAU.

2. The Umbral Dots

Some attempts have already been made to investigate multicomponent umbrae. In 1963, Makita postulated a two-component umbra in order to explain the observation of both Fe I and Fe II lines. Obridko later considered the effect on the mean umbral opacity of patches of photospheric matter occupying less than 10% of the spot area.

Both Makita and Obridko appear to envisage hot vertical columns extending through a height of order 1000 km. However, it is probably intuitively obvious that a hot column of diameter 200 km, height 1000 km, and effective temperature 5800 K cannot exist in a region of temperature 4000 K for periods of up to one hour without having a considerable warming effect on its surroundings. This is a problem in three-dimensional transfer theory and I recently carried out an analysis of some possible models (Wilson, 1969). It was shown that, because of these three-dimensional effects, the postulated surface area of a bright feature in the umbra places a constraint on the depth of the feature. By considering various models, it was possible to show that those with surface diameters 300 km or less could not extend through depths greater than 300 km below the surface.

This is of some significance because, if convection (i.e. the upward motion of hot plasma along the field lines) is to be considered as the explanation of the dots, they must extend to considerable depths in the umbra since most umbral models do not show a sharp temperature increase until depths of order 800 km are reached. Because models with surface diameters of 300 km or less cannot extend through depths greater than 300 km, it would appear that they are inconsistent with convection as the energy source. Although a model of diameter 500 km could not be limited in this way, the computed emergent intensities (when integrated over a seeing function) were inconsistent with the direct optical data of Beckers and Schröter. Of the smaller models considered, one, with diameter 200 km and depth 250 km, was found to be consistent with these data. Although the data are too uncertain to warrant any strong claims for it, it will serve as a working model for this discussion.

3. Departures from Radiative Equilibrium

The method of solving the transfer equation involves postulating models for the emission of non-radiative energy into the region of the dot. A similar method was used in investigating granulation models.

For model 1 it was shown that a maximum emission of 2.5×10^4 erg s^{-1}/cm^{-3} is required. If convection is ruled out as the source of this energy, the magnetic data, which indicate large horizontal field gradients in the dot region, suggest Joule heating as a likely source and this may be estimated from $E = (c/4\pi \, \text{curl} \, H)^2/\sigma$. Mogilevsky's observations suggest that curl $H = \partial H_z/\partial x$ might be of order 20 G/km and, taking the conductivity $\sigma = 10^{10}$ esu, this yields only 20 erg cm^{-3}s^{-1}, a value consistent with estimates of 0.4–40 erg cm^{-3}/s^{-1} given by Kopecký and Obridko (1968).

Beckers (private communication) suggested that it was perhaps unfair to consider the maximum energy emission required. However, the model derived from the transfer analysis gives the dimensions of the region of abnormal emission (i.e. diameter 200 km and depth 50 km) and if the actual non-radiative emission is averaged over this volume, the energy required is 4×10^3 erg cm^{-3} s^{-1}, which is still two orders of magnitude greater than the best which might be obtained from Joule heating. Having made considerable but unsuccessful efforts to bridge this gap, I feel that it is necessary to look for another source of non-radiant energy.

4. Alfvén Waves

Both Musman (1967) and Savage (1969) have investigated the possibility of energy transport through the umbra by hydromagnetic waves. Both use linear stability analysis which must be somewhat suspected in view of the complex nature of sunspot fields revealed by these data. However, the boundary conditions also appeared to present considerable difficulties. Musman, using open boundary conditions at the upper boundary of the superadiabatic layer, found that so much energy was carried away by the hydromagnetic waves that all the oscillatory perturbations were quickly damped. Savage, however, found that allowing for the density decrease in the stable layer permits sufficient reflection of these waves to reduce the threshold for the emission of hydromagnetic waves from these layers. Using a sunspot model of Yun, Savage suggests that hydromagnetic waves can be maintained and, calculating the energy flux from

$$F = \tfrac{1}{2} \varrho v_{max}^2 v_a$$

(v_a is the Alfvén velocity), he finds a flux of 2×10^{10} erg cm^{-2} s^{-1}. Thus he is able to account for perhaps $\tfrac{1}{3}$ of the entire sunspot flux deficit (for umbral area 3×10^{18} cm^2 this gives 6×10^{28} cf 4×10^{29} erg/s). Savage considers the possibility that umbral dots are due to hydromagnetic waves being absorbed in these layers but discards it because the frictional damping length, as given by the formula $\zeta_{frict} = (v_A/\omega^2)(\varrho_n + \varrho_i/\varrho_n \tau_n)$, is of order 10^7 km which is far too large.

However, these calculations assume vertical parallel fields whereas the evidence of the large gradients in these regions suggest that the field lines are folded over and under these conditions the frictional damping length must be greatly reduced.

Consider an umbral model in which field lines are vertical outside a region of diameter 200 km and carry a hydromagnetic wave flux of 2×10^{10} erg cm^{-2} s^{-1}. Within this region the field lines are folded over a height which, according to our working model, is of order 50 km. Consider the case in which all the hydromagnetic energy flux is converted into thermal energy by frictional damping throughout the volume. Then the energy released per unit volume and second is

$$E = F/h_c.$$

If $h_c = 50$ km, then $E = 4 \times 10^3$ erg cm^{-3}/s^{-1}.

This agreement with the energy required is too good to be true. However, it does suggest that this mechanism might be worth further investigation.

5. Discussion

It is important to point out that a field reversal in the neighborhood of the dot is not a necessary condition for the mechanism, although it would probably be sufficient. Any large local gradient in the field would probably be adequate to disrupt the flow of Alfvén energy but whether it would be reflected back downwards or dissipated locally would depend on the particular configuration considered.

Zwaan's observation of a particularly dark region of uniform field without dots and a more normal umbral region in which dots and field reversals can be detected is of considerable significance. If a sunspot is cool because the deep magnetic field inhibits the convection, then undoubtedly there will be some equipartition of magnetic and thermal energy in this region which will enhance the field strength. Gough (1966) has shown that the most unstable mode of convection in an umbral region is a deep narrow cell. When fully established, eddy viscosity changes the most likely configuration to a broader eddy structure, but where the magnetic field prevents the flow around the eddy, the field may be deformed into a shape typical of the long thin mode rather than those which exist in established convection. In these regions the field would exhibit a pronounced fine structure and even a reversal of polarity over a small horizontal scale without altering the average overall field strength.

If this takes place well below the visible region, one would expect that the umbra should be dark and that the field should be uniform, with no evidence of reversal. However, if in parts of the umbra this interaction occurs closer to the visible region, then some folding (and hence reversal) of the field may still exist in the visible layers and thus give rise to the umbral dots.

These considerations suggest that it would be an observational point of some interest to know whether dots are always seen provided the umbra is sufficiently over-exposed, or whether there are some spots in which no umbral structure is observable. If over-exposure increasingly reveals the presence of more dots, one would like to know whether they become everywhere dense or whether at a given depth in the umbra they can be regarded as individually isolated features.

References

Beckers, J. M. and Schröter, E. H.: 1968, *Solar Phys.* **4**, 303.
Gough, D. O.: 1966, thesis, University of Cambridge.
Kopecký, M. and Obridko, V.: 1968, *Solar Phys.* **5**, 354.
Makita, M.: 1963, *Publ. Astron. Soc. Japan* **15**, 145.
Mogilevski, E. I., Demkina, L. B., Ioshpa, B. A., and Obridko, V. N.: 1968, in K. O. Kiepenheuer (ed.), 'Structure and Development of Solar Active Regions', *IAU Symp.* **35**, 215.
Musman, S.: 1967, *Astrophys. J.* **149**, 201.
Obridko, V. N.: 1968, *Bull. Astron. Inst. Czech.* **19**, 186.
Savage, B. D.: 1969, *Astrophys. J.* **156**, 707.
Wilson, P. R.: 1969, *Solar Phys.* **10**, 404.
Zwaan, C. and Buurman, J.: 1971, this volume, p. 220.

Discussion

Rösch: On the photographs which have been shown, more penumbral filaments appear, less bright, on the inner side of the penumbra, when the aperture increases, and finally, umbral dots appear to fill the entire umbra. So it seems more realistic to imagine that the limit between penumbra and umbra is not a vertical cylinder, but a dish, through the content of which one sees less and less bright penumbral filaments on the edges, and umbral dots in the center; these umbral dots could be just the tips of filamentary features of the same nature as the so-called penumbral ones. The smaller the diameter of the dish, the more transparent its central part, so that umbral dots appear brighter in small spots than in large ones. Of course, one should try to explain how such structures could fit with all that is known about the three dimensional magnetic velocity fields.

Wilson: You have raised several interesting questions of umbral and penumbral structure. However it would require much more time than we have now to reply to them adequately. I do think that the question of the intrusion of penumbral filaments into the umbra and the form of the umbral-penumbra boundary is of great importance.

Kuperus: It seems to me that you treat the convective motions and the Alfvén waves rather separately. There is not really a need for doing so since it seems that both types of motions can be strongly coupled with a high degree of efficiency. For rotational waves (internal gravity waves) in stable layers this coupling has been demonstrated by M. J. Lighthill.

Wilson: I did not intend to imply that they should be treated separately – far from it. I think that the inhibition of convection and the generation of Alfvén waves at some depth below the surface of the penumbra must be intimately connected. The equipartition of energy between convective and magnetic modes may give rise to the enhancement and bending of the field in these deep regions. Some of these bends in the field may also occur in the visible layers and give rise to the dots. In umbral regions where no dots are seen the magnetic-convective interaction may occur at great depths; where many dots are observed it may be much closer to the surface.

Athay: I find it very difficult to have any degree of confidence in quantitative estimates of mechanical energy requirements in sunspots. Such estimates imply that you have both a good radiative equilibrium model and a good actual model for the spot, and I doubt that we have either.

Wilson: The model does not involve a discussion of the mechanical energy requirements of sunspots as a whole. The calculations concern only the non-radiative energy required to produce this particular model of a bright dot isolated in a surrounding medium which has temperature density and pressure parameters of a typical umbra.

Musman: I do not believe that the amplitude of Alfvén waves would be large enough to cause reversals in the sign of the magnetic field. My theory predicted that disturbances would most easily escape at small amplitudes.

Wilson: I do not suggest that the Alfvén wave motions cause the field reversals. Rather that the 'observed' reversals, caused possibly by 'would be' convective motions will cause dissipation of energy carried by the Alfvén waves. However, I agree that there is no quantitative theory for this process as yet.

STATISTICAL MODEL OF SMALL SCALE DISCRETE STRUCTURE OF MAGNETOPLASMA IN ACTIVE REGIONS OF THE SUN

E. I. MOGILEVSKY

Academy of Sciences of the U.S.S.R., Institute of Terrestrial Magnetism, Ionosphere and Radio Wave Propagation, Dept. of Solar Physics, Moscow, U.S.S.R.

Abstract. A possible statistical model of the solar magnetoplasma in solar active regions consisting of a totality of small-scale current vortex plasma elements ('subgranules') is discussed. Some results are given which naturally follow from such a model: the small value of effective conductivity in the macro-structures of the plasma (hence, the possibility of a greater mobility and changeability of field and plasma for a short time), the formation of filamentary-structural elements, the oscillating regime of the magnetoplasma in an active region. A principal scheme is given of the experiment on a solar magnetograph with very high resolution, which is able to reveal discrete fields of the subgranules. The dispersion equation is derived for macro-magnetic oscillations of the statistical ensemble of magnetized subgranules. The possibility is noted to reveal macro-magnetic oscillations and waves during observations of low-frequency modulation in the solar radio-emission. It is shown that at solar radioburst propagation in the corona, (according to the model under consideration, some regularly located plasma inhomogeneities should be present in the active region) Bragg's diffraction takes place. Typical properties of complex radiobursts may be received (extreme narrowness of the band, short lifetime, directivity).

Recent observations of magnetic fields on the Sun have shown (Severny, 1965; Beckers, 1968), that the solar magnetoplasma (where $8\pi/H^2 \times (P + \frac{1}{2}\varrho u_\perp^2) \ll 1$) is characterized by a small-scale ($\simeq 3$–5×10^7 cm) structure. This fundamental property of the solar magnetoplasma inherent in the terrestrial and cosmic plasma as well, determines the main macroscopic characteristics of solar activity phenomena. With a large scale characteristic of solar phenomena, the great dynamic behavior of the activity in the magnetized plasma is incompatible with the practically infinite conductivity of the solar atmosphere. The stable filamentary structure of the cosmic plasma, where a longitudinal section is much greater than a transverse one, cannot be explained with the theory of stability of a laboratory plasma (Kadomzev, 1963).

A number of other peculiarities are observed in solar activity phenomena, and for their explanation some independent hypothesis is necessary. These peculiarities are: variations of the magnetic flux balance in an active region; the problem of transport of energy, field and matter in the magnetized solar plasma; special features of the spectra of flares and prominences; a long time-scale generation of subthermal particles in active regions; two component structure of magnetic fields in the spots, and others.

In our paper (Mogilevsky, 1968) an attempt has been made to show that the above mentioned properties of the solar magnetoplasma may result from the observed fine structure of the field and plasma. We try to give a description of the magnetoplasma as a statistical ensemble of discrete small-scale elements.

In theoretical works by Ermakov (1969, 1970) some static properties of individual plasma clouds with current vortices have been considered. He has theoretically proved the possibility of the existence of discrete excited elements of plasma with their own magnetic fields in the potential field, those are known as 'subgranules' or 'gyrons'. The analytical presentation of field and plasma during the process of the streamlining of the element (without a boundary discontinuity) has been obtained for a spheroid. The geometry of the element may be different as well. Such a solution for the stable subgranules is possible, when there is minimum energy of a system with a current vortex, i.e. a spontaneous appearance of subgranules may occur. However, aside from a 'diamagnetic' model, a model with a field discontinuity on the surface, i.e. that with the boundary current, is possible. A stable axi-symmetric subgranule of that type may be estimated. At least two types of subgranule models possibly exist: a mixed model and such a type of model in which a transformation of a 'diamagnetic' one into a 'paramagnetic' and that of the reverse action, may occur. The determination of models is done on the basis of corresponding experimental data.

The problem of the mechanism and place of generation of individual subgranular decay has not yet been formulated. However the properties of a statistical ensemble of interacting 'collisionless' subgranules in the external homogeneous (or quasi-homogeneous) potential field H may be discussed. For the description of that anisotropic statistical ensemble a distribution function $f(q, v)$ in phase space (q, v) may be obtained from Vlasov's (1966) kinetic equation for space-limited structures:

$$\frac{\partial f}{\partial t} + \text{div}_q \, vf + \text{div}_v \langle \dot{v} \rangle f = 0 \tag{1}$$

together with an integral equation

$$\int \int f(q, v, t) \, dq \, dv = F(\theta_m, \varphi_m, H), \tag{2}$$

where θ is the parameter of the energy of sporadic subgranular movement (the 'magnetic temperature');

$$\varphi_m(q) = \int f(q, v) \, dv \text{ is the space concentration of subgranules.}$$

In the most simple axi-symmetric case (Z is the axis of symmetry) the distribution function is:

$$f = A\varphi_m(z - ut) \exp\left(-\frac{\varepsilon_m}{\theta_m} + \alpha I\right), \tag{3}$$

$$I = r\left(u + \frac{e}{mc} A_\varphi\right) \tag{4}$$

where A and α are constants, ε_m is the density of the subgranular kinetic energy, A_φ is the component of the vector potential of the magnetic field, e and m are the total charge of polarization and subgranular mass, u is the macrovelocity of the ensemble. Knowing the distribution function macro properties of that plasma may be determined.

So, for example, the transference of the magnetic field in the active region from the photosphere into the chromosphere and the corona may be the result of sub-granular diffusion. It is possible to estimate the value of the effective conductivity in the macro volume ($V \gg V_s$ the subgranular volume)

$$G_{ef} = G_o R_m^{-1} \tag{5}$$

R_m is Reinold's magnetic number (for the solar atmosphere it is $\gtrsim 10^4-10^5$). Physically it means that the conductivity is sharply reduced in the macro volume due to the frequent scattering of electrons in the external subgranular fields, while the common values of the plasma conductivity inside the subgranules are not changed. The sharp difference between turbulent and molecular viscosity may be an analog to the above-mentioned effect. Hence, during macro events in the solar magnetoplasma with discrete magnetized subgranules, the time scale of the field dynamics will be essentially different from that which it should be according to the classical values of the plasma conductivity. It is shown in (Mogilevsky 1968) that if the subgranular model considered is used, the above mentioned characteristic properties of the solar magnetoplasma may be explained (at least qualitatively) from a *unified point of view*. In our work (Mogilevsky *et al.*, 1968) the data and our explanation of the observed quasi-pi-component of Zeeman's splitting in the sunspot umbras testify to the correctness of the model. The two component model of the field and the plasma of the spots proves the truth of that idea as well (Mogilevsky *et al.*, 1968; Obridko, 1968).

At present the urgent necessity appears to be to obtain direct experimental data not only about the existence of discrete small-scale structures, but to determine characteristic parameters of a subgranule. There are great difficulties in carrying out field measurements for elements, the characteristic dimensions of which are of the order of the spatial resolution of modern solar telescopes. However, we may try to undertake such a task, using a special method of observation by means of our solar magneto-graph. As a matter of fact the idea of statistical Fourier spectrometry may be used. Instead of the usual optical light modulator with an ADP crystal, a light modulator employing photo elasticity is used. In the water or crystal of the modulator a frequency-operated ultrasonic space grating is generated by a piezocrystal. The phase difference Γ of the wave λ with an optical path in the modulator $1' = 1\Delta n$, where Δn is the difference arising from the refraction constant, due to photo elasticity (it depends on the modulation wave phase), will be

$$\Gamma = \frac{2\pi}{\lambda} 1\Delta n\,(t). \tag{6}$$

Evidently, it is possible to select such a condition in the modulator, when the maximum $\Gamma_m = \lambda/4$. Then along the entrance slit of the magnetograph a modifying polarizing 'mosaic' will appear. A number of strips and their dimensions may be changed by frequency variations in the modulator. For example for the ultrasonic frequency $\Omega = 10^7$ cps the strip width (resolving capacity) of the IZMIRAN tower telescope is $0''.1$. If the magnetic field along the whole entrance slit is homogeneous, the magnitude of a magnetic field signal would not depend on the modulator frequency. If the slit crosses, for example, N discrete magnetic elements, then the strongest signal will be when $N = 2K(d/\Lambda)$, where d is the modulator height Λ is the length of the ultrasound wave, and K is a constant. Thus, the light is modulated in even harmonics Ω, it is changed in amplitude, and the number of strips in the modulator corresponds to the magnetic element number on the slit. Using Fourier analysis for the field intensity change during the frequency variation of Ω modulation, a motion about the average picture of the field distribution within the limits of similar discrete subgranules is formed. The disturbing influence of the atmosphere (vibration) may be reduced by the following:

(a) by means of putting into operation the phase detecting of the magnetic field signal, synchronized with the frequency of the modulator Ω and carrying out measurements on both frequencies of modulation (2Ω and 4Ω).

(b) carrying out some additional measurements (synchronous with the magneto-graph) on the same spectrograph in a neighboring nonmagnetic line of the atmosphere scintillations with leading in its spectrum (through the reverse Fourier filter) in the magnetograph channel.

The estimations show sufficient effectiveness of such a method.

Important information about properties of the considered statistical model of the magnetoplasma may be obtained from the analysis of its wave characteristics. Since a definite value of the magnetic moment μ is connected with discrete subgranules, some space regulation of the magnetic moments arises in the quasihomogeneous external potential magnetic field $\mathbf{H_0}$. This effect is broken by disturbances of individual subgranules. These disturbances may be of a non-coherent character, then they determine the macromagnetic noise (numerically it is characterized by parameter θ_m). They also may be of a coherent character as well, resulting in macromagnetic waves. For their description a mathematical formalism of a phenomenological description of spin waves may be used (Achiezer $et\ al.$, 1967). In our case the magnetic energy ω is in a single volume $V_0 > V_s$

$$\omega = F\left(\mu, \frac{\partial \bar{\mu}}{\partial q}\right) + \frac{1}{8\pi}(\tilde{H}^m)^2 - \bar{\mu}\tilde{H}_0, \tag{7}$$

where F is potential energy of subgranular interaction, $\tilde{H}^m = \langle \sum_i \tilde{H}_i^e \rangle$ is the magnetic field of the subgranular ensemble, averaging in Volume V_0, $\bar{\mu}$ is the average magnetic moment in a volume V_0. The time variations of the magnetic moment are described by a vector equation

$$\frac{\partial \bar{\mu}}{\partial t} = C[\bar{\mu} \times \tilde{H}], \tag{8}$$

where C is a constant proportionality factor, and the effective magnetic field $\tilde{H}$ is determined by a functional derivative

$$\tilde{H} = -\frac{\partial \omega}{\partial \mu}. \tag{9}$$

Equation (8) may be linearized, if the following approximate expressions are used for the magnetic moment $\bar{\mu}$ and the field $\bar{H}^m$

$$\bar{\mu}(q, t) = \bar{\mu}_0 + \bar{m}(q, t),$$
$$\bar{H}^m(q, t) = \bar{H}_0^m + h(q, t),$$

(10)

where $\bar{\mu}_0$ and $\bar{H}_0^m$ are the corresponding equilibrium values, around them occur regular variations of the moment and the magnetoplasma field, considered here. Then from (8) we shall obtain a linearized equation

$$\frac{\partial \bar{\mu}}{\partial t} = C[\bar{\mu}_0 \times \tilde{H}],$$

(11)

where

$$\tilde{H} = h - \beta \bar{m} + \alpha \Delta \bar{m},$$

(12)

β and α are constants.

For low frequencies (less than the ion gyrofrequency w_i) Maxwell's equations for h and $\bar{m}$ will be

$$\text{rot}\, h = 0$$
$$\text{div}\, h = 4\pi \,\text{div}\, \bar{m}.$$

(13)

In the common case, using Fourier's presentations* of vectors h and m we shall obtain

$$\bar{m} = \int m_0 e^{i(kq - wt)} dw$$

$$h = \int h_0 e^{i(kq - wt)} dw.$$

(14)

Then the Equations (13) give

$$\bar{k} \times h = 0$$
$$\bar{k}h = -4\pi \bar{k}\bar{m},$$

(15)

or

$$h(kw) = -ik\psi(kw)$$

(16)

where ψ is the Fourier component of a magnetic potential. On the other hand, using (14) in (8), we obtain

$$\bar{m}_i(kw) = \chi(kw) h_i(kw),$$

(17)

where χ is a tensor of the oscillating magnetic receptivity. Using (16) in (17), we obtain

$$\{k^2 + 4\pi k_i k_j \chi(kw)\} \psi(kw) = 0$$

(18)

or

$$\{k^2 + 4\pi k_i k_j \chi(kw)\} = 0.$$

(19)

Equation (19) represents the dispersion law of macro-magnetic waves, showing the

* In our case this expression may be confined by a finite row of Fourier harmonics.

connection between the frequency w and the wave vector k. Here it is important that for low frequencies w (where $w < w_i$) a dependence of χ (or μ) on frequency appears. And this result is important and may give the essential effects in wave propagation of the considered discrete elementary magnetoplasma (for example, the interaction between the macromagnetic and the low frequency plasma waves). Discovery and research of these low frequency oscillations ($10^{-1} < w < 10^3$ cps) in the solar plasma would give some new information about its structure.

The discrete structure of the magnetoplasma may bring up a number of interesting peculiarities in the metric and decametric radiowave propagation through the corona*. Let us assume that the subgranular structure exists in the active regions of the corona as well. Then we shall consider the radiowave propagation in frequencies $w \lesssim w_0$, where w_0 is the plasma frequency (the nature of these wave generations is not important in this case). A relatively regular space distribution of plasma clouds should be as a space diffraction grating. During the radiowave propagation with an angle to a chain of subgranules (the angle corresponds to Bragg's diffraction angle) the radio-emission intensity of a wave λ in the first side maximum will be determined by an expression:

$$I_{-1} = I_0 \sin^2 \frac{\pi \Delta n l}{\lambda},\tag{11}$$

where Δn is the difference of the refraction index in the unit path, l is the optical wave in the 'grating'. Let us take an example for 25 Mc/s (the characteristic dimensions of subgranules in the radiofrequency accepted by us) the deviation angle will $\sim 18°$ and $\sim 10\%$ of the emission will be directed to this side. A small space angle of deviation, determined by the length of a wave and 'grating' parameters causes a noticeable effect due to the static 'grating'. 'Statics of the grating' comes true because of the fact that macrowaves propagate with a velocity $V \lesssim V_A$, that is much lower than the group velocity of radiowaves in the corona. However, relatively long radio bursts may drift (to one or the other side), passing through the swinging space 'grating'. Those bursts are actually observed (Ellis, 1969; Markeev and Chernov, 1970). The narrow strip, the drift velocity and the burst duration will be more distinct for low frequencies than for high ones. This effect is actually revealed if we compare similar events on frequencies in the region of both 200 MHz and $\simeq 40$ MHz (Ellis, 1969; Markeev and Chernov, 1970).

References

Achiezer, A. I., Bazjachtar, V. G., and Peletminsky, S. V.: 1967, *Spin Waves*, Nauka, Moscow.
Beckers, J. M.: 1968, *Solar Phys.* 3, 258.
Beckers, J. M.: 1968, *Solar Phys.* 4, 303.
Ellis, G. R. A.: 1969, *Australian J. Phys.* 22, 177.
Ermakov, F. A.: 1969, *Geomagnetizm i Aeronomiya* 9, 593.
Ermakov, F. A.: 1970, Preprint IZMIRAN, (Rus.).
Kadomzev, B. B.: 1963, in *Some Questions of Plasma Theory*, Vol. 2, Gosatomizdat, Moscow.
Markeev, A. K. and Chernov, G. P.: 1970, *Astron. Zh.* 47, 1044.

* As is seen from estimation, for the shorter radiowaves the effect is insignificant.

Mogilevsky, E. I.: 1968, in *5th Consult. on Heliophysics and Hydromagnetics in Potsdam*, Geodat. Geophys. Veröffentl., Berlin, p. 95.

Mogilevsky, E. I., Demkina, L. B., Ioshpa, B. A., and Obridko, V. N.: 1968, in K. O. Kiepenheuer (ed.), 'Structure and Development of Solar Active Regions', *IAU Symp.* **35**, 215.

Obridko, V. N.: 1968, *Bull. Astron. Inst. Czech.* **19**, 183; 186.

Severny, A. B.: 1965, *Astron. Zh.* **42**, 217.

Vlasov, A. A.: 1966, *Statistical Function Distribution*, Nauka, Moscow.

Discussion

Sweet: What is the size of the subgranules, and how could they influence the non-radiative flux from the photosphere?

Mogilevsky: In order to determine the subgranule sizes it is necessary to assume a definite mechanism of dissipation for the current fields of subgranules. For admitted assumptions, estimations give the subgranule size value of the order of hundreds of km. It is close to the observed sizes of 'magnetic dots'.

If the magnetic field gradient exists then according to a model adopted by us the diffusion of subgranules and their macro-oscillations determine a non-radiative flux of the photosphere. This influence is determined with the variations of the plasma temperature and density which are put into the 'background' plasma by the subgranules.

EVOLUTION OF TURBULENT MAGNETIC FIELDS – APPROACH
TO A STEADY STATE

S. NAGARAJAN

MATSCIENCE, Madras 20, India, and
Université Libre de Bruxelles, Brussels, Belgium

Abstract. The dynamical evolution of a weak, random, magnetic excitation in a turbulent electrically-conducting fluid is examined under varying kinematic conditions. It is found that the results of an earlier paper (Kraichnan and Nagarajan, 1967) can be reliably extended to a stage of evolution wherein the magnetic spectrum has reached local equipartition with the velocity. The transfer of the magnetic energy to smaller wavenumbers (larger scales) is considerable and significant. This result is highly pertinent to the *turbulent dynamo* question, which has been variously investigated recently. The relevance of the coupling of the rms magnetic field to the magnetic modes of all scales in deciding the efficiency of this transfer is discussed.

1. Introduction and Review

In a number of recent investigations, (Parker, 1970; Moffatt, 1970; Parker, 1969; Krause, 1968; Rädler, 1968; Steenbeck *et al.*, 1966; Steenbeck and Krause, 1966, 1967; Krause and Rädler, 1971; Fitremann and Frisch, 1969; Vainshtein, 1970), the question of regeneration of a magnetic field, by turbulent motions has been reconsidered, under a variety of kinematic assumptions about the turbulence. In an earlier paper (Kraichnan and Nagarajan, 1967), we have reviewed the previous work on this subject in great detail and found that simple intuitive statistical arguments like equipartition, or analogical and heuristic kinematic considerations like the vorticity analogy are highly inadequate in resolving this question. In a recent paper, Kraichnan (1970) has considered the analogous question of the growth and propagation of the deviations between the point-to-point velocity fields in two flow systems, which are statistically identical. Here again, one finds that the ultimate evolution depends on the quantitative competition between the local-enhancement and sweeping-away processes in the wave-number domain. One needs a considerable amount of knowledge of the internal dynamics and characteristic times, and assertions of kinematic nature based on universal equilibrium hypotheses are highly inadequate.

In our paper referred to earlier, we could not carry our calculations very much forward in time, because we had no reliable information about the internal time structure of the combined fields of velocity and magnetic field, at that time. In a more recent paper (Nagarajan, 1971), we have investigated the internal structure of the steady state spectra on the basis of a detailed *dynamical* theory. In this, we have also reviewed the relevance of the ideas of Kolmogorov to the hydromagnetic case, keeping in mind the Galilean non-invariance of the hydromagnetic equations to a random constant magnetic field transformation. The cascade of energy in the hydromagnetic case is not strictly local in the wave number domain. A large scale rms magnetic field presents the possibility of Alfvén wave propagation along it and thus provides a significant dynamical coupling between magnetic fields of large and small

Howard (ed.), Solar Magnetic Fields, 487–504. All Rights Reserved.
Copyright © 1971 by the IAU.

scales. Our steady state considerations provide us with the necessary information about the local internal relaxation features and their relative magnitudes, so much so we plan to extend our earlier study of evolution of weak magnetic fields – to a stage in which the spectrum of the magnetic field has evolved sufficiently to a point of dynamical feedback to the velocity field and consequently a statistical steady-state.

And since we are basing our calculations on a well-considered dynamical theory of turbulence, we will be able to throw some light on the nature of the transfer of energy in the magnetic spectrum: in particular, without using either oversimplifications or idealisations of the characteristic length and time scales of the magnetic field and turbulence as have been done by Moffatt (1970), Parker (1969), Fitremann and Frisch (1969) or Vainshtein (1970).

2. The Dynamical Model

We start with a steady turbulence with an extended inertial range. The choice of the kinematic parameters and the wave number range is made suitably, so that we can talk of an extended equilibrium range, without worrying about the sources of input of energy into the system from the geometric range. Further, there exists a sufficiently noticeable dissipative tail to the spectrum at the high wave number end. The form of the spectrum and parameters are chosen so as to be compatible with the asymptotic requirements of the direct interaction approximation of Kraichnan (1958, 1959, 1965, 1966), with suitable modifications to reproduce Kolmogorov scaling.

A disturbance which is localized in the wave number range of the magnetic spectrum is introduced at time $t=0$.

Following the notations of our earlier papers (Kraichnan, 1958; Kraichnan and Nagarajan, 1967; Nagarajan, 1971), we can write the equation for the secular evolution of the two spectra for times >0 as

$$\left(\frac{\partial}{\partial T} + 2vk^2\right) E^V(k; T)$$

$$= \iint \frac{k}{2pq} \, dp \, dq \, [\{k^2 a_{kpq} E^V(p; T) \, E^V(q; T) \, \theta_{kpq}^{VVV}$$

$$- p^2 b_{kpq} E^V(k; T) \, E^V(q; T) \, \theta_{pqk}^{VVV}\}$$
$$+ \{k^2 a_{kpq} E^M(p; T) \, E^M(q; T) \, \theta_{kpq}^{VMM}$$
$$- p^2 C_{kpq} E^V(k; T) \, E^M(q; T) \, \theta_{pqk}^{MMV}\}] \tag{1}$$

$$\left(\frac{\partial}{\partial T} + 2\lambda k^2\right) E^M(k; T)$$

$$= \iint_{\Delta} \frac{k}{2pq} \, dp \, dq \, [k^2 d_{kpq} E^M(p; T) \, E^V(q; T) \, \theta_{kpq}^{MMV}$$

$$- p^2 h_{kpq} E^M(k; T) \, E^V(q; T) \, \theta_{pqk}^{MVM}$$

$$- p^2 j_{kpq} E^M(k; T) \, E^M(q; T) \, \theta_{pqk}^{VMM}] \tag{2}$$

The spectral functions $E^V(k; T)$ and $E^M(k; T)$ are connected to the velocity and magnetic fields as follows:

$$W^V(k; t, t') = (2\pi)^{-3} \int d^3(\mathbf{x} - \mathbf{y}) \langle \mathbf{U}(\mathbf{x}; t) \cdot \mathbf{U}(\mathbf{y}; t') \rangle e^{i\mathbf{k}\cdot(\mathbf{x}-\mathbf{y})}$$

$$W^M(k; t, t') = (2\pi)^{-3} \int d^3(\mathbf{x} - \mathbf{y}) \langle \mathbf{W}(\mathbf{x}; t) \cdot \mathbf{W}(\mathbf{y}; t') \rangle e^{i\mathbf{k}\cdot(\mathbf{x}-\mathbf{y})}$$

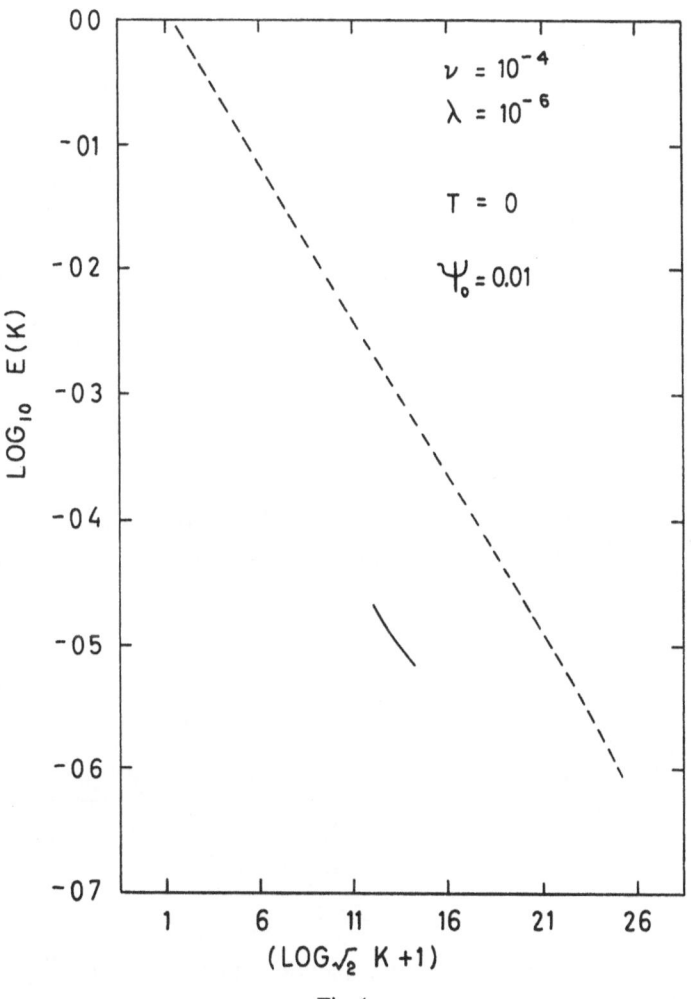

Fig. 1.

where $\mathbf{U}(\mathbf{x}, t)$ is the fluid velocity and $(4\pi\mu\varrho)^{1/2} \mathbf{W}(\mathbf{x}, t)$ is the magnetic induction field, ϱ is the fluid density, μ the magnetic susceptibility of the fluid, v and λ are the kinematic viscosity and magnetic diffusivity respectively.

We assume the turbulence to be homogeneous and isotropic

$$\tfrac{1}{2}W^V(k; t, t') = (4\pi k^2)^{-1} E^V\left(k; \frac{t+t'}{2}\right) R^V(k; t-t').$$

$$\tfrac{1}{2}W^M(k; t, t') = (4\pi k^2)^{-1} E^M\left(k; \frac{t+t'}{2}\right) R^M(k; t-t')$$

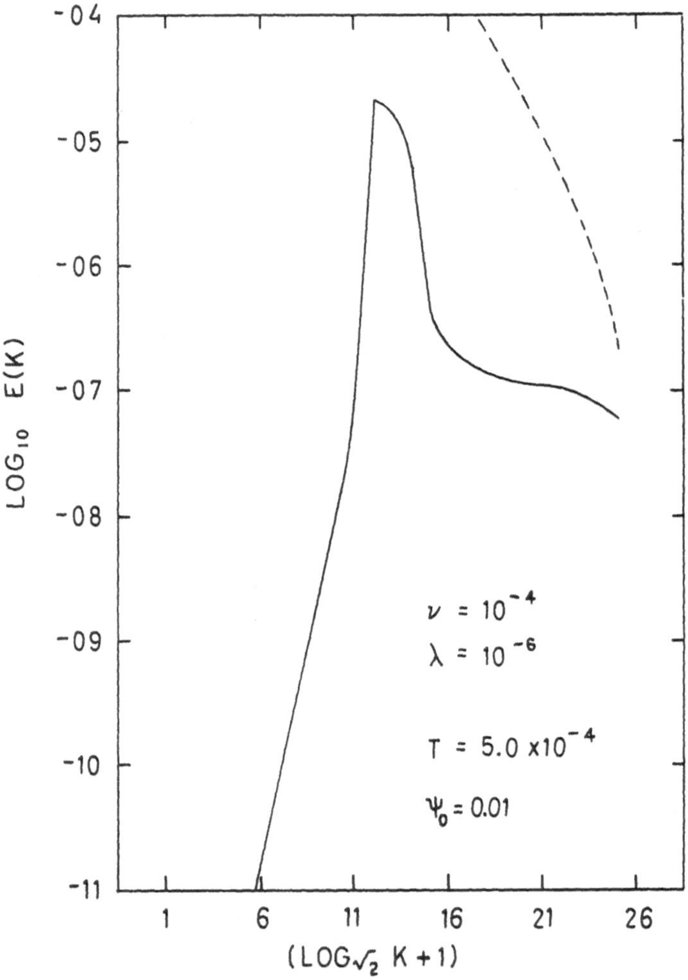

$\nu = 10^{-4}$

$\lambda = 10^{-6}$

$T = 5.0 \times 10^{-4}$

$\psi_0 = 0.01$

Fig. 2.

where $W^V\{\ \}$ and $W^M\{\ \}$ are energy functions and $R^V\{\ \}$ and $R^M\{\ \}$ are modal correlation functions.

The θ'-s which appear in Equations (1) and (2) are the effective memory times of the interaction between the three respective wave numbers. They are given by

$$\theta_{lmn}^{abc}(T) = \int_{-\infty}^{\infty} G_l^a(T-s) R_m^b(T+s) R_n^c(T+s)\, ds$$

(where a, b, $c = V$ or M) and $G^V(k; T)$ and $G^M(k; T)$ are the averaged response functions of the velocity and magnetic fields for the given wave number respectively.

In a general turbulent system in which a weak macroscopic (i.e. geometric range) disturbance in the magnetic spectrum is introduced at time $t = 0$, the θ'-s will be very complicated functions of the correlation and response features of the turbulence

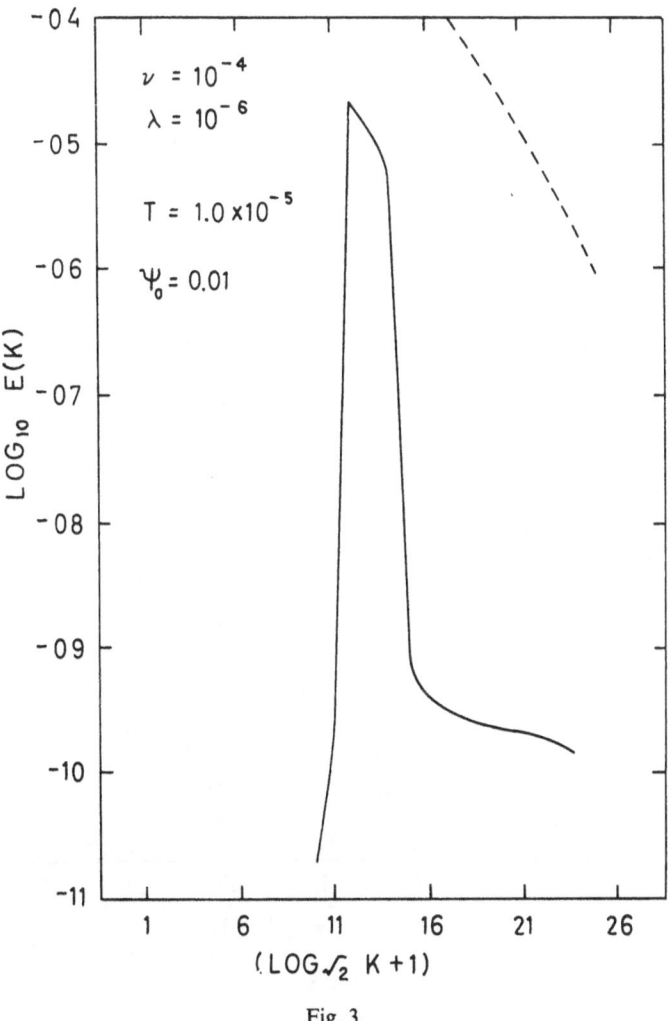

Fig. 3.

and initial magnetic field. But if we assume that the weak magnetic excitation is sufficiently localized in the inertial range, the secular time dependence of the θ'-s can be ignored. This point has been discussed in detail by Kraichnan (1959) in the hydrodynamic context. In the magnetic situation also much of the argument goes through unaltered.

We choose a form for the correlation and relaxation functions and the θ's from Nagarajan, 1971.

$$R^a(k; t) = \exp\left\{-\tfrac{1}{4}\pi\left(\zeta_a(k)\, t\right)^2\right\}$$
$$G^a(k; T) = \exp\left\{-\tfrac{1}{4}\pi\left(\eta_a(k)\, t\right)^2\right\}$$

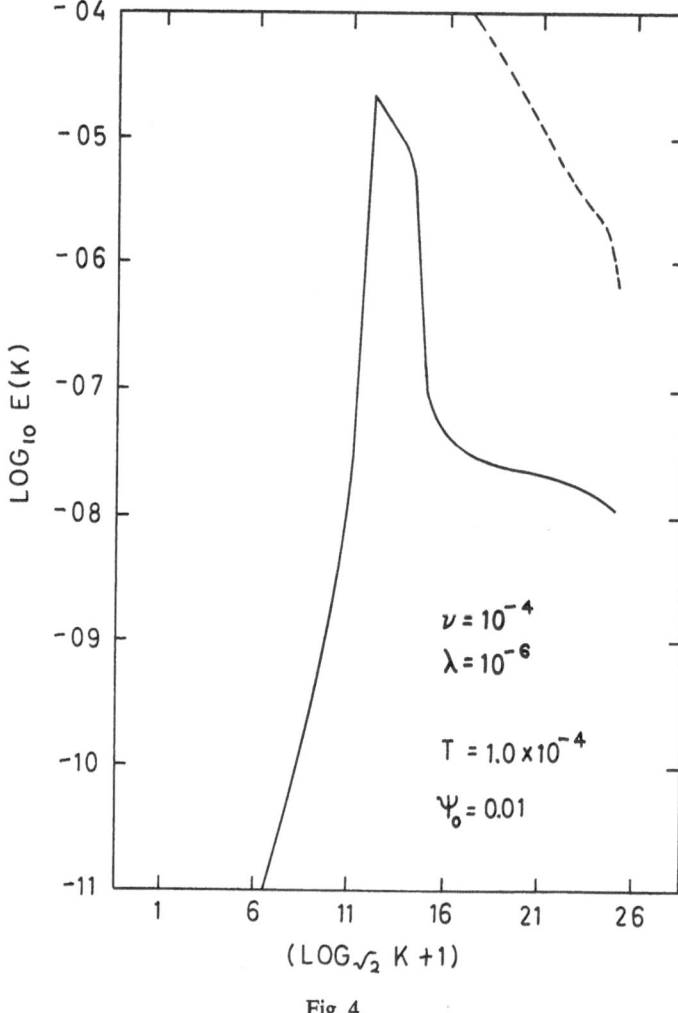

Fig. 4.

which gives for θ

$$\theta_{kpq}^{abc} = \left[\{\eta_a(k)\}^2 + \{\zeta_b(p)\}^2 + \{\zeta_c(q)\}^2\right]^{-1/2}.$$

Our elaborate study of the various extreme considerations of Galilean-invariance and Kolmogorov's arguments on the one hand and Galilean non-invariant Eulerian solutions on the other in the steady-state case (Nagarajan, 1971) convinces us that in so far as energy transfer information is concerned, the details of the internal correlation times are not very important. Using the results of this study, we evolve a quasi-Lagrangian scheme. We take the velocity correlations and relaxations to be Kolmogorovian i.e. decided by the local parameters of the position in the wave number spectrum. The magnetic terms are modulated by energy range parameters as in the unmodified direct interaction approximation of Kraichnan (1959, 1965). With

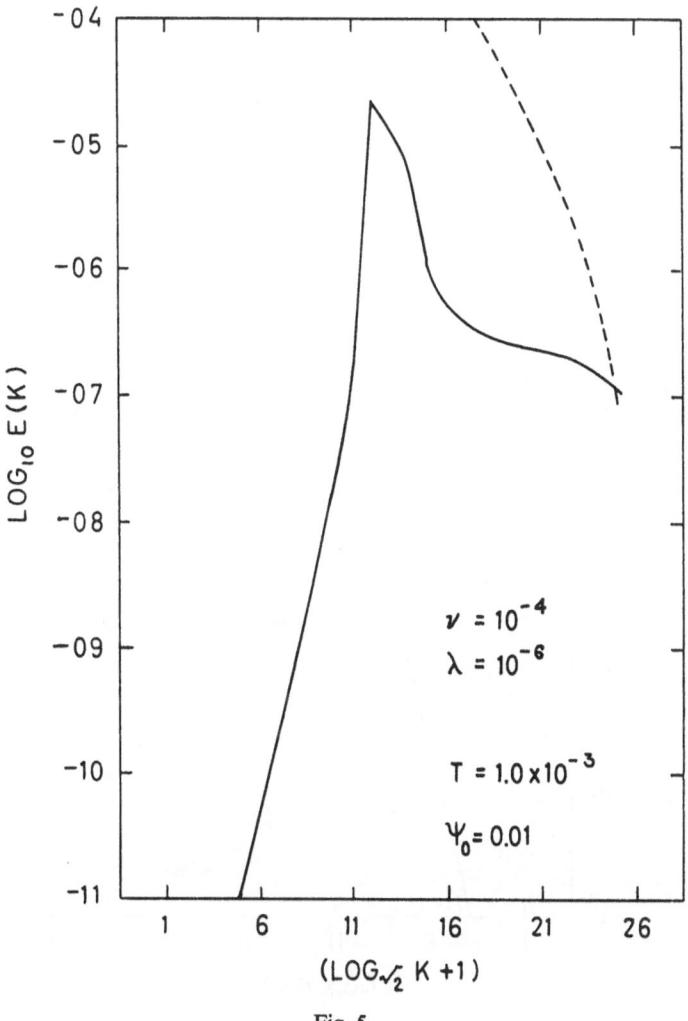

Fig. 5.

these preliminaries one can write

$$\zeta_V(k) = [E^V(k; T) k^3]^{1/2}$$
$$\eta_V(k) = [\{\zeta_V(k)\}^2 + (vk^2)^2]^{1/2}$$
$$\zeta_m(k) = (v_0 k)$$
$$\eta_m(k) = [\{\zeta_m(k)\}^2 + (\lambda k^2)^2]^{1/2}.$$

Here v_0 is the rms velocity in the energy range. (It will be apparent that this energy-range mixing was the reason why we chose the initial magnetic excitation to be localized in the inertial range. But for that the results of the hydrodynamic case or even the steady-state study will be inapplicable.) We choose a convenient unit of wave numbers and time scales such that $v_0 = 1$.

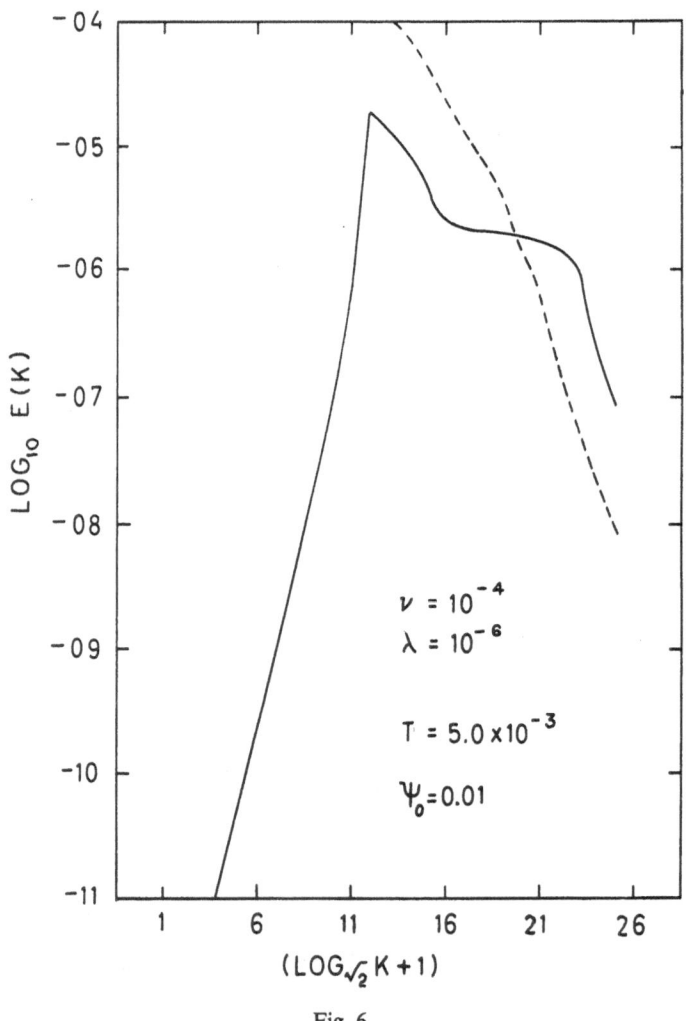

$\nu = 10^{-4}$

$\lambda = 10^{-6}$

$T = 5.0 \times 10^{-3}$

$\Psi_0 = 0.01$

Fig. 6.

3. Evolution Study

Now that all the quantities in Equations (1) and (2) are completely defined, we integrate them forward in time. In time, they have the character of a set of non-linear coupled differential equations. But for each time value there is an integral to be performed over the contributions from various regions of wave number space. We discretise the wave number region into twenty-five logarithmic half-octave intervals.

The details of this procedure are much the same as in an earlier paper (Nagarajan, 1971). We perform the time integration using a fourth-order variable-step Runge-Kutta Scheme. The details of the numerical scheme are given elsewhere (Nagarajan,

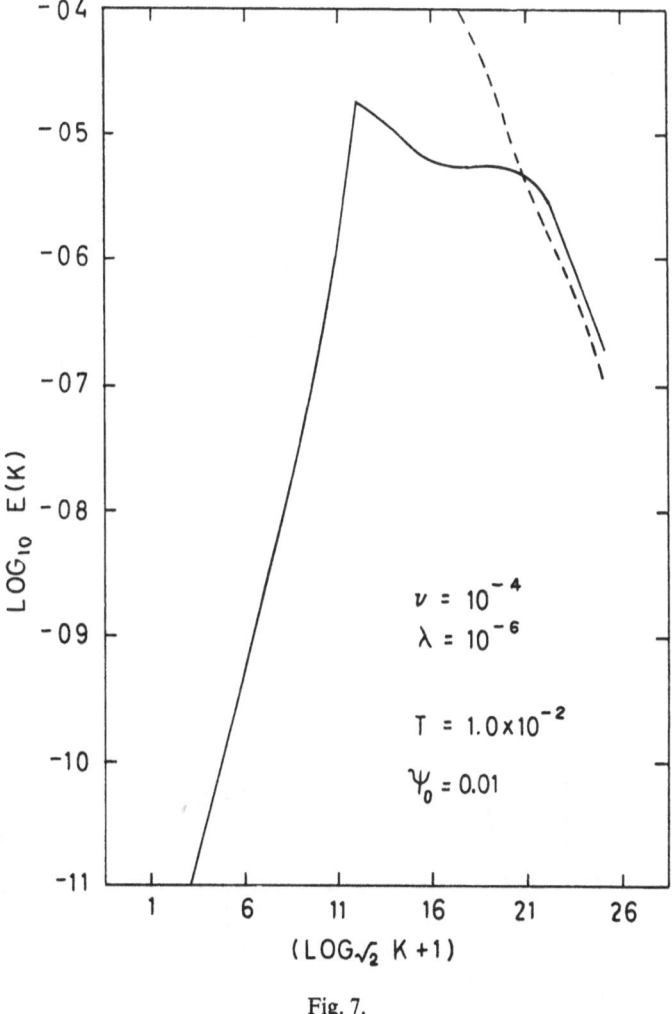

$$\nu = 10^{-4}$$
$$\lambda = 10^{-6}$$
$$T = 1.0 \times 10^{-2}$$
$$\psi_0 = 0.01$$

Fig. 7.

1970). We shall here consider only the results and their astrophysical implications.

Figure 1 shows the initial spectral disposition in one of the runs. The dotted line gives the velocity spectrum, and the continuous line, the magnetic disturbance. ψ_0 is the value of the initial ratio of the magnetic spectrum to the velocity spectrum at nonzero points, which is a parameter of the run. Though we are going to display here only initial disturbances which have the same spectral shape as the velocity and are localised in wave number space in a delta-function way, we had performed a number of runs with a variety of initial shapes $ak^n \exp(-bk^m)$ and initial ratio ψ_0. There was no pathological feature arising from the initial choice either numerically or otherwise.

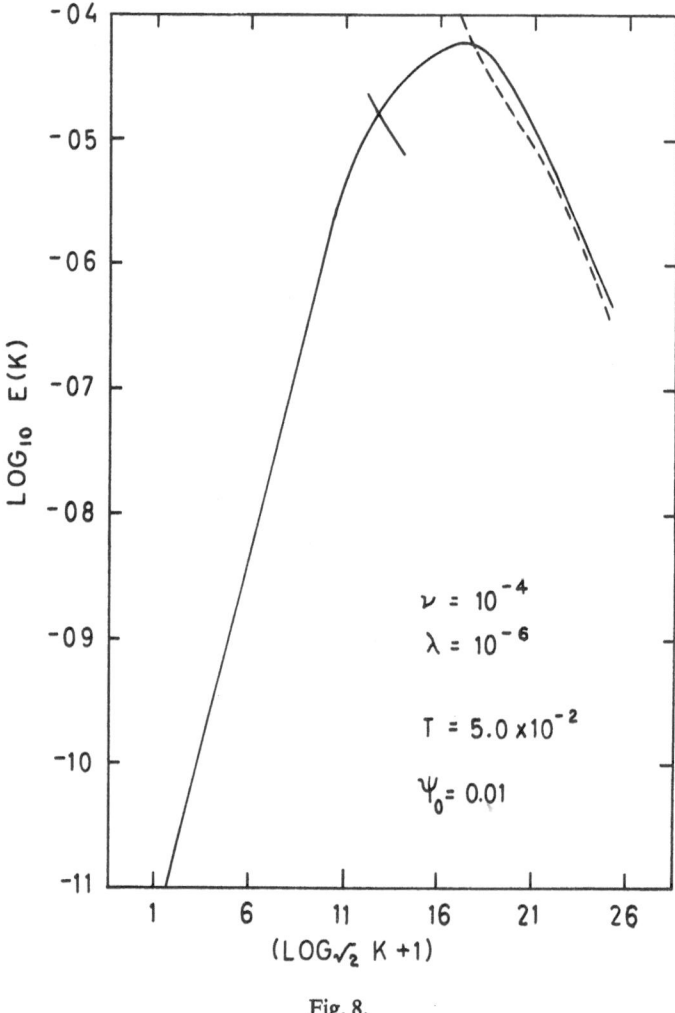

$$\nu = 10^{-4}$$
$$\lambda = 10^{-6}$$
$$T = 5.0 \times 10^{-2}$$
$$\psi_0 = 0.01$$

Fig. 8.

Figures 2 and 3 give the spectra at characteristic times $t=1.0 \times 10^{-5}$ and $t=1.0 \times 10^{-4}$. These time scales are so normalised that they are unity for the largest wave numbers in our system. The noteworthy feature of the curves is that the energy has now moved both to higher and lower wave numbers. The rate of transfer to lower wave numbers is essentially smaller than the rate of transfer to higher wave numbers, because the characteristic times of transfer are of the order of the internal times of the given scale.

Figures 4 and 5 give the spectra at $t = 5.0 \times 10^{-4}$ and 1.0×10^{-3}. Already, within a time of the order of the local eddy-circulation time in the largest wave numbers, the magnetic spectrum has wrapped up sufficiently to almost equality with the velocity

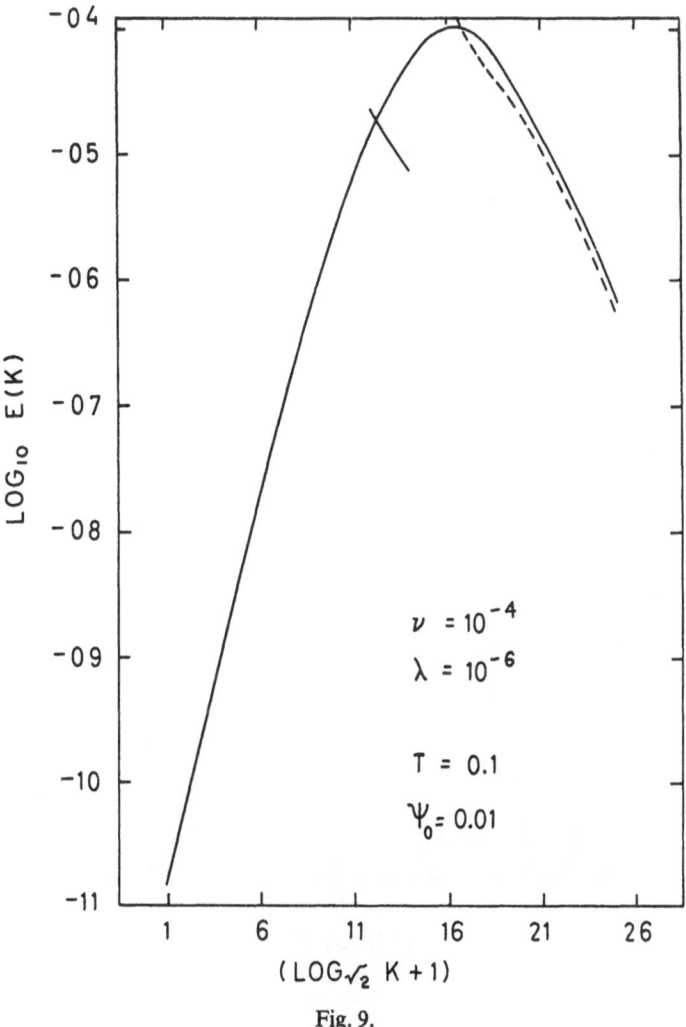

Fig. 9.

spectrum at the highest wave number. Figure 5 to some extent and Figure 6, in a more profound way show that the magnetic spectrum has overshot significantly above the velocity at lowest scales. This arises because of two reasons: (1) The choice of kinematic parameters v and λ. In this run λ is very much smaller, so much so the magnetic spectrum has a *longer* dissipative tail. (2) The second reason for the overshooting is the fact that the form of the spectrum is still non-equilibrium so much so the approach to local equipartition is in an overstable way.

Figure 7 and more prominently Figure 8 show how the feature of equipartition is transferred to smaller wave numbers, much in the same way as argued by Biermann and Schlüter (1951). By now the evolution has reached a stage in which any peculiar

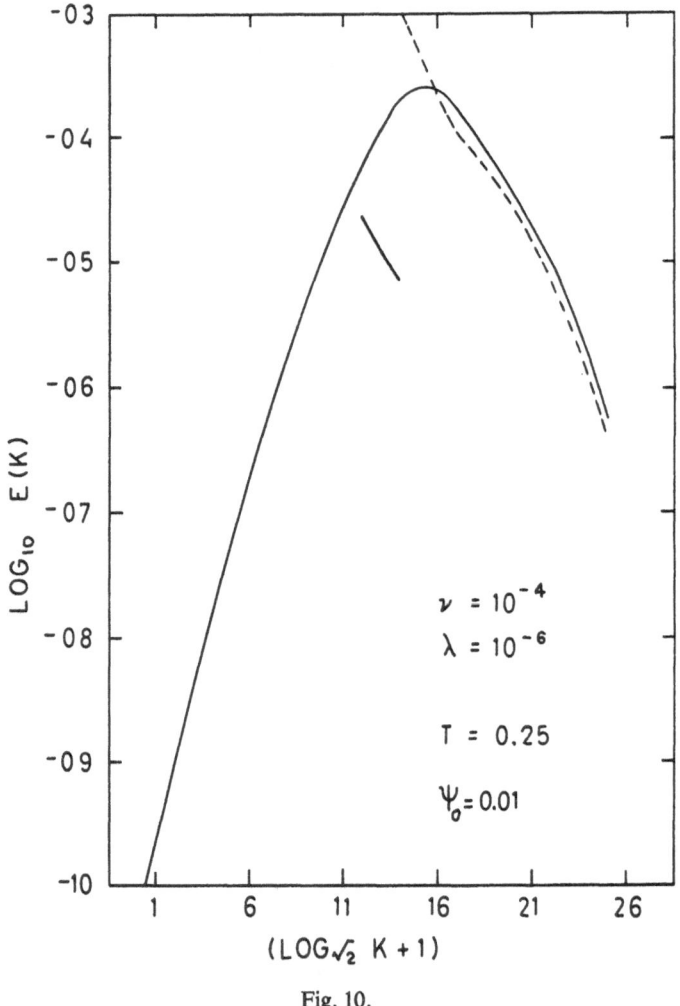

$$\nu = 10^{-4}$$
$$\lambda = 10^{-6}$$
$$T = 0.25$$
$$\Psi_o = 0.01$$

Fig. 10.

dependence on choice of initial form has been completely lost. Figures 9 and 10, which are for the same run for times $t=0.1$ and 0.25 show that by now the evolution has reached a stage when one can safely conclude about ultimate features. The numerical integration times involved at this stage are so large that one stops the calculations because no new features are likely to evolve from further evolution study.

Figures 11, 12 and 13 feature the final and initial spectra for a few other runs which start different initial ratios and kinematic parameters. These are meant for the purist to show that pathological features are not included in the choice of initial assumptions.

In all these runs, at a fairly advanced evolution, the spectral shape reaches an approximate form $A(t) k^4 \exp\{-B(t) k^2\}$. Thereafter the integral features of the spectrum evolve more or less without change of form.

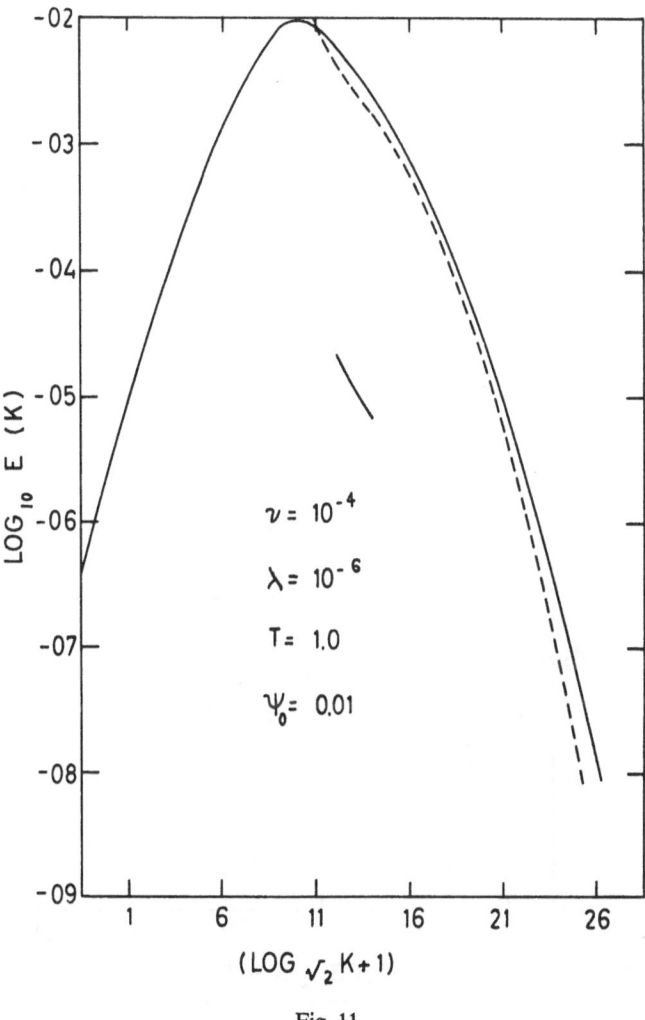

Fig. 11.

4. Conclusions

Apart from the fact that this evolution study fills many a gap in our earlier study, this proves more or less conclusively that there is no reason to expect, in evolving non-equilibrium hydromagnetic turbulence, that the transfer will take place only to larger wave numbers. In fact, the transfer to smaller wave numbers is significant and this can provide just the missing link in the *turbulent dynamo* problem. The regeneration of larger magnetic loops through a co-operative interaction of the velocity fluctuations of all scales and magnetic fluctuations of smaller scales is not only feasible but very significant. In our study, we find that this is facilitated by two dynamical requirements. Firstly, the non-equilibrium feature of the magnetic spectrum: the ultimate steady-

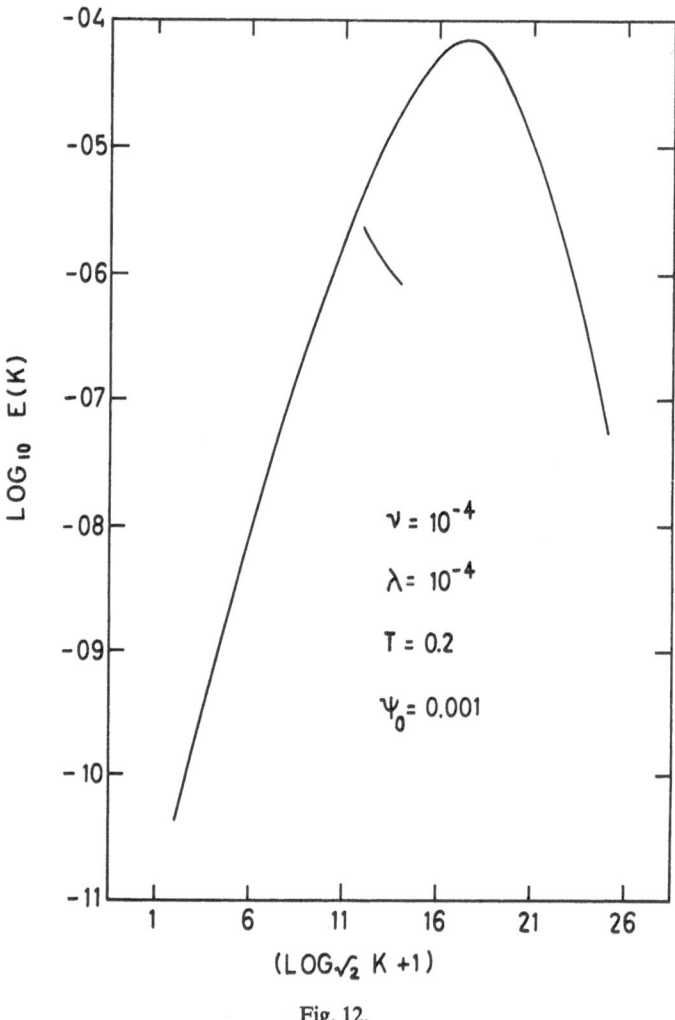

Fig. 12.

state magnetic spectrum will be in equipartition with the velocity in all scales other than the ones where either the inputs of energy from external sources of the train of energy through molecular dissipation depresses or raises either of them. Any other form of the spectral ratio is not an invariant form which will be left invariant by the non-linear interaction. The non-linear interaction will change the ratio to get into the equilibrium form. Secondly, the Galilean non-invariance: The fact that a magnetic field cannot be gauged out makes a profound modification in the internal dynamics. Here probably one can stretch our comparison a bit with other recent studies. Krause (1968), Rädler (1968), Steenbeck *et al.* (1966), Steenbeck and Krause (1966, 1967), Krause and Rädler (1971) and Moffatt (1970) have considered the α-effect of regenera-

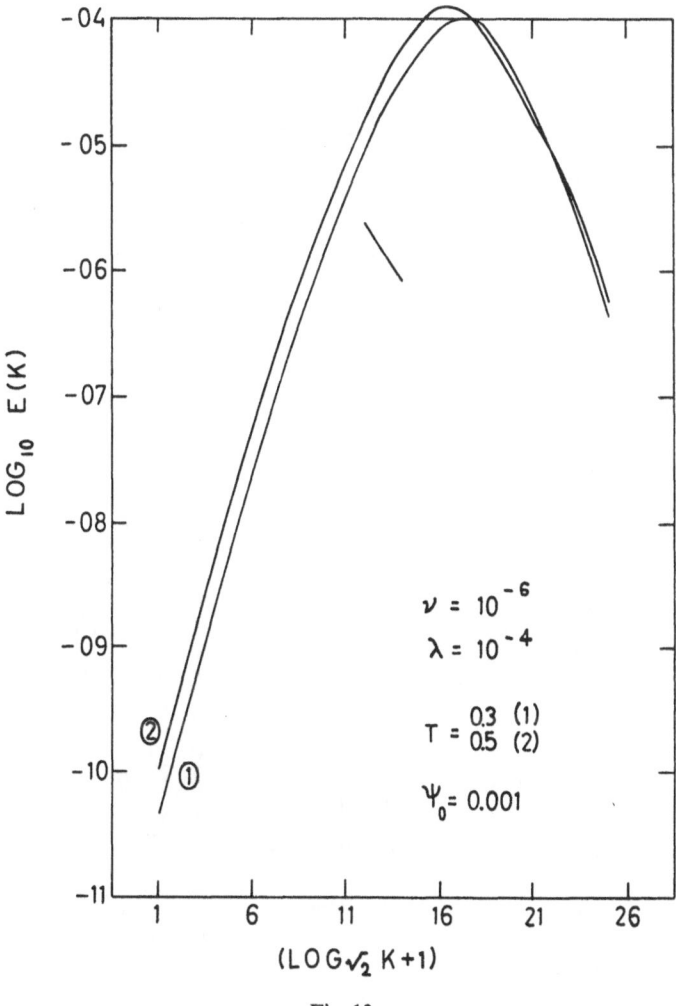

Fig. 13.

tion in great detail. A certain aspect of the α-effect is included in our Galilean non-invariance picture, because a larger magnetic loop, when it is impressed on a system of smaller magnetic and velocity fluctuations, introduces a condition of reflectional non-invariance. Beyond this point one cannot carry the analogies because their inferences about the values of the α-effect are based on equilibrium transfer theory, which as our study has clearly shown, are inapplicable.

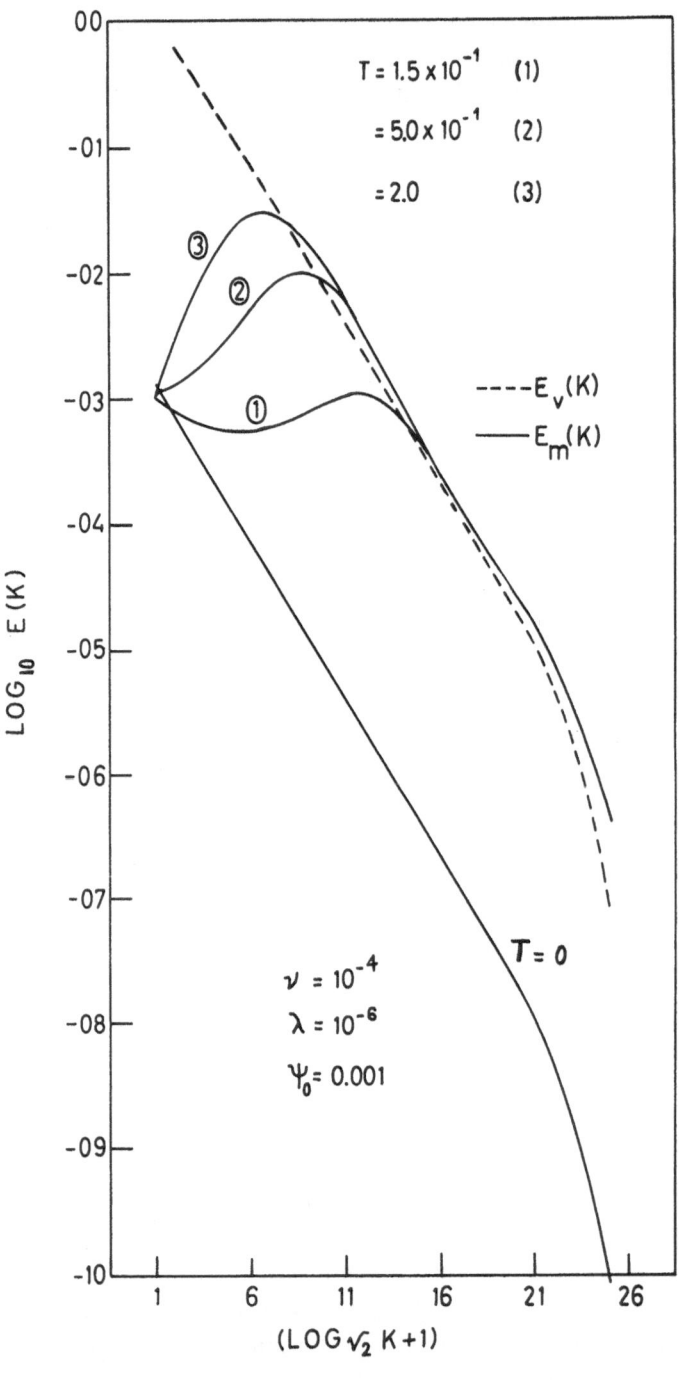

Fig. 14.

Parker (1969) and Vainshtein (1970) have asked much the same question, as we have, but since they had to invoke some extreme idealisations to get their results, the physical validity of their conclusions is in doubt. Qualitatively, our results corroborate theirs.

Robinson and Rusbridge (1971), in a study of Plasma turbulence in the Zeta plasmas, have found that plasma turbulence seems to resemble fluid turbulence except that the turbulent elements are enlarged along the mean magnetic field to form rolls and suggest that an appropriate comparison would have to explain the existence of significant transfer to large scales from small-scales, as against isotopic hydrodynamic theory, which will not permit this. One hopes that it will not be too presumptuous to believe that the effect, they find is contained in our procedure. Further the importance of this to heat transfer in the presence of magnetic turbulence is also very tempting.

References

Biermann, L. and Schlüter, A.: 1951, *Phys. Rev.* **82**, 863.
Fitremann, M. and Frisch, U.: 1969, *Compt. Rend. Acad. Sci. Paris* **268**, 705.
Kraichnan, R. H.: 1958, *Phys. Rev.* **109**, 1467.
Kraichnan, R. H.: 1959, *J. Fluid Mech.* **5**, 497.
Kraichnan, R. H.: 1965, *Phys. Fluids* **7**, 1723.
Kraichnan, R. H.: 1966, *Phys. Fluids* **8**, 575.
Kraichnan, R. H.: 1966, *Phys. Fluids* **8**, 1385.
Kraichnan, R. H.: 1970, *Phys. Fluids* **13**, 569.
Kraichnan, R. H. and Nagarajan, S.: 1967, *Phys. Fluids* **10**, 859.
Krause, F.: 1968, *Z. Angew. Math. Mech.* **48**, 333.
Krause, F. and Rädler, K.-H.: 1971, this volume, p. 770.
Moffatt, H. K.: 1970, *J. Fluid Mech.* **41**, 435.
Nagarajan, S.: 1970, *Proc. of the Symposium on Computing and Applications*, Institute of Mathematical Sciences, Madras.
Nagarajan, S.: 1971, *Phys. Fluids*, in press.
Parker, E. N.: 1969, *Astrophys. J.* **157**, 1119 and 1129.
Parker, E. N.: 1970, *Astrophys. J.* **160**, 383.
Rädler, K. H.: 1968, *Z. Naturforsch.* **23a**, 1841.
Robinson, D. C. and Rusbridge, M. G.: 1971, *Phys. Fluids*, in press.
Steenbeck, M., Krause, F., and Rädler, K. H.: 1966, *Z. Naturforsch.* **21a**, 369.
Steenbeck, M. and Krause, F.: 1966, *Z. Naturforsch.* **21a**, 1285.
Steenbeck, M. and Krause, F.: 1967, *Magn. Gidrodyn.* **3**, 19.
Vainshtein, S. I.: 1970, *Zh. Eksperim. i Teor. Fiz.* **58**, 153 = *Soviet Phys. JETP* **31**, 87.

Discussion

Nakagawa: What is your assumption concerning the initial velocity and magnetic field spectra?

Nagarajan: The initial velocity is in quasi-equilibrium with an extended inertial range. The magnetic spectrum is localized in the middle of the inertial range in all but one of the runs, with a level of excitation very much lower than the velocity.

Weiss: After equipartition has been achieved for intermediate wave numbers, is your steady energy spectrum maintained over periods comparable with the resistive decay time for the smallest wave numbers?

Nagarajan: Yes. We follow the time evolution until the initial form dependence is washed out. Essentially this turns out to be larger than the resistive time scale of the initial specimen. But after that time, the further buildup of the spectrum – even towards smaller wave numbers – takes energy from the velocity spectrum. This time-invariant self-preserving form with the tail in steady-state with the velocity, keeps growing in over-all energy and extent. This may look like a violation of simple physical and statistical requirements. But it is not.

Cowling: In many ways the assumptions made (nature of background fields, motions, statistical assumptions) appear to be as important in the theory of magnetohydrodynamic turbulence as the detailed theory.

Nagarajan: True: statistical description does not in any sense minimize the number of necessary assumptions. But the statistical theory has an advantage in that one requires only on-the-average features. So many of the phasing requirements are weakened. But the main feature of this investigation has been to show that the back-transfer in wave-number spectrum is significant, which can have truly deep conceptual consequences.

DISTRIBUTION OF THE MAGNETIC FORCE IN THE SURFACE LAYERS OF SUNSPOTS

J. JAKIMIEC

Astronomical Institute, Wrocław University, Poland

Abstract. At the beginning the problem of constructing the three-dimensional magnetohydrostatic models of the photospheric layers in sunspots is discussed in some detail. It is pointed out that the construction of such models by solving the set of equations of magnetohydrostatics cannot be effectively carried out.

In order to solve the difficulties a suitable method of determining the distribution of the magnetic force in sunspots from measurements of the magnetic field has been worked out. Tentative results of the computations are presented.

General features of the distribution of the magnetic force in the photospheric layers of stable sunspots are discussed. It is pointed out that significant magnetic forces are necessary in the penumbra; they secure its transversal equilibrium, but are rather unimportant for its vertical structure. And it is quite probable that the magnetic field in the umbras of stable spots is nearly potential or force-free down to the photospheric level.

There are many observational data relevant to the photospheric layers of sunspots. Nevertheless, it is not easy to construct a complete model of these layers.

Such a complete model must be a three-dimensional one, because physical quantities in a sunspot vary in a horizontal as well as in the vertical direction. At present we are able to deal with the simplest such model only, viz., with the one which has the following properties:

(1) it is an average model, i.e. fluctuations connected with fine structures of sunspots are neglected;

(2) it is an axially symmetric model, i.e. the considerations are restricted to the sunspots of the most regular structure;

(3) it is a static model, i.e. systematic motions (Evershed effect) are ignored – estimations show that the observed velocity field is of secondary importance for the over-all equilibrium of the considered layers of a sunspot (Jakimiec, 1965), and therefore a static model should provide a good approximation.

Figure 1 is intended to visualize the peculiar geometry of the considered region of a sunspot. Transverse dimensions of the region are much larger than its vertical thickness (in the figure the z-scale is even stretched). Moreover, the detailed shape of the region is not known from observations; only an estimation of the depression Δz at the umbral-penumbral boundary can be obtained, from analysis of the Wilson effect. Thus the complete model of the photospheric layers of a sunspot must determine not only the distribution of the physical quantities in the region but also the geometry of the region. And the consistency between the geometry and the distribution of forces must be secured in the model.

It is rather obvious that the construction of such a model should be based on the ample experience gained already in constructing the average models of the undisturbed

Howard (ed.), Solar Magnetic Fields, 505–511. All Rights Reserved.
Copyright © 1971 by the IAU.

photosphere. It is known that the standard procedure in the case of the undisturbed photosphere falls into two main stages:

(1) Determination of the empirical temperature distribution $T(\tau)$, from the measurements of the limb darkening.

(2) Determination of the pressure distribution by integrating the equation of hydrostatic equilibrium:

$$dp/dz = -\varrho g.$$

Introducing the optical depth τ as a new independent variable,

$$dp/d\tau = g/\kappa,$$

and making use of the previously obtained $T(\tau)$ the integration can be effectively performed.

It will be easily understood that such a division into two stages should be retained also while constructing the model of the photosphere in a sunspot. From very precise

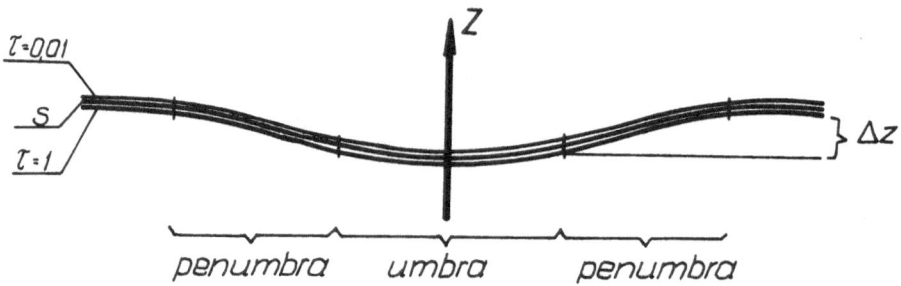

Fig. 1.   Geometry of the photospheric layers in a sunspot, schematically. The probable shape of the surfaces $\tau = $ const is shown. S is the surface, to which the measured magnetic field relates.

spectrophotometric observations the empirical temperature distributions $T(\tau; r)$ for individual vertical columns of the photosphere in a spot could, in principle, be obtained. (Accuracy of the present observations is not as yet sufficient for that.)

The integration of the equations of magnetohydrostatic equilibrium to obtain the pressure distribution (together with the distribution of the magnetic field) would be now a natural generalization of stage 2) for our case. The equations form a set of non-linear partial differential equations; necessary boundary conditions should be supplied by observational data. Unfortunately, this direct way of determining the pressure distribution cannot be utilized effectively. The main reasons for that, without going into details, are as follows:

(a) in our three-dimensional case the change of variable $z \to \tau$ cannot be effectively carried out and therefore the temperature distribution $T(\tau; r)$ cannot be directly used in the integration;

(b) the available observational data fail to provide appropriate boundary conditions;

(c) the shape of the surface S, to which the boundary conditions (observations) relate, is not known *a priori*.

Moreover, the theory of the boundary value problems for such non-linear elliptical systems of equations is not yet worked out in detail.

Because of this state of affairs, the present author undertook the investigation of the problem of determining precise values of the magnetic force in sunspots from measurements of the magnetic field. It can be easily understood that the possibility of determining the distribution of the magnetic force, i.e. magnetic terms in the equations of magnetohydrostatic equilibrium, with an appropriate accuracy would open the way for determining the pressure distribution in the model of the photosphere in a sunspot.

As it is, a promising method of determining magnetic forces in sunspots has just been worked out (Jakimiec, 1970). It is mathematically expedient to consider the problem in terms of determining electric currents in a sunspot; then the main difficulty consists in determining the azimuthal component j_φ of the current which is of the prevailing importance for our problem. The currents are determined by solving a boundary value problem for differential equations. The measured magnetic field is used as the boundary condition $(\mathbf{H})_S$.

The idea of detecting the currents situated in the region above S can be summarized as follows: The presence of the currents in the region will have such an effect on the measured field $(\mathbf{H})_S$ that it will not fit to any potential field, and from the deviations of $(\mathbf{H})_S$ from the potential case inferences about the currents can be drawn. In order for the solution of the problem to be determined uniquely, use is made of the fact that in the considered region the gas pressure, and therefore also magnetic forces, decrease steeply with height. (For mathematical details see the paper cited above.)

The method has, *inter alia*, the following advantage: It allows to solve relatively easily the difficulty that the detailed shape of the surface S, to which the measurements of the magnetic field relate, is not known from observations. The shape is determined as an additional unknown function $z_0(r)$ by making use of the fact that it must be closely related to the distribution of the magnetic forces on the surface: The greater the slope of the surface S at a given point, the greater, on the average, should be the radial pressure gradient $\partial p/\partial r$, and therefore also the radial magnetic force F_r, at the point. In the computations actually performed this relation was approximated by the proportionality:

$$dz_0(r)/dr = K(F_r)_S,$$

where $z_0(r)$ describes the shape of the surface S.

The detailed calculations have been carried out making use of the observational data of Stepanov and Gopasjuk (1962). Figure 2 shows the obtained distribution of the radial component F_r of the magnetic force across the sunspot. The obtained shape of the surface S is shown in Figure 3. It is seen that it agrees with the traditional picture of the depression in a sunspot (gradual slope in the region of the penumbra and flat bottom in the umbra).

Because present-day measurements of the magnetic field in sunspots may be subject to large systematic errors (see, e.g., Severny, 1967), the results should be considered as rather tentative ones. It can be estimated that in order to obtain reliable determination of the magnetic forces of the relevant order of magnitude it is necessary to have measurements of the individual components of the magnetic field with an accuracy of ± 50 G. The requirements do not seem to be excessive ones and they probably will be met in the near future.

Thus, there is, indeed, a real possibility of determining reliably the regular magnetic forces in the surface layers of sunspots from the measurements of the magnetic field, provided that main systematic errors of the measurements will be eliminated.

However, basing on the experience gained in the above calculations, some conclu-

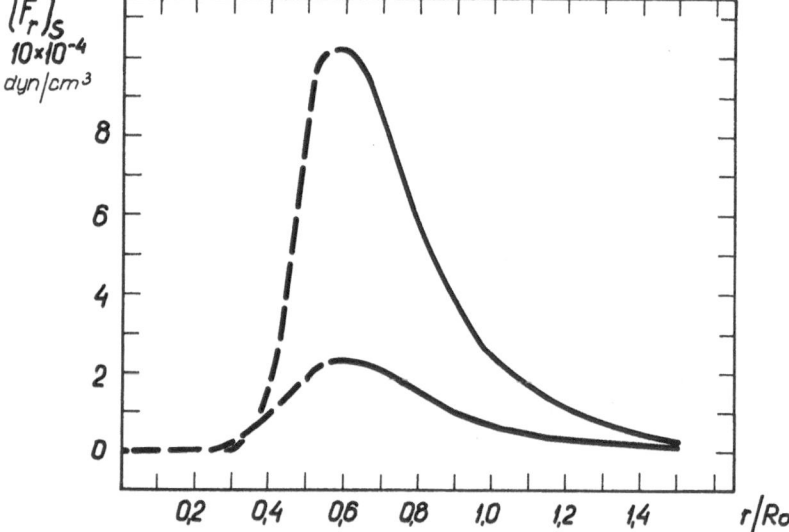

Fig. 2.   The obtained distribution of the radial magnetic force F_r on the surface S in the investigated spot. Two variants of the computations corresponding to extreme values of the rate of decrease of the magnetic force with height are shown. The dashed lines are drawn for the region of the umbra where measurements of the magnetic field are the most uncertain. R_0 is the radius of the spot.

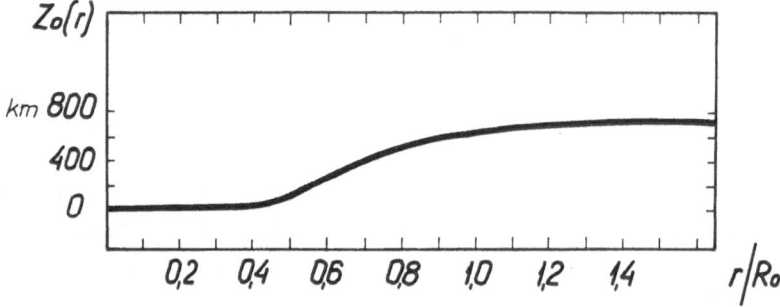

Fig. 3.   The shape of the surface S obtained in the computations. (The z-scale has been stretched.)

sions about the distribution of the magnetic forces in the layers of sunspots can as-
suredly be drawn also at present.

First of all, if we start with the rather well-established fact that in a large spot the
photosphere at the umbral-penumbral boundary is depressed some 700–1000 km
below the undisturbed photosphere, we are forced to conclude that significant radial
magnetic forces $F_r > 0$ are necessary for the equilibrium of the penumbral photosphere.
Moreover, from this fact reliable estimation of the mean value of F_r in the penumbra
can be made. The relevant equations are:

$$F_r = \frac{\partial p}{\partial r}; \quad \frac{\partial p}{\partial r} = -\frac{\partial p}{\partial z} \tan \zeta,$$

where ζ is the angle between the z-direction and the vector $(-\operatorname{grad} p)$ at a given point.

Characteristic values of $\partial p/\partial z$ at the level of line formation in the penumbra can be
taken from the empirical model of the penumbral photosphere, calculated on the
assumption of hydrostatic equilibrium (see remarks below): $|\partial p/\partial z| \sim 10^{-3}$ dyn cm^{-3}.
(The sub-hydrostatic model of Makita (1963) yields in fact the same order of magni-
tude of this quantity – cf. Jakimiec, 1970.)

As a characteristic value of ζ the mean slope of the solar surface in the penumbra
can be taken:

$$\tan \zeta \approx \frac{\text{depression } \Delta z \text{ in the sunspot}}{\text{width of the penumbra}} \sim 10^{-1}.$$

Hence

$$(\overline{F_r})_S \sim 10^{-4} \text{ dyn cm}^{-3}.$$

It is worthwhile to note that the values of the force F_r derived from the above mentioned measurements of the magnetic field are of the correct order of magnitude (see
Figure 2).

As to the forces F_r in the umbra, we can say only that the usually accepted shape of

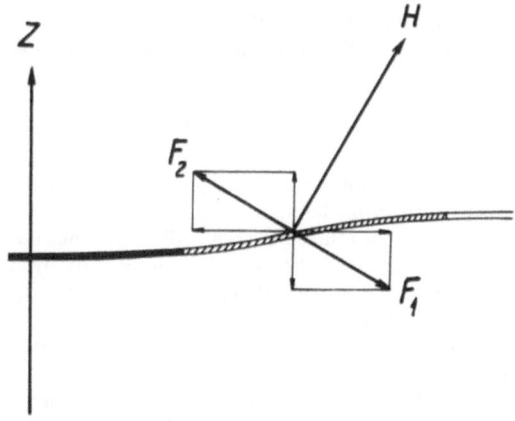

Fig. 4. Schematic diagram for discussing the direction of the vertical magnetic force F_z
in the penumbra.

the depression in a sunspot, with flat bottom in the umbra, implies absence of significant regular radial forces F_r there, i.e. the distribution of F_r being of the type as shown on Figure 2.

Now several comments on the vertical component F_z of the magnetic force in the penumbra are in order. On the schematic diagram (Figure 4) the typical inclination of the magnetic field **H** in the penumbra, as given by observations, is shown. The magnetic force **F** must be perpendicular to **H**, so its projection on the plane of the diagram will be directed along F_1 or F_2. On the other hand, it has been shown in the previous considerations that the radial component F_r of the magnetic force in the penumbra should be directed outwards $(F_r > 0)$. Thus the case F_1 should take place and consequently the vertical component F_z should be directed downwards $(F_z < 0)$.

We see that it is rather difficult to expect regular magnetic forces supporting the solar atmosphere in the penumbra. The forces tend to increase effective gravitation, i.e. to compress the atmosphere. Nevertheless, the calculations show that F_z is rather small in comparison with gravity, so that the deviations of the pressure from the distribution corresponding to the hydrostatic equilibrium do not exceed 10–20%.

To summarize, the best overall self-consistent model of the photosphere in a sunspot, which can be proposed at present, would be the following one: Significant magnetic forces are present in the penumbra, they secure its transversal equilibrium, but are rather unimportant for its vertical structure. Hence also in the penumbra the vertical gas colums should be in nearly hydrostatic equilibrium. And it is quite probable that in the umbras of stable spots there are no significant regular magnetic forces, neither vertical, nor transversal ones. In other words, the smoothed magnetic field in there would be nearly potential or force-free down to the photospheric level.

References

Jakimiec, J.: 1965, *Acta Astron.* **15**, 145.
Jakimiec, J.: 1970, *Astron. Zh.* **47**, 520.
Makita, M.: 1963, *Publ. Astron. Soc. Japan* **15**, 145.
Severny, A. B.: 1967, *Izv. Krymsk. Astrofiz. Obs.* **36**, 22.
Stepanov, V. E. and Gopasjuk, S. I.: 1962, *Izv. Krymsk. Astrofiz. Obs.* **28**, 194.

Discussion

Musman: What in your model is the difference in gas pressure between the center of a sunspot and the surrounding photospere at the same level.

Jakimiec: The depth of the sunspot is taken to be 700–1000 km. Thus the total transverse pressure differences at the level of the umbral photosphere are of the order of 10^6 dyn cm^{-2}. The essential point here is that the pressure differences are equilibrated by magnetic forces in the region of the penumbra (mainly below the visible penumbra – cf. Figure 1).

Maltby: Could you comment on the stability of your sunspot model. Does not the energy equation enter in your discussion at all?

Jakimiec: Concerning the first question: the considerations concern only the thin layer of a sunspot (cf. Figure 1) and the problem of stability cannot be investigated for such a restricted region of the sunspot. But we need not trouble ourselves very much about the stability problem in the above considerations, because: (1) all the considerations concern the most stable, very slowly varying sunspots,

(2) the stability of a sunspot as a whole is determined by the situation in the deeper layers and it is quite improbable that inaccuracies of a model of the considered photospheric layers could have any significant effect on the stability of the whole sunspot.

Concerning the second question: the equation of energy does not enter here because we use an empirical temperature distribution $T(\tau, r)$, which presents a ready (approximate) solution of the problem of the energy transport in the layers considered. This is quite analogous to the case of the empirical model of the undisturbed photosphere.

OBSERVATION OF SOLAR FLARE TYPE PROCESSES IN THE LABORATORY

W. H. BOSTICK, V. NARDI, L. GRUNBERGER and W. PRIOR

Stevens Institute of Technology, Hoboken, N.J., U.S.A.

Abstract. A filamentary magnetic structure is produced on the plasma current sheath of a coaxial accelerator operated with deuterium. Space and time analysis of X-ray, neutron and visible-light emission indicates that the magnetic energy of the filaments is transferred to the plasma during a process of decay of the filaments. X-ray photographs show very localized regions (diameter <0.5 mm) of strong emission. Some of these regions are also located where the plasma is not subject to a maximum of compression. Similar bright spots (Hβ) are observed by 5 ns image converter photographs. The detailed analytical description of the self-consistent fields is deduced. The localized regions of strong emission may well correspond with the explosive onset of an instability at a point on a filament (single filament decay) or at a point where two filaments with opposite fields coalesce with magnetic field annihilation. The similarities with solar flares are considered.

1. Experimental Technique and Observations

The idea proposed by several authors (Gold and Hoyle, 1960; Alfvén and Carlquist, 1967), that a substantial amount of energy can be transferred from magnetic field to plasma at a very high rate (as in solar flares) is tested with laboratory data.

In order to produce the plasma with a suitable magnetic structure, we have used a coaxial accelerator (CA) with electrode radii 5 cm and 1.7 cm – center electrode positive – connected at one end (breach) to a 45 μF capacitor bank initially at a potential $V = 13$ kV. This device has been extensively described in the literature (Marshall, 1960); some data are recalled here. The cylindric coaxial electrodes (see Figure 1) are in a vacuum system containing deuterium at a well defined initial pressure p. With the onset of the discharge at the breach, a current sheath $(CS) \approx 1$ cm thick carrying most of the current ($\approx 90\%$) through the plasma, moves between the electrodes with a velocity $u_0 = u_0(p, V, ...)$ from the breach to the other end (muzzle) of CA. CS is driven in the axial direction by the force density $\mathbf{j} \times \mathbf{B}$; $\mathbf{j}$ is the current density through the plasma on CS and $\mathbf{B}$ is given essentially by the magnetic field $B_\theta \approx |\mathbf{B}|$ in the plasma region behind CS due to the current I in the electrodes.

By magnetic probe measurements and image-converter (IC) photographs taken at specific times, it is verified that the thin layer (≈ 3 mm thick) which carries about half of all the current on the sheath, contains the plasma front ≈ 1 mm thick which strongly radiates in the visible spectrum. By CS, we indicate from now on specifically this luminous layer.

The fine structure of CS is better observed by choosing the initial value of the CA parameters $(p, V, \text{etc.})$ in a specific interval. Our CA is usually operated with deuterium $p = 8$ torr, peak value of the electric current across CA $I_{max} \approx 0.5$ Ma. Then $CS(u_0 \approx 4.0 \times 10^6$ cm/s) is observed to be composed largely of luminous plasma filaments (diam 0.3–0.5 mm) oriented in the radial direction, i.e. orthogonal to B_θ

direction (see Figure 2; optimum conditions for filament observation are found at $V = 11$–13 kV).

By decreasing p (e.g. down to 0.6 Torr) at a constant $V = 13$ kV, IC photographs of CS do not readily show a filamentary structure by visible light whereas simultaneous observations of the electron density gradient $|\nabla \varrho_-|$ by Schlieren shadowgrams show a

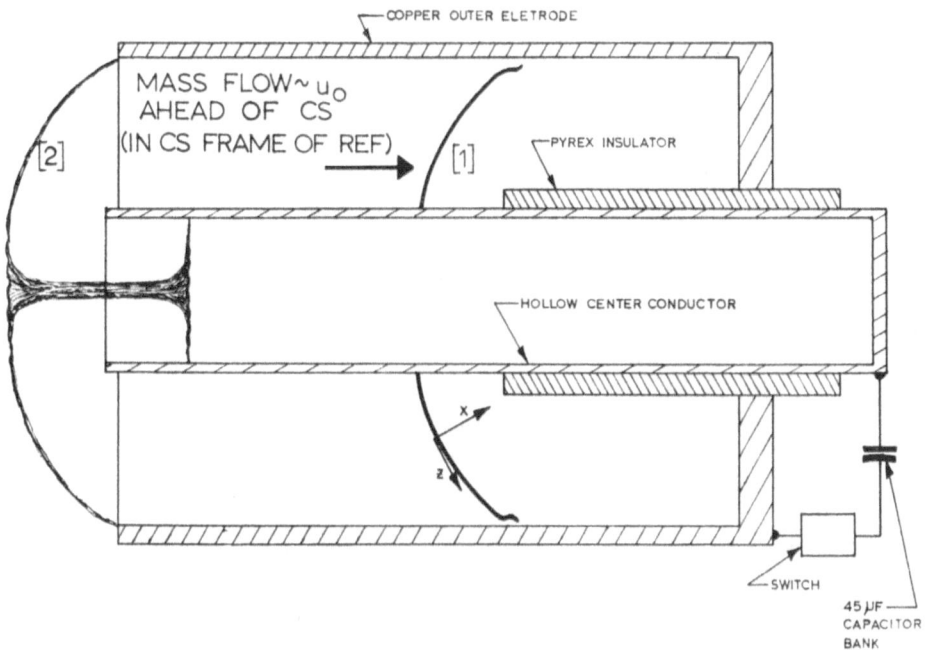

Fig. 1.   Cross-sectional view of the CA. Current sheath between the electrodes [1] and at the later stage of axial collapse [2].

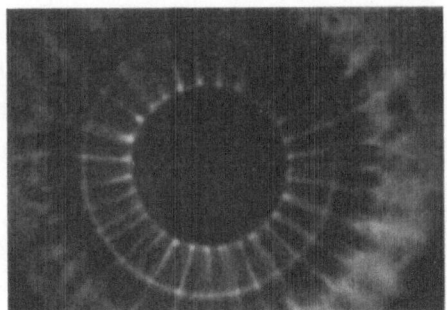
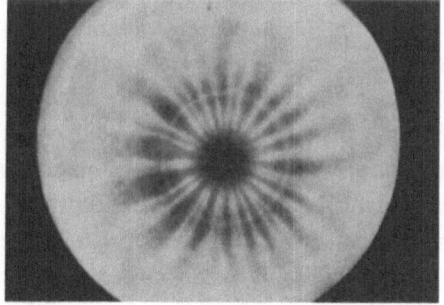

Fig. 2.   Left: Front view of CS with filaments rolling off the end of the center electrode. Image-converter photos by visible light; 5 ns exposure (at $t = -120 \pm 20$ ns before time of maximum axial compression). The circular edge of the center electrode (diam 3.4 cm) is visible in the background (D_2 initial pressure 8 Torr; peak value of electric current 0.5 Ma; applied potential ~ 13 kV). Right: Same conditions but 100 ns exposure. A relevant fact is the absence of azimuthal motion of the filaments on CS. No blurring of the filament profile is observed with the long exposure if compared with 5 ns exposure. This fact demonstrates the permanent, stable character of the filamentary structure of CS.

filamentary pattern. An apparently uniform CS by visible light is also observed by increasing V to sufficiently high values. A statistical analysis on many discharges indicates that the average number of radial filaments for a given p increases by increasing V (see Table I). CS and filament arrangement on IC photos remain substantially unchanged by using an interference filter with a 100 Å band width and transmission peak near the Hβ line. Spectrograph data confirm that a major percentage of the visible light is emitted near Hβ. The surface Σ, defined as the foremost luminous face of CS, is a direct source of information on flow and current distribution on CS. As a consequence of the observed nonplanarity and specific curvatures of Σ, one can easily prove by using MFD shock equations that vorticity $\boldsymbol{\omega}=\tfrac{1}{2}\nabla\times\mathbf{u}$ is different from zero in the filament region immediately behind Σ and is oriented in the same direction as the filament axis (Bostick *et al.*, 1969). It is observed that the plasma density, plasma flow density and magnetic fields are concentrated in these filaments (vortex filaments) (Bostick *et al.*, 1966, 1970). The electron density by Hβ Stark broadening in the filaments appears greater by a factor ≈ 3 than in the near luminous region.

TABLE I

Operating initial conditions	Average number of radial filaments before CS collapse	Average number of radial filaments after CS collapse
9 kV, 8 mm D_2	28	16
11 kV, 8 mm D_2	38	23
13 kV, 8 mm D_2	44	34

Probe data show that B_θ vanishes ahead of CS which coincides with the region of quick transition to high B_θ (B_θ increases to $\approx 1.4\times 10^4$ G by crossing CS, up to $\approx 2\times 10^4$ G behind the whole current layer). CS has a magnetic structure much more complicated than the simple B_θ field.

By different magnetic probe measurements, it is possible to detect a strong component ≈ 2500 G of the magnetic fields along the filament axis. This magnetic field component vanishes outside the filaments and usually has an opposite orientation for two neighboring filaments (Bostick *et al.*, 1966). The preference sometimes shown by two neighboring filaments for a parallel arrangement (instead of the radial, spokelike arrangement which is suggested by the axial symmetry of the whole system) directly shows that a pairing effect exists between two filaments. A pair of parallel vortex filaments normally keeps an equilibrium distance from one another for a relatively long time (Figure 2). When this equilibrium distance is upset (for example by a shock wave which occurs in the medium at a specific stage of the discharge) the vortex pair is observed to coalesce (presumably with appreciable magnetic field annihilation), the region of combination traveling along the vortex filaments with a speed ≈ 2–5×10^7 cm/s. When CS reaches the muzzle, it rolls off the electrodes, comes around the edge of the center electrode, proceeds across the front face with a radial velocity

$\approx 7.5 \times 10^6$ cm/s (radial and axial velocities of CS are found statistically by plotting the observed positions versus time for a number of discharges) and collapses on the CA axial region in a cylindric column (diam ≈ 4–5 mm) collinear with the electrode axis. The number of filaments decreases during this stage approximately by a factor $\frac{2}{3}$ (see Table I). This occurs mainly by two-by-two-filaments coalescence processes which ordinarily begin at the radially innermost point of the combining pair.

In correspondence with a coalescence process, a typical fork configuration of the filaments can be observed on the moving off-axis part of CS where resolution of the structure remains possible. At the stage of maximum axial compression, the remaining filaments partially blend in the cylindrical pattern on the axial region (dense plasma focus) but in the off-axis part of CS filaments can still be observed quite clearly by visible light (see Figure 3).

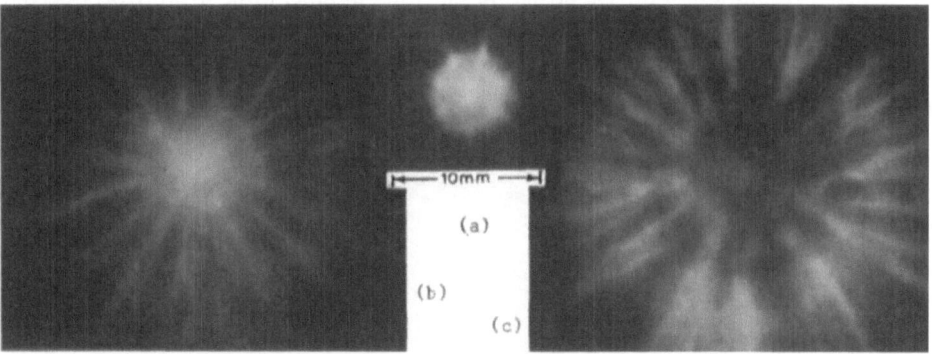

Fig. 3. 5 ns photos at different times from the time of maximum axial compression. (a) $t \sim 30$ ns; (b) $t \sim 100$ ns; (c) $t \sim 180$ ns. (a), (b) show clearly bright spots where filaments coalesce. Neutron production starts in (a); reaches its peak value in (b) and is still high ($\sim \frac{1}{3}$ of the peak value) in (c) where no plasma concentration is shown by visible light in the axial region.

The stage of maximum compression, as shown by visible-light photos, occurs on the axial region starting from the end of the center electrode at a time coincident with a typical well-defined negative peak in the $\partial I/\partial t$ curve given by an oscilloscope trace. This peak time is conventionally taken as $t=0$. At $t=0$, a bright spot (BS) appears on the front face of the center electrode and apparently moves on the axial column with a velocity of ≈ 2–5×10^7 cm/s. The off-axis part of CS branches off immediately ahead of BS. A dark region in the column follows behind BS which appears the region where the filaments coming together, start to coalesce and/or decay by high rate processes. The production of both X-rays (≈ 1–10^3 keV; $\approx 95\% \lesssim 15$ keV) and neutrons starts almost simultaneously at $t \approx 20$–40 ns. The neutrons produced by $D+D \rightarrow He^3 + n + 3.2$ MeV are detected in two ways, viz. by a silver activation counter (standard type developed at Los Alamos Sci Lab.) which gives a measure of the total number of neutrons per discharge and by a fast plastic scintillator (NE-102) which is connected via photomultiplier (RCA 5819) to an oscilloscope (rise time of whole detection system ≈ 20 ns). A similar scintillator (but with a thinner plastic disk) is used for X-rays. The small thickness in this case insures against response to neutrons. Other methods based on neutron time-of-flight and on X-ray absorbing foils (aluminum) are used to separate neutron and photon signals also for calibration procedures. X-ray-time-integrated images of the off-electrode plasma region are also recorded by pin-hole camera photographs. About 90% of the X-ray signal appears to originate within 7.5 mm of the anode and are mainly due to the anode bombardement and anode vapor. The amplitude of this soft X-ray ($\lessgtr 10$ keV) signal can be decreased to $\approx \frac{1}{10}$ by replacing the solid with a hollow center electrode (a pipe with a 3 mm thick wall). Unless specifically indicated, we will not refer to these electrode X-rays which can be easily discriminated from the X-ray produced in the deuterium plasma. Most of the X-rays which originate out in the plasma, away from the anode, have energies $\gtrless 10$ keV far in excess of the electrode potential. The first X-ray burst is mainly due to this hard component. Since our CA is operated near neutron threshold for a better resolution of the CS structure, the total number of neutron N produced in each discharge can fluctuate up to a factor 10^4. The mean value is 10^8 with an observed maximum of $\approx 4 \times 10^8$.

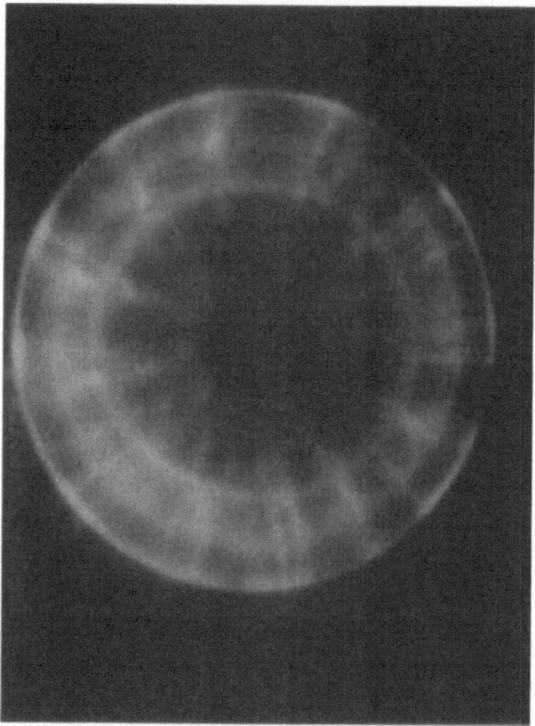

Fig. 4. Same conditions as Figure 2; 5 μs exposure at $t=250\pm 20$ μs. The glowing of the radial filaments in the off-axis region preceeds the decay induced by the radial shock. This correlated (shock-induced) decay of many filaments resembles features of the sympathetic flare phenomenon. The formation of a new kind of filament (rings) parallel to B_θ (related also to the slowing-down of CS) is evident.

No appreciable variations in the neutron production are introduced by using the hollow anode (in this case CS after collapsing in the axial region propagates inside the anode, with a filamentary structure quite similar to that occurring outside the anode; see Figure 1). In coincidence with the *en mass*-combination and decay of the vortex filaments in the axial column, the first $\Delta N/\Delta t$ pulse rises and reaches the peak value in ≈ 40–60 ns, i.e. at the time $t = 60$–100 ns in which BS has reached the end of the column. A clear indication exists that the dark region behind BS where complete decay of the filament occurs, coincides with the region of X-ray and neutron production since this starts only when the dark region begins to grow between BS and the center electrode.

Neutrons and X-rays are frequently emitted also in a later pulse at $t \approx 200$ ns apparently in coincidence with the observed vortex filament coalescence (by pairs) and decay (by single filament processes) in the off-axis region of CS when a disruptive shock wave, partially due to plasma ejected from the axial region, invests CS and starts to propagate in the radial direction across CS (see Figure 4). Overall, there is thus evidence indicating that the X-ray and neutron production is due to the decay of the filamentary magnetic structure of CS, in particular to the magnetic field annihilation associated with the combining of vortex filaments. This last decay process is essentially the solar flare process as proposed by Gold and Hoyle (1960).

2. Particle Acceleration Mechanism

Production of high energy electrons and ions and so of X-rays and neutrons at the focus is frequently explained in the literature (Mather, 1965) as the result of the plasma radial compression in the axial column due to the inward motion of CS and to pinch effect (CA is operated with I reaching I_{max} at the time of CS collapse). Specifically two models which ignore the filamentary structure have been proposed to account for the neutron yield (Patou *et al.*, 1969). It is now generally acknowledged in the literature that both those models cannot consistently explain the measured anisotropies of the neutron emission and the center of mass velocity $(1.2$–$2) \times 10^8$ cm/s of the reacting deuterous as given by the neutron energy spectrum. (Patou *et al.*, 1969). The resolution of the CS structure in our experiments yield the information necessary for a more refined analysis of the focus stage. To summarize the essential role of the magnetic structure decay in the particle acceleration and ultimately in the plasma heating, we single out the following experimental observations:

(a) The different neutron pulses (frequently from one to three pulses are detected on an interval of time between 0.06 to 0.6 μs from $t = 0$) cannot be explained simply in terms of successive pinchings of the plasma column. Second and third $\partial I/\partial t$ minima are occasionally observed. These minima are often – not always – followed by X-ray and neutron bursts but temporal correlation between all three signals cannot be established at the late $\partial I/\partial t$ peak times as it is for the first (compression) peak at $t = 0$. The time separation between different neutron pulses is in a better agreement with the different times of decay of the filamentary structure observed in different regions on the plasma column and on the off-axis region of *CS*. These decay times can be obtained by observing the intense glowing of decaying filament segments (immediately before their disappearance) and of the forks in coalescence processes. The experimental evidence indicates that the glowing of a filament region is a precursor phenomenon ≈ 20 to 100 ns ahead of the complete decay of the filamentary structure in that region.

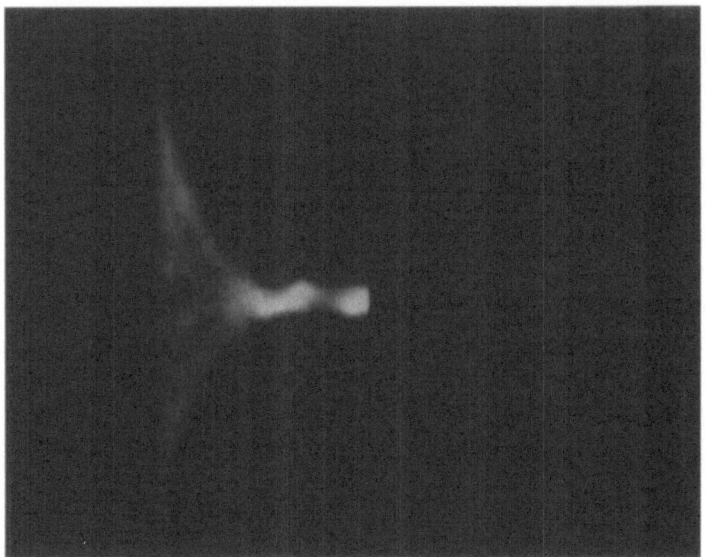

Fig. 5.   Same conditions as Figure 2. Side view at $t = 40 \pm 20\ \mu s$.

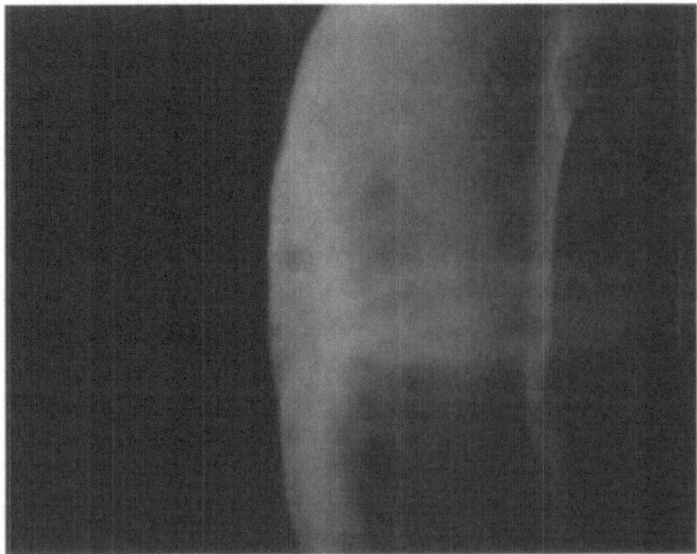

Fig. 6.   Same conditions as Figure 2 but for the introduction of an axial magnetic field ($\approx 10^3$ G) in the anode-front region. Side view at $t = 100 \pm$ ns. The external axial field reduces the neutron yield by a factor $\approx 10^{-2}$.

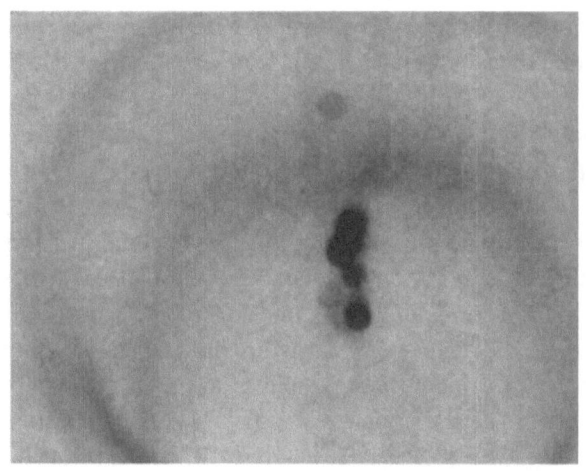
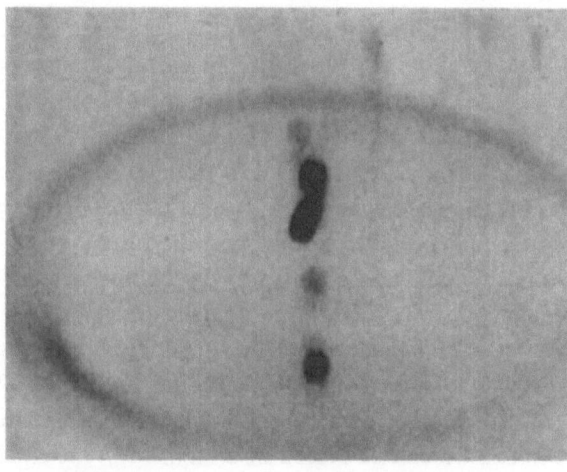
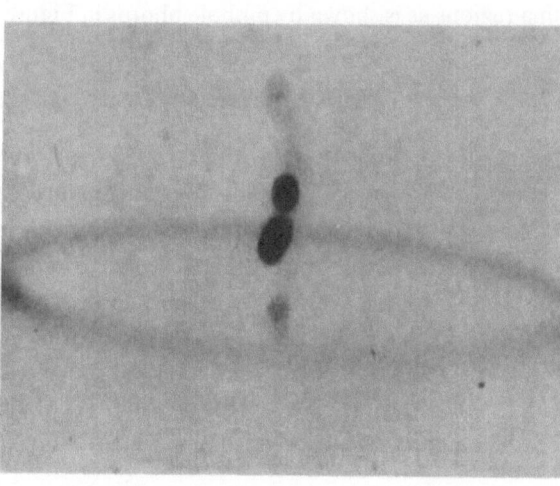

Fig. 7. X-ray pin-hole camera photos (time integrated) of the same discharge taken at 85°, 50°, and 15° with respect to the axis of the gun. In particular, localized regions of high intensity (X-ray dots with diameter < 0.5 mm) can be observed both on and off the axis. Pin-hole diameter 0.5 mm in 0.25 mm thick copper foil (pin-hole 10 cm distance from the center of anode face; film 5 cm from pin-hole). The X-ray arcs on the internal surface of the anode are due to electrons ejected tangentially to CS for short periods of time Δt_x. From CS speed and arc width in the axial direction we have $\Delta t_x \stackrel{\approx}{<} 10^{-8}$ s.

(b) Space-time resolution of the X-ray source is obtained by a collimator (1.5 mm pass-through gap) orthogonal to the axial column. The hard X-ray generating region is apparently moving along the column with a velocity $\sim 3 \times 10^7$ cm/s. This is shown by the relative delay of the X-ray peak signals for two different locations (e.g. 13.5 and 26 mm) of the collimator gap along the axial column. This velocity agrees with the velocity of BS and with the apparent axial velocity of the neutron-generating region as determined by collimated-neutron measurements from other laboratories (Bernstein et al., 1969). Since a clear space-time correlation is so established among X-ray, neutrons and the filament decay precursor, it seems reasonable to conclude that the filament decay is the direct cause of X-ray and neutron production.

(c) The importance of the filamentary structure in particle acceleration is also indicated by X-ray ($\gtrsim 15$ keV) pinhole photographs of CS (front view, time integrated) which show a filamentary structure similar to visible light and Schlieren images (Lee et al., 1968). The fact that the filamentary structure exists in the axial column, where optical resolution is usually more difficult than for the previous stages of CS, is indicated by inhomogeneity and irregular shape of BS on the column. Sometimes BS is confined to a part of the column diameter (see Figure 5). The same irregular pattern is shown by side-view-X-ray pinhole photos. More specifically the existence of the filaments in the column can be proved by slowing down the CS collapse in the axial region. This is accomplished by introducing an external magnetic field up to $\approx 10^3$ G confined to the axial region in the direction of the electrode axis. The reaction of this field to the compression due to the collapsing CS, results ultimately in a greater diameter of the column at maximum compression so that the resolution of the column structure in terms of filaments remains clear also by IC photos (see Figure 6). Our conclusion is that the magnetic structure decay propagates along the filaments with a velocity $2-5 \times 10^7$ cm/s. The corresponding inductive fields eject particles both in the radial and axial directions with a high energy which accounts for the $1.2-2 \times 10^8$ cm/s center-of-mass velocity of the reacting deuterons. The magnetic structure decay can be of an explosive type in some localized region of a filament in the axial column as well as in the off-axis part of CS. This is proved by the localized intense X-ray sources located in different plasma regions as is shown by pinhole photos in Figure 7.

3. Self-Consistent Fields, Structure and Decay

The density in phase-space $f_\pm$ for ions and electrons satisfies $df_\pm/dt = S_\pm$ where $d/dt = \partial/\partial t + \mathbf{v} \cdot \nabla \pm e/m(\mathbf{E} + \mathbf{v} \times \mathbf{B}) \cdot \nabla_v$ is the Liouville-Vlasov operator. The source term S accounts for production and disappearance of ions and electrons by ionization and recombination phenomena. According to S, a particle can be removed from an orbit, which is defined in terms of the constants of the motion as in the Vlasov theory ($S=0$), also by radiation and collision phenomena. Electric and magnetic fields $\mathbf{E} = -\nabla \phi$, $\mathbf{B} = \nabla \times \mathbf{A}$ satisfy Maxwell equations with charge and current densities $e(\varrho_+ - \varrho_-)$, $\mathbf{j} = e(\mathbf{u}_+\varrho_+ - \mathbf{u}_-\varrho_-)$ given by $f_\pm$. Exact solutions of this nonlinear system of integro-differential equations have been obtained for stationary conditions

$(\partial/\partial t = 0)$ on the *CS* frame of reference (moving with a velocity u_0 with respect to the laboratory) by assuming constraints on ϱ, $\mathbf{u}$, $\mathbf{B}$, $\mathbf{E}$, and so on $S_\pm$, $f_\pm$, which are suggested by the experimental evidence (Nardi, 1970). Specifically the space variations along the filament axis are ignored (the filaments are considered essentially as parallel cylinders, $\partial/\partial z \approx 0$) and in this direction $u_{z\pm} = -\varrho_{v\pm} c_0^2/\varrho_\pm c_\mp$ where c_0 is the speed of light and $\varrho_{v\pm}$ is given by the orbit-controlled component f_v of $f = f_v + f_s (df_v/dt = 0$; the *x*-axis is taken $\perp$ to *CS*, *z*, *y* on *CS* (see Figure 1) and so $\mathbf{B}_y \approx \mathbf{B}_\theta$). Since by definition $|c_\pm| = |\int f_{v\pm} v_z \, dv/\varrho_{v\pm}| = $ constants $< c_0$, the contribution ϱ_s to ϱ from the source-controlled part f_s must satisfy $\varrho_\pm = \varrho_{v\pm} + \varrho_{s\pm} > \varrho_{v\pm} c_0/|c_\mp|$. This choice of u_z is based on the criteria: (I) of coupling the mean (macroscopic) velocity of one species of particles with the other species with a maximum of simplicity and (II) to obtain self-consistent solutions of great generality, enough to match all constraints suggested by the experiments. Since $u_{z-} < 0$ for a positive center electrode, $c_+ > 0$; $u_{z+} > 0$ (for both polarities of the center electrode, see Bostick *et al.*, 1970) so $c_- < 0$. The self-consistent fields, flows, and densities are derived for $\varrho_{s+} = \varrho_{s-}$, $f_{v\pm} = \varrho_{0\pm} (\alpha_\pm m_\pm/2\pi)^{3/2} \exp[-\alpha_\pm (\varepsilon_\pm + c_\pm p_{z\pm} + m_\pm c_\pm^2/2)]$, $\varepsilon_\pm$, $p_{z\pm} = m_\pm v_{z\pm} \pm eA_z$ are the energy and *z*-momentum of a particle. Our solutions are (emu):

$$A_z = [e(c_- - c_+)]^{-1} \ln(F_1^{1/\alpha_+} F^{1/\alpha_-}) \Big\}$$
$$\phi = c_+/e(c_+ - c_-) \ln(F_1^{c_-/c_+ + \alpha_+} F_2^{1/\alpha_-}) \Big\}$$
$$\text{where} \quad F_1 = [4/\alpha_+ K^2] |\partial g_1/\partial \eta|^2 [1 + |g_1|^2]^{-2},$$

g_1 is an arbitrary (non-constant) function of the complex variable $\eta = x + iy$, $K^2 = -e^2 c_0^2 \varrho_{0+} + (c_+ - c_-) 2\pi/c_-$; by replacing in F_1 the constants $\alpha_+, \varrho_{0+}, c_+, c_-$ respectively with $\alpha_-, \varrho_{0-}, c_-, c_+$ and g_1 with any non-constant g_2 we have F_2. The functions $= \pm e(\phi + c_\pm A_z) \geq 0$ have the role of effective potentials in $f_{v\pm}$ which represent jets of particles affecting the self-consistent field without directly entering short-range collision and ionization phenomena. The parameters $\alpha_\pm, \varrho_{0\pm} > 0$ and $c_\pm$ play an essential role in the stability and decay of the filament system. Since the plasma is far from thermodynamical equilibrium, a temperature can be defined by $T_\pm = \text{trace} \int f_\pm (\mathbf{v} - \mathbf{u}_\pm)(\mathbf{v} - \mathbf{u}_\pm) \, d\mathbf{v} m_\pm/3k\varrho_\pm$ rather than by $(k\alpha_\pm)^{-1}$ ($k = $ Boltzmann const.). Charge neutrality may or may not be imposed everywhere. If we take for simplicity $\varrho_{v+}/\varrho_{v-} = $ const., i.e. $F_1 \sim F_2$, $\phi \sim A_z$ then the choice $g_1 = g_2 = a + (1 + a^2)^{1/2} e^{mn}$ gives $F \sim (\cosh mx + r\cos my)^{-2} = \tau^{-2}$, $r = a(1 + a^2)^{-1/2}$, which accounts for: (I) periodic arrangement of the filaments on the *y*-axis (simple period $= 2\pi/m$); (II) ionization high on *CS*, low far from *CS*, so $\varrho_\pm$ has maximum values on *CS* at $x = 0$, $\varrho_\pm (x \to \pm \infty) \to 0$; (III) sharp increase of B_y across *CS*, etc. Since $\partial/\partial z \approx 0$, A_z determines completely B_x, B_y i.e. $\mathbf{B}_\tau \equiv \mathbf{B} - \mathbf{B}_z$. The velocity fields for electrons and ions in a filament region $(\tau < 1 + r)$ are given by $\mathbf{u}_- = \lambda(x, y) \mathbf{B}$, $\mathbf{u}_{\tau+} = \beta(x, y) \mathbf{B}_\tau$. The first condition accounts for the small electron inertia and r_e ($=$ electron Larmor radius) $< \lambda_-$ ($=$ electron mean free path) so that the electrons are tied to $\mathbf{B}$ (Lighthill, 1960). According to $\mathbf{u}_+ (x \to -\infty) \to \mathbf{u}_0$ the ions have a kinetic energy which is mainly directed $(kT_+/\bar{\varepsilon}_{+k} < 1; \bar{\varepsilon}_{+k} = \int (\varepsilon_+ - e\phi) f_+ \, dv/\varrho_+)$ on *CS* and for $x < 0$. In spite of $r_+ > \lambda_+$, we can consider then the mass-flow lines on CS nearly coincident with the (non-adiabatic) orbit of an ion unaffected by collisions. The condition $u_{\tau+}$ collinear with B_τ is then a consequence of: (I) the observed invariance of the (equilibrium) filament configuration respect to sign $(u_\tau \times B_\tau)$ as it is shown by the apparently identical configurations of two filaments with opposite B_z and so opposite u_τ: (II) the fact that any equilibrium with $u_\tau \times B_\tau = 0$ for which this invariance is automatically satisfied, is the simplest to describe analytically. The equations for λ, β are given by the x, y components of $\nabla \times B = j$ (which are invariant for a simultaneous change of the signs of λ, β) with the solution $\lambda = \varrho_v - Q(A_z)$, $\beta = \lambda + (c_0^2/ec_+) \, d(1/Q)/dA_z$ where Q is an arbitrary (non-constant) function of A_z.

The field structure of two filaments with opposite B_z components is represented in Figure 8. Each filament is a bundle of helical B-lines with a pitch increasing from the periphery to the axis of the filament (B_z, $\varrho_\pm$ maximum on the axis). It is clear that other equilibrium configurations can be easily obtained by a different choice of g_1, g_2. If B, density, optical pattern, etc. from the experimental observation can be fitted by more than one solution, the choice of one self-consistent solution can be based on further constraints, e.g. the condition of greatest stability against modes growing exponentially with time. If we consider our solutions as the most stable against this

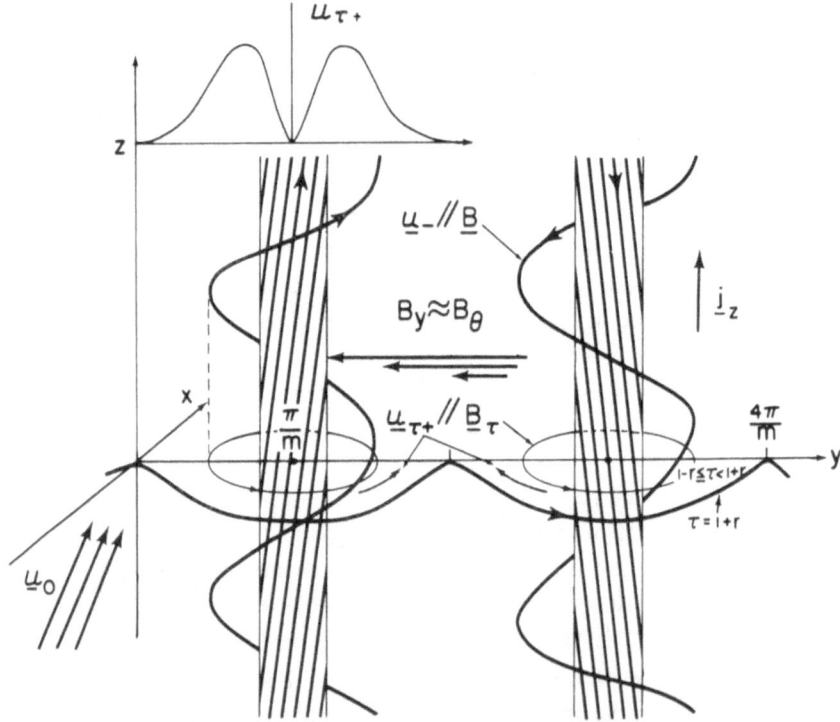

Fig. 8.   Magnetic field lines and their projection ($A_z = \text{const.}$) on the x, y plane in the filament region $\tau < 1 + r$. The magnitude u_τ of the ion mean velocity on the same region of the x, y plane is plotted in the upper diagram (arbitrary scale). The mass flow ($\sim u_0$) far ahead of CS is parallel to the x, z plane.

kind of (secular) variations, we can still analyze the filament decay on a shorter time scale defined by some interval $\Delta t \ll T \approx 8 \ \mu s \ (= CA$ current period). Plasma and CS are not an isolated system but the dynamic behavior on Δt can be considered as determined mostly by the local plasma conditions. As an example, the collapse of CS on $\Delta t_1 \approx 0.1$–$0.2 \ \mu s$ introduces time variations which cannot be considered of the same secular type as the CA current period. This collapse as well as the related slowing-down of CS in the off-axis region is responsible for flow and density rearrangements which trigger the decay (sometime explosive) of the filamentary structure. Explosive onsets of instabilities on a $\Delta t_2 \ll \Delta t_1$ can be predicted by considering a whole series of continuous configurations of equilibrium (a Poincaré linear series, see Jeans, 1928) obtained by varying one of the parameters $\alpha_{\pm}, \varrho_{0\pm}, c_{\pm}$.

These variations of the parameters on a time interval Δt_1 can be associated with the rearrangements due to collapse and slowing-down of CS. We have analized more in detail the linear series corresponding to different values of c_+. There is a critical value of c_+, depending also on the choice of Q, beyond which no stable configuration exists. This kind of instability could be related to the localized X-ray sources (see Figure 7) and to the appearance of bright spots on the filaments during the filament decay (corresponding to the flash phase of the flare). A detailed picture is supplied by our equilibrium configuration. Even a slight alteration in the collinearity of $\mathbf{u}_-$ and $\mathbf{B}$ (eventually induced by an incoming jet of particles) can interrupt the j_z current at a point of a filament. The consequence is the same as if the plasma there suddenly becomes essentially non conducting. The whole magnetic energy of the filament tends to be dissipated at that point so that the filament explodes (Alfvén and Carlquist, 1967).

A major part of the energy stored in the filamentary structure can be released by some other decay process (in correspondence with the main phase of the flare). The cyclotron resonant transfer of kinetic energy from filament-axis direction into the Larmor rotation (and vice-versa) can be considered as such a process (Nardi, 1970) along with the magnetic field annihilation between neighboring vortex filaments (Bostick et al., 1966; Gold and Hoyle, 1960) and other cooperative effects.

4. Laboratory Experiment and Solar Flares

The basic ingredients to produce the filamentary structure in the laboratory are: (I) a flow of dense highly-ionized gas (10^{17}–10^{18} particles/cm^3) with a mean velocity $u_0 \approx 10^6$–10^7 cm/s from a region of low (or vanishing) magnetic field toward a region with a strong field $B_\theta (B_\theta \approx 10^4$–$10^5$ G); (II) a steep gradient of this background field $B_\theta (\approx 10^4$–10^5 G/cm between the two regions) in the direction of the velocity $\mathbf{u}_0$; (III) $\mathbf{u}_0$ nearly orthogonal to $\mathbf{B}_\theta$. In this case, a filamentary structure appears in the region of sharp transition; each filament axis is orthogonal to $\mathbf{B}_\theta$ and in the plane containing $\mathbf{u}_0$. We recall that in the case, not considered here, of high field but small gradient (Bostick et al., 1966) or for relatively smaller values of u_0 (see Figure 4), filaments can be produced with the axis in the same direction of the background field B_θ. The self-consistent field resulting from the interaction of the plasma with B_θ (in the transition region) has a fine structure which is underlined by the (visible light) filament arrangement. This self-consistent field region rests upon the high-B_θ region (without great variations with time) until an instability is triggered by some disturbance, e.g. a modulation in the incoming particle flow or a localized disruption. The decay of the fine-structure field with luminosity increase and particle acceleration can leave the background field B_θ substantially unchanged. Similar features can be found in solar flare phenomena. The field connecting solar regions (spots) of different polarity may correspond to our B_θ. If an inward flux of particles flows over a region of predominantly horizontal field (i.e. near the null line of the observed longitudinal field) the formation of filaments can be expected. These fine-structure filaments (orthogonal to the background field) will appear in general as parallel to the isogauss lines (of the background longitudinal field). This seems to be in agreement with flare observations (Howard and Harvey, 1964). The inward motion can be observed by Balmer-line profile asymmetry (red wing brighter than blue wing in 90% of flares, Smith and Smith, 1963). Other observations by Severny (1958) indicate that field gradients near the neutral point are quite steep. The idea that only the filamentary fine-structure of the field decays during a flare is supported by the observation of flares for which no changes in the isogauss maps could be detected before and after the flare (Howard and Harvey, 1964). The flare X-ray spectrum in the range 20–100 keV can be represented by a power law $N_f(\varepsilon) \sim \varepsilon^{-n}$, $3 \leq n \leq 5$, (Fichtel and McDonald, 1967) which is essentially the same as the power law $N(\varepsilon) \sim \varepsilon^{-4 \pm 1}$ for the range 100–500 keV from the plasma focus (Lee *et al.*, 1970). These observations are perhaps sufficient to point out the useful role of laboratory experiments in the study of solar flares. The use of similarity transformations to gain a better insight in solar flare phenomena requires further work. It is a well-known fact that arc-plasma does not obey the same similarity laws available for other kinds of discharges.

Acknowledgements

Work supported by U.S. Air Force under Grant AFSOR-70-1842. (U.S.A.) and by Istituto Avogadro di Tecnologia, Roma C.P. 10757 (Italy).

References

Alfvén, H. and Carlquist, P.: 1967, *Solar Phys.* 1, 220.
Bernstein, M. J., Meskan, D. A., and van Paassen, H. L. L.: 1969, *Phys. Fluids* 12, 2193.
Bostick, W. H., Prior, W., Grunberger, L., and Emmert, G.: 1966, *Phys. Fluids* 9, 2078.
Bostick, W. H., Grunberger, L., Nardi, V., and Prior, W.: 1969, *Proc. 9th Int. Conf. Ionized Gases*, Bucharest, p. 66.
Bostick, W. H., Grunberger, L. Nardi, V., and Prior, W.: 1970, *Proc. 4th Int. Conf. Thermophys. Properties*, Newton, Mass., p. 495.
Fichtel, C. E. and McDonald, F. B.: 1967, *Ann. Rev. Astron. Astrophys.* 5, 531.
Gold, T. and Hoyle, F.: 1960, *Monthly Notices Roy. Astron. Soc.* 120, 89.
Howard, R. and Harvey, J. W.: 1964, *Astrophys. J.* 139, 1328.
Jeans, J.: 1928, *Astronomy and Cosmogony*, Cambridge University Press., Cambridge, England, Ch. 7.

Lee, J. H., Conrads, H., Williams, M. D., Shomo, L. P., and Hermansdorfer, H.: 1968, *Bull. A.P.S.* **13**, 1543.

Lee, J. H., Loebbaka, D. S., and Roos, C. E.: 1970, *Bull. A.P.S.* **15**, 642.

Lighthill, M. J.: 1960, *Phil. Trans.* **A252**, 397.

Marshall, J.: 1960, *Phys. Fluids* **3**, 134.

Mather, J. W.: 1965, *Phys. Fluids* **8**, 366.

Nardi, V.: 1970, *Phys. Rev. Letters* **25**, 718.

Patou, C., Simmonet, A., and Watteau, J. P.: 1969, *Phys. Letters* **29A**, 1.

Severny, A. B.: 1958, *Izv. Krymsk. Astrofiz. Obs.* **20**, 22.

Smith, H. J. and Smith, E. P.: 1963, *Solar Flares*, Ch. 6.

Discussion

Brueckner: What is the density ratio in the center of the arc prior to and during the collapse?

Nardi: Our initial D_2 density was 2.8×10^{17} cm^{-3}; 10^{19} cm^{-3} can be accepted as the correct order of magnitude of the maximum density of electrons ($\cong$ ion density) after current sheath collapse during the neutron pulse. Different methods (Schlieren, best fit to the bremsstrahlung, etc.) point to this value. A maximum value $> 4 \times 10^{18}$ cm^{-3} for electrons has been obtained just before the beginning of the neutron pulse, by measurement of 90° cooperative scattering of a ruby-laser beam by the plasma (coaxial accelerator and operating conditions were comparable to ours; J. P. Baconnet *et al.*, *Proc. Ninth Int. Conf. Ionized Gases*, Bucharest 1969, p. 643). Our measurements (by Hβ Stark broadening) of the electron density in the filaments, before collapse of the current sheath, gives $1-2 \times 10^{18}$ cm^{-3}, i.e. $\sim$ three times the initial density.

Wilcox: Referring to the slide (Figure 2) why do the radial filaments pair up with the current and field in opposite directions in each member of a pair?

Nardi: The ions can move across the magnetic field with a Larmor radius of the same order as the filament optical radius, whereas the electrons are tied to the field. Therefore a distinction must be made about the different components of current density j. The current carriers are different in different directions. The z-component j_z, along the filament axis, is mainly carried by the electrons (this is indicated by specific experimental data, see W. H. Bostick *et al.*, *Proc. Fifth Symp. Thermophysical Properties*, Newton, Mass. 1970, Amer. Soc. Mech. Eng. edit., p. 495) and has the same direction in all filaments with a positive or a negative value depending only on the electrode polarity. The corresponding (B_y, B_x) field lines in a filament region are closed loops with center on the filament axis, having a common direction defined by sign (j_z). On the other hand, in the (x, y) plane, orthogonal to the current sheath driven by the electrode applied field, the ions are the current carriers. This becomes clear by considering: (I) the electron inertia as negligible if compared with ion inertia and (II) net charge density as very small compared with the electron density. Then it can be easily shown (M. J. Lighthill, *Phil. Trans. Roy. Soc.* **A252**, 397 (1960)) that a low rate of loss of electron momentum, per unit volume, by collision with ions implies that the magnetic lines of force 'move with' or are 'frozen into' the electron gas rather than 'with' or 'into' the flowing mass. This is our case and it can be considered an accurate picture whenever the magnetic field is so high that the Hall-current effect is much more important than the electrical resistivity (i.e. the gyro-frequency of electrons greatly exceeds their collision frequency).

Since the electrons are tied to the magnetic field lines, in the current sheath frame of reference the ions carry the current in any direction orthogonal to these lines (i.e., essentially in the (x, y) plane since B_z is by far the largest component in a filament region). For our self-consistent solutions the projection on the plane (x, y) of the mass flow lines corresponding to the vortex structure of each filament, are closed loops nearly, or exactly, collinear (depending on our choice) with the (B_x, B_y) field loops. Solutions exist corresponding to both parallel or antiparallel motions with respect to these field loops. The mass flow which minimizes the shear, however, corresponds to a vorticity with opposite directions in neighboring filaments and is, likely, the preferred one by the plasma. Opposite rotational velocity of the ions implies opposite currents in the (x, y) plane and so opposite components B_z of the self-consistent field. The paired arrangement of the filaments can be considered then as corresponding to the most stable configuration with a minimum shear in the velocity field.

INTERACTION OF CORONAL MATERIAL
WITH MAGNETIC FIELDS

G. W. PNEUMAN and ROGER A. KOPP

High Altitude Observatory, National Center for Atmospheric Research* Boulder, Colo., U.S.A.

Abstract. The optical resolution in recent eclipse photographs is now sufficient to illustrate clearly many qualitative features of gas-magnetic field interactions in the solar corona. Close to the Sun where the field is strong, the coronal gas over certain regions of the solar surface can be contained within closed loop structures. However, since the field strength in these regions declines outward rapidly, the pressure and inertial forces of the solar wind eventually dominate and distend the field outward into interplanetary space. The overall configuration is determined by a complex interplay of inertial, pressure, gravitational, and magnetic forces. The present work is oriented toward the understanding of this interaction.

Of central importance to this general problem is the typical 'helmet' streamer generally overlying a large diffuse bipolar region on the solar surface. A helmet streamer consists of closed magnetic loops near the Sun and open field lines adjacent to and above the loops. The configuration possesses a 'cusp-type' neutral point (to be differentiated from a y-type neutral point) at the top of the closed loops. In addition to volume currents, a single sheet current separates the regions of opposite polarity above the neutral point. Below the neutral point, sheet currents also are present between the open and closed regions. These sheet currents are necessary to support the discontinuity in gas pressure which arises since the open region is expanding and the closed region is not. The neutral point problem is discussed in some detail and sample numerical examples are presented.

The interaction of coronal material with the magnetic fields of the Sun is quite evident in recent white light eclipse photographs of the inner corona. For example, large, closed magnetic loops are visible over extensive regions which presumably completely inhibit coronal expansion. Beyond about 2 $R_\odot$, however, the field lines are all open, allowing streaming out into interplanetary space.

This observed picture is consistent with two theoretical concepts. Firstly, due to the high electrical conductivity and the large scale of coronal structures, the magnetic Reynolds numbers appropriate for these conditions are enormous. Consequently, the magnetic fields are effectively frozen into the material. Secondly, the ratio of magnetic pressure to dynamical pressure evidently varies from a value somewhat greater than one at the coronal base to values much less than one at large distances, with the two being equal somewhere near the tops of the closed loop systems.

In this work, we attempt to describe within the context of a theoretical model this complex interplay of inertial, pressure, gravitational, and magnetic forces. In order to simplify the problem as much as possible without undermining the relevant physics, we will consider the corona as being isothermal and symmetric about the rotation axis. Also, we neglect the effects of solar rotation.**

---

* The National Center for Atmospheric Research is sponsored by the National Science Foundation.
** Although rotation introduces a toroidal magnetic field in the corona, it can be shown to have a negligible influence on the poloidal components of field and velocity which are of interest here.

Howard (ed.), Solar Magnetic Fields, 526–533. All Rights Reserved.
Copyright © 1971 by the IAU.

1. The Helmet Streamer

Of central importance to the general problem of the interaction of the solar wind with the coronal magnetic fields is the so-called 'helmet' streamer. This structure is of special interest since, by definition, it represents the transition from closed to open field lines and requires special considerations which distinguish it from totally open magnetic configurations. These distinguishing considerations are the neutral point existing at the top of the closed loops and the sheet currents both above the neutral points where the field changes polarity and below the neutral point on the interface between the open and closed regions. This helmet configuration is schematically sketched in Figure 1.

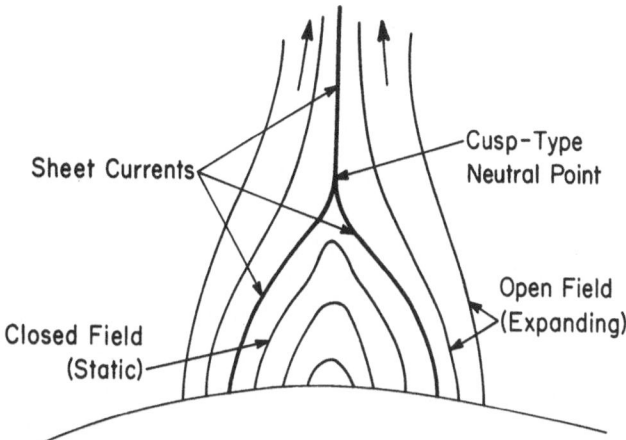

Fig. 1.   Schematic diagram of a helmet streamer configuration.

Since motion of material across field lines is essentially prohibited under coronal conditions, we can safely assume that coronal expansion is not taking place in the closed magnetic region (if, indeed, the field lines themselves are stationary). Likewise, conductive losses across the field are insignificant. If radiative losses are negligible, conditions should be almost isothermal there and the gas pressure determined by the equation of hydrostatic equilibrium applied *along* the field.*

In the region of open field lines, outward heat conduction and expansion are taking place. These processes cause the temperature to decline outward there. This decline results in more rapid pressure drop with height in the open region than in the closed region and is directly responsible for the large observed outward increase of density enhancement in helmet streamers (Pneuman and Kopp, 1970).

Since the gas pressure is falling faster in the open region than in the immediately adjacent closed region, there is a pressure discontinuity at the interface separating the regions. The quantity $p + B^2/8\pi$, however, must be continuous across the interface. Hence, there must be a jump in $B^2/8\pi$ equal and opposite to the jump in p. In the limit

* By this, we mean that the pressure at any given point in the closed region is related to the base pressure only on the same field line and not, for example, to the pressure directly below.

of infinite electrical conductivity, a sheet current must exist at the boundary to produce this jump in field strength. There must also be a sheet current at the neutral sheet above the neutral point separating the regions of opposite magnetic polarity.

The field configuration in the vicinity of the neutral point can theoretically be of three types (Sturrock and Smith, 1968), two of which can be ruled out by physical considerations. The 'y-type' neutral point can be ruled out since, for this case, the field vanishes when approaching the neutral point from all directions and there is consequently no way to balance the pressure jump between the open and closed regions. In the 'inverted T-type' neutral point, the field vanishes in the open region but not in the closed region and the jump in B is of the wrong sign. The only type of neutral point consistent with the forces involved is the 'cusp-type' geometry in which the field goes to zero approaching the neutral point from the inside, but does not vanish on open field lines. The condition to be met there can, therefore, be mathematically stated as

$$p_c = p_s + B_s^2/8\pi \tag{1}$$

where the subscript c refers to the closed region and the subscript s to the streaming region.

Finally, we point out another interesting property of the neutral point. If the sonic point lies well above the cusp (this is probably the case under actual coronal conditions), then it can be shown that the pressure difference between the open and closed regions at the neutral point is about equal to the kinetic energy density in the solar wind just outside the closed region (Pneuman and Kopp, 1971). Since this pressure difference at the cusp must also balance the magnetic pressure just outside (see Equation (1)), we have

$$\tfrac{1}{2}\varrho_s V_s^2 \approx B_s^2/8\pi$$

or

$$V_s \approx B_s/\sqrt{4\pi\varrho_s}\,.$$

Hence, for the expansion just outside the closed region. *the Alfvénic point occurs at the cusp*.

2. Numerical Method of Analysis

Having discussed the physical aspects of closed and open magnetic structures, we now turn to the general problem of numerically determining the properties of a model corona subject to specified boundary conditions at the coronal base. The necessary base conditions are density, temperature, and normal component of the magnetic field. Also, some appropriate model of coronal heating must be incorporated.

To solve the relevant differential equations, we employ an iterative technique. The iterative loop consists of three distinct steps, each of which is manageable using an electronic computer. A detailed description of the iterative method appears elsewhere (Pneuman and Kopp, 1971). We only briefly describe the contents of the three steps here.

The first step consists of calculating the magnetic field configuration for a prescribed distribution of volume currents **J** and sheet currents **J*** over all space as well as the specified normal component of the magnetic field at the solar surface. Since $\nabla \cdot \mathbf{B} = 0$, we employ the vector potential **A** in which

$$\nabla \times (\nabla \times \mathbf{A}) = 4\pi \mathbf{J}. \tag{2}$$

This differential equation for **A** is of the elliptic type and is integrated using standard relaxation methods. For axially symmetric distributions, both **A** and **J** have only azimuthal components, A and J.

In this analysis, we treat only field configurations which are symmetric about the equator. The conditions to be met at the neutral point in this case are that

$$\frac{\partial}{\partial r}(Ar\sin\theta) = \frac{\partial^2}{\partial r^2}(Ar\sin\theta) = 0,$$

where θ is the colatitude and r the radial distance from the solar center. For a given set of electric currents, these two conditions can be simultaneously satisfied at only one point, thus providing a unique solution.

The magnetic field configuration determined in step 1 defines a streamtube geometry in which the solar wind equations may be integrated outward along each field line. Solution of these equations in this geometry is greatly facilitated by the fact that there are no magnetic forces along the field. On closed flux tubes, hydrostatic equilibrium holds along the field and the velocity is taken to be zero. On open field lines, we find the so-called 'critical' solution in which the solar wind starts subsonically near the Sun, passes smoothly through the critical point, and is supersonic at large distances.

The integration of the equations for conservation of momentum and mass along the field in step 2 determines all the necessary physical variables. However, the equilibrium of forces *normal* to the field has not yet been imposed. In step 3, we use this condition to determine new electric currents. In other words, we determine an electric current **J** which in conjunction with the magnetic field **B** determined in step 1, produces a $\mathbf{J} \times \mathbf{B}$ force necessary to balance the components of the pressure, inertial, and gravitational forces normal to the field.

Having found new expressions for the current density in step 3, the iterative loop is complete and we can return to step 1 continuing the process. The iteration is considered converged when the field on a given iteration differs from that of the previous iteration by some specified small amount.

3. A Numerical Example

The field line configuration of a sample converged solution using the iterative technique of Section 2 is shown in Figure 2. Here, the normal component of the field at the surface is that of a dipole of strength 1 G at the pole. The gas pressure at the surface is taken to be independent of latitude and equal to the surface magnetic pressure at the pole. The coronal temperature is chosen to be 1.56×10^6 K and also independent of

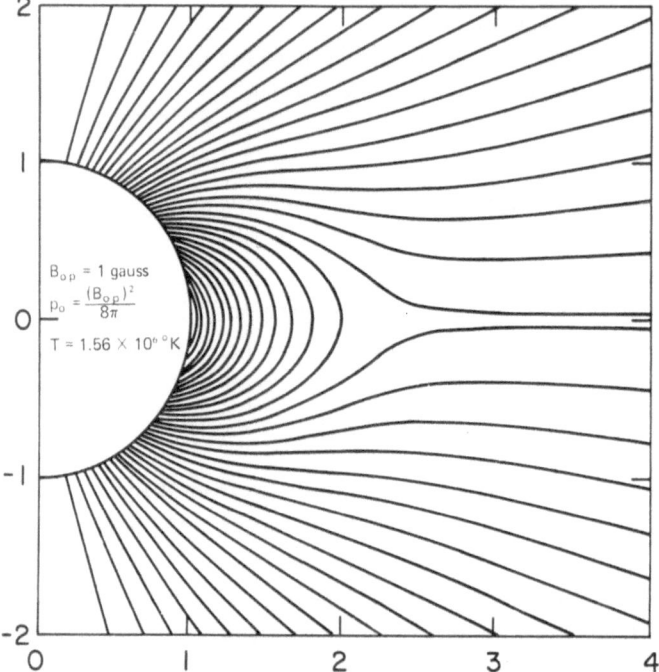

Fig. 2.   Field and streamline distribution corresponding to base conditions in which the normal component of the magnetic field is that of a dipole and the reference density is independent of latitude.

latitude. The computed configuration is similar in many respects to the large helmet streamers observing during solar eclipses. Closed loops extend out to about 2.5 $R_\odot$, the field everywhere being distended outward by the solar wind. The height at which the field lines open at the equator is most sensitive to coronal temperature due to its influence on the scale height.

A more instructive illustration than Figure 2 is Figure 3, in which the same solution is plotted in a different coordinate space. The ordinate here is $\cos\theta$ and the abscissa $R_\odot/r$. The advantage of this system is that the solution over all space can be shown in a finite box. The top and bottom boundaries represent the poles, the right-hand boundary the solar surface, and the left-hand boundary infinity. We see that the field becomes asymptotically radial at large distances. The nearly vertical dashed curve depicts the sonic surface and the solid curve the Alfvénic surface. The sonic surface is at 4 $R_\odot$ at the pole and moves out to 5 $R_\odot$ near the equator. The Alfvénic surface, on the other hand, is at about 5 $R_\odot$ at the pole and moves in to the cusp at the equator ($\approx 2.5\ R_\odot$), as expected from the discussion in Section 1.

The variation of solar wind velocity with latitude is shown at two different heights in Figures 4 and 5. Figure 4 depicts the flow at a typical height below the top of the helmet with the region of zero velocity indicating the region of closed field lines. The velocity on open field lines is seen to increase from the pole toward the equator. In Figure 5, representing a level above the neutral point, the variation of velocity with

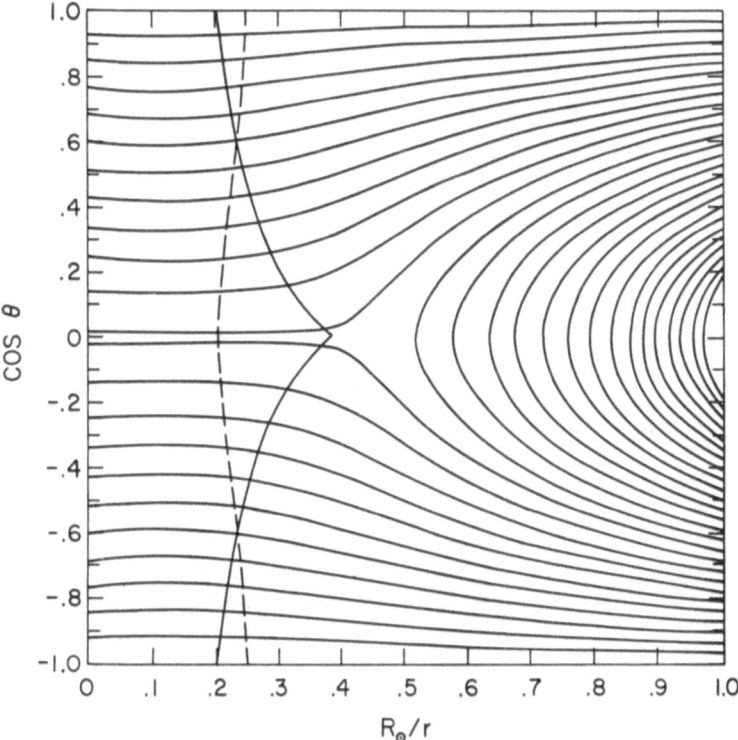

Fig. 3.   The same field configuration as in Figure 2 but plotted with $\cos\theta$ as the ordinate and $R_\odot/r$ as the abscissa. The vertical dashed curve represents the sonic surface and the solid curve the Alfvénic surface.

latitude is the opposite, with the lowest velocity occurring at the equator. This qualitative difference in velocity distribution is due to the behavior of the flux tube cross section. Below the neutral point, the cross section expands more rapidly near the equator than at the pole. Above the neutral point, the opposite is the case.

4. Conclusions

The results of Section 3, in spite of applying only to an extremely simple configuration symmetric both to the rotational axis and the equator, nevertheless demonstrate the possibility of numerically modeling the structure and physical properties of the solar corona and interplanetary medium as a function of measured quantities at the coronal base. The obvious extensions which must be made are the inclusion of the energy equation within the framework of the model as well as relaxing the symmetry conditions imposed here. When this is done, more realistic surface distributions of magnetic field and density can be treated.

Acknowledgements

The authors wish to offer special thanks to Nancy Werner who did the computer programming for this work.

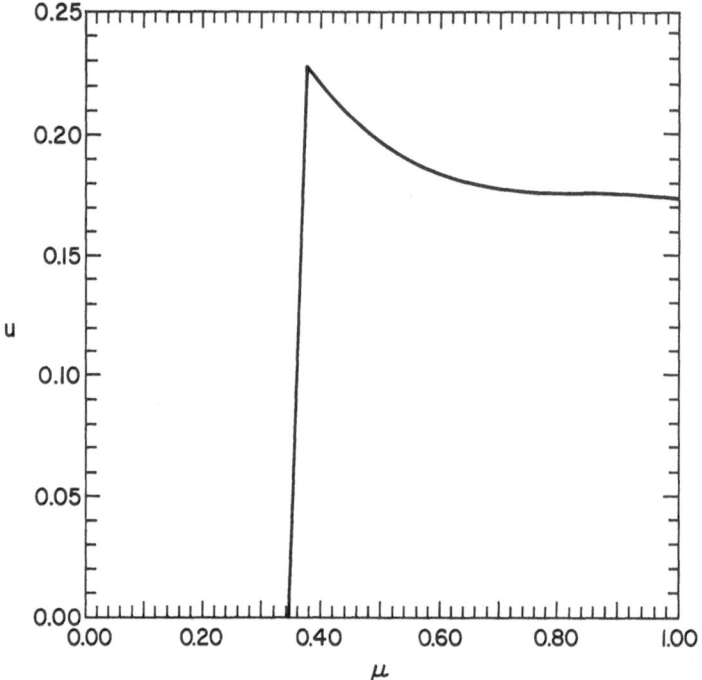

Fig. 4. Variation of velocity with latitude at a typical level below the top of the helmet. Here $u = V/\sqrt{2}\, V_s$, where V_s is the sound speed, and $\mu = \cos\theta$. The region of zero velocity is the region of closed field lines.

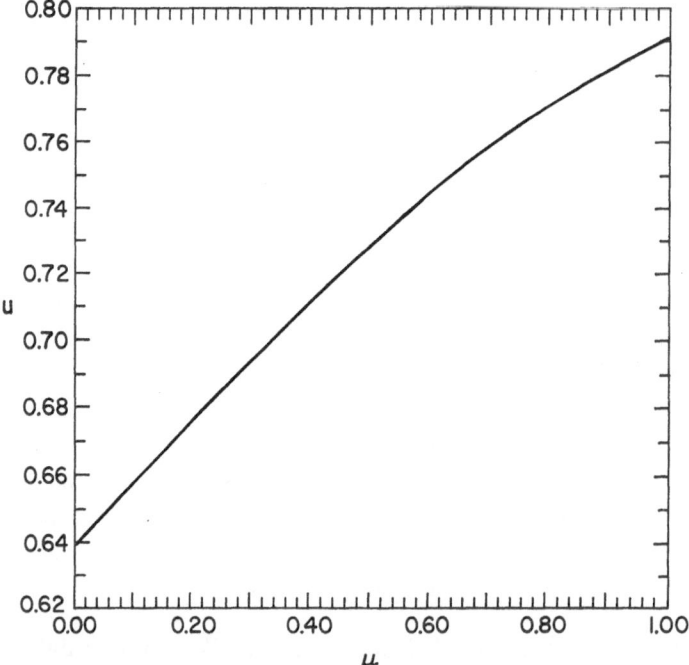

Fig. 5. Variation of velocity with latitude at a typical level above the neutral point. Note that, as opposed to Figure 4, the velocity here decreases from pole to equator.

References

Pneuman, G. W. and Kopp, R. A.: 1970, *Solar Phys.* **13**, 176.
Pneuman, G. W. and Kopp, R. A.: 1971, *Solar Phys.* **18**, 258.
Sturrock, P. A. and Smith, S. M.: 1968, *Solar Phys.* **5**, 87.

Discussion

Kuperus: The neutral sheet type magnetic field configuration which you used in your model has the complication of being unstable due to a magnetohydrodynamical instability called the tearing mode instability. The result of this instability is that field and matter might be dragged towards the neutral sheet where the field is partly annihilated. The first remark is that this effect might considerably change your field topology close to the 'neutral' sheet. My second remark is that due to the enhanced annihilation of the magnetic field in the sheet layer an extra heating term enters your energy equation thereby modifying the expansion of the outer corona.

Pneuman: I don't know whether the neutral sheet in this configuration is stable or unstable. A stability analysis has not been carried out to answer this question. If such an instability exists, it is probably relevant only in the near vicinity of the sheet and will not affect the overall structure which is of interest here. That this must be true is borne out by the observation that these helmet streamers appear very stable for several rotations and seem to evolve only as the underlying fields change their configuration.

In answer to your second remark, the example shown here is for an isothermal corona and energy considerations have not yet been included in the analysis. The point you raise, however, is an interesting one and may indeed be an important consideration in the central regions of streamers.

THE POSSIBILITY OF MAGNETIC FIELD ORIGIN IN FINE STRUCTURE ELEMENTS OF SOLAR FEATURES

M. KOPECKÝ

Astronomical Institute of the Czechoslovak Academy of Sciences, Ondřejov, Czechoslovakia

and

G. V. KUKLIN

Siberian Institute of Terrestrial Magnetism, Ionosphere and Radio Propagation, Irkutsk, U.S.S.R.

Abstract. It is demonstrated that in principle the magnetic field may originate in the fine structure elements of the photosphere and the sunspots as a result of lack of coincidence of isobaric and isothermic surfaces in those elements.

In recent years a large number of papers have appeared in which the fine structure of solar features was considered. Fine structure elements are observed in the photosphere, in sunspots, in faculae etc. They are connected closely with the structure of magnetic fields. The estimations of the magnetic field decay time show that the time scale is comparable with the lifetime of these elements. Naturally the magnetic field must be regenerated, or else the magnetic fields of larger scale may decay during the time interval of the same order, if we consider them as only chance collections of the fine structure elements.

Among various conceivable mechanisms for magnetic field origin conforming to the fine structure elements, an assumption about the magnetic field origin in the stars proposed by Biermann and Schlüter is the most attractive one. If surfaces of equal electronic pressure and concentration do not coincide then a priming magnetic field may appear and may be amplified by any mechanism, for instance by turbulence. Therefore in any inhomogeneous layer of the Sun where suitable conditions exist, the origin of the magnetic field is possible in principle. Maybe this mechanism works in the subphotospheric layers as well as in the outer solar atmosphere.

Here we shall estimate the possibility of magnetic field origin in the fine structure elements of the photosphere and of the sunspots. The cause of currents generating the magnetic field is an electric field connected with the pressure gradient. It is very difficult to consider the whole evolution of the magnetic field because we must solve the full system of nonlinear equations of hydromagnetics. This is accompanied by great computational difficulties and with uncertainties in values of a number of the physical parameters and in the fine structure element models. Earlier we obtained the equations for the electric field caused by the pressure gradient (Kopecký and Kuklin, 1967). One may conclude that in the presence of a magnetic field, there is no full compensation of the pressure gradient drift by the electric field because in a coordinate system connected with different components of the plasma, uncomparable expressions for the electric field are obtained.

So we shall try to estimate the magnetic field origin rate at an initial moment when

Howard (ed.), Solar Magnetic Fields, 534–541. All Rights Reserved.
Copyright © 1971 by the IAU.

there is no magnetic field either in the fine structure elements or in the environment. Then the expression for the electric field is independent of the coordinate system,

$$\mathbf{E}^* = \frac{1}{en_e}(\varepsilon \mathbf{G} - \nabla p_e). \tag{1}$$

The initial equation of the magnetic field dynamics

$$\frac{\partial \mathbf{H}}{\partial t} = c \, \text{rot} \, \mathbf{E}^* + \text{rot} \, (\mathbf{v} \times \mathbf{H}) + \frac{c^2}{4\pi\sigma} \Delta \mathbf{H} + \frac{c^2}{4\pi\sigma^2} (\nabla \sigma \times \text{rot} \, \mathbf{H}) \tag{2}$$

becomes

$$\frac{\partial \mathbf{H}}{\partial t} = \frac{c}{e} \, \text{rot} \, \frac{\varepsilon \mathbf{G} - \nabla p_e}{n_e} \tag{2'}$$

where

$$\mathbf{G} = \xi_n \nabla (p_e + p_i) - \xi_i \nabla p. \tag{3}$$

Here $\xi_i \approx 1 - \xi_n$ is the relative ion mass and ξ_n is the relative neutral mass. Assuming

$$T_e = T_i = T_n = T = \frac{5040}{\theta}, \quad n_e = n_i = \frac{p_e\theta}{5040 \, k}, \quad p_e = p_i = sp$$

we obtain

$$\frac{\partial \mathbf{H}}{\partial t} = -5040 \, \frac{ck}{e} \, \text{rot} \left[\frac{\varepsilon(1 - \xi_n)}{\theta p_e} \nabla p + \frac{1 - 2\varepsilon}{\theta p_e} \nabla p_e \right] \tag{4}$$

or introducing

$$\psi_0 = \frac{\varepsilon(1 - \xi_n)}{s\theta}, \quad \psi_2 = \frac{1 - 2\varepsilon}{\theta}$$

$$\frac{\partial \mathbf{H}}{\partial t} = 5040 \, \frac{ck}{e} \, [\nabla \ln p \times \nabla \psi_0 + \nabla \ln p_e \times \nabla \psi_2]. \tag{4'}$$

The auxiliary values may be computed with the help of the following formulae

$$1 - \xi_n = \frac{sM}{1 - 2s}, \quad \varepsilon = (1 + Cr_{in}^2)^{-1}, \quad C = \frac{42.69}{r_{en}^2} \sqrt{\frac{\mu_i}{1 + M}},$$

where $M = \mu_i/\mu_n$ and μ_i, μ_n are the ion and neutral masses. Let the fine structure element have an axial symmetry and then considering ψ_0 and ψ_2 as functions of θ, $\lg p$ and $\lg p_e$ we obtain

$$\frac{\partial H_\varphi}{\partial t} = 10^8 \left[\frac{\partial \psi_0}{\partial \theta} \left(\frac{\partial \theta}{\partial r} \frac{\partial \lg p}{\partial z} - \frac{\partial \theta}{\partial z} \frac{\partial \lg p}{\partial t} \right) + \right.$$

$$\left. + \frac{\partial \psi_2}{\partial \theta} \left(\frac{\partial \theta}{\partial r} \frac{\partial \lg p_e}{\partial z} - \frac{\partial \theta}{\partial z} \frac{\partial \lg p_e}{\partial r} \right) \right] =$$

$$= 10^8 \left[\frac{\partial \psi_0}{\partial \theta} (\nabla \lg p \times \nabla \theta)_\varphi + \frac{\partial \psi_2}{\partial \theta} (\nabla \lg p_e \times \nabla \theta)_\varphi \right]. \tag{5}$$

This expression may be transformed also in such a way

$$\frac{\partial H_\varphi}{\partial t} = 10^8 \, \psi \, (\theta, \, p) \, (\nabla \lg p \times \nabla \theta)_\varphi \tag{5'}$$

where

$$\psi \, (\theta, \, p) = \frac{\partial \psi_0}{\partial \theta} + \frac{\partial \psi_2}{\partial \theta} \left(1 + \frac{\partial \lg s}{\partial \lg p}\right) - \frac{\partial \psi_2}{\partial \lg p} \frac{\partial \lg s}{\partial \theta}. \tag{6}$$

In order to make numerical estimations we need information about the geometry of the fine structure elements, their models and tables of functions ψ_0, ψ_2, and ψ. We shall not discuss defects of existing models or an accurate form of these elements because we are interested in making estimations with an accuracy of an order of magnitude, solving principally to see if magnetic field origin is possible in such conditions. Thus we selected the inhomogeneous photosphere model by de Jager (1959) and the inhomogeneous sunspot model by Obridko (1968) assuming the hot elements to be vertical cylindrical columns in the cold medium which are some hundreds of km in length. The adopted geometry of the physical parameter distribution is presented in Figure 1. If we are wrong then the errors in the gradient computation do not exceed one order and most probably we are close to the lower limit of the values.

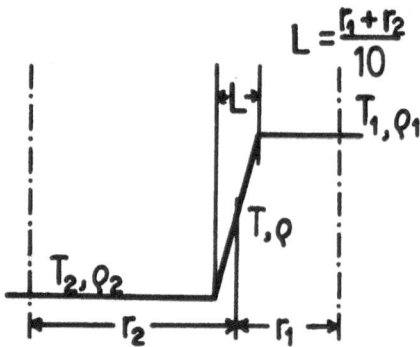

Fig. 1.   The adopted distribution of physical parameters in hot and cold elements of the fine structure.

We have computed tables of ψ_0, ψ_2, and ψ using the data published in our previous papers (Kuklin, 1966; Kopecký and Kuklin, 1969) for the same composition of the solar atmosphere. Only the selection of the ion-neutral interaction cross-section value r_{in}^2 is uncertain. We made computations of tables for r_{in}^2 equal to 10^{-16}, 10^{-14} cm^2 but all estimations of the magnetic field origin rate are made for $r_{in}^2 = 10^{-15}$ cm^2.

We used all 3 variants of the inhomogeneous sunspot model by Obridko and by this for the initial model of Michard we considered the value α in computations of dark element sizes assuming for bright elements $r_1 = 150$ km. In the case of the initial model by Fricke-Elsässer we took a distance between centers of bright and dark elements $r_1 + r_2 = 1000$ km. The results of our computations are presented in Figure 2 (1: the initial model by Fricke-Elsässer; 2: initial model by Michard,

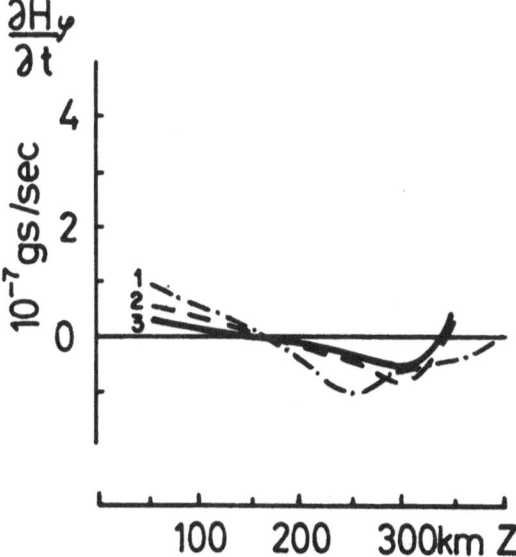

Fig. 2. The rate of magnetic field origin in the fine structure elements of the two-component umbra models: (1) Initial model by Fricke and Elsässer; (2) Initial model by Michard $\alpha = 0.10$; (3) Initial model by Michard $\alpha = 0.05$.

TABLE I

Values of $1 - \xi_n$

θ / $\lg p$	0.8	1.1	1.4	1.7
3.0	3.57×10^{-3}	1.057×10^{-3}	3.49×10^{-4}	7.76×10^{-5}
4.5	1.16×10^{-3}	7.04×10^{-4}	1.23×10^{-4}	3.33×10^{-5}
6.0	6.85×10^{-4}	2.64×10^{-4}	4.58×10^{-5}	9.41×10^{-6}

TABLE II

Values of ε

r_{in}^2	θ / $\lg p$	0.8	1.1	1.4	1.7
10^{-16}	3.0	1.45×10^{-1}	1.22×10^{-1}	1.30×10^{-1}	1.26×10^{-1}
	4.5	1.30×10^{-1}	1.22×10^{-1}	1.26×10^{-1}	0.97×10^{-1}
	6.0	1.16×10^{-1}	1.17×10^{-1}	0.93×10^{-1}	0.67×10^{-1}
10^{-15}	3.0	1.66×10^{-2}	1.38×10^{-2}	1.47×10^{-2}	1.42×10^{-2}
	4.5	1.48×10^{-2}	1.37×10^{-2}	1.42×10^{-2}	1.06×10^{-2}
	6.0	1.29×10^{-2}	1.31×10^{-2}	1.02×10^{-2}	0.71×10^{-2}
10^{-14}	3.0	1.69×10^{-3}	1.39×10^{-3}	1.49×10^{-3}	1.44×10^{-3}
	4.5	1.50×10^{-3}	1.39×10^{-3}	1.44×10^{-3}	1.07×10^{-3}
	6.0	1.31×10^{-3}	1.32×10^{-3}	1.03×10^{-3}	0.72×10^{-3}

TABLE III
Values of ψ_0

r_{in}^2	lgp \ θ	0.8	1.1	1.4	1.7
10^{-16}	3.0	1.82×10^{-1}	1.78×10^0	1.78×10^0	1.45×10^0
	4.5	2.85×10^{-1}	2.18×10^0	1.72×10^0	7.92×10^{-1}
	6.0	7.18×10^{-1}	1.99×10^0	9.67×10^{-1}	4.26×10^{-1}
10^{-15}	3.0	2.09×10^{-2}	2.00×10^{-1}	2.01×10^{-1}	1.63×10^{-1}
	4.5	3.24×10^{-2}	2.41×10^{-1}	1.94×10^{-1}	8.68×10^{-2}
	6.0	8.02×10^{-2}	2.23×10^{-1}	1.06×10^{-1}	4.53×10^{-2}
10^{-14}	3.0	2.12×10^{-3}	2.03×10^{-2}	2.04×10^{-2}	1.65×10^{-2}
	4.5	3.28×10^{-3}	2.44×10^{-2}	1.97×10^{-2}	8.77×10^{-3}
	6.0	8.11×10^{-3}	2.25×10^{-2}	1.07×10^{-2}	4.56×10^{-3}

TABLE IV
Values of $10\psi_2$

r_{in}^2	lgp \ θ	0.8	1.1	1.4	1.7
10^{-16}	3.0	8.89	6.86	5.29	4.40
	4.5	9.24	6.88	5.34	4.74
	6.0	9.60	6.96	5.81	5.09
10^{-15}	3.0	12.08	8.84	6.93	5.72
	4.5	12.13	8.84	6.94	5.76
	6.0	12.18	8.85	7.00	5.80
10^{-14}	3.0	12.46	9.07	7.12	5.87
	4.5	12.46	9.07	7.12	5.87
	6.0	12.47	9.07	7.13	5.87

TABLE V
Values of $\partial\psi_0/\partial\theta$

r_{in}^2	lgp \ θ	0.8	1.1	1.4	1.7
10^{-16}	3.0	9.42	1.95	-1.27	-0.24
	4.5	12.37	1.34	-3.37	-1.77
	6.0	11.15	-1.13	-4.15	2.09
10^{-15}	3.0	1.050	0.223	-0.139	-0.038
	4.5	1.341	0.161	-0.366	-0.240
	6.0	1.257	-0.133	-0.481	0.244
10^{-14}	3.0	0.1063	0.0226	-0.0141	-0.0037
	4.5	0.1357	0.0163	-0.0371	-0.0246
	6.0	0.1273	-0.0135	-0.0477	0.0248

TABLE VI

Values of $\partial\psi_2/\partial\theta$

r_{in}^2	θ / $\lg p$	0.8	1.1	1.4	1.7
10^{-16}	3.0	−0.723	−0.612	−0.424	−0.159
	4.5	−0.915	−0.656	−0.362	−0.032
	6.0	−1.246	−0.574	−0.253	−0.282
10^{-15}	3.0	−1.376	−0.822	−0.485	−0.363
	4.5	−1.401	−0.828	−0.477	−0.347
	6.0	−1.422	−0.818	−0.464	−0.380
10^{-14}	3.0	−1.457	−0.847	−0.491	−0.389
	4.5	−1.459	−0.848	−0.490	−0.387
	6.0	−1.464	−0.847	−0.489	−0.390

TABLE VII

Values of $\psi(\theta, p)$

r_{in}^2	θ / $\lg p$	0.8	1.1	1.4	1.7
10^{-16}	3.0	9.30	1.46	−1.59	−0.34
	4.5	12.06	0.84	−3.60	−1.77
	6.0	10.57	−1.50	−4.28	1.98
10^{-15}	3.0	0.508	−0.436	−0.482	−0.335
	4.5	0.633	−0.485	−0.715	−0.522
	6.0	0.438	−0.695	−0.848	0.026
10^{-14}	3.0	−0.490	−0.655	−0.358	−0.326
	4.5	−0.621	−0.645	−0.399	−0.342
	6.0	−0.719	−0.597	−0.443	−0.205

TABLE VIII

Values of $10^7 \, \partial H\varphi/\partial t$ G/s

z	Initial sunspot umbra models by			Photosphere model by
	Michard $\alpha = 0.05$	Michard $\alpha = 0.10$	Fricke-Elsässer	de Jager
50	+0.29	+0.53	+0.94	
100	+0.17	+0.32	+0.54	
150	+0.06	+0.12	+0.17	−0.05
200	−0.12	−0.16	−0.40	+0.50
250	−0.31	−0.43	−0.98	+0.64
300	−0.49	−0.78	−0.59	
350	+0.49	+0.30	−0.39	
400			+0.09	

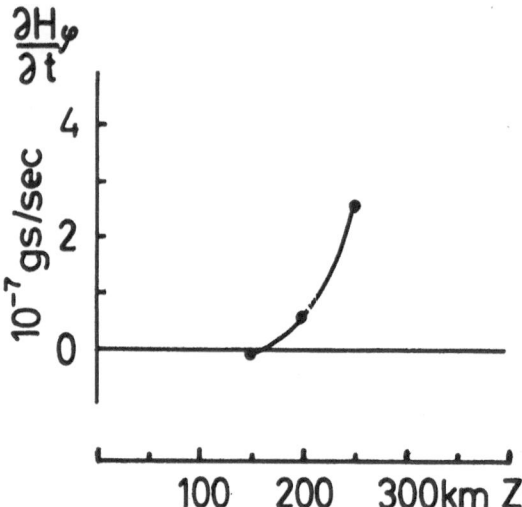

Fig. 3.   The rate of the magnetic field origin in the fine structure elements of the inhomogeneous
photospheric model by de Jager.

$\alpha = 0.10$; 3: $\alpha = 0.05$). We used here the expression (5′). In the case of the inhomogeneous photosphere model by de Jager we took $r_1 = 0''8 \approx 580$ km for hot elements and $r_2 = 0''4 \approx 290$ km for cold elements. In computations the expression (5) was used but strictly speaking this is not sufficiently correct because the solar atmosphere composition differs from that for which $\partial\psi_0/\partial\theta$ and $\partial\psi_2/\partial\theta$ are computed. For estimations with accuracy up to an order of magnitude we can neglect this contradiction. The results are given in Figure 3.

One may see that according to Figure 2 and 3 and Table VIII at the initial moment when the external magnetic field is absent in the fine structure elements, the toroidal magnetic field originates with a rate close to 10^{-7} G/s. Usually above the depth level 150 km its direction is opposite to that under this level. For rough estimation, using the life time of the order of 10^3 s, we obtain a priming magnetic field of the order of 10^{-4} G which is sufficient for the action of the amplifying mechanisms. It is difficult to estimate how long this effect will work because as the magnetic field arises so the complex processes of plasma component diffusion and magnetic field dissipation begin to act. Then the velocity field is important which may be neglected if $H = 0$. Returning to the sunspot models we may find that the difference of the gas pressures in dark and bright elements is of such an order that the corresponding difference in the magnetic field intensities is of the same order. Naturally then our computations made for boundary layers without magnetic field are very conditional ones. Finally according to Wilson (1969) and Teplitzkaya (1970) the hot fine structure elements cannot stretch deeper than 230–250 km. Therefore with the adopted distribution geometry at the bottom of the elements we reach the maximum values 3–10 times larger than at 150 km level.

In spite of these objections, we presume to declare that in general the magnetic field may originate in the fine structure elements of the photosphere and the sunspots as a result of lack of coincidence of isobaric and isothermic surfaces in those elements. In the future we hope to succeed in numerical simulation of the evolution of magnetic fields appearing in such a way.

Acknowledgements

The authors are grateful to Drs R. B. Teplitzkaya and S. J. Vainstejn for discussions.

References

de Jager, C.: 1959, *Handbuch der Astrophysik*, Bd. **52**, p. 83.
Kopecký, M. and Kuklin, G. V.: 1967, *Bull. Astron. Inst. Czech.* **18**, 77.
Kopecký, M. and Kuklin, G. V.: 1969, *Solar Phys.* **6**, 241.
Kuklin, G. V.: 1966, *Rezultaty nabl. i issled. v period MGSS, vyp. I 'Nauka' Moskva*, p. 17.
Obridko, V. N.: 1968, *Bull. Astron. Inst. Czech.* **19**, 183.
Teplitzkaya, R. B.: 1970, Private Communication.
Wilson, P. R.: 1969, *Solar Phys.* **10**, 404.

ANISOTROPY OF ELECTRIC CONDUCTIVITY AND DISSIPATION OF MAGNETIC FIELDS

M. KOPECKÝ

Astronomical Institute of the Czechoslovak Academy of Sciences, Ondřejov, Czechoslovakia

and

V. KOPECKÝ

Institute of Plasma Physics, Czechoslovak Academy of Sciences, Prague 9, Czechoslovakia

Abstract. The Joule dissipation of magnetic fields in the highest photosphere and spot layers can be considerably accelerated by the anisotropy of electric conductivity.

Most papers on magnetic field dissipation in the solar photosphere have assumed isotropic electric conductivity. Neglecting motion of the plasma, the magnetic field variation has then been described by the equation

$$\frac{\partial \mathbf{H}}{\partial t} = \frac{c^2}{4\pi\sigma} \Delta \mathbf{H},\tag{1}$$

where σ is the isotropic plasma conductivity. It is known, however, that a magnetic field may change a plasma into an anisotropic medium, and the conductivity anisotropy in a sunspot can be rather great depending upon the spot model, the intensity of the magnetic field, and the optical depth as was shown by Kopecký and Kuklin (1966, 1969). In considerations of the magnetic field dissipation, it is therefore necessary to utilize the equation that takes into account the influence of a magnetic field on conductivity. This fact has been noted already by Altschuler (1967), who came to the conclusion that the influence of the conductivity anisotropy is negligible. However, this conclusion holds only for the deeper layers of the photosphere and spots.

In a weakly ionized plasma $(n_e/n_n \ll 1)$, which the photosphere and spot may be considered to be, a change of the magnetic field is described by the equation derived by Cowling (1957). For the case considered it may be approximately written in the form

$$\frac{\partial \mathbf{H}}{\partial t} = \frac{c^2}{4\pi} \operatorname{rot} \left\{ \frac{1}{\sigma_0} \operatorname{rot} \mathbf{H} + \frac{1}{c^2 \alpha_{in}} \left[\mathbf{H} \times [\operatorname{rot} \mathbf{H} \times \mathbf{H}] \right] + \frac{1}{ecn_e} \left[\operatorname{rot} \mathbf{H} \times \mathbf{H} \right] \right\},\tag{2}$$

where σ_0 is the parallel conductivity and α_{in} is the coefficient of friction between ions and neutral atoms. At the same time we assume that $v=0$ and that the terms containing the pressure gradients may be neglected.

Using (2) we obtain for a change of energy of the magnetic field

$$\frac{\partial}{\partial t} \int \frac{H^2}{8\pi}\, dV = \frac{-1}{4\pi} \int dV \left\{ \frac{c^2}{4\pi\sigma_0} |\operatorname{rot} \mathbf{H}|^2 + \frac{1}{4\pi\alpha_{in}} |[\operatorname{rot} \mathbf{H} \times \mathbf{H}]|^2 \right\}.\tag{3}$$

Howard (ed.), Solar Magnetic Fields, 542–544. All Rights Reserved.

In the case of a force-free magnetic field it is obvious that the conductivity anisotropy does not play a role because electric currents flow along the magnetic field and hence the field dissipates according to Equation (1). In the opposite case (rot $\mathbf{H} \times \mathbf{H} \neq 0$) the dissipation increases approximately α times, where

$$\alpha \equiv \frac{\sigma_0}{\sigma_\perp} = 1 + \frac{H^2 \sigma_0}{c^2 \alpha_{in}}. \tag{4}$$

For the solar photosphere and spots the coefficient of anisotropy α takes values much greater than unity only in the highest layers as may be seen in Table I, where the values of α are determined for the chemical composition used by Zwaan (1965), and the effective cross sections are those considered in the paper by Kopecký and Kuklin (1969) at $r_{in}^2 = 10^{-15}$ cm^2.

TABLE I

Approximate values of the anisotropy coefficient α
of electric conductivity

| $H =$ | 300 | 1000 | 3000 |
$\log P_g$			
3.0	9.02	1.13×10^2	1.01×10^3
4.5	1.01	1.11	2.03
6.0	1.00	1.00	1.00

In these highest layers of the spots and photosphere the anisotropy of electric conductivity can thus essentially lessen the time of the Joule dissipation of magnetic fields up to one or two orders. Therefore, in these highest photospheric layers, magnetic fields which are not force-free should dissipate more rapidly than they have been assumed to do until now.

In the first place this effect should appear in a relatively rapid dissipation of inhomogeneities of the magnetic field within sunspots. Thus, for example, for bright points in a large umbra with observed lifetimes of 10^3 to 10^4 s, the time for Joule dissipation is 10^3 to 10^5 s at $\sigma_0 = 10^9$ to 10^{11} CGSE (Kopecký and Obridko, 1968). If the electric conductivity is anisotropic with $\alpha = 10$ to 10^2, their Joule dissipation is thus completely comparable with the observed lifetime.

We can see to what extent σ_0 can differ from $\sigma_\perp$ in some spot models from Figure 1, where the dependence of $\log \sigma_0$ (solid lines) and $\log \sigma_\perp$ at $H = 3000$ G (dashed lines) on the optical depth τ is given for the case of the spot models by Van 't Veer (1963) with an umbral area $A = 50$, 80, and 110 millionths of the Sun's hemisphere (Kopecký and Kuklin, 1966). Figure 1 shows, first of all, a considerable dependence of the anisotropy of electric conductivity, and therefore the characteristic dissipation time, on the physical conditions in a spot as a consequence of a change of its area.

The Joule dissipation of magnetic fields in the highest photosphere and spot layers can thus be considerably accelerated by the anisotropy of electric conductivity. This again gives evidence for the fact that the dissipation processes cannot be completely neglected in hydromagnetic considerations of processes in the observed layers of the

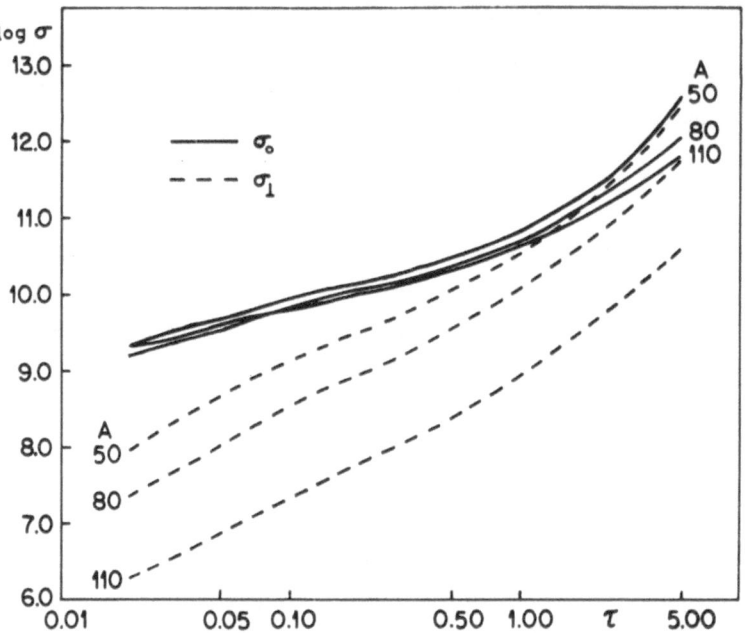

Fig. 1. The dependence of $\log \sigma_0$ ——————— and $\log \sigma_\perp$ - - - - - - - - at $H = 3000$ G on the optical depth τ for the case of the spot models by Van 't Veer with the umbra area $A = 50$, 80 and 110 millionths of the Sun hemisphere.

Sun. Above all the dissipation processes cannot be neglected in considerations of the generation of magnetic fields as, for example, in the theory by Getling and Tversky (1968), where invariably currents with a component perpendicular to the magnetic field are generated, so that the generated fields are not force-free, and therefore characteristic dimensions of the field generation and dissipation processes are comparable.

References

Altschuler, M. D.: 1967, *Solar Phys.* **1**, 377.
Cowling, T. G.: 1957, *Magnetohydrodynamics*, Interscience Publishers, London.
Getling, A. V. and Tversky, B. A.: 1968, *Astron. Zh.* **45**, 606.
Kopecký, M. and Kuklin, G. V.: 1966, *Bull. Astron. Inst. Czech.* **17**, 45.
Kopecký, M. and Kuklin, G. V.: 1969, *Solar Phys.* **6**, 241.
Kopecký, M. and Obridko, V.: 1968, *Solar Phys.* **5**, 354.
Zwaan, C.: 1965, *Rech. Astron. Observ. Utrecht* **12**, (4).

PART VI

OPTICAL AND RADIO OBSERVATIONS OF LARGE SCALE MAGNETIC FIELDS ON THE SUN

LARGE SCALE SOLAR MAGNETIC FIELDS AND
THEIR CONSEQUENCES

GORDON NEWKIRK, JR.

High Altitude Observatory, National Center for Atmospheric Research*, Boulder, Colo., U.S.A.

Abstract. The general properties of large scale solar magnetic fields are reviewed. In order of size these are: (1) Active region, generally bipolar fields with a lifetime of about two solar rotations. These are characterized by fields of several hundred G and display differential rotation similar to that found for the photosphere. (2) UM regions which appear to be the remnants of active region fields dispersed by the action of supergranulation convection and distorted by differential rotation. These are characterized by fields of a few tens of gauss and have lifetimes of several solar rotations. (3) The polar fields which are built up over the solar cycle by the preferential migration of a given polarity towards the poles. The poloidal fields are of a few gauss in magnitude and reverse sign in about 22 yr. (4) The large scale sector fields. These appear closely related to the interplanetary sector structure, cover tens of degrees in longitude, and stretch across the equator with the *same* polarity. This pattern endures for periods of up to a year or more, is *not* distorted by differential rotation, and has a rotation period of about 27 days. The presence of these long enduring sector fields may be related to the phenomenon of active solar longitudes. The consequences of large scale fields are examined with particular emphasis on the effects displayed by the corona. Calculated magnetic field patterns in the corona are compared with the density structure of the corona with the conclusion that: (1) Small scale structures in the corona, such as rays, arches, and loops, reflect the shape of the field and appear as magnetic tubes of force preferentially filled with more coronal plasma than the background. (2) Coronal density enhancements appear over plages where the field strength and presumably the mechanical energy transport into the corona are higher than normal. (3) Coronal streamers form above the 'neutral line' between extended UM regions of opposite polarity. The role played by coronal magnetic fields in transient events is also discussed. Some examples are: (1) The location of Proton Flares in open, diverging configurations of the field. (2) The expulsion of 'magnetic bottles' into the interplanetary medium by solar flares. (3) The relation of Type IV radio bursts to the ambient field configuration. (4) The guiding of Type II burst exciters by the ambient magnetic field. (5) The magnetic connection between widely separated active regions which display correlated radio bursts.

1. Introduction

The suspicion that the Sun maintained a general magnetic field was first stimulated by the similarity of the shape of the solar corona at sunspot minimum to that of the field lines around a bar magnet (Bigelow, 1889; Störmer, 1911). A brief history of our knowledge of large scale fields shows that this suspicion was to remain unsubstantiated for many years. Following his pioneering work (Hale, 1908) which demonstrated the presence of magnetic fields in sunspots, Hale (1913) turned his newly developed equipment to the detection of a dipole field of the Sun. In spite of great care taken to avoid systematic errors and the averaging of a great number of photographic observations of magnetically sensitive lines, the derived value of 50 G for the field at the poles was, as we now know, erroneous (Stenflo, 1970). In fact, subsequent observations at Mt. Wilson gave values of the polar magnetic field between zero and 50 G and, thus, cast doubt on the existence of such a field.

---

* The National Center for Atmospheric Research is sponsored by the National Science Foundation.

Howard (ed.), Solar Magnetic Fields, 547–568. All Rights Reserved.
Copyright © 1971 by the IAU.

Following the application of photoelectric techniques to the problem, several investigators (von Klüber, 1951; Thiessen, 1952; Kiepenheuer, 1953; Babcock and Babcock, 1953) were able to show that the Sun did, indeed, have a polar field of magnitude 1–2 G. Continuous observation (Babcock and Babcock, 1955) of the magnetic field distribution over the entire disk of the Sun for several years not only confirmed the existence of a general polar field but also revealed two general types of low latitude fields: the bipolar regions (BMR) associated with active regions and the unipolar regions (UM), which were believed to be the long sought 'M' regions responsible for recurrent geomagnetic storms. The well documented weakening and subsequent reversal of the polar field during the sunspot maximum of 1958-59 (Babcock, H. D., 1959) led H. W. Babcock (1961) to formulate a qualitative model of the solar cycle which was based on a suggestion of Cowling (1953). The success of this model is well known to us all.

2. General Description of Large Scale Fields

The past decade has brought a remarkable increase in the sensitivity of solar magnetographs and a subsequent revision of our earlier ideas of large scale magnetic structures (Howard, 1967). Bumba and Howard (1965) found that weak magnetic fields pervade nearly the entire surface of the Sun and that this background field contains a persistent pattern of UM regions which lasts for many rotations. These extended regions (Figure 1) appear to consist of the weak, expanded fields of old active regions (Howard, 1967) and their shape is largely determined by the shearing produced by differential rotation. Leighton (1964) has suggested that the initially compact fields of an active region are gradually eroded by the motions within the constantly changing pattern of supergranulation cells and spread to widespread areas of the solar surface. The random dispersal of these fields, coupled with the differential rotation, produces, in his model, a shape for a UM region quite similar to that observed (Figure 2). The preferential transport of following polarity toward the pole accounts for the reversal of the polar fields.

Leighton (1969) has expanded his investigation to produce a model of the solar cycle in which the presence and dispersal of large scale magnetic fields play a crucial role. This model extends the earlier work of Babcock (1961) and Leighton (1964) to include a semi-quantitative treatment of the amplification, eruption to the surface, and spreading of magnetic field. Since the analysis requires the *ad hoc* assumption of several parameters, such as the critical field magnitude which causes eruption of a flux tube and the depth dependence of differential rotation, this work must be a considered midway between Babcock's pioneering qualitative model and a full scale solution of the hydromagnetic equations. One of the first attempts to formulate the full solution will be presented at this meeting by Nakagawa (1970).

Thus, many of the features of large scale solar magnetic fields are well observed and appear to be understandable in terms of theoretical models which are simple only in comparison to the complexity of the problem. However, recently we have become aware

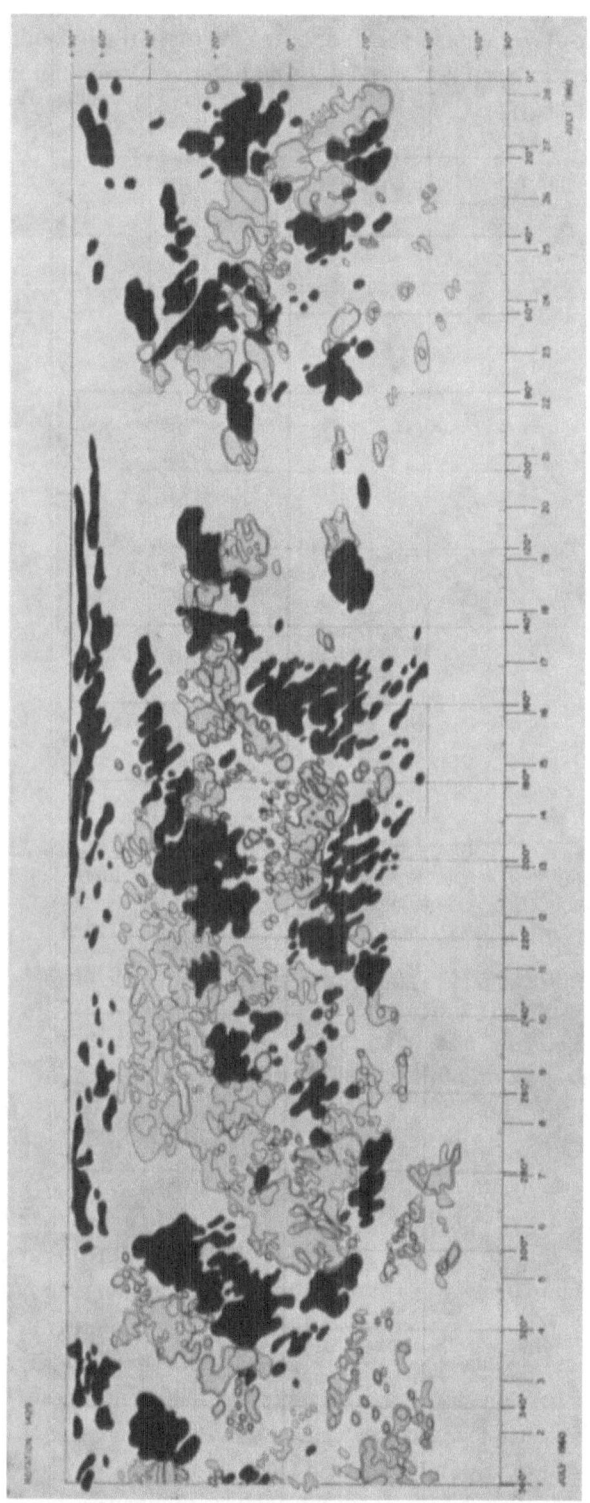

Fig. 1. A magnetic map of the entire solar surface dramatically displays the extended UM regions which result from the dispersal of the compact regions associated with active regions (courtesy Howard et al., 1967).

of some other unsuspected aspects of large scale fields. Most outstanding is the existence of a gross pattern in the field which appears to rotate rigidly with a period of
approximately 27 days rather than partake in the differential rotation characteristic
of the directly observed surface layers.

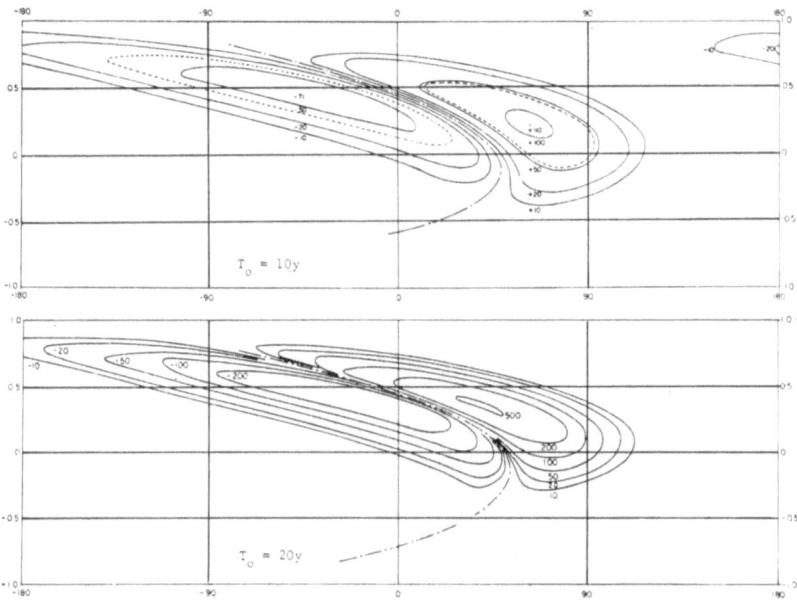

Fig. 2. Calculated isogauss maps of a simulated bipolar source influenced by dispersal and differential rotation ½ yr after its birth. The quantity T_0 is the fundamental decay time of the field. Compare
the shape of these regions to those observed (from Leighton, 1964).

3. Rotation and Persistent Patterns of Large Scale Fields

The synodic period of 27 days for recurrent geomagnetic activity has long been interpreted simply as a reflection of the rotation period of the Sun for the mean latitude
of active regions. As we shall see, this may represent an oversimplification. The first
concrete suggestion that the Sun has a rigidly rotating core with measurable effect at the
photosphere came from the statistical evidence for the existence of remarkably active
Carrington longitudes (Warwick, 1965; Dodson-Prince and Hedeman, 1968; Sawyer,
1968; Švestka, 1968; Van Hoven et al., 1969). These papers noted that certain solar
longitudes are particularly favorable for the production of various symptoms of solar
activity such as flares, cosmic ray events on the Earth, etc. Although not *prima facie*
evidence that there are large scale features of solar magnetism which do not share the
differential rotation, these phenomena are suggestive. The connection with the large
scale magnetic field is made more secure when we realize that active regions frequently
erupt within previously existing and long-lived UM regions.

A second line of evidence has come from the analysis of magnetic fields measured in interplanetary space. Early investigations of this field (Ness and Wilcox, 1965) showed that it had, at least during periods of low solar activity, a remarkably simple sector structure (Figure 3) with the field generally directed toward or away from the Sun over sectors of 60° to 90° in longitude. This discovery was quickly followed (Ness and Wilcox, 1966; Wilcox and Howard, 1968) by the realization that the polarity of the interplanetary field corresponded rather well to that of the large scale photospheric field which has passed central meridian $4\frac{1}{2}$ days earlier. The concept was extended by Schatten *et al.* (1969) who suggested that the field in the solar corona below a 'source sphere' with a radius of 1.6 to 2.5 R_0 was essentially the potential field

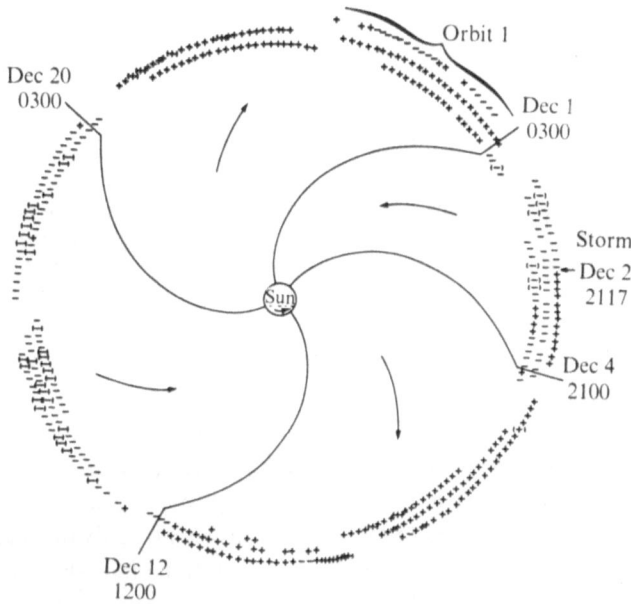

Fig. 3. The interplanetary sector structure observed by IMP-I is representative of one of the largest patterns of organization of the solar magnetic field (from Ness and Wilcox, 1965).

distribution, while the field pattern at the 'source sphere' was mapped by the solar wind out into interplanetary space.

In an attempt to discover the solar origin of the interplanetary field, Wilcox and Howard (1970) determined the rotation rate of the photospheric fields detected on Mt. Wilson magnetograms and found that at low latitudes the rotation rate of the magnetic patterns is the same as that of sunspots. At higher latitudes, the pattern rotates at a rate consistent with that found for prominences and the corona (Hansen *et al.*, 1970). However, we must conclude that these patterns are not primarily responsible for the interplanetary sector structure. The surface fields associated with the sector boundary projected back on the Sun appear to rotate rigidly and have the *same* polarity north and south of the equator (Schatten *et al.*, 1969) (Figure 4). Apparently, the interplanetary field is dominated by very large scale, weak fields which were not

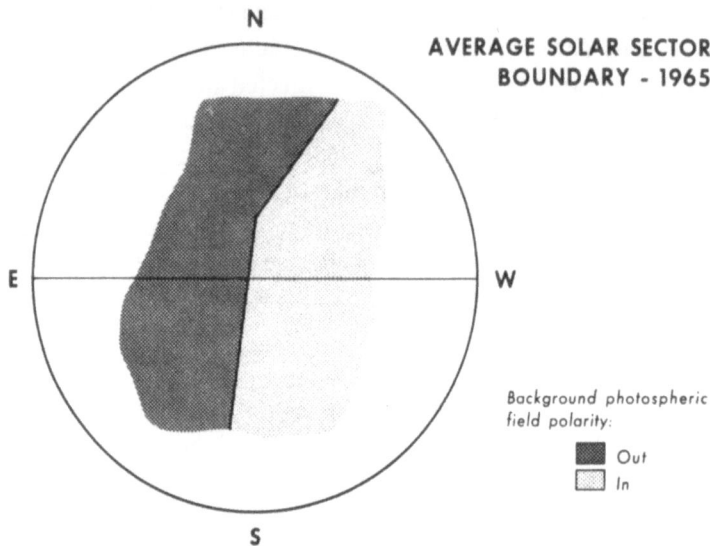

Fig. 4.   The average background photospheric field associated with the sector boundaries observed during 1965 and projected back on the solar disk. On each side of the boundary the weak photospheric background field is predominantly of the same polarity. This weak pattern extends across the equator and does not appear to be sheared by differential rotation (from Schatten, *et al.*, 1969).

dominant in the Wilcox and Howard analysis. (They did not distinguish rotation rate according to field strength or the scale of the pattern.)

These very large scale, weak fields do show up in the hemispheric average of the field, which is closely related to the interplanetary field (Wilcox *et al.*, 1969; Severny *et al.*, 1970). They also appear in long term synoptic observations as shown by the analysis of Bumba and Howard (1969), which reveals not only the familiar, transient active regions which evidence the differential rotation, but also extended regions which occupy ten's of degrees in longitude and which have a rotation period of about 27 days. These extended regions persist for many months or years and stretch up to 20° on either side of the equator with the *same polarity*. A recent, statistical analysis of an equally long time series (Wilcox *et al.*, 1970) confirms these characteristics of the persistent, large scale pattern of the weak (~ 1 G) background field (Figure 5) as does the harmonic analysis (Altschuler *et al.*, 1971) to be described later today.

4. Consequences of Large Scale Solar Magnetic Fields

We have already touched on probably the most profound aspect of large scale fields – their central role in the solar cycle. Of course, these fields make their presence known by other ways than on our magnetographs. The extended weak field regions can be outlined by the practiced eye on calcium spectroheliograms (Howard and Harvey, 1964; Veeder and Zirin, 1970). They are undoubtedly responsible for the appearance of white light faculae over the poles during activity minimum (Waldmeier, 1955; Howard, 1959). The occurrence of filaments along the interface between regions of opposite polarity has been well documented (Howard, 1959). Of course, magnetic fields in the corona are, by and large, dominated by the large scale surface fields and we shall devote the remainder of the discussion to various aspects of coronal fields.

A. CORONAL MAGNETIC FIELDS

Unlike other areas of solar physics the study of coronal magnetic fields has had to rely almost exclusively upon calculation. Perhaps a brief review of such calculations

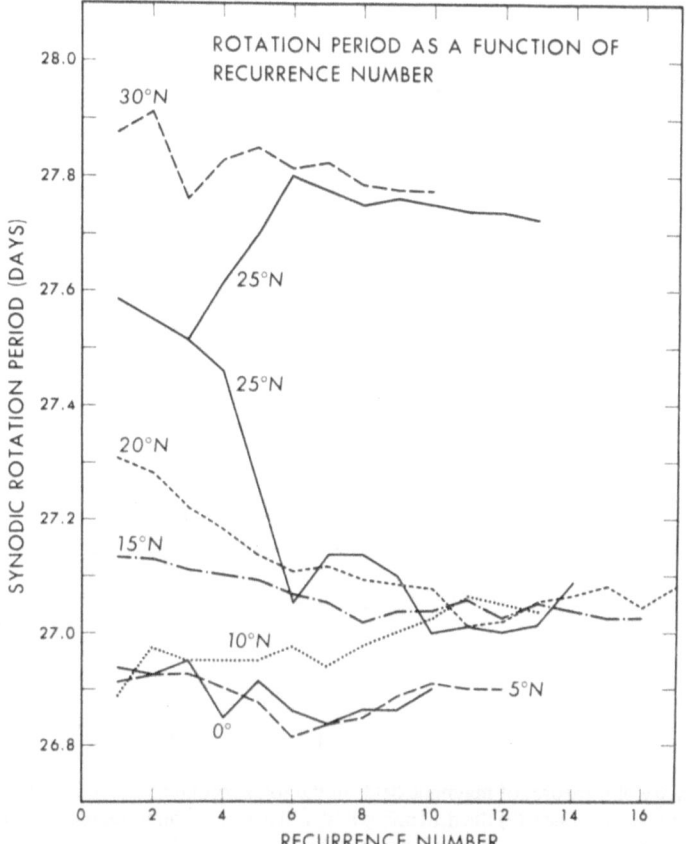

Fig. 5. The synodic rotation rate of solar magnetic fields from an autocorrelation analysis of eight years of Mt. Wilson data. Individual curves are for particular latitude zones. The very persistent features which can be followed through many recurrences have a period of 27 days with little indication of differential rotation (from Wilcox *et al.*, 1970).

is in order. They all begin with measures of the photospheric field and then, either in rectangular coordinates (Schmidt, 1964) over a volume small with respect to the Sun, or in spherical coordinates over a large volume (Newkirk *et al.*, 1968; Schatten *et al.*, 1969; Altschuler and Newkirk, 1969) calculate the potential field present. Without discussing the differences in the various mathematical techniques used in such calculations, we note that they share a common deficiency in either ignoring the electric currents which occur in the solar wind or incorporating them only by means of the rather crude technique of a zero potential surface or source surface. Moreover, the calculations requiring measurements of the field distribution over the entire photosphere are forced to use time averages of unknown accuracy.

Methods of checking these calculations are unfortunately scarce (Newkirk, 1967; Takakura, 1966) except where direct detection of the Zeeman splitting in prominences can be made (Severny and Zirin, 1961; Rust, 1966, Harvey, 1969; Tandberg-Hanssen, 1970). Analyses of the Razin effect in a single radio burst (Boischot and Clavelier,

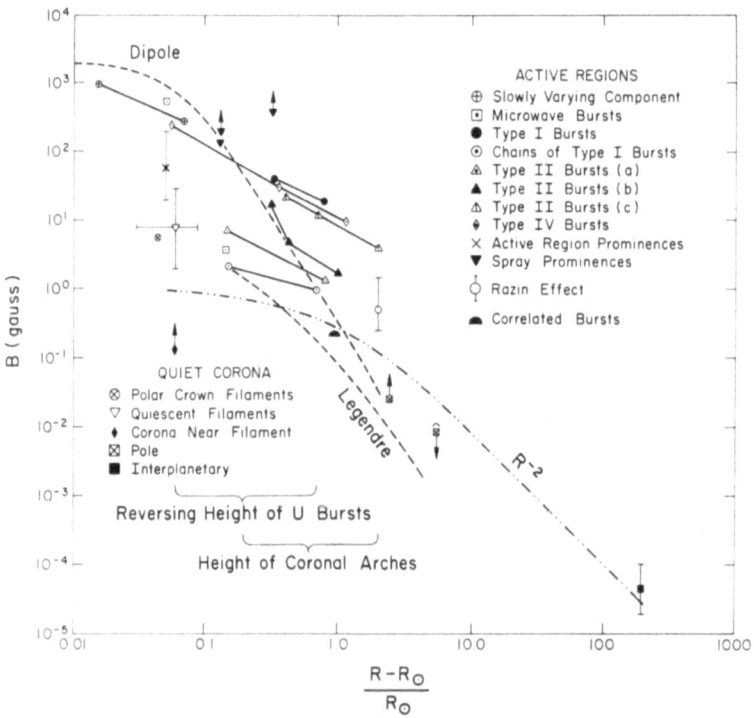

Fig. 6. Summary of measures of magnetic fields in the solar corona compared to (1) R^{-2} extrapolation from interplanetary space, (2) the field above a typical active region using the harmonic expansion method with $N=9$, and (3) a simple dipole potential model for an active region. Except for the Razin effect and the correlated burst measurements, all other references are to be found in Newkirk, 1967.

1967; Ramaty and Lingenfelter, 1968; Bohlin and Simon, 1969) and of the weak polarization in some correlated bursts (Kai, 1969a) have, in the last few years, added a few more measures (Figure 6) of the magnitude of the field at $\sim 2 R_\odot$. In the absence of any event by event comparison between observed and calculated coronal fields we compare the observations with three simple models: (1) a R^{-2} extrapolation from interplanetary space, (2) the Legendre polynomial field above a plage for the surface fields of November 1966, and (3) a potential dipole model of a plage region. This comparison suggests two cautions: (1) the Legendre approximation will not yield accurate results near active regions (a fact well known) and (2) radio bursts at $\sim 2\, R_\odot$ may well represent events in which a transient field disturbance is ejected into the corona and may be unsuitable as a measure of the quiet (i.e. current-free) magnetic field.

In fact, the only comparison between observed and calculated fields now at our disposal is in prominences (Harvey, 1969; Rust, 1966) which show, in general, an agreement between the shapes of the fields and currently accepted ideas for the occurrence of prominences within the fields. However, in active prominences, particularly, the measured fields are in excess of those calculated by a factor of five. The origin of this discrepancy is unknown. Similarly, comparison of the shapes of bright coronal emission regions and those of the calculated magnetic fields (Rust, 1970) gives some confidence that the potential field is a good first approximation and that the coronal loops delineate magnetic tubes of abnormally high material density.

B. MAGNETIC FIELDS AND THE SHAPE OF THE CORONA

This brings us to the important question of the influence of large scale magnetic fields on the shape of the solar corona. Although magnetic structures as small as 30000 km may affect such features as polar plumes (Saito, 1965; Newkirk and Harvey, 1968; Ivanchuk, 1968), in general only the extended fields will have major influence in the corona. The investigation of the relation between the magnetic field and the density structure of the corona has followed two lines. One is to compare the calculated fields with the known shape of the corona (Newkirk *et al.*, 1968; Bohlin, 1968; Schatten, 1968; Altschuler and Newkirk, 1969; Newkirk *et al.*, 1970; Newkirk and Altschuler, 1970). Whether the comparison is intended as a prediction (Schatten, 1970a) or as a *post facto* analysis, the method is basically the same. The second line of attack is to use a simplified distribution of the field and to solve the hydromagnetic and solar wind equations simultaneously (Pneuman, 1968; Pneuman, 1969; Pneuman and Kopp, 1970) to determine the resultant distribution of material, the field, the velocity structure, and the energy flow in the modified corona. We shall discuss some examples of both approaches.

To begin, we first examine the pattern of calculated coronal fields present during a typical period as seen against an Hα spectroheliogram (Figure 7). The magnetic fields may be conveniently divided into (1) Diverging Fields which are found in close association with plages, (2) Low Magnetic Arcades and (3) High Magnetic Arcades. Perhaps, most striking is the existence of magnetic arches connecting widely separated active regions. Such arches may well be the lines of communication which give rise to nearly simultaneous radio bursts in separated active regions (Wild, 1969a). In view of the close correlation between the positions of plages and coronal density enhancements, it is not surprising to find a similar correlation between such enhancements and the Diverging Field patterns. Comparing the overall structure of the corona with the field as in Figure 8, we find that coronal streamers appear to form over the High Magnetic Arcades. This is illustrated in Figure 8 by the superposition of the K-coronameter isophotes of a streamer, identified on the 12 November 1966 eclipse photograph

on the coronal magnetic map. This substantiates the idea long used in theoretical
models (Kuperus and Tandberg-Hanssen, 1967; Pneuman, 1968, 1969) *that streamers
develop above the neutral line separating regions of opposite polarity.*

An examination of the relationship between the shapes of small scale features in
the corona and of the magnetic field lines is almost inevitably restricted to an evaluation of their *projected* positions and appearances. Returning to the 1966 eclipse (Fi-

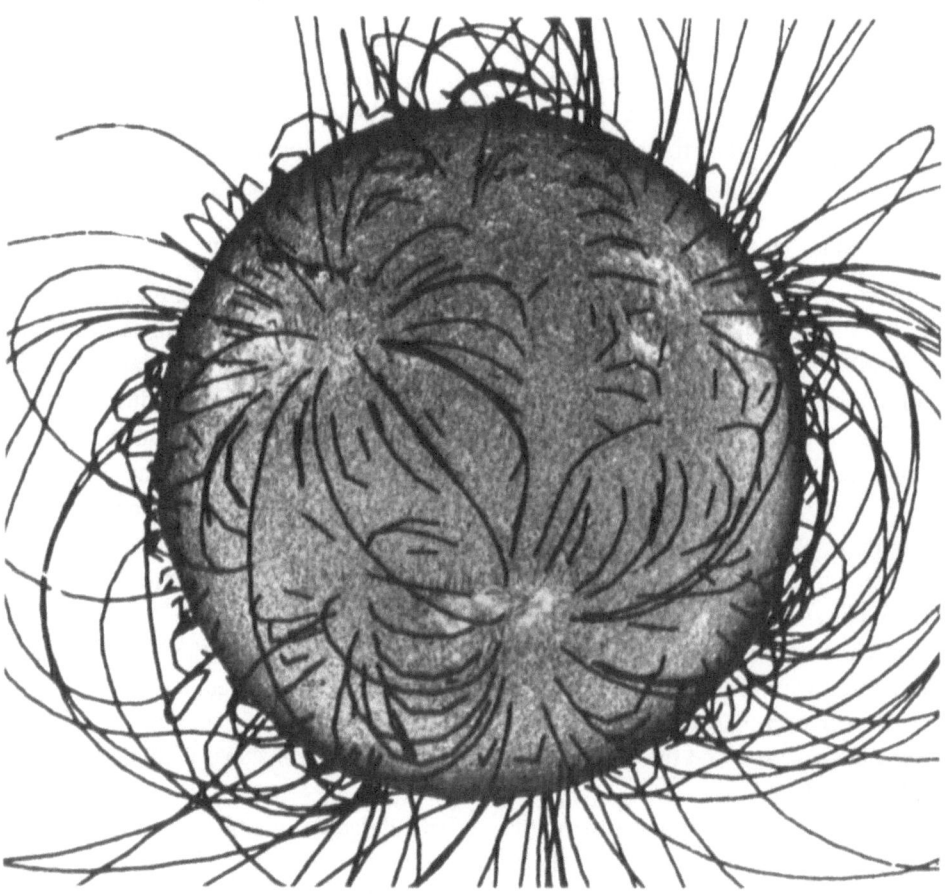

Fig. 7.   Superposition of coronal field map (least-mean-square fit to B_L, $R_W = \infty$, corrected for
magnetograph saturation) and the corresponding Hα filtergram (Sacramento Peak Observatory).

gure 8), we find that the agreement is quite good – we find open rays, arches, loops,
etc. in the corona where they are indicated in the field. A similar conclusion is reached
by examination of the most recent eclipse (Figure 9) as well as the 1965 eclipse (Figure
10). *Thus, we conclude that much of the fine structure visible in the corona is simply a
mapping of magnetic tubes in the approximately potential field above the photosphere.*

As an example of the more theoretical approach I cite the work of Pneuman (1968,
1969) and Pneuman and Kopp (1970). They assume a simple distribution of field as

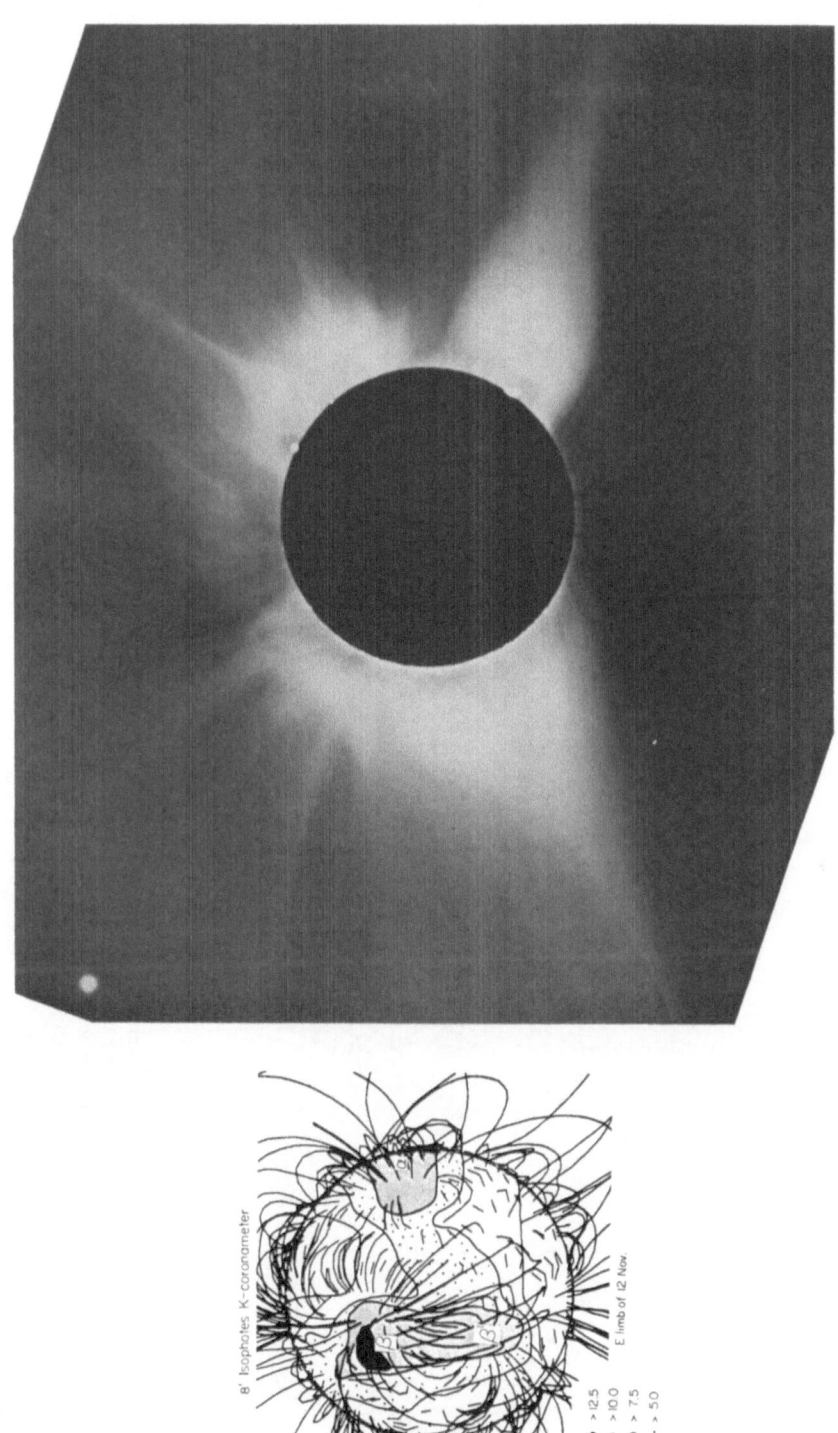

Fig. 8. Comparison of K-coronameter isophotes at 1.5 $R_\odot$ and a coronal magnetic field map (left) with the eclipse corona of 12 November 1966 (right). The central meridian of the magnetic map corresponds to the east limb (left) at the time of the eclipse and the line-of-sight proceeds from right to left across the map. Corresponding arches and rays can be easily located in the field and in the corona.

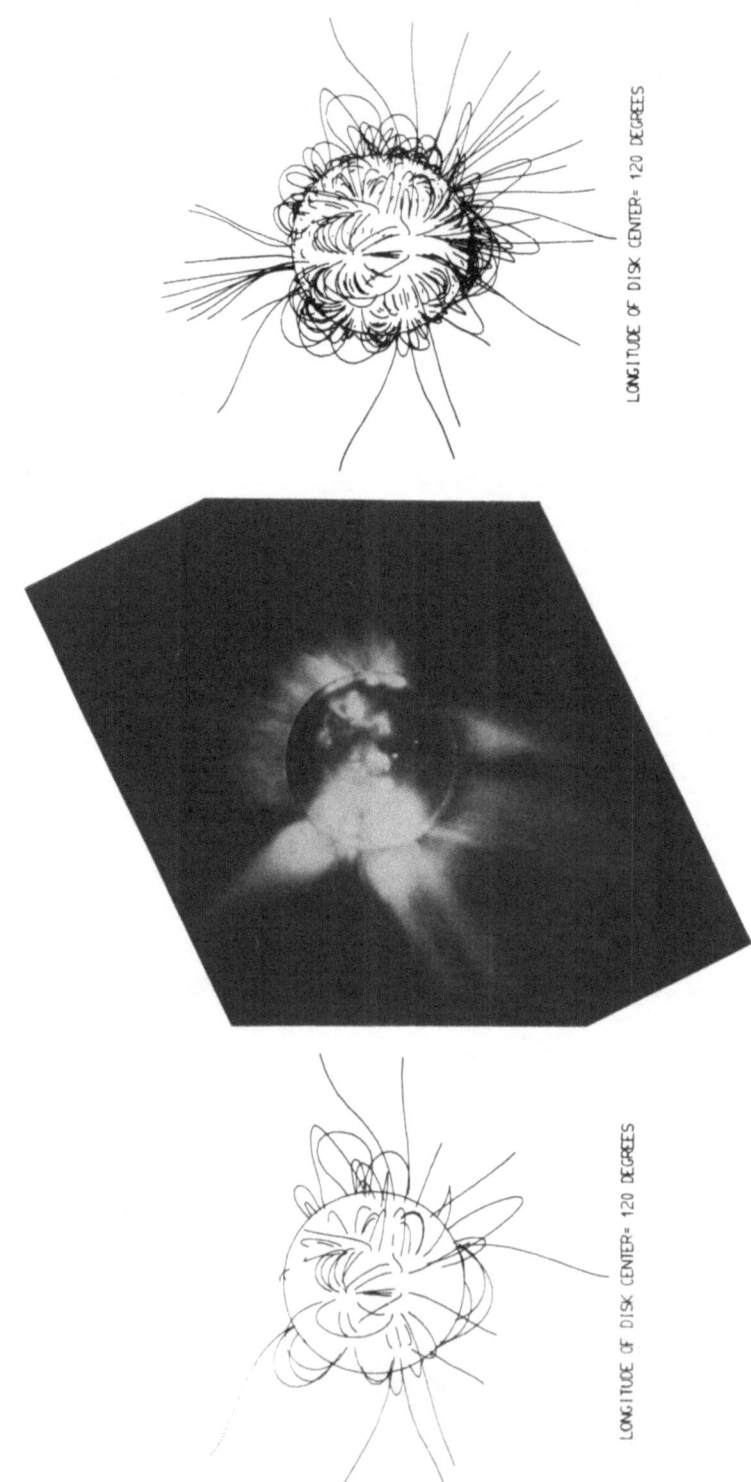

Fig. 9.   Comparison of the solar corona of 7 March 1970 (outer corona HAO; X-ray corona seen on the disk courtesy Vaiana *et al.*, 1970, American Science and Engineering) with the corresponding coronal magnetic maps. In this and subsequent coronal magnetic displays, the weak field map on the right shows field lines originating at foot points where $B_L \geq 0.16$ G while in the strong field map at the left only field lines originating where $B_L \geq 10\%$ of the maximum line-of-sight field present at the surface are displayed.

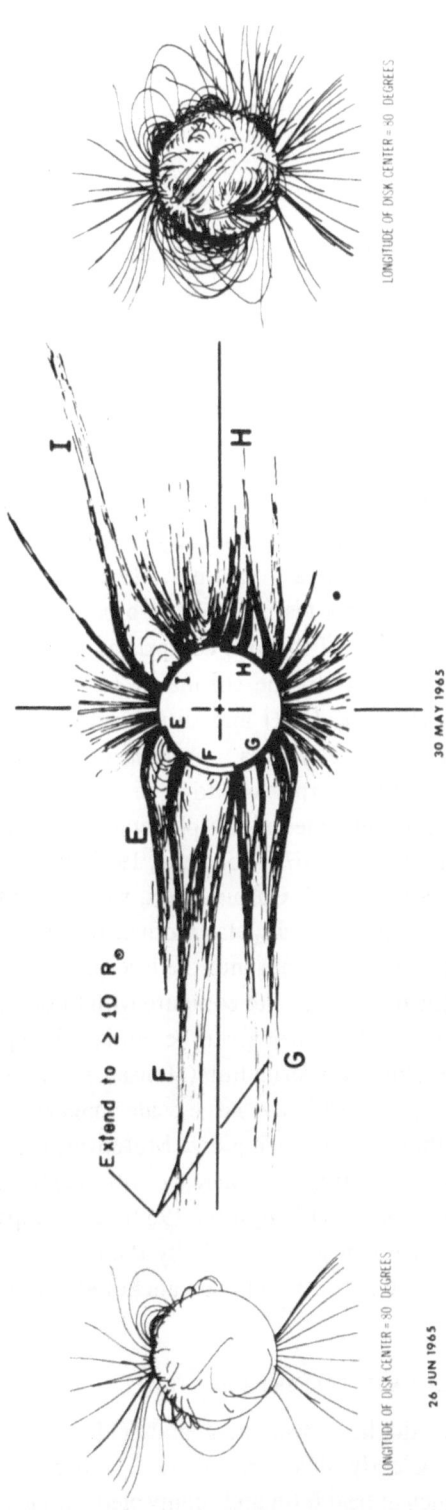

Fig. 10. Comparison of the solar corona of 30 May 1965 (drawing from Bohlin, 1968) with the corresponding magnetic maps. Note particularly the similarity between (1) the magnetic and coronal arches in streamer I and (2) the polar magnetic field and polar plumes (see caption Figure 9).

well as pressure equilibrium at the base of the corona. The hydromagnetic and energy transport equations are then solved iteratively to arrive at a model for the streamer which includes such parameters as:

(1) the profile of the boundary between streamer and interstreamer;
(2) the axial enhancement;
(3) the temperature profile; and
(4) the velocity structure.

At least for those parameters which can be measured, the agreement between the model and the structure of the corona is impressive. Such models are important because they allow us to see how the visible structures in the corona are molded by the magnetic field and how they influence the structure and dynamics of the interplanetary medium.

C. INFLUENCE ON CORONAL ROTATION

In addition to influencing the distribution of material and the expansion velocity of the solar corona, large scale magnetic fields clearly determine the rotation and the angular momentum of the interplanetary medium. Here we must take care to distinguish between the corotation of a feature such as a coronal streamer or a sector boundary and the angular velocity of the ions comprising the feature. Observational evidence for the tangential velocity of the corona at 1 AU comes to us from the orientation of comet tails (Brandt, 1967) and direct detection from space probes (Hundhausen, 1968). Both techniques yield a tangential velocity of 4–10 km/s, which would require rigid rotation of the corona out to $\sim 15\,R_\odot$ if conservation of angular momentum were to hold in the remainder of interplanetary space. Theoretical analyses (Pneuman, 1966; Weber and Davis, 1967; Modisette, 1967; Brandt *et al.*, 1969) show this concept to be vastly oversimplified. Coronal ions, while lagging behind the solar surface at all heights, receive significant angular momentum from the solar magnetic field from the surface to large distances out into the interplanetary medium. We have no data on the rotation of the inner corona to compare with these calculations.

The rotation of *structures* in the corona can be entirely independent of the motions of the individual ions. Present information (Hansen *et al.*, 1970) shows that the low coronal enhancements rotate with the large scale magnetic structures (Wilcox and Howard, 1970) rather than with active regions. Moreover, these data suggest that, at a given latitude, the rate of rotation may *increase* with height in a manner similar to that found in the photosphere (Livingston, 1969). This apparently anomalous phenomenon may be explained (Pneuman, 1971) by the confinement of coronal gas within magnetic loops which have their foot points anchored at different latitudes with different rates of rotation.

D. CORONAL MAGNETIC FIELDS AND TRANSIENT EVENTS

Thus far we have discussed the large scale magnetic field and its influence as if the field were constant in time. Clearly, this is not the case, and we now examine several types of transient events which appear intimately connected with magnetic fields.

One such phenomenon is associated with solar cosmic rays, which imply (1) a more or less direct channel for the escape of the particles from the flare region into inter-planetary space and (2) storage and/or continuous generation of particles at the Sun for a period of many days (Fan *et al.*, 1968). An examination of the coronal magnetic field associated with a proton flare (Valdez and Altschuler, 1970) (Figure 11) suggests that the channel of direct escape may be found in the Diverging Fields asociated with every active region and that storage may occur in some of the closed loops connected with most active regions. That a proton flare may be associated with a permanent dis-ruption of the large scale fields is shown by comparing Figures 11 and 12. Figure 12

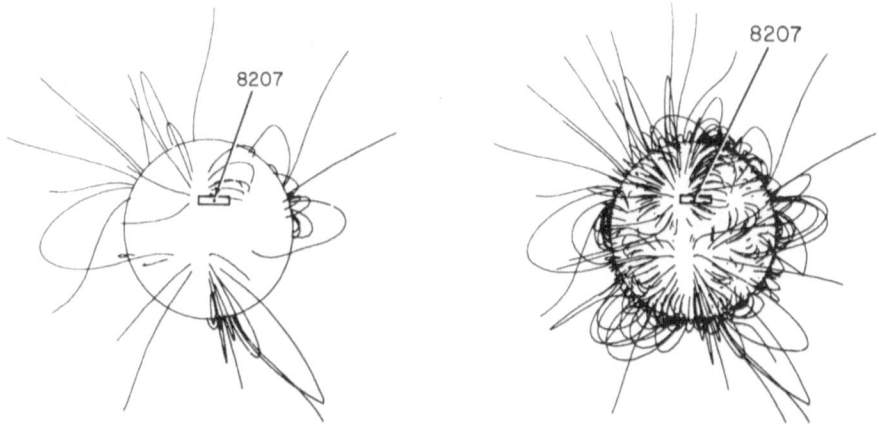

Fig. 11.   Coronal magnetic maps for the Proton Flare of 16 April 1966 (Valdez and Altschuler, 1970) based on surface data taken *before* the occurrence of the flare. The location of the flare is marked with a rectangle (see caption Figure 9).

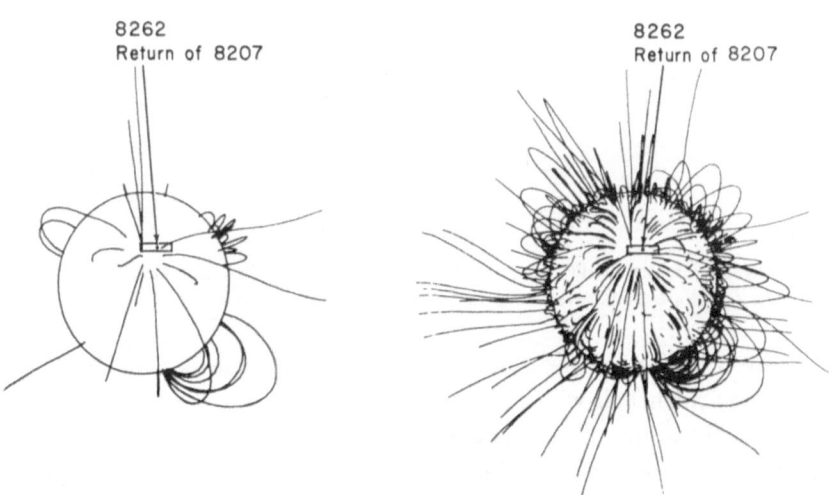

Fig. 12.   Coronal magnetic maps for the Proton Flare of 16 April 1966 (Valdez and Altschuler, 1970) based on surface data taken *after* the occurrence of the flare. The location of the flare is marked with a rectangle (see caption Figure 9).

shows the same region as Figure 11 but one solar rotation later, after a proton flare. Note that the previously closed magnetic loops are open after the event. The fact that the open field lines appear in the current-free approximation indicates that a readjustment of the surface fields has occurred.

Radio occultation observations, either of natural sources (Dennison, 1970) or of satellite-borne transmitters (Levy *et al.*, 1969) give evidence for impulsive changes in

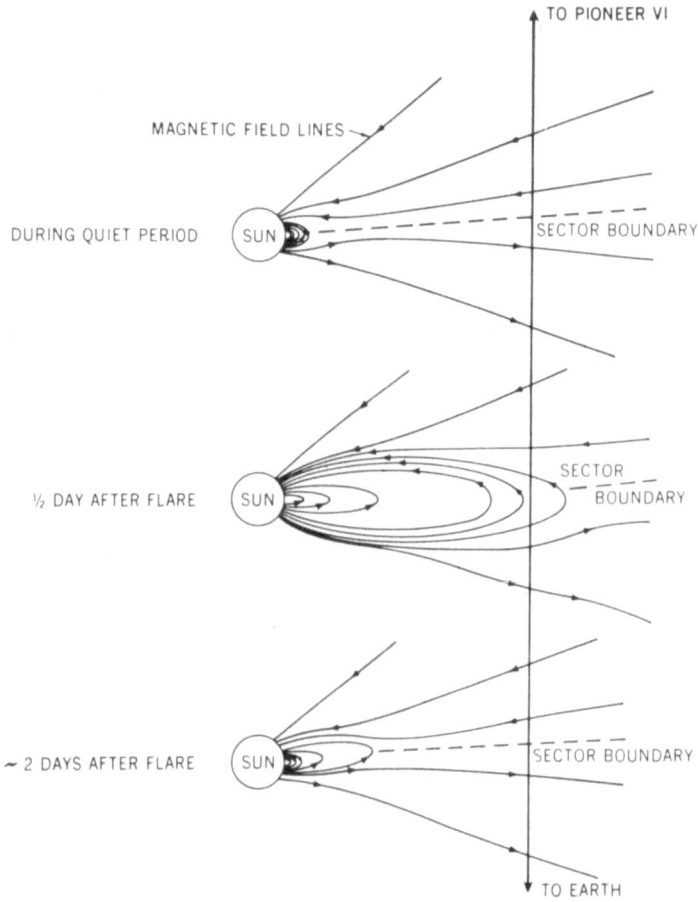

Fig. 13.   Inferred geometry of the 'magnetic bottle' envisioned by Schatten (1970b) to account for the post flare transient change in Faraday rotation observed by Pioneer VI.

the large scale magnetic field near the Sun. Transient changes in the angular broadening of the occulted Crab nebula with a time scale down to minutes may be due to streamers or other coronal features of magnetic origin intruding into the line-of-sight. A comparison of such observations with the calculated coronal fields has yet to be made. In the case of a spacecraft transmitter (Schatten, 1970b) (Figure 13), transient changes in the Faraday rotation were interpreted as a 'magnetic bottle' intruding into the line-of-sight at $10 R_{\odot}$ after ejection by an observed flare. The observed direction of

Faraday rotation was found to be consistent with the calculated fields in the lower magnetic arch which presubably expanded into the line-of-sight. Because the disturbing density and field both exceeded the ambient value by an order of magnitude, we appear to have an example of a true intrusion of a magnetic bottle rather than a minor perturbation of the previously existing corona.

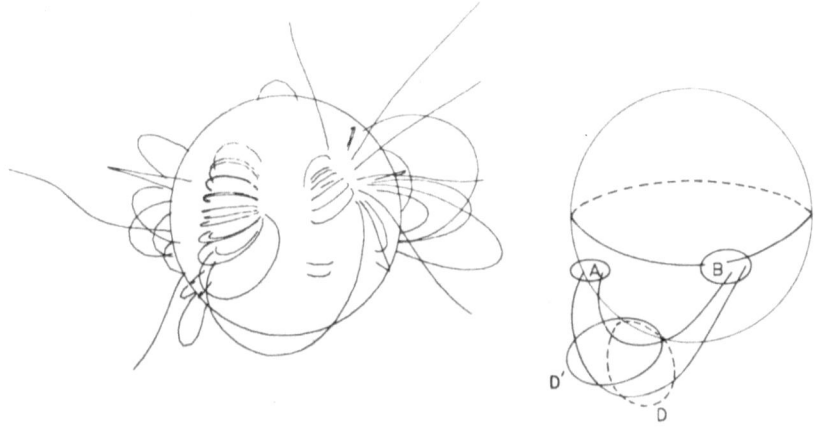

LONGITUDE OF DISK CENTER = 100 DEGREES

Fig. 14.   Comparison of the magnetic arch inferred by Dulk (1970) from the radio bursts of 29–30 August 1969 (right) with the calculated magnetic map (left). Only field lines originating at foot points where $B_L \geqslant 10\%$ of the maximum line-of-sight field present at the surface are displayed.

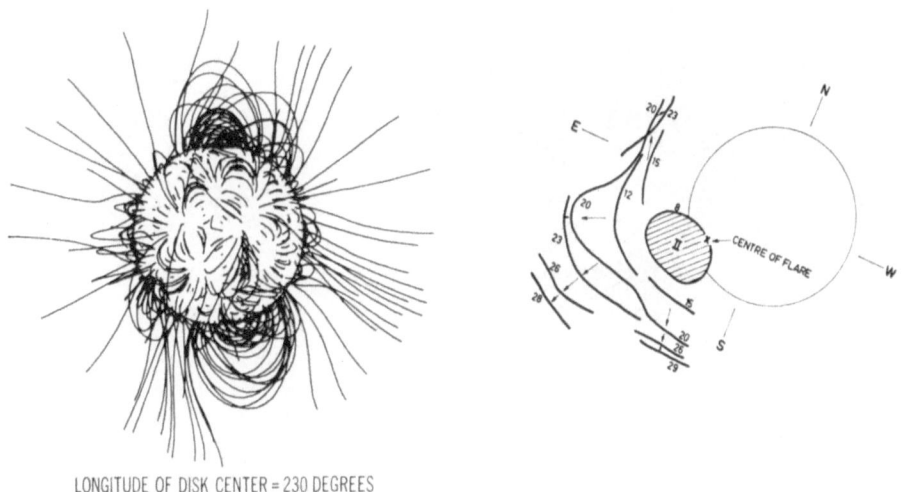

LONGITUDE OF DISK CENTER = 230 DEGREES

Fig. 15.   Comparison of the successive positions of a rising Type IV burst (Kai, 1969b) with the corresponding coronal map. Field lines originating at foot points where $B_L \geqslant 0.16$ G are displayed.

Comparisons of radioheliograph observations of various radio bursts with corresponding magnetic maps have been made only recently and thus no detailed analyses can be reported. However, a brief examination of some of the data suggests that we have some exciting discoveries in store for us.

One of the most energetic of radio events is the Type IV burst believed to be due

LONGITUDE OF DISK CENTER = 170 DEGREES

Fig. 16.   Channeling of a directed shock, which gave rise to a complex of Type II, Type IV and Type III radio events and a disappearing prominence (Kai, 1969a), by the coronal magnetic field. Field lines originating at foot points where $B_L \geq 0.16$ G are displayed.

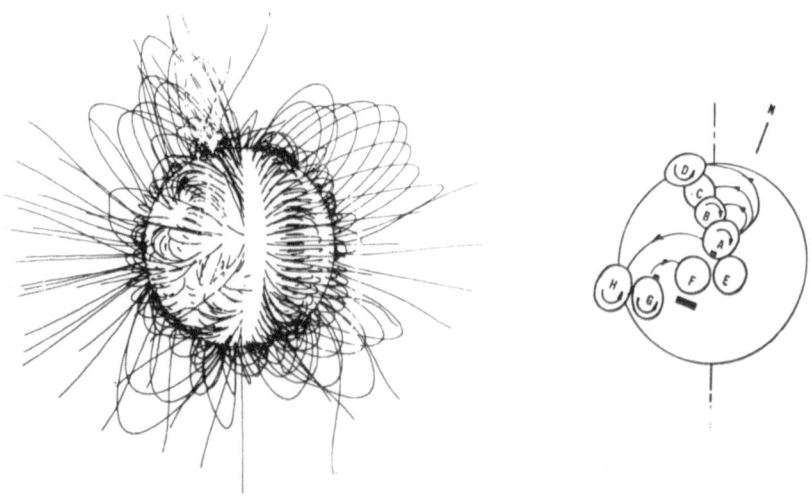

LONGITUDE OF DISK CENTER = 190 DEGREES

Fig. 17.   Correlated radio bursts (Wild, 1970) between various centers (A, B, etc.) connected by curved lines in the radioheliograms (right) appear to be connected by magnetic arches in the magnetic map (left). Field lines originating at foot points where $B_L \geq 0.16$ G are displayed.

to synchrotron radiation from mildly relativistic electrons. A comparison between radioheliograph observations (Wild, 1970) of several of these bursts and the calculated magnetic fields in the corona (Smerd and Dulk, 1970; Newkirk, 1970) shows that various subclasses of the Type IV bursts appear to be influenced by the field in different ways. In one subclass (see Smerd and Dulk, 1970, Figure 12) we see a loop of Type IV emission (Wild, 1969b), which was associated with the flare (X) under the low magnetic loops. Smerd and Dulk (1970) interpret this event as the expansion of the low magnetic loops in which high energy particles are trapped. Alternatively, we might imagine a shock disturbance propagating along the Diverging Fields which proceed to the east from the general region of the flare; however, the slow source speed (<400 km/s) and high polarization are drawbacks to this interpretation.

From the detailed analysis of another burst Dulk (1970) has inferred the presence (Figure 14, right) of a magnetic arch in which the particles are trapped to produce the Type IV burst (D) at the top. That such a loop is a fairly permanent feature of the field is suggested by the comparison with the magnetic map (Figure 14, left).

In another expanding Type IV burst (Kai, 1969b) (Figure 15) the expansion appears to have occurred outward *along* the field lines. Here we may have evidence that the initiating shock wave, which may be responsible for the acceleration of the particles, has been guided by the field and that the moving burst is really the moving shock front. However, since we have not established the coincidence of the burst and the field in three dimensions, this conclusion may be premature (Smerd and Dulk, 1970).

A final moving Type IV (Riddle, 1970) was accompanied by the expulsion of a spray prominence (McCabe and Fisher, 1970). Both the prominence and the burst appear to have been conducted out along the magnetic field (see Smerd and Dulk, 1970, Figure 14). Riddle has suggested that this particular radio event may represent a vortex ejected from below.

Magnetic fields also appear directly responsible for guiding other types of radio disturbances. Kai (1969a) has reported a Type II burst which proceeded from a flare in only one direction and was followed by the disruption of a filament and a slowly moving Type IV. Inspection of Figure 16 strongly suggests that the channeling of the disturbance was, indeed, magnetic.

It has been suggested (Wild, 1970) that correlated radio bursts occur when high magnetic arches connecting widely separated regions conduct a triggering disturbance back and forth between the burst locations. Figure 17 suggests that this is actually the case, although channels for the southern group of correlated bursts are not nearly so obvious as those for the northern group.

5. Prospects and Problems

We have seen that the consequences of large scale solar magnetic fields appear in a variety of forms. Such fields are of primary importance in understanding the fundamental mechanisms underlying the solar cycle. Moreover, they appear to determine the density structure and rotation of the corona as well as its projection out into the interplanetary medium. It is only the large scale features of the field which have an influence on the solar wind as it eventually reaches the Earth. In addition, our present information suggests that it is the large scale fields which act as the guiding force for a variety of impulsive events in the outer solar atmosphere such as radio bursts and cosmic rays and may well act as the container for the storage of cosmic rays over many days.

Many of these suggestions are speculative to say the least and much work remains to be done. However, many questions regarding the basic nature of large scale fields also persist. Their meridional structure and apparent rigid rotation does not fit comfortably into any of the theoretical models now receiving current attention. The evolution of these large, weak regions of field is known only in its bare outlines as is their connection to the density, velocity and magnetic structure of the corona and interplanetary medium.

I suggest that one of the most critical needs at the present time is for more, daily, accurate full disk magnetograms so that the growth and evolution of the large scale structures may be followed for many years. Also, we need observations relating directly to magnetic fields in the outer corona as well as synoptic observations of the corona itself and the transient events which penetrate it. Observations of the magnetic field in the lower corona and in prominences are also needed. In the theoretical area we require the diagnostic tools which utilize the observations and yield information on magnetic fields – I mention coronal emission line polarization and radio burst excitation and polarization as two examples. Finally, the existence and behavior of the large scale fields must ultimately have a theoretical interpretation which not only describes what we observe but explains their origin.

Note added in proof: The coronal magnetic field maps presented in this paper are based on photospheric magnetograph data furnished by R. Howard (Hale Observatories) and obtained in a program supported in part by the Office of Naval Research under contract NR 013-230, N000 14-66-C-0239.

References

Altschuler, M. D. and Newkirk, G., Jr.: 1969, *Solar Phys.* **9**, 131.
Altschuler, M. D., Newkirk, G., Jr., Trotter, D. E., and Howard, R.: 1971, this volume, p. 588.
Babcock, H. D.: 1959, *Astrophys. J.* **130**, 364.
Babcock, H. W.: 1961, *Astrophys. J.* **133**, 572.
Babcock, H. W. and Babcock, H. D.: 1953, in *The Sun*, (ed. by G. P. Kuiper) p. 704.
Babcock, H. W. and Babcock, H. D.: 1955, *Astrophys. J.* **121**, 349.
Bigelow, F. H.: 1889, *The Solar Corona*, Smithsonian Institute, Washington D.C.
Bohlin, J. D.: 1968, Ph.D. Thesis, Department of Astro-Geophysics, University of Colorado.
Bohlin, J. D. and Simon, M.: 1969, *Solar Phys.* **9**, 183.
Boischot, A. and Clavelier, B.: 1967, *Astrophys. Letters* **1**, 7.
Brandt, J. C.: 1967, *Astrophys. J.* **147**, 201.
Brandt, J. C., Wolff, C., and Cassinelli, Joseph: 1969, *Astrophys. J.* **156**, 1117.
Bumba, V. and Howard, R.: 1965, *Astrophys. J.* **141**, 1502.
Bumba, V. and Howard, R.: 1969, *Solar Phys.* **7**, 28.
Cowling, T. G.: 1953, in *The Sun*, (ed. by G. P. Kuiper) p. 565.

Dennison, P. A.: 1970, Private Communication.
Dodson-Prince, H. W. and Hedeman, E. R.: 1968, in K. O. Kiepenheuer (ed.), 'Structure and Development of Solar Active Regions', *IAU Symp.* **35**, 56.
Dulk, G. A.: 1970, submitted to *Australian J. Phys.*
Fan, C. Y., Pick, M., Pyle, R., Simpson, J. A., and Smith, D. R.: 1968, *J. Geophys. Res.* **73**, 1555.
Hale, G. E.: 1908, *Astrophys. J.* **28**, 315.
Hale, G. E.: 1913, *Astrophys. J.* **38**, 27.
Hansen, R. T., Hansen, S. F., and Loomis, H. G.: 1970, *Solar Phys.* **10**, 135.
Harvey, J. W.: 1969, Ph.D. Thesis, Department of Astro-Geophysics, University of Colorado.
Howard, R.: 1959, *Astrophys. J.* **130**, 193.
Howard, R.: 1967, *Ann. Rev. Astron. Astrophys.* **5**, 1.
Howard, R., Bumba V., and Smith, S. F.: 1967, *Atlas of Solar Magnetic Fields*, Carnegie Institute of Washington, Publication 626.
Howard, R. and Harvey, J. W.: 1964, *Astrophys. J.* **139**, 1328.
Hundhausen, A. J.: 1968, *Space Sci. Rev.* **8**, 690.
Ivanchuk, V. I.: 1968, *Problems in Cosmic Physics*, p. 129.
Kai, K.: 1969a, *Proc. Astron. Soc. Australian* **1**, 186.
Kai, K.: 1969b, *Solar Phys.* **10**, 460.
Kiepenheuer, K. O.: 1953, Proc. **9**, Volta Conference, Rome, 1.
Krieger, A. S., Vaiana, G. S., and Van Speybroeck, L. P.: 1971, this volume, p. 397.
Kuperus, M. and Tandberg-Hanssen, E.: 1967, *Solar Phys.* **2**, 39.
Leighton, R. B.: 1964, *Astrophys. J.* **140**, 1547.
Leighton. R. B.: 1969, *Astrophys. J.* **156**, 1.
Levy, G. S., Sata, T., Seidel, B. L., Stelzried, C. T., Ohlson, J. E., and Rusch, W. V. T.: 1969, *Science* **166**, 596.
Livingston, W. C.: 1969, *Solar Phys.* **9**, 448.
McCabe, M. K. and Fisher, R. R.: 1970, *Solar Phys.* **14**, 212.
Modisette, J. L.: 1967, *J. Geophys. Res.* **72**, 1521.
Nakagawa, Y.: 1971, this volume, p. 25.
Ness, N. F. and Wilcox, J. M.: 1965, *Science* **152**, 161.
Ness, N. F. and Wilcox, J. M.: 1966, *Astrophys. J.* **143**, 23.
Newkirk, G. Jr.: 1967, *Ann. Rev. Astrophys. J.* **5**, 213.
Newkirk, G., Jr.: 1970, in Macris (ed.), *Physics of the Solar Corona*, NATO Advanced Study Institute on Physics of the Solar Corona, Greece.
Newkirk, G. Jr., Altschuler, M. D., and Harvey, J.: 1968, in K. O. Kiepenheuer (ed.), 'Structure and Development of Solar Active Regions', *IAU Symp.* **35**, 379.
Newkirk, G. Jr. and Altschuler, M. D.: 1970, *Solar Phys.* **13**, 131.
Newkirk, G. Jr. and Harvey, J. W.: 1968, *Solar Phys.* **3**, 321.
Newkirk, G. Jr., Schmahl, E. J., and Deupree, R. G.: 1970, *Solar Phys.* **15**, 15.
Pneuman, G. W.: 1966, *Astrophys. J.* **145**, 242.
Pneuman, G. W.: 1968, *Solar Phys.* **3**, 578.
Pneuman, G. W.: 1969, *Solar Phys.* **6**, 255.
Pneuman, G. W.: 1971, *Solar Phys.*, in press.
Pneuman, G. W. and Kopp, Roger: 1970, *Solar Phys.* **13**, 176.
Ramaty, R. and Lingenfelter, R. E.: 1968, *Solar Phys.* **5**, 531.
Riddle, A. C.: 1970, *Solar Phys.* **13**, 448.
Rust, D.: 1966, Ph.D. Thesis, Department of Astro-Geophysics, University of Colorado.
Rust, D. M. and Roy, J.-R.: 1971, this volume, p. 569.
Saito, K.: 1965, *Publ. Astron. Soc. Japan* **17**, 1.
Sawyer, C.: 1968, *Ann. Rev. Astron. Astrophys.* **6**, 115.
Schatten, K. H.: 1968, *Nature* **220**, 1211.
Scahtten, K. H.: 1970a, *Nature* **226**, 251.
Schatten, K. H.: 1970b, *Solar Phys.* **12**, 484.
Schatten, K. H., Wilcox, J. M., and Ness, N. F.: 1969, *Solar Phys.* **6**, 442.
Schmidt, H. U.: 1964, in *NASA Symposium on Physics of Solar Flares*, p. 107.
Severny, A., Wilcox, J. M., Scherrer, P. H., and Colburn, D. S.: 1970, Univ. Calif, Berkeley, Preprint Series 11, Issue 36.

Severny, A. and Zirin, H.: 1961, *Observatory* **81**, 155.

Smerd, S. F. and Dulk, G. A.: 1971, this volume, p. 616.

Stenflo, J. O.: 1970, *Solar Phys.* **13**, 42.

Störmer, K.: 1911, *Compt. Rend. Acad. Sci. Paris* **152**, 425.

Švestka, Z.: 1968, *Solar Phys.* **4**, 18.

Takakura, T.: 1966, *Space Sci. Rev.* **5**, 80.

Tandberg-Hanssen, E.: 1971, this volume, p. 192.

Thiessen, G.: 1952, *Z. Astrophys.* **30**, 185.

Valdez, J. and Altschuler, M. D.: 1970, *Solar Phys.* **15**, 446.

VanHoven, Gerard, Sturrock, P. A., and Switzer, Paul: 1969, AAS Meeting, Pasadena, California.

Veeder, G. J. and Zirin, H.: 1970, *Solar Phys.* **12**, 391.

Von Klüber, H.: 1951, *Observatory* **71**, 9.

Waldmeier, M.: 1955, *Z. Astrophys.* **38**, 37.

Warwick, C. S.: 1965, *Astrophys. J.* **141**, 500.

Weber, E. J. and Davis, L. Jr.: 1967, *Astrophys. J.* **148**, 217.

Wilcox, J. M. and Howard, R.: 1968, *Solar Phys.* **5**, 564.

Wilcox, J. M. and Howard, R.: 1970, *Solar Phys.* **13**, 251.

Wilcox, J. M., Schatten, K. H., Tanenbaum, A., and Howard, R.: 1970, Univ. Calif. Berkeley, Preprint Series 11, Issue 22.

Wilcox, J. M., Severny, A., and Colburn, D. S.: 1969, *Nature* **224**, 353.

Wild, J. P.: 1969a, *Proc. Astron. Soc. Austral.* **1**, 181.

Wild, J. P.: 1969b, *Solar Phys.* **9**, 260.

Wild, J. P.: 1970, *Proc. Astron. Soc. Austral.* **1**, 365.

Discussion

Schatten: Jan Stenflo and I discussed the fact that if the photospheric fields are measured to be 2 or 3 times too small, then a 'source surface' or zero potential surface more in accord with your observations of 2.5 or 3 $R_\odot$ would be appropriate. This would not change the calculated polarities much but would substantially weaken the 'source surface' field relative to the photospheric field and so could bring them into agreement with the larger photospheric fields suggested.

Newkirk: Then apparently we are in agreement that the source surface should be at 2 to 2.5 $R_\odot$ during 1966 although its height undoubtedly depends upon the solar cycle.

CORONAL MAGNETIC FIELDS ABOVE ACTIVE REGIONS

DAVID M. RUST

Sacramento Peak Observatory, Air Force Cambridge Research Laboratories, Sunspot, New Mex., U.S.A.

and

J.-RENÉ ROY

Dept. of Astronomy, University of Western Ontario, London, Ontario, Canada

Abstract. We have made detailed comparisons of coronal structure, as photographed at $\lambda 5303$ Å, with magnetic lines of force, as computed from measurements of the longitudinal component of the underlying photospheric magnetic fields. Coronal fields were computed under the assumption that the space above the active regions is current-free. Out of 36 regions for which there exist both magnetic and coronal data at the Sacramento Peak Observatory, we found only four suitable for analysis. Using a light-pen attachment to a digital computer, we were able to choose lines of force whose footpoints in the photosphere best matched those of the lines of force which were suggested by local intensity variations in the corona. The lines of force thus computed give excellent agreement with the apparent height and shape of coronal fieldlines. In particular the magnetic field in the loop structures of 2 November 1969 and 18–19 November 1968 falls from several hundred gauss at 10000–20000 km above the photosphere to 2–20 G at heights of 100000–150000 km. These computed values are in general agreement with other measurements. Spectra of the loops reveal that the direction of the line-of-sight component of the motion agrees with that anticipated from a comparison of the computed fieldline orientation and the observed motion in the plane of observation. We conclude that coronal magnetic fields above slowly changing active regions are nearly the same as the vacuum potential fields derived from underlying photospheric field sources (principally the larger sunspots in a region) and that calculations of this type may be relied upon to give the magnetic fields for studies of coronal dynamics.

Above active centers on the Sun, the solar corona is a luminous forest of loops, arcs and rays. Because of the strong magnetic fields known to exist in the underlying photosphere and because of the suggestive loops and arches, we expect that there are intense, complex magnetic fields in the corona over active centers. The only magnetograph measurements of these magnetic fields have been made by Harvey (1969), but because of the very low signal-to-noise ratio he encountered, he had to integrate the field measurements over areas much larger than that of typical coronal fine structure. Harvey's measurements indicated average coronal fields of a few gauss. Hyder (1964) inferred field intensities of 60 to 80 G from his measurements of linear polarization in Hα loops at heights of between 5000 and 45000 km above the solar limb. Measurements in quiescent prominences, which, in contrast to loop prominences, are invisible in photographs taken with the coronal emission lines, may not be reliable indicators of the surrounding coronal fields. Correll *et al.* (1956), Correll and Roberts (1958), Warwick (1957), Bumba and Kleczek (1961) and Hyder (1966) have inferred either coronal field configurations or limits on coronal field intensities from the motions and shapes of solar prominences and flare sprays, but none of these indirect approaches allows one to specify the fields completely. Newkirk *et al.* (1968) and Altschuler and Newkirk (1969) used fictitious multipoles at the Sun's center to compute the complete magnetic field of the solar corona. They assumed that the corona is free of large scale

Howard (ed.), Solar Magnetic Fields, 569–579. All Rights Reserved.
Copyright © 1971 by the IAU.

currents and that the coronal field depends only upon the photospheric fields from which their multipoles are derived. They compared their computed field configuration with eclipse photographs of the white light corona and found good agreement between fieldlines and large scale structures such as streamers, domes over quiescent prominences, and arches linking active centers.

In 1963 Schmidt (1964) suggested that one may be able to neglect currents in the lower corona, i.e., between 10000 km and 150000 km above the limb, and compute the field directly from a potential derived from magnetograph measurements of the radial component of photospheric magnetic fields. Schmidt identified the measurements of the normal component of the field at disk center with fictitious magnetic monopoles. He then computed the vacuum field above the plane of observation, which he assumed to be flat. Harvey (1969) and Rust (1966, 1970) used the computer program devised by Schmidt to compute fields for comparison with coronal structure and with observed fields in prominences. In the work we are describing here, we extended Rust's and Harvey's studies to loop prominences and to some less distinct coronal structures above active regions. We used magnetograms from the Sacramento Peak non-sa-

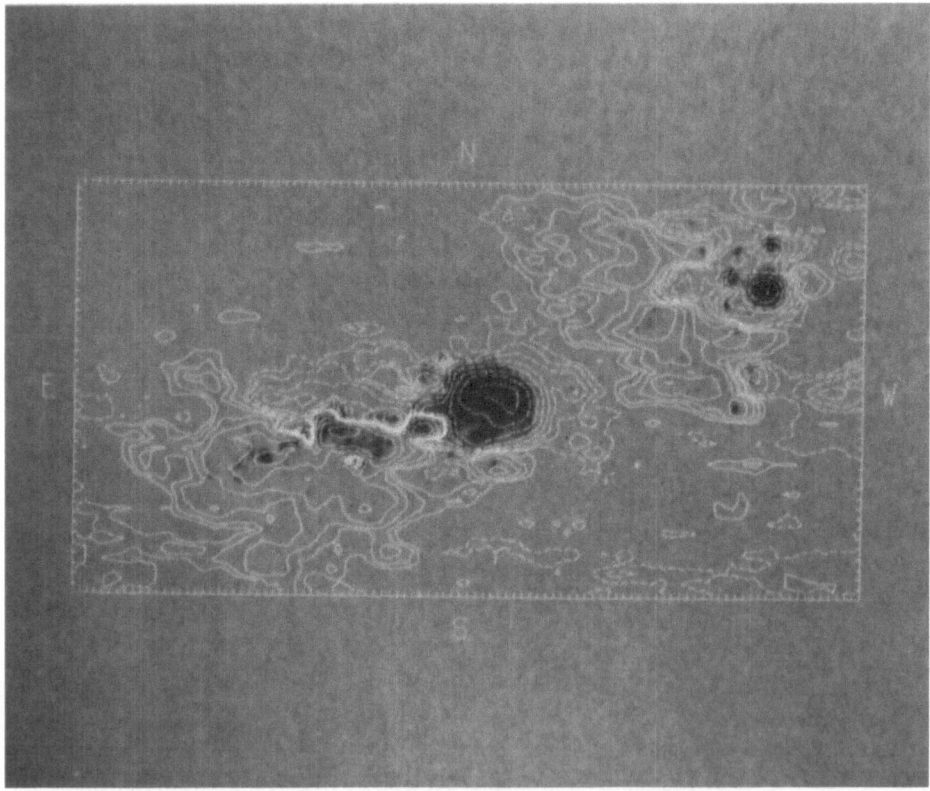

Fig. 1. Contours of longitudinal magnetic field intensity and the sunspots of Mt. Wilson spot regions 17533, 17544, and 17535. Time: 1500 UT, 24 October 1969. Dashed contours indicate negative fields, solid contours indicate positive fields. Contour levels are ± 10, ± 20, ± 40, etc.

turating Doppler-Zeeman Analyzer. The instrument was described by Dunn in an earlier session of this symposium. Monochromatic pictures of the corona at $\lambda 5303$ Å and spectra and slit-jaw photographs in Hα obtained at the Observatory during the years 1968 and 1969 were available for this study. Dr. Robert Howard provided us with copies of the visual sunspot field measurements from the Mount Wilson Observatory for comparison with the measurements from the Doppler-Zeeman Analyzer.

From the rich store of data available to us, we found 36 active centers for which there exist disk-center magnetic maps and green-line photographs of the corona as

Fig. 2.   Hα loops seen above the region of Figure 1, when it was 12° beyond the west limb on 2 November 1969. Feet of the loops are hidden behind the chromosphere.

the center passed over the limb. The first step in our analysis was to eliminate from further consideration all centers that underwent substantial evolution from disk center to limb passage. Furthermore, the quality of some magnetograms was too poor for quantitative work, and other magnetograms were rejected because they did not include important regions of magnetic flux near the active center under study. We were left with only four relatively stable active centers for analysis. We have found that, in each of these cases, the potential fields derived from photospheric observations

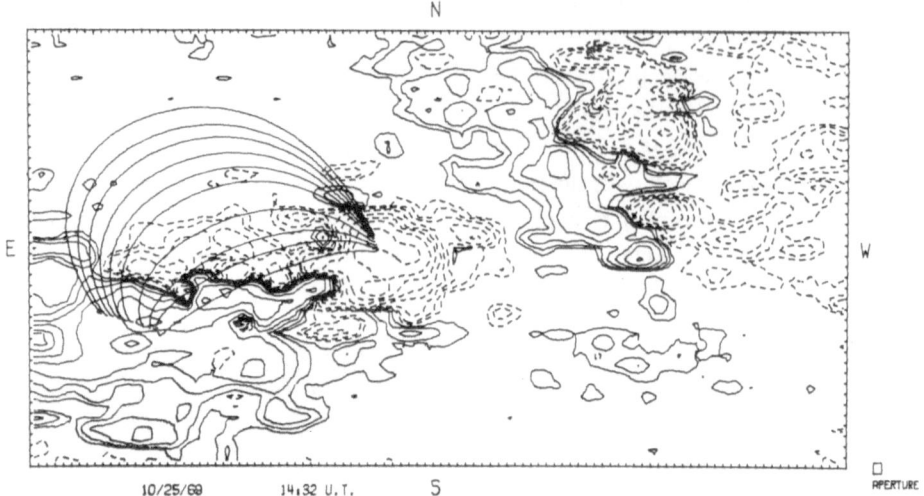

Fig. 3.   Contours of longitudinal magnetic field one day later than Figure 1. The smooth, thin lines are the fieldlines which give the best fit to the loops seen in the preceding figure. Notice how the left-most footpoints of the fieldlines form a ribbon parallel to the dividing line between regions of opposite polarity. The other footpoints concentrate in a small area of the strong negative fields in the big sunspot (see Figure 1).

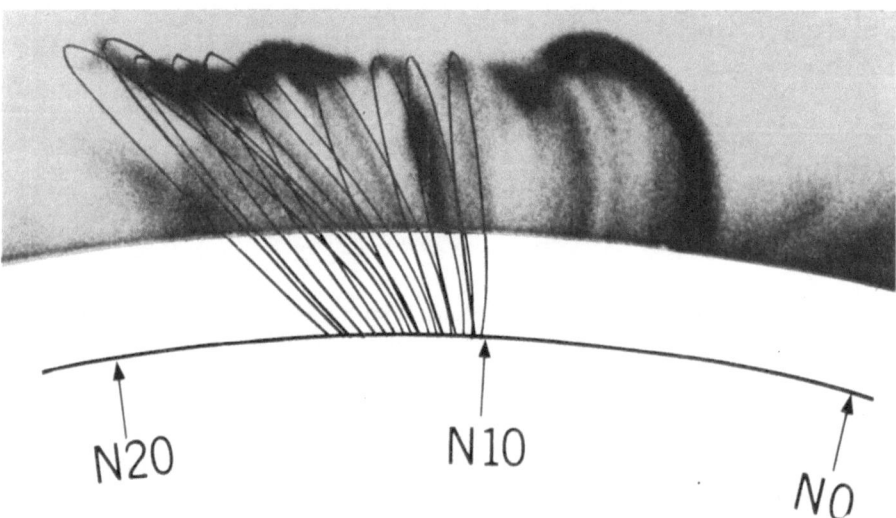

Fig. 4.   Computed fieldlines for the loops of 2 November 1969. The loops are seen here on an over-occulted negative photograph of the corona at $\lambda 5303$ Å. The lower arc indicates the level of the chromosphere. These coronal loops are nearly identical to the Hα loops of Figure 2, which were photographed at about the same time.

match the coronal structure very closely. No data which met the selection criteria were eliminated from study once we started to compute the magnetic fields.

Figure 1 shows the magnetic-field contours in Mount Wilson spot regions Numbers 17533, 17544 and 17535. The lowest contour level is 10 G and the highest field measured was in the large spot at the center and was equal to 1800 G. Visual measurements of line splitting, made at Mount Wilson, indicated a maximum field of 2100 G. After several comparisons of this sort we concluded that, although our magnetic observations were made in the temperature-sensitive line of iron at 5250 Å, the magnetic fields recorded by the Doppler-Zeeman Analyzer are not less than about 70% of the true fields. When the photograph of Figure 1 was taken the largest region in the picture was 20° east of the central meridian. On the following day, 25 October 1969, we obtained another magnetogram. The indicated fields were very similar to those seen in Figure 1. When the easternmost of the regions shown was 12° over the west limb, there occurred in the region a very large flare, producing the largest X-ray burst of the year and a spectacular series of loops, as shown in the hydrogen-alpha photograph (Figure 2). This region has a fairly featureless and apparently unchanging corona on the 17th and 18th of October, when it was on the east limb, and on the 30th of October and the 1st of November, as it passed over the west limb. The loops photographed in Figure 2 provide us with a distinct structure to compare with computed fieldlines. Figure 3 shows the fieldlines that fit the loops. We would like to remark here that, from the work of Bruzek (1964), we don't think that the appearance of loops where there was previously an amorphous structure indicates a change in the coronal fields. Loops only show where material ejected during the flare is condensing in the corona. Furthermore, it is well known that there are no changes in the major sunspots or magnetic fields at the time of a flare. Whatever changes occur in association with flares take the form of minor adjustments in the field intensity or in the position of small magnetic features, such as may be seen next to the large spot in Figure 1. The shapes of the fieldlines shown here are little affected by any but the largest magnetic poles in the region. Figure 4 shows the loops as they appeared at $\lambda 5303$ Å. The computed fieldlines shown are the same as those in the previous figure, but they are seen at the position they took when the region was 12° over the west limb at about 1500 UT on 2 November 1969. Both sets of loops are 108 000 km above the underlying photospheric surface. The field intensity at the loop tops varies from 6 G to 16 G and increases to hundreds of gauss at about 25 000 km above the large spot. The flare associated with these loops was a major producer of protons, and we note that the footpoints of the loops penetrate the umbra of the major spot (Figure 3). Bruzek has shown that the footpoints of loop systems fall along the two ribbons of large Hα flares. Since proton flares are known for the fact that the Hα emission of one of the ribbons penetrates the umbra of major spots, the appearance of the computed footpoints in the umbra further confirms the correctness of the computed fields.

Examination of the spectra of the loops shows that the legs of the loops which are more inclined to the solar surface are blue-shifted along the upper half of the loops. Since we are looking at the loops nearly edge-on, the more inclined, blue-shifted loops must be the closest to us. The computed fieldlines also show the same mixture of inclinations and the more inclined loops are also closer to the observer. A side view of the computed loops suggests that if the material in the loops is moving along the fieldlines everywhere with the same velocity, one should see red shifts in the rearward legs of the loops which are greater than the blue shifts coming from the forward legs. An examination of the spectra confirmed that this is the case.

In two details the computed loops do not fit the observed loops. First, the computed

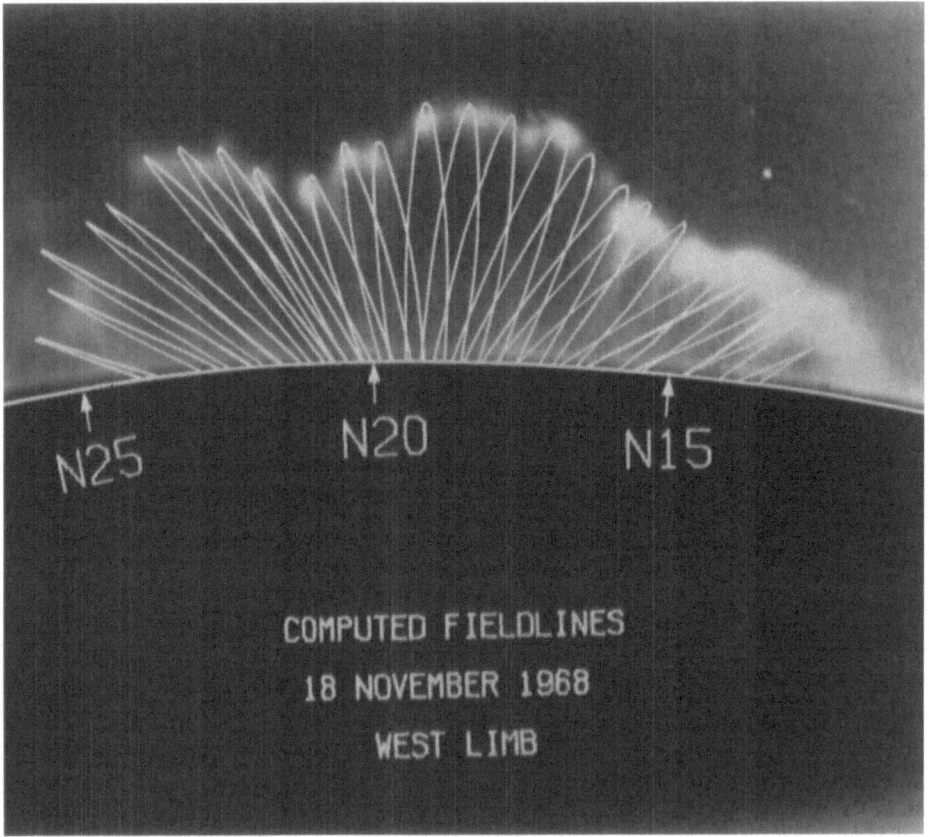

Fig. 5.   Loops seen in the coronal $\lambda 5303$ Å emission with computed fieldlines superposed upon them. Numbers indicate heliographic coordinates.

fieldlines lean slightly more to the side. This may probably be explained as an edge effect, arising from the fact that outside the plane of observation, we have assumed that there is no field in the photosphere. The exact configuration of leaning loops is sensitive to small magnetic features near the edges of the active center. The second detail in which our computed loops do not fit the observations concerns our inability to fit the large loop on the right. We think there may have been some rotation of the following spots about the leader – this is often observed. A change of this sort between the disk observation of 25 October and the coronal observation of 2 November could produce the needed field configuration.

Figure 5 shows the computed fieldlines and observed loops for 18 November 1968. The fit is excellent everywhere, except for a small difference in the inclination of the loops, as observed and as calculated, on the edge of the frame. The loops are about 90 000 km high at the center. The field there equals 11 G, while the field intensity at the tops of the side loops is 4 G. The loops observed on this date as well as the loops shown in the earlier figures are typical in height and configuration. They are two-to-three times the height of the loops measured by Hyder. When lower, but similar, loops

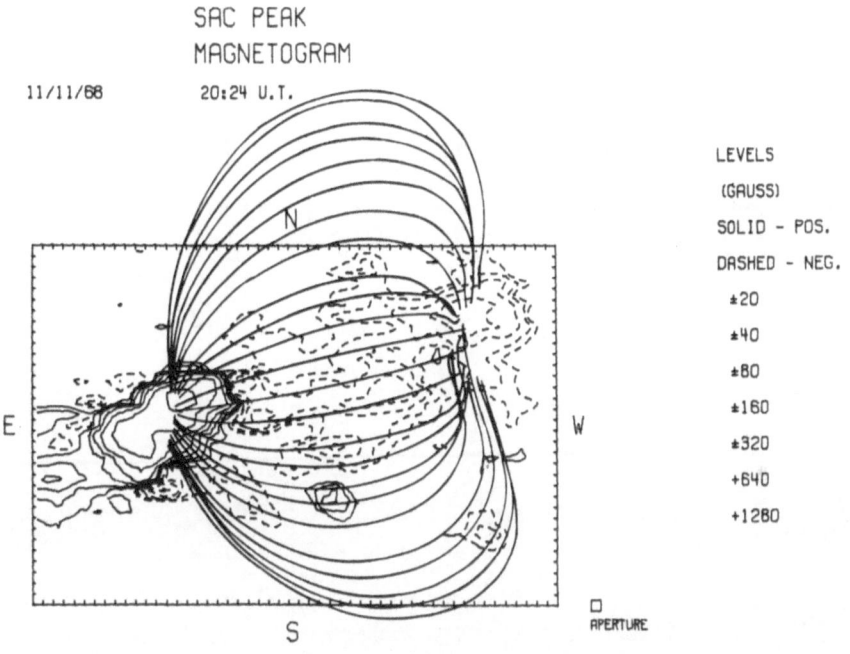

Fig. 6. Fieldlines of Figure 5 as seen from directly above the active center.

are computed for the regions studied here, the resultant field intensities at the loop tops agree well with those measured by Hyder.

Figure 6 shows the loops as seen from directly above the active center. At the top of the picture, the loops swing out above the region included in the magnetogram. If there had been some positive field there, instead of the zero field we assumed, those lines of force wouldn't have been so inclined toward the surface, and they would have fit the observations better. We feel that this failure to fit the observed structures exactly derives from this limitation in the measurements, and it is not due to currents in the corona. Notice that the lines of force fall in two ribbons which flank the line dividing magnetic polarities. We expect that these ribbons reveal the location of the Hα flare associated with the loops.

At the lower half of Figure 6, the fieldlines pass over two small concentrations of field. The effect of these fields is to add several inflection points along the edge loops. A large-scale motion picture of the region taken in the D_3 emission of helium shows that falling material apparently follows fieldlines with inflection points similar to those we found in our calculations. Inflection points of this sort may imitate twisted fieldlines and suggest the presence of currents. We feel that current-free fields are adequate to explain the apparent twisting of the fields observed here.

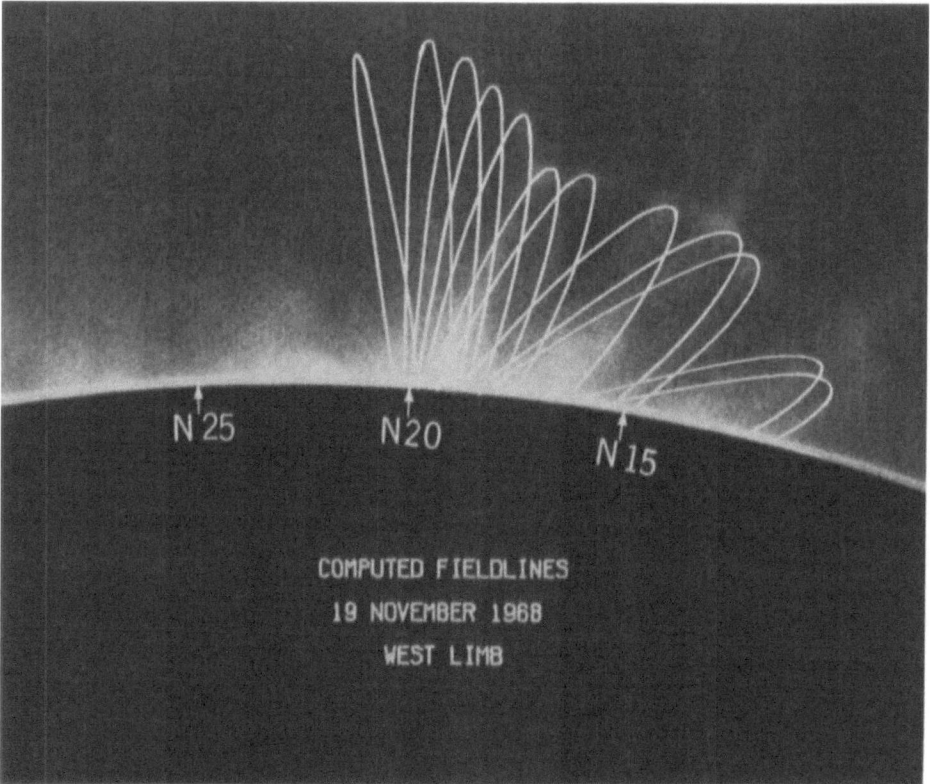

Fig. 7.   Computed fieldlines and observed coronal structure for 19 November 1968. The brilliant loops of the previous day (Figure 5) have faded to a higher and barely detectable veil.

Figure 7 shows how the loops appeared on the next day. They were higher and much fainter. This time they were visible only in the corona. Except for some edge effects, the computed loops fit the observations very well. At the loop tops, from 80000 km to 145000 km high, the field intensity is 2 to 4 G. The footpoints form two ribbons in the photosphere somewhat farther apart than those for the brighter, lower loops of the previous day.

Spectra of the loops show more blue-shifted material than red-shifted material. A view of the loops face-on readily explains this result since both legs of the loops are inclined toward the observer. Only along a small segment of the visible loops does downfalling matter have a component of motion away from the observer.

So far, we have discussed only loop-prominence observations. Figure 8 illustrates the success enjoyed in fitting vacuum fields to the coronal condensation of 27 March 1969. Since it is impossible to tell from the coronal pictures whether the fieldlines are closed or open at the top, we calculated both kinds of lines. The difference in open and closed lines is only a few seconds of arc in the position of the footpoints away from the center-line of the underlying fields. The fields at 70000 to 100000 km are 7 to 4 G respectively. Another case of fitting to a coronal condensation is illustrated in Figure 9. As in the previous case, the indistinctness of the corona limits the impressiveness of the correlation.

Fig. 8. Computed fieldlines and coronal rays above an east limb active center.

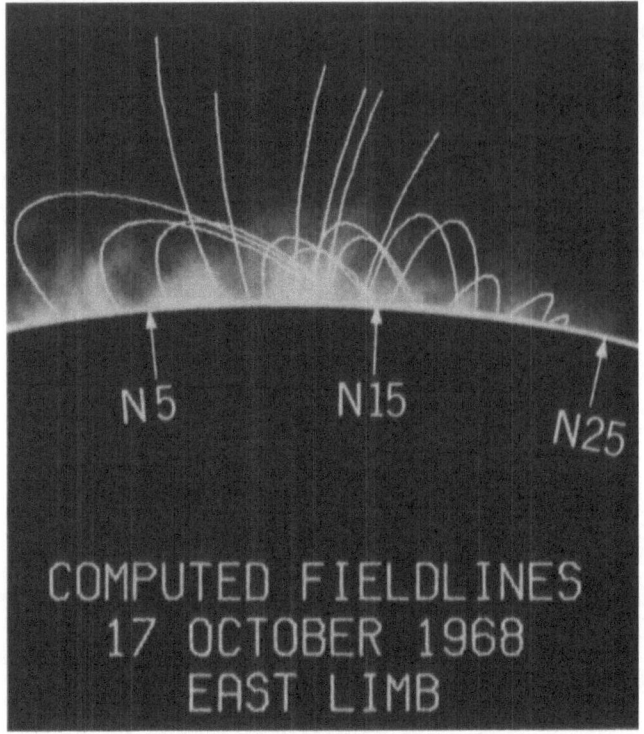

Fig. 9.   Less distinct coronal structure and computed fieldlines for an east limb condensation. The
original negatives reveal a background of rays paralleling the computed fieldlines.

Summary

We have found that vacuum potential fields, computed from disk-center observations
of slowly changing active centers, seem to fit coronal structures very closely. The
computed fieldlines apparently reveal the direction of the fields in the corona. The
magnitude of the fields we compute agrees well with observations and probably is too
low by only about 30% due to the difficulty of measuring the true photospheric fields
with the 5250 line. Where movies and spectra are available, it should be possible,
with the aid of the Schmidt program, to derive complete three-dimensional maps of
the magnetic and velocity fields in the corona above active centers. We plan to extend
our study into this area to better understand the dynamics of the corona.

Acknowledgements

We thank Howard DeMastus and Lewis Gilliam of the Sacramento Peak Observatory
for lending us the coronal and the D_3 films. They obtained these excellent photographs
through a program of careful observations whenever the weather would allow. We
are also grateful to Dr Charles Hyder and to Dr William Wagner for the use of the
Hα pictures and for the spectra of the loops.

References

Altschuler, M. D. and Newkirk, G. A.: 1969, *Solar Phys.* **9**, 131.

Bruzek, A.: 1964, *Astrophys. J.* **140**, 746.

Bumba, V. and Kleczek, J.: 1961, *Observatory* **81**, 141.

Correll, M., Hazen, M. and Bahng, J.: 1956, *Astrophys. J.* **124**, 597.

Correll, M. and Roberts, W. O.: 1958, *Astrophys. J.* **127**, 726.

Harvey, J. W.: 1969, Thesis, University of Colorado.

Hyder, C. L.: 1964, *Astrophys. J.* **140**, 817.

Hyder, C. L.: 1966, *Z. Astrophys.* **63**, 78.

Newkirk, G. A., Altschuler, M. D., and Harvey, J. W.: 1968, in K. O. Kiepenheuer (ed.), 'Structure and Development of Solar Active Regions', *IAU Symp.* **35**, 379.

Rust, D. M.: 1966, Thesis, University of Colorado.

Rust, D. M.: 1970, *Astrophys. J.* **160**, 315

Schmidt, H. U.: 1964, in *AAS-NASA Symp. Phys. Solar Flares*, (ed. by W. N. Ness), NASA SP-50, p.107.

Warwick, J. W.: 1957, *Astrophys. J.* **125**, 811.

Discussion

Leighton: How does the Sun decide which lines of force to condense matter upon, and how do you choose which lines to compute?

Rust: I don't know the answer to the first question. As for the way I choose the lines to draw, I take my cue from the Sun. There is an infinite number of possible lines to draw, but all but a few of them are irrelevant to the comparisons shown.

Severny: My question has much in common with that of Dr. Leighton. So I will phrase the question by asking how many 'sources' or equivalent dipoles do you include in your calculations of lines of force? I found that your calculated field has only two equivalent $\pm$ and $-$ 'poles' or places of divergence and convergence of lines of force. Meanwhile on the maps you showed there are many magnetic poles of equal strength in between these particular places. Generally the problem of calculation of lines of force in such circumstances is equivalent to n-body problem with dipole-type interaction. In the Crimea we usually include in the computer program not less than four or five equivalent dipoles to find the actual behavior of lines of force and this behavior appears to be very peculiar sometimes, more complex than you showed.

Rust: At each point along the field lines, we take into account the influence of all 5000 or so points of the magnetogram. If we drew the lines of force which lie close to the photosphere, we would get more complex shapes, but the loops are simple structures because they are so high.

Schatten: Rather than being critical of Dr. Rust's choice of the footpoints of the field lines, I am impressed that there are some field lines which agree as well as they do! Furthermore, in some of my field calculations I observe the highest closed arches above active regions to be about 1/10–1/5 of a solar radius. This occurs often when an active region is surrounded by a strong background field of one polarity. It is possible that these field lines that you have chosen are the highest closed lines and that coronal material is 'raining' down on these lines. Were these the biggest closed field lines in your calculation?

Rust: No. There were other, higher lines that closed.

Jordan: You say that your calculated loop structure agrees with that observed in the active regions. However, you have compared only with the Fexiv line. The results from the 1970 eclipse (Speer *et al.*, 1970, *Nature*) show that the loop structure is quite different in lines formed in different temperature regions.

Rust: The green-line observations are the only ones available to us, at least on a regular basis. It would be interesting, but also difficult, to gather the data for the more complete comparison you mentioned.

EXPERIMENTAL STUDY OF THE ORIENTATION
OF MAGNETIC FIELDS IN THE CORONA

P. CHARVIN

Observatoire de Paris, Meudon, France

Abstract. We present polarization measurements obtained in 1970 in the green coronal line with a new coronameter located at the Pic du Midi. The analysis of these data has been conducted with the theory given by the writer in 1964 and 1965. It appears that magnetic field orientations in the Corona can be deduced from the above measurements. First results showing large scale magnetic structures are presented.

1. Introduction

The writer theoretically studied in 1963 and 1964 the polarization of coronal forbidden lines, and showed that the azimuth of the magnetic field could be determined from the polarization measurements of certain lines, particularly the 5303 green line (Charvin, 1964a, b, 1965).

Since then, several eclipse experiments provided experimental results. The most important contribution has been obtained by Hyder *et al.* (1968). They have shown:

(1) the validity of the general predictions of the theory (6374 line not polarized; increasing polarization of the 5303 line with the distance from the limb);

(2) moreover, that the polarization of the 5303 line generally lies within numerical values calculated in several simple models;

(3) finally, that the direction of the major electric vector is not radial everywhere, as can be expected from the theory if the magnetic coronal fields are not also radial everywhere.

Unfortunately, their data were not accurate enough to allow the determination of magnetic field orientations.

Preliminary measurements made at Meudon in 1963 (Charvin, 1965) had shown that the sensitivity of the monochromatic coronameter (Charvin, 1963) should be great enough to allow the measurement of the coronal green line polarization without eclipse. Therefore, we rebuilt this instrument to get improved facilities. It is now located at the Pic du Midi Observatory.

This new coronameter is equipped with a 16 cm diameter coronagraph lens. Specially designed for polarization measurements, it works on-axis, on a field less than 1 min of arc. Instrumental polarization can be neglected. The disturbing effects due to the sky background and to the K-Corona are compensated with an accuracy which can reach 10^{-4}. In good observing conditions, this instrument can detect the 5303 green line up to an altitude higher than 1 $R_\odot$. Its sensitivity is about $0,01 \times 10^{-6}$ $B_\odot$ (1 Å). The following work has been done with this instrument during the last 1970 summer.

Howard (ed.), Solar Magnetic Fields, 580–587. All Rights Reserved.

2. Measurements

The Corona is photoelectrically observed, point by point, with a 51 or 29 s of arc aperture. The green line intensity, its polarization and the angle between the major electric vector component of the light and a fixed direction are simultaneously measured. In general, each measurement lasts between 2 and 4 min.

The degree of observed polarization varies within a wide range. As a general rule, it increases with the distance from the limb, and it is higher above polar regions (weak electronic density) than above active regions. The minimum and maximum values measured during these first observations were respectively less than 0,4% (near the limb), and close to 30% (at 1 $R_\odot$).

Generally, the percent polarizations measured near the solar limb are about 3 or 4 times weaker than the maximum values that were computed in several models with radial magnetic field (Charvin, 1965). On the contrary, far from the limb, the observed values are frequently similar to the computed ones.

In addition to the depolarizing effects due to the inclination of the magnetic field, the theory provides different possible explanations for these results: for instance, use of underestimated collisional cross-sections; inhomogeneities in the Corona (for example, filamentary structure); overlapping of several magnetic structures along the line of sight.

Indeed, the first magnetic effects play a quite important role, since one often observes important departures from radial configurations. Even, in some cases, one finds the major electric vector parallel to the solar limb. Then, one obtains higher values for the degree of polarization when, using the magnetic maps finally obtained, one takes in account the angular effects. Thus, the agreement with the theoretical computed values frequently becomes much better.

Figure 1 gives an example of the polarization directions observed on July 26, 1970, in the vicinity of Northern solar Pole. During this four-hour observation, the sky brightness generally stayed between 25 and 50×10^{-6} $B_\odot$. The accuracy of the angular measurements ranges from about 1° in best cases, to about 12° for weak polarizations observed under poor conditions (high and variable sky brightness).

One can distinguish regions where the directions of polarization are homogeneous, and regions where the polarization azimuth changes much from point to point. In these apparently inhomogeneous regions, the observed degree of polarization is likely to be frequently diminished by unresolved angular effects in the observed element (here, 51 s of arc).

3. Obtaining the Magnetic Field Orientations

A. PROBLEMS

The above mentioned theory connects in a simple way the polarization of the light emitted by a small volume element to the magnetic field orientation inside this element. Polarization does not depend on the magnetic field intensity; its direction is only dependent on the angular parameters involved in this problem (Charvin, 1964b, 1965). As a final result, and given the values of these angles, the direction of the major electric vector gives, either the azimuth of the magnetic field (i.e. the direction of the magnetic field as projected on the plane normal to the line of sight), or the perpendicular direction. This alternative is governed by strict rules; in many cases, it can be solved in

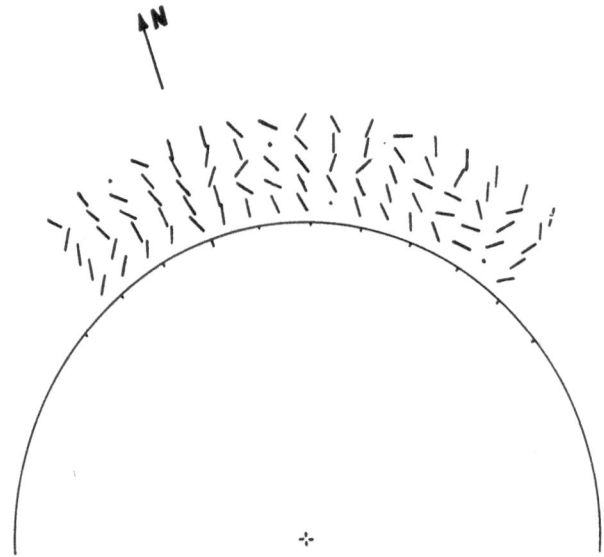

Fig. 1.   Observation of the azimuth of the green coronal line between 1 and 5 min of arc as measured during July 26, 1970 (from 6.50 UT to 10.45 UT; aperture field: 51 s of arc). The direction of bars gives the direction of the major electric vector component of the light. The dots indicate points where polarization has not been observable.

a fully objective way. However, there are some cases when the choice of the right determination raises a problem. This last ambiguity may be solved by means of internal consistency considerations. This is the first problem.

Let us go now through the second one. Considering the characteristics of the measuring system on one hand (diameter of the observed field, spacing of the observed points, minimum duration necessary to establish the map of a significant part of the Corona), the line of sight effects on the other hand, the existence of large scale magnetic structures in the major part of the Corona seems to be necessary. It has to be that way, so that, more particularly:

(1) the percent polarization has a measurable value on most of the observed elements;

(2) the resulting data have a clear physical meaning. In this respect, the maximum consistency is obtained when polarization has constant magnitude and orientation along the line of sight (Charvin, 1965);

(3) the polarization observed on near elements are (except on both sides of boundary lines) truly connected to the same general magnetic field. Then, the ambiguities previously mentioned can be correctly removed.

In fact, this assumption that large scale structures do exist in the major part of the Corona, seems to be generally well-founded. For instance, it is well known that structures visible on K-Corona and monochromatic photographs may be, without contradiction, explained as resulting from magnetic structures, and that they seem generally to possess the expected characteristics, i.e. scale and stability. Furthermore, we must emphasize the fact that our polarization data frequently had significant values, and that in addition, equal polarization directions were observed in several cases during 2 or 3 successive days in the same coronal regions. These last points give an experimental support to the existence of such structures. Nevertheless, in some parts of the Corona, the danger still exists that measurements concerning, in fact, 2 or 3 different structures could be improperly ascribed to the same magnetic pattern. This risk could be greatly reduced by increasing the density of the measured points.

B. OBTAINING MAGNETIC MAPS

The achievement of the right determination of magnetic orientations (magnetic field azimuth parallel or normal to the measured direction), and the drawing of the maps are then performed both together. However, for better comprehension, we present Figure 2 as well as Figure 3.

In order to check our results, up to a certain extent, we proceeded as follows:

First, we outlined a map of the field orientation, taking only in account the values of the measured angles and the general properties of magnetic fields (continuity, conservation). This leads to the identification of most of the main structures. At the end of this step, the arches 1, 2, 4 and 5, the polar features 3 and the regions 6 and 8 are, in the example given in Figure 3, clearly distinguished.

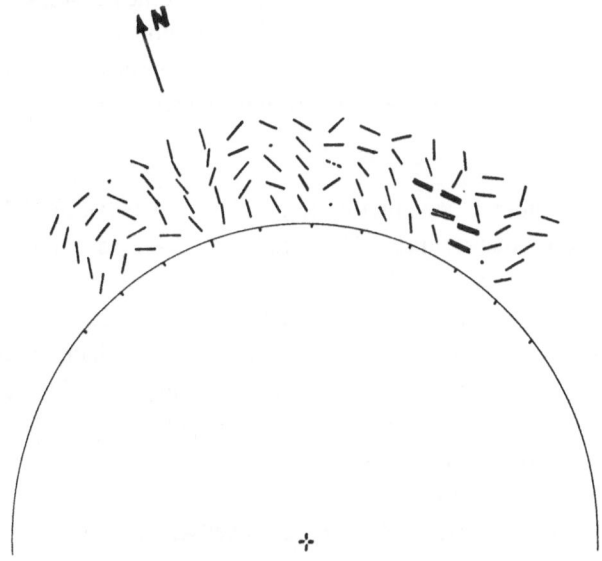

Fig. 2.   Diagram of magnetic field orientations as projected on the plane normal to line of sight. The double bars indicate points where the magnetic field has a strong component along the line of sight.

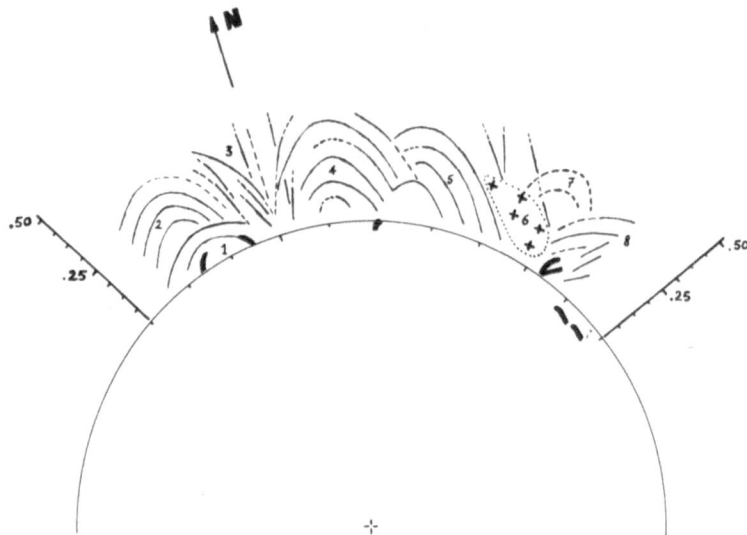

Fig. 3.   Schematic map of the coronal magnetic field (July 26, 1970). The prominences and filaments
observed on spectroheliograms are shown on the map.

Second, in order to check and to refine these preliminary results, we made use of the
degrees of polarization and of the isophotes. Thus, we are able to fix in a better way
the apparent boundaries of the different structures. In difficult cases, the close connections often existing between the green line intensities and the magnetic structures
make some extrapolations possible. Thus, the existence of structure 7 is suggested by
the isophotes.

An ultimate analysis could be undertaken if the electronic densities were known with
precision. It would consist of the computation of models concerning the studied
regions, and of their comparison to the observations. This last step would, particularly
make possible the estimation of the inclination of the structures with the line of sight.

C. FIRST RESULTS

When considering Figure 3*, one first notices a similarity between such a map and the
appearance of numerous eclipse photographs as well as monochromatic ones.
Differences should also exist, at least those resulting from some properties of our
instrument (weak space resolution, linear response), and from the different ways in
which the polarizations of monochromatic emissions on one hand, and the white light
or monochromatic intensities on the other hand, should be weighted along the line of
sight. Let us mention that this similarity is also visible on the other maps we drew
above active regions (see the footnote).

More precisely, Figure 3 shows:

– polar features 3 which rise from a very restricted area;

* Several other maps have been presented at this Symposium. They will be published later.

– arch structures 1, 2 and 4 which connect the Northern Pole to lower latitude regions;

– feature 6, whose main characteristic is a strong magnetic component along the line of sight;

– the relations between the quiescent prominences, the filaments and the coronal magnetic structures. These relations, also clearly visible on the other maps, indicate that the magnetic field must be parallel to the apparent axis of the prominences, i.e. to the long axis of the filaments. This seems to give a new support to the Kippenhahn-Schlüter theory (1967).

A detailed study of the connection between coronal matter and magnetic structures will be published later.

4. Conclusions

The experimental study of the coronal magnetic field orientations by means of forbidden lines polarization provides a new method in solar research. In fact, the observations which have been presented here are the first accurate observations of these phenomena made without eclipse. They are also the first ones which allow, through specific optical measurements, the drawing of magnetic maps in the Corona. Hence, it is easy to realize that the right use of such observations still requires additional work. Otherwise, our main purpose, when making these first observations, was primarily to test the method and the reducing procedures. At this point, a comparison of our results with those provided by other methods seems to be important. In this respect, a comparison with pictures taken in the green line, and also with structures computed with the method developed by Newkirk and Altschuler (1969) still remains to be done.

However, the following points should be already noticed:

(1) The large scale magnetic structure of the Corona has been observed during the course of these first observations. The actual measurements of significant degrees of polarization, often close (after correction) to theoretical values, provide first evidence of this. These measurements also afford a good estimate of the extent and stability of these structures. The existence of several structures visible at the same apparent place during 2 or 3 successive days, produce one more proof. The identification of some peculiar features, as the diverging appearance of the polar magnetic field, or the relations between the coronal magnetic field and prominences, seems to be a third remarkable point.

These results substantiate the general acceptance that structures visible on eclipses and monochromatic pictures correspond to lines of force of the coronal magnetic field.

(2) The good internal consistency of the results here obtained gives a second verification of the theory, more achieved than the verification obtained by Hyder *et al.* (1968).

(3) We think the magnetic field maps are already obtained in a rather satisfying way, even if some parts or details have not yet the required degree of certainty. The main difficulties we met do not arise from the 90°-ambiguity existing for certain angular values, but much more from a too weak density of the observed elements (compared

to the extent and complexity of actual coronal structures). We think a systematic use of a lower field aperture (about 30 s of arc), along with a doubling of the observed point density would strongly reduce these difficulties. Particularly, we think that the degree of subjectivity yet existing in the interpretation of data could be practically nullified.

Some improvements of coronameter should allow this doubling without a correlative increase of the total observing time.

(4) We think these first results should stimulate further eclipse observations of coronal line polarization. The method we are using has in fact two limitations: its weak space resolution, and its long scanning time. It does not seem possible, even in the best case, to measure in a whole day more than 350 or 400 points with this instrument.

(5) Finally, we think this method should yield new valuable information in the field of coronal physics. Besides the study of magnetic structures in themselves and in connection with photospheric fields, their knowledge should allow us to interpret, much better than now, measurements of white light and monochromatic intensities. New evidence concerningin homogeneities in the Corona as well as magnetic field intensities is likely to be obtained.

Acknowledgements

I thank Prof. J. Rösch, Director of the Pic du Midi Observatory, whose hospitality allowed me to set up the coronameter in the best conditions. I am grateful to Mr J. Arnaud for invaluable help during the course of observations.

This research is partly supported by the 'Direction des Recherches et Moyens d'Essais'.

References

Altschuler, M. D. and Newkirk, G., Jr.: 1969, *Solar Phys.* **9** 131.
Charvin, P.: 1963, *Compt. Rend. Acad. Sci.* **256**, 368.
Charvin, P.: 1963, *Compt. Rend. Acad. Sci.* **256**, 1078.
Charvin, P.: 1964a, *Compt. Rend. Acad. Sci.* **258**, 1155.
Charvin, P.: 1964b, *Compt. Rend. Acad. Sci.* **259**, 733.
Charvin, P.: 1965, *Ann. Astrophys.* **28**, 877.
Hyder, C. L., Mauter, H. A., and Shutt, R. L.: 1968, *Astrophys. J.* **154**, 1039.
Kippenhahn, R. and Schlüter, A.: 1957, *Z. Astrophys.* **43**, 36.

Discussion

Severny: How could you discriminate between the polarization due to magnetic field and just due to the scattering effect in the line examined by Lyot (observationally) by Thiessen and others and which is essentially nonmagnetic? Why do you ascribe all the effect to magnetic field?

Charvin: The polarization I am observing is not due to the magnetic field. In fact, this polarization is due to a diffusion process similar to that which, as you point out, has been observed by Lyot in the prominence lines. It should occur even without magnetic fields, because it is due to the anisotropy of the radiation field producing the line. However, the effects of this type are magnetic field dependent. In the present case, the magnetic field acts by reducing the percent polarization and by tilting the polarization plane according to the theoretical rules I have given. I make use of these last effects for the determination of the magnetic field in the corona.

Beckers: Can you explain why the polarization is parallel or at right angles to the field lines? From the contribution by Dr. House I understand that the angle of polarization depends also on the amount of precessing of the bound electron in the magnetic field.

Charvin: Quantum mechanics gives the answer. As the strength of the magnetic field is certainly higher than 10^{-5} or 10^{-6} G in the regions of the corona I am observing, we have not taken into account precession effects in the case of forbidden coronal lines. In this case, the polarization is either normal to the field or parallel to the field as projected on the plane normal to the line of sight. Let us consider a small volume element in the corona. The angle between the normal to the Sun, which is the symmetry axis of the radiation field, and the magnetic field, which is the quantization axis, gives the sign of the polarization. When this angle increases, the polarization decreases, then cancels for Van Vleck's value (about 55°), and finally becomes negative. Roughly speaking, because of line of sight effects, the direction of the field is directly given by the observation for angles observed between 35° and 55°. For the observed values lower than 35° the ambiguity is not present because of the magnitude of the percent polarization.

Lamb: It should be borne in mind that in the analysis described by Charvin it is not necessary to consider level crossing-interference, and hence the dependence of the plane of polarization on field strength described by Dr House does not need to be taken into account.

TIME EVOLUTION OF THE LARGE-SCALE SOLAR
MAGNETIC FIELD

MARTIN D. ALTSCHULER, GORDON NEWKIRK, JR.,
and DOROTHY E. TROTTER

High Altitude Observatory, National Center for Atmospheric Research*, Boulder, Colo., U.S.A.

and

ROBERT HOWARD

Hale Observatories, Pasadena, Calif., U.S.A.

Abstract. The six years of data from the Mt. Wilson Magnetic Atlas were analyzed in terms of surface harmonics. Between 1959 and 1962 the dominant harmonic corresponded to a dipole lying in the plane of the equator (2 sectors). There was also a significant zonal harmonic in which both solar poles had the same magnetic polarity, opposite to that at the equator. From the end of 1962 through 1964, the harmonic corresponding to 4 sectors was dominant. In 1965 and 1966, the harmonic of the north-south dipole became significant.

The Mt. Wilson magnetic atlas (Howard *et al.*, 1967) provides about six years of data of the line-of-sight component of the photospheric magnetic field. For any 28-day period (1 solar rotation), the data can be expanded in surface harmonics of the form $P_n^m(\theta) \cos m\phi$ and $P_n^m(\theta) \sin m\phi$. The coefficients of the Legendre series expansion are chosen by a least-mean-square fit to the given line-of-sight magnetic field data (Altschuler and Newkirk, 1969). We normalize the surface harmonics by the technique of Schmidt (1935) as in geomagnetic work (see Chapman and Bartels, 1940) because then the magnitude of the coefficient of a surface harmonic reflects the importance of that harmonic in representing the data. A normalization which makes the coefficient precisely proportional to the importance of the harmonic (in the sense that the average square value of the surface harmonic over the sphere is unity) (Schmidt, 1895) would give lower values for the coefficients of the large n harmonics than would the Schmidt (1935) normalization. Thus the Schmidt (1935) normalization tends to over-emphasize the importance of the harmonics of large n. We compared data samples using both the Schmidt (1935) and Schmidt (1895) normalization and found a rather small difference. A more detailed analysis will use a Schmidt (1895) normalization to evaluate the relative importance of the surface harmonics.

The geometrical interpretation of the spherical harmonics is simple. The harmonic $P_n^m(\theta) \cos m\phi$ (or $P_n^m(\theta) \sin m\phi$) is zero on $2m$ different meridians equally spaced in longitude; a plot of the harmonic around a parallel of latitude will show $2m$ zeroes. If $m \neq 0$, there are $2m$ sectors, m positive and m negative. In the case $m=0$, the harmonic has a constant value along a parallel of latitude and the harmonic is called a zonal harmonic. The poles of the Sun can have a uniform magnetic polarity if and only if $m=0$. The integer $n-m$ is the number of times the surface harmonic goes

---

* The National Center for Atmospheric Research is supported by the National Science Foundation.

to zero between (but not at) the poles. Along a meridian of longitude there are $n-m$ zeroes of the surface harmonic excluding those at the poles. Thus there are $n-m+1$ zones. If $n-m$ is an odd integer, the harmonic goes to zero at the equator, and the magnetic polarity changes when we cross the equator. If $n-m$ is an even integer, the region about the equator has a definite magnetic polarity. If $n-m=0$, there is a pure sector structure; that is, there are no zeroes along a meridian of longitude. The integer n gives the index of the multipole; $n=0$ is a monopole, $n=1$ is a dipole, $n=2$ is a quadrupole, $n=3$ is an octopole, and $n=N$ is a '2^N-pole'. In our procedure we assume there is no monopole component for the solar field, and the data is uniformly calibrated accordingly (Altschuler and Newkirk, 1969).

As an example, Figure 1 (top) shows the three surface harmonics $(n, m)=(7, 0)$, $(n, m)=(7, 5)$, and $(n, m)=(7, 7)$. The first is a pure zonal harmonic $(m=0)$ with 7 zeroes along a meridian connecting the poles. Thus there are 8 zones. The second is a tesseral harmonic with $n-m+1=3$ zones and $2m=10$ sectors, which divide the spherical surface into $2m(n-m+1)=30$ separate surface regions of alternating

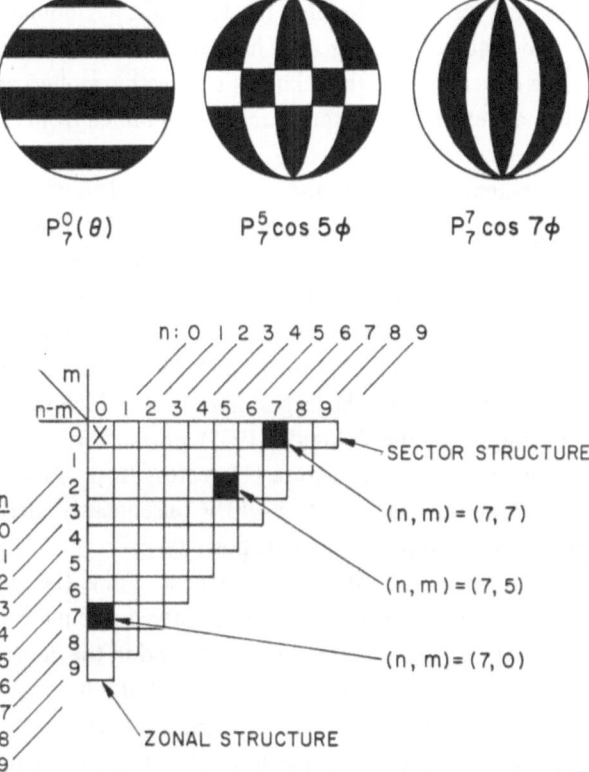

Fig. 1. Geometrical interpretation of surface harmonics. The integer n is the index of the multipole. The integer m is half the number of (longitude) meridians on which the surface harmonic is zero. The integer $n-m$ is the number of (latitude) parallels between (but not including) the poles on which the surface harmonic is zero. If $n-m=0$, there is a pure sector structure. If $m=0$, there is a pure zonal structure.

magnetic polarity. The third is a pure sector structure harmonic $(n-m=0)$ with 14 zeroes along a parallel of latitude and 14 sectors (since $2m=14$). Figure 1 (bottom) also shows the location of these harmonics on a graph of $n-m$ vs m. Since all 3 of these harmonics were chosen with $n=7$, they all lie along a diagonal of constant n.

To test the reliability of our procedure, we compared the harmonic analysis of the Mt. Wilson magnetic data with that of the Kitt Peak data for the same date 7 March 1970 (corresponding to an eclipse). The Kitt Peak magnetograph does not saturate even at sunspot field strengths whereas the Mt. Wilson magnetograph saturates at about 80 G. Although the Mt. Wilson data should be corrected for saturation, this was not done for the present comparison. Nevertheless, the agreement is remarkably good. Figure 2 (top) identifies the 5 largest harmonics (black squares) and their rank in magnitude (numbers in the squares). Actually, we omit the phase information and rank the magnitudes of $S_n^m = (g_n^m)^2 + (h_n^m)^2$ where g_n^m is the coefficient of $P_n^m(\theta) \cos m\phi$ and h_n^m is the coefficient of $P_n^m(\theta) \sin m\phi$. The zonal and sector boundaries could be obtained by considering g_n^m and h_n^m separately; this has not yet been done.

Of the five largest harmonics in each data set, four agree in their (n, m) identity. The other large harmonic had the same n in both data sets but differed in the value of m by $\Delta m=1$. Figure 2 (bottom) identifies the 11 largest harmonics (black and stippled squares). We find that 8 of the largest 11 harmonics have the same (n, m) identity in both data sets. We conclude that the 5 largest surface harmonics are characteristic of the photospheric magnetic field and can be reliably obtained from existing magnetograph data.

The photospheric magnetic field for the eclipse date 7 March 1970 is interesting because there was no significant dipole component $(n=1)$. Both the north-south dipole component $(n=1, m=0)$ and the dipole component in the plane of the equator $(n=m=1)$ were relatively insignificant. There is a relatively weak quadrupole component $(n=2)$ and a strong octopole component $(n=3)$. The weakness of the dipole and quadrupole components means that the solar field decreased rather rapidly with radial distance on the eclipse date. The largest harmonic (which was twice as large as the next highest harmonic in the Schmidt (1935) normalization) corresponded to $n=m=8$, a very high order sector structure (16 sectors, 8 positive and 8 negative). There was no zonal structure (for which $m=0$). Consequently, neither pole of the Sun was characterized by a dominant magnetic polarity.

The six years of data from the Mt. Wilson magnetic atlas (Howard et al., 1967) were analyzed. Surface harmonic expansions were derived at equally spaced intervals (roughly every half rotation), except where gaps occurred in the data. Between 20 and 30 such harmonic expansions, spanning a time interval of about a year were considered as a single data period. We analyzed 6 data periods between 1959-1966. Figure 3 summarizes the harmonic structure for the 6 data periods. Each period is labelled with the last 3 digits of the first and last Carrington rotation and with the longitude at the center of the first and last data sample. For example, 417/170–431/0 means the period runs from (McMath) rotation 1417 centered at Carrington longitude 170° to rotation 1431 centered at Carrington longitude 0°. The number of 'samples' is the

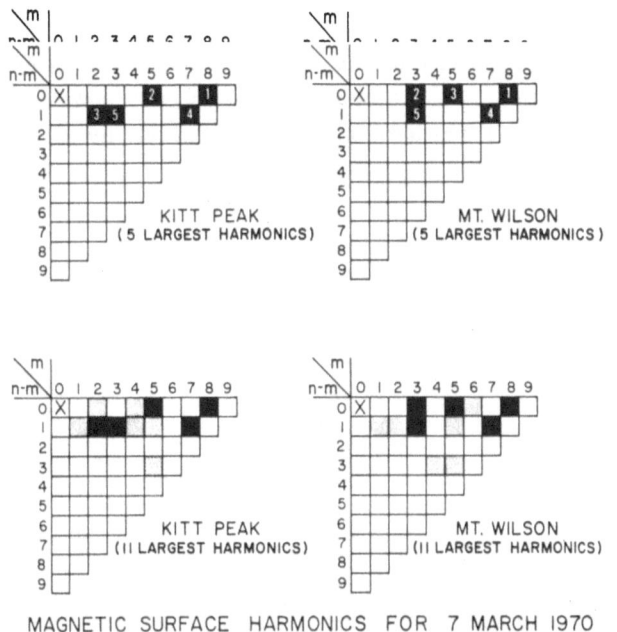

MAGNETIC SURFACE HARMONICS FOR 7 MARCH 1970
COMPARISON OF DATA FROM DIFFERENT MAGNETOGRAPHS

Fig. 2. Comparison of analysis using data from different magnetographs. The largest five surface harmonics are ranked in importance.

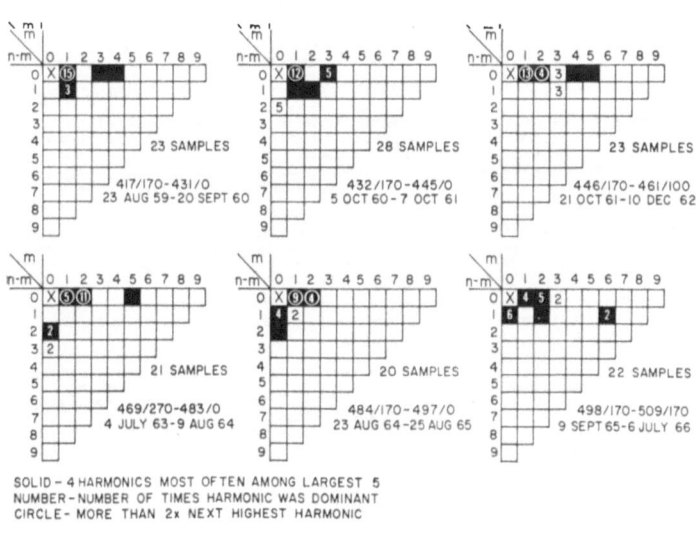

DOMINANT SURFACE HARMONICS FOR SOLAR MAGNETIC FIELD 1959-1966

Fig. 3. Roughly six years of dominant surface harmonics. Black boxes are harmonics which appear most often in top five in importance. Numbers in boxes give number of times harmonic was dominant. Circle means harmonic was at least twice as large as the next largest harmonic on at least two occasions.

number of harmonic expansions we computed for the data period. These samples
are essentially uniformly spaced in time. In this analysis, the data were corrected for
magnetograph saturation by a procedure (Altschuler and Newkirk, 1969) based on
the ideas of Stenflo (1967). The harmonic shaded black are those which appear most

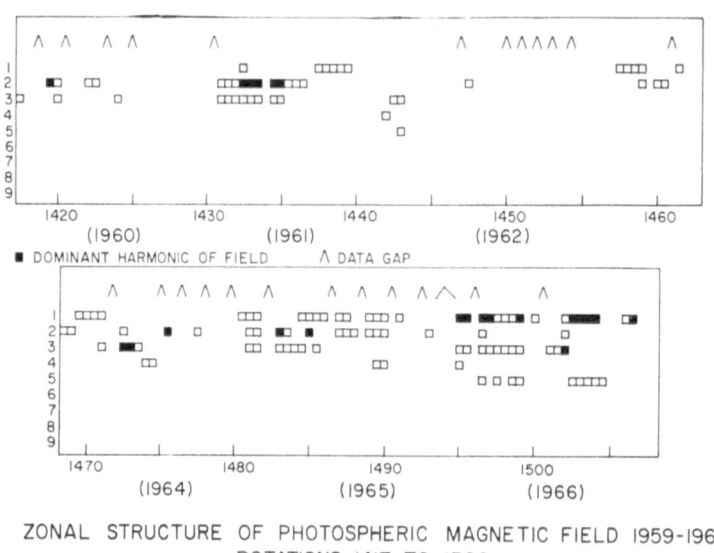

Fig. 4.   Time variation of zonal harmonics. Plot of *n-m* vs time. Blank box means harmonic was
among the largest five harmonics. Black box means harmonic was the dominant harmonic.

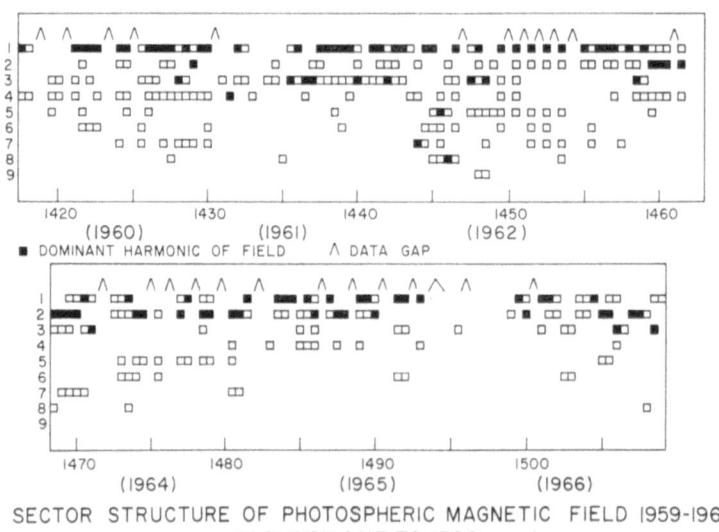

Fig. 5.   Time variation of sector structure. Plot of *m* vs time. (Same code as in Figure 4.)

frequently among the largest 5 harmonics. The number in the box of a particular (n, m) harmonic is the number of times the harmonic was the dominant (largest) harmonic of a data period. A circle around a number means that for at least two data samples in the data period the harmonic was at least twice as large as the next largest harmonic.

For most of the time between 1959 and 1962, the dominant harmonic was a dipole lying in the plane of the equator $(n=m=1)$. This harmonic corresponds to the simplest sector structure, with one hemisphere of positive polarity and the other hemisphere of negative polarity; the two sector boundaries are meridians of longitude 180° apart. There was also a significant zonal harmonic $(n=2, m=0)$ in which both solar poles had the same magnetic polarity, opposite to that at the equator. At the end of 1962 and lasting through 1964, the sector structure $(n=m=2)$ with 4 sectors (2 positive and 2 negative alternating in longitude) became the dominant harmonic largely because the dipole harmonic $(n=m=1)$ decreased in importance. The harmonic $(n=m=2)$ corresponds to the sector structure described by Wilcox and Ness (1965). With our technique we can establish that the sector structure of 1963 corresponded to $(n=m=2, n-m=0)$ and not to $(n=4, m=2, n-m=2)$. Both of these patterns are similar near the equator. In 1965 and 1966 in the post-minimum part of the solar cycle, the north-south dipole component $(n=1, m=0)$ became significant. Figures 4 and 5 show the pure zonal and sector structures respectively as a function of time.

We point out that even during times of reliable data, there were rapid changes in the surface harmonic 'spectrum' of the photospheric magnetic field. Often significant changes in the dipole $(n=1)$ and quadrupole $(n=2)$ harmonics occurred in times of less than half a solar rotation. The low n harmonics represent very large scale structures in the solar magnetic field. Rapid changes in such harmonics indicate the possible occurrence of very large and rapid changes in the energy of the solar magnetic field. Moreover, the coronal magnetic field at large distances would be most affected by the lowest n harmonics. Therefore, any rapid changes in the low n harmonics of the photospheric field would almost certainly affect the magnetic environment of the Earth.

In summary, we believe that our technique for examining all the magnetic surface harmonics should be useful in tracing the solar cycle.

Acknowledgement

The Atlas of Solar Magnetic Fields, which was our primary source of data, was funded in part by the Office of Naval Research (USA).

References

Altschuler, M. D. and Newkirk, Jr., G.: 1969, *Solar Phys.* **9**, 131.
Chapman, S. and Bartels, Jr.: 1940, *Geomagnetism*, Clarendon Press, Oxford.
Howard, R., Bumba, V., and Smith, S. F.: 1967, *Atlas of Solar Magnetic Fields 1959–1966*, Carnegie Institute of Washington, Publication No. 626.

Schmidt, A.: 1895, *Abhandl. Bayer. Akad. Wiss. Munchen, II. Classe* **19**, 1.

Schmidt, A.: 1935, *Tafeln der Normierten Kugelfunktionen, Sowie Formeln zur Entwicklung*, Gotha, Engelhard-Reyer.

Stenflo, J. O.: 1967, *Acta Univ. Lund. Sect. II*, No. 35.

Wilcox, J. M. and Ness, N. F.: 1965, *J. Geophys. Res.* **70**, 5793.

Discussion

Wilcox: The rapid change with time of the number of sectors found in your analysis is consistent with the observations of the interplanetary magnetic field showing that a new sector may be formed within a time interval of a few days. For example this might change the number of sectors from two to four. From our investigations it seems that the formation of a new sector may not be related to a flare.

Altschuler: Thank you.

Schatten: I think your analysis is very interesting. However, I would like to suggest that other orthonormal sets might more closely represent (and shed additional light) on the physical processes occurring in the photosphere. As Dr Newkirk showed, from the model of Leighton, there are backwards 'C' shaped structures on the Sun. This was observed by Dr Bumba and Dr Howard. This field would be represented by a series of Legendre polynomial terms. Perhaps it would be possible to combine the terms and thus form a new set that more closely resembles physical processes.

Altschuler: The Legendre polynomials are the natural orthonormal polynomials for a spherical surface. We can determine the sector and zonal structures, the magnetic polarity of the poles of the Sun, the multipole structure, and how the field decays with radial distance. Using the Legendre surface harmonic expansion, we have drawn many maps of the coronal field, and we feel confident that our method can delineate the photopheric boundaries between different magnetic polarities.

Ward: The axes of magnetic regions shift longitude as a function of latitude. Why did you not attempt to fit functions to the magnetic data which better represented this known feature of their appearance?

Altschuler: What we have tried to do is to summarize the photospheric magnetic field pattern in terms of a few parameters. The present Legendre polynomial procedure gives an excellent 'fit' to the available magnetic data and does it in an impartial way. Any other set of orthonormal polynomial functions should also work, but the geometrical interpretation might be more difficult.

Simon, M.: What is the ratio of the amplitude of the first five largest tesseral components, to the next five largest components? That is, thinking of your analysis as a power spectrum analysis in spherical harmonics, how rapidly does the 'power' decrease toward the higher spatial frequencies?

Altschuler: The most important surface hormonic (i.e. the surface harmonic with the largest coefficient) might have a value of $S_n{}^m$ about an order of magnitude greater than that of the sixth most important surface harmonic, and about two orders of magnitude greater than the eleventh ranking harmonic. The steepness of drop-off (i.e. the sharpness of the spectrum) is greatest during quiet years and is least during active years, probably because there is more randomness in the magnetic field during active years.

Vaiana: Do you have a map of the sectors for the March 7, 1970 eclipse?

Altschuler: We know the sectors and their phase relations but we have not made a sector map. We do have a map of the coronal magnetic field.

THE MAGNETIC FIELD STRUCTURE IN THE
ACTIVE SOLAR CORONA

KENNETH H. SCHATTEN

Laboratory for Extraterrestrial Physics, NASA-Goddard Space Flight Center,
Greenbelt, Md. 20771, U.S.A.

Abstract. The structure of the magnetic field of the active solar corona is discussed with reference to optical and radio observations of the solar atmosphere. Eclipse observations provide evidence of fine scale structures in the solar atmosphere that appear to relate to the coronal magnetic field. The coronal magnetic field used for comparison is that field calculated from potential theory: the influence of solar activity upon the potential theory field is discussed with reference to observations of the Faraday rotation of a microwave signal from Pioneer 6 as it was occulted by the solar atmosphere. Evidence has been found suggesting the existence of expanding magnetic bottles located at $10\,R_\odot$ above flaring active regions. The dynamics of these events is discussed. It is further suggested that these magnetic bottles are an important component in the solar corona.

I would like to discuss optical and radio observations of the corona and their relation to the magnetic field in the active corona.

First I would like to mention briefly the inactive corona for reference. Gordon Newkirk has described the research relating the computed coronal magnetic field to the observed features in the visible corona at the time of an eclipse.

Some of the basic physics involved in the structuring processes of the solar corona may be understood by referring to Figure 1. This figure illustrates the energy densities of various components of the solar atmosphere as a function of distance above the photosphere. The energy curves shown are to be interpreted from a somewhat qualitative viewpoint in that temporal and spatial variations may significantly alter these curves. Nevertheless the curves do show the relative importance of various components of the corona. Close to the Sun both the total magnetic field and the transverse magnetic field predominate indicating a force-free field configuration results. Beyond about $0.6\,R_\odot$ the plasma thermal energy density supersedes the transverse field energy density. This allows the plasma to stream away from the Sun carrying the magnetic field. The supersonic point is about $3\,R_\odot$. The Alfvén point is between 20 and $30\,R_\odot$. This is where the flow energy density exceeds the field energy density and thus the flow is super-Alfvénic and escape of the plasma is inevitable. This point will be important later.

The second figure is a schematic showing the magnetic field in the solar corona as it extends outward to 1 AU. Region 1 represents the photosphere where the magnetic fields are essentially held rigidly due to the large amount of plasma present and rotate with the Sun. Region 2 represents the inner corona where the magnetic field energy density supersedes the plasma energy density. Thus the magnetic field determines the geometry and thus the plasma rotates with the field lines. The large-scale field lines obey the current-free field equations and thus form arches and rays. At about $0.6\,R_\odot$ the plasma energy density supersedes the transverse field energy and thus currents flow which cause the magnetic field to form an open field configuration and allow the plasma to escape. This field is then stretched into an Archimedes spiral. This was predicted by Parker and observed on many spacecraft experiments.

Comparisons of field calculations with observations of coronal structure from eclipse photographs adds support to these ideas. Figure 3 shows a comparison of a drawing based upon field calculations with a photograph of the November 12, 1966

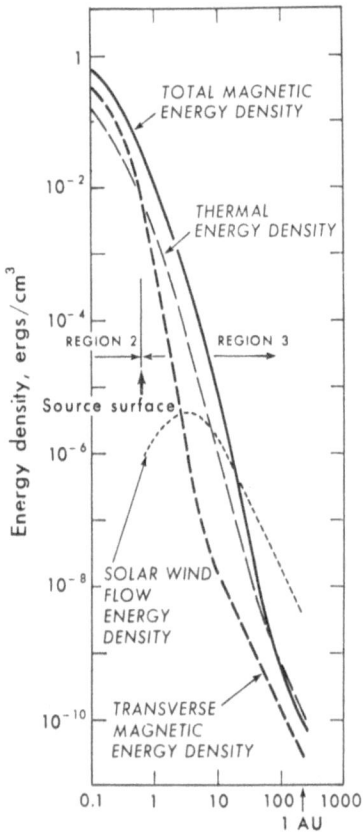

Fig. 1. The coronal energy density of the total magnetic field, transverse magnetic field, thermal motion, and solar wind flow versus distance above the photosphere. In Region 2 the magnetic field dominates the structure. On the source surface currents flow which allow the field to be transported by the solar wind.

eclipse. The Sun was not yet very active. There is fairly good agreement of many of the large scale features.

I would now like to discuss one aspect, perhaps the most significant of the influence of solar activity upon coronal magnetic field structure. This is the expulsion of magnetic flux from the inner corona by flare activity. The influence of a flare upon the coronal field occurs primarily from the creation of a hotter denser plasma expanding outward

from the flare region. We shall therefore discuss the manner in which the coronal
magnetic field reacts to this change.

A unique experiment conducted by Gerry Levy and others at the Jet Propulsion
Laboratory allowed observations of the coronal magnetic field from 4 to 16 $R_\odot$ that
enabled an interpretation of the interaction. Figure 4 shows the geometry involved in
the experiment. Levy *et al.* (1969) measured the Faraday rotation of the microwave
signal transmitted by Pioneer 6 as it passed through the solar corona to Earth.

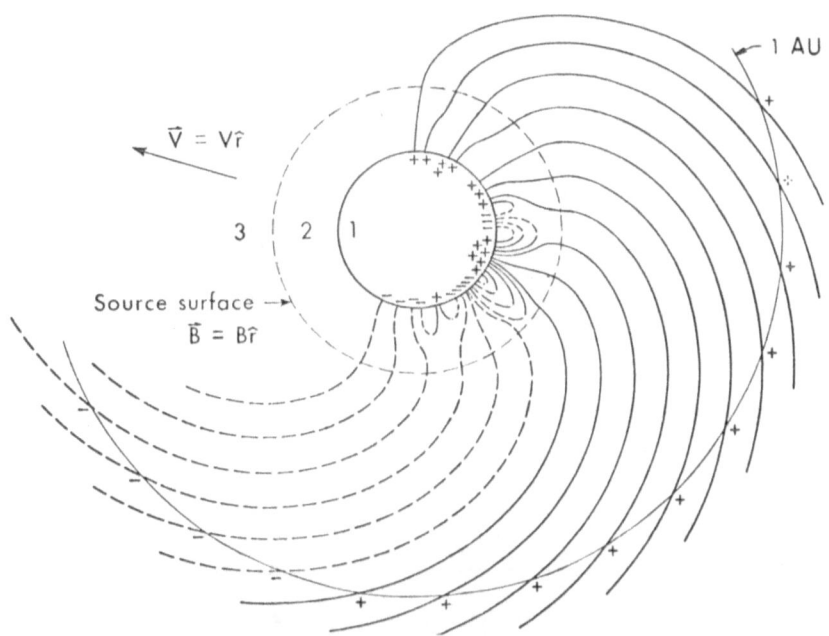

Fig. 2.   Schematic representation of the source surface model. The photospheric magnetic field is
measured in Region 1 at Mount Wilson Observatory. Closed field lines (loops) exist in Region 2. The
field in this region is calculated from potential theory. Currents flowing near the source surface
eliminate the transverse components of the magnetic field, and the solar wind extends the source
surface magnetic field into interplanetary space. The magnetic field is then observed by
spacecraft near 1 AU.

In this figure the undisturbed coronal magnetic field pattern is shown as suggested
by the coronal models of Altschuler and Newkirk (1968), and Schatten *et al.* (1968).
Solid lines indicate away from the Sun magnetic field and dashed lines indicate toward
the Sun field. Close to the Sun (within approximately a solar radius), the magnetic
field may form closed loops but beyond this distance a Parker type Archimedean
spiral geometry is thought to occur on average. In the absence of any disturbance the
signal to Earth from Pioneer 6 passes through the open field geometry.

This Faraday rotation experiment provides a measure of the line integral of the
electron density times the component of the magnetic field along the line of sight
from the spacecraft to Earth. Levy *et al.* (1969) report three transient phenomena with

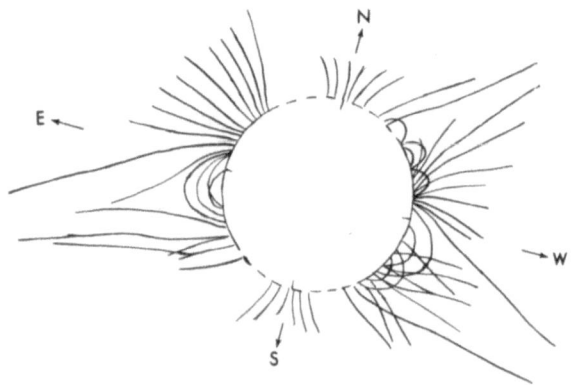

Fig. 3.   Photograph of the corona of November 12, 1966 by S. Smith – NASA Ames Research Center (top). Drawing of the magnetic structure of the corona from source surface potential theory (bottom).

Faraday rotations on the order of 40° with a duration approximately two hours. These Faraday rotation signals were observed when the distances from the Sun to the Pioneer 6-Earth line of sight were 6, 9 and 11 $R_\odot$.

A possible model for producing the Faraday rotation observed by Pioneer 6 but allowing the interplanetary sector pattern to remain intact is shown in Figure 5.

A flare of importance 1 or a subflare occurs in the active region resulting in the heated coronal plasma expanding to produce the magnetic field configuration shown here. This field configuration is similar to that proposed by Gold (1959) for a solar outburst reaching 1 AU. In this case the heated plasma expands the looped coronal magnetic field past the Pioneer 6 Earth line of sight at about 10 $R_\odot$. The tension in the magnetic field, however, may prevent the coronal plasma from escaping further into interplanetary space. Let us now examine the observation of Levy et al., that suggest this hypothesis.

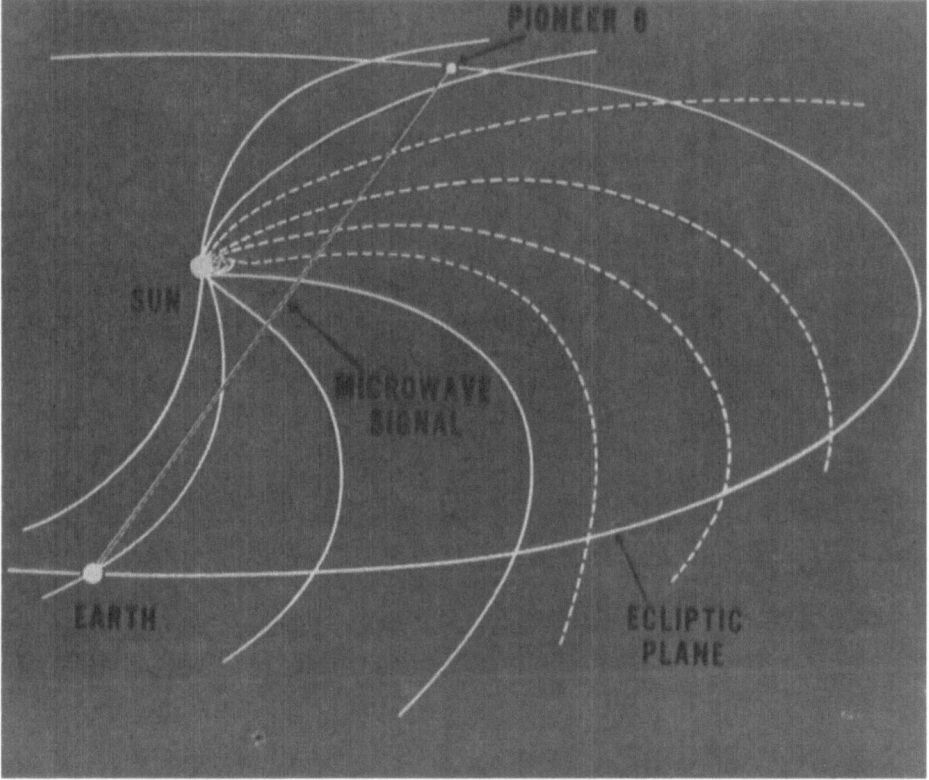

Fig. 4.   Sketch of coronal and interplanetary magnetic field and the geometry V involved in the Pioneer 6 Faraday rotation experiment. See text.

Figure 6 from Levy et al., shows the polarization angle of the radio signal in degrees versus universal time on November 4, 1968. The spacecraft was transmitting its signal at a polarization angle of 90°. The background Faraday rotation is due to the effects of the ionosphere and the corona. Approximate velocities of transport of the coronal material are calculated by dividing the distance from the Sun to the Pioneer 6-Earth line-of-sight by the time between the prior flare and the observed Faraday event. Velocities of about 200 km/s are obtained. Using this speed and the average 2 hr duration of the observed Faraday rotation effect one can find an approximate value for the dimension of the region moving past the line of sight. A value of two solar radii is thus obtained for the thickness of the region. It is interesting that J. P. Wild has reported at this meeting large-scale flare ejections using observations from the Culgoora radioheliograph.

Further insight into the dynamics of the phenomena associated with these coronal disturbances may be obtained by considering their effect on the interplanetary medium.

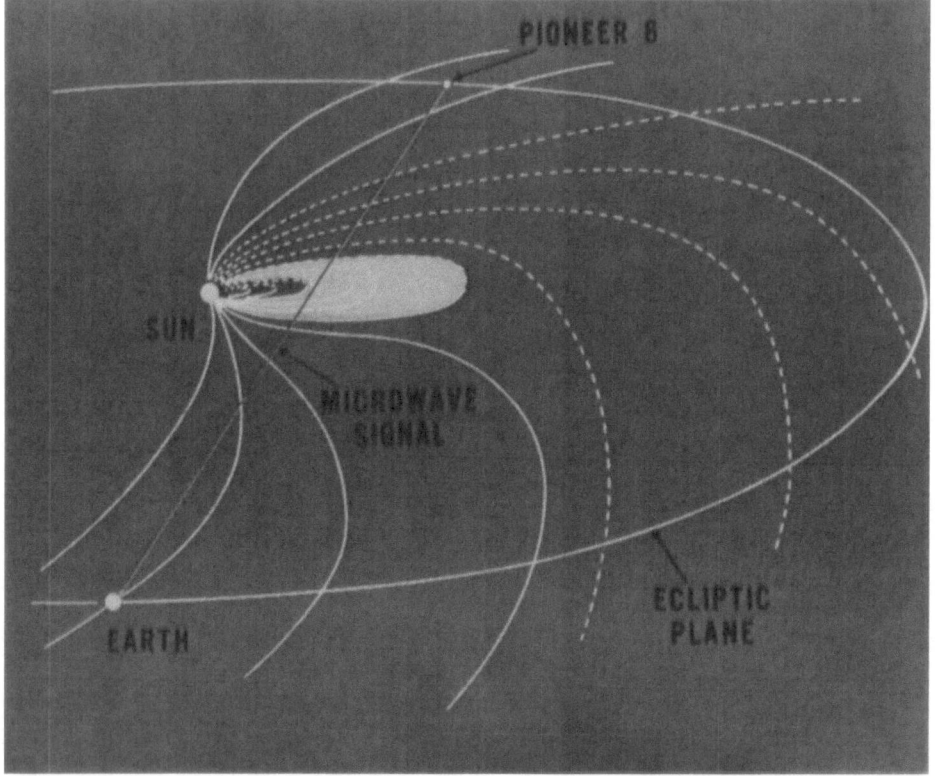

Fig. 5.   Sketch showing an interpretation by Schatten of the magnetic field for the events observed on November 4, November 8, and November 12, 1968. The light shaded looped field region indicates the area affecting the Faraday rotation measurements for the 3 events.

The sector pattern, it appears, relates more to the background field and is usually not affected by most solar flares occurring within active regions. This was shown by Wilcox and Severny (1969) in their solar field interplanetary field correlations. Another example to show this is that an average of 17 flares of importance 1 or greater occurred on the Sun with 30 degrees of central meridian per 27 day solar rotation during the last quarter of 1968. Whereas only a few disruptions in the sector pattern each 27 day period are noted for this time. Thus it is expected that events of the type observed with Pioneer 6 at about $10 R_\odot$ do not usually affect the interplanetary magnetic field pattern at 1 AU. In addition, the Faraday rotation returns to background level after each event has ended suggesting no permanent influence on the field beyond about $10\,R_\odot$.

Figure 7 shows magnetograms of the Sun obtained from the Mount Wilson Observatory 7 days prior to each of the three events. Thus due to the effects of solar rotation, central meridian in these magnetograms corresponds to the west limb at the three events. The three regions thought responsible for the events are labelled by their McMath plage number. Regions 9747 and 9754 were simple bipolar magnetic regions oriented with the positive flux (solid contours) following (eastward of) the negative flux. This orientation is of the proper sense to produce the Faraday rotation observed,

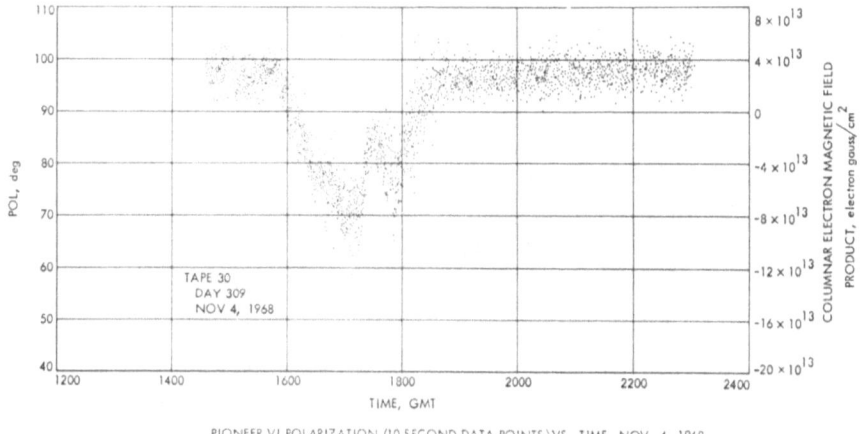

PIONEER VI POLARIZATION (10 SECOND DATA POINTS) VS. TIME, NOV. 4, 1968

Fig. 6. Observations of Levy *et al.* (1969) of the Faraday rotation of the microwave signal observed from Pioneer 6 on November 4, 1968. The signal was transmitted at a 90° angle from the ecliptic. The background effect is due to the steady-state corona and the Earth's ionosphere.

if the model previously outlined is correct. Region 9740 is a magnetic region with a more complex field configuration. The field arrangement does have positive flux following negative flux and hence could also produce the observed Faraday rotation in accordance with the model.

These regions contain approximately 3×10^{21} G cm^2 of magnetic flux. If this flux is spread through a meridional cross-section perpendicular to the flux loop one can estimate the field strength in the region where the Faraday rotation occurs. A 0.02 G field strength at $10\,R_\odot$ results. A proton and electron density of 2.0×10^4/cm^3 is assumed. Enhanced coronal densities near this value are found near solar maximum at these distances from the Sun. The coronal plasma at $10\,R_\odot$ would thus have an Alfvén velocity of 330 km/s. This is slightly greater than the 200 km/s speed of transport of the plasma calculated from delay times. Thus the magnetic field can transport the force necessary to retain the expanding coronal plasma.

The expected Faraday rotation can then be calculated. A degree of rotation results from about 4×10^{12} eG/cm^2. Allowing an azimuthal extent of 30°, the line integral

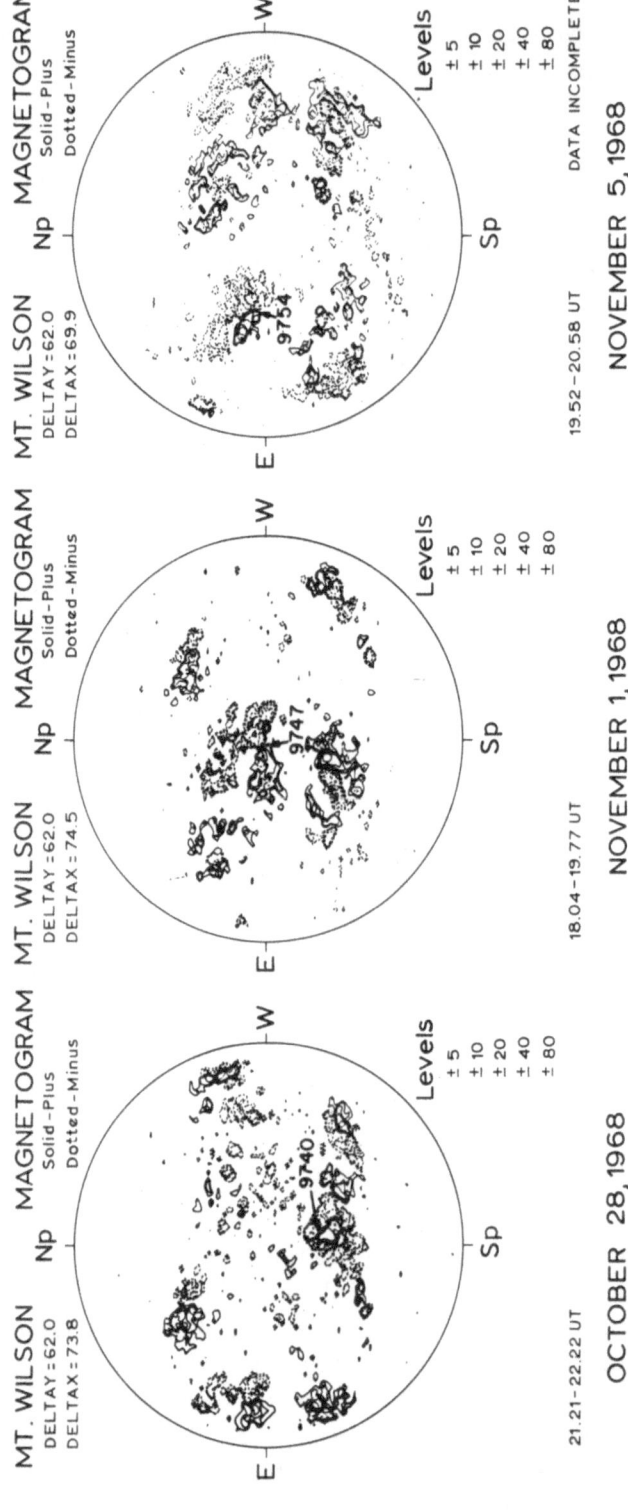

Fig. 7. Magnetograms of the Sun on October 28, November 1 and November 5, 1968. Solid lines indicate away from the Sun magnetic field and dashed lines indicate toward the Sun magnetic field. Contour levels are 5, 10, 20, 40 and 80 G.

of $N\mathbf{B} \cdot \mathbf{dl}$ is approximately 1.3×10^{14} G/cm². Hence approximately 33° of Faraday rotation should occur as the signal from Pioneer 6 passes through the solar corona. The observed Faraday rotation is from 30 to 40° in each of the three events. Thus the proposed model is consistent both in sign with the observed photospheric field and in magnitude with the observed Faraday rotation.

If an adiabatic expansion of the flare heated coronal gas is allowed, one can consider the variation in gas pressure relative to magnetic pressure as expansion occurs to gain a qualitative understanding of the dynamics. The adiabatic gas pressure varies as $P_G \propto V^{-\gamma} = V^{-5/3} \propto L^{-5}$, where V represents volume and L is a characteristic dimension of the system. The magnetic pressure varies as $Pm \propto B^2 \propto L^{-4}$ if the field varies as an inverse square with distance as it would if the plasma expands radially into a conical shape. Thus the gas pressure falls off more rapidly than the magnetic tension and at some distance the field can retain the plasma. If, however, coronal heating occurs thereby increasing the plasma pressure, or the hot coronal plasma expands past the Alfvén point which is about 20–30 $R_\odot$, the expansion can continue to 1 AU. This would be the case in large flares associated with terrestrial effects. In the current investigation it is suggested that although the coronal plasma is expanding with a supersonic velocity, the Alfvén velocity exceeds the expansion rate and escape can thus be prevented by the magnetic field.

After a balance between plasma pressure and magnetic pressure develops, the coronal gas cools by radiative losses and conduction, mostly in the lower corona, and the field configuration slowly returns to its initial state.

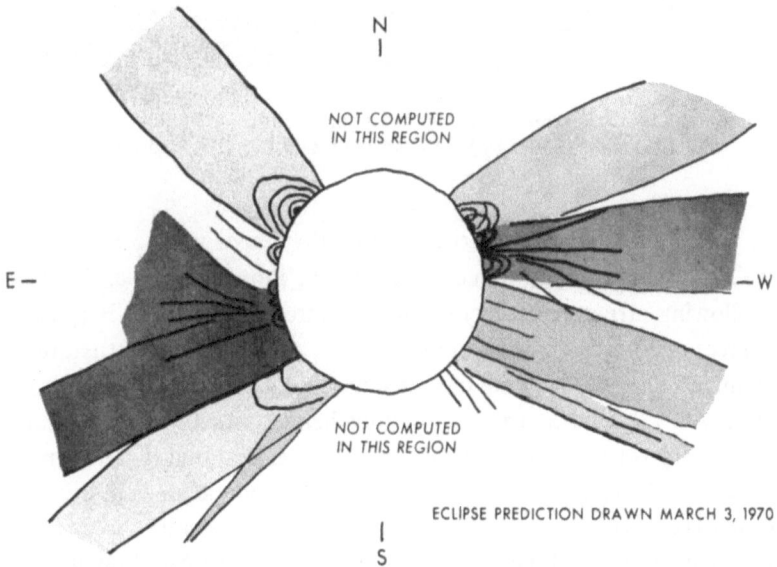

Fig. 8.   Prediction of the coronal magnetic field structure for the March 7, 1970 solar eclipse. Compare with Figures 9 and 10.

Thus it appears that even moderate solar activity may influence the coronal magnetic field structure. We shall now examine the recent March 7, 1970 solar eclipse (total over Mexico, the United States and Canada) for effects of solar activity upon coronal field structure.

Figure 8 shows a prediction of the structure of the corona at the March 7, 1970 eclipse and Figure 9 shows a photograph of the corona during the eclipse. The photo was obtained with a radial transmission filter by Sheldon Smith. Waldmeier (1970) as well as Smith and Schatten (1970) compared the prediction with observations of coronal

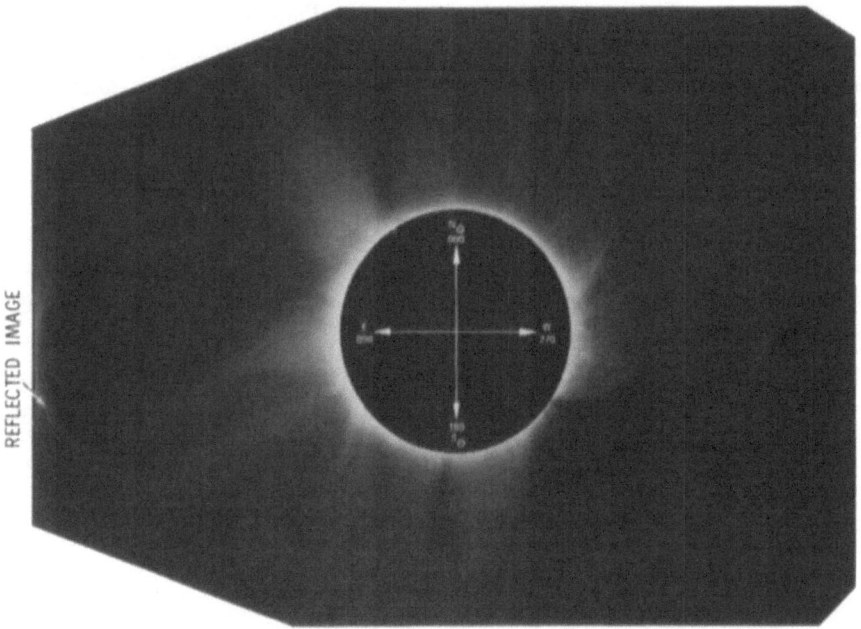

Fig. 9.   Photograph of the corona during the total solar eclipse of March, 7, 1970 using a radial transmission filter by Smith and Weinstein. See also Figure 10.

structure. In both findings comparisons show that there were certain features that agreed well and others that disagreed. Some of the more obvious areas of agreement are the following structures: the long helmet streamer in the NE (position angle 30–70, degrees counterclockwise from the north), short ray open structure in the SW (position angle 210–230), a system of nested arches located above the western equator (position angle 292); and a streamer without helmet structure located south of the eastern equator (position angle 100). Waldmeier notes that that region of most serious discrepancy is in the southwest quadrant. The photospheric fields in this region were not well observed prior to the eclipse due to inclement weather at Mt. Wilson. In addition new activity developed there just prior to the eclipse. Martin *et al.* (1970) report in *Nature* that on the SW limb an active region began developing on the preceding day around 1700 UT.

Smith and Schatten point out that the regions of disagreement are ususally associated with fan shaped structures. These are located near the equator on the east and west limbs (position angle 109–122, 268–283). These fan shaped structures could be a visible manifestation of the flare ejected plasma just discussed. The structure of the regions is that of a concave outward series of rays emanating from a small region near the limb. This is just the shape that would characterize flare ejected field and plasma.

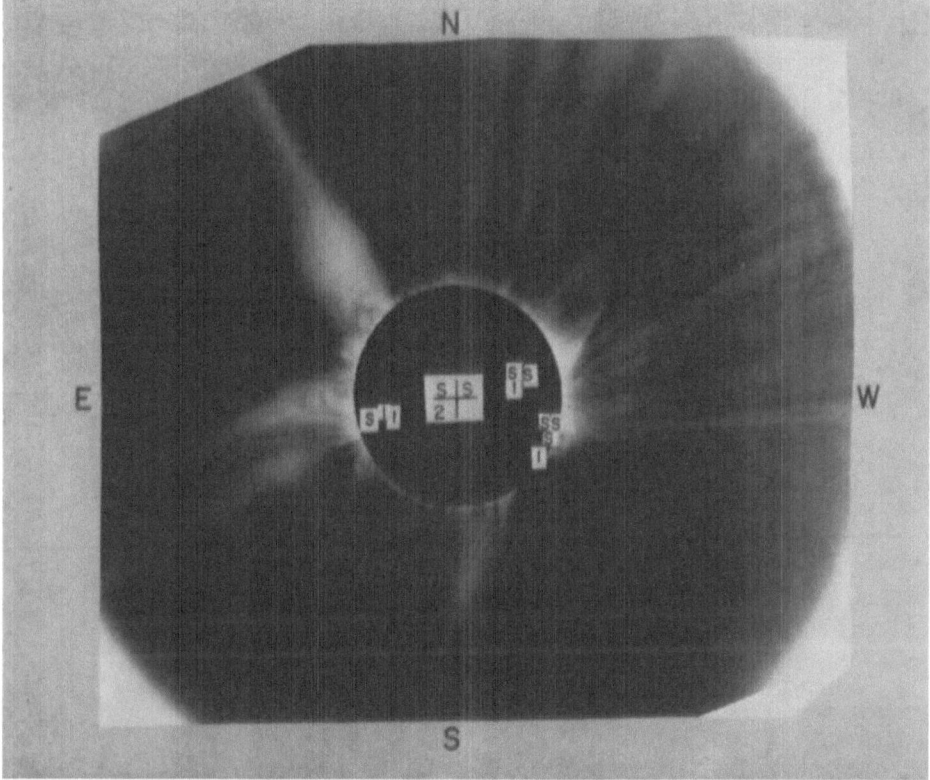

Fig. 10.   Photograph of the March 7, 1970 solar eclipse by Laffineur and Koutchmy. Superposed are the flares that occurred on the visible side of the Sun 12 hr prior to the solar eclipse. The letter 'S' indicates a subflare, a 1 indicates an importance 1 flare and a 2 indicates an importance 2 flare.

Figure 10 shows superposed upon a photo of the corona by Serge Koutchmy all the flares and subflares listed in the *ESSA Bulletin on Solar Geophysical Data* 12 hours prior to the solar eclipse. If the flare ejected plasma emanates radially, the eastern fan structure may be explained very well. The fan structure on the west limb also appears close to active regions recently flaring. In fact it may be the region Martin *et al.* observed.

I would now like to present evidence that these coronal magnetic bottles produced by small flares are not uncommon. First I shall note a few relevant observations. One feature of the active corona we have seen is the large number of apparently open-rayed structures. The interplanetary magnetic field near the ecliptic does not increase in magnitude from solar minimum to solar maximum as shown in Figure 11. This figure shows the magnetic field magnitude and direction as observed by Ness near solar minimum (top) and near solar maximum (bottom). Note that the field magnitude (the top graph) is near five gamma at both times (1 gamma equals 10^{-5}G). This indicates that roughly the same number of field lines are leaving the outer corona and extending to 1 AU at solar minimum and at solar maximum. The photospheric field varies considerably from solar minimum to solar maximum. There is a substantial increase in the photospheric field strength from minimum to maximum due to the presence of active regions. The predominantly open structure of the inner corona from 1 to 4 $R_\odot$ from the eclipse observations indicates most of these aditional field lines

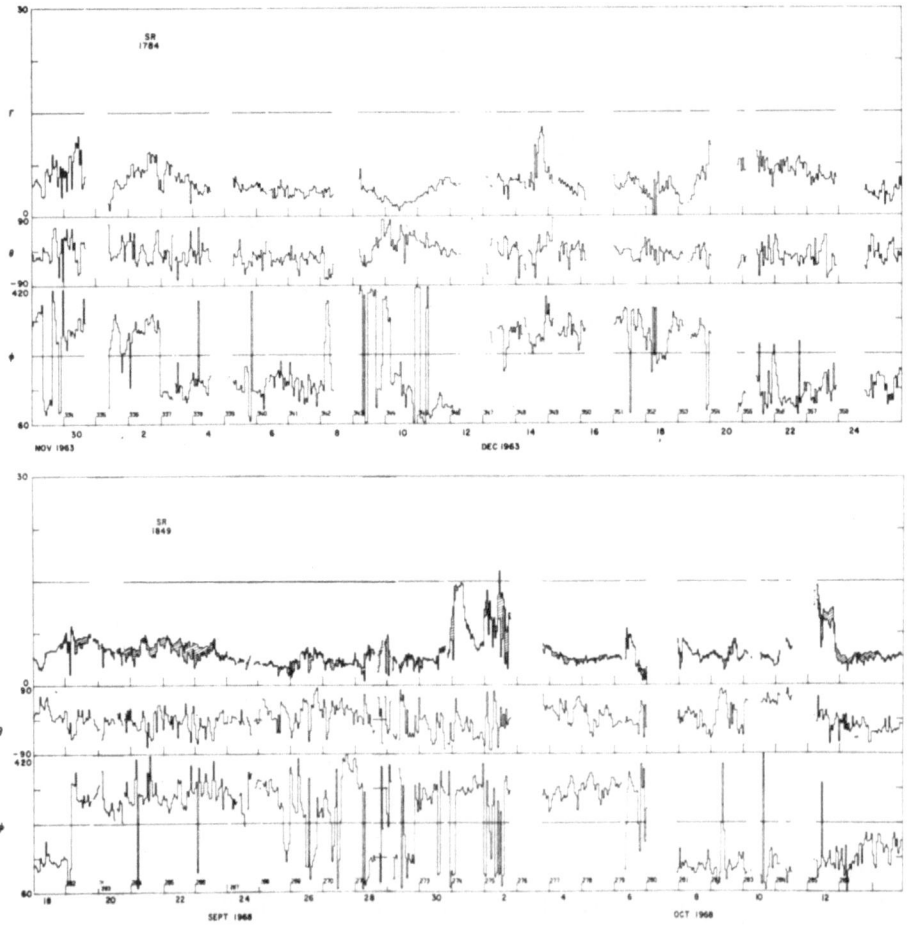

Fig. 11.   The observed interplanetary magnetic field at 1 AU for 1963 (top) and 1968 (bottom). The first graph is field magnitude in gammas (1 gamma equals 10^{-5} G). Beneath this are the solar ecliptic field direction angles. Note that field magnitude shows very little change with solar cycle.

leave the inner corona; the constant interplanetary field magnitude throughout the solar cycle indicates the additional field lines do not reach 1 AU and in fact do not reach the Alfvén point at 20–30 $R_\odot$ as they would then be convected to 1 AU by the solar wind where they are not seen. Thus much of the additional field at solar maximum may reside in magnetic bottles located at 10 to 20 $R_\odot$. Flares may be responsible for this new field configuration. A suggested field topology of the active solar corona is thus illustrated in Figure 12 in a logarithmic polar coordinate graph.

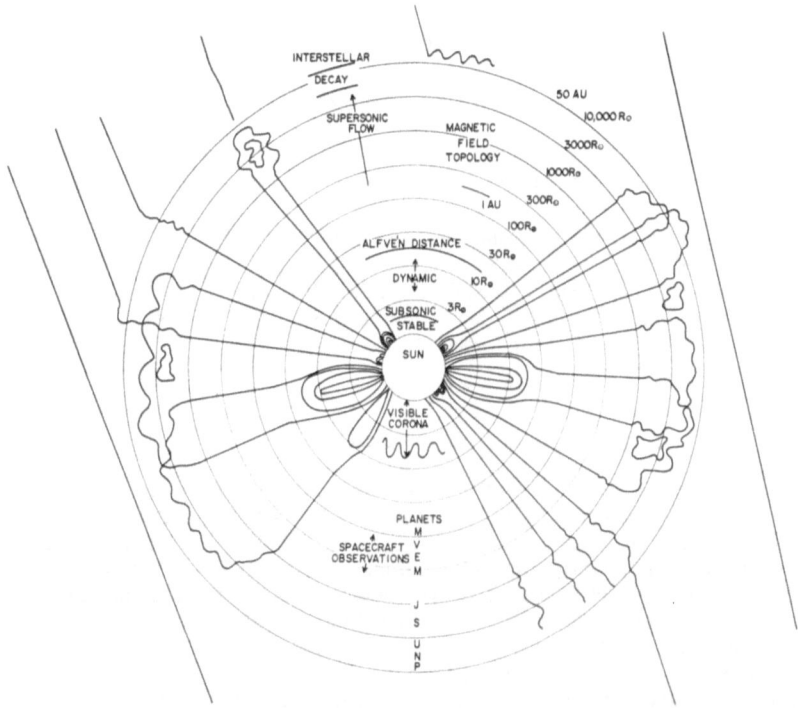

Fig. 12.   Magnitude field topology in the solar system. Stable loops form above magnetic regions beneath 2 $R_\odot$. Flare ejected loops exist below 20 $R_\odot$. The field is then convected out to the heliospheric boundary at about 50 AU. The sector structure and the spiral structure are not shown for clarity.

Mostly we are concerned with the central region. Close to the Sun, the corona is stable and inactive coronal magnetic loops may form in accordance with the magnetic field calculations. These loops rotate rigidly with the Sun. Larger field loops are ejected by small flares. The inner portion of these loops is in the visible corona and appears as radial rays emanating from a common location. This region is labeled Dynamic as these bottles expand when flare energy is released and contract when cooling. The bottle may extend out to anywhere between 5 and 20 or 30 $R_\odot$. This outer portion of the bottle in general would not be observed. Beyond 20 $R_\odot$ the field lines are open and form Archimedes spirals. These, as well as the sector structure, are not shown for clarity.

Occasional field loops will emanate from the Sun and exist in this region but they will quickly be convected out by the supersonic solar wind. At about 50 AU the field lines presumably merge and the local instellar field predominates.

Acknowledgements

I would like to acknowledge critical discussions and other forms of support which led to the formulation of the ideas in this paper by John Wilcox, Robert Howard, Norman Ness and Gordon Newkirk.

References

Altschuler, M. D. and Newkirk, G., Jr.: 1969, *Solar Phys.* **9**, 131.
Gold, T.: 1959, *J. Geophys. Res.* **64**, 1665.
Laffineur, M., Burnichon, M.-L., and Koutchmy, S.: 1969, *Nature* **222**, 461.
Levy, G. S., Sato, T., Seidel, B. L., Stelzried, C. T., Ohlson, J. E., and Rusch, W. V. T.: 1969, *Science* **166**, 596.
Martin, D. C., Smith, S. F., and Chapman, G. A.: 1970 *Nature* **226**, 1138.
Schatten, K. H.: 1970, *Solar Phys.* **11**, 236.
Schatten, K. H., Wilcox, J. M., and Ness, N. F.: 1969, *Solar Phys.* **6**, 442.
Severny, A., Wilcox, J. M., Schener, P. H., and Colburn, D. S.: 1971, *Solar Phys.* (submitted).
Smith, S. M. and Schatten, K. H.: 1970, *Nature* **226**, 1130.
Waldmeier, M.: 1970, *Nature* **226**, 1131.

Discussion

Meyer: One is worried a little bit about the lateral equilibrium in your model of a magnetic bottle. When such a bottle is pulled out radially with a velocity of 200 km/s, but on the other hand conserves an inside Alfvénic velocity of 300 km/s, the configuration would have time enough to set itself into equilibrium with the weak lateral pressures of the normal solar wind magnetic field, which are an order of magnitude smaller than the field strengths you infer for your bottle. Thus one should expect a considerable lateral expansion together with the radial motion. Consequently, the fields inside the bottle should in effect be much smaller than would be inferred from purely radial motion, in fact of the order of the field strength in the surrounding ordinary solar wind.

Schatten: Your argument is interesting. However, one may allow the tension in the magnetic bottle to provide the force that prevents lateral expansion. In carrying out the calculations, I used a 30 degree lateral spread. This would be correct to a factor of 3 in any case.

Maltby: Is there any indication of Faraday depolarization in the radio observation. If so an interpretation with magnetic field perturbation may be more feasible.

Schatten: Gerry Levy *et al.* measured the depolarization Doppler shift, angle polarization, and other quantities that relate to the signal. It is the Faraday rotation of the angle of polarization that relates directly to the field component along the line of sight. Weakening of the signal and other processes other than Faraday rotation do not change the direction of polarization. They tracked the signal from 16 to 4 $R_\odot$ at which time the noise from the Sun obscured the signal. The angle of polarization was adequately tracked *throughout* the three events so as to eliminate the possibility of loss of signal as the cause of the change.

OPTICAL AND RADIO OBSERVATIONS OF
LARGE SCALE MAGNETIC FIELDS ON THE SUN

G. DAIGNE, M. F. LANTOS-JARRY, and M. PICK

Meudon-Nançay Observatory, France

Abstract. It is possible to deduce information concerning large scale coronal magnetic field patterns from the knowledge of the location of radioburst sources.

As the method concerns active centers responsible for corpuscular emission, the knowledge of these structures may have important implications in the understanding of corpuscular propagation in the corona and in the interplanetary medium.

The object of this paper is to show it is possible to deduce information concerning the large scale coronal magnetic field patterns indirectly from knowledge of the location of radioburst sources.

It is now well known that noise-storms, type I bursts and continuum are associated with active centers and the location of emission sources is closely dependent upon the magnetic structure; thus the emission regions of type I bursts and continua may be simple or double or complex (Clavelier, 1968; Daigne, 1968). For several cases the type I bursts source has been found to be bipolar (Kai, 1970). Kai has advanced a qualitative model in which electrons with energies of a few keV are trapped in the magnetic field of a pair of sunspots of inverse polarity.

Recently, an active region giving rise to a noise storm center for several successive days has been studied by Lantos-Jarry (1970). In this case a global and systematic displacement of the mean distribution of the continuum and type I burst sources was observed. This displacement was explained in terms of the coronal magnetic field structure. This structure is illustrated in Figure 1. It represents an arch connecting two regions of opposite polarity, widely separated at the surface of the solar disk. In the present case, one of these regions is the active center giving rise to the noise storm and the other is a large and stable region of opposite polarity, characterized by a faint thermal radio enhancement observed in the corona at centimetric wavelengths.* Very

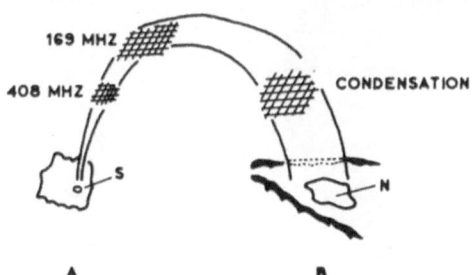

Fig. 1. Coronal magnetic field structure (from Lantos-Jarry, 1970).

* The metric thermal emission is observed inside the streamer associated with the presence of filaments.

Howard (ed.), Solar Magnetic Fields, 609–615. All Rights Reserved.
Copyright © 1971 by the IAU.

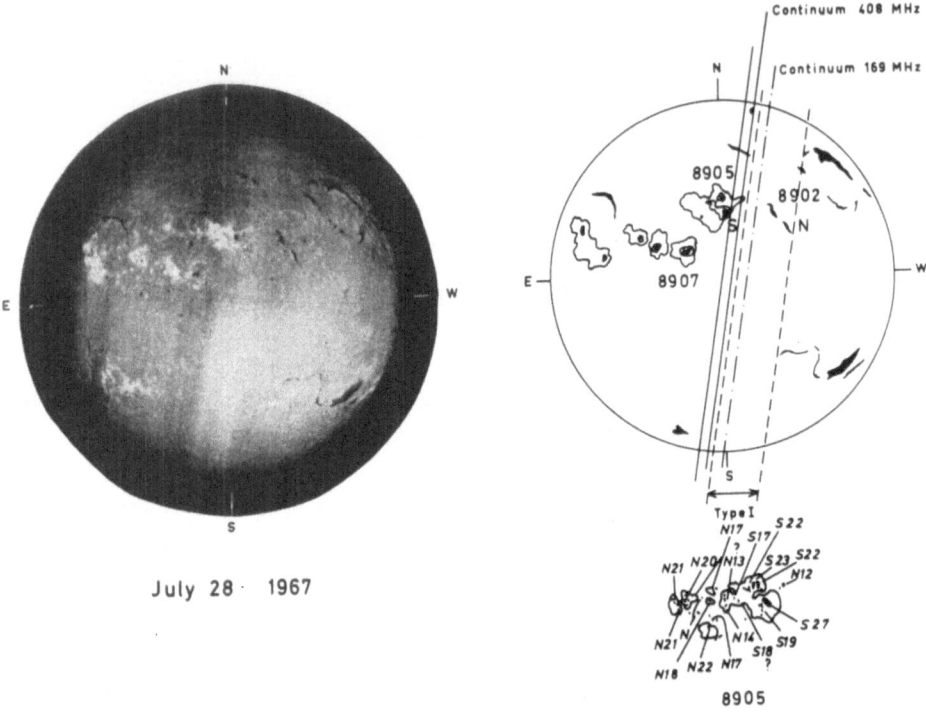

Fig. 2.   July 28, 1967, – Hα spectroheliogram (from Meudon Observatory); – schematic representation of the sunspots observed on the spectroheliogram. Mean positions of the continuum storms observed at 169 MHz (–·–·–) and 408 MHz (————); the two broken lines indicate the range inside which the type I bursts are located; – sunspot polarities from Crimea Observatory.

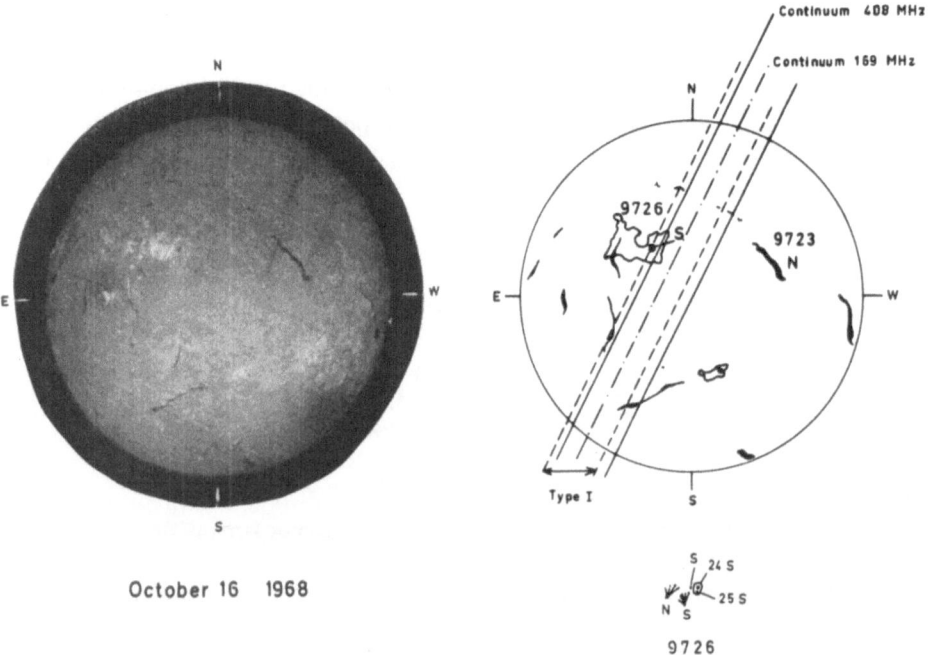

Fig. 3.   October 16, 1968, same caption as Figure 2.

likely, this arch is similar to the low or high magnetic arcades as defined by Newkirk and Altschuler (1970) when coronal magnetic fields are calculated.

In this paper, this conclusion will be developed further with examination of several similar sets of observations. The radio-burst positions are obtained with an accuracy of about 1' of arc from the Nançay interferometer at 408 MHz and from the Nançay East-West radioheliograph at 169 MHz.

In Figures 2, 3, 4, and 5 are summarized observations of active centers for which a displacement of the associated noise storm source has been observed during several successive days. On each figure, the following observations have been shown:

– The Hα spectroheliogram for the day when the active center concerned is near the central meridian.

– A schematic representation of the sunspots observed on K_1 spectroheliograms, the quiescent filaments observed in Hα. On the same diagram are indicated the mean positions of the continuum storms observed at 169 MHz and 408 MHz. Two broken lines indicate the range inside which type I bursts are located.

– The sunspot polarities of the active centers published by the Crimean Observatory.

From Figures 2 through 5, we note the following common features.

(1) The observed mean position of the continuum and type I bursts measured at 169 MHz is systematically displaced relative to the position of the optical active region by about 10° to 20° in the East-West direction, the sense and magnitude depending on the storm center observed. In each case, the type I burst positions are widely spread.

(2) This displacement is always oriented toward a region of polarity opposite to that of the main spots of the eruptive structure giving rise to the noise-storm.

(3) This region of opposite polarity is stable and most frequently delimited by a large quiescent filament. As is well known, filaments represent the limit of two regions of opposite polarity. This region is an old calcium plage persisting on several solar rotations and associated with a thermal enhancement observed on centimetric wavelengths. A thermal metric emission is observed inside the streamers associated with the presence of filaments.

(4) A striking feature is the absence of any other plage between the active center and the region which is at the other foot of the inferred connecting magnetic arch. This property is visible on the Hα spectroheliograms.

We conclude that there may exist large scale coronal magnetic structure between two regions of opposite polarity when they are at the limit of a large region free of any magnetic field. On the other hand, the case of several neighbouring plages will lead to local magnetic field structures with magnetic field lines confining themselves to closed regions.

As an example, some interesting observations from July 1967 may be shown. The Hα spectroheliogram reproduced on Figure 2 shows that there existed for this period several active centers on the Sun: Center 8905, located at 28° N, passed through central meridian passage on July 28 and gave as we have already discussed a noise center systematically shifted westwards. A large number of flares are located in the leading

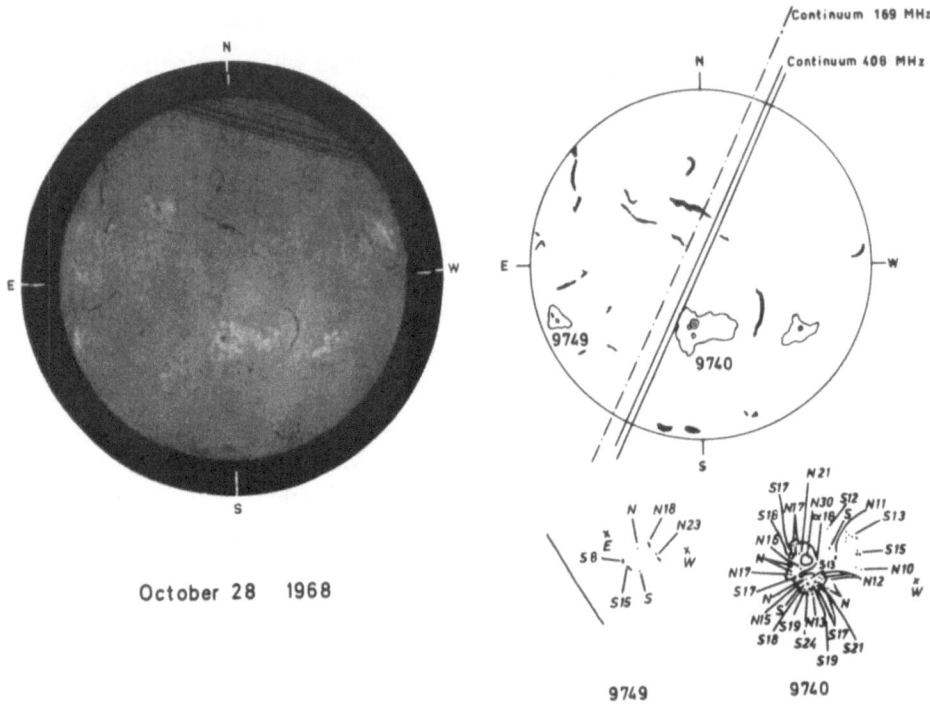

Fig. 4.   October 28, 1968, same caption as Figure 2.

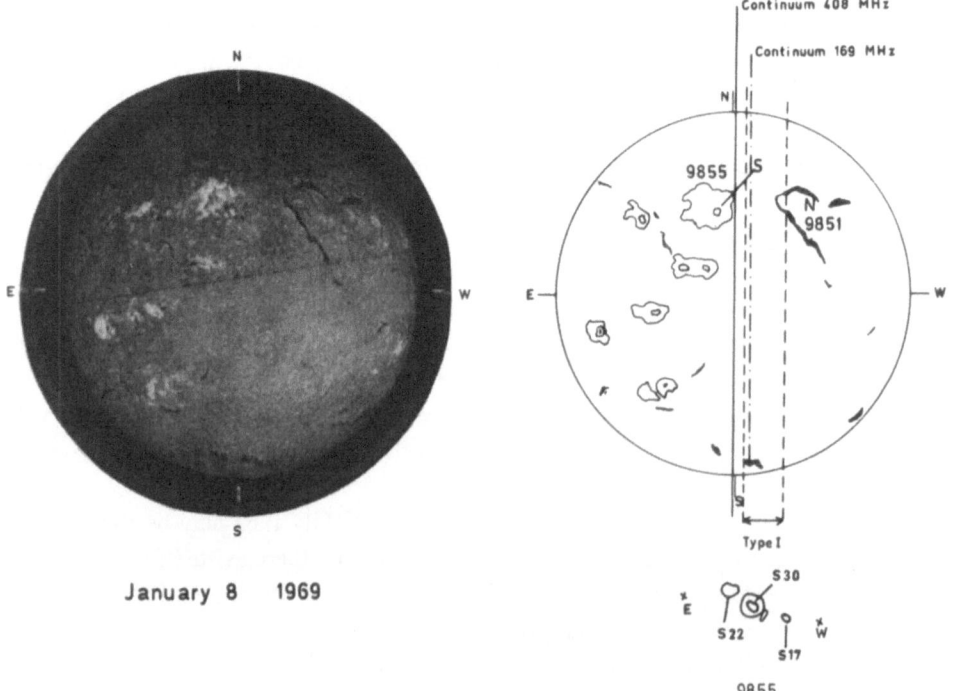

Fig. 5.   January 8, 1969, same caption as Figure 2.

part of the active center, whose main polarity is South. A magnetic arch probably exists between Center 8905 and the old calcium plage 8902, barely visible on the Hα spectroheliogram. Center 8907, whose latitude is 15° North also produced a noise storm observed at 408 MHz and 169 MHz on July 29 and 30. The location of the radio sources is shown in Figure 6. In this case, no displacement in the radio position is observed: note that flares are located in the trailing part of the active center. In fact this trailing part is formed by 2 or 3 regions: the noise storm center is located above this part. Like Center 8905, the leading spot is of South polarity but probably does not play a role in the formation of a large scale magnetic field structure, in which electrons whose acceleration is related to eruptive structure would be able to propagate.

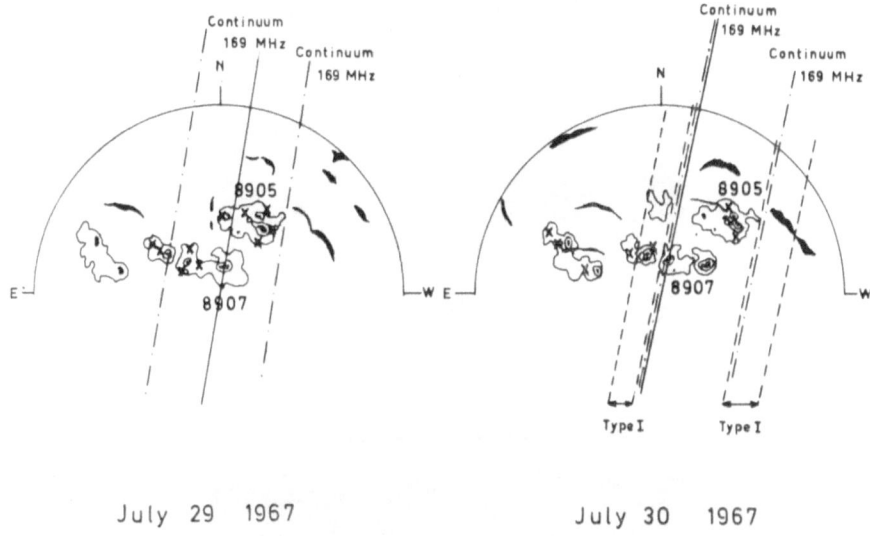

July 29 1967   July 30   1967

Fig. 6.   July 29, 30 1967, same caption as Figure 2. Crosses indicate the position of the flares.

According to Lin (1970) solar X-ray bursts of energies ≥ 20 keV were associated with both Regions 8905 and 8907. But only for Region 8905 were electron events observed near the Earth. Lin concludes that the coronal structure of Region 8907 may differ from Region 8905 in that electrons are trapped at a lower height (below 1 $R_\odot$) while 8905 coronal structure provides our open path from the acceleration region to the interplanetary field lines.

The same kind of magnetic connection as we have just described may exist between two active centers. In this case, the two feet of the magnetic loop would be situated at sunspots: this is the case for 1968 October 28, for which the arch connects sunspots of opposite polarity widely separated at the solar surface. It is now known that we may observe correlated activity above the two active centers. Figure 7 gives an example of this kind of correlation observed on 1969 July 8 with the Nançay radioheliograph. The two active centers were separated by 70° of longitude.

In summary, radio observations give the possibility of indirectly detecting large scale coronal magnetic field patterns. This is consistent with the results of Altschuler and Newkirk (1969–1970). The same kind of observations also gives an estimation of the stability of these structures which may last for up to a few days. Because this method concerns active centers responsible for corpuscular emission, knowledge of these structures may have important implications in the understanding of problems

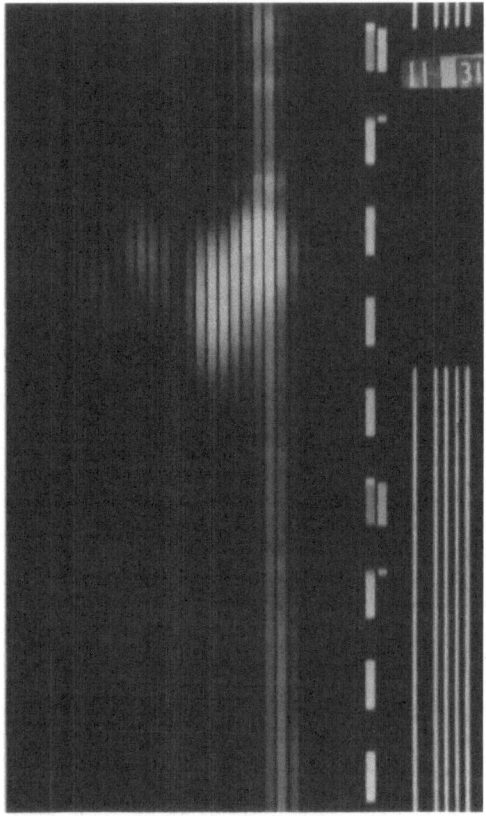

Fig. 7.   July 9, 1969, correlated activity between two active centers. The distance between two successive strips is 2′ of arc in the East-West direction. The time scale is indicated by the vertical lines. One single line corresponds to one second.

relative to corpuscular propagation in the corona and in the interplanetary medium: indeed sometimes both electrons and protons start to reach the Earth very quickly even when they are emitted from an Eastern flare. Transverse diffusion is unable to account for these transit times. It is possible to invoke magnetic guidance giving the possibility of the corpuscules reaching the region 'connected' with the Earth. Further-more, Martres et al. (1970) have recently shown that large coronal condensations, such as those just associated with the foot of magnetic archs, are the main sources from which the interplanetary magnetic field would emanate. In the same plage region, there are some field lines connecting with the active center and some passing into interplanetary space, and small scale diffusion of energetic particles may be expected. Particles would propagate into the solar wind more easily from these regions. This rough model finds some confirmation in the recent study by Mc Donald and Desai (1970): they show that particles escape preferentially from a region preceding the active center associated with their production. The maximum of corpuscular emission occurs when the active center is still east of the central meridian.

We shall not discuss the implications of these observations for the emission mechanism.

Acknowledgements

We thank G. Servajean and J. Geoffroy for their help and the members of the Radioheliograph Group for the observations.

References

Altschuler, M. D. and Newkirk, G., Jr.: 1969, *Solar Phys.* **9**, 131.
Clavelier, B.: 1968, in K. O. Kiepenheuer (ed.), 'Structure and Development of Solar Active Regions', *IAU Symp.* **35**, 556.
Daigne, G.: 1968, *Nature* **220**, No. 5167.
Kai, K.: 1970, *Solar Phys.* **11**, 456.
Lantos-Jarry, M. F.: 1970, *Solar Phys.* **15**, 40.
Lin, R. P.: 1970, University of California, Berkeley.
Martres, M.-J., Parkes, G. and Pick., M.: 1970, *Solar Phys.* **15**, 48.
McDonald, F. B. and Desai, U. D.: 1971, *Recurrent Solar Cosmic Ray Events and Solar M. Regions*, in press.
Newkirk, G. N., Jr. and Altschuler, M. D.: 1970, *Solar Phys.* **13**, 131.
Wild, J. P.: 1968, *Proc. ASA* **1**, 137.

80 MHz RADIOHELIOGRAPH EVIDENCE ON

MOVING TYPE IV BURSTS AND CORONAL MAGNETIC FIELDS

S. F. SMERD

Division of Radiophysics, C.S.I.R.O., Sydney, Australia

and

G. A. DULK

Division of Radiophysics, C.S.I.R.O. and Department of Astro-Geophysics, University of Colorado, Boulder, Colo., U.S.A.

Abstract. The characteristics of 12 moving type IV bursts observed with the 80 MHz radioheliograph at the Culgoora Observatory between February 1968 and April 1970 are summarized.

Three classes of moving sources can be recognized; they are described as: (1) Expanding arch; (2) Advancing front; (3) Isolated source.

The first class has been identified (Wild, 1969) with the expansion of a magnetic arch or loop; the second class is here identified with an advancing MHD disturbance which may accelerate the radiating electrons *in situ* when moving at greater than Alfvén speed; the third with solar ejecta in the form of magnetized plasma clouds, or plasmoids. In all cases the radiation mechanism is probably synchrotron radiation from mildly relativistic electrons; energies in the range ~ 0.1 to ~ 1 MeV could account for the observed strong circular polarizations.

With an expanding magnetic arch, source and magnetic-field movement are inseparable; the field remains a closed loop throughout the event. The MHD front probably moves largely along and the plasmoids between the open magnetic-field lines of unipolar regions or helmet structures. In the latter case it is the internal magnetic field – possibly toroidal – of the moving plasmoid that determines the polarization of the synchrotron radiation. A preliminary comparison of moving type IV sources with Newkirk-Altschuler maps of coronal magnetic fields shows suitably located closed loops for 2 events identified as expanding magnetic arches and unipolar open field lines along the path of a moving source identified as a plasmoid.

1. Introduction

Type IV bursts were first recognized and classified as a separate type by Boischot (1957), largely on the basis of the long-lasting, smooth character of their radiation and the substantial movement of the burst source as observed with the 169 MHz interferometer at Nançay. Subsequently, the type IV label was applied to all long-lasting, fairly smooth and wide-band continua at all wavelengths, even though most such radiation came from stationary sources. To distinguish the original type IV burst from other metre-wave bursts with similar spectral characteristics, Weiss (1963) introduced the terms 'moving' and 'stationary' type IV bursts. Characteristically, the moving type IV burst occurs at metre and longer wavelengths where it is probably the rarest type of burst. Wild *et al.* (1959) using swept-frequency interferometry showed that the radiation at all frequencies in the range 40–70 MHz came from the same source region, as would be expected on Boischot and Denisse's (1957) hypothesis that synchrotron radiation is emitted by a moving source.

In Section 2 of this paper, we summarize the 80 MHz radioheliograph observations of 12 moving type IV bursts observed in the period February 1968 to April 1970.

Howard (ed.), Solar Magnetic Fields, 616–641. All Rights Reserved.

Three bursts, which we consider to be representative of three phenomenologically different classes of moving type IV bursts, are described in greater detail. In Section 3, we discuss the physical implications of the observed properties of the bursts; this leads to a physical classification of moving type IV bursts. In Section 4 we relate the observed and derived characteristics of the bursts to the magnetic-field structure in the corona.

2. The Observations

2.1. GENERAL

Table I summarizes the characteristics of 12 moving type IV bursts and their association with other radio and optical activity on the Sun. Some of the parameters in Table I result directly from the radioheliograph observations; others are derived, using an additional assumption or supplementary data. For six events the derived distances and speeds are given by bracketed values in the Table. Of these, four events (1968 Feb. 25, 1969 Mar. 1–2, 1969 Oct. 10, 1970 Mar. 21) are 'limb events' in which the associated flare or eruptive prominence was located close to the limb and hence the observed distances were probably within 10% of the actual distances. Two events (1968 Nov. 22, 1970 Apr. 29) were associated with flares at intermediate longitudes and no supple-

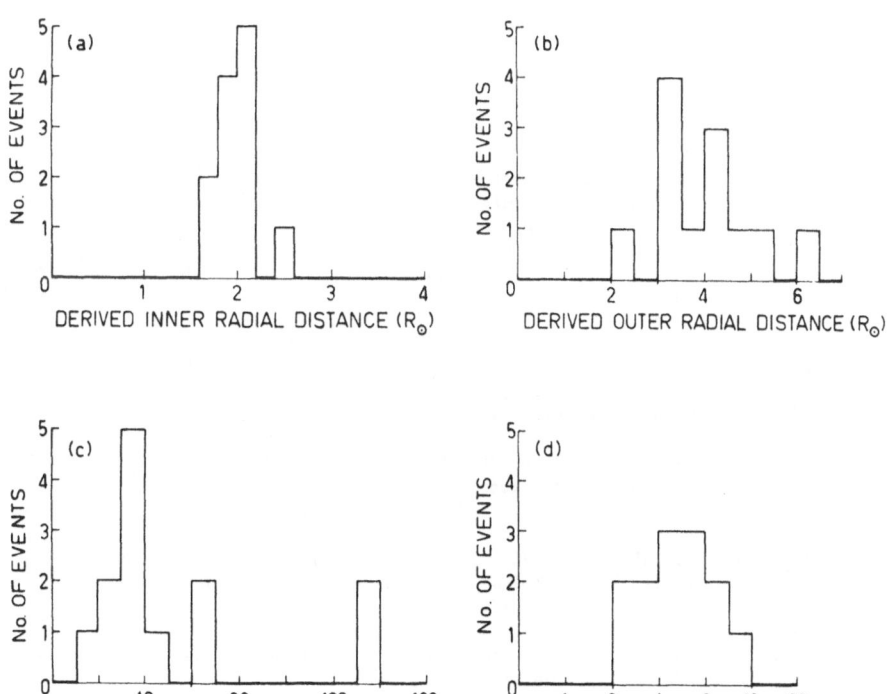

Fig. 1.   Histograms of the derived inner (a) and outer (b) radial distances, duration (c), and derived source speed (d) for the 12 moving type IV bursts observed with the 80 MHz radioheliograph between February 1968 and April 1970.

Characteristics of the twelve moving type-IV bursts observed with

Date	Time (UT) of Start h m	Time (UT) of Finish h m	Associated events Radio	Associated events Optical	Distances form centre ($R_\odot$) Projected First	Distances form centre ($R_\odot$) Projected Last	Distances form centre ($R_\odot$) Derived First	Distances form centre ($R_\odot$) Derived Last	Velocity Projected Speed (km/s)	Velocity Projected PA (°)	Velocity Derived Speed (km/s)	Velocity Derived Angle from line of sight (°)
1968 Feb. 25	04 05	04 30	III, II (+II?) μ-burst (C)	1B, 03ʰ48ᵐ S30, W17; EPL	1.6	2.0	(1.6)	(2.0)	400	120	(400)	(90)
1968 Sep. 4	00 59	01 22	III, 2 II's, IVS μ-burst (C)	1N, 00ʰ28ᵐ N13, W14; DF	0.6	1.4	2.5	4.0	500	90	760	41
1968 Oct. 23–24	00 06	00 20	III, 2 II's IVS? μ-burst (C)	2B, 23ʰ56ᵐ S12, E59; ADF	1.5	2.9	2.0	3.6	1200	50–140 Max.displ. 105	(1400)	(57)
1968 Oct. 26	01 20	02 20	III, 2 II's, IVS μ-burst (C)	1N, 01ʰ09ᵐ S20, E32; ADF, DSD	0.9	2.1	2.0	4.3	250	110	480 (Identified with speed of second type II)	30
1968 Nov. 22	01 00	01 45	IVS μ-burst (C)	1N, 00ʰ56ᵐ S16, E39; APR	1.2	2.0	(1.9)	(3.2)	210	115	(334)	(39)
1969 Mar. 1–2	< 23 16	01 31	III, II μ-burst (C)	1N, 21ʰ04ᵐ N08, W89; EPL 21ʰ36ᵐ	2.1	5.25	(2.1)	(5.3)	270	− 90	(270)	(90)
1969 Mar. 21	01 52	02 28	III, 3 II's, IVS μ-burst (C)	2N, 01ʰ42ᵐ N19, E16; ADF	1.1	2.4	1.6	4.8	470	+ 10 to − 25	980 (Ident. with speed of type II bursts)	30
1969 Aug. 29–30	23 30	00 00	III, II, 2 IVS's μ-burst (S)	1B, 23ʰ04ᵐ N04, E38; ADF	1.1	1.8	2.0	3.2	370	140	480	50
1969 Oct. 10	03 48	04 22	None (Initiating event probably behind Sun, W-hemisphere)	Unknown	1.8	3.4	(1.8)	(3.4)	650	− 135	(650)	(90)
1969 Dec. 17	00 50	01 20	III, IVS μ-burst (S-C)	2B, 00ʰ31ᵐ N09, E36;	1.2	1.4	1.7	2.0	500	80→40	700 ≈	45
1969 Dec. 17	01 16	02 20	III, IVS μ-burst (S-C)		1.2	1.7	1.9	3.3?	150	70 → 50	260	35
1970 Mar. 21	01 12	03 30	III, 2 II's, IVS μ-burst (S-C)	2N, 00ʰ23ᵐ N20, E69; EPL 00ʰ30ᵐ	1.9	6.0	(1.9)	(6.0)	340 (For most distant fragment)	40	(290)	(90)
1970 Apr. 29	01 00	01 30	III, IVS? μ-burst (S-C)	2N, 00ʰ47ᵐ S10, E46; ADF	1.4	3.0	(2.1)	(4.2)	580	85	(820)	(46)

In the 'Associated Radio Events' μ-burst is used to denote burst activity (S-simple, C-complex) reported at cm and dm wavelengths. In the 'Associated Optical Events' filament and prominence activity is indicated by the abbreviations of the ESSA 'Preliminary Reports – SDFR', namely: APR – active prominence; EPL – eruptive prominence on the limb; ADF – active dark filament; DSD – dark surge on disk; DF is used to denote the disappearance of a filament.

I

the 80-MHz radioheliograph, February 1968–April 1970

Source Structure		Descriptive Class	Remarks	Reference
Spatial	Polarization			
Single, elliptical	Unknown	Isolated source	Most movement late in life (after 04^{h}18^m); associated with triggered prominence eruption	Wild *et al.* (1968)
Single, elliptical	40→ > 85 %, LH	Isolated source	Same speed and direction as second type II	Kai (1969a, b)
110°-arch	10 – 20 % LH	Advancing front	Same speed and general direction as second type II	Kai (1970)
Jet	50→ ≲ 100 % LH	Expanding arch (edge on?)	Accompanied for half-hour by weak, RH polarized source moving from (0.8 $R_\odot$, 140°) to (1.1 $R_\odot$, 160°)	Stewart *et al.* (1970); Stewart (1971)
Three sources	Apex: unpolarized feet: opp.polarized	Expanding arch	Attributed to expanding magnetic arch	Wild (1969)
Single (A)→ Double (A, B)→ Single (A)	A:0→ > 70 % RH B:0→ 60 % LH	Isolated source	In direction of eruptive prominence, delayed ≈ 1^h	Riddle (1970)
Three sources (M_1M_2 M_3) within arch	M_1:unpolarized, M_2:20→70 %RH, M_3:≈ 15 % LH	Expanding arch	II and IV attributed to same shock wave (expanding magnetic arch?)	Stewart and Sheridan (1970)
Inverted U	0→80 % bi-polar	Expanding arch	Bi-polar ∩-shaped apex accompanied by two stationary feet on disk	Dulk (1970, 1971)
Single (M_1)→ Double (M_1 M_2)→ Triple ($M_1M_2M_3$)	M_1:0→ ≈ 50 % LH M_2:0→ ≈ 30 % LH M_3:40→ > 80 % LH	Isolated Source	Initial movement southward; sources diverge while fading rapidly	Smerd (1971)
Single	0→90 % RH	Isolated source	Peripheral movement; remained at end point for ≈ 1^h	Dulk *et al.* (1971)
Single	> 90 % LH	Isolated source	Slightly curved path, ≈ 45° to radial. Both sources expand while fading	
Single→ Four fragments	30→ > 90 % RH	Isolated source	Centroid velocity 290 km/s; from 02^{h}30^m source expands at 90 km/s and fragments	Sheridan (1970)
Single	< 10→ > 70 % LH	Isolated source	Intermittent III bursts from 10 m before to end of event	Dulk and Altschuler (1971)

The 'Projected Direction' is given by the Position Angle measured eastward (+) and westward (−) from the north of the projected rotational axis of the Sun. In the 'Derived Distances from Centre' and in the 'Derived Speed and Direction' the bracketed values were derived on the assumption of radial propagation; a 'Derived Direction' of (90°) indicates that the event is considered a limb event.

mentary evidence on their directions of motion was available; in these cases, the motion was assumed to be radial.

With due allowance for the uncertainty in the initial and final heights of a few events,
Table I and Figure 1a show that characteristically the sources were first seen at a radial
distance of about 2 $R_\odot$. In a coronal model with twice the densities of Newkirk's (1961)
active region (hereafter designated our standard density model), this initial height
corresponds to the 44 MHz plasma level, 0.4 $R_\odot$ above the 80 MHz plasma level.
Similarly, Table I and Figure 1b show that the final radial distance averages 3.9 $R_\odot$,
which is approximately the 20 MHz plasma level.

The burst durations, shown in Figure 1c, average 51 min. The sample suggests, but
is too small to establish, that the distribution is bi-model. The two 'long-duration'
events are also the events with the greatest outer radial distances.

Figure 1d shows that the average derived speed is 602 km/s. The two 'long-duration
and distant' limb events were relatively slow moving, at about half the average speed
(270 and 290 km/s respectively). Figure 2 presents the same data as a comparison of
burst duration and radial distance moved.

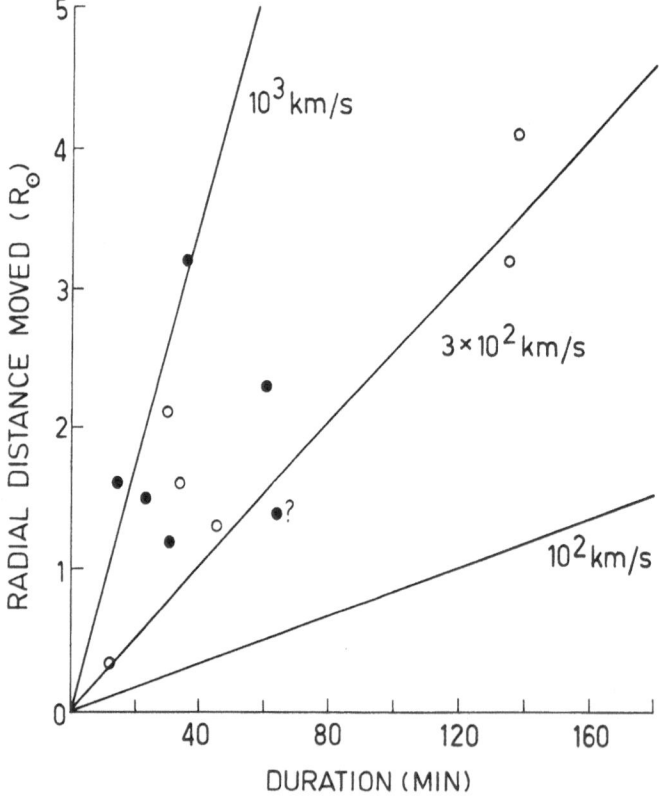

Fig. 2.   The radial distance moved as a function of the duration for 12 moving type IV bursts. The
filled circles refer to events for which the direction of motion was derived from supplementary data,
the open circles to events where the motion was assumed to be radial. Three lines of constant speed
are shown for comparison.

At the time of the first event (Feb. 1968), polarization was not measured by the radioheliograph; of the other 11 events, 9 attained circular polarization greater than 70%. In all these events the polarization increased with time, in some cases from initially zero polarization. In some events multiple sources existed from the beginning (1968 Nov. 22; 1969 Aug. 29–30) while in other events, a single source either fragmented into several components or new components appeared nearby (1969 Mar. 1–2, Mar. 21, Oct. 10; 1970 Mar. 21). In one case (1969 Mar. 1–2) there were two sources with opposite circular polarizations. In three cases, (1968 Nov. 22; 1969 Mar. 21 and 1969 Aug. 29) two oppositely polarized sources were accompanied by, or overlapped into, a third unpolarized source. In two cases (1969 Oct. 10; 1970 Mar. 21), the three or four moving sources all had the same sense of circular polarization.

The associated optical flares were not remarkable. The flare importance ranged from 1N to 2B, 6 flares were of importance 1 and 5 of importance 2, 7 were of normal brightness and 4 were listed as bright. In one case, 1969 Oct. 10, the flare probably occurred beyond the west limb and was not observed.

The outstanding optical feature related to the bursts is the almost invariable association with active or disappearing filaments (disk events) and active or eruptive prominences (limb events).

Ten of the events were preceded by type III bursts and 8 by type II bursts, 6 of which were multiple type II bursts. The absence of type II observations in some of the remaining events may be linked with the much higher sensitivity of the radioheliograph compared with the Culgoora spectrograph. Many of the moving type IV events were quite weak ($\lesssim 10^{-21}$ Wm^{-2} Hz^{-1}); indeed two of the events (1968 Nov. 22; 1969 Oct. 10) were not detected by the spectrograph. We conclude, as did Boischot and Denisse (1957), that there is a strong association between type II and moving type IV bursts. It may be significant that in several cases (e.g. 1968 Sept. 4, Oct. 23–24; 1969 Mar. 21) the moving type IV sources were associated with a slower-moving, second or third component of a multiple type II burst.

Following Wild (1970a), we distinguish three types of continuum emission associated with these events: (1) 'flare continuum', which appears soon after the flare starts, (2) moving type IV continuum, and (3) 'storm continuum' or the stationary type IV phase, which follows the moving type IV phase. Of the twelve events described here, at least 4 and possibly 7 included a flare continuum phase. Typically, the flare continuum comes from a large source ($\simeq 15'$ diam) and is unpolarized. Nine events included a storm continuum phase, typically with a small source ($\simeq 5'$ diam) and strong circular polarization.

2.2. INDIVIDUAL EVENTS

In this section we describe four events in some detail; three illustrate fairly distinct classes of moving type IV bursts, while the fourth may fit either of two classes.

Example 1: Isolated Source of 1970 March 21

Sheridan (1970) reported a moving type IV source which moved outward to a radial distance of more than 6 $R_\odot$. The event started at 00^h23^m with a 2N flare at N20, E69 and a spectacular eruptive prominence (Figure 3). The prominence material moved with a speed of about 260 km/s at position angle (P.A.) 36° and was last seen at a radial distance of 1.3 $R_\odot$ at 00^h54^m. Approximately 20 min later the moving type IV source became visible at 1.9 $R_\odot$ and at the same position angle, whence it moved

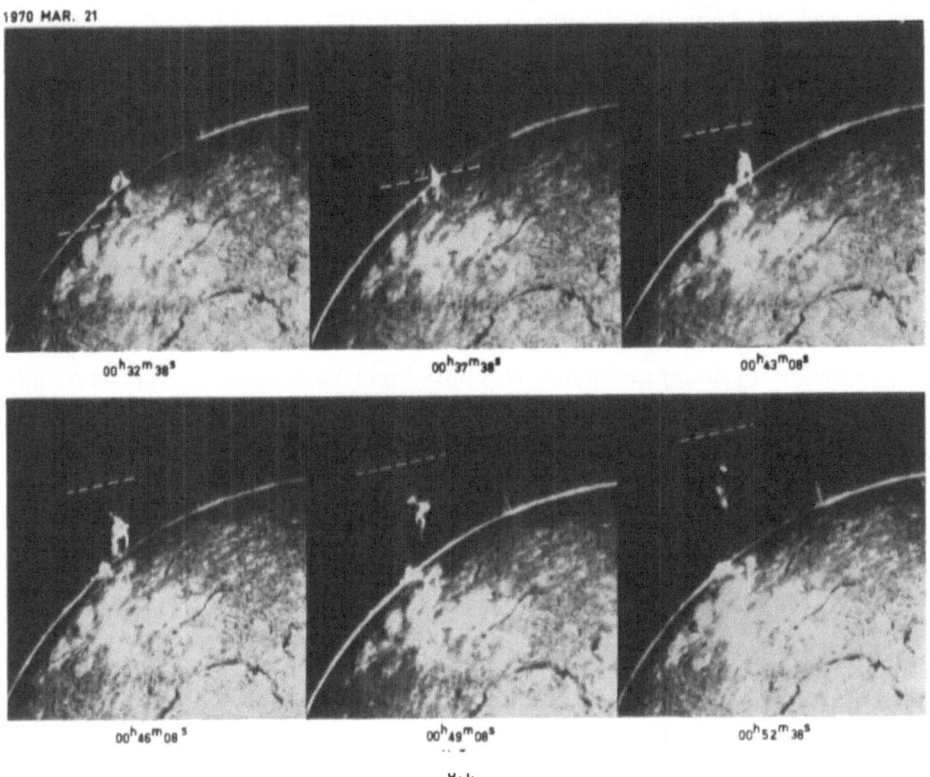

Fig. 3.   Hα filtergrams of a spectacular prominence eruption that preceded the far-moving type IV burst of 1970 March 21. The broken lines move out from the apparent starting point (marked X at the top left) at 290 km/s, the speed of the radio source (Sheridan, 1970).

outward at about 290 km/s for about 90 min. The evolution of the source is illustrated by the contour diagrams of Figure 4. The radio source seen at 00^h55^m (2nd from top) is the early, stationary flare continuum. It is distinct from the moving source which first appeared in the same vicinity at about 01^h12^m. The moving source remained single and expanded slightly until 02^h30^m, at 3.7 $R_\odot$, when it broke up. The four major fragments moved apart at about 90 km/s while the centroid of the configuration continued outward at 290 km/s. During the outward progress the polarization increased from 30% RH to more than 90% RH, with all fragments polarized to the same high degree and in the same sense. A storm continuum developed after 02^h00^m and remained visible until 03^h30^m, again at a radial distance of about 1.9 $R_\odot$. This source also showed RH polarization, which gradually increased from 30% to 70%.

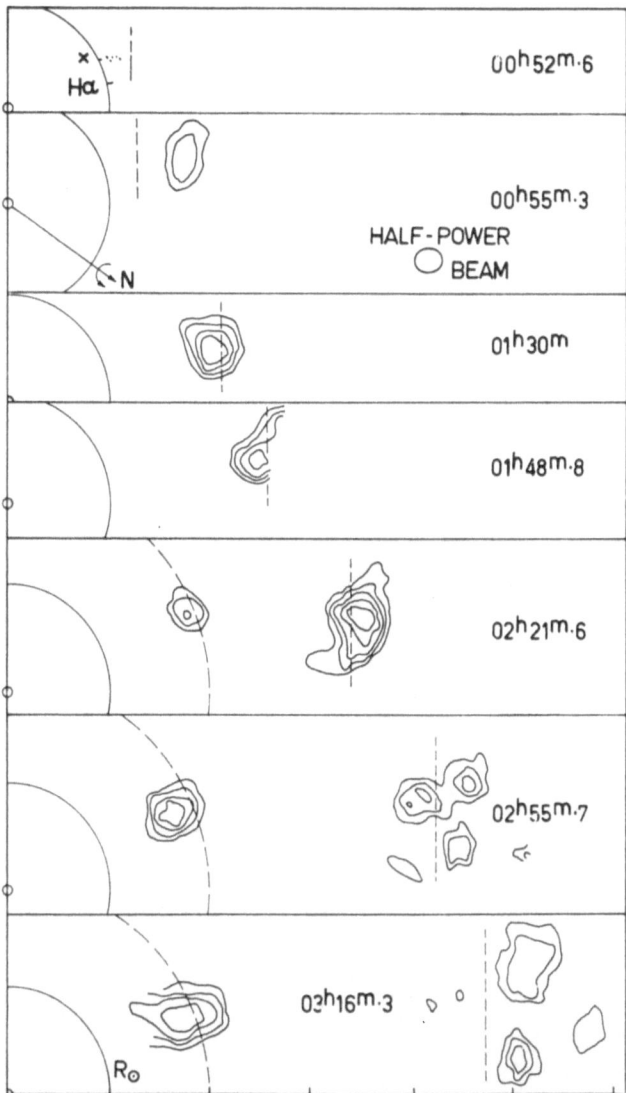

Fig. 4.   80 MHz contours illustrating the outward movement and fragmentation of the 'isolated source' of March 21 (adjacent contours are in the ratio $\sqrt{2}:1$). The broken lines denote a disturbance speed of 290 km/s. The top picture shows schematically a late phase of the eruption observed in Hα (see Figure 3) (Sheridan, 1970).

The spectrum showed that the type IV continuum was preceded by two type II bursts. The velocity of the second and stronger type II shock wave, deduced from the frequency drift and the standard density model, was >1500 km/s. In view of the large and irreconcilable difference between the speeds of the type II shock front (>1500 km/s) and of the moving type IV source (<300 km/s), it is clear that, in this case, the type II and the moving type IV bursts did not result from the same outward-moving disturbance.

Example 2: Expanding Arch of 1968 November 22

The evolution of this event (Wild, 1969) is illustrated in Figure 5, which shows the development of an unmistakable arch structure. At times during the early phases, the arch became strongly circularly polarized, LH on the east side, RH on the west.

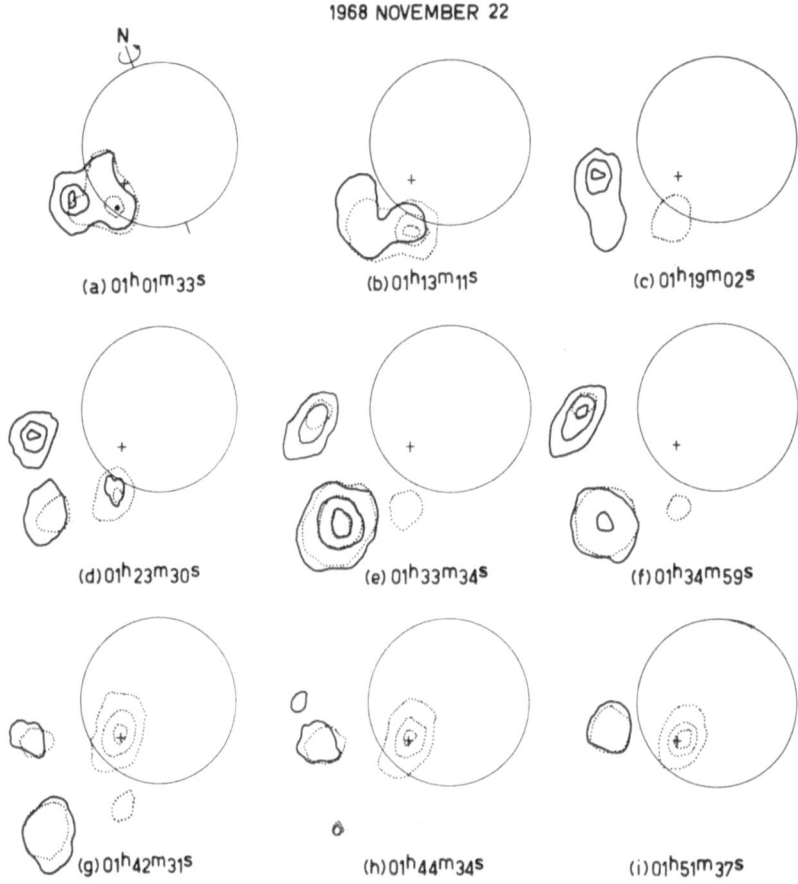

Fig. 5. 80 MHz contours illustrating the development at the indicated times of the 'expanding magnetic arch' of 1968 November 22. Full-line and broken-line contours represent LH and RH polarization respectively. The three contours represent 15, 50 and 85% of peak brightness. The circles in this and later Figures indicate the optical disk of the Sun (After Wild, 1969).

As the arch expanded, the emission concentrated in three sources; two oppositely polarized sources near the feet of the arch and an unpolarized source (at times with oppositely polarized edges) at the top. During the development of the arch the projected rate of expansion decreased from about 300 km/s to 100 km/s. The unpolarized source reached a projected distance of $2\,R_\odot$ before the three sources began to fade. While the arch sources faded, a storm continuum, strongly RH polarized, appeared above the flare region.

The spatial distribution and polarization of the sources strongly suggest an expanding magnetic arch (Wild, 1969), in which the unpolarized source resulted from synchrotron emission from a concentration of energetic electrons trapped in circular orbits near the top of the arch and the sources near the feet of the arch resulted from plasma radiation from electrons whose mirror points lay below the 80 MHz plasma level.

Example 3: Advancing Front of 1968 October 23–24

Kai (1970) reported a complex event which was associated with a 2B-flare and filament activation. The radio spectrum showed numerous type III bursts, two type II bursts and a moving type IV burst, and a final phase which may be best described as a storm continuum. Figure 6 illustrates the various burst positions at different times during the event; we note that the type II source and the type IV movement are in the same direction from the flare. All parts of the moving type IV front were slightly ($<20\%$) LH-polarized. The possible storm continuum, which settled down over the flare region near the end of the event, was 30–40% LH-polarized.

The derived velocities of the two II shock fronts were 3100 and 1400 km/s respectively. The projected velocity of the advancing type IV front was 1200 km/s; correcting for a line-of-sight component (assuming radial propagation) the velocity becomes

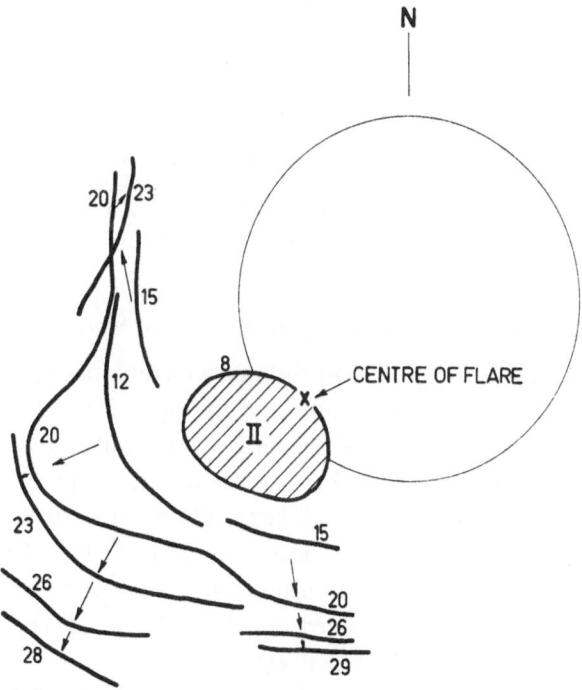

Fig. 6.   Schematic diagram illustrating the evolution of the 'advancing front' of 1968 October 23–24. The thick lines represent the fronts along which radiation was concentrated. The numbers give the time in minutes from the start of the flare (Kai, 1970).

1400 km/s. Figure 7 illustrates the excellent agreement and continuity between the heights and speeds of the second type II burst and the movement of the type IV source. Of the twelve moving type IV bursts listed in Table I, this event shows the closest relationship between the type II and type IV bursts. It is unique for the large peripheral extent and the high speed of the outward-moving front.

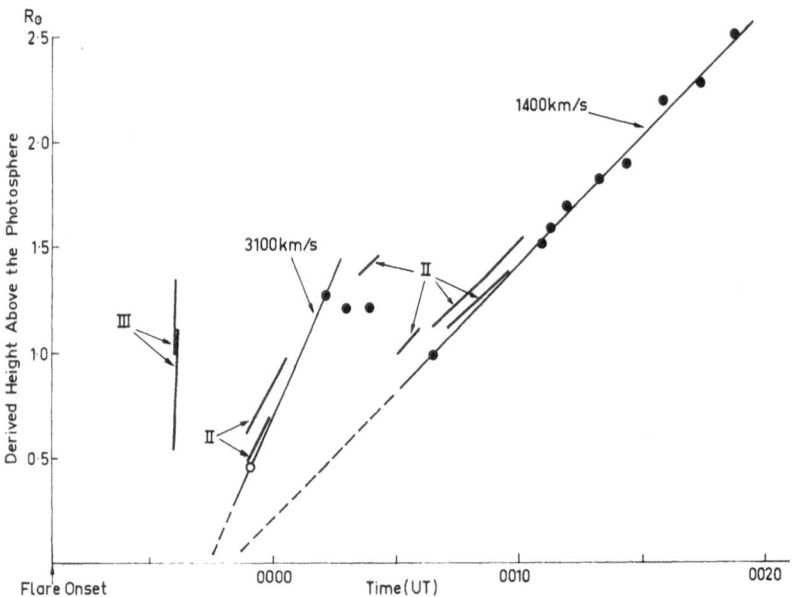

Fig. 7.   Height-time plot for the type III and type II bursts and for the 'advancing front' type IV source of 1968 October 23–24. All heights were derived on the assumption that the sources lay radially above the flare. The type III and type II lines were derived from the dynamic spectra and the standard coronal model (see text); the open and filled circles were derived from the source positions observed with the radioheliograph. The filled circles beyond 1.5 $R_\odot$ relate to the advancing front (Kai, 1970).

Example 4: Expanding Arch or Advancing Front of 1969 March 21

Stewart and Sheridan (1970) reported an outburst with multiple type II and moving IV bursts which combined some of the features of Examples 2 and 3 above. The type II bursts implied a disturbance moving outwards with a radial velocity of 980 km/s. When the type II disturbance reached a distance of 3 $R_\odot$ from the Sun's centre (as derived from the spectrum), the source of the type IV burst started its outward movement at a projected speed of 470 km/s. The movement, the spatial structure and the polarization structure of this burst are illustrated by the contour diagrams of Figure 8. The central part of the source remained unpolarized; the adjacent regions became 70% RH and 20% LH polarized respectively.

Stewart and Sheridan have interpreted this event as one in which type II and the type IV bursts originate in a common advancing shock front. By identifying the type II and type IV source velocities they deduced that the type IV source moved at an angle of 30% to the line of sight as illustrated in Figure 9. On this basis, the event is in the same class as Example 3.

Alternatively, this event may be interpreted as an expanding magnetic arch, such as that of Example 2, where the moving source is associated with the apex of the arch.

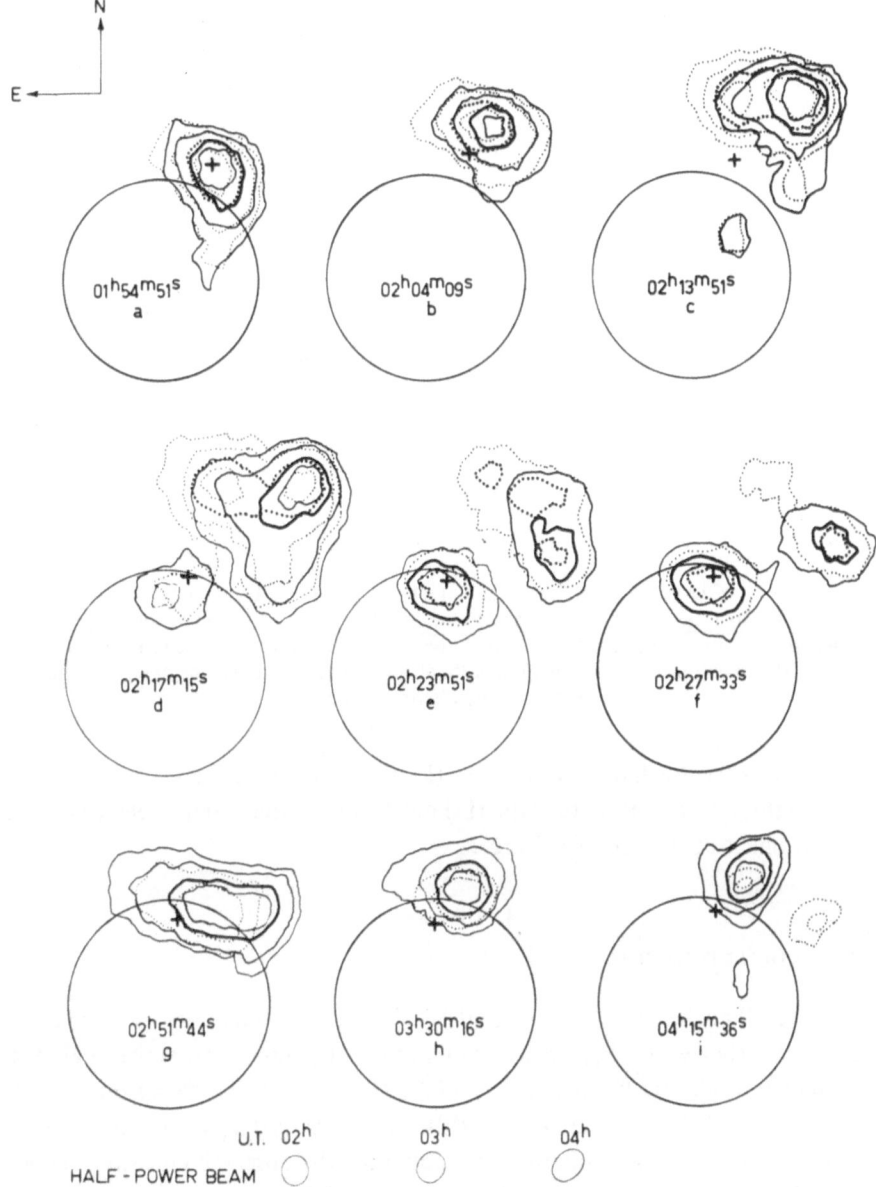

Fig. 8.   80 MHz contours illustrating the advancing front or 'expanding arch' of 1969 March 21. Full-line and dotted-line contours represent LH and RH polarization respectively. Adjacent contours have a brightness ratio $\sqrt{2}:1$. The crosses indicate centroid positions as measured with the 158 MHz interferometer (Stewart and Sheridan, 1970).

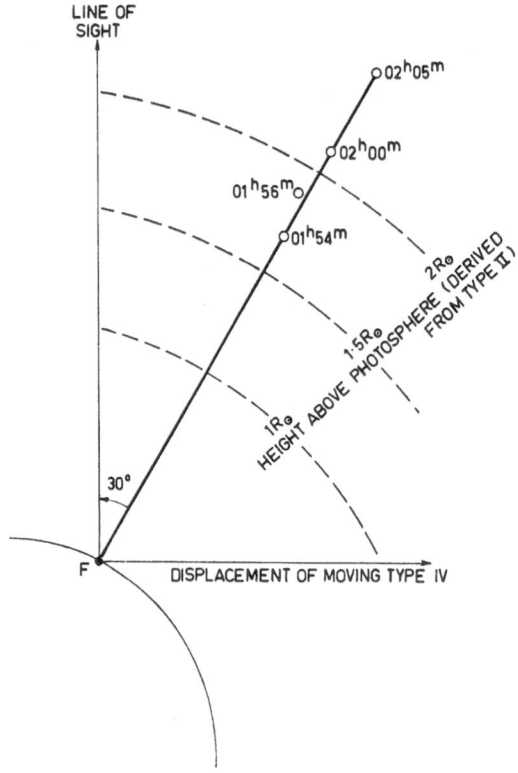

Fig. 9.   The event of 1969 March 21 (see Figure 8) interpreted as an advancing front, with a common shock front exciting type II and moving type IV radiation. The speed of the moving type IV source is assumed to be identical with the shock speed derived from the spectrum of the last and most intense of three type II bursts. The required direction of the shock front is 30° to the line of sight, as illustrated (Stewart and Sheridan, 1970).

The source near the north point of the disk (Figure 8d, e, f) then represents one foot of the arch. In this interpretation the type II shock front is unrelated in speed or direction to the movement of the type IV source.

3. Discussion

3.1. RADIATION MECHANISM

We have seen that the type IV sources at 80 MHz moved outwards from an average initial radial distance of $2\,R_\odot$. In the standard density model (an active region) this corresponds to a plasma frequency $f_{p,A} = 44$ MHz, while in the background coronal plasma, $f_{p,B} \approx 15$ MHz (Van de Hulst, 1950). The ratios $(f/f_p)_A = 1.8$ and $(f/f_p)_B = 5.3$ are far too high for plasma radiation to account for the observed emissions. 80 MHz plasma radiation could occur at a radial distance of $2\,R_\odot$ only if the density were 6.7 times that in Newkirk's model of an active region.

Gyromagnetic radiation in the extraordinary (x) mode and at the fundamental gyro-frequency, f_B, has been proposed (Fung, 1969) as a possible radiation mechanism.

With this mechanism, the wide-band nature of the emission depends on a wide range of viewing angles. Narrow-band bursts should be common and large magnetic fields must necessarily be frozen into the moving plasma, e.g. $B=29$ G for $f_B=80$ MHz. Another severe restriction is the requirement that $f_p/f_B \lesssim \frac{1}{3}$. Little is known about the value of f_p/f_B or its changes in the moving source. However, as Fung (1969) pointed out, in the expanding source model used, for instance, by Zheleznyakov and Trakhtengerts (1965), $f_p/f_B \propto L^{1/2}$ – where L is the linear dimension of the moving plasma. Therefore, gyro-radiation, even if possible at the lowest levels, would soon cease as the moving cloud expanded.

Synchrotron radiation at the higher harmonics of the relativistic and Doppler-shifted gyro-frequency has been invoked from the earliest interpretations of moving type IV bursts (Boischot and Denisse, 1957). The energy range of the radiating electrons is probably 100 keV $<E<1$ MeV because the bandwidth is often less than 100 MHz and because the polarization is often strongly circular. A high energy cutoff of about 1 MeV is also indicated by the condition $E>\frac{1}{2}f/f_p$ under which synchrotron radiation in a medium is thought to be suppressed (Ginzburg and Syrovatskii, 1965). The combination of suppressant medium and outward-moving source should make the onset of the burst drift from high to low frequencies in the spectrum; it has not been possible to check this feature for our bursts because the moving type IV component was usually weaker than the other spectral features with which it was intermingled.

3.2. INTENSITY AND POLARIZATION CHANGES

In Section 2 we found that the burst evolution usually involves increasing polarization and decreasing intensity. In an attempt to explain these features, Dulk (1970), using Wild's (1970b) much simplified formulae for the synchrotron emissivity of mildly-relativistic electrons, has evaluated the intensity and polarization of the radiation emitted by such electrons. The electron energy (E) and pitch angle (ϕ) distributions were assumed to be of the form: $n_{E,\phi}(t)=n_E(t) n_\phi(t)$. The power-law energy distribution used was $n_E(t)=n_{E_0}(t) (E/E_0)^{-\alpha} \exp (-E/E_1)$, with $E_1 \gg E_0$ (electrons with $E<E_0$, the low-energy cutoff, radiate negligible power; the high energy cutoff E_1 eliminates any small contribution from high-energy electrons with $\beta \gtrsim 0.95$, $E \gtrsim 1.2$ MeV where Wild's approximations are invalid). Two pitch-angle distributions were used: $n_\phi =$ constant, and $n_\phi = |\sin 2\phi|$ (the latter corresponds to a deficiency of electrons at $\phi=0$ and $\pi/2$). The results, plotted as a function of emission angle (with respect to the magnetic field), for the isotropic and anisotropic pitch-angle distributions are shown in Figures 10 a and b respectively. The intensity peaks at $\theta=90°$ for isotropic pitch angles and at $\theta \approx 60°$ and $120°$ for anisotropic pitch-angles. The intensity is widely distributed in angle, with half-power widths $\approx 80°$ and $\approx 110°$ respectively. With $f/f_B=10$ the degree of polarization at the half-power points is about 60% in the isotropic case and 75% in the anisotropic case. The circular polarization is right-handed for $\theta < \pi/2$, left-handed for $\theta > \pi/2$. The power and the degree of polarization decrease with increasing f/f_B, i.e. with increasing harmonic number.

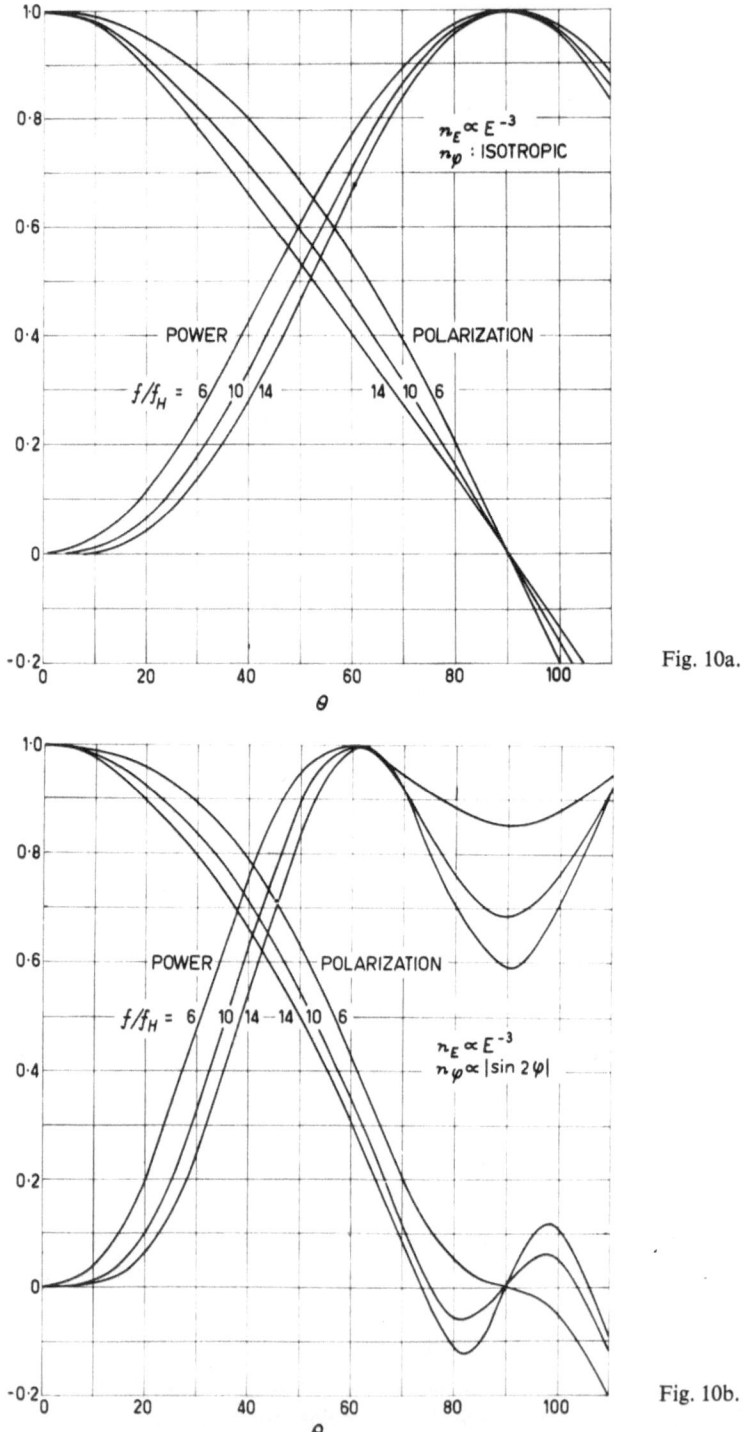

Fig. 10a.

Fig. 10b.

Fig. 10a–b. The power and polarization polar diagrams of synchrotron radiation from mildly relativistic electrons with a powerlaw energy distribution and an (a) isotropic and (b) anisotropic pitch-angle distributions. φ represents the angle between the magnetic field direction and the observer (Dulk, 1970).

The decrease in the magnetic field as the source moves outwards and expands causes the electrons to experience betatron deceleration. Since the time-scale of the field changes ($\sim 10^2$s) is long compared with the gyro-period ($\sim 10^{-6}$s), the magnetic moment remains adiabatically invariant and field changes are accompanied by energy changes such that $E_\perp/B$=constant. The deceleration of the electrons is accompanied by a decrease in the emitted intensity, partly because as f_B decreases, a higher harmonic is needed to radiate a given frequency f, and partly because the number of electrons with $E > E_0$ decreases as $n_{E_0}(t) = [B(t)]^\alpha$. Also, since the deceleration affects only $E_\perp$, it decreases the number of electrons with large pitch angles ($\phi \approx \pi/2$). Electrons with $\phi \approx 0$ are also lost because they travel along magnetic field lines without mirroring. The anisotropic distribution $n_\phi = |\sin 2\phi|$ is intended to simulate these conditions. These effects may partially explain the rapid fading and increasing polarization observed in the late stages of moving type IV sources.

3.3. ASSOCIATION WITH OTHER METRE-WAVE CONTINUA

We found in Section (2.1) that the fading of the moving type IV source is frequently accompanied by the brightening of one or two storm continuum sources (stationary type IV sources). Characteristically, the storm continuum is situated near the flare, possibly in a region of strong magnetic field. In terms of the above interpretation of the fading phase of a moving source, the storm-continuum source may sometimes be related to the 'dumping' of mildly-relativistic eletrons from the moving source.

We also noted in Section (2.1) that in some events the moving type IV burst was preceded by a stationary, unpolarized flare continuum. The onset of a striking example of a flare continuum above the limb, consisting of a rapidly expanding source, was attributed (Smerd, 1970) to the injection of fast electrons (possibly during the flash phase) into a bi-polar closed-field region. The mirroring of such particles can produce opposing electron streams in the trapping region which, in the manner described by Zheleznyakov and Zaitsev (1968) for type V bursts, emit unpolarized, second harmonic ($f \approx 2f_p$) radiation over an appreciable range of frequencies. The persistence of the flare continuum beyond several minutes may be due to repeated electron injections or to electron acceleration in the trapping region following the impact of the shock wave.

3.4. ASSOCIATION WITH TYPE II BURSTS

The present observations show a fairly strong association between moving type IV bursts and type II bursts. A case for a common shock wave as the source of both type II and moving type IV bursts has been made by Stewart *et al.* (1970) for three of the present events (1968 Oct. 23–4, Oct. 26; 1969 Mar. 21). The first of these, which we consider to show the closest relationship between the two bursts, has been described as Example 3 in Section 2.2.

Shock waves, not necessarily associated with type II bursts, have been involved as the sources of the moving bursts from the earliest interpretation (Boischot and Denisse, 1957). One moving type IV burst observed at Nançay with one-dimensional resolution at 169 and 408 MHz (Boischot and Clavelier, 1967, 1968; Boischot and Daigne, 1968; Lacombe and Mangeney, 1969) was explained in terms of synchrotron radiation from electrons which were accelerated to relativistic energies within the advancing shock front. The explanation invoked a theory limited to weak (magnetic Mach number $M_A \lesssim 2$), collisionless shocks perpendicular to the magnetic field to provide a dissipative shock wave. The dissipative shock process provides the first stage of electron acceleration as had previously been suggested for type II bursts by Pikelner and Ginzburg (1963). Therefore, it seems that the physics of weak, collisionless, perpendicular plasma shocks – as understood at present – might account for the generation of both type II and type IV radiation by the same shock as it advances through the corona.

If we identify the disturbance with a weak, perpendicular shock moving at $M_A = v/v_A = 2$ (where the Alfvén velocity $v_A = 7 \times 10^3 \, f_B/f_p$ km/s) then $v = 2v_A = 1.4 \times 10^4 \, (f_B/f_p)_{amb}$ (where the subscript amb denotes the ambient medium through which the shock moves). In Section 2 we found that the synchrotron radiation of harmonic number $f/f_B \approx 10$ can best explain the burst characteristics. Therefore, for 80 MHz radiation from inside the shock, we have $f_{B, shock} \approx 8$ MHz and $B \approx 3$ G. We also found that the initial height of moving type IV bursts was near the 44 MHz level in the standard model; we assume that this represents the enhanced plasma frequency within the shock, i.e., $f_{p, shock} = 44$ MHz. Using these values, together with the relation $(f_B/f_p)_{shock} = 1.54 \, (f_B/f_p)_{amb}$ (Zaitsev, 1968), we find $(f_B/f_p)_{shock} = 0.182$, $(f_B/f_p)_{amb} = 0.118$, and $v = 1650$ km/s.

This velocity is close to the observed velocity of the advancing front in Example 3. On the other hand, a shock moving at about 600 km/s (the average source speed found earlier) is not likely to be a moving type IV source because high harmonic numbers (≈ 30) are required. Also, a shock moving at, say, 3000 km/s would require very low harmonic numbers (≈ 5) and correspondingly high magnetic fields.

In the case of strong shocks ($M_A \gg 1$) moving perpendicular to the magnetic field, we can use the relation $(f_B/f_p^2)_{shock} = (f_B/f_p^2)_{amb}$ (Helfer, 1953). The limiting value as $M_A \to \infty$ is $f_{p, shock} = 2 f_{p, amb}$. Writing $v_{shock} = M_A v_A$ and substituting for v_A in terms of $(f_B/f_p)_{amb}$ we obtain

$$(f_B/f_p)_{shock} = \frac{v_{shock} \, (\text{km/s})}{3.5 \times 10^3 \, M_A}.$$

Choosing as before, $f_{p, shock} = 44$ MHz and $(f/f_B) = 10$, this gives $v_{shock} \, (\text{km/s}) = 1.27 \times 10^3 \, M_A$. Since, by definition, $M_A \gg 1$, this condition can only be satisfied for shock velocities that are faster (> 6000 km/s, say) than any source velocities yet observed. In the case of oblique shocks, only $B_\perp$ is involved in the jump conditions and the velocity of the strong shock is even higher.

On the other hand, Warwick (1965, 1968) identifies the moving type IV source with a disturbance moving parallel to the magnetic field and attributes the observed synchrotron radiation to relativistic electrons encountering the disturbance as they diffuse out along magnetic field lines from an accelerating region low in the Sun's atmosphere. He suggests that the disturbance, a large amplitude Alfvén wave or soliton, will degenerate into a shock as it moves in the corona; as such, it could also be a source of type II radiation. Longitudinal, collisionless shocks are known to exist, for example, in interplanetary space. Their large scale features have been successfully explained by aerodynamic methods (e.g. Sonnett, 1969); however, the wave-particle interactions that allow the wave to become a dissipative shock remain unspecified. Without this knowledge, we cannot discuss the radio emission from such waves.

These considerations lead us to conclude the moving type IV sources cannot always be identified with shock fronts of accompanying type II bursts. Does then the observed close association between type II and type IV bursts have some other physical significance? We suggest that an impinging shock front may be responsible for the expansion of a magnetic arch or that it may initiate or trigger the outward movement of the plasma-field configuration which is observed as a moving type IV source. We note that in such cases the speed and direction of the moving type IV source is not necessarily the same as that of the type II disturbance.

The almost invariable association of moving type IV bursts with active or disappearing filaments or prominences suggests that prominence material and magnetic field may be ejected from the low corona (perhaps by a shock wave) and later appear as a radio source. In some cases, the field structure may remain tied to the photosphere, resulting in an expanding arch. In other cases, magnetic reconnection may occur, resulting in a self-contained plasmoid which appears as an isolated source. In all cases, some of the material must be accelerated to relativistic energies to account for the observed radio emission.

3.5. PHYSICAL CLASSIFICATION OF THE BURSTS

We now attempt to use the results of the preceding discussion to provide a physical basis for the classification of the bursts as isolated sources, expanding arches, or advancing fronts.

The most clear-cut class is the expanding arch which, following Wild (1969), we identify with an expanding magnetic arch. It is clear that, depending on the orientation and location of such an arch on the Sun, the appearance of the arch as observed by the radioheliograph may differ considerably from event to event.

We identify the advancing front with the propagation through the corona of a MHD disturbance, either a soliton or a shock front. It is clear from example 3 of Section 2.2 that agreement between the type II speed (derived from the spectral drift rate) and the type IV speed (from projected positions) can be convincingly demonstrated from the observations only if the source movement is mainly transverse to the line of sight.

The isolated source illustrated by Example 1 was probably produced by solar ejecta in the form of a magnetized plasma cloud, part of the preceding eruptive prominence. A similar situation prevails in the event of 1969 March 1–2, the other 'long, distant and slow' event that seems to fit the same classification fairly unambiguously. However, it is important to recognize that in some circumstances an expanding arch or an advancing front might appear as an 'isolated source', e.g. an expanding arch seen edge-on or an advancing front intersecting a streamer structure.

In Table II we attempt to classify the 12 events into the proposed three types of moving-IV source.

TABLE II

Classification of 12 moving type IV bursts

Model A: Isolated source

1968 Feb. 25 (Models B and C cannot be ruled out because of short duration and small distance moved).
1968 Sept. 4 (Close relationship with type II burst leaves Model C possible).
1969 Mar. 1–2 (Large distance moved and source appearance strongly suggest this model).
1969 Oct. 10 (Advancing front intersecting a streamer structure, a variant of Model C, is possible).
1969 Dec. 17 (Two moving sources; peripheral movement of source 1 may fit variant of Model C).
1970 Mar. 21 (Large distance moved and source appearance suggest this model; same sense of polarization of all fragments may be more easily interpreted by Model B, expanding arch, seen edge-on).
1970 Apr. 29 (Appearance favours this model).

Model B: Expanding magnetic arch

1968 Oct. 26 (Strong polarization and jet-like appearance suggest arch seen edge-on).
1968 Nov. 22 (Appearance and polarization strongly suggest this model).
1969 Mar. 21 (Stewart and Sheridan's interpretation as an advancing front cannot be ruled out).
1969 Aug. 29–30 (Appearance and polarization support this model; advancing front cannot be ruled out).

Model C: Advancing front

1968 Oct. 23–24 (Source appearance and close relation with type II burst strongly suggest this model).

---

4. Moving Type-IV Sources and Coronal Magnetic Fields

4.1. SOURCE MOVEMENT AND CORONAL FIELD

In the case of an expanding magnetic arch the source movement and magnetic-field movement are inseparable. The magnetic loop links weak fields of opposite polarity in two regions as illustrated in Figure 11a. When the force which caused the expansion has spent itself the loop may contract to its original, much lower, position, it may remain as an 'exchange loop' at a high level (and may be carried out by the solar wind), or similar loops may be regenerated by photospheric currents.

The relation between an advancing front and the coronal magnetic field is more ambiguous. The 'slow' MHD wave (at $v \approx V_T \approx 170$ km/s) is probably not involved in moving type IV bursts. Both the remaining magnetohydrodynamic wave modes propagate in the corona at about the Alfvén velocity; the Alfén wave propagates along the field only, the 'fast' wave in all directions. Meyer (1968) has found that the combined effects of the two wave modes make the direction of the magnetic field the preferred direction for coronal disturbances to propagate because it is the direction of least dispersion. Thus the advancing front type of moving type IV burst should move along the magnetic fields illustrated in Figure 11b. However, as noted in Section 3.4 shock-wave theories for type-II and moving type-IV bursts have been formulated only for perpendicular shocks.

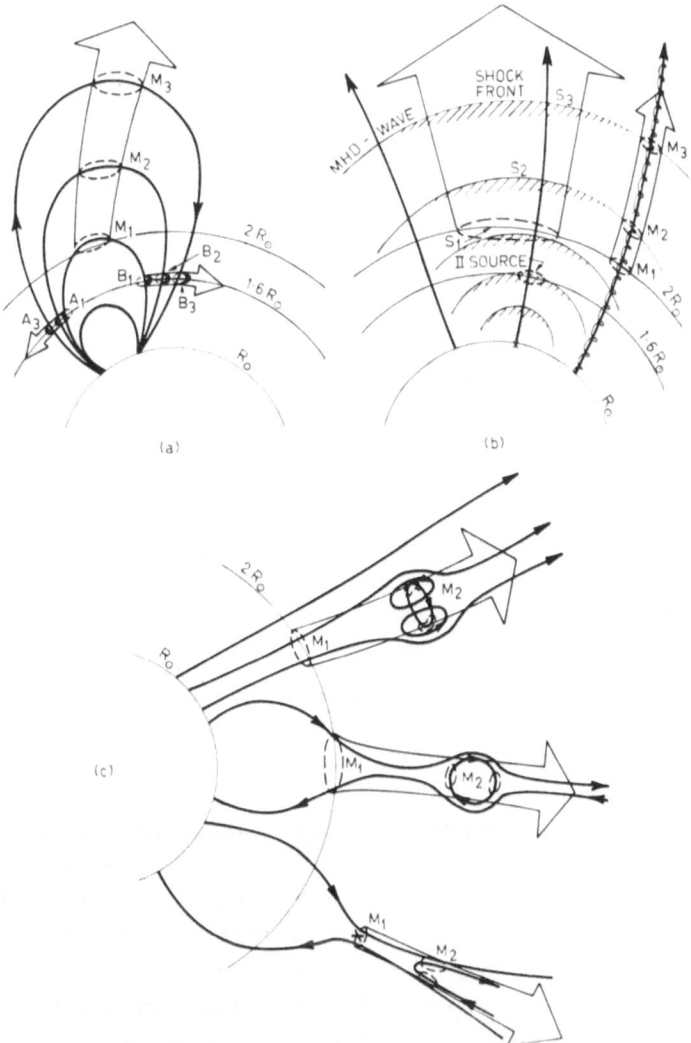

Fig. 11. Schematic diagrams of the magnetic-field structure in three models of moving type IV sources. The magnetic field lines are shown as thick arrowed lines, the paths of the moving IV sources by broad arrows. The optical disk, the 80 MHz plasma level and the average starting height of the moving IV sources are shown by thin arcs at $R_\odot$, 1.6 $R_\odot$ and 2 $R_\odot$ respectively. The subscripts 1, 2, 3, refer to the source positions at three successive times. (a) An 'expanding magnetic arch' with A and B representing plasma level sources and M representing the moving source near the top of the arch. (b) An 'advancing front' of an MHD-wave moving along the magnetic field from an explosion near the surface. In the center and left, we depict a shock wave which generates type II radiation as it passes through the plasma level and synchrotron radiation beyond 2 $R_\odot$. On the right, relativistic electrons are shown traveling along the field line; they emit synchrotron radiation as they pass through the 'soliton' which propagates outwards at the much lower Alfvén velocity. (c) 'Isolated sources' moving out between open field lines. On top, the moving IV source is shown as a toroid with a poloidal stabilizing component moving in a unipolar field region. In the middle, the toroidal type IV source moves outwards between the opposing field lines of a helmet structure. At the bottom the moving IV source is at the end of an open-ended loop which retracts outwards after magnetic re-connection at X.

In the case of isolated moving type IV sources due to ejecta from flare explosions, the motion, possibly at the Alfvén velocity, occurs along or between open field lines. Two open-line systems suggest themselves: those of a helmet structure where opposing field lines are separated by neutral sheets (see Figure 11c, middle), and the aligned 'radial' field lines high above unipolar regions (see Figure 11c, top).

A special case involving a helmet structure is Wild's (1970a) adaptation of Sturrock's (1966) model for the high-energy phase of a flare, illustrated in Figure 11c, bottom. The opposing fields, after reconnection across the neutral sheet, retract outwards and inwards from the place of reconnection. Wild suggests that the outward-moving, open-ended loop may explain the isolated source in Exemple 1 and the other distant event of 1969 Mar. 1–2.

4.2. Polarization and Source Fields

In Section 3.2 we found that the general pattern of the evolution of moving type IV sources – decreasing intensity and increasing polarization – is compatible with the consequences of a gradual decrease of the magnetic field as the source moved outwards. We now discuss the observed polarization structure in relation to the magnetic field at the source and the wave-mode in which the radiation is emitted.

Wild (1969) has shown that the polarization structure on 1968 Nov. 22 is compatible with the emission of ordinary (o) mode radiation from the two regions where the arch intersects the 80 MHz plasma level and synchrotron radiation in the extraordinary (x) mode from electrons at the apex of the arch, high above the plasma level. This situation is schematically illustrated in Figure 11a.

The same interpretation may apply to Example 4, the event of 1969 Mar. 21, and to the other events listed as possible expanding magnetic arches in Table II. The event of 1968 Oct. 26, when interpreted as an arch seen edge on, suggests that the field had a line-of-sight component towards the observer at the plasma level and away from the observer at the synchrotron heights. In the event of 1969 Aug. 29–30 the synchrotron source at the top of the arch showed both polarizations, implying a field component towards the observer on the west side and in the opposite direction on the east side of the arch. The plasma-level source on the west side of the arch was polarized and the one on the east side of the arch was unpolarized; this may indicate that the field on the east side was weak or that it was nearly perpendicular to the line of sight.

The polarization of the event of 1968 Oct. 23–24, the advancing front, was weakly LH all along the extended arch. For the x-mode to be LH the field must have a component away from the observer: the weak degree of polarization suggests that the field at the source was directed approximately 100°–120° from the observer. If the field lines were everywhere aligned along the advancing front, the line of sight component would necessarily reverse direction at some point along the arc. However, if the field lines were directed nearly radially inwards in the vicinity of the advancing front (which moved at an angle of approximately 60° to the line of sight), the observed polarization could be explained. This suggests that the front advanced along the magnetic field.

The remaining events in Table II are classified as isolated sources.

Polarization was not measured for the event of 1968 Feb. 25. For two events – 1968 Sept. 4, and 1970 Apr. 29 – the moving source remained single from beginning to end with the polarization increasing from 40% to >85% LH in the former and from <10% to >70% LH in the latter event. The increasing polarization fits the pattern discussed in Section 3.2. Again, for x-mode radiation to be LH, the magnetic field within the source must have had a component away from the observer; the high degree of polarization makes it likely that at least late in the events, the sources contained few electrons in near-circular orbits and that the sources were observed at moderate angles (≈ 20–$40°$) with respect to the magnetic field.

In another event, that of 1969 Dec. 17, the two separate moving sources had different initial positions, different starting times, different speeds and moved in different directions. The first source moved peripherally and its polarization increased from 0 to $\approx 90\%$ RH. Again we attribute the increasing polarization to the changing magnetic-field effects described in Section 3.2; the RH polarization suggests a magnetic-field component towards the observer. The second moving source was strongly LH polarized from the beginning; the movement was approximately rectilinear until late in the event. The sense of polarization requires a magnetic-field component towards the observer. The high degree of polarization requires a deficiency of near circular-orbit electrons throughout the event and a viewing angle of no more than $30°$ with respect to the field.

The three remaining events – 1969 March 1–2, 1969 October 10 and 1970 March 21 – have intriguing polarization structures; the moving type IV phase began with a single source which later was accompanied by, or fragmented into, several moving sources.

On 1969 March 1–2 (Riddle, 1970) the initial source A, moving at ≈ 270 km/s and unpolarized in the early stages, was joined after ≈ 1 hr by another source B about $11'$ arc towards the NNE. Source B faded from view after about 45 m; source A persisted for 40 m after B had disappeared. The polarization of source A started to increase 20 m before source B appeared and reached 70% RH. The polarization of source B increased, though not monotonically, from zero to 60% LH. Riddle suggested that the two sources formed part of the same structure, possibly ring-shaped, with a predominantly toroidal field stabilized by a poloidal component. Viewed side-on the two bright regions are those of greatest longitudinal extent near the extremities of the toroid where the field points towards and away from the observer respectively. This model could account for the observations with the field towards the observer at the leading edge. The model is illustrated at the top of Figure 11c, where the source is identified with an ejected 'plasmoid' moving out between the aligned open field lines above a unipolar region.

In the event of 1969 October 10 (Smerd, 1971), the original moving source M_1 fragmented into or was accompanied by source M_2 which lay about $12'$ east of M_1; late in the event a third source M_3 appeared about $5'$ east of M_1. The sources moved apart at about 300–400 km/s while the centroid motion continued at ≈ 650 km/s.

M_1 was initially unpolarized, but after fragmentation all sources became LH polarized. In the final, fading stages the polarization reached $\approx 50\%$ for M_1, $\approx 30\%$ for M_2 and $> 80\%$ for M_3.

The event of 1970 March 21 (Sheridan, 1970) was very similar in many respects. The initially single source, moving at 290 km/s, broke into 4 parts which moved apart at ≈ 90 km/s. The polarization increased from 30% to $> 90\%$, all parts having the same RH sense of circular polarization. Sheridan notes that a toroid cannot account for the observed polarization structure and considers Wild's (1970a) suggestion of an outward-moving, open-ended loop (the result of re-connection across a neutral sheet) illustrated at the bottom of Figure 11c. However, this model can produce multiple sources with like polarization only if viewed within a very small range of viewing angles near 90°, in which case the degree of polarization is low.

The results of Section 3.2 suggest that high polarization in several sources can be observed only if the field direction is about 20° (or 160°) from the line of sight. One plasma-field configuration meeting these geometrical requirements is a twisted loop (like a Figure 8, though without intersecting lines) radiating at its extremities. Another possibility is a dipole field with a radiation belt. However, we know of no evidence that solar ejecta have such field configurations.

Wild (private communication) also suggested the model of an expanding magnetic arch seen nearly edge-on. We consider that this model provides the simplest explanation of the polarization of the events of 1969 Oct. 10 and 1970 March 21, providing that there is no reconnection of the dragged-out field lines, and providing that the breakup into several subsources can be explained.

4.3. COMPARISON WITH CORONAL MAGNETIC-FIELD MAPS

Altschuler and Newkirk (1969) described a technique for deriving the coronal magfield structure from Mount Wilson observations of the photospheric field. Using magnetometer observations taken at Mt. Wilson over a 27 day period, together with an approximate simulation of the effects of the solar wind, they calculate the three-dimensional structure of the coronal field and display the results in stereo projection. As part of a joint investigation, they have produced coronal field maps pertaining to the times of several radioheliograph events (see also Newkirk, 1971). We now make a preliminary comparison of the radio burst positions observed in three events with the corresponding coronal field maps.

(1) Expanding arch of 1969 November 22 (Example 2 in Section 2.2). Figure 12 shows the expanding-arch positions at various times superimposed on one map of a stereo pair. The flare X occurred underneath or at the edge of a low arch tunnel; the activated filament Y lay at the far end of the tunnel. We suggest that the flare disturbance initiated an expansion of a portion of the magnetic arch tunnel. After the arch rose above the 80 MHz level, the radio arch became visible. The polarization of the two halves of the arch in the early stages and the isolated 'feet' in later stages corresponds to the ordinary mode in the field configurations of the maps. This is consistent with the

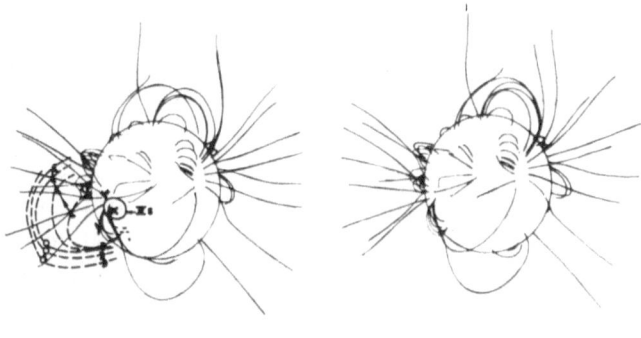

1968 NOVEMBER 22

Fig. 12. The 'expanding magnetic arch' of 1968 November 22 (Wild, 1969) plotted schematically on one of a stereo pair of coronal magnetic field maps (by courtesy of Newkirk and Altschuler, High Altitude Observatory, Boulder). The flare (X) and active filament (Y) seem to be located at the edge and at the far end of an arch tunnel respectively. The symbols (,), and $\bigcirc$ denote LH, RH, and unpolarized sources respectively. $+$ and $-$ magnetic source regions are denoted. The expanding arch traced out by the radio event is compatible with an expansion of a portion of the magnetic arch tunnel.

hypothesis that electron streams moving along the arch set up plasma oscillations and that the extraordinary mode radiation has difficulty in escaping (Wild, 1969). In later stages, the unpolarized central source presumably resulted from synchrotron emission from a concentration of electrons near the top of the arch. Finally, the location of the RH polarized storm continuum over a region of $+$ field (N-polarity) implies that emission in the extraordinary mode is involved. This is opposite to the general rule, which implies ordinary-mode emission on the 'leading spot' hypothesis.

(2) Expanding arch or advancing front of 1969 March 21 (Example 4 in Section 2.2). Figure 13 shows the motion of the type IV burst super-imposed on the appropriate coronal field map. The flare X and disappearing filament Y occurred under high

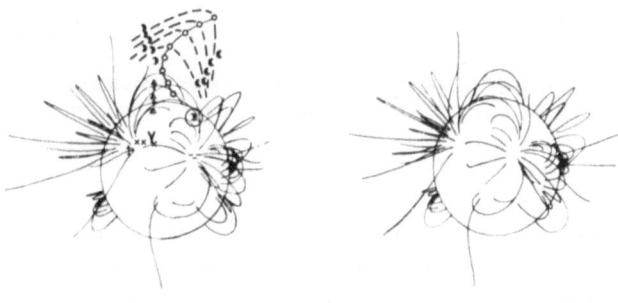

1969 MARCH 21

Fig. 13. The complex moving type IV event of 1969 March 21 plotted as in Figure 12. The flare (X) and disappearing filament (Y) seem to have occurred underneath high magnetic loops. If the event is interpreted as an 'expanding magnetic arch' it is found to be compatible with an expansion of these magnetic loops.

magnetic loops. If, as suggested by Stewart and Sheridan (1970), the moving type IV burst was associated with an advancing shock front moving predominantly along the line of sight, then the maps give little information because no field lines are shown in this vicinity above about 1.5 $R_\odot$. Alternately, we can interpret the event as an expanding arch as suggested earlier. In this case, the field maps relate closely to the type IV source. We suggest that the disturbance from the flare or disappearing filament caused the arch structure to move outward at an angle of $\approx 60°$ to the line of sight. The radiation from the top of the magnetic arch is unpolarized; the regions on either side show opposite polarizations because of the oppositely directed magnetic-field components in the line of sight. The polarity of the radiation corresponds to the extraordinary mode in the field configuration of the maps. The polarization of the storm continuum source (possibly representing one foot of the arch), corresponds to the ordinary mode. Thus the polarization data and field maps are consistent with synchrotron emission from the moving source and with plasma radiation from the stationary source.

(3) Isolated source of 1969 March 1–2. Figure 14 shows the path of the isolated source of March 1–2 and the preceding eruptive prominence superimposed upon the corresponding coronal field map. Note that there were two oppositely polarized sources

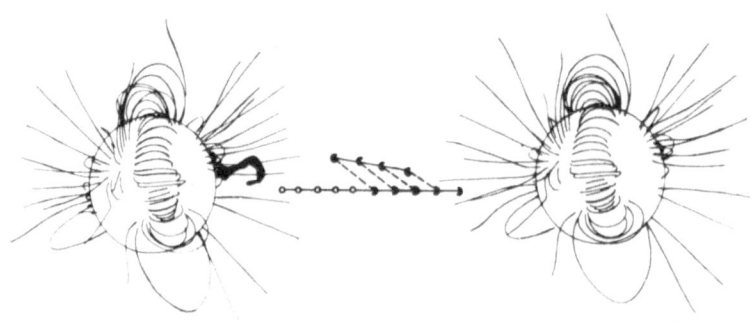

1969 MARCH 1-2

Fig. 14. The far-moving 'isolated source' type IV event of 1969 March 1–2 plotted as in Figure 12. The radio source seems to have moved out between open field lines of north (positive) polarity.

during the intermediate stages of the event. The coronal field lines in the region of interest are predominantly open, thus providing an open pathway for the moving source. The field lines are directed outwards, which results in RH polarization for the extraordinary mode. This agrees with the observed polarization of the dominant component of the burst. However, the LH polarization of the accompanying source, and probably the RH polarization of the dominant source as well, are presumably determined by the field structure carried out within the plasmoid.

References

Altschuler, M. D. and Newkirk, G. Jr.: 1969, *Solar Phys.* **9**, 131.

Boischot, A.: 1957, *Compt. Rend. Acad. Sci. Paris* **244**, 1326.

Boischot, A. and Clavelier, B.: 1967, *Astrophys. Letters* **1**, 7.

Boischot, A. and Clavelier, B.: 1968, *Ann. Astrophys.* **31**, 445.

Boischot, A. and Daigne, G.: 1968, *Ann. Astrophys.* **31**, 531.

Boischot, A. and Denisse, J. F.: 1957, *Compt. Rend. Acad. Sci. Paris* **245**, 2194.

Dulk, G. A.: 1970, *Proc. Astron. Soc. Austral.* **1**, 372.

Dulk, G. A.: 1971, *Australian J. Phys.* **24**, 217.

Dulk, G. A. and Altschuler, M. D.: 1971, submitted to *Solar Phys.*

Dulk, G. A., Stewart, R. T., Black, H. C., and Johns, I. A.: 1971, *Australian J. Phys.* **24**, 239.

Fung, P. C. W.: 1969, *Can. J. Phys.* **47**, 179.

Ginzburg, V. L. and Syrovatskii, S. I.: 1965, *Ann. Rev. Astron. Astrophys.* **3**, 297.

Helfer, H. L.: 1953, *Astrophys. J.* **117**, 177.

Kai, K.: 1969a, *Proc. Astron. Soc. Austral.* **1**, 189.

Kai, K.: 1969b, *Solar Phys.* **10**, 460.

Kai, K.: 1970, *Solar Phys.* **11**, 310.

Lacombe, C. and Mangeney, A.: 1969, *Astron. Astrophys.* **1**, 325.

Meyer, F.: 1968, in K. O. Kiepenheuer (ed.), 'Structure and Development of Solar Active Regions', *IAU Symp.* **35**, 485.

Newkirk, G. Jr.: 1961, *Astrophys. J.* **133**, 983.

Newkirk, G. Jr.: 1971, this volume, p. 547.

Pikelner, S. B. and Gintsburg, M. A.: 1963, *Astron. Zh.* **40**, 842 = *Sov. Astron.* **7**, 639, 1964.

Riddle, A. C.: 1970, *Solar Phys.* **13**, 448.

Sheridan, K. V.: 1970, *Proc. Astron. Soc. Austral.* **1**, 376.

Smerd, S. F.: 1970, *Proc. Astron. Soc. Austral.* **1**, 305.

Smerd, S. F.: 1971, *Australian J. Phys.* **24**, 229.

Sonett, C. P.: 1969, *Comments Astrophys. Space Phys.* **1**, 178.

Stewart, R. T.: 1971, *Australian J. Phys.* **24**, 209.

Stewart, R. T. and Sheridan, K. V.: 1970, *Solar Phys.* **12**, 229.

Stewart, R. T., Sheridan, K. V., and Kai, K.: 1970, *Proc. Astron. Soc. Austral.* **1**, 313.

Sturrock P. A.: 1966, *Nature* **211**, 695.

Van de Hulst, H. C.: 1950, *Bull. Astron. Inst. Neth.* **11**, 135.

Warwick, J. W.: 1965, in *Solar System Radio Astronomy*, ed. by J. Aarons, p. 131, Plenum Press, N.Y.

Warwick, J. W.: 1968, *Solar Phys.* **5**, 111.

Weiss, A. A.: 1963, *Australian J. Phys.* **16**, 526.

Wild, J. P.: 1969, *Solar Phys.* **9**, 260.

Wild, J. P.: 1970a, *Proc. Astron. Soc. Austral.* **1**, 365.

Wild, J. P.: 1970b, *Proc. Astron. Soc. Austral.* **1**, 348.

Wild, J. P., Sheridan, K. V., and Kai, K.: 1968, *Nature* **218**, 536.

Wild, J. P., Sheridan, K. V., and Trent, G. H.: 1959, in *Paris Symp. Radio Astron.* (ed. by R. N. Bracewell), p. 176, Stanford University Press.

Zaitsev, V. V.: 1968, *Astron. Zh.* **45**, 766; 1969, *Sov. Astron.* **12**, 610.

Zheleznyakov, V. V. and Trakhtengerts, V. Yu.: 1965, *Astron. Zh.* **42**, 1005 = *Sov. Astron.* **9**, 775, 1966.

Zheleznyakov, V. V. and Zaitsev, V. V.: 1968, *Astron. Zh.* **45**, 19; 1968, *Sov. Astron. – A. J.* **12**, 14.

ACTIVE REGIONS AT MILLIMETER WAVELENGTHS AND THE MEASUREMENT OF MAGNETIC FIELDS

M. R. KUNDU

Astronomy Program, University of Maryland, College Park, Md., U.S.A.

Abstract. Some properties of solar active regions at 9 and 3.5 mm wavelengths are discussed. The regions have excess brightness temperatures of up to 1000 and 700 K at 9 and 3.5 mm wavelengths respectively. The background radiation at 3.5 mm is often seen to be 'absorbed' in regions closely coincident with Hα dark filaments on the disk. Interpretation of this 'absorption' as due to the large optical thickness of the overlying filamentary material leads to an estimate of electron density in the filaments. The active regions at millimeter wavelengths show almost one-to-one correspondence with the Ca-plage regions as well as with the regions of longitudinal magnetic fields on Mt. Wilson magnetograms. A comparison of the mm-λ maps with the magnetograms 'smoothed' with the beams of mm observations shows this correspondence in a striking manner. This relationship suggests the possibility of measuring chromospheric magnetic fields from the measurement of polarization at millimeter wavelengths.

In this paper, I shall discuss some high resolution observations of solar active regions at millimeter wavelengths and their possible usefulness in the measurement of chromospheric magnetic fields. The observations discussed here were taken at 9 and 3.5 mm wavelengths with the 36-ft radio telescope at the National Radio Astronomy Observatory, situated at Kitt Peak in Tucson, Arizona. The half-power beamwidths of the telescope are 3.5 and 1.2 arc respectively at 9 and 3.5 mm wavelengths.

The observing procedure has been discussed previously (Kundu, 1970); it consists of scanning the Sun in right ascension at the rate of 1° per minute, the successive scans being separated by 1' of arc in declination. A complete solar map consists of a 45' by 37' arc grid of data points, centered on the Sun. Typical maps at 9 and 3.5 mm wavelengths are shown in Figure 1. The active regions were scanned repeatedly over a RA range of 35' arc and at declination intervals of 1' arc over a suitable declination range in which one or more dominant active regions are located. These scans are repeated at time intervals of approximately d minutes, where d is the declination range in arc minutes. A typical set of active region maps at 3.5 mm successively taken over a period are shown in Figure 2.

The maps at 3.5 and 9 mm are quite similar, although the maps at 3.5 mm show considerably more structure than the 9 mm maps, obviously due to higher angular resolution. The maps at 9 and 3.5 mm both show sources whose peak excess brightness temperatures can reach up to 1000 K and 700 K respectively, representing about 10–15% of the quiet Sun radiation at these wavelengths. The regions at 9 and 3.5 mm are distributed practically all over the disk. Such a distribution is typical of Calcium plage regions. Indeed the 9 and 3.5 mm regions have very close correspondence in size and intensity with the plage regions. The intense sources at both 9 and 3.5 mm usually appear in pairs; these correspond to bipolar regions on magnetograms. A complex region at 3.5 mm consisting of three or more components usually corresponds

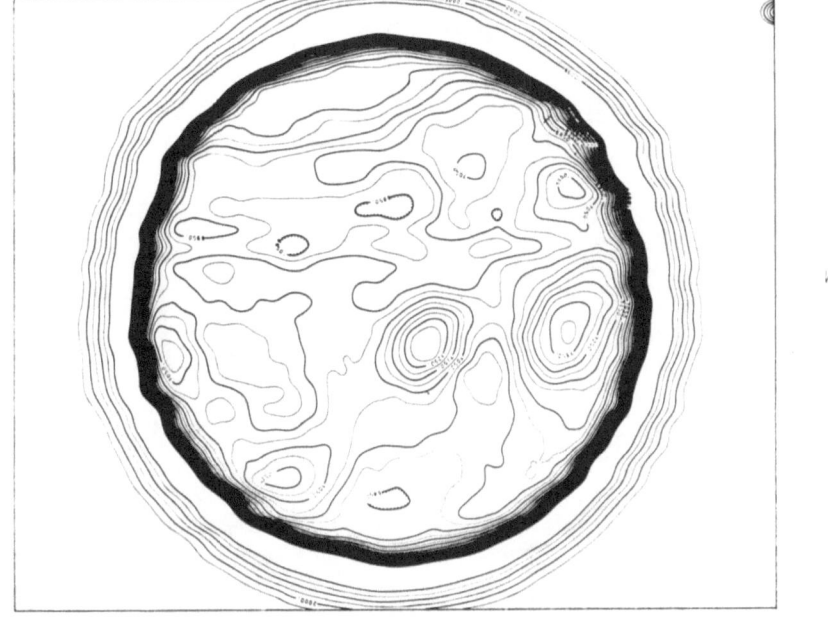

Fig. 1a. Solar Map at 9 mm-λ for Sept. 17, 1969. The grid used was 45' by 37' in RA and δ at intervals of 1' arc in δ. The regions marked by small dashes indicate temperature depressions (from Kundu, 1970).

to an equally complex region on magnetograms. Such complex structure is greatly smoothed out on the 9 mm maps because of lower resolution, although the existence of several components can often be discerned. Figure 3 shows the magnetogram which should be compared with the 3.5 and 9 mm maps of Figures 1a and 1b. The very close correspondence between the mm regions and the magnetic regions is quite clear. Detailed studies (Kundu, 1970) indicate that (a) the size of the mm regions is very closely related to the Ca-plage regions as well as the magnetic regions; (b) the excess brightness temperatures of the 3.5 and 9 mm regions is proportional to the magnetic field strength of the plage regions; and (c) the excess brightness temperatures at 3.5 and 9 mm are very poorly correlated with magnetic fields and areas of associated sunspots.

One of the most important characteristics of most 3.5 mm and some 9 mm maps is the existence of 'temperature depressions' or 'absorption features'. These features usually correspond to dark disk filaments observed in Hα. As we know, the dark filaments form in regions of zero longitudinal magnetic field; the filamentary material is supported by a magnetic field parallel to the surface of the Sun. It is relevant to note that the temperature contours between the two components of a mm-λ double source

NORTH

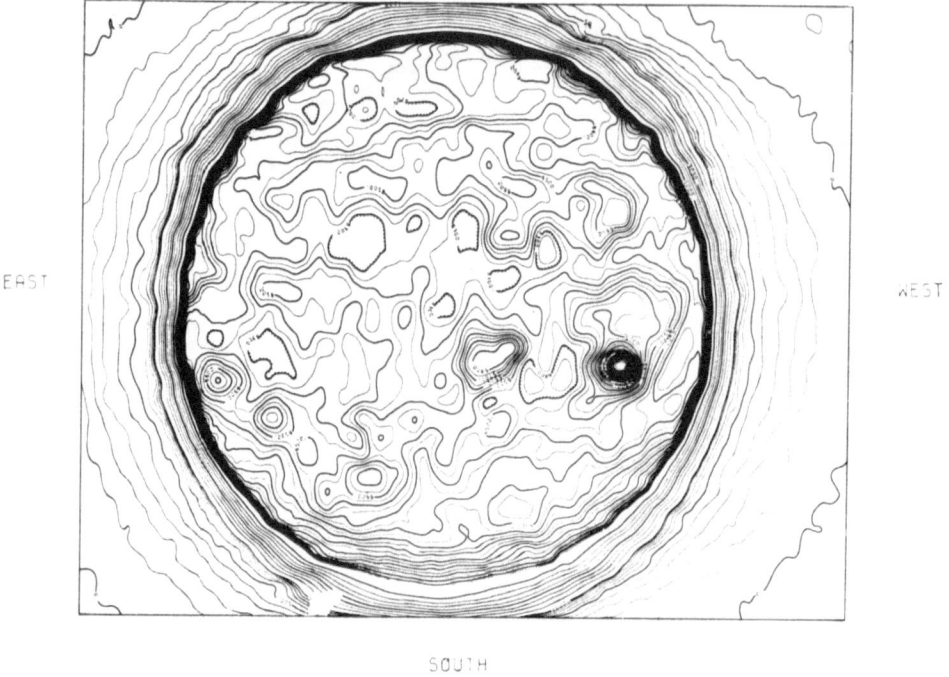

EAST

WEST

SOUTH

Fig. 1b.   Solar Map at 3.5 mm-λ for Sept. 17, 1969. The grid used was 45′ by 37′ in RA and δ at intervals of 1′ arc in δ. The regions marked by small dashes indicate temperature depressions (from Kundu, 1970).

very often delineates the boundary of the neutral region of the magnetic bipolar region even when the absorption feature is not so obvious. If we interpret the absorption as being caused by free-free absorption of the radio radiation by the filamentary material having high electron density and lower temperature, then we may have a method of estimating the electron density in the filament from the observed amount of absorption at 3.5 mm wavelength. The mean optical depth τ is given by $\tau = Kl$, K being the absorption coefficient and l the dimension of the filamentary material through which the radio radiation passes. Since K depends upon the wavelength λ, electron density N_e and electron temperature T_e, therefore assuming an appropriate value of T_e, one can determine N_e. If one can measure the absorption at a second wavelength with comparable angular resolution, one can estimate both N_e and T_e. The measurement of absorption at a number of millimeter wavelengths should lead to a fairly good estimate of the height of the disk filaments, since at a certain wavelength the absorption will cease to be observable.

The properties of the slowly varying radiation at millimeter wavelengths seem to be consistent with the thermal origin of this radiation in the higher than normal density regions at chromospheric temperatures. Indeed, the spectrum of brightness

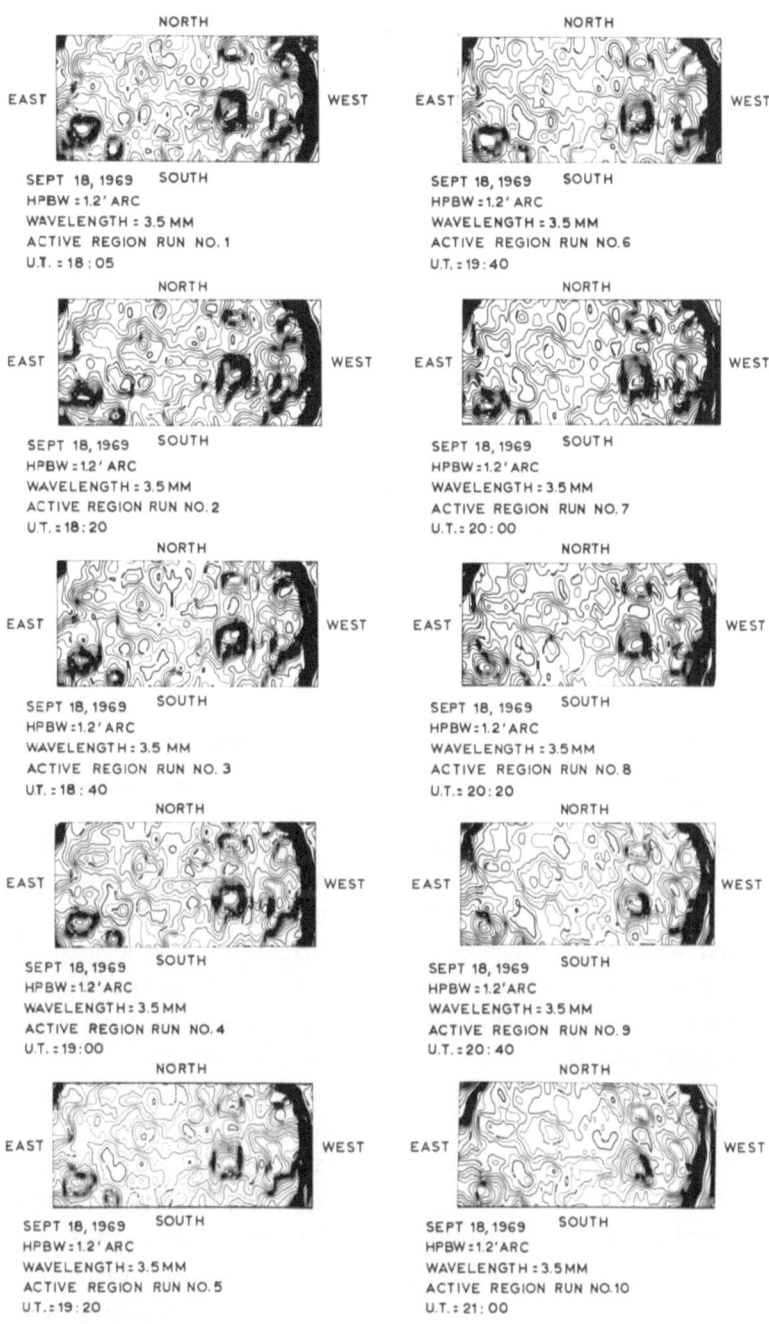

Fig. 2.   A typical set of active region maps at 3.5 mm wavelength, illustrating the time
evolution of active region.

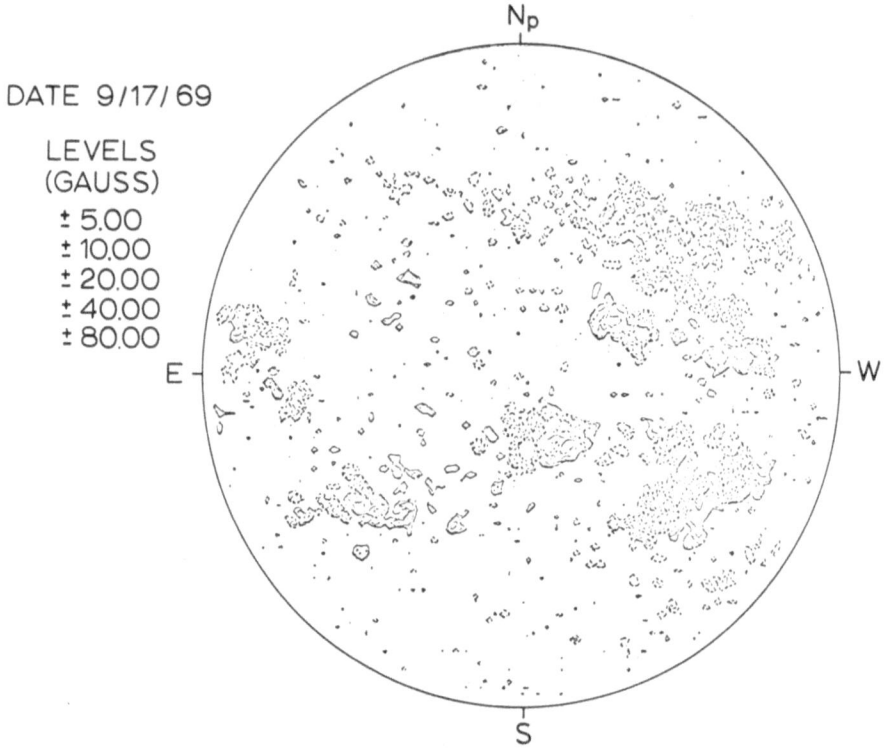

Fig. 3.   Mount Wilson magnetogram for Sept. 17, 1969. The angular resolution used was 17″ arc.

temperature which decreases with decreasing wavelength is compatible with this explanation. It is quite obvious that the gyroresonance absorption which is invoked to explain the slowly varying component at centimeter wavelengths cannot play any role at millimeter wavelengths, since the required magnetic field is at least an order of magnitude stronger than observed. This interpretation is strengthened by the fact that there is poor correlation between brightness temperature and sunspot magnetic field. The thermal radiation at millimeter wavelengths in the presence of a magnetic field should be circularly polarized; the degree of circular polarization can be used to measure the longitudinal fields in the chromosphere. Indeed for quasi-longitudinal propagation of radio waves in the magneto-ionic medium of the chromosphere, the degree of polarization is related by a simple expression to the magnetic field (see for example, Kundu, 1965). As we have noted earlier, the millimeter wavelength regions show very good correspondence with the regions of longitudinal magnetic fields on magnetograms; however, the angular resolution used for magnetograms is 17″ as against 3′.5 and 1′.2 for 9 and 3.5 mm maps. In order to compare the two kinds of observations in a more meaningful manner, we have smoothed the magnetogram shown in Figure 3, with equivalent beams corresponding to those of 3.5 and 9 mm observations. The resulting 'smoothed' maps are shown in Figures 4 and 5. A compa-

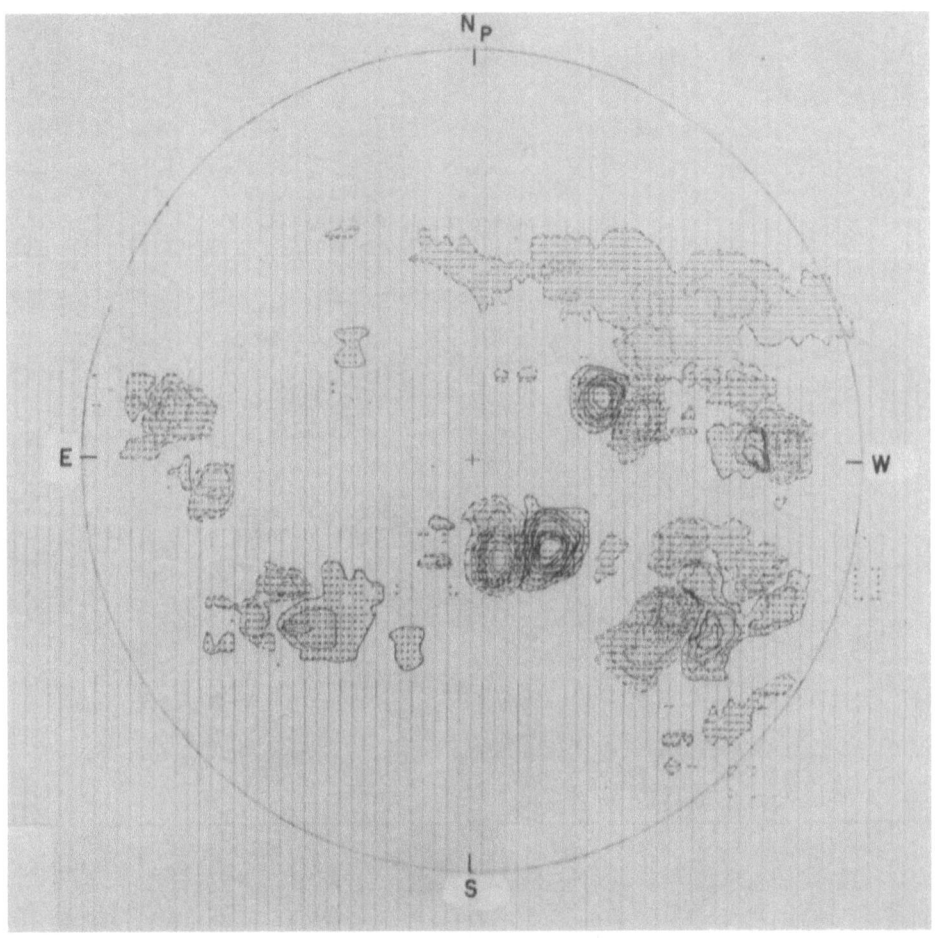

Fig. 4a.   Magnetogram of Sept. 17, 1969 (shown in Figure 3), smoothed by a pencil beam of 1'.2 arc. The regions delineated by continuous and broken lines correspond to north and south polarities respectively. The outermost contour corresponds to a magnetic field of 0.03–3 G; the successive inner contours correspond to magnetic fields of 3–9, 9–15, 15–21, 21–27, 27–33, 33–39, 39–45 and > 45 G respectively.

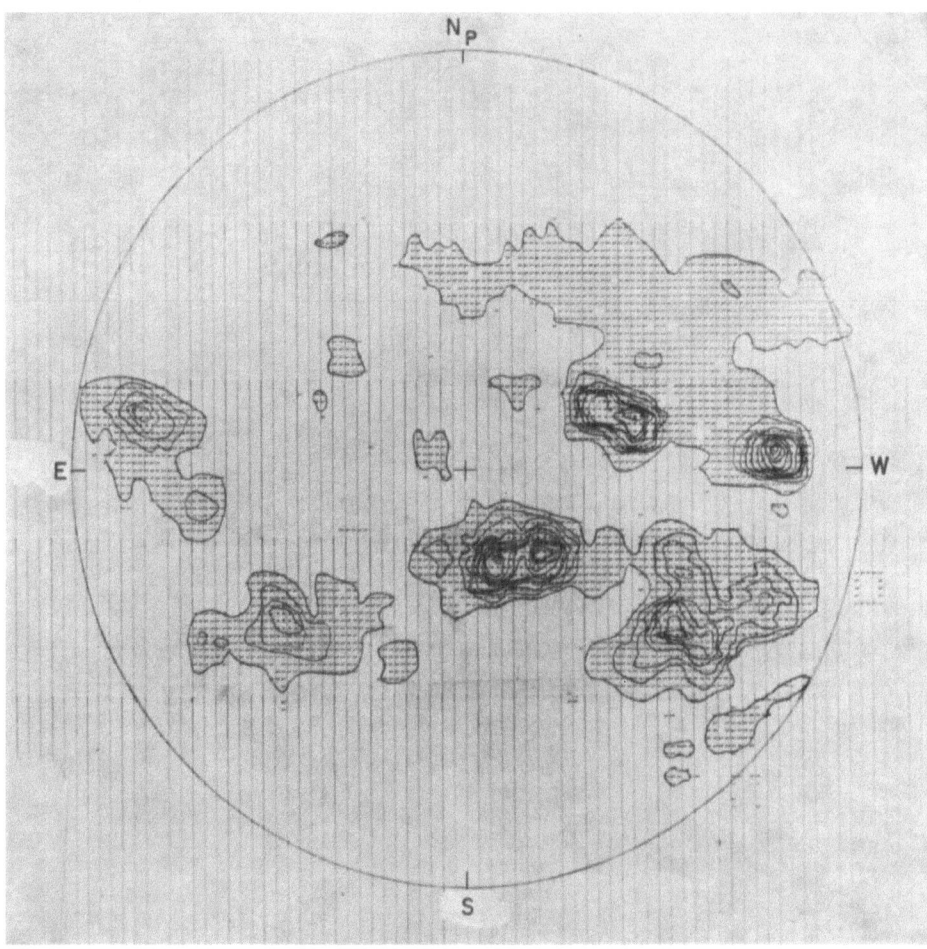

Fig. 4b. Same as Figure 4a, except that opposite polarities are not distinguished here, for direct comparison with the 3.5 mm map. The outermost contour corresponds to a magnetic field of 0.03–3 G; the successive inner contours correspond to magnetic fields of 3–9, 9–15, 15–21, 21–27, 27–33, 33–39. 39–45, 45–51, 51–57, 57–63, 63–69, and 69–75 G respectively.

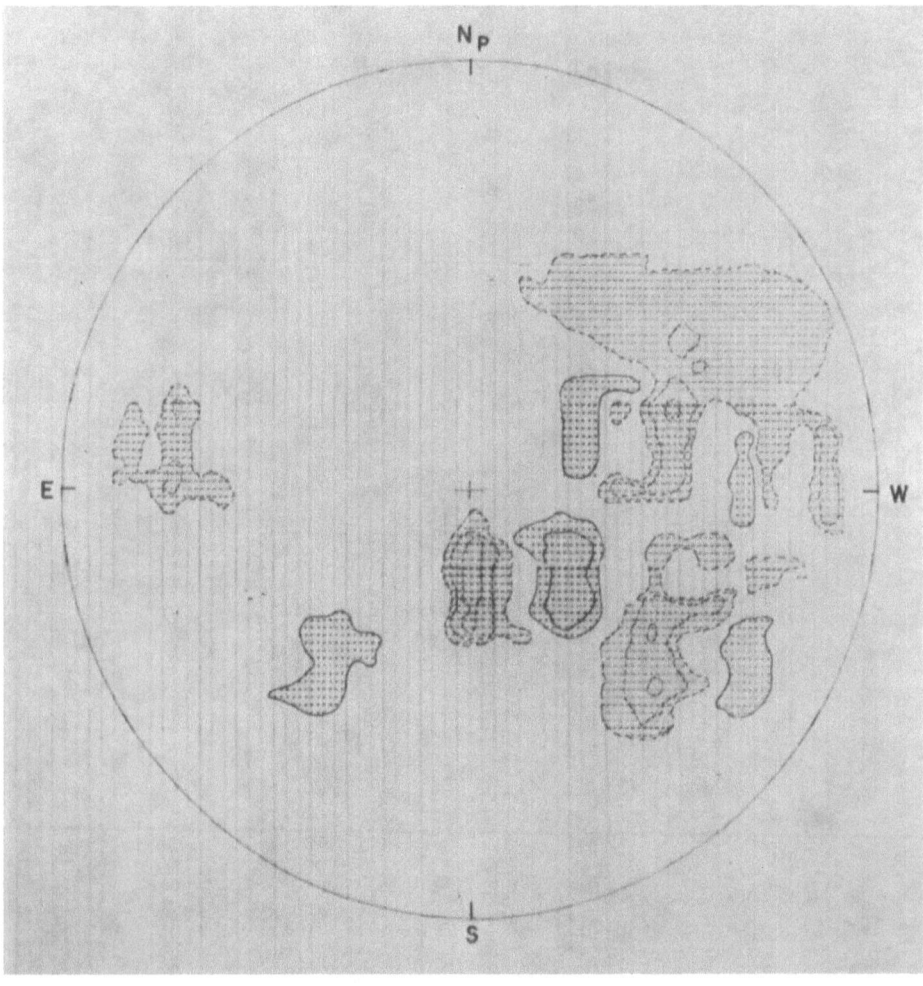

Fig. 5a.   Magnetogram of Sept. 17, 1969 (shown in Figure 3), smoothed by a pencil beam of 3'.5 arc. The regions delineated by continuous and broken lines correspond to north and south polarities respectively. The outermost contour corresponds to 0.03–3 G; the successive inner contours correspond to 3–9, 9–15 and 15–21 G respectively.

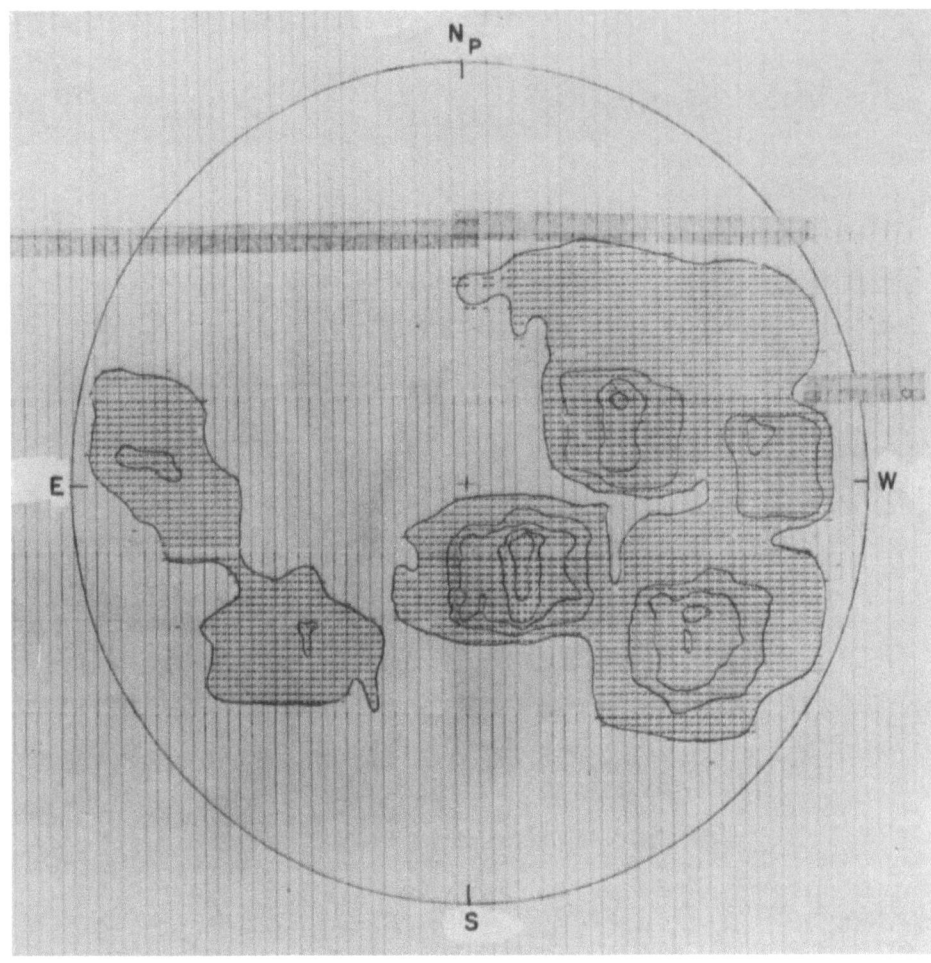

Fig. 5b.   Same as Figure 5a except that opposite polarities are not distinguished here, for direct
comparison with the 9 mm map. The outermost contour corresponds to 0.03–3 G; the successive
inner contours correspond to 3–9, 9–15, 15–21, 21–27, 27–33 and 33–39 G respectively.

rison of these maps with those of Figure 1 establishes the one-to-one correspondence between the 3.5 and 9 mm active regions on the hand, and the magnetic regions on magnetograms on the other. Thus, the assumption of quasi-longitudinal propagation seems to be justified. Consequently, if one can measure the degree of circular polarization at 3.5 and 9 mm-λ, one should be able to get fairly good estimates of longitudinal magnetic fields at corresponding heights in the chromosphere.

References

Kundu, M. R.: 1965, *Solar Radio Astronomy*, John Wiley-Interscience Publishers, p. 234.
Kundu, M. R.: 1970, *Solar Phys.* **13**, 348.

RADIO-ASTRONOMICAL EVIDENCE FOR MAGNETO-
HYDRODYNAMICAL PULSATIONS IN THE CORONA

HANS ROSENBERG

Observatory 'Sonnenborgh', Utrecht, The Netherlands

At Utrecht Observatory a 60-channel solar radiospectrograph has been in regular operation since the fall of 1968, in the bandwidth 160–320 MHz, with a time resolution of $0.^{s}03$ (De Groot and Van Nieuwkoop, 1968). In addition to all generally known and classified emissions as noise storms, type IV bursts, type III bursts, type IV continua, a number of unclassified emissions have been observed. The spectrograph is particularly well suited to study short time scale fluctuations, even of small amplitude. One striking feature is an often recurring broad band (> 80 MHz) weak, quasiperiodic fluctuation superimposed on a type IV-like continuum (Figure 1). Since no interferometric measurements were available, it is not known whether it is a stationary or a moving type IV burst. However, due to the sometimes very long duration, we expect it to be a stationary type IV.

We have suggested (Rosenberg, 1970) that the fluctuation in the synchrotron radiation causing the type IV emission is due to the fluctuation of the local magnetic field strength. The time scale of such a fluctuation will in general be determined by the Alfvén-velocity and the characteristic length of the magnetic field inhomogeneities, in the direction perpendicular to the magnetic field. Assuming field strengths in the order of 3 G and local densities of 10^8 cm^{-3}, while using a characteristic time of $0.^{s}3$, the scale of the inhomogeneities equals 1000 km, in accordance with optical and radio measurements (Hewish and Symonds, 1969; Dennison, 1969; Van de Hulst, 1950).

Often a low-frequency cutoff is observed around 220 MHz, which has also been noted by other observers (Takakura, 1963; Takakura and Kai, 1961) and has been explained by Ramaty and Lingenfelter (1967) and others as being a suppression of the synchrotron emission due to the influence of the ionized medium (Razin effect). In some cases, however, the ridges bend towards later times at the lower frequencies (Figure 2). We think this to be due to group retardation effects as proposed by Jaeger and Westfold (1950) when they tried to explain type III bursts by this process. This will be investigated in more detail.

On some occasions type I storms have been observed, during which the type I bursts did not occur in chains (as they generally do), but were more situated on the intensity maxima of the underlying pulsating structure (top of Figure 1).

Finally in a few examples we found groups of type III bursts starting from a pulsating structure, where again every type III burst seemed to be associated with a maximum in the pulsating structure.

We therefore propose the following model, which we are studying in more detail at present. The pulsating structure is due to synchrotron radiation in a force-free

Howard (ed.), Solar Magnetic Fields, 652–655. All Rights Reserved.
Copyright © 1971 by the IAU.

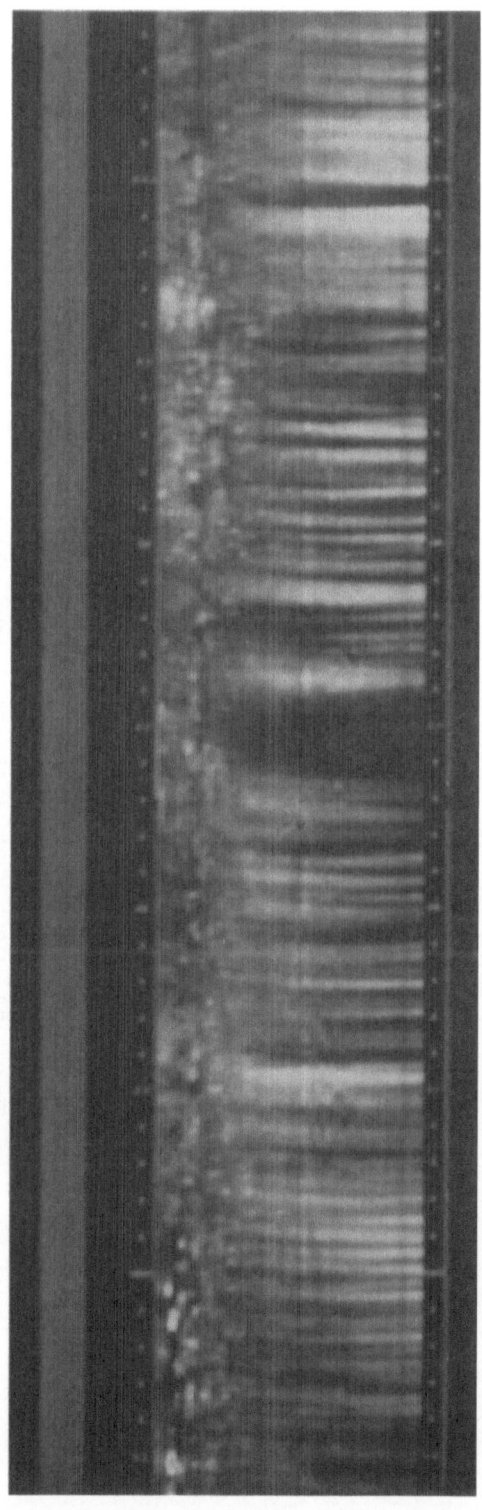

Fig. 1. 25 February 1969; $09^h08^m54^s$–$09^h09^m34^s$; 160 MHz at top, 320 MHz at bottom of the picture; distance between two dots equals one second.

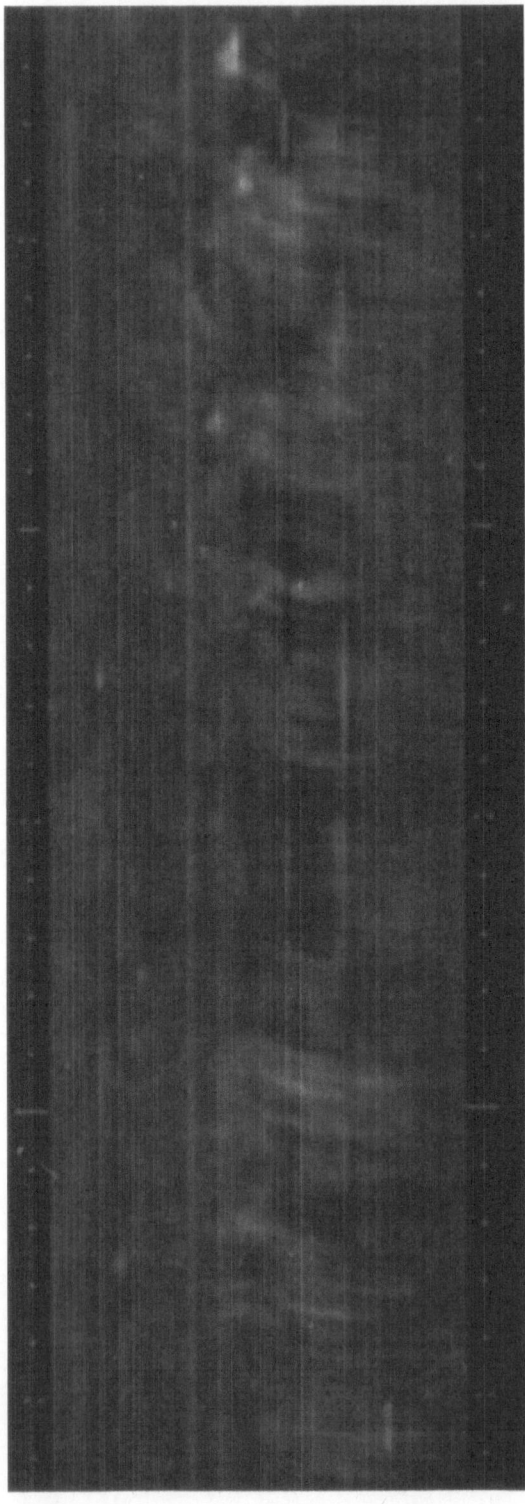

Fig. 2. 10 June 1969; $15^h03^m54^s–15^h04^m19^s$; curved pulsating structure.

magnetic flux tube which may reach quite high into the corona. A compression of the tube, thus enhancement of the field strength, corresponds to a maximum in intensity, an expansion to a minimum. The mechanism responsible for the type I radiobursts seems to be preferentially active at the maximum field strength.

The release of fast charged particles, causing type III bursts, also seems to occur at the moment of highest intensity of the flux tube. These fast particles are apparently stored in the flux tube and can be released periodically at the moments of highest compression.

The stability of such a flux tube has been studied in recent years by a number of people (Callebaut and Voslamber, 1962; Anzer, 1968), finding seemingly contradicting results. It is quite certain that a very long and thin flux tube will be unstable. However the critical numbers for which stability is obtained are not available yet. Furthermore the influence of the bending of a straight flux tube (with a helical field) into a semicircle on the stability, and the effect of the solar wind has to be taken into account. This programme is at present being carried out at the Utrecht Observatory.

References

Anzer, U.: 1968, *Solar Phys.* **3**, 298.
Dennison, P. A.: 1969, *Planetary Space Sci.* **17**, 189.
Groot, T. de and van Nieuwkoop, J.: 1968, *Solar Phys.* **4**, 332.
Hewish, A. and Symonds, M. D.: 1969, *Planetary Space Sci.* **17**, 313.
Hulst, H. C. van de: 1950, *Bull. Astron. Inst. Neth.* **11**, 150.
Jaeger, J. C. and Westfold, K. C.: 1950, *Australian J. Sci. Res.* **A3**, 376.
Ramaty, R. and Lingenfelter, R. E.: 1967, *J. Geophys. Res.* **72**, 879.
Rosenberg, H.: 1970, *Astron. Astrophys.* **9**, 159.
Takakura, T.: 1963, *Publ. Astron. Soc. Japan* **15**, 327.
Takakura, T. and Kai, K.: 1961, *Publ. Astron. Soc. Japan* **13**, 94.
Voslamber, D. and Callebaut, D. K.: 1962, *Phys. Rev.* **128**, 2016.

Discussion

Boischot: How intense are the fluctuating structures compared to the intensity of the continuum?

Rosenberg: The amplitudes are very variable. During the short ($\tau \sim 0^s.3$) pulsations they may be only a few percent, but we have observed long pulsations ($\tau \sim$ a few seconds) with amplitudes of 20–30%.

ON THE ORIENTATION OF MAGNETIC FIELDS IN
QUIESCENT PROMINENCES

ULRICH ANZER* and E. TANDBERG-HANSSEN

*High Altitude Observatory, National Center for Atmospheric Research**, Boulder, Colo., U.S.A.*

1. The Observations and Their Reduction

The longitudinal component of the magnetic field in quiescent prominences has been measured directly with magnetographs using the Zeeman effect on selected spectral lines (Rust, 1966; Ioshpa, 1968; Harvey, 1969). We know that as a general rule the magnetic field enters the, largely-vertical, sheet-like quiescent prominence on one side and exits on the other. The field traverses the prominence plasma with components both along and at right angles to the long axis of the prominence. It is the purpose of this paper to describe observations that may indicate the relative importance of the two components of the magnetic field, and to derive a distribution function for the magnetic field vectors.

The data for the longitudinal component of the magnetic field, $B_\parallel$, are based on observations of about 70 quiescent prominences studied with the Climax magneto-graph during the period 1968–69 (see the paper Tandberg-Hanssen: 'Observations of Magnetic Felds in Quiescent Prominences', this volume, p. 192). We included only those prominences for which the approximation of a plane-parallel sheet seemed satisfactory. Let us consider an orthogonal coordinate system with x-axis horizontal and perpendicular to the (idealized) plane prominence sheet, y-axis horizontal and along the filament and z-axis perpendicular to the photosphere. The observations of a prominence on the solar limb near the equator and in the plane of the sky give $B_\parallel = B_x$, see Figure 1, prominence 1. In the more general case, as a prominence at latitude ϕ and not in the plane of the sky move onto the disk and toward the central meridian, we have measured the angle θ_1 that the assumed vertical plane filament makes with the north-south direction on the Sun, see Figure 1, prominence 2. Figure 2 shows how the observed longitudinal magnetic field, $B_\parallel$, for a given prominence depends on this angle between the y-axis and the north-south direction. We emphasize the uncertainties involved in determining this angle for any filament, but taken at face value Figure 2 indicates that $B_\parallel$ increases with increasing angle θ_1; in other words $B_y > B_x$.

The procedure outlined to determine the angle θ_1 is valid only if the filament does not change in shape or orientation during the time from its limb passage until observed near the central meridian. Furthermore, this measurement is only good up to a certain latitude, as the following discussion will show.

Let θ_1 be the observed angle at solar latitude ϕ. Then, the corresponding angle θ,

---

* On leave from Max-Planck-Institut für Physik und Astrophysik, München.
** The National Center for Atmospheric Research is sponsored by the National Science Foundation.

Howard (ed.), Solar Magnetic Fields, 656–662. All Rights Reserved.
Copyright © 1971 by the IAU.

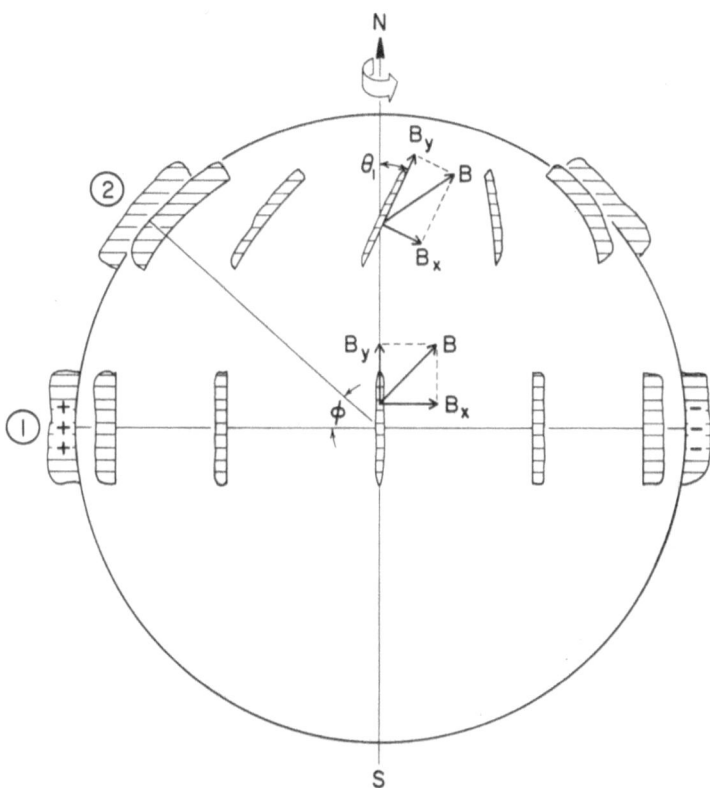

Fig. 1. Parameters defining (a) prominence in the plane of the sky on the solar equator (Prominence 1) and (b) a prominence at latitude φ and not in the plane of the sky (Prominence 2).

which one would measure if the prominence were on the equator, is given by

$$\tan \theta = \tan \theta_1 \cos \phi . \tag{1}$$

For high-latitude prominences (the polar crown, $\phi > 60°$) the measured angle θ_1 differs significantly from the angle θ. Since we know ϕ from the observations, we could, in principle, correct our data for this effect. But for ϕ and θ_1 close to 90° the value of θ varies strongly with θ_1. Since θ_1 cannot be determined very accurately, the calculated θ would be much less accurate. Consequently, we did not correct θ according to Equation (1), but treated the observations of polar-crown filaments as a separate set of data.

2. Relation Between the Observed Field Distribution and the True Angular Distribution

From the observations, see Figure 2, we can calculate a mean value, $\bar{B}_{\parallel}$, for each θ, and in principle we can describe the relationship by a function, $F_{obs}(\theta)$ say. However, of primary interest is the angle, α, between the filament and the magnetic field, see Figure 3.

To proceed we make a number of assumptions:

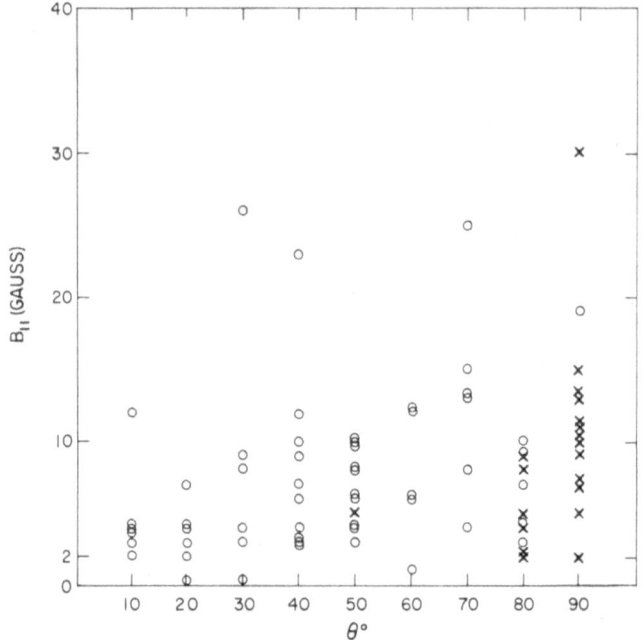

Fig. 2. The observed longitudinal component of the magnetic field, $B_{\parallel}$, plotted against the angle θ between the north-south direction on the Sun and the long axis of the prominence. The crosses indicate polar crown prominences.

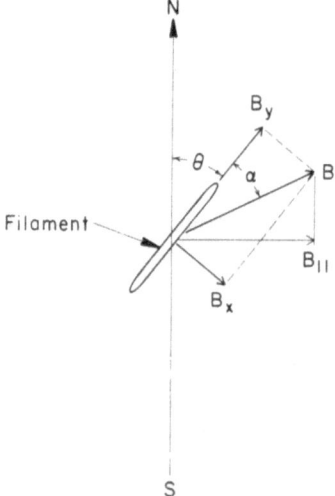

Fig. 3. The relationship between the angles α and θ, and the components of the magnetic field vector B.

(i) The magnetic field is horizontal (a possible z component may be added without changing the results). Then

$$B_{\parallel} = B_0 |\sin(\alpha + \theta)| \qquad (2)$$

where α and θ can take values $0 \leqslant \alpha \leqslant \pi$, and $-\pi/2 \leqslant \theta \leqslant +\pi/2$.

(ii) The horizontal magnetic field vectors in quiescent prominences may be described by a distribution function, viz.

$$\frac{N}{N_{\text{tot}}} dB_0 \, d\alpha = f \, (B_0, \, \alpha) \, dB_0 \, d\alpha, \tag{3}$$

where N is the number of filaments having a field vector of strength B_0, making an angle α with the y-axis, and N_{tot} is the total number of filaments. Equations (2) and (3) provide the average line-of-sight field

$$\bar{B}_{\parallel} \, (\theta) = \int_0^\infty \int_0^\pi B_0 \, |\sin \, (\alpha + \theta)| \, f \, (B_0, \, \alpha) \, d\alpha \, dB_0 \,. \tag{4}$$

(iii) The distribution function is separable, i.e.

$$f \, (B_0, \, \alpha) = f_1 \, (B_0) \, f_2 \, (\alpha). \tag{5}$$

This is not correct if the field strength and the angle α are correlated. To the best of our knowledge, however, such a correlation has not been shown. With Equation (5) we find

$$\bar{B}_{\parallel} \, (\theta) = \int_0^\infty B_0 f_1 \, (B_0) \, dB_0 \int_0^\pi |\sin \, (\alpha + \theta)| \, f_2 \, (\alpha) \, d\alpha = \bar{B}_0 F \, (\theta). \tag{6}$$

Now, if $F(\theta) = \overline{\bar{B}_{\parallel} \, (\theta)/B_0}$ is given from the observations, one has to determine $f_2 \, (\alpha)$ such that

$$F \, (\theta) = \int_0^\pi |\sin \, (\alpha + \theta)| \, f_2 \, (\alpha) \, d\alpha \,. \tag{7}$$

(iv) The distribution is symmetric around $\alpha = \pi/2$, or $f_2(\pi - \alpha) = f_2(\alpha)$. Then $F(\theta)$ is symmetric with respect to $\theta = 0$ and

$$F \, (\theta) = 2 \left(\cos \theta \int_\theta^{\pi/2} \sin \alpha f_2 \, (\alpha) \, d\alpha + \sin \theta \int_0^\theta \cos \alpha f_2 \, (\alpha) \, d\alpha \right). \tag{8}$$

To invert this integral equation for $f_2 \, (\alpha)$ we note that

$$\tfrac{1}{2} F' \, (\theta) = - \sin \theta \int_\theta^{\pi/2} \sin \alpha f_2 \, (\alpha) \, d\alpha + \cos \theta \int_0^\theta \cos \alpha f_2 \, (\alpha) \, d\alpha, \tag{9}$$

and

$$\tfrac{1}{2} F'' \, (\theta) = - \cos \theta \int_\theta^{\pi/2} \sin \alpha f_2 \, (\alpha) \, d\alpha + \sin^2 \theta f_2 \, (\theta) +$$

$$+ \sin \theta \int_0^\theta \cos \alpha f_2 \, (\alpha) \, d\alpha + \cos^2 \theta f_2 \, (\theta), \tag{10}$$

whence

$$f_2(\alpha) = \tfrac{1}{2}[F''(\alpha) + F(\alpha)]. \tag{11}$$

Equation (9) shows that

$$F'(0) = 0 = F'\left(\frac{\pi}{2}\right) \tag{12}$$

for all functions f_2 which do not contain δ-functions at $\alpha=0$, $\alpha=\pi/2$. Therefore, only functions $F(\theta)$ which fulfill Equation (12) can be used in Equation (11).

3. Numerical Examples

It would seem a reasonable approach to try to approximate the observations by simple analytic curves $F(\theta)$ which fulfill conditions (12), and then calculate $f_2(\alpha)$ from Equation (11). However, for polynomials $F(\theta)$ in reasonable agreement with observations the calculated $f_2(\alpha)$ turns out to be not positive definite over the whole range $0 \leqslant \alpha \leqslant \pi/2$. This means that we have to find an analytic function $F(\theta)$ which describes the observations well enough, and at the same time gives a positive definite distribution function $f_2(\alpha)$. Since our observations do not define $F_{\text{obs}}(\theta)$ with sufficient accuracy, this procedure amounts to trying any possible function $F(\theta)$ and calculate the corresponding $f_2(\alpha)$ until one finds a solution which fulfills $f_2(\alpha) \geqslant 0$ for $0 \leqslant \alpha \leqslant \pi/2$.

Instead, we assume some simple distribution functions $f_2(\alpha)$, then calculate the $F(\theta)$'s and compare these functions with observations. We normalize $f_2(\alpha)$ to give $\int_0^\pi f_2(\alpha)d\alpha = 1$, and use Equation (8) to calculate $F(\theta)$.

Case 1 $f_2(\alpha) = \dfrac{1}{\pi}$.

This case gives $F(\theta) = 2/\pi$.

Case 2 $f_2(\alpha) = \tfrac{1}{2}[\delta(\alpha - \alpha_0) + \delta(\alpha - (\pi - \alpha_0))]$,   with $0 \leqslant \alpha_0 \leqslant \dfrac{\pi}{2}$.

This leads to

$$F(\theta) = \begin{cases} \cos\theta \sin\alpha_0 & \text{for } 0 \leqslant \theta \leqslant \alpha_0 \\[2mm] \sin\theta \cos\alpha_0 & \text{for } \alpha_0 \leqslant \theta \leqslant \dfrac{\pi}{2}. \end{cases}$$

Special cases are: $\alpha_0 = 0$ (the field is parallel to the filament), and $\alpha_0 = \pi/2$ (the field is along the x-axis, i.e. the Kippenhahn-Schlüter (1957) configuration).

Case 3 $f_2(\alpha) = \dfrac{2n+1}{2}\left(\dfrac{2}{\pi}\right)^{2n+1}\left(\dfrac{\pi}{2} - \alpha\right)^{2n}$.

With increasing n this gives higher and higher concentration close to $\alpha=0$. We then find

$$F(\theta) = \frac{2n+1}{2}\left(\frac{2}{\pi}\right)^{2n+1}\left[\cos\theta \int_0^{\pi/2-\theta} \alpha^{2n}\cos\alpha\, d\alpha + \sin\theta \int_{\pi/2-\theta}^{\pi/2} \alpha^{2n}\sin\alpha\, d\alpha\right].$$

These integrals may be solved by a recursion procedure which leads to

$$F_{2n} = \frac{2n+1}{(\pi/2)^2} \left[\frac{\pi}{2} \left(1 - \frac{\theta}{\pi/2} \right)^{2n} + 2n \sin\theta - 2nF_{2(n-1)} \right]$$

and $F_0 = 2/\pi$.

4. Interpretations

From the observations, Figure 2, the average field strength for each 10-degree interval in θ is calculated together with the root mean squares of the deviation, leaving out the four points which lie far above the bulk of the observed values. The results are plotted in Figure 4.

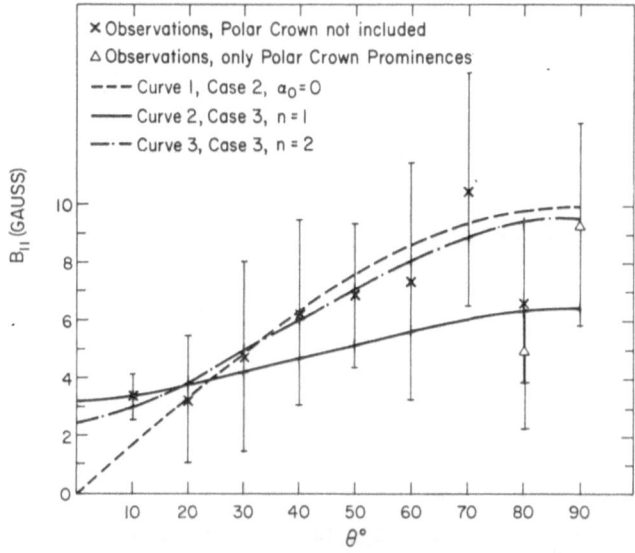

Fig. 4. Different functions, $F(\theta)$, from Equation (8) compared to the observed function $F_{obs}(\theta)$.

The observations immediately rule out a distribution function which is independent of α (Case 1, i.e. $F(\theta) = const$), as well as the configuration where the field is strictly perpendicular to the y-axis (Case 2, with $\alpha_0 = \pi/2$, i.e. $F(\theta) \propto \cos\theta$). The curves drawn in Figure 4 are (1) the field parallel to the y-axis (Case 2, with $\alpha_0 = 0$, i.e. $F(\theta) \propto \sin\theta$), (2) a quadratic distribution function (Case 3, with $n=1$), and (3) a fourth order distribution function (Case 3, with $n=2$).

Curve (1) seems to be in disagreement with the observations at small angles, $\theta \approx 10°$. Here the observed $\bar{B}_{\parallel}$ is about twice as large as given by curve (1). This means that the field in general is not exactly parallel to the filament. Curve (2) is much too flat, i.e. the ratio $F(\pi/2)/F(0)$ is about a factor of two too small. The curve which fits best is curve (3), calculated from the distribution function

$$f_2(\alpha) = \frac{5}{2} \left(\frac{2}{\pi} \right)^5 \left(\frac{\pi}{2} - \alpha \right)^4, \tag{13}$$

shown in Figure 5.

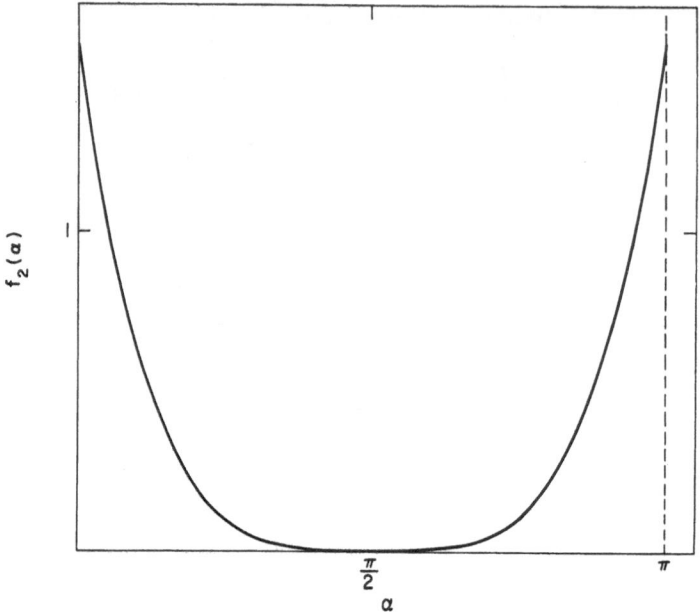

Fig. 5.   The distribution function, $f_2(\alpha)$ from Equation (13).

Using the distribution function given by Equation (13), we can calculate the mean angle, $\bar{\alpha}$,

$$\bar{\alpha} = \int\limits_0^{\pi/2} f_2(\alpha)\,\alpha\,\mathrm{d}\alpha \bigg/ \int\limits_0^{\pi/2} f_2(\alpha)\,\mathrm{d}\alpha = \frac{1}{6}\frac{\pi}{2},$$

or $\bar{\alpha} \approx 15°$. This indicates that the magnetic field traverses the quiescent prominence sheet under a small but finite angle. From curve (3) of Figure 4 and Equation (6) we find that the average total field strength $\bar{B}_0$ is about 10 G.

References

Harvey, J.: 1969, Ph.D. Diss., Univ. of Colorado.
Ioshpa, B.: 1968, *Results of Researches on the International Geophysical Projects, Solar Activity*, No. 3, Nauka, Moscow, p. 44.
Kippenhahn, R. and Schlüter, A.: 1957, *Z. Astrophys.* **43**, 36.
Rust, D.: 1966, Ph.D. Diss., Univ. of Colorado.

OBSERVATIONS OF THE CORONAL NETWORK

GEORGE W. SIMON

Sacramento Peak Observatory, Sunspot, N. Mex. 88349, U.S.A.

and

ROBERT W. NOYES

Harvard College Observatory, Cambridge, Mass. 02138, U.S.A.

Abstract. Spectroheliograms of quiet regions obtained by the OSO-G satellite in chromospheric, transition zone, and coronal lines indicate that the chromospheric network as seen in Ca K can be resolved despite the low resolution (30 arc s) of OSO-G. The network pattern becomes more diffuse with increasing height, indicating that the magnetic field pattern spreads with height.

We wish to present a few preliminary results obtained from the Harvard experiment on the OSO-G satellite. These observations were made in a number of EUV lines

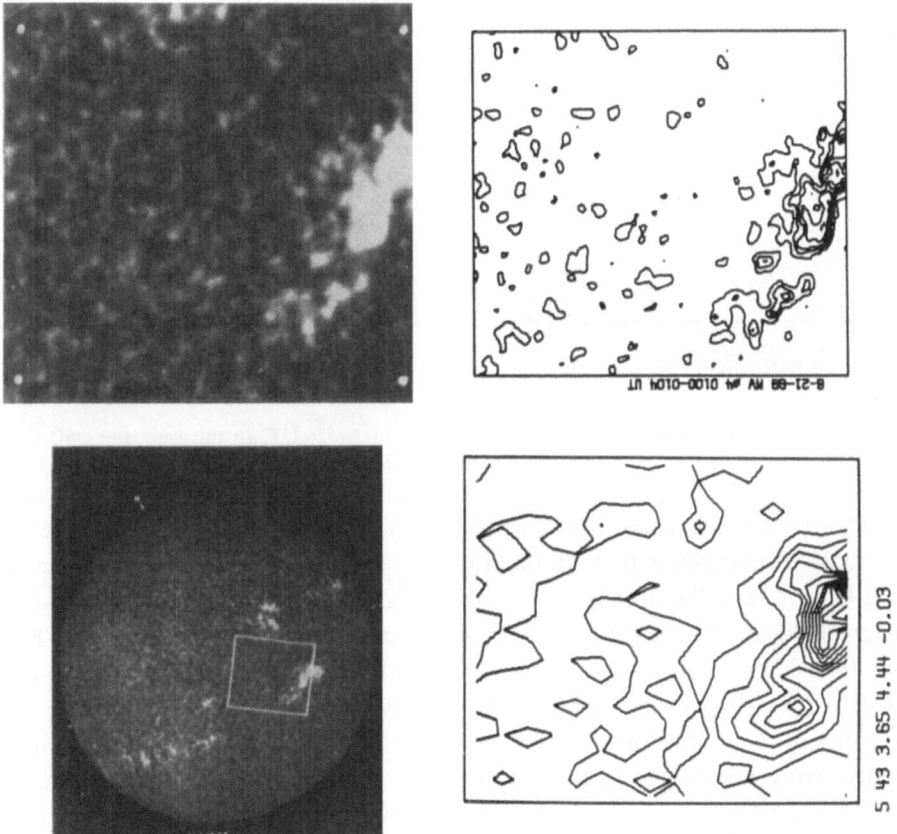

Fig. 1.   Simultaneous Mt. Wilson Ca II 3934 and OSO-G satellite Ly Cont 897 spectroheliograms of 21 August 1969, 0100 UT. *Lower left:* Mt. Wilson spectroheliogram showing area observed by OSO-G. *Upper left:* Enlargement in Ca II of observed area. *Upper right:* Contour map of Ca II spectroheliogram. *Lower right:* Contour map of OSO-G spectroheliogram.

Howard (ed.), Solar Magnetic Fields, 663–666. All Rights Reserved.

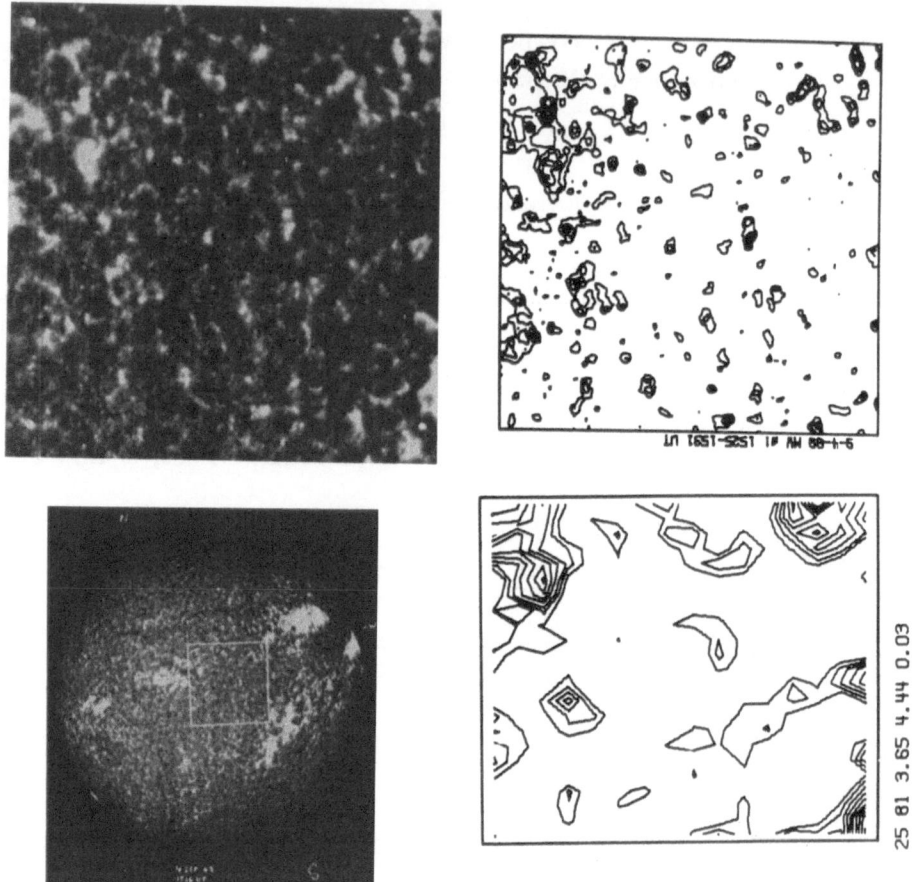

Fig. 2.   Similar to Figure 1, taken on 4 September 1969, 1526 UT. In this case, the OSO-G data
are of C$_{II}$ 1335.

which are formed between 2000 and 15000 km above the photosphere, that is, from
the high chromosphere, through the transition zone, into the low corona. Since it is
well-known that there exists a strong spatial correlation between photospheric
magnetic fields and the chromospheric calcium K brightness network, it was hoped
that these higher temperature observations would enable us to extend this correlation
upward into the corona. However, there exist two difficulties. First, the photon
statistics in the coronal lines were not sufficient to draw convincing conclusions, so
that we were able only to go to the height of oxygen VI, which is formed in the transition zone at a temperature of 300000K, 5000 km above the photosphere, that is,
about 3000 km above calcium K. Second, we were trying to observe the quiet-sun
network pattern which has a typical scale of 30000 km, using a telescope with an
entrance aperture almost as large, namely 22000 km (30 arc s) on a side. Thus there were
doubts whether the network structure could even be resolved with such a large
aperture.

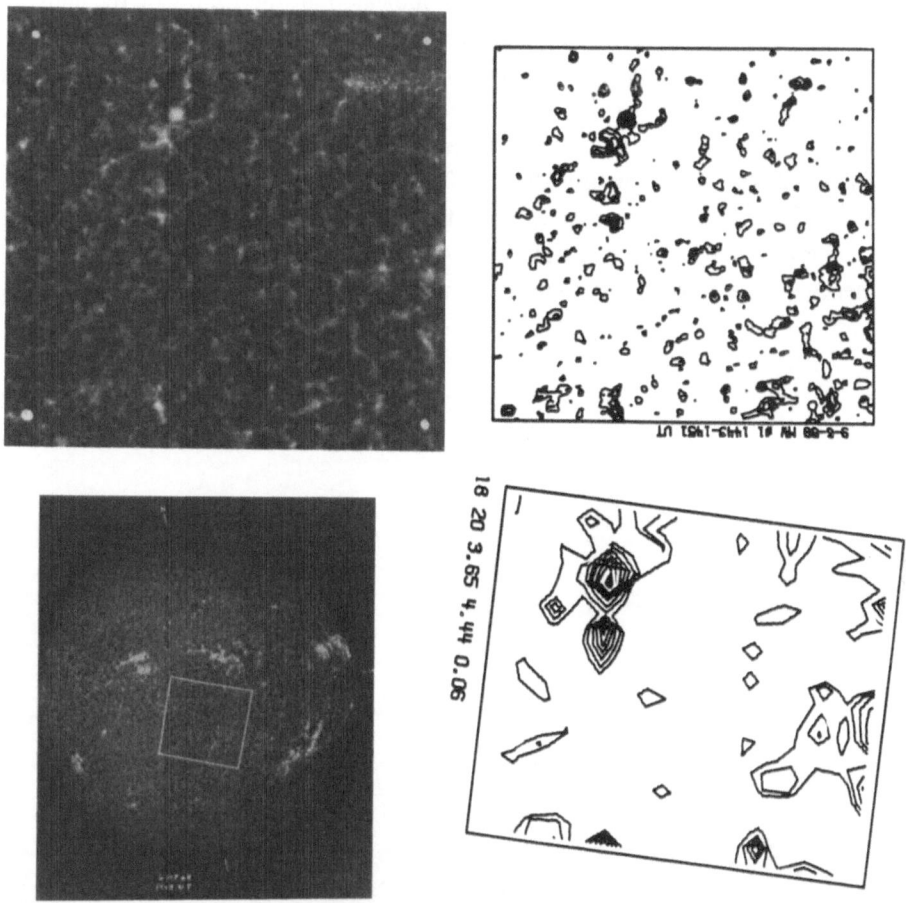

Fig. 3. Similar to Figure 1, taken on 6 September 1969, 1445 UT, with the OSO-G data in the line C II 1335.

Keeping this in mind, we will now show you some qualitative results. The lower left hand part of Figure 1 shows a calcium spectroheliogram taken at Mt. Wilson simultaneously with the satellite observation. (We wish to thank Dr. Robert Howard and his staff for their cooperation in this program.) The marked area shows the size of the satellite raster, which is 300 000 km on a side. In the upper left you see an enlargement of the pertinent area of the calcium spectroheliogram, to the right of it a contour map of the spectroheliogram with 3 arc s resolution, and at the bottom right a contour map of the satellite observation, in this case obtained in the Lyman continuum at 897 Å, which comes from a height very close to that of the calcium K. Note that many of the bright points in the calcium do indeed appear in the satellite observation, despite the low resolution. This correlation shows up even more clearly, if the two contour maps are superimposed, rather than shown side by side as in this figure. Two more such examples are shown in Figures 2 and 3.

On a quantitative basis, we have obtained auto-correlation curves in a number of these EUV lines, and the preliminary results indicate that the widths of the emission features increase with increasing height up to a width of 17000 km at a height of 5000 km. We thus have the following picture of the network, and thus hopefully also of the magnetic pattern: In the photosphere we find widths under 3000 km; in Hα at a height of 1000 km we find 5000 km widths; in calcium K at 2000 km height, the width is 10000 km; and in oxygen VI at 5000 km height, the width is 17000 km.

We conclude that, as one might expect, the magnetic field continues to spread out with increasing height, and will at some still greater height in the corona, merge together to form a more or less continuous magnetic sheath covering the solar surface.

PRELIMINARY RESULTS OF SPECTROSCOPIC
DETERMINATION OF THE CORONAL ROTATION

V. E. STEPANOV and N. F. TJAGUN

Siberian Institute of Terrestrial Magnetism, Ionosphere and
Radio Wave Propagation, Irkutsk, U.S.S.R.

Abstract. Coronal rotation is determined by means of the spectroscopic method. The mean value of the rotation rate vs. the heliographic latitude is found. At high latitudes the corona rotates much faster than the underlying photosphere. This fact confirms Waldmeier's and Billing's hypothesis that the high-latitude phenomena should depend on the magnetic fields of low latitudes and that the interchange of the matter between the active zone and polar coronal region along the field lines should take place.

1. Introduction

Determinations of the coronal rotation have been made for the first time from observations of single long-lived coronal regions. Waldmeier (1950) studied an emission ($\lambda 5303$) region at 55°, which was identified by him for almost seven rotations; he determined its rotation rate from the successive observations. The values of this rate fit very closely the curve found by d'Azambuja (1948) from the rotation of filaments. The most extensive study of the coronal rotation from monochromatic observations was carried out by Trellis (1957). He also used successive east and west passages of bright $\lambda 5303$ regions. The difference between the coronal rotation rate and the Carrington rate for latitudes from 0° to 70° was determined. He found that the corona rotates a little slower than the photosphere at latitudes below 35° and more rapidly at latitudes above 35°. The difference becoming significant at high latitudes.

Cooper and Billings (1962) from 29 successive rotations of an emission region at 65° have determined a rotation rate higher than the rotation found by Waldmeier (1950) and Trellis (1957). Hansen *et al.* (1969) from autocorrelation analyses of K-coronameter observations have established average yearly rotation rates of coronal features as a function of latitude. At low latitudes the corona was found to rotate at the same rate as sunspots but at higher latitudes it rotates much faster than the underlying photosphere. The white light corona in 1964–1967 rotated much faster than the green corona in 1943–1955 as reported by Trellis. Nevertheless the high rates found from the recurrence of two stable high latitude coronal regions (Waldmeier, 1950; Cooper and Billings, 1962) are confirmed by the results of Hansen *et al.* (1969). However the problem of the variations of coronal rotation with latitude cannot be solved only from observations of successive limb passages of bright emission regions. In addition there may be systematic differences between the rotation of the bright regions usually used for such studies and the rotation of the coronal matter.

In this paper we investigate the coronal rotation, using the Doppler displacements of the coronal emission line $\lambda 6374$ Å.

Howard (ed.), Solar Magnetic Fields, 667–671. All Rights Reserved.
Copyright © 1971 by the IAU.

2. Data and Results

During the time interval from March 1968 until September 1969 a special program was undertaken with the 53 cm coronagraph of the Sajan Observatory (Siberia). The optical system of the coronagraph gives us a final solar disk image of 129 mm (Nikolsky, 1966) at the slit of the spectrograph.

The spectra of the corona used in this study were obtained in the wave length region of $\lambda\lambda 6350$–6385 Å in second order at a dispersion of 1 mm per Å with a curved slit at height intervals of 20000–40000 km above the photosphere. The microphotometer of Sibizmir was used to obtain microphotometer traces. Altogether 600 traces were done along the dispersion. Line shifts of the line $\lambda 6374$ Å due to coronal motion were measured relative to the Fraunhofer lines in the scattered light background. In Table I the wavelengths of the lines used, are given. The accuracy of the derived coronal rotation rates with the help of the spectroscopic method is limited by the presence of macroscopic motions within the corona, by the method of measurement of lineshifts, and by the width of the coronal emission line $\lambda 6374$ Å. The measurement of the positions of the Fraunhofer lines at the disk center showed that it is necessary to make a correction which is equal to 0.0165 Å ± 0.0016 Å at the Sun's equator. The correction diminishes with increase of latitude according to the law of differential rotation. This displacement of the Fraunhofer lines in the background is caused by scattering of light in the instrument because of the solar limb close to which the coronal line is being photographed. The strong coronal line profiles differ very little from a Gaussian curve; most profiles of the weak lines are asymmetric. Most of those are distorted by film grain and by the presence of motions which are not randomly distributed. Furthermore, since the coronal lines are quite broad, a precise determination of the wavelength position of the intensity maxima is difficult. With this in mind we determined the position of the 'center of gravity' of the coronal lines and of the Fraunhofer lines in the scattered light background:

$$x_c = \frac{\int xI \, dx}{\int I \, dx}$$

were I is the intensity or depth of the lines. The height of the scanning aperture in the direction perpendicular to the dispersion was $8\overset{''}{.}45$ of arc; and the width was 0.05 Å. Traces were obtained at interval of $1°$–$2°$, at the solar limb. In Table II the number of the positions for different latitudes at the east and west solar limb is given. To obtain a quantitative change of the rotation rate as a function of heliographic latitude, the northern and southern hemisphere and east and west limb coronal observations are analyzed together, and the average rotation rates were taken for each latitude.

Figure 1 (see Table III) shows the rotation rate in km/s as it depends on solar latitude. To compare the rates obtained from other solar traces including K-corona, regions of bright green coronal emission and photosphere, Figure 2 is given. At lower latitudes

TABLE I

Characteristics of the base Fraunhofer lines

Wave-length/Å	Equivalent width mÅ	Solar identi-fication	Low EP line	Note
6358.687 S	82	FeI	0.86	reliable
6366.491 m	26	NiI	4.17	unreliable
6371.355 m	35	SlII	8.12	unreliable
6378.256 S	27	NiI	4.15	reliable
6380.750 S	40	FeI	4.19	reliable

TABLE II

The number of contours for the east and west limb

$\varphi°$	0–10	11–20	21–30	31–40	41–50	51–60	61–70	71–80	81–90
East	74	65	54	26	15	14	21	27	12
West	54	61	34	16	11	2	10	19	9

our rates are much less than those found by Hansen and Loomis (1969) from analyses of K-coronameter observations, and within an uncertainty of 1% to 2% have the same rotation rate for latitudes up to about 25° as determined by Howard (1970) from observations of the photosphere Doppler shifts.

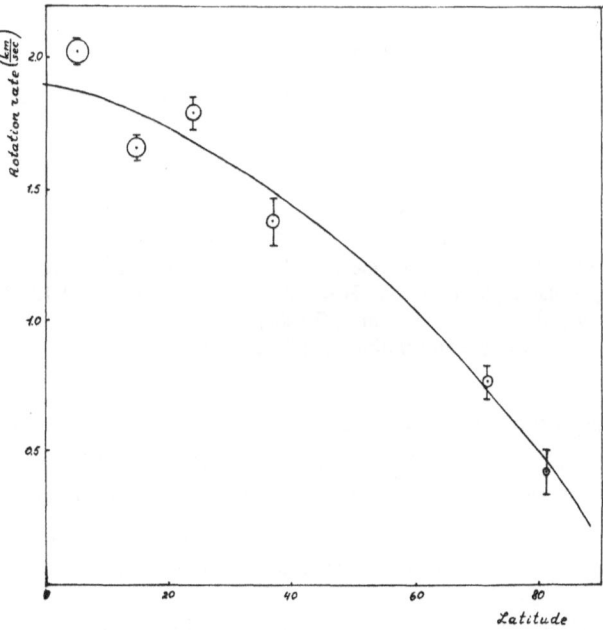

Fig. 1. Latitude dependence of coronal rotation rate, determined by the spectroscopic method. Dimensions of the open circles correspond to the number of data used for determination of the point. Relative errors are shown as vertical bars.

V. E. STEPANOV AND N. F. TJAGUN

TABLE III

Results presented in Figure 1

Heliographic latitude φ°	The number of points	Doppler displacements Å	Rotation rate km/s	Errors
5.4	139	0.043	2.02	±0.193
14.9	113	0.035	1.66	±0.198
23.9	91	0.038	1.79	±0.236
38.3	78	0.029	1.38	±0.287
71.5	56	0.016	0.763	±0.264
81.2	31	0.009	0.424	±0.344

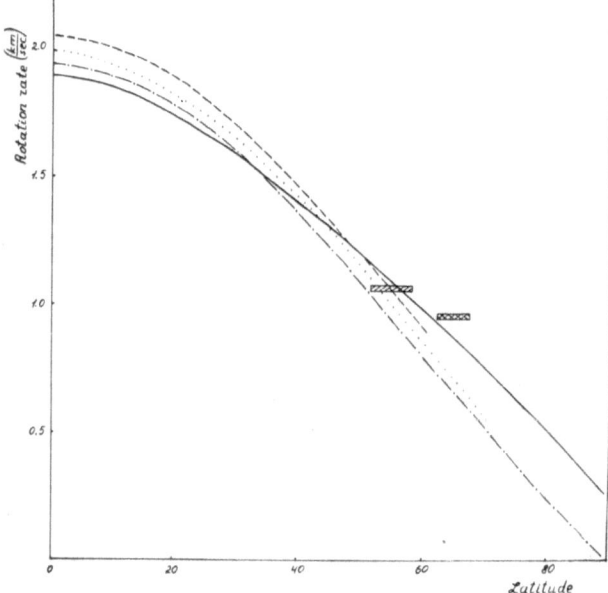

Fig. 2. Comparison of average rotation rate according to the spectroscopic method (solid curve) with those for the photosphere (dotted curve, Howard, 1970); K-corona at 1.125 Ro (dashed curve, Hansen *et al.*, 1969); Green Corona (points, Trellis, 1957); Waldmeier (shaded rectangle, 1959); Cooper and Billings, 1962 (blacked rectangle).

At high latitudes the corona rotates substantially faster than the underlying photosphere. Also our rates are generally faster than reported by Trellis for the green corona in 1945–1955 and for the K-corona (Hansen *et al.*, 1969). However the coronal rate at 30°–55° of the green corona (Trellis, 1957) of the K-corona (Hansen *et al.*, 1969) and the fast rate of the stable high latitude coronal regions are confirmed by our results.

At low latitudes the general agreement is remarkable on the one hand between the average photosphere and coronal rates determined through spectroscopic methods, and on the other hand between rates determined with the help of spots, filaments and coronal bright emission regions. At high latitudes however the difference between all measurements becomes great.

The differences are obviously due to systematic errors taking place when the measurements of rates are made at high latitudes. But there can be no doubt that at high latitudes the corona apparently rotates substantially faster than the underlying photosphere.

In earlier discussions of the difference between coronal rotation rates and polar faculae (Billings, 1966; Hansen *et al.*, 1969) suggested that high latitude phenomena should depend on magnetic fields at low latitudes and that the interchange of matter between the active zone and the polar coronal region along force-lines should take place. Our results prove this hypothesis.

References

d'Azambuja, M. and d'Azambuja, L.: 1948, *Ann. Obs. Paris* **6**, No. 7.
Billings, D. E.: 1962, *A Guide to the Solar Corona*, Academic Press, New York.
Cooper, R. H. and Billings, D. E.: 1962, *Z. Astrophys.* **55**, 28.
Hansen, R. T., Garcia, C. I., Hansen, S. F., and Loomis, H. G.: 1969, *Solar Phys.* **7**, 417.
Howard, R.: 1970, *Solar Phys.* **12**, 23.
Nikolsky, G. and Sazonov, T.: 1966, *Astron. Zh.* **43**, 868.
Trellis, M.: 1957, *Ann. Astrophys. Suppl. Ser.* **5**, 81.
Waldmeier, M.: 1950, *Z. Astrophys.* **43**, 29.

PART VII

THE POLAR FIELDS OF THE SUN AND THE MAGNETIC ACTIVITY CYCLE

THE POLAR FIELDS AND TIME FLUCTUATIONS OF
THE GENERAL MAGNETIC FIELD OF THE SUN

A. B. SEVERNY

Crimean Astrophysical Observatory, Nauchny, Crimea, U.S.S.R.

Abstract. In an attempt to summarize the present knowledge on the general magnetic field (gmf) of the Sun we pointed out the fine structure and the statistical nature of the gmf as one of its most important properties. The dipole-like behaviour of the mean polar field strengths is combined sometimes (since 1964) with a bias of the S-polarity flux for both poles. Highly uneven distribution of gmf with latitude and longitude, the disappearance of gmf at the South pole for months, and short period, almost synchronous at both poles, variations in the sign of gmf are pointed out. The fluctuations with time of the mean magnetic field of the Sun seen as a star (as well as mf at different latitudes) shows periodicity connected with the rotation of the Sun and very close agreement with the fluctuations of the interplanetary field (sector structure). The effect of faster rotation of N-polarities as compared with S-polarities as well as the bias of mean solar as well as interplanetary S-polarity fields are also pointed out. The possibility of short time-scale (hours) intrinsic changes in the local pattern of gmf is demonstrated.

1. Introduction

Starting from Hale's *et al.* pioneer work (1918), earlier observations (Langez, 1936; Adams, 1934, 1949; Babcock, 1948; Thiessen, 1946; Von Klüber, 1951; Beggs and

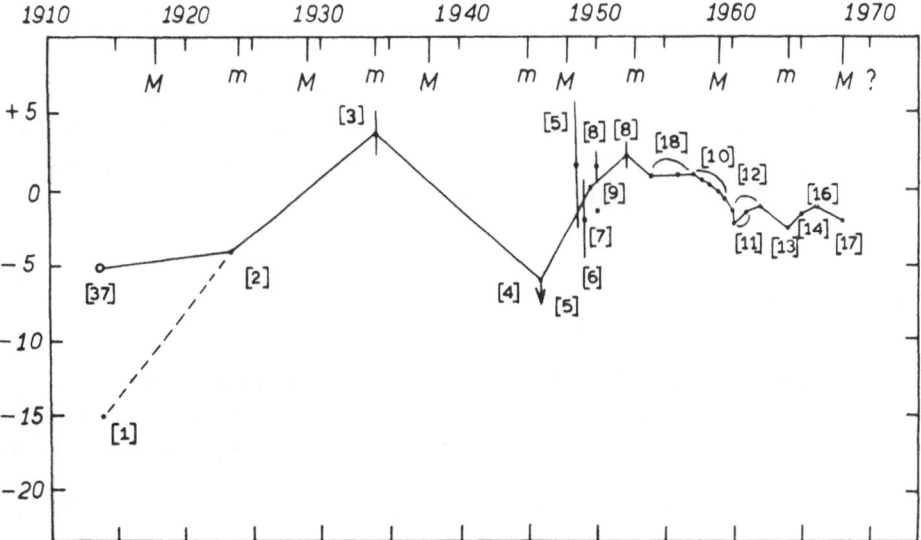

Fig. 1. Representing on a graph the separate determinations of the polarity and magnitude of the general magnetic field of the Sun. [1]=Hale *et al.*, 1918; [2]=Langez, 1936; [3]=Adams, 1934; [4]=Babcock, 1948; [5]=Thiessen, 1946, 1952; [6]=Adams, 1949; [7]=Von Klüber, 1951; [8]=Babcock and Cowling, 1953; [9]=Kiepenheuer, 1953; [10]=Babcock, 1959; [11]=Howard, 1965; [12]=Von Klüber, 1965; [13]=Severny, 1966; [14]=Severny, 1967; [16]=Stenflo, 1968; [17]=Stenflo, 1968; [18]=Babcock and Babcock, 1955; [37]=Stenflo, 1970*.

* As you will see from the following contribution of Dr. Stenflo the circle [37] on Figure 1, representing remeasured by him observations by Hale *et al.*, should be put somewhere near the zero line.

Von Klüber, 1964; Babcock and Cowling, 1953; Kiepenheuer, 1953) of the general magnetic field of the Sun compiled on a graph (see Figure 1) permit one to suspect the existence of secular variations of the polarity and magnitude of the general field. (Here the ordinate is the difference: field strength at N-pole minus the same at S-pole). We observe that polarity reversals coincide more or less exactly with the epochs of maximum activity (M) while the minima of activity are approximately in phase with the strongest negative or positive fields, (except for the first determination by Hale (1918)). The graph includes also more recent observations in 1964–66 made in the Crimea (Severny, 1966, 1967; Stenflo, 1966a, 1968a) and 1968 simultaneous observations at Mt. Wilson and the Crimea (Stenflo, 1968a) showing the same negative polarity at N-heliographic pole and positive at S-pole, as it has had since 1959. The expected interchange of polarities between S and N poles during the maximum of activity has not appeared so far.

2. Statistical Properties of the General Magnetic Field of the Sun

Until 1952 the observations were based on measurements of Zeeman-shifted spectral lines in spectra of polar regions. These are of limited accuracy, and hence the errors were comparable with the magnitude to be determined. The photoelectric method introduced in 1952 (Babcock and Babcock, 1955) increased the sensitivity at least by 10 times, and the accuracy was determined only by the noise level. To increase the signal to noise Babcock and Babcock (1955) used a long slit and image-slicer collecting the light from an area $40'' \times 70''$. Although many important results were obtained, this brought a great deal of averaging and led to an overestimated role of the large scale magnetic fields, because the signal of the magnetograph, having an entrance area S and compensated for brightness fluctuations,

$$\delta i_\parallel \sim \sum_{i=1}^{n} \frac{S_i}{S} h_i \qquad (1)$$

is proportional to the area S_i and field strength h_i of a magnetic feature.

If the general magnetic field has fine structure (with dimensions $d_i < \sqrt{S}$), small elements $(S_i/S \ll 1)$ do not contribute to the signal even if h_i is appreciable. As the signal:

$$\delta i_\parallel \quad \text{must be} \quad \geqslant \delta_i (\text{noise}) \qquad (2)$$

to be recorded, we can easily find (Severny, 1967) (using (1), the data about the noise level, and the weighted mean h, see below) that at a resolution of $23'' \times 23''$ (usual for Mt. Wilson) we lose information about all magnetic features with dimensions $\leqslant 8''$ ($\sim 40\%$ of total number, see histograms below) while at a resolution of $2\overset{.}{''}5 \times 9''$ (Crimea) the corresponding loss is for sizes $\leqslant 3''$, ($\sim 10\%$ of total number).

If we increase the resolution, the amplitude (maximum field strength) of a given magnetic element (at a given scan) can increase by ~ 2.0–3.0 times, while the mean strength over an extended area (e.g. polar cap) also increases but only by ~ 30–50%

(Stenflo, 1966b; Severny, 1967). This uneven increase is obviously due to (1) the fact that at low resolution the contributions into the sum (1) from opposite polarities are partly cancelled, and (2) the mean size (area) of magnetic features decreases with increase of resolution but slower than the field strength*. These effects are shown on Figure 2.

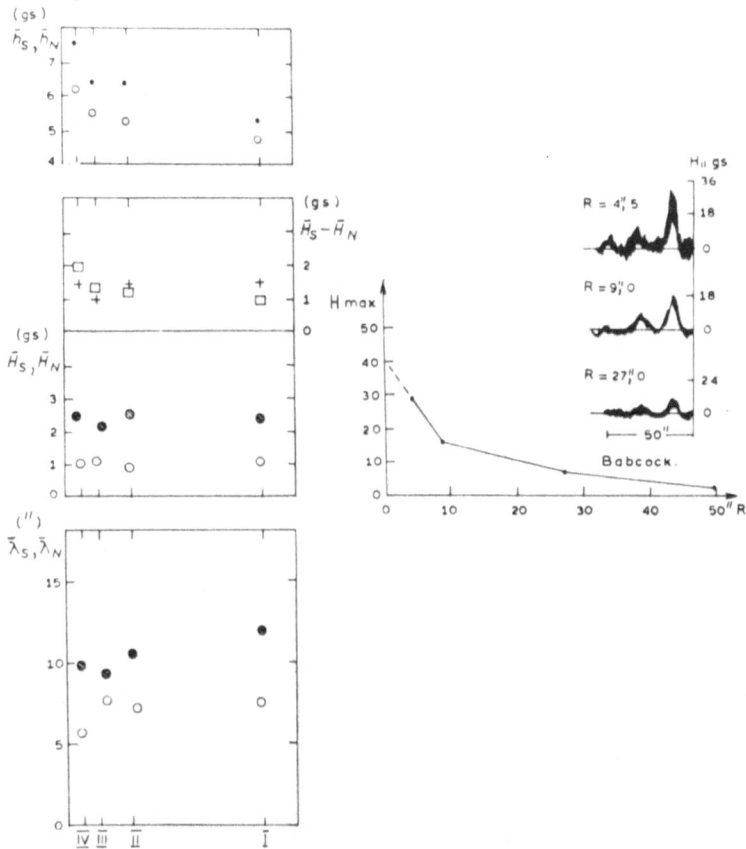

Fig. 2.   Illustrating the influence of the resolution on the maximal field strength of a given element (right) and on the mean values of field strength h, mean flux H of S-(solid dots) and N-(open circles) polarity, and on the mean size of magnetic elements λ. The roman numbers correspond to different slit heights: $I = 27''$, $II = 9''$, $III = 4''.5$, $IV = 2''.25$ while the slit width remains the same $= 2''.0$.

We can see that the mean size at our highest resolution (2".25) does not fall below 5".

The problem arises: what are the smallest elements of the general magnetic field? The answer depends on the highest resolution, available at the present time only at Kitt Peak where it is 560 km ($\approx 0''.65$) at good seeing, and Livingston (1968) found that while brightness and velocity exhibit a spatial fine structure down to this highest resolution, *no similar* fine structure is found in the magnetic field. The smallest

---

* Of course, the net flux must remain the same at different R provided that we scan a given area without gaps and overlappings, cf. arguments between Stenflo (1966b) and Howard (1966).

elements seen on magnetic maps with $R = 930$ km square at Kitt Peak are $2''$ while at our successive maps with $R = 2''3 \times 2''25$ the smallest cells persistent and reproducible on every map are $\simeq 2''5$ (see Figure 3). We think, therefore that *the smallest magnetic cells in quiet regions are about $2''$ in size.*

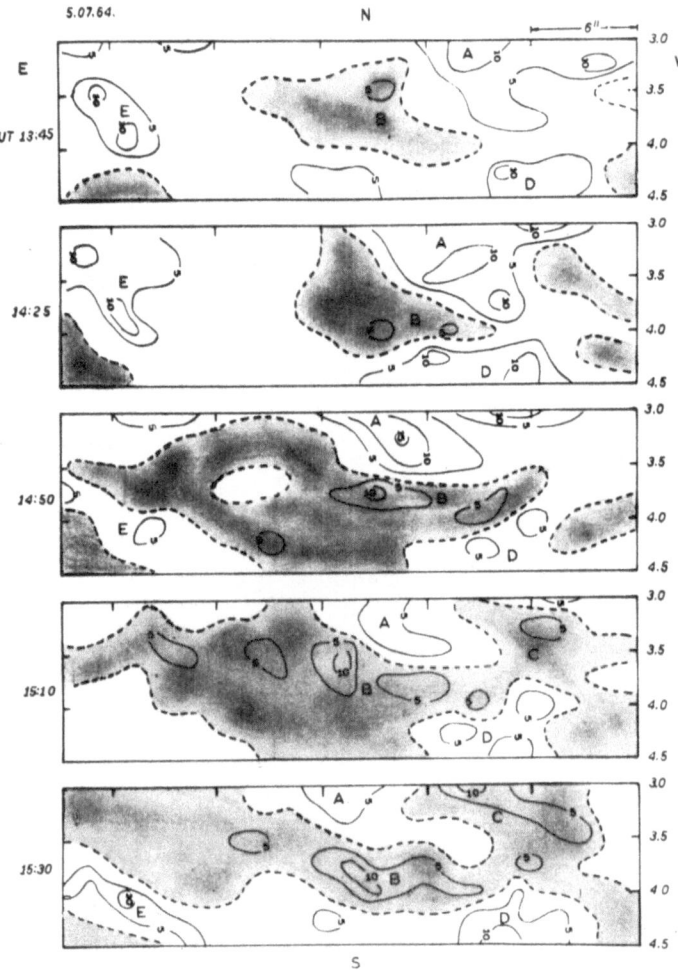

Fig. 3.   High resolution ($2''.5 \times 2''.0$) successive magnetic maps of a quiet region on the disk, showing the growth of opposite polarity in the previous region occupied by N-polarity.

All said above illustrates the fundamental role of resolution in examination of the nature of the general magnetic field.

The picture of the general magnetic field in quiet and polar regions looks like a carpet consisting of a large number of small elements, cells of different polarity, strength, and area mixed sometimes at random as Figure 4 shows. Extreme inhomogeneity and rapid time variations in this picture makes inadequate and accidental some old measurements of the general magnetic field at some separate fixed points (Adams, 1934, 1949; Babcock, 1948, 1959; Thiessen, 1946).

The effect of averaging at low resolution can also seriously distort the pattern obtained with the magnetograph (Babcock and Babcock, 1955) by producing so called 'UM-regions' – unipolar large-scale regions. These regions being unipolar at low resolutions appear as multipolar at higher resolution (see examples in Severny (1965)), and neither Crimean nor Kitt Peak magnetograms show such solid unipolar regions.

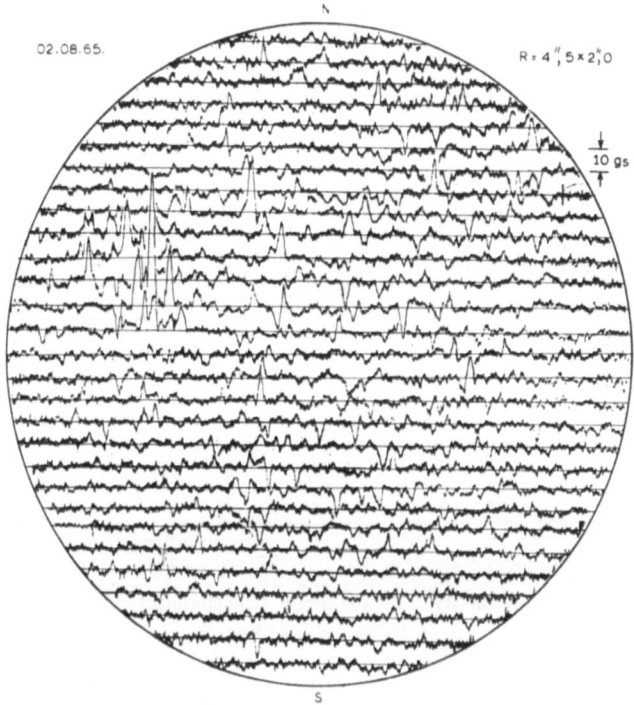

Fig. 4.   The Crimean records of the magnetic field over the whole disk.

We think that the word 'unipolar' means only (the predominance) *the bias* in integrated flux of elements of a given polarity.

For quiet regions *at the center* of the disk we have the following values typical for the fine structure under consideration; while: $H_S = H_N = 3.74$ with peaks 20 G, (Crimea (Severny, 1967), $R = 2''.5 \times 9''$, 1964–65), rms$(H) = \pm 4.9$ G, (Kitt Peak, $R = 10'' \times 10''$, (Livingston, 1966) and ± 2.8 G with peaks $\simeq 10$ G, ($R = 2'' \times 2''$ the same source).

The autocorrelation for magnetic features gives characteristic half-widths

$$r_m(0.5) = \frac{8''.5 \quad \text{(Crimea (Severny, 1967))}}{8''.7 \quad \text{(Livingston (1966))}}$$

which is, by the way, *three times* as large as half-widths of the corresponding curve for velocities (Severny, 1965). Neither at Kitt Peak nor at Crimea has any correlation between magnetic fields and velocities been found: the coefficient of cross correlation never reached 0.2 for different latitudes. Meanwhile this coefficient is, as a rule, larger than 0.5 for simultaneous magnetic records at two different levels (corresponding to $\lambda 5250$ and $\lambda 6103$). These records also show an increase of autocorrelation radius and a decrease of the mean strength outside. Simultaneous records in $\lambda 5250$ and $H\alpha$ show a decrease of mean field strength by 1.5 times from photosphere to chromosphere. The corresponding decrease found at Kitt Peak is $2.8/1.8 = 1.55$ (Livingston, 1966).

Before passing to polar regions we wish to emphasize that statistics for these regions differ but little from that just described for the quiet center of the disk. Histograms of

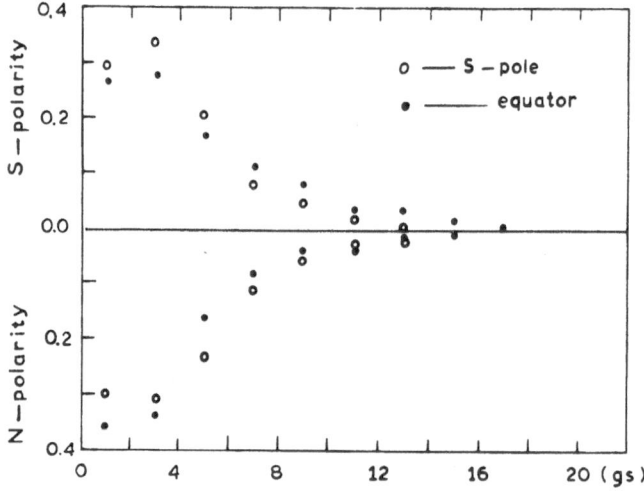

Fig. 5.   Showing the similarity of histograms of field strengths at the polar and central regions of the disk.

field strengths on Figure 5 (Severny, 1966) are practically the same for polar and central regions, except for a slight increase of half width at the equator (see also Stenflo, 1968a). Livingston (1966) also writes that "magnetic structure shows no center-to-limb variation. Since at the center we see 'longitudinal fields' while at the limb 'transverse fields', we conclude the field distribution is isotropic". Histograms of sizes of magnetic cells in Figure 6 are also quite similar (the mean for the whole of 1965). As the weighted mean field at the center is a little stronger and not so concentrated (more fragmentary) as at the poles the distribution of field directions is rather semi-isotropic, with little bias of radial component as compared with transversal one. However this can also be due to the effect of projection near the border of the disk.

On the other hand, the histogram of sizes (Figure 6) as well as autocorrelation curves of magnetic elements do not show well defined characteristic dimensions ascribed to supergranulation (Leighton, 1965). Slight secondary maxima we have near 12″, 24″ and 48″ as if they were due to successive modes of standing surface waves on the Sun, (the same is also true for the 1964 distribution). If, according to Simon and Leighton (1964), strong fields correspond to the common boundaries of supergranules (fields higher than, say, 10 G) the weak fields belonging to the general magnetic field "are elements which have escaped the concentrating action of supergranular flow", (Livingston, 1968). The above mentioned absence of the correlation between magnetic and velocity structures is also suggestive in this respect because supergranulation is essentially a velocity structure.

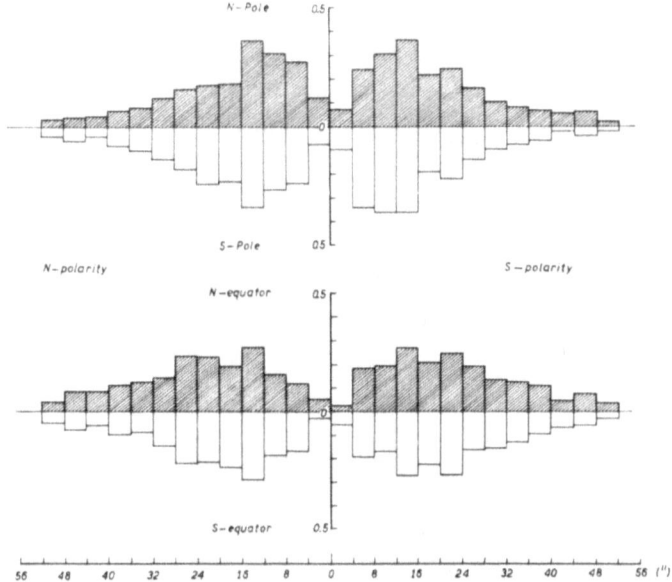

Fig. 6.   The histograms of the sizes of 'magnetic elements' in polar and equatorial regions.

The mean statistical properties of polar field on the basis of more or less regular high-resolution records were followed in 1964 (27 days (Severny, 1966), 1965 (33 days (Severny, 1967)), 1966 (32 days (Stenflo, 1968)) and 1968 (Stenflo, 1968). At each day of observation the polar cap (or area 50°–75° in φ and $\pm 20°$ in λ) was scanned ($R = 4\overset{''}{.}5 \times 2\overset{''}{.}3$) and weighted over the number of elements. The mean strength

$$\bar{H} = \sum n_i H_i / \sum n_i$$

on the basis of the histogram was determined for each polarity and each heliographic pole separately. Also the mean flux was determined,

$$\bar{F} = \int H_{\parallel} \, dl,$$

by planimetry of different scans in the area (also separately in the same way). The results are summarized in Table I, showing that the resulting field H_N–H_S was invariably

TABLE I

Resulting longitudinal field strength (weighted mean)

	1964 (I)	1964 (II)	1965	1966	1968
Pole N	-2.09	-1.09	-0.63	-1.02	-0.7
Pole S	$+0.25$	$+0.98$	$+0.79$	$+1.11$	$+1.44$

negative at the N-pole and positive at the S-pole and was of about the same value as if it were in the mean the field of a dipole, except for the first half of 1964, when the field practically disappeared at the S-pole. (It should be noted that the 1966 values of Stenflo (1968) were obtained by simultaneous records of a double magnetograph, and his results of 1968 (Stenflo, 1968) obtained at Mt. Wilson are not the weighted mean but the values of F averaged over a 27-day period of rotation.)

An essentially different pattern is displayed by the net flux F_N-F_S: it showed permanently through the years 64–65 and 68–69 (see below) a bias of S-flux or 'magnetic asymmetry', see Table II (for 64–65):

TABLE II

Ratio of magnetic fluxes F_S/F_N

	Ratio $F_S:F_N$	
Year	N-Pole	S-Pole
1964	2:0.5	1:1
1965	3.2:0.6	1.5:1

The effect was checked by direct planimetry of isogauss-maps and by comparison of the mean dimension of S- and N-polarity elements showing $\lambda_S/\lambda_N \cong 1.5$. This bias is exclusively due to the larger area occupied by the S-polarity. This magnetic asymmetry was reflected by the predominance of north polar flocculi according to Howard (1965) and accompanied by enhancement of all other activity (sunspots, K-plages, coronal, etc.) in the northern hemisphere. We will see that this 'monopole' – like behavior is also reflected by the interplanetary magnetic field.

The distribution of mean (averaged over $\pm 90°$ of longitude) field strength with latitude φ in polar regions shows the run which is rather opposite to that expected for a dipole, (Severny, 1967). Moreover if we pass now to the whole disk records, considering the same dependence on latitude we find no simple regularity: the mean field changes its sign very rapidly with latitude, several times between the N- and S-heliographic poles, see Figure 7 from Severny, 1967. The first change of polarity (moving from the poles) appears at latitudes 60–70°, and Stenflo at Mt. Wilson (1968) observed these first changes at $+70°$ and $-55°$, and correlated them to the zones of polar prominences. Some peaks in this latitude distribution of the general magnetic

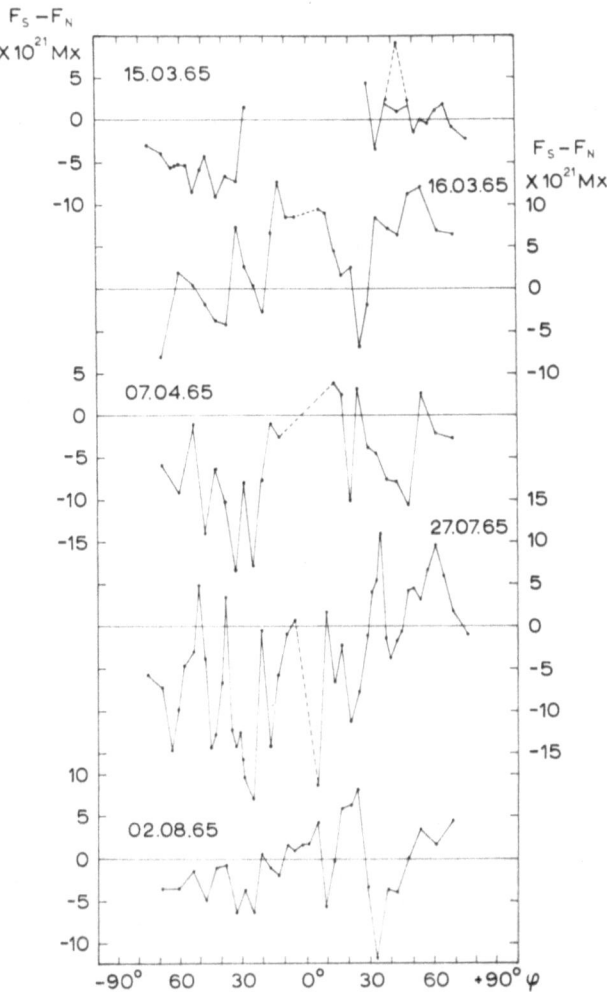

Fig. 7. The examples of the distributions with latitude of the mean (averaged over longitude) flux $F_S - F_N$ of the general magnetic field ($H_\| < 20$ G). (Vertical scales are units of 10^{+21} Mx.)

field can be identified from day to day, shifting gradually towards the equator (with a velocity 1–6 km/s). These results together with the differential rotation of the Sun point to the principal possibility when *one polarity overtakes the other polarity* thus giving rise to some kind of dynamo action (see below). It seems also impossible to bring such an asymmetrical (with respect to the equator) distribution of polarities into agreement with any dynamo-theory, even with one such as that of Steenbeck and Krause (1969), assuming quadrupole-like magnetic fields with S-magnetic pole at the equator.

This latitude-distribution also suggests that the high-latitude zones ($\varphi > 30$–40) can bring into the fluctuating net flux of the whole disk a *contribution comparable* with that for the low-latitude ($\varphi < 30\text{--}40$) zones (see e.g. the distribution for 7/04/65 where the flux $+7.88 \cdot 10^{21}$ Mx is mostly due to the high latitudes).

3. The Time Fluctuations of the General Magnetic Field

The most remarkable thing about the general magnetic field is, in our opinion, its *rapid fluctuations with time*. Babcock and Babcock (1955) were the first who observed that "irregular fluctuations in magnetic flux in the vicinity of the heliographic poles have been so large as to defy satisfactory explanations". They first noted the 'unprecedented' *disappearance* of the general magnetic field at the S-pole for 13 days (from July 29, 1954) and they claimed that they found annual variation of polar fluxes in

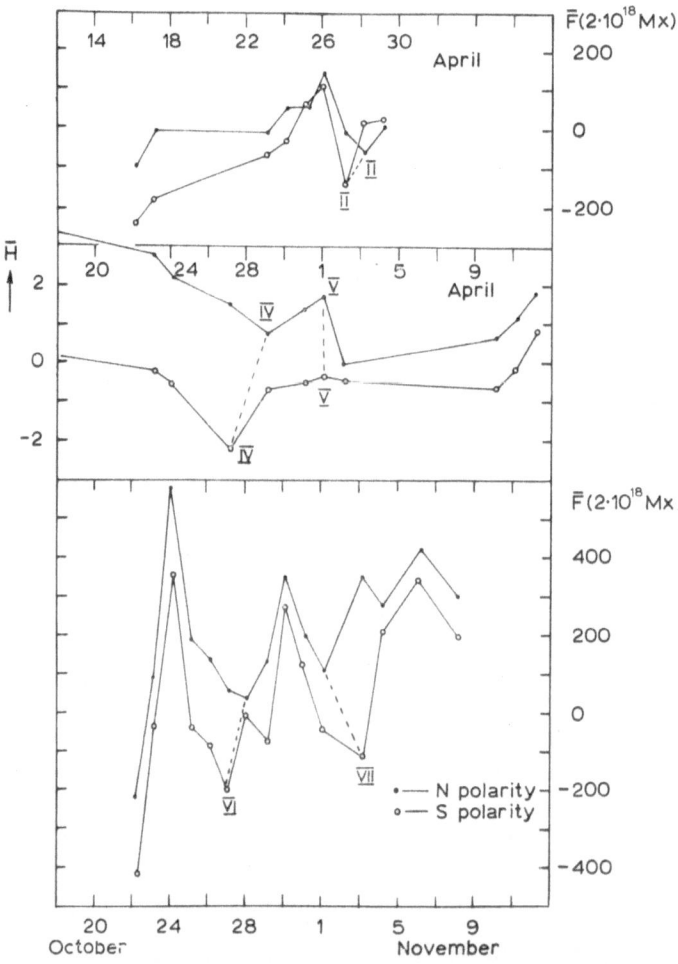

Fig. 8.   Short time-scale fluctuations of the mean polar fields of the Sun recorded in August 1965 at Crimean Observatory (in the middle) and the same fluctuations according to data provided by Mt. Wilson Observatory (top and bottom) for April and October – November 1965.

phase with the annual variation of the heliographic latitude of the Earth, B_0. The simultaneous appearance of the *same* positive (N)-*polarity at both helio-poles* for a period of more than one year (1957.5–1958.7) was first observed by Babcock (1959), who also demonstrated the lack of any correlation between long term variations of the general magnetic field at the poles and the heliographic latitude of the Earth.

The next disappearance of the general magnetic field at the S-pole was observed starting from March 1961 until the end of 1962 according to Howard (1965). He did not detect it in summer 1963 at Mt. Wilson and we could not detect it in the fall of 1963 as well as until September 1964 in the Crimea. Meanwhile at the N-pole we had quite measurable negative (S) polarity. Our more or less regular observations during 1965 showed several times (in March, July, September, October) the simultaneous appearance of negative (S) polarity general magnetic field at both poles. These 1965-

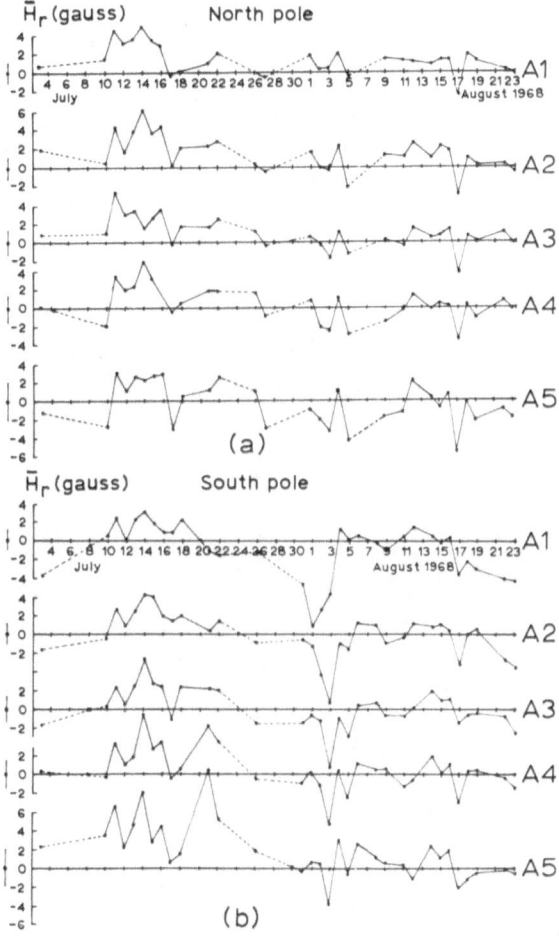

Fig. 9.   Short time-scale fluctuations of the polar fields observed in 1968 at Mt. Wilson by Stenflo (1968b).

fluctuations not only did not show any correlation with the sine-like run of B_0, but were rather of the opposite character.

Even more striking are *short time-scale fluctuations* (of the order of one day) illustrated in Figure 8 from Severny (1967). Remarkable here are the almost synchronous appearance of peaks at *both* poles, including the reversal of polarity. In 3 cases out of 4 the minima-peaks are about 1 day *earlier* at the S-pole than at the N-pole as if the positive (N) polarity tried to overtake the negative one. This looked very unlikely, but in 1968 Stenflo at Mt. Wilson (1968) observed the same strange simultaneous appearance of peaks at both poles (Figure 9), including also the reversals of polarity, (but no delay). In both cases Crimean and Mt. Wilson observations, the effect cannot be ascribed to the shift of the zero, because the corresponding records in non-magnetic lines do not reveal any such systematic error. Moreover Stenflo also found a highly uneven distribution of polar fields according to *longitudes* – very far from being rotationally symmetric and rather in a manner of sectors, (positive, negative then positive and negative again, see figure 10 from Stenflo (1968b). So that one can get a wrong estimate of the average polarity if the measurements do not cover one complete rotation at least.

The important question arises whether the time fluctuations specific for the polar fields are also specific for the whole solar disk or for the Sun as a star? Our high resolution ($2.3'' \times 9''$) full disk records showed a variation of the net total flux in a

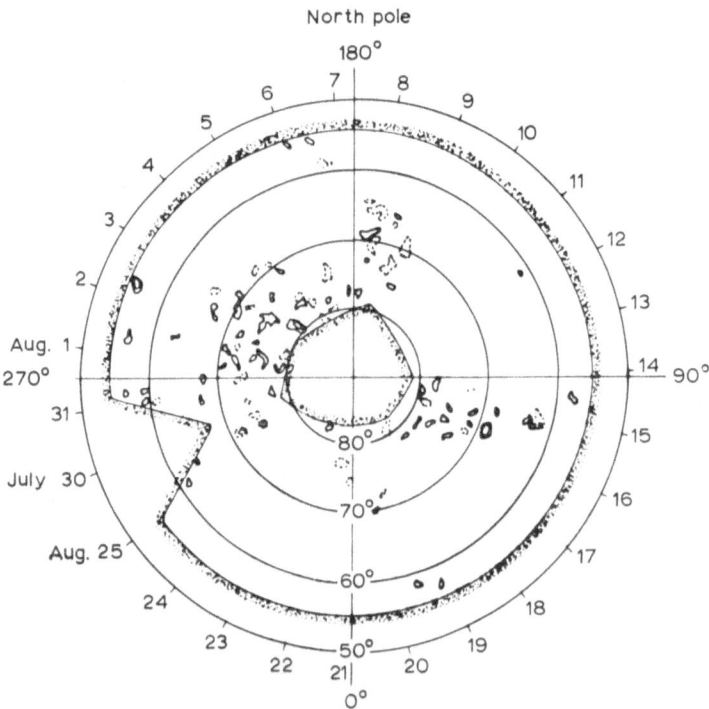

Fig. 10. The uneven distribution of polar fields according to longitudes according to Stenflo (1968b).

matter even of a day (from -0.6 to $+0.5$ G for the magnitude of $F_{net}/4\pi R_\odot^2$), see Severny (1967). The disadvantage of such a record is that it takes almost a whole clear day – the time interval during which intrinsic changes of the general magnetic field can happen (see below). The low resolution ($23'' \times 23''$) full disk magnetic charts for seven rotations (starting in April 1961), being planimetered, showed much slower variations of net flux which remained always positive and repeating, with some time lag, the curve of daily sunspot numbers, (Bumba *et al.*, 1967). Disadvantages of such records are (1) the loss, as we said above, of information about the field of $\sim 50\%$ of the magnetic features, (2) the main contribution in the flux here was, according to the authors, from fields with a strength of 16 G (after proper calibration) which corresponds to Ca-plages and (3) all the fields around the border ($>60°$) were disregarded although, as we showed above, they can contribute a flux comparable with the central part of the disk.

Being incapable of increasing considerably the brightness of the solar image in order to reduce the time needed for high resolution scanning of the whole disk, we started, in the beginning of 1967, the more or less regular measurements of the magnetic field in the line 5250 of the whole solar disk using, instead of the solar image, a *parallel beam* from the coelostat mounting falling on the magnetograph slit of the Crimean tower telescope Severny (1969). This method provides immediately the mean (averaged over all elements of the image with the distribution of the brightness over the image as a weighting function) longitudinal field strength of the Sun seen as a star. As this strength is small – fractions of a gauss and the maximum amplitude of the noise can reach 3 G, we accumulate the weak signal during 15^m–20^m, leading to an accuracy of ± 0.15 G, (after subsequent planimetry of the records, all procedure is ~ 100 times shorter than full disk scanning measures). The most important point is to fix the position of the zero for the field strength which is made by similar records in the non-magnetic line ($\lambda 5123$) and the magnetic field strength is the difference between the mean reading for $\lambda 5250$ and that for $\lambda 5123$. (If we use the instrumental zero (dark current) instead of the mean signal from the non-magnetic line we can get an error of 300% and even the wrong polarity). Figure 11 plots these values for March–June 1968. As sunspots can in principle contribute to this field the total flux from all spots H_S (divided by the visible area of the solar disk) is plotted also according to routine observations of the solar patrol, (more precisely we plot the value $H_S = 0.15 \sum s_i H_i / \pi R^2$ where S_i is the area of the ith spot, including penumbra, having a maximum absolute value of field strength H_i inside the umbra; 0.15 is the mean ratio of umbral to penumbral area).

We see *a periodicity of fluctuations*: twice in the course of one rotation (27^d) we have positive and negative polarity, the mean time interval between $+$ and $-$ peaks is almost half of the rotation period. There can hardly be any doubt that these changes of the flux from the whole disk are due to rotation, and the peculiar behavior of the Sun is similar to a rotating quadrupole. The magnitude and the sign of the mean solar field can change very rapidly in a matter of one day and the change can be as large as 1 G per day. This is quite consistent with the above mentioned results (by Severny

(1967) and Stenflo (1968)) about the rapid changes of the mean polar fields and it implies the idea that a wide range of latitudes is involved in these changes. An inspection of the Mt. Wilson synoptic magnetic charts by Wilcox and Howard (1968) shows that during several rotations we have more or less a clear demarcation in longitude between opposite polarities.

However the most remarkable point is the very close agreement between the mean solar field and the longitudinal component of the interplanetary field plotted in the same Figure 12, found by Wilcox *et al.* (1969), taking into account the transit time of solar wind plasma from Sun to Earth$=4\frac{1}{4}$ days. I will not enter into the details of this our common work with Dr. Wilcox on the comparison of the mean solar and interplanetary field which is in progress now. I wish only to make the following comments from the 'solar side' of the problem:

(1) From Figure 11 we see that the contribution of sunspots to the mean field can usually be disregarded. Moreover sunspot flux is either in antiphase with the mean solar field or disappears. Further if we plot (see Figure 12) also the latitude of the main sources contributing to the net flux which are usually preceding spots (leaders), and suppose that the behavior of the Sun as a magnetic star is determined only by sunspot magnetic fields, we find that the source determining the polarity of this magnetic star would pass *periodically* from N-hemisphere to S-hemisphere and back in the course of rotation. This periodicity follows also from the earlier work of Grotrian and Kunzel (1950). Meanwhile Wilcox and Howard (1968) and Shatten *et al.* (1969) demonstrated the existence of a correlation between the sign of the solar field and of the interplanetary field for a wide range of latitudes:* the peaks at ~5 day lags are

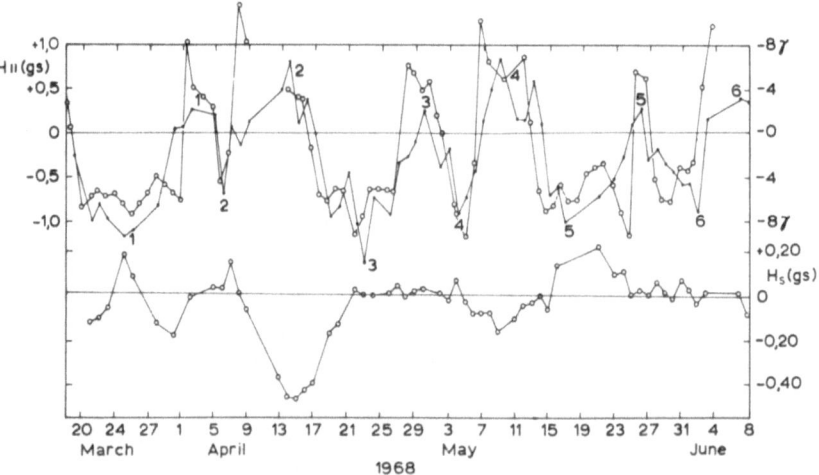

Fig. 11.   The fluctuations of the mean magnetic field of the Sun seen as a star (solid dots, top) compared with the same fluctuations of the mean longitudinal component of interplanetary magnetic field (open circles, top). The fluctuations of the total flux of sunspots is plotted at the bottom.

* They used the synoptic chart method in which only the central (± 12 h) zone near the meridian is supposed to be responsible for producing the magnetism of interplanetary space.

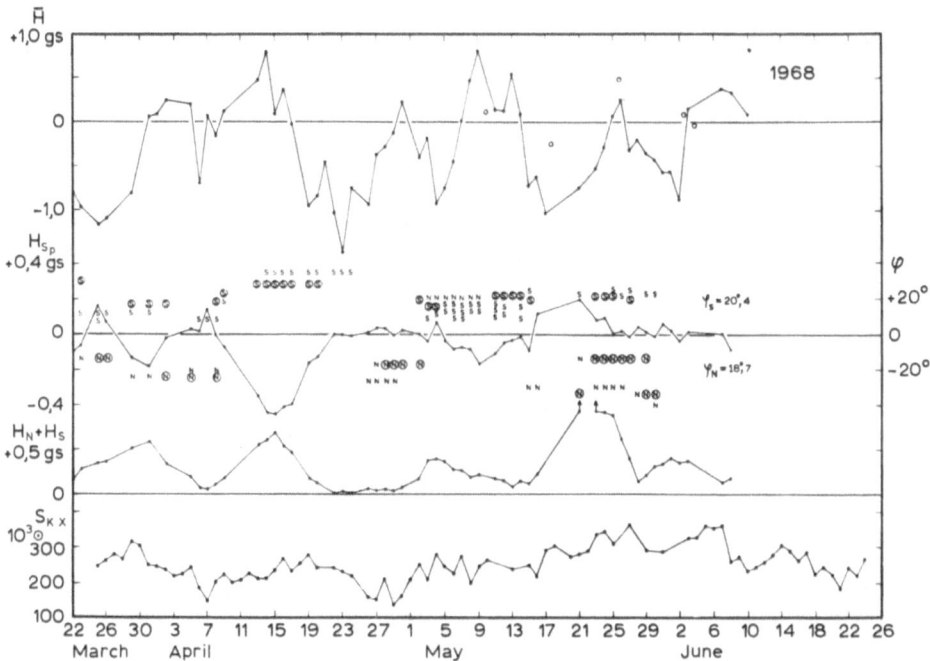

Fig. 12. The same as on Figure 11 fluctuations of the mean solar field (top) and of all sunspots Hs_p (solid line, second from the top) compared with the fluctuations in the position (latitude) of the principal contributor to the net sunspot flux (circles with the letter S or N), and with the area of Ca^+ − plages (bottom).

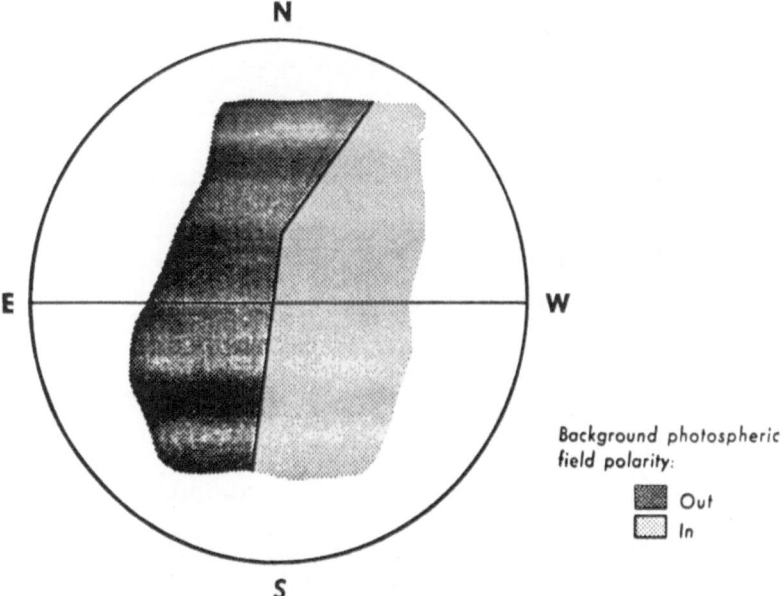

Fig. 13. The average solar sector boundary (1965) of the solar source of interplanetary magnetic field, see Wilcox *et al.*, 1969.

clearly expressed for latitudes from $-40°$ to $+40°$, pointing again to the demarcation line close to the NS direction on the solar surface and to the longitudinal sector structure of the solar source of the mean field (Figure 13).

(2) The solar source of the interplanetary field could hardly be connected with active regions: good correspondence (within a factor less than 2) between the magnitudes of the solar and interplanetary fields shows that the solar wind is capable of dragging out the very photospheric fields with mean strength $\sim 1-2$ G which can hardly be possible in active regions with strong fields where usually $V_A > V_S$, especially if we take Livingston's correction for the flattening of the line profile. This follows clearly from Shatzman (1962) and Mestel's (1968) considerations of the origin of the stellar wind. Now in the same Figure 12 we also plotted the total area of calcium plages S_p which characterizes the total magnetic flux of both polarities from active regions. There is no clearly defined correlation of this amount with the mean magnetic field we are measuring. Finally when we look carefully at cross correlations between polarities of solar and interplanetary fields (Schatten *et al.*, 1969) for different latitudes we can find that the height of the cross correlation peaks increases with increasing latitude and they are highest at $\varphi = \pm 40°$, i.e., when we are outside of the zone of solar activity, (Figure 14).

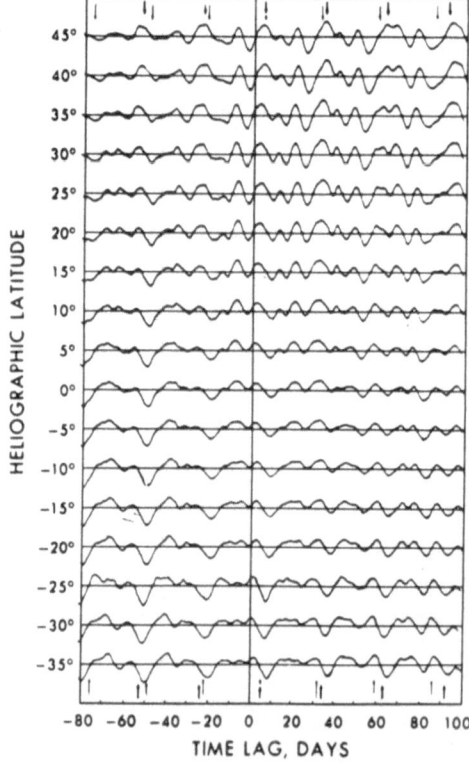

Fig. 14.   The cross-correlations solar-interplanetary magnetic fields according to Schatten *et al.* (1969).

(3) If, further, we take the integral over time of the mean field fluctuations, we find the mean value to be -0.29 G, that is, the effect of magnetic asymmetry described earlier and existing since, probably, 1961. Now, the reality of this phenomenon is supported by the agreement between solar and interplanetary fields, and we can see for example in Figure 15 the same predominance in time of negative polarity in the interplanetary field during the whole of 1968. We think that well-known statement by Prof. Alfvén (1967) that magnetographic measurements are inadequate and lead to absurd results as regards the general magnetic field can now be laid to rest. The magnetic asymmetry is a real phenomenon and the nature of such a predominance of one polarity is one of the most challenging problems in solar physics.

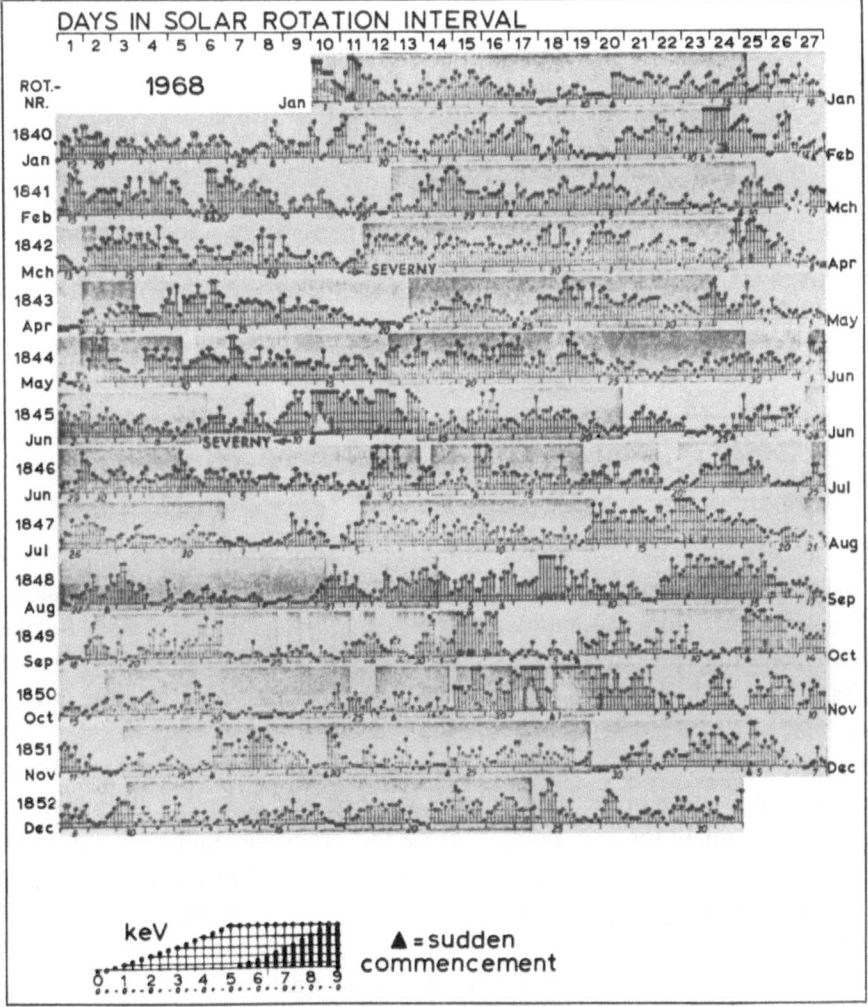

Fig. 15.   Interplanetary magnetic sector structure (1968) overlayed on a chart of planetary 3-hr range indices Kp. It is seen the predominance of dark shading during 1968 pointing on the predominance on the field toward the Sun.

(4) With the permission of Dr. Wilcox I wish also to make an additional remark: that 6-month's observations of mean solar field made in 1969 are also in agreement with the results just described, based on 1968 observations of solar and interplanetary fields.

(5) At first sight the periodic variations of the mean magnetic field of the Sun seen as a star, having two maxima and two minima during one rotation, show only accidental fluctuations of period (from 24 to 28.5 days). However if we compare (for 1968 and also 1969) the mean period between successive maxima (S-polarity) and minima (N-polarity) on the graph of $H_{\parallel}$ versus time we find

$$P_{max}(S) = 27\overset{d}{.}60 \quad (10)$$
$$P_{min}(N) = 26\overset{d}{.}8 \quad (9)$$

The difference is $0\overset{d}{.}8$ which can be interpreted as that, in the mean, N-polarity is rotating faster than S-polarity. Or, if we use Newton and Nunn curve for long-lived sunspots, it means that N-polarity is concentrated very near to the solar equator while the 'center of gravity', so to say, of S-polarity is near 20° lat. A similar difference of periods we obtain also if we overlap all 27-day period variations on one graph, making coincident all maxima, and then making coincident all minima. The third piece of evidence that such a difference can exist for the period 1968–1969 comes from a comparison of the mean latitude of sunspots giving the main contribution to the total flux from sunspots: we have

	1968	1969
φ(N-pol. spots) =	18.7 (24)	11.9 (94)
φ(S-pol. spots) =	20.4 (59)	16.1 (57)

The inspection of the Mt. Wilson synoptic charts of photospheric magnetic field for the period 1959–1966 also shows that sometimes (e.g. in 1959–1960) the blue areas of N-polarities are more concentrated near the equator than the red areas of S-polarity which spreads out up to the polar regions. This agrees also with the predominance of S-polarity flux in higher latitudes which we found for the years of solar minimum 1964–1965.

It would be interesting to repeat the investigation made by Wilcox and Howard of differential rotation of the photospheric magnetic field using Mt. Wilson synoptic charts but separately for each polarity. If we find from the autocorrelation analysis for some periods a real difference in rotation periods for S and N-polarity it would open a new interesting possibility for the dynamo mechanism, arising from the over-taking of one polarity in the course of rotation and thus, leading to the formation of toroidal fields connected with the torsion of poloidal (or meridional) fields in a way different from that proposed by Babcock.

(6) The fluctuations of the general magnetic field we have discussed so far were mostly interpreted as due to *rotation* of the Sun, having probably a stationary large

scale four sector magnetic structure. The purpose of further investigations is to examine more closely this structure. However, besides these fluctuations and supposed 22-year-cycle of fluctuations of the general magnetic field with which we started our talk, there exist *intrinsic* short-time-scale fluctuations manifesting the hydromagnetic activity of the quiet solar surface (magnetic activity connected with active regions is, of course out of the scope of the present review). The large scale semiregular pattern of mean field is changing but slowly during several rotations showing attraction of features of the same polarity and repulsion of features of opposite polarity, as shown by Bumba and Howard* (1965). They found that sometimes weak features covering a large area apparently disappear over a period of few rotations. They found also, as did Meudon astronomers as well the continuing development of *new* magnetic regions *inside* the regions occupied by *the old* magnetic region. Some idea about how rapid this process can be in quiet regions is given in Figure 3. These charts, obtained in the Crimea by repeated frequent scanning of the same small area of the quiet Sun (resolution $2''.3 \times 4''.5$) during several hours, show clearly the process of emergence and growth of one polarity, B, inside the region primarily occupied mainly by the opposite polarity, A. The new-born polarity, B, pushes the 'old' one a little away, and the corresponding graph of magnetic flux shows that the original unbalance of fluxes for the area considered tends to disappear. All processes take about 2.5 hr. In other cases we have observed that a previously disappeared magnetic hill appears again and the whole process looks like a very slow oscillation on a time scale ~ 2–3 hr. If the process we are talking about is characteristic for all quiet areas on the Sun and not an oscillatory one, the pattern of the general magnetic field on the disc can be renewed during a quarter of a day. These intrinsic changes of the general magnetic field coupled with those due to rotation of the Sun should present an extremely complicated pattern and only very high resolution records of the whole disk combined with 'zero-resolution' measures of the mean field of the Sun as a star can shed some light on the problem.

References

Adams, W. S.: 1934, *Annual Rept. Mt. Wilson Obs., C.I.W. Yearbook*, p. 138.
Adams, W. S.: 1949, *Annual Rept. Mt. Wilson Obs., C.I.W. Yearbook*, p. 12.
Alfvén, H.: 1967, *Bjerkeland Symposium*, Sandefjord, Norge, Stockholm.
Babcock, H. D.: 1948, *Publ. Astron. Soc. Pacific* **60**, 244.
Babcock, H. D.: 1969, *Astrophys. J.* **130**, 364.
Babcock, H. W. and Babcock, H. D.: 1955, *Astrophys. J.* **121**, 349.
Babcock, H. W. and Cowling, T.: 1953, *Monthly Notices Roy. Astron. Soc.* **113**, 357.
Bumba, V. and Howard, R.: 1965, *Astrophys. J.* **141**, 1502.
Bumba, V., Howard, R., and Smith, S.: 1967, in *Magnetic and Related Stars* (ed. by Robert C. Cameron), Mono Book Corp. Baltimore, p. 131.
Grotrian, W. and Kunzel, H.: 1950, *Z. Astrophys.* **28**, 28.

---

* It remains unclear how large is the influence of the behavior of active regions on the conclusions drawn in Bumba and Howard (1965), because at the resolution of $23'' \times 23''$ used by these authors the signal from one element with size $2''$ and with the strength 10^3 G (Schröter-Beckers, or Sheely-elements) is equivalent to that from a uniform background field spread over the area $23'' \times 23''$ with strength 4 G minimal strength recorded on the charts in Stenflo (1970).

Hale, G.: 1915, *Nature* **136**, 703.

Hale, G., Sears, F., Van Maanen, A., and Ellerman, F.: 1918, *Astrophys. J.* **47**, 206.

Howard, R.: 1965, in R. Lüst (ed.), 'Stellar and Solar Magnetic Fields', *IAU Symp.* **22**, 129.

Howard, R.: 1966, *Observatory* **86**, 73.

Kiepenheuer, K. O.: 1953, *Astrophys. J.* **117**, 117.

Langez, R.: 1936, *Publ. Astron. Soc. Pacific* **48**, 208.

Leighton, R.: 1965, in R. Lüst (ed.), 'Stellar and Solar Magnetic Fields', *IAU Symp.* **22**, 158.

Livingston, W.: 1966, Private letter.

Livingston, W.: 1968, *Astrophys. J.* **153**, 929.

Mestel, L.: 1968, *Monthly Notices Roy. Astron. Soc.* **138**, 359.

Severny, A.: 1965, in R. Lüst (ed.), 'Stellar and Solar Magnetic Fields', *IAU Symp.* **22**, 238.

Severny, A.: 1966, *Izv. Krimsk. Astrophys. Obs.* **35**, 97.

Severny, A.: 1967, *Izv. Krimsk. Astrophys. Obs.* **38**, 3.

Severny, A.: 1969, *Nature* **224**, Oct. 4.

Schatten, K., Wilcox, J., and Ness, N.: 1969, *Solar Phys.* **6**, 442.

Shatzman, E.: 1962, *Ann. Astrophys.* **25**, 1.

Simon, G. W. and Leighton, R.: 1964, *Astrophys. J.* **140**, 1120.

Steenbeck, M. and Krause, F.: 1969, *Astron. Nachr.* **291**, Heft 2.

Stenflo, J. O.: 1966a, *Arkiv for Astronomi*, Band 4, nr. 13, 173.

Stenflo, J. O.: 1966b, *Observatory* **86**, 73.

Stenflo, J. O.: 1968a, *Acta Univers. Lund. Sectio II*, 1968, N. 1, 5.

Stenflo, J. O.: 1968b, *The Polar Magnetic Fields of July and August* 1968, preprint.

Stenflo, J. O.: 1970, *Hale's Attempt to Determine the Sun's General Magnetic Fields*, preprint.

Thiessen, G. J.: 1946, *Ann. Astrophys.* **9**, 101.

Thiessen, G. J.: 1952, *Astrophys.* **30**, 185.

Von Klüber, H.: 1951, *Monthly Notices. Roy. Astron. Soc.* **111**, 2. See also Beggs, D. and Von Klüber, H.: 1964, *Monthly Notices Roy. Astron. Soc.* **127**, 153.

Von Klüber, H.: 1965, in R. Lüst (ed.), 'Stellar and Solar Magnetic Fields', *IAU Symp.* **22**, 144.

Wilcox, J. and Howard, R.: 1968, *Solar Phys.* **5**, 564.

Wilcox, J., Severny, A., and Colburn, D.: 1969, *Nature* **224**, 353.

Discussion

Altschuler: At this conference we have now heard at least three different groups state that they have seen rapid changes in the large scale photospheric magnetic field. You have seen rapid changes in the polar field, Wilcox has seen sudden changes in the sector boundaries, and we have evidence of rapid changes in the surface-harmonic spectrum of the photospheric magnetic field. The theoretical question that must now be posed I think is how such rapid changes in the large scale photospheric magnetic field can occur.

Cowling: You state that your observations cast doubt on the dynamo theory of solar fields. Babcock's theory began as a semi-empirical interpretation of observations. Theoreticians would be grateful if observers would agree what are the observations they ought to interpret. My second point is one which I think was made by Gold at Rottach-Egern. He pointed out that the structure of the solar plumes is remarkably constant, despite the rapid fluctuations in the photospheric field below. Are we to understand that plumes can rise indiscriminately from regions of north or south polarity, or that the tangles of opposite polarity do not extend more than a short distance above the photosphere?

Severny: I don't think we are now in a position to reject Babcock's theory which was, by the way, based on empirical grounds relating mainly to the strong sunspot field and was influenced to a larger extent by ideas of a dipole field, which is, if consistent at all, a very restricted meaning. If as we have shown, one polarity can overtake the other, the new possibility is offered: namely of the torsional dynamo effect because of differential rotation of different polarities. As to the second point the most plausible picture seems to me that proposed by Schatten *et al.* when the stable pattern of magnetic fields carried out with the solar wind should exist only outside of active regions where the Alfvén velocity is large compared with the sound velocity.

Nagarajan: The observations of the large scale field, its time structure and the sector structure explain why the classical dynamo theory is inapplicable. The hydro magnetic approximation has to be thrown away and one has to look at the possibility of having a nonstationary dynamo – presumably oscillating with complex multiple periodicities. One also has to treat the gas as in a non-equilibrium kinetic stage. Perhaps convection plays a larger role in the dynamo mechanism than just presenting a passing battery scheme, as, for example in the Biermann generator. Though these added complications make the models mathematically more complicated, they make it physically more reasonable.

Brueckner: Are the Mt. Wilson and Crimean observations of the simultaneous rapid changes of the 'general magnetic field of the Sun' in phase?

Severny: They probably can not be due to the difference in longitude, if we speak about short period fluctuations. For larger time-scales they agree quite satisfactorily, as Dr. Stenflo will probably speak about.

Pick: You have observed a good correlation between the mean photospheric field and the interplanetary magnetic field. In some cases, there seems to exist an anticorrelation between the interplanetary field and the main polarity of active centers. Is this true? And in this case, do you think that there may exist some reconnection between magnetic field lines originating from active centers and magnetic field lines originating away from the mean photospheric field? Thus this field reconnection would enable a sector to enlarge as it is seen during the increasing part of the cycle.

Severny: Yes it is true, and sometimes, as you have seen on my graph of mean solar field, we have just the opposite run on the overall flux from sunspots as the mean field in question. The kind of interaction between the mean photospheric field and the field of active centers should exist and I think that one of such suggestions of interaction as the one about which Prof. Tuominen will speak at this session.

OPPOSITE POLARITIES IN THE DEVELOPMENT OF SOME REGULARITIES IN THE DISTRIBUTION OF LARGE-SCALE MAGNETIC FIELDS

P. AMBROŽ and V. BUMBA

Astronomical Institute of the Czechoslovak Academy of Sciences, Ondřejov, Czechoslovakia

R. HOWARD

Hale Observatories, Carnegie Institution of Washington and
California Institute of Technology, Pasadena, Calif., U.S.A.

and

J. SÝKORA

Astronomical Institute of the Slovak Academy of Sciences, Skalnaté Pleso, Czechoslovakia

Abstract. When viewed separately in the two polarities, the large-scale pattern of magnetic fields on the Sun gives an appearance reminiscent of a very large cellular structure. These structures and their relation to 'active longitudes' are discussed.

1. Introduction

Recently there have been several attempts to study the characteristics and motions of large-scale solar magnetic fields (Bumba and Howard, 1965a; Bumba and Howard, 1969a, b; Bumba, 1970, 1971). The suggestion was made that giant regular structures, formed from many active regions in complexes of activity, could be due to very large scale convective elements. In one study (Bumba, 1970) some indication was found that opposite polarities, either plus and minus or leading and following, behaved differently. In order to follow up this suggestion and study in more detail large-scale patterns and apparent regularities, we have undertaken this investigation, which utilizes more recent observational data which is of relatively high quality.

2. The Observational Data

The basic observational data are those of the *Atlas of Solar Magnetic Fields 1959–1966* (Howard *et al.*, 1967), supplemented by draft copies of magnetic synoptic charts of subsequent rotations also constructed from the daily magnetograms of the Mount Wilson Observatory. The data used extended through the end of 1969.

In addition we used synoptic charts of solar activity drawn from the *Daily Maps of the Sun* published by the Fraunhofer Institute in Freiburg, Germany, and the *Daily Geomagnetic Character Figures C9* edited by the Institut für Geophysik, Göttingen, Germany.

Because of the inhomogeneity of some of the magnetic synoptic charts and in order to cancel some rapid changes in time, the maps were integrated by overlaying two or three sequential maps. For example in a series of synoptic charts we would compare: $n, n+1; n+1, n+2; n+2, n+3$; etc.

3. Two Types 'Active Longitudes'

A. 'ACTIVE LONGITUDES' ROTATING WITH A 27-DAY PERIOD

In earlier studies (Bumba and Howard, 1969a, b) it was convenient to define 'active longitudes' as the locations where large-scale patterns in the magnetic fields recurred for many rotations. Features of the background field pattern may be traced for 10 or more rotations on synoptic charts mounted in chronological order and confined to strips including only certain latitude zones. Very often individual 'rows' may be associated with either one of two individual 'streams' which may be observed to persist for several years. The inclination of such rows and streams in Carrington coordinates for latitudes between $\pm 20°$ shows a small spread around a value corresponding to a rotation period of 27 days. A certain fine structure is noted in these inclinations (Bumba, 1971).

B. 'ACTIVE LONGITUDES' ROTATING WITH A 28-DAY PERIOD

During periods of relatively high solar activity in the equatorial strips of the synoptic charts, successive rows and streams of activity may be seen to be intensified in succession, giving the appearance on the large scale of a different rotation period. This is illustrated in Figure 1.

Another manifestation of this type of 'active longitude' is seen when comparing the equatorial strips with $20°-40°$ strips in either hemisphere. Individual rows are seen later displaced at higher latitudes in such a way as to give the impression of a different rotation period. This effect is best seen on the Zürich *Heliographic Maps of the Photosphere* which are available for four cycles of solar activity (Bumba, 1971). These synoptic charts represent photospheric faculae, which themselves represent well the distribution of magnetic field strength.

The (synodic) period represented by both effects mentioned above is close to 28 days, sometimes slightly less (Švestka, 1968), sometimes greater (Toman, 1967; Wolfer, 1897). One of us (Sýkora, 1971) has recently found a 28-day rotation for 'active longitudes' formed from the green corona maximum. These results are illustrated in Figures 2 and 3. If we compare the position of Sýkora's 28-day coronal 'active longitudes' with the magnetic and facular synoptic charts, we can see that the 28-day features coincide very well on both charts.

4. The Role of 'Active Longitudes' in the Development of Large-Scale Activity

A. COMPLEXES OF ACTIVITY

Although the boundaries of a complex of activity are in general not clearly defined, on the average the expansion during the early development of a complex proceeds at a rate of about 20 m/s to the west and about 50 m/s to the east. These velocities are

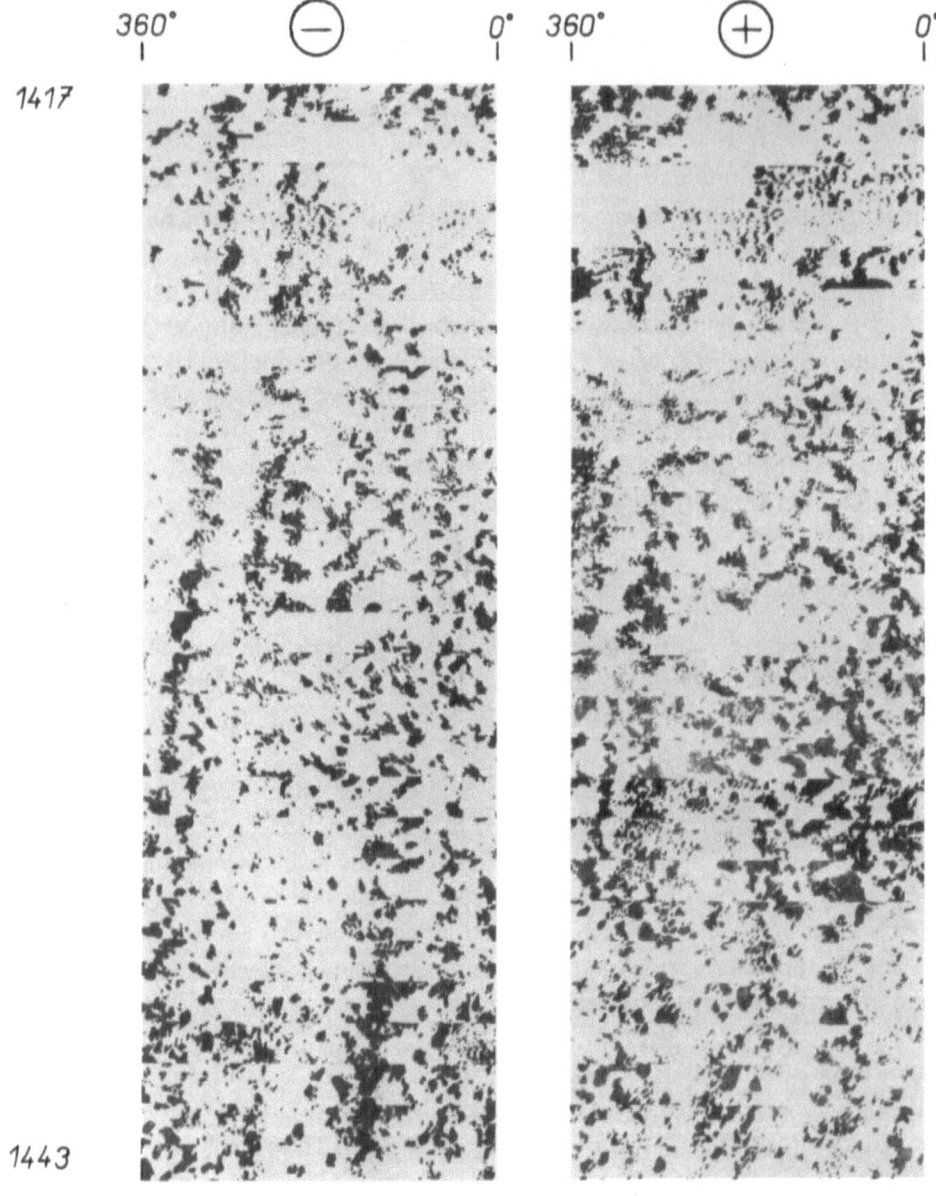

Fig. 1.   Magnetic synoptic charts cut into equatorial strips ($\pm 20°$) and mounted in chronological order for rotations Nos. 1417–1443 (1959–1961). Both polarities are drawn separately. Minus polarity is on the left, plus on the right of the figure. The main inclination of 27 days can clearly be seen in both pictures. On the minus polarity drawing may be found the intensification of subsequent individual magnetic rows, starting in the upper left-hand corner and going in a regular succession, more inclined in the opposite direction than the 27-day rows, toward the lower right-hand corner, forming the 28-day active longitude.

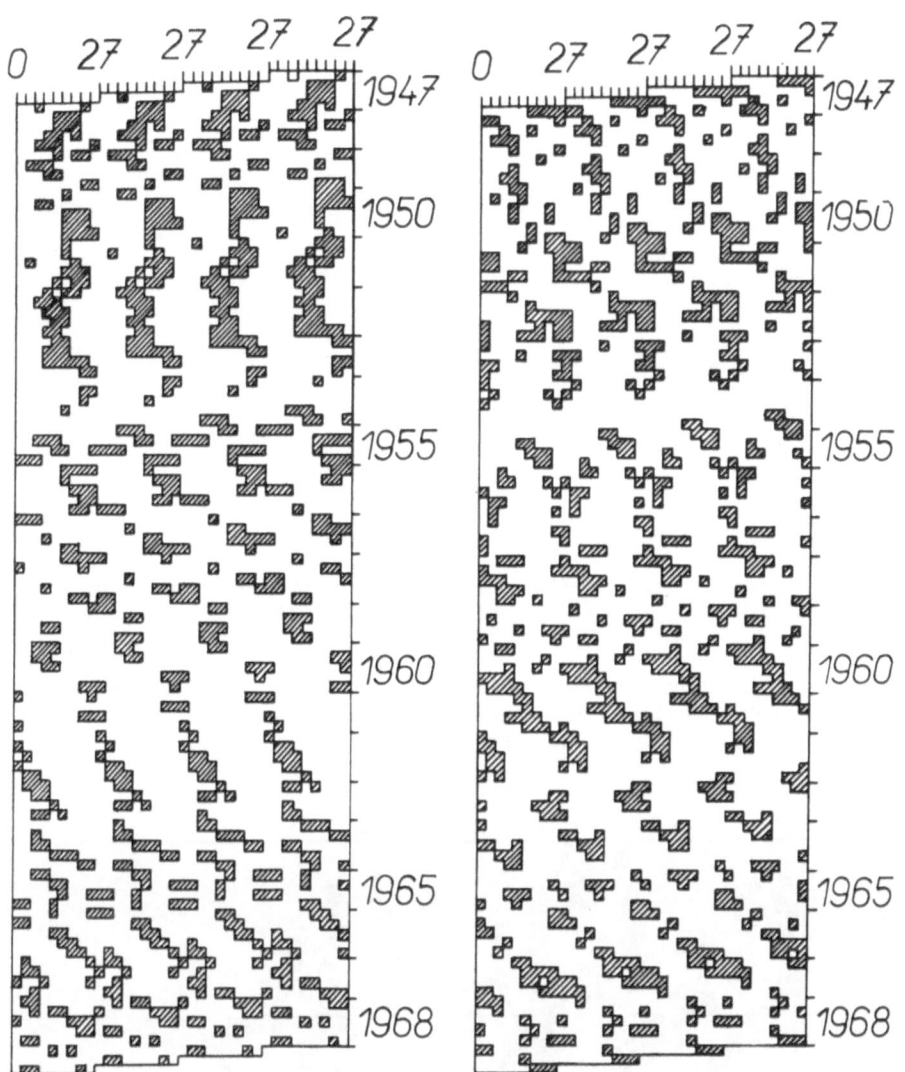

Fig. 2. Synoptic charts of green corona maxima for highest latitude intervals (to the left $+40°-+60°$, to the right $-40°--60°$), repeated four times, drawn in Bartel's coordinates (27-day rotation). From normalized and integrated coronal intensity for three rotations only the highest values were drawn in the form of shaded areas. During certain phases of the cycle, maxima appear with a 27-day rotation period, during the other phase maxima with 28-day rotational values are seen.

based on the Carrington period of 27.3 days and thus represent a period of 27 days at the west boundary and 28 days at the east boundary (Bumba *et al.*, 1968; Šykora, 1969).

Švestka in a study of long-term forecasts of proton flares followed 81 different proton-flare regions from their appearance on the disk to their eventual decay and disappearance. He has plotted these groups in several figures. These figures show two main longitude zones of increased flare activity with interesting fine structure: there

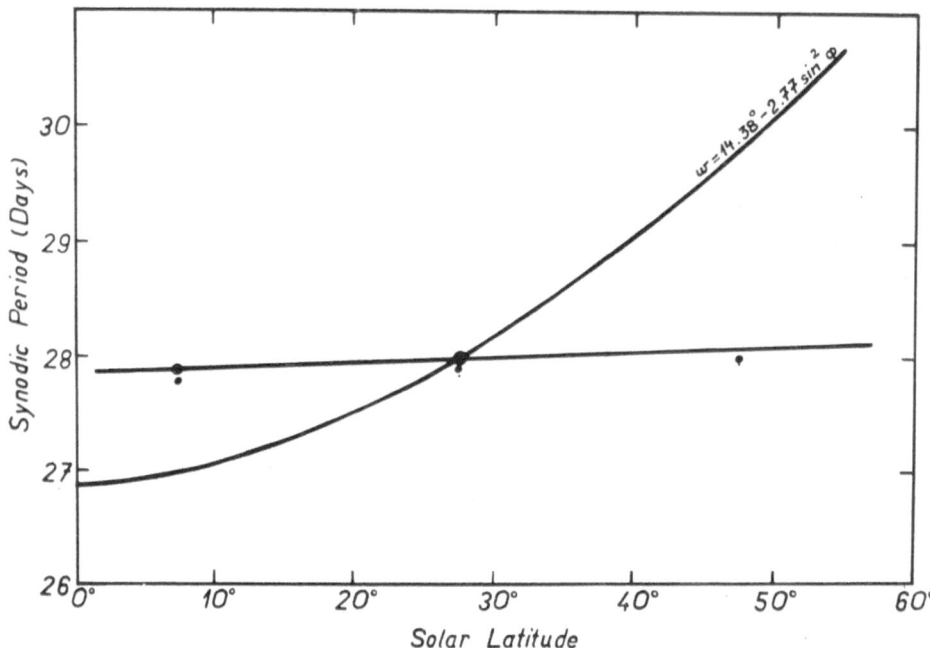

Fig. 3.   The very small change of the rotational value of the 28-day active longitude in green corona in dependence on heliographic latitude (estimated for three latitude strips (0–20°, 20°–40°, 40°–60°). For comparison the differential rotation of sunspots is given.

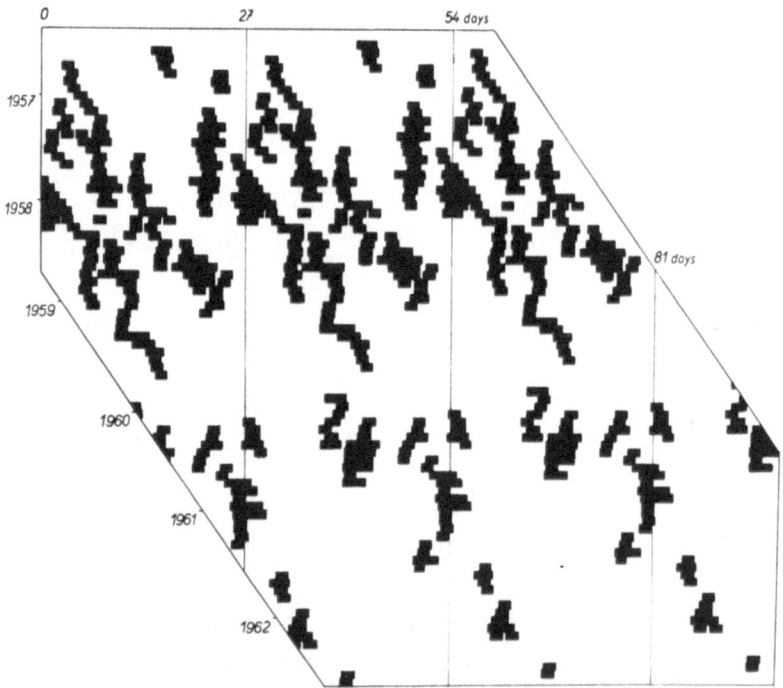

Fig. 4.   Švestka's Figure 3 (1968a) showing his complexes of proton-flare activity in the southern solar hemisphere. Bartel's solar rotation periods (27 days are plotted on the horizontal axis and the subsequent solar rotation on the vertical one. The situation is repeated three times in order to stress the sequences of shorter 27-day strips forming the 28-day active longitudes.

are individual proton-flare regions showing a 27-day recurrence, and these relatively short-lived rows form together a longer sequence which is inclined as are the 28-day 'active longitudes'. Švestka estimates their rotation period to be 27.8 days. See Figure 4. In the northern more active hemisphere the 27-day features are more pronounced, and in the south the 28-day features are more clearly visible.

B. 'SUPERGIANT' FEATURES

From the definition of the 28-day 'active longitudes' it follows that the maximum in the development of both the 27-day and 28-day features occurs when these features cross on the multiple synoptic chart figure. Such maxima are formed from very large roughly elliptical structures with high latitude tails formed apparently by differential rotation. Because of the similarity of these features in appearance to granular cells, we call these 'supergiant' structures. They are best seen separately in the negative polarity magnetic fields. They are related to the development of complexes of activity, and they develop through consecutive enhancement of individual patterns of the background field as the boundary of the complex expands. Such a huge structure develops and disappears in one 27-day stream, then again in the next one (to the east), and so on as they are successively crossed by the 28-day 'active longitude'. During the decay of one structure the following one may already be forming. In about nine rotations such 'supergiant' structures develop to their maximum size – starting about the size of an active region and expanding until the size reaches about one half of the area in the active longitudes, or even more because of the tail that extends to high latitudes (55°–60°). This effect is illustrated in Figure 5.

'Supergiant' structures which develop at different times in different longitudes are very similar. Even the structures developed in opposite polarities may be very similar, as we shall see. They also have their own internal structure in which during the periods of best visibility we may recognize the 'giant' structures (30°–35°) and their clusters (about 60°).

5. The Differences Between the Polarities

A. DIFFERENCES IN THE LARGE-SCALE DISTRIBUTION

During certain time intervals the concentration of features of each polarity into one of the two principal 27-day streams, and the clear separation of the features of both polarities may be seen. The best example of this type of large-scale configuration may be seen during rotations 1434–1443 (see Figure 6). During this interval there were two principal 27-day streams. The 28-day 'active longitude' appeared to cross one of

→

Fig. 5.   Magnetic synoptic charts for rotations Nos. 1417–1443, constructed separately for opposite polarities. The minus polarity is on the left and the plus on the right of the figure. For integration three individual charts always overlap. Three or four sequences of minus polarity 'supergiant regular structure' development may be found. In the positions of minus polarity 'supergiant' bodies, certain maxima in the plus polarity charts can be seen. Two types of cellular features especially in the plus polarity maps may be followed: sone of them rotate with a 27-day period and some of them have a constant Carrington longitude (27.3 days).

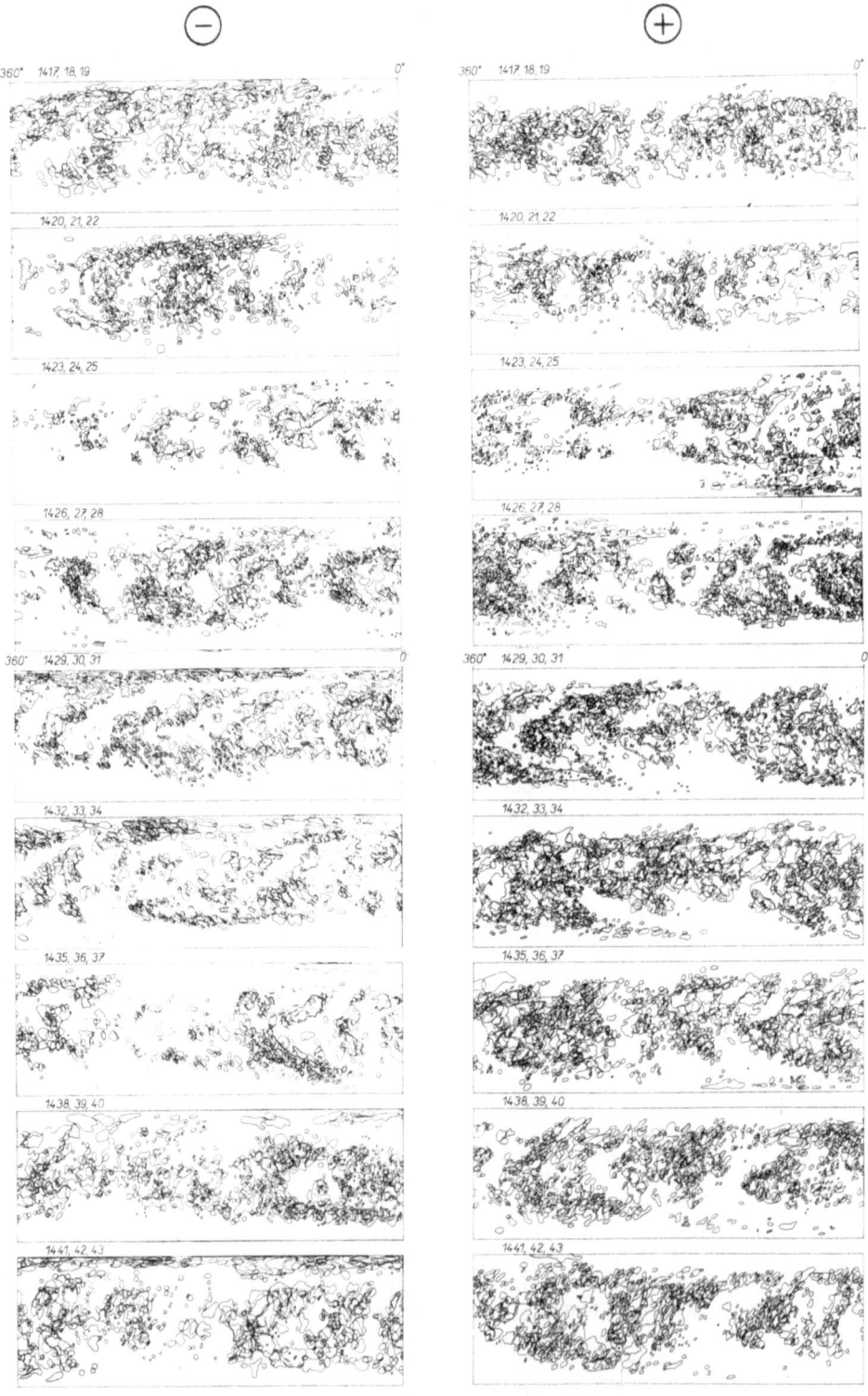

Fig. 5.

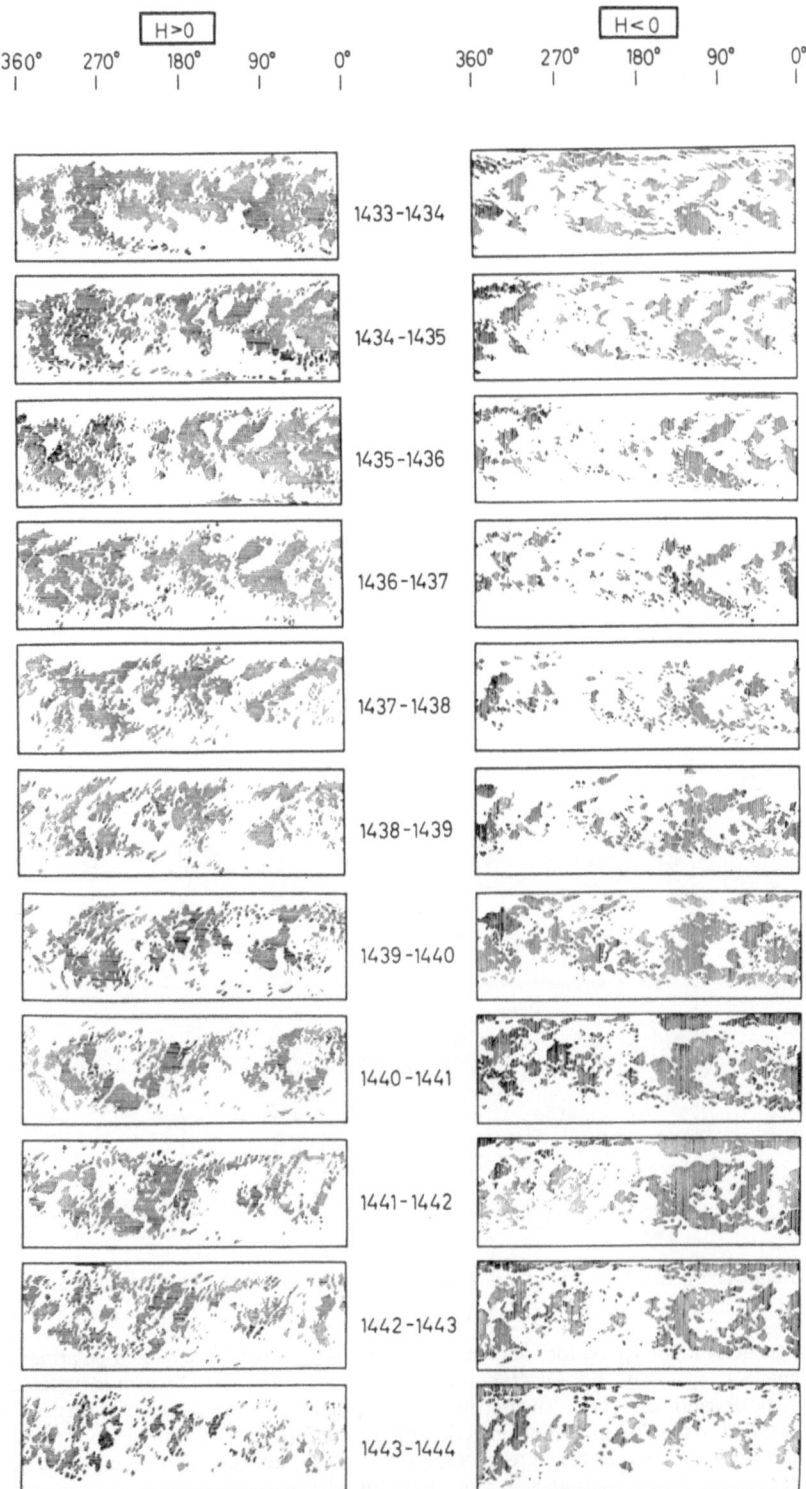

Fig. 6.

the streams. In this stream the negative polarity 'supergiant' structure developed, and the positive polarity 'supergiant' structure developed practically simultaneously in the other stream which was 160° west of the first stream. The negative polarity feature, which was leading in the southern hemisphere, formed a tail at high southern latitudes, and the positive polarity feature formed a tail at high northern latitudes. The similarity

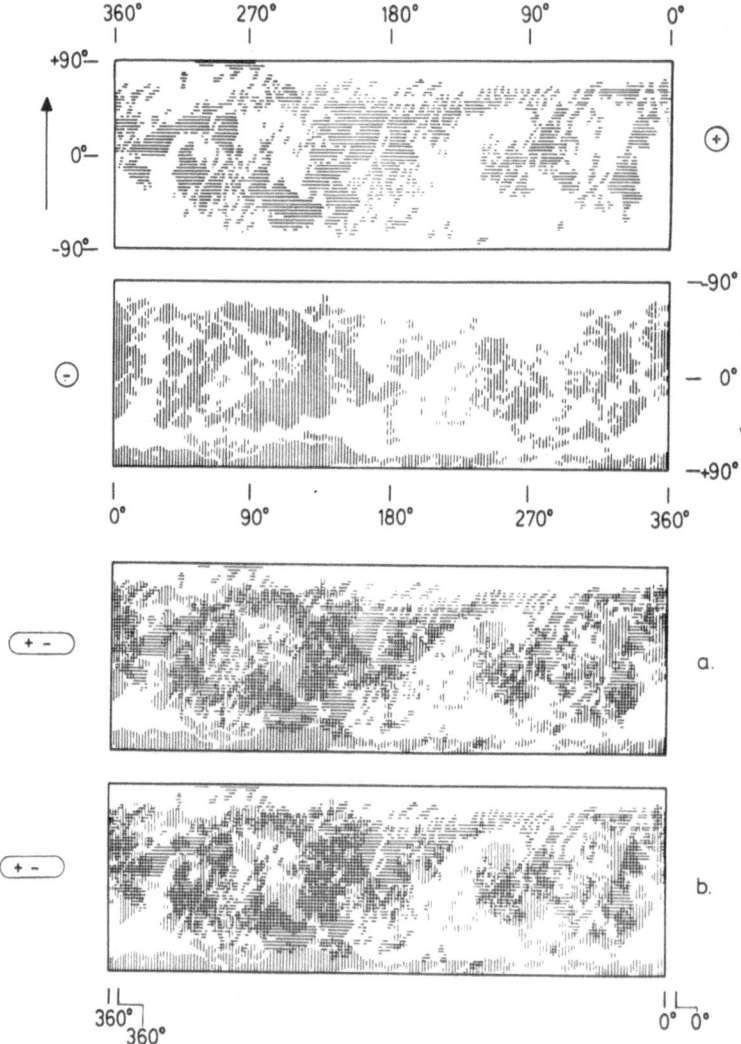

Fig. 7.   Synoptic charts demonstrating the degree of symmetry of both polarity 'supergiant' structures.

Fig. 6.   Magnetic synoptic charts for rotations Nos. 1434–1444 drawn in separated polarities (plus to the left, minus to the right). For integration two consecutive maps, one of which is repeated, are overlapped. The development of two 'supergiant' regular structures is shown, each of them occupying practically one half of the Sun in both main active longitudes, formed from opposite polarities. The regularity and repeated interval structure of 'supergiant' bodies and the 27-day rotational period of both active longitudes may be seen.

of both these enormous features is so striking that for a period of time one can overlay the two and get almost perfect agreement in the position of features over nearly half the solar surface. This is illustrated in Figure 7.

On features of both polarities the characteristic structure of the 'supergiant' features may be seen during this interval. The smaller cell-like feature in the western part of the main body has a diameter of about 30°–35°. The larger cell-like feature near the center of the 'supergiant' structure has a diameter of about 60°–70°. The length of the whole structure is nearly 180° not counting the tail.

During this time interval and viewed in this manner (polarities examined separately)

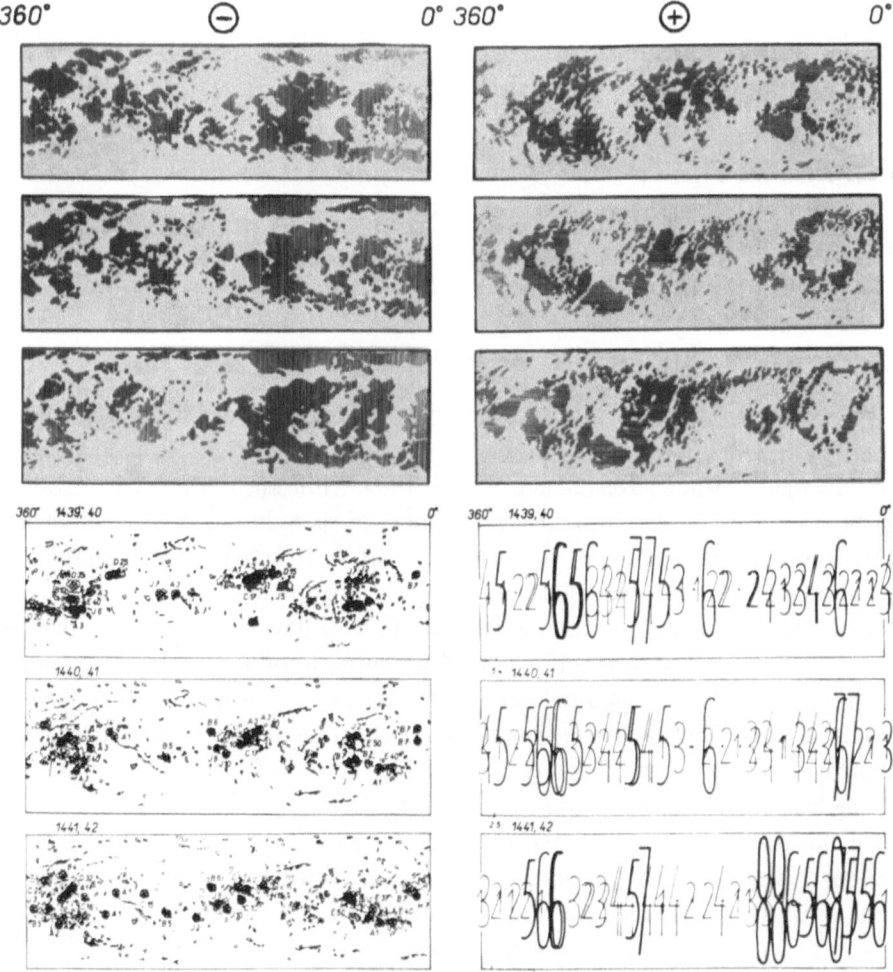

Fig. 8.   Magnetic synoptic charts for rotations Nos. 1439–1442, drawn as in Figure 6, compared with the large-scale solar activity distribution (Fraunhofer Institute, Freiburg) and the geomagnetic activity distribution (Institut für Geophysik, Göttingen) for the same time interval. (The four days, needed by the particles to arrive at the Earth, are taken into account). The close relation of younger minus fields to the actual activity, as demonstrated by calcium plages, and the relation of areas of older plus fields to the geomagnetically disturbed days may be seen.

the large-scale distribution resembles a huge bipolar sunspot group with the two polarities enhanced on either side of the Sun.

A concentration of one polarity in one stream may be observed relatively often, although the distance separating two main streams is not always 160°.

B. CORRELATION WITH SOLAR AND GEOPHYSICAL ACTIVITY

The frequent concentration of opposite polarity fields in streams on opposite sides of the Sun is not the only difference between the polarities. If we compare the distribution of fields of the two polarities with the distribution of activity as seen in the synoptic charts made from the Freiburg daily solar maps, we may see that concentrations of activity inevitably coincide with concentrations of negative polarity fields on the large scale. In addition a comparison of daily geomagnetic disturbances, shifted four days to account for the travel time of particles from the Sun to the Earth, with the magnetic synoptic charts shows a fairly close correlation of the positive polarity fields with positive polarity magnetic features. This is seen in Figure 8.

This remarkable distinction between the polarities can nearly always be seen in plots of separate polarity synoptic charts. Moreover there appears to be no change with the new cycle, although a larger time interval in this present cycle is needed to be certain of this. See Figure 9.

6. Discussion

Regularities in the distribution of sunspots were noted as early as 1863 by Carrington. Recently more attention has been paid to this problem (Dodson-Prince and Hedeman, 1968; Švestka, 1968a, b; Švestka and Simon, 1969; Vitinskij, 1969a, b). It is still not possible to say, of course, how such recurrences of spots in the same location may result from the mechanism of the generation of solar activity. In the large-scale distribution of weak magnetic fields some of the large patterns give the appearance of some sort of convective origin and rotate in 27.3 days. Some cell-like patterns rotate in 27.0 days. The form and mode of development of 'supergiant' structures is striking: an ellipse with a high-latitude tail stretched by differential rotation. It grows from an original single active region as the result of a complex of activity, with an expanding boundary, and reflects to some extent the polarity alternation of the pre-existing background field pattern.

It is not now possible to say whether the characteristic enormous pattern which develops is a natural consequence of the normal redistribution on the solar surface of the magnetic fields of active regions which are formed in more or less regular patterns across the Sun, or whether the large-scale weak pattern which we observe is

→

Fig. 9.   The same comparison of different polarity distributions with solar activity middle part for rotations Nos. 1545–1547 from the present cycle. As in the preceding cycle, the correlation of minus polarity (upper part) with solar activity may again be found. The comparison of plus polarity and geomagnetic activity distribution for this time interval was not possible because of many geomagnetic events with sudden commencements during the interval.

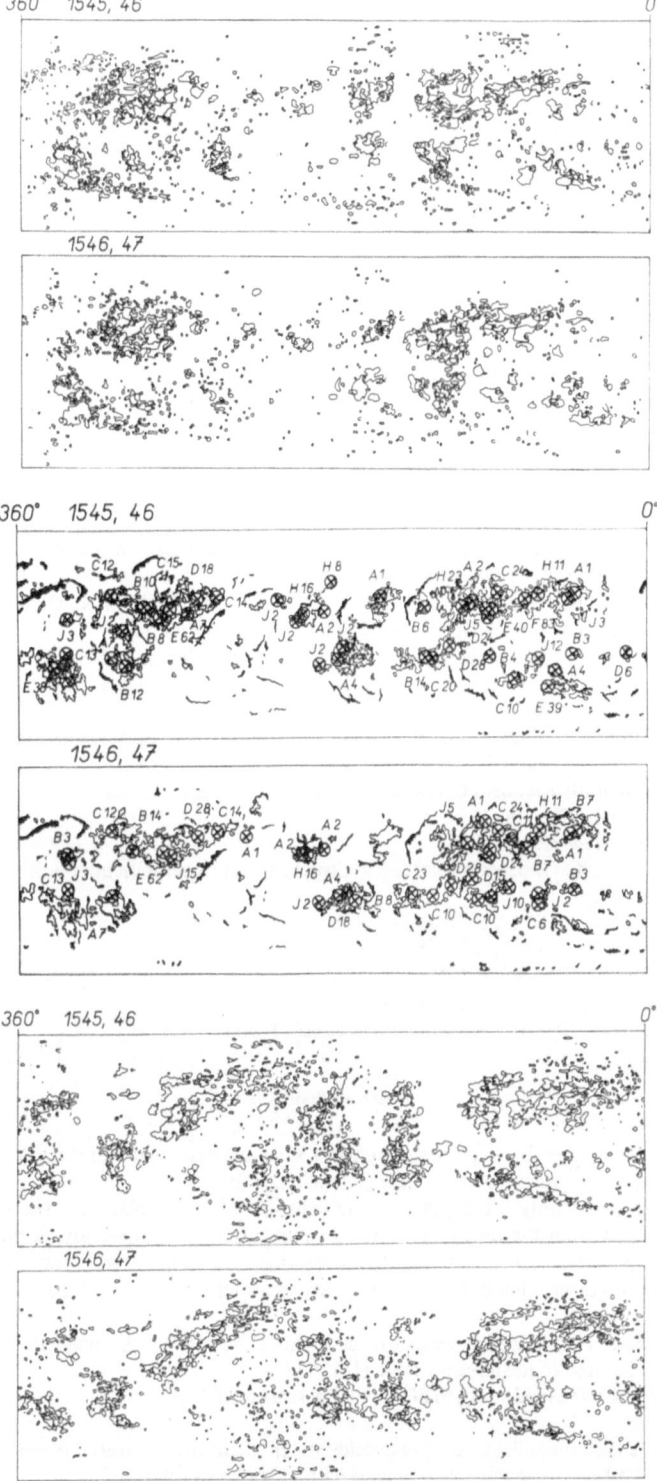

Fig. 9.

a direct reflection of large-scale subsurface fields which themselves produce active regions at the surface in a preferential pattern.

Clearly more work is needed in the study of large-scale patterns. In particular more full-disk data of high quality are needed covering various phases of the solar cycle.

Acknowledgements

The later magnetograph observations and reductions were supported in part by the Office of Naval Research through contract No. N00014-66-C-0239.

References

Bumba, V.: 1970, *Solar Phys.* **11**, 111.
Bumba, V.: 1971, The Proceedings of the Leningrad Symposium on Solar-Terrestrial Physics, in press.
Bumba, V. and Howard, R.: 1965a, *Astrophys. J.* **141**, 1502.
Bumba, V. and Howard, R.: 1965b, *Astrophys. J.* **141**, 1492.
Bumba, V. and Howard, R.: 1969a, *Solar Phys.* **7**, 28.
Bumba, V. and Howard, R.: 1969b, in *Solar Flares and Space Research* (ed. by C. de Jager and Z. Švestka), North-Holland, Publ. Co., Amsterdam, p. 387.
Bumba, V., Howard, R., Martres, M. J., and Soru-Iscovici, I.: 1968, in K. O. Kiepenheuer (ed.), 'Structure and Development of Solar Active Regions', *IAU Symp.* **35**, 13.
Carrington, R. C.: 1863, *Observations of the Spots on the Sun*, Williams and Norgate, London, p. 246.
Dodson-Prince, H. W. and Hedeman, E. R.: 1968, in K. O. Kiepenheuer (ed.), 'Structure and Development of Solar Active Regions', *IAU Symp.* **35**, 56.
Howard, R., Bumba, V., and Smith, S. F.: 1967, Carnegie Inst. of Washington Publ. No. 626, Washington.
Sýkora, J.: 1969, *Bull. Astron. Inst. Czech.* **20**, 70.
Sýkora, J.: 1971, *Solar Phys.* **18**, 72.
Švestka, Z.: 1968a, *Solar Phys.* **4**, 18.
Švestka, Z.: 1968b, in K. O. Kiepenheuer (ed.), 'Structure and Development of Solar Active Regions', *IAU Symp.* **35**, 287.
Švestka, Z. and Simon, P.: 1969, *Solar Phys.* **10**, 3.
Toman, K.: 1967, *J. Geophys. Res.* **72**, 5570.
Vitinskij, Ju. I.: 1969a, *Solar Phys.* **7**, 210.
Vitinskij, Ju. I.: 1969b, *Soln. Dann.* No. 4, 88.
Wolfer, A.: 1897, *Publ. Sternw. Eidg. Polytechn. Zürich*, **1**, XII.

Discussion

Severny: Am I correct that you find the same effect I have just described that there is an overtaking of the south polarity by the north polarity of the general magnetic field?

Bumba: There is certainly a difference in the behavior of both polarities. The south structures change more rapidly than the north structures. In our graphs we can see also a phase shift in the maximum development of south and north structures which can make several rotations.

Severny: Did you discriminate strong fields of active centers from general magnetic fields in your analysis?

Bumba: We do not exclude the sunspot fields because the Mt Wilson magnetic maps do not see them practically, because of the large aperture ($17'' \times 17''$).

Tuominen: What is the field strength of these large scale fields?

Bumba: 4–6 G.

Tuominen: Are then these fields, the polar fields which are in the equatorial plane, more important than the 'classical' fields where the poles coincide with the poles of rotation?

Bumba: I do not know.

Deubner: One must admit that it is very impressive to look at the large ring-like structures you showed. But could not these structures be built up from the large ⊂ shaped flux distributions, generally present on the solar surface, together with some additional flux at the open side of these features just by chance sometimes?

Bumba: I do not think that the regularities are due to chance. They are regular not only in longitude but also in time. They repeat too often to be due to chance and you may observe their development from small to very large areas.

MAGNETIC FIELDS IN POLAR PROMINENCES

G. Y. SMOLKOV

Siberian Institute of Terrestrial Magnetism, Ionosphere and Radio Propagation, Irkutsk, U.S.S.R.

Abstract. The results of longitudinal magnetic field measurements in polar prominences (PP), obtained by Rust during 1964–1965 and by the author during 1966, 1969 and 1970, are considered. The magnetic field configuration in PP is similar to that of an arc. In opposite branches of such an arc the magnetic field has different polarities that do not coincide with the magnetic field polarity in current cycle BMR's. The magnetic field value increases as the angle between the line of sight and filament long axis grows. PP's reflect the behaviour of the border line between the magnetic fields of preceding and current cycles.

Polar prominences (PP), less than other prominences, are subjected to the influence of active regions. Therefore they are the most stable prominences (filaments). Study of the basic properties of PP inevitably leads to the conclusion that PP are located at borders between following parts of the current sunspot cycle BMR's and the same parts of the preceding sunspot cycle BMR's. In connection with this fact magnetic fields in PP's must reflect the general regularities of solar magnetic fields.

Only a relationship between their polarities may serve as a criterion of the correlation of a PP magnetic field with the underlying photospheric magnetic fields. The magnetic field value is not significant in this case. Its determination sufficiently depends on the prominence orientation and the magnetograph sensitivity. It does not change monotonically with height and along a prominence. To use this characteristic one needs measurements of the magnetic field intensity vector.

As result of the factors mentioned above, the polarity of the field in a PP must be inverse to that which is observed in BMR's of the current sunspot cycle.

During 1964–1965 Rust measured the magnetic field longitudinal component at 35 points of six PP's with a 20″ aperture (1966, 1967). These measurements made it possible to find only the tendency of the PP magnetic fields to have a polarity inverse to the polarity of the magnetic field of the current cycle BMR's. In one PP the magnetic field was found equal to 0 ± 2 G. In two others the magnetic field polarity coincided with the polarity of the current cycle BMR magnetic field. In each of them the magnetic field was measured only at one point. But the magnetic field measurement in one of them (also a single measurement) two days earlier showed the opposite polarity of the magnetic field.

The magnetic field measurements in separate points of a prominence do not permit us to obtain the magnetic field distribution picture. Therefore the presence of both polarities in one of the prominences led Rust to the conclusion of an extremely complicated picture of the spatial distribution of magnetic fields in prominences.

Independent measurements of magnetic fields in quiet prominences (QP) with the help of the SibIZMIR magnetograph (Kuznetzov *et al.*, 1966) were started by the author in January, 1966 at Sayan Mountain Observatory (Smolkov, 1966). These observations made it possible to obtain a two-dimensional continuous distribution of the longitudinal component of the magnetic field vector in a whole prominence (Smolkov, 1967a, b). The results of measurements in 1966 showed that the magnetic field configuration in QP's and PP's may be that of an arc. The magnetic field polarity in opposite branches of such an arc may or may not coincide with the magnetic field polarity in a current cycle BMR. An example of such magnetic field distribution is given in Figure 1.

The unipolar distribution of the QP magnetic fields corresponds to cases when the magnetic field of one of the polarities is behind the limb or at the Sun's disk or when the resulting field of one polarity is measured.

Magnetic fields in PP's were measured by the author additionally in 1969 and 1970. The total number of PP's measured is equal to 12, the number of days on which magnetic field measurements were made is equal to 25 (in 1966 correspondingly 7 and 13). In the majority of cases PP's were scanned completely one or several times a day. In some PP's the magnetic field distribution was recorded during several days one after the other or was repeated at the other limb or in another solar rotation.

The magnetic field measurements in PP's made by Rust and the author cover nearly all years from the minimum to the maximum of the current cycle. To consider the magnetic field polarity question we included in addition high-latitude old quiet

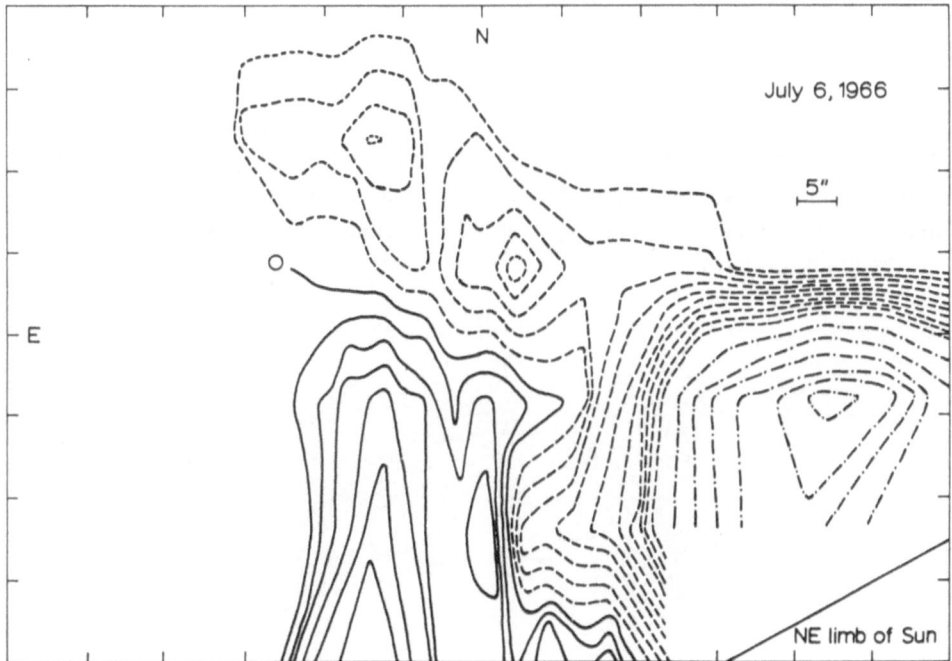

Fig. 1.   Magnetogram, obtained for the Polar Prominence of July 6, 1966. Prominence coordinates 61°N 83°E. Dashed contours enclose negative fields, solid contours enclose positive fields. Isolines on the magnetogram are at 3, 6, 9... G levels. Dash-dot isolines corresponds to interval of 9 G. Thick line means zero field. Standard error is close to 1 G. Scanning aperture is 0″.6 × 5″. The prominence orientation at the moment of magnetic field measurements conforms to the western part of the filament on the diagram of Figure 2.

prominences (OQP). The total number of prominences considered is equal to 27 and the number of days with magnetic field measurements is equal to 47. The total number of cases of lack of magnetic field polarity coincidence with the BMR magnetic field polarity of the current cycle before discussion was equal to 77% of the number of observation days and 76% of the number of all prominences.

Each of these prominences was considered using the Meudon synoptic charts of the solar chromosphere and the Atlas of solar magnetic fields (Howard *et al.*, 1967). Unfortunately the Atlas covered generally only latitudes between −40° and +40° and only the years 1959–1966. A minimum value of the magnetic field on these maps is equal to 6 G. It is not possible to define the photospheric magnetic field distribution for all the prominences considered. However the overwhelming majority of the prominences indicate a scheme of the magnetic field polarities in a OQP which is discussed below. The measurements of the magnetic fields in prominences in 1969 and 1970 conform to this scheme generally. But because just now the author has no synoptic maps of the chromosphere and of the photospheric magnetic fields these measurements require additional study.

The polarity of the recorded magnetic field in high-latitude prominences depends on the long-axis orientation of a prominence (filament) in respect to the line of sight. Filaments which have a relatively large life time stretch along the heliographic latitude. At the later stage of development of filaments their eastern parts become inclined to the equator like rounding the large-scale magnetic field features.

The scheme of a developed OQP orientation in the northern hemisphere is presented in Figure 2. Measuring the magnetic fields in the western part of the filament we obtain the magnetic field polarity coinciding with the magnetic field polarity of the current cycle BMR. In the highest latitude point of filament *A* the longitudinal magnetic field may be small and less than the magnetograph sensitivity level or may be absent altogether. In the eastern part the orientation of the photospheric magnetic

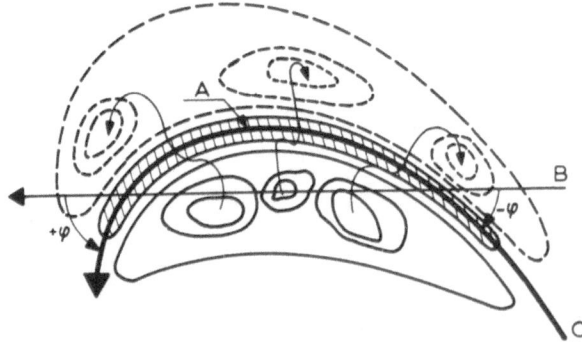

Fig. 2. Diagram of magnetic field polarity in old quiescent filaments for the northern hemisphere. Filaments divide the following patches of BMR preceding and current sunspots cycles. Dashed contours enclose negative fields, solid contours enclose positive fields. φ is angle between the line of sight and the filament long axis. A negative angle means the coincidence of the polarity of the magnetic field of old quiescent filaments and of a current cycle BMR, a positive angle is of the opposite sense. The point *A* dividing the filament in the eastern and western parts is closest to the pole. The letter *B* means the line of sight. The letter *C* means the filament long axis.

field force lines becomes inverse. Therefore the author in the case of such a filament orientation recorded a magnetic field polarity which does not coincide with the magnetic field polarity in the current cycle BMR's, as did Rust.

For PP's having a primary orientation along a heliographic latitude, such conditions of the magnetic field measurements arise as a result of the inclination of the Sun's rotation axis.

The magnetic field in two prominences of 1964–1965 (December 17, 1964 and September 27, 1965) and in one prominence of 1966 (April 2, 1966) deviated from this rule. In all these cases owing to the development of new BMR's, local temporary inhomogeneities of the magnetic field were formed which changed its polarity into an inverse one temporarily. In both cases of the first set, the photospheric magnetic field distribution had been restored by the next solar rotation. The filaments were changed after this. The prominence of 1966 disappeared soon.

All the rest of the prominences of 1964–1966 had a magnetic field polarity strictly corresponding to Figure 2.

The inversion of the magnetic field polarity when passing the point A (Figure 2) and the null value of the magnetic field in such parts of the filament testify to the predominance of the magnetic field directed for the most part perpendicularly to the long axis filament. During the measurements in the same prominence, the magnetic field value increases as the angle between the sight line and its long axis grows. These facts show that OQP's have no magnetic fields of their own. Using the magnetograph measurements one records the magnetic field caused by photospheric sources which penetrate into the prominence. This is in agreement with the theory of filaments by Kippenhahn and Schlüter (1957).

The results obtained confirm the close connection of OQP's with the photospheric magnetic fields. One may consider that PP's reflect the behaviour of the line of division of the magnetic fields of preceding and current cycles. This was shown theoretically by Leighton (1969). The coincidence of the differential rotation velocity of the photospheric magnetic fields with the filament rotation velocity found by Wilcox and Howard (1970) may serve as an additional reason.

References

Howard, R., Bumba, V., and Smith, S. F.: 1967, Carnegie Institution of Washington Publication, No. 626.

Kippenhahn, R. and Schlüter, A.: 1957, Z. Astrophys. 42, 36.

Kuznetzov, D. A., Kuklin, G. V., and Stepanov, V. E.: 1966, 'Results of Observations and Researches in Period IQSY. Siberia and Far East. I', Solar Physics and Cosmic Rays, p. 80.

Leighton, R. B.: 1969, Astrophys. J. 156, 1.

Rust, D. M.: 1966, Ph.D. Thesis, Univ. of Colorado.

Rust, D. M.: 1967, Astrophys. J. 156, 313.

Smolkov, G. Y.: 1966, Researches on Geomagnetism and Aeronomy p. 189.

Smolkov, G. Y.: 1967a, 'Results of Observations and Researches in Period IQSY. Siberia and Far East, II', Solar Physics and Cosmic Rays, in press.

Smolkov, G. Y.: 1967b, 'Results of Observations and Researches in Period IQSY. Siberia and Far East, II', Solar Physics and Cosmic Rays, in press.

Wilcox, J. M. and Howard, R.: 1970, Solar Phys. 13, 251.

OBSERVATIONS OF THE POLAR MAGNETIC FIELDS

JAN OLOF STENFLO

Astronomical Observatory, Lund, Sweden

Abstract. Results from observations of the polar magnetic fields made in 1968 with the Mount Wilson magnetograph are reported. The field was of south polarity near the heliographic north pole and of north polarity near the south pole. The inversion line of the field was at latitude $+70°$ in the north hemisphere and $—55°$ in the south, which coincides with the position of the polar prominence zones at that time. Observations made simultaneously at the Crimean Astrophysical Observatory are in good agreement with the Mt. Wilson data and confirm the latitude variation of the field.

A few hundred of Hale's plates from 1914, which he had used to determine the Sun's general magnetic field, have been remeasured. The plates were automatically scanned with a digitized microphotometer, and the treatment of the data was made with a computer. While the visual measurements by Van Maanen had given a strong general field similar to that of a dipole, the computer reductions did not reveal any significant field (stronger than 5 G) at any latitude.

1. Introduction

The various theories of the solar cycle make different predictions about the magnetic fields at high heliographic latitudes. Although the polar fields are so important for tests of the solar cycle theories, our knowledge of the behaviour of the polar fields is very limited. It is difficult to observe the polar fields because they are very weak and always situated close to the limb.

The first attempts to determine the average properties of the Sun's magnetic field were made by Hale (Hale, 1912, 1913; Hale *et al.*, 1918; Seares *et al.*, 1918, 1919). He found a latitude variation of the field between latitudes $-50°$ and $+50°$ that corresponded to that of a dipole-type field. The polarity of the solar magnetic field was the same as that of the Earth. Assuming a dipole-type field for the Sun Hale could calculate what the field strength would be at the poles. The observed field was however found to depend very much on the spectral line used; the field was stronger in weaker lines. This was interpreted in terms of a depth-dependence of the field. Accordingly the polar field strength would vary from about 50 G in the lower photosphere to zero in the upper photosphere. This extremely rapid variation of the field with height is however quite inconsistent with the assumption of a dipole-type field and implies that the field lines should be almost parallel to the solar surface (Rosseland, 1925; Stenflo, 1970b).

Determinations of the inclination i of the magnetic axis to the rotational axis and the period of rotation P of the magnetic pole around the axis of rotation were also presented (Seares *et al.*, 1918, 1919). The results were: $i = 6°0 \pm 0°4$ and $P = 31.52 \pm 0.28$ days. The magnetic pole was on the central meridian at time $t = $ June 25.38 ± 0.42 (UT), 1914.

It has long been a riddle what these early determinations of the Sun's general magnetic field really mean, because later observations have given entirely different results. Hale tried to make a new determination of the Sun's general field in 1931–1932

Howard (ed.), Solar Magnetic Fields, 714–724. All Rights Reserved.
Copyright © 1971 by the IAU.

but obtained zero results. This made him suspect that the solar magnetic field could vary in time with an unknown period (*Mount Wilson Observatory Year Book* **33**, 1933–1934). It is interesting to note that the determinations of the general magnetic field by Hale in 1912–1914 were made during a period of minimum solar activity, which was the lowest minimum we have had since 1810. It was preceded by the lowest maximum there has been since 1816.

The first extensive photoelectric observations of the fields at high heliographic latitudes were made by Babcock and Babcock (1955). The observed longitudinal field was of the order of 1 G and showed opposite sign in the two hemispheres. The polarity of the field was opposite to that of the Earth. In the spring of 1957 the field near the south pole was observed to reverse its sign (Babcock, 1959). The field seemed to be of the same polarity (directed outwards) at both poles for $1\frac{1}{2}$ yr, until it suddenly, in November 1958, reversed sign at the north pole. It was suggested by Waldmeier (1960) that this difference in time of the field reversals was related to the difference in phase between the activities in the two hemispheres.

On the basis of these observations of the reversal of the polar fields Babcock (1961) proposed a theory for the solar cycle according to which the field will reverse its sign around each period of maximum solar activity. This theory has been significantly developed by Leighton (1959), who however assumes infinite electrical conductivity and thereby avoids the problem of the maintenance of the field. Steenbeck and Krause (1966a, b, 1967, 1969) succeeded to build up a solar cycle theory that is based on a dynamo mechanism for the maintenance of the field and which qualitatively yields the same results as the phenomenological theory by Babcock. All these theories predict a polarity of the solar field in the period 1912–1914 which is opposite to that found by Hale.

2. Observations of the Polar Fields near the Last Maximum of Solar Activity

As an actual reversal of the polar fields has only been observed once, in 1957–1958, it is of considerable interest to study the polar fields around the activity maximum in 1968–1969. I will here report on systematic observations that were carried out with the Mt Wilson magnetograph during July and August 1968 (Stenflo, 1970a). The north and south polar regions were scanned almost daily during a period covering approximately two solar rotations. The scanning aperture was $5'' \times 5''$, and the spectral line Fe I 5250.2 Å was used.

A computer program was written to plot synoptic charts of the polar regions. Synoptic charts for August 1968 are presented in Figure 1. These maps show isogauss contours for $H_{\parallel}/\cos\theta$, where $H_{\parallel}$ is the observed longitudinal field and θ the heliocentric angle. $H_{\parallel}/\cos\theta$ is the total field strength if the field is assumed to be directed along the solar radius. It is often quite misleading to present observations of polar magnetic fields in terms of $H_{\parallel}$ only, since the observations are often made at various heliocentric angles. Close to the limb the magnetic field should be mainly transversal.

N polarity elements seem to predominate in both polar regions. Accordingly one

North pole

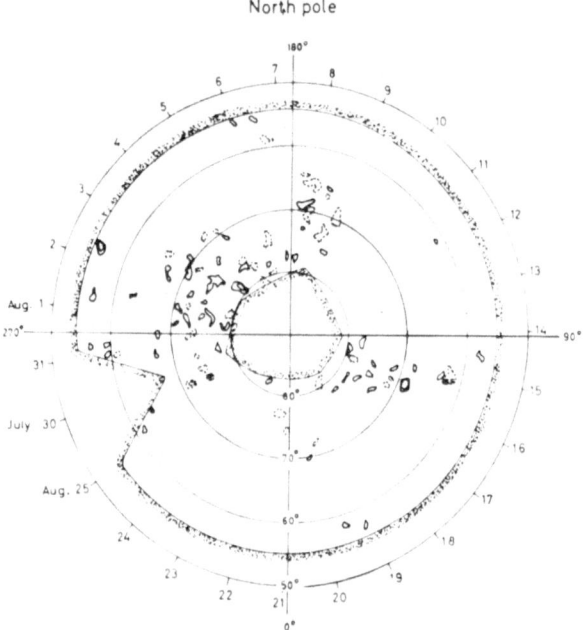

Fig. 1a.

South pole

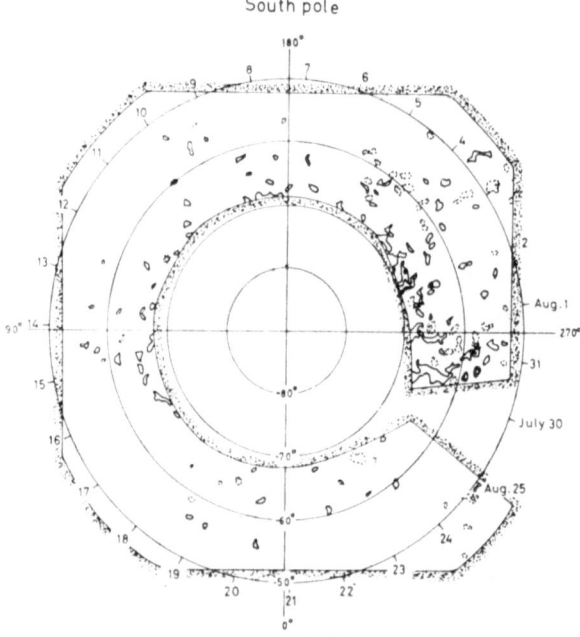

Fig. 1b.

Fig. 1a–b. Synoptic charts of the polar magnetic fields $H_{\parallel}/\cos\theta$. Solid lines represent N polarity, dashed lines S polarity. The isogauss levels are 10 and 20 G. The areas covered by the observations are enclosed by curves with shadings on the outside. (a) North pole. Observations from August 1– August 23, 1968. (b) South pole. Observations from July 31–August 23, 1968.

could get the impression that the average magnetic field was directed outwards from both poles, a situation similar to that observed by Babcock (1959) for the field in 1957–1958. A more careful analysis shows however that this is not quite so.

The structure of the polar fields is very complicated and is far from rotationally-symmetric. Some average properties may be seen, however, by averaging the radial field $H_{\parallel}/\cos\theta$ over all longitudes and studying the latitude variation of this field. Figure 2 shows the latitude variation of the field for July–August 1968. H is the observed average radial field multiplied by a factor of 3 to account for the weakening of the 5250 Å line in magnetic regions (Harvey and Livingston, 1969). It is hoped that H by this correction will be a good approximation of the true average field strength. The synoptic charts on the other hand have not been corrected for line weakenings.

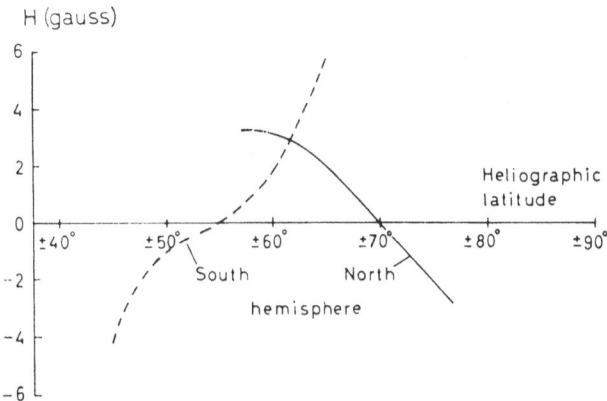

Fig. 2.   The average radial magnetic field of July–August, 1968, corrected for line weakenings in magnetic regions.

The field in the north hemisphere varies in strength from about $+3$ G at lat. $+60°$ to -2 G at $+75°$, with polarity reversal at lat. $+70°$. In the south hemisphere it varies from -4 G at lat. $-45°$ to about $+6$ G at $-65°$, with polarity reversal at lat. $-55°$. The field is of N polarity in both polar regions between lat. $55°$ and $70°$. This is the reason why one gets the impression that the field is directed outwards from both poles.

The mean latitudes of the polar prominence zones in 1968 were $+72°$ and $-58°$ (Waldmeier, 1969). This coincides within the limits of observing accuracy with the latitudes of polarity reversals in Figure 2. Such a relation between inversion lines of the average magnetic field at high latitudes and the polar prominence zones was first suggested by Waldmeier (1960). It is also of interest to compare our result with that of Bumba et al. (1969) that quiescent prominences and filaments at high latitudes coincide in position with minima in the radio emission at 1420 MHz and with minima in the brightness of the green ($\lambda\,5303$ Å) corona.

The asymmetry or phase-shift between the two hemispheres may be explained in the theory of Steenbeck and Krause (1966a, b, 1967, 1969) in terms of a constant quadrupole-type field that is superposed on a variable dipole-type field. When the dipole-type field disappears, only the quadrupole-type field, which has the same polarity at both poles, is seen.

3. Comparison of Observations at Mt Wilson and Crimea

It is important that such delicate observations as those of the polar fields can be verified by other observers, if possible by simultaneous observations. On August 17 and 20 the polar fields were recorded at Crimea and Mt Wilson with a time difference of only about 10 hr between the recordings at the two observatories (Kotov and Stenflo, 1970). The position of the regions scanned is illustrated in Figure 3. The characteristics of

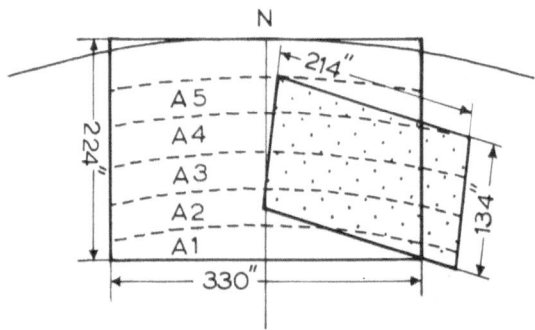

Fig. 3.   Illustration of the north polar regions scanned at Mt Wilson and Crimea (with dots inside). For the reduction, the scanned regions were divided in smaller areas A1–A5, in which the average properties of the field were determined.

TABLE I

Data for the recordings with the Mt Wilson and Crimean magnetographs

Observatory	Mt Wilson	Crimea
Scanned area	330″ × 224″	214″ × 134″
Slit size	5″ × 5″	9″ × 2″.5
Scanning velocity v	5″.9/s	1″/s
Time constant τ	0^s.4	2^s.5
R.m.s. noise	3 G	1.7 G
Position of exit slits	7–87 mÅ	27–90 mÅ

the two magnetographs are compared in Table I. In Figure 4 two isogauss maps covering the same region on the Sun are shown. The six largest magnetic elements in the Crimean map could all be identified in the Mt Wilson map. Due to the time difference the Crimean elements are displaced in the westward direction by about 6° relative to the Mt Wilson elements. These magnetic features obviously have lifetimes exceeding 10 hr. The other isogauss maps do not show such a good agreement as the maps in Figure 4.

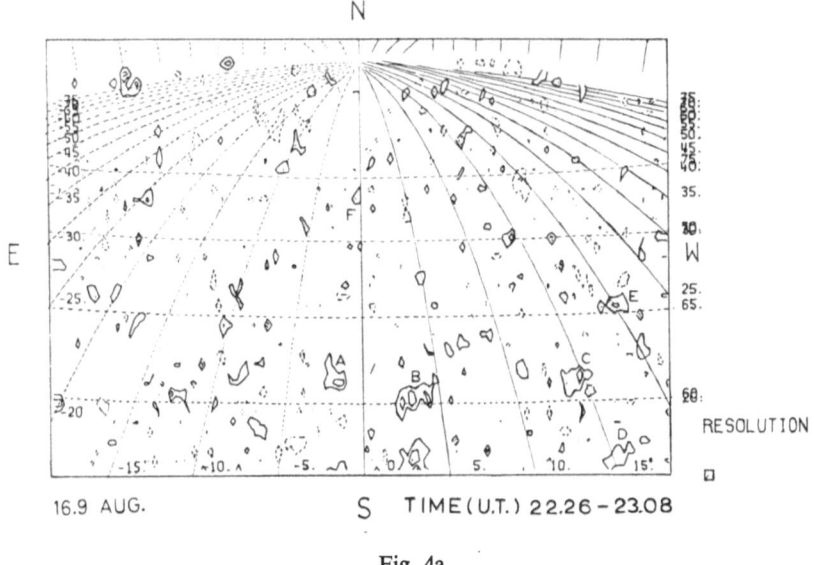

Fig. 4a.

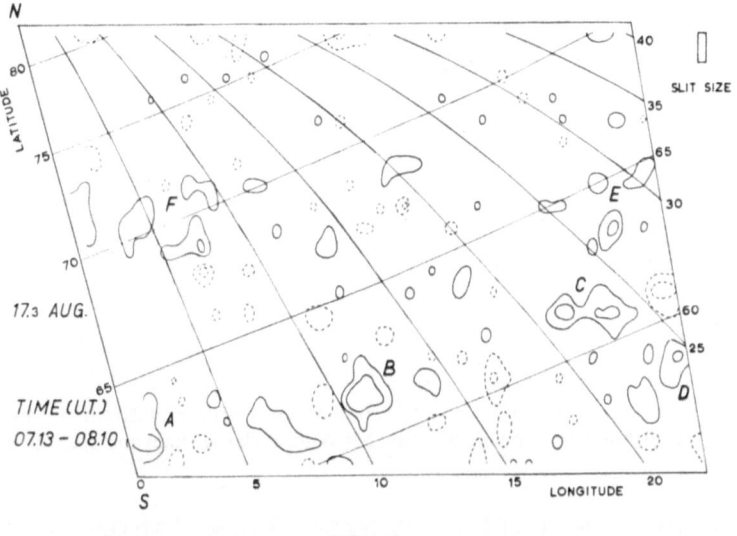

Fig. 4b.

Fig. 4a–b.   Isogauss maps for the north pole obtained on August 16 at Mt Wilson (a) and on August 17 at Crimea (b). Solid lines represent N polarity, dashed lines S polarity. The isogauss levels are 5, 10 and 20 G.

A more quantitative comparison is presented in Figure 5, in which the average radial field $\bar{H}_r = \overline{H_\parallel / \cos\theta}$, the average longitudinal N- and S-polarity fields, $\bar{H}_N$ and $\bar{H}_S$, and the percentage of the scanned area covered by N-polarity fields, $100\,N_N/N$, are given as functions of latitude. N_N is the number of observation points with fields of N polarity, while N is the total number of points. Corrections have been made for the asymmetric position and time difference of the Crimean recording relative to the Mt Wilson one, so that the values in Figure 5 refer to the same area on the Sun. Only the results from August 20 are shown in Figure 5; the results from August 17 are qualitatively the same.

Considering the many factors that may influence recordings of the polar fields, the agreement between the Crimean and Mt Wilson results can be regarded as satisfactory. The correlation coefficient between the Crimean and Mt Wilson values of $\bar{H}_r$ is 0.7 for the north and 0.5 for the south hemisphere. The conclusions on the latitude variation of the polar fields that were based on the Mt Wilson recordings are thus supported by the Crimean data.

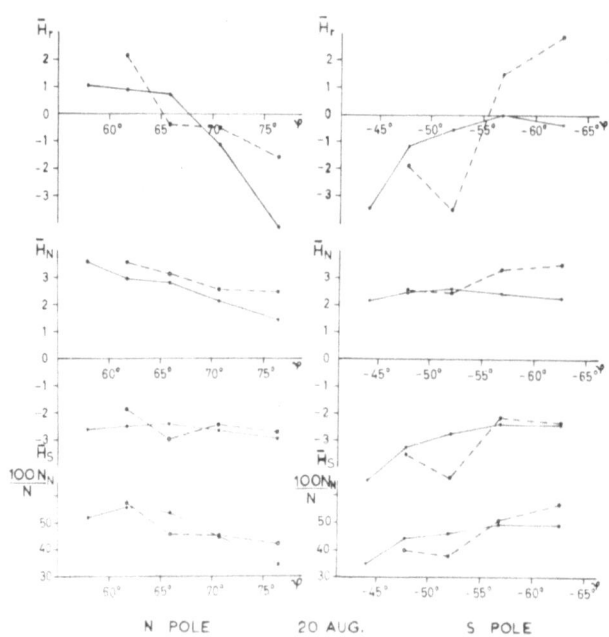

Fig. 5.   The latitude variation of $\bar{H}_r$, $\bar{H}_N$, $\bar{H}_S$(G) and 100 N_N/N as determined with the two magnetographs for August 20. Solid lines and filled circles: Mt Wilson. Dashed lines and open circles: Crimea.

4. Remeasurements of Hale's Plates for the General Magnetic Field

Hale's early results on the Sun's general magnetic field contradict the results of modern photoelectric observations. Practically all the plates used by Hale in the investigation of the general field were reduced by Van Maanen by visual settings with a tipping glass-plate micrometer.

To free the reductions from any possible personal bias, a few hundred of the plates from the 1914 series were chosen at random and taken by the author to the Sacramento Peak Observatory, where they could be automatically scanned with a digitized microphotometer. The line profiles were recorded on magnetic tape, and the reductions were made with a CD 3600 in Uppsala (Stenflo, 1970b).

The same plates had been measured by Van Maanen. His reduction book with all his micrometer settings was available. This material had then been used for the determination of the inclination and period of rotation of the magnetic pole relative to the Sun's rotational axis.

Each plate is composed of 2 mm wide strips caused by the compound $\lambda/4$ plate. Successive strips are alternately right- and left-handed circularly polarized. On each plate there is a 'marked strip'. The adjoining strips are circularly polarized in the left-handed direction.

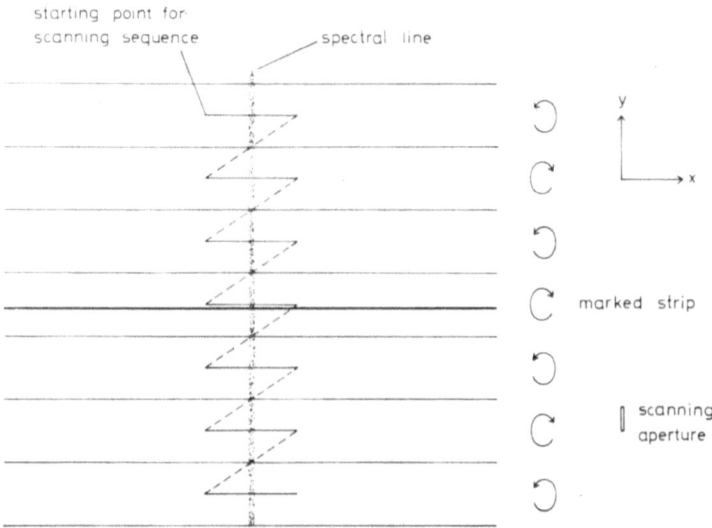

Fig. 6. Illustration of the procedure for automatic scanning of the Hale plates.

The scanning sequence is illustrated in Figure 6. The position (mean x-coordinate) of the line in each strip was computed. Strips with defects (specks in the emulsion etc.) were skipped. The marked strip was always skipped in the computations. In Figure 7 are illustrated the line displacements $\bar{X}_k$ relative to strip no. 1. A least-mean-square fit of a second-order curve has been made to the values for strips of *odd* ordinal number, i.e. strips of the same polarization as strip no. 1 (filled circles in Figure 7). The displacement caused by a magnetic field is defined as the difference between the $\bar{X}_k$ in strips of *even* ordinal number (open circles in Figure 7) and the curve. The correct sign of the Zeeman displacement depends on (1) if the marked strip and strip no. 1 are of the same or opposite polarization, and (2) if the compound $\lambda/4$ plate was in the normal or inverted position.

The average line displacements are given in Figure 8 as a function of heliographic latitude. The same spectral lines as those used by Van Maanen, the Cr I lines $\lambda\lambda\,5247.57$, 5300.75 and 5329.15 Å were used. The results obtained for each line separately are very similar. Accordingly only the results for all three lines together are given in Figure 8. Using the effective Landé factors of the lines we can express the displacements in G. The scale to the right in Figure 8, in which 1 μ corresponds to 4 G, represents an average for the three lines.

Van Maanen's results for the same plates as were remeasured are shown in Figure 8a. The measured field shows a neat variation with heliographic latitude, with the solar equator forming a sharp demarcation line between the opposite polarities in the two hemispheres. The field strength attains a maximum of about 11 G at lat. + and $-45°$.

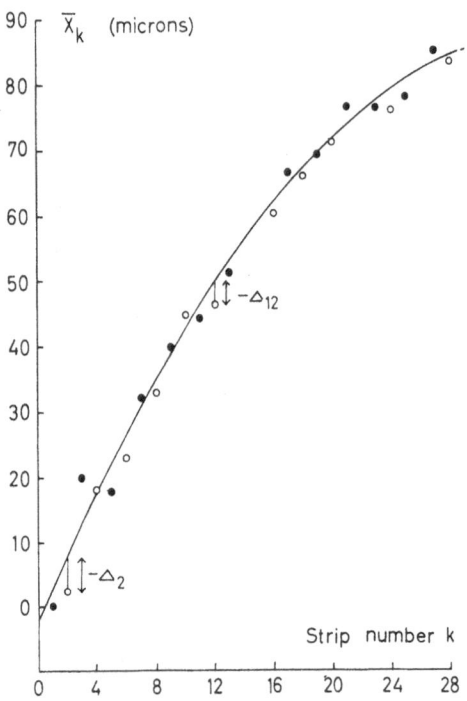

Fig. 7. Displacement $\bar{X}_k$ of the line in strip No. k relative to the line in strip No. 1. The filled circles represent values in strips of *odd* numbers (of the same polarization as strip No. 1), while the open circles represent values in strips of *even* numbers. The curve represents a second-order least-mean-square fit to the filled circles. The Zeeman displacement Δ_k is defined as the distance (with sign) of an open circle from the curve as illustrated in the figure.

In contrast, the computer reductions do not show any significant field at any latitude. It was not clear if north was 'up' or 'down' on the plates, and therefore two separate reductions were made to cover both these cases. Figure 8b gives the results for the assumption that north was in the direction of decreasing strip numbers, while Figure 8c is based on the opposite assumption. An approximate upper limit for the average field strength is 5 G for all latitudes. There is no correlation between the remeasured displacements and Van Maanen's values.

The standard deviations in the mean values are also shown in Figure 8. Van Maanen's standard deviations are about 3 times smaller than ours. His standard deviations in *one* value are only 1–2 μ, which is quite remarkable, considering that the width of the spectral line was about 500 μ, that the exposure time was about 20 min and that the standard deviations should include the influence of local magnetic fields and Doppler shifts that are not coherent over successive strips. Our standard deviations in *one* value are 3–6 μ.

It is interesting to compare the attempts to determine the Sun's general magnetic field with the determinations of proper motions in the spiral galaxy M33. Van Maanen found the proper motions to be directed along the spiral arms (Lundmark, 1927).

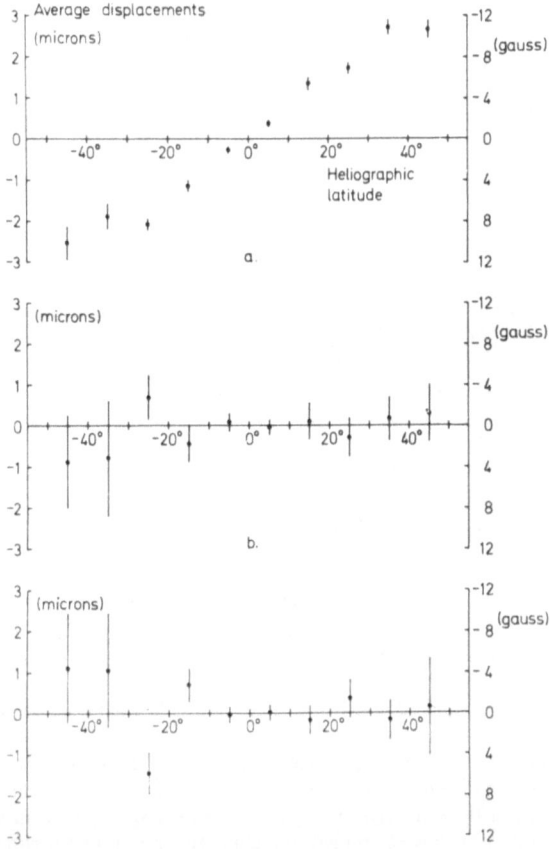

Fig. 8.   Average Zeeman displacements expressed in microns and gauss as a function of heliographic latitude. The material has been grouped in the latitude intervals $-50°$–$-40°$, $-40°$–$-30°$,..., $+40°$–$+50°$. The standard deviations in the mean value are shown. (a) Van Maanen's results; (b) New results with the assumption that north is in the direction of decreasing strip numbers; (c) New results, with the assumption that north is in the direction of increasing strip numbers.

The proper motions were largest at the edge of M33 and decreased towards the centre of the galaxy, where they reversed direction. The rotation period found by Van Maanen was 230000 yr, which is smaller by a factor of 1000 than the period of rotation of our own galaxy. At that time these results were strong evidence against the idea that the so-called spiral nebulae were 'island universes' at large distances from our Milky Way system (Lundmark, 1927).

According to our remeasurements of Hale's plates from 1914 there is no contra-diction between the field observed at that time and the magnetic field that is presently observed with photoelectric techniques.

References

Babcock, H. D.: 1959, *Astrophys. J.* **130**, 364.
Babcock, H. W.: 1961, *Astrophys. J.* **133**, 572.
Babcock, H. W. and Babcock, H. D.: 1955, *Astrophys. J.* **121**, 349.
Bumba, V., Kleczek, J., Olmr, J., and Ruzickova-Topolova, B.: 1969, *Bull. Astron. Inst. Czech.* **20**, 67.
Hale, G. E.: 1912, *Terrest. Magn. Atmospheric Electr.* **17**, 173.
Hale, G. E.: 1913, *Astrophys. J.* **38**, 27.
Hale, G. E., Seares, F. H., Van Maanen, A., and Ellerman, F.: 1918, *Astrophys. J.* **47**, 206.
Harvey, J. and Livingston, W.: 1969, *Solar Phys.* **10**, 283.
Kotov, V. A. and Stenflo, J. O.: 1970, *Solar Phys.* **15**, 265.
Leighton, R. B.: 1969, *Astrophys. J.* **156**, 1.
Lundmark, K.: 1927, *Uppsala Astron. Obs. Medd.* No. 30.
Rosseland, S.: 1925, *Astrophys. J.* **62**, 387.
Seares, F. H., Van Maanen, A., and Ellerman, F.: 1918, *Proc. Nat. Acad. Sci.* **4**, 4.
Seares, F. H., Van Maanen, A., and Ellerman, F.: 1919, *Proc. Nat. Acad. Sci.* **5**, 242.
Steenbeck, M.: 1966a, *Wissenschaft und Menschheit* **2**, 315, Urania-Verlag Leipzig/Jena/Berlin.
Steenbeck, M. and Krause, F.: 1966b, *Z. Naturforsch.* **21a**, 1285.
Steenbeck, M. and Krause, F.: 1967, Gustav-Hertz-Festschrift, Akademie-Verlag, Berlin.
Steenbeck, M. and Krause, F.: 1969, *Astron. Nachr.* **291**, 49.
Stenflo, J. O.: 1970a, *Solar Phys.* **13**, 42.
Stenflo, J. O.: 1970b, *Solar Phys.* **14**, 263.
Waldmeier, M.: 1960, *Z. Astrophys.* **49**, 176.
Waldmeier, M.: 1969, *Astron. Mitt. Zürich* Nr. 288.

Discussion

Livingston: Did you compare or try to correlate your Mt Wilson polar fields with white light polar faculae or other brightness pictures? I am thinking that if no facular correspondence is seen the fields might be diffusely weak and 5250 might not be thermally weakened (so your field strength as shown might be correct).

Stenflo: I have not compared my magnetic-field observations with brightness pictures.

Rust: A look at your last slide makes me think that a least-squares fit to your reduction of the data would imply polar fields of the same sign, if not of the same magnitude, as in the earlier reduction. A slide that Prof. Severny just showed seemed to include this point, and it showed that the polarity of the Sun's dipole was opposite to that to be expected from extrapolation of modern observations back to that time. Does your conclusion agree with this?

Stenflo: The conclusion I can draw from my reductions is that there is no significant field in excess of 5 G at any latitude. It is not possible to determine the average polarity of the field from my data.

A NUMERICAL STUDY OF THE SOLAR CYCLE

Y. NAKAGAWA

High Altitude Observatory, National Center for Atmospheric Research*, Boulder, Colo., U.S.A.

Abstract. Models relevant to numerical studies of the solar cycle are reviewed briefly with discussions of pertinent physical mechanisms. It is suggested that the observed surface activities are secondary in nature and an example of possible non-axisymmetric steady state solutions is given, together with the results of preliminary numerical computations.

1. Introduction

In the model study of the solar cycle, it is necessary that a model satisfies various well-established laws of observations. These laws are: (1) Hale's law of polarity of sunspots (Hale, 1908, 1913), namely, the preceding sunspot has opposite polarity in the northern and southern hemispheres and their polarities reverse from one 11-yr cycle to the next, (2) Spörer's law of zone of activity (Carrington, 1858; Spörer, 1894), i.e., at the beginning of activity the sunspots appear near latitude 30° and the zone of activity gradually shifts towards the equator in 11 yr, and (3) the reversal of polar magnetic field (Babcock, 1959) with approximately 11-yr period.

Models which satisfy some or all of these laws of observations have been proposed by a number of authors on the basis of somewhat diversified physical mechanisms. Among the models, notably, the topological model of Babcock (1961) has combined the observed features of the 22-yr cycle of the Sun in a consistent manner, and most other models have been based on this topological model in one way or the other. The first quantitative examination of the Babcock model (1961) has been given by Leighton (1964). Leighton (1964) showed that the redistribution and annihilation of magnetic flux of bipolar sunspots by supergranulation could account for the development of the bipolar magnetic regions which lead to the final reversal of the polarity of polar magnetic field.

A more extended quantitative examination of the Babcock model has been presented by Nakagawa and Swarztrauber (1969). Starting with the initial state of axisymmetric quasi-dipole magnetic field they have numerically integrated the hydromagnetic equations within the context of axisymmetric incompressible hydromagnetics. Their results showed in agreement with the Babcock model, (1) the development of the toroidal magnetic field, (2) the gradual drift of zone of activity (Spörer's law), and (3) Hale's law of polarity. However, the assumption of axisymmetry limited their study only to a single 11-yr cycle. Recently a different type of approach to the model of the solar cycle has been suggested by Leighton (1969), in the form of a 'magneto-kinematic' model based on a set of synthesized equations which govern the magnetic fields. It was mentioned that these equations account for the various physical processes needed at

---

* The National Center for Atmospheric Research is sponsored by the National Science Foundation.

Howard (ed.), Solar Magnetic Fields, 725–736. All Rights Reserved.
Copyright © 1971 by the IAU.

the different stages of development of the solar cycle as postulated in the Babcock model. However, it is rather difficult to justify his basic equations from the ordinary hydromagnetics.

Apart from these studies, in a series of papers, Krause (1967) and his collaborators have developed a model of the solar cycle, emphasizing the effect of hydromagnetic turbulence resulting from the subphotospheric convection. Their model utilized an induced quadrupole type magnetic field first developed near the surface which migrates towards the equator to account for Spörer's law of zone of activities and towards the interior to achieve the polarity reversal of initial dipole-like polar magnetic field.

In this paper, the basic concept of the model of an the solar cycle under the assumption of non-axisymmetric hydromagnetics is discussed and some preliminary results of a non-axisymmetric hydromagnetic model study of an incompressible, viscous medium of finite electrical conductivity are presented.

2. Basic Concept of the Model

In constructing a model of the solar cycle, the question must be focused on the physical mechanisms which are pertinent to produce the observed phenomena. First, we may follow the Babcock model (1961) to examine the various physical mechanisms postulated in his model. Babcock (1961) divided his model into five successive stages of development each characterized by the dominance of different physical mechanisms.

The initial stage (stage 1) corresponding to the state of solar minimum was assumed to be dominated by an approximately axisymmetric dipole-like magnetic field with the interior flux lying in a relatively thin submerged layer, while the surfaces of constant angular velocity (isotachs) cut more deeply into the Sun than the magnetic lines of force. This assumption led to stage 2 of the generation and amplification of toroidal magnetic field from the initial poloidal magnetic field by the differential rotation. This was followed by stage 3, the emergence of the amplified toroidal magnetic field as bi-polar sunspots due to a postulated instability associated with the magnetic buoyancy. Stage 4 was characterized by the redistribution and annihilation of the emerged magnetic flux which eventually led to the reversal of polarity of the polar magnetic field. The final stage 5 was assumed to be the solar minimum reached after 11-yr similar to stage 1 but with the dipole polarity reversed.

In Leighton's model (Leighton, 1964), the redistribution and cancellation of bipolar magnetic flux (the stage 4 of Babcock model) was shown to be the result of a two-dimensional random walk process due to the supergranulation in the presence of a differential rotation. Leighton (1964) utilized an important result of observation that the bipolar sunspots emerge with a slight difference in the latitude governed by Hale's law, i.e., the preceding spot appears closer to the equator. The use of this result led the polarity of reversed polar magnetic field to be that of the following spot.

In the hydromagnetic model of Nakagawa and Swarztrauber (1969), the law of isorotation (Ferraro, 1937) was assumed at the initial stage. As a consequence, the angular velocity of their model decreased with depth. However, in the presence of viscous dissipations (which could be identified with the granulation as well as super-granulation), a net angular velocity transfer towards the equator was required to maintain the faster equatorial rotation. The angular momentum transfer was achieved by a meridional circulation directed towards the equator near the surface. This meridional circulation produced a concentration of the interior magnetic lines of force near the surface similar to the Babcock model of stage 1, and led to the subsequent generation and amplification of toroidal magnetic field of stage 2. The manner of distortion of the initial poloidal magnetic lines of force then successfully accounted for Hale's law. Spörer's law of the equatorial migration of the zone of activity was satisfied by the gradual equatorial drift of the concentrated toroidal magnetic field by meridional circulation. In contrast to this model, Krause (1967) and Leighton (1969) have assumed a differential rotation increasing with depth.

The initial state of Krause's model (Krause, 1967) was characterized by a prevailing dipole-like magnetic field which cut deeply into the bottom of the convection zone contrary to the shallow submergence assumed by Babcock (1961). Also, as stated above, it was assumed that the core rotates faster than the outer surface in his model. The assumed differential rotation induced a toroidal magnetic field similar to stage 2 of the Babcock model, however, the hydromagnetic turbulence coupled with the density stratification led to the production of quadrupole type poloidal magnetic field with the polarity opposite to the initial dipole-like field. The subsequent emergence of the quadrupole type magnetic field was identified as the solar activity, and the final polarity reversal of the initial dipole-like magnetic field was achieved by a gradual migration of the induced quadrupole type magnetic field towards the interior.

Apart from the difficulties of justification of the basic equations, the 'magneto-kinematic' model by Leighton (1969) has been most successful, as this model has reproduced the observed solar cycle over three 22-yr cycles including fluctuations. Therefore, it may be worthwhile to summarize some of the assumptions and conclusions relevant to the present discussions. Those are: (1) a faster interior rotation is preferred, (2) the differential rotation confined close to the equator is also preferred, (3) the formation of bipolar sunspots can be considered as a random (eruption) process, (4) the emerged bipolar sunspots have the latitudinal difference given by Hale's law, and (5) the redistribution and annihilation of erupted magnetic flux, including the reversal of polarity of the polar magnetic field, are the result of a two-dimensional random walk process due to supergranulation in the presence of a differential rotation.

In view of the quasi-regularity of the solar cycle, we may consider that the observed surface phenomena could be perturbations (finite amplitude or otherwise) over a possible, long-time averaged, basic state of approximately axisymmetric equilibrium. Then, it may be worthwhile to examine the physical characteristics of the state of axisymmetric hydromagnetic equilibria. It has been shown (Woltjer, 1959a, b; Mestel, 1961; Nakagawa, 1970b) that for a perfectly conducting medium, the meridional velocity field must be similar to the poloidal magnetic field. Hence, for a convective circulation confined within a spherical shell, the poloidal magnetic field must also be confined within the same shell. Further, it has been shown that if the boundary of the shell is formed by a line of force of the poloidal magnetic field, the velocity of rotation must be a constant. Table I presents a summary of the principal characteristics of axisymmetric equilibria deduced from the integrals of hydromagnetic equations. It can be seen from Table I that with a meridional circulation the possible surface rotation is at most a constant. In this connection, it is interesting to note that Wilcox (1971) found that the large scale magnetic pattern of the Sun rotates with practically a constant angular velocity of rotation, while the movement of sunspots and the Doppler shift observations showed the faster equatorial differential rotation.

TABLE I

Summary of principal physical characteristics of axisymmetric hydromagnetic equilibria

Case *A*	I	Constant surface rotation
	II	No surface rotation
	III	No rotation (anywhere)
Case *B*	I	Constant surface rotation
	II	Surface differential rotation possible
	III	Surface differential rotation possible with magnetic field extending outside

Case *A* meridional circulation, Case *B* no meridional circulation, Subcase *I* poloidal and toroidal magnetic field, Subcase *II* purely toroidal magnetic field, Subcase *III* purely poloidal magnetic field.

In view of this conclusion, it appears logical to consider that the observed differential rotation of the Sun could be a locally confined phenomenon (Leighton, 1969). Also, it appears reasonable to assume that the general surface magnetic fields are the results of leakages or eruptions of interior magnetic fields as postulated by Babcock (1961) and Leighton (1969) rather than a systematic large scale field. Then we may consider that the first task of the hydromagnetic model study should be directed to the determination of an interior self-consistent systematic large scale magnetic field and convective flow.

3. Basic Equations and Boundary Conditions

The basic equations of the problem could be divided into two groups, i.e., the equations governing the hydromagnetics and the equations governing the thermodynamics of the problem. In the present study, we consider only the hydromagnetics of the problem and in particular, an incompressible medium so that the successive future improvements of the model such as the introduction of compressibility of medium and thermodynamics of the problem would provide us with the understanding of the importance of these effects.

For an incompressible medium, the appropriate equations of hydromagnetics can be written in the following forms, in a system of coordinates rotating with a constant velocity Ω around the polar axis:

$$\frac{\partial \mathbf{v}}{\partial t} = (\nabla \times \mathbf{B}) \times \mathbf{B} - (\nabla \times \mathbf{v}) \times \mathbf{v} - \nabla(\tfrac{1}{2}|\mathbf{v}|^2) -$$
$$- 2\Omega \times \mathbf{v} - \nabla\Pi - \nu\nabla \times (\nabla \times \mathbf{v}), \quad (1)$$

$$\frac{\partial \mathbf{B}}{\partial t} = \nabla \times (\mathbf{v} \times \mathbf{B}) - \eta\nabla \times (\nabla \times \mathbf{B}), \quad (2)$$

$$\nabla \cdot \mathbf{v} = 0, \quad (3)$$

$$\nabla \cdot \mathbf{B} = 0, \quad (4)$$

where $\mathbf{v}$ is the velocity, $\mathbf{B}$ the magnetic field measured in terms of velocity unit $[\mathbf{B} = \mathbf{B}^*(4\pi\varrho)^{-1/2}$; $\mathbf{B}^*$ the strength of magnetic field induction, ϱ the density], Π a scalar defined by $\Pi = p\varrho^{-1} - \tfrac{1}{2}\tilde{\omega}^2|\Omega|^2 + \Phi$ (p the pressure, $\tilde{\omega}$ the radial distance from the rotating axis, Φ the gravitational potential), ν the kinematic viscosity, and η the magnetic diffusivity. In deriving Equations (1) and (2), ν and η are assumed constants.

The basic equations of the present numerical study consist of a set of seven equations: three components of Equation (1), three components of Equation (2) and an equation for Π which is obtained by taking the divergence of Equation (1), i.e.,

$$\nabla^2\Pi = \nabla[(\nabla \times \mathbf{B}) \times \mathbf{B}] - \nabla[(\nabla \times \mathbf{v}) \times \mathbf{v}] - \nabla^2(\tfrac{1}{2}|\mathbf{v}|^2) - 2\nabla(\Omega \times \mathbf{v}), \quad (5)$$

where

$$\nabla^2 = \frac{1}{r^2}\frac{\partial}{\partial r}\left(r^2\frac{\partial}{\partial r}\right) + \frac{1}{r^2 \sin\theta}\frac{\partial}{\partial\theta}\left(\sin\theta\frac{\partial}{\partial\theta}\right) + \frac{1}{r^2 \sin^2\theta}\frac{\partial^2}{\partial\phi^2}, \quad (6)$$

and r, θ, ϕ, denote the radius, co-latitude and longitude, respectively. These equations are currently solved numerically subject to the boundary conditions described below.

In the model, the convection zone of the Sun is represented by a spherical shell between radius $r = 1$ and $r = \beta$ ($\beta < 1$). Then, in order to simulate the effect of convection, which could be interpreted in terms of the differential rotation (Kato and Nakagawa, 1969; Busse, 1970), we shall assume that: (1) The observed differential rotation is maintained at the outer surface ($r = 1$); (2) at the bottom (the core) surface ($r = \beta$), a constant rotation is maintained; (3) the fluid motions are confined within the spherical shell and (4) the magnetic fields are also confined within the spherical shell.

The last condition is assumed in consideration of the physical characteristics of the averaged state of axisymmetric equilibria. In other words, the present study is focused on the investigation of the possible recycling of the subsurface flow and magnetic field consistent with the basic hydromagnetic equations subject to the boundary conditions representing the averaged state.

At the beginning of our computation, we select an arbitrary set of approximate steady state non-axisymmetric solutions in order to facilitate the study. The choice of such solutions is designed to introduce certain longitudinal perturbations which otherwise cannot be generated from the equations. The possible choice of a set of non-axisymmetric solutions follows from the fact that a divergence free vector can be represented by two scalar functions, say, $P(r, \theta, \phi)$ and $T(r, \theta, \phi)$. For example, we can express the velocity field by (e.g. Chandrasekhar, 1961),

$$\mathbf{v} = \nabla \times \nabla \times (P\mathbf{1}_r) + \nabla \times (T\mathbf{1}_r) =$$

$$= \left[-\frac{\Lambda^2}{r^2} P \right] \mathbf{1}_r + \left[\frac{1}{r} \frac{\partial^2 P}{\partial r \partial \theta} + \frac{1}{r \sin \theta} \frac{\partial T}{\partial \phi} \right] \mathbf{1}_\theta +$$

$$+ \left[\frac{1}{r \sin \theta} \frac{\partial^2 P}{\partial r \partial \phi} - \frac{1}{r} \frac{\partial T}{\partial \theta} \right] \mathbf{1}_\phi, \tag{7}$$

where

$$\Lambda^2 = \frac{1}{\sin \theta} \frac{\partial}{\partial \theta} \left(\sin \theta \frac{\partial}{\partial \theta} \right) + \frac{1}{\sin^2 \theta} \frac{\partial^2}{\partial \phi^2} \tag{8}$$

and $\mathbf{1}_r$, $\mathbf{1}_\theta$, and $\mathbf{1}_\phi$ denote unit vectors in the three principal directions.

In the present problem, we may write the scalar functions in terms of spherical surface harmonics,

$$Y_l^m(\theta, \phi) = P_l^{|m|}(\cos \theta) e^{im\phi}, \tag{9}$$

where $P_l^m(\cos \theta)$ is the Legendre function (e.g. Janke and Emde, 1945). Introducing the notations U, V, and W for the radial, latitudinal, and longitudinal velocity components, and choosing

$$P = {}^{(I)}\Phi(r) \, {}^{(I)}Y_l^m(\theta, \phi), \quad T = {}^{(II)}\Phi(r) \, {}^{(II)}Y_l^m(\theta, \phi), \tag{10}$$

we obtain

$$U = \frac{1}{r^2} {}^{(I)}\Phi(r) \, l(l+1) \, {}^{(I)}Y_l^m(\theta, \phi), \tag{11}$$

$$V = \frac{1}{r} \frac{d}{dr} {}^{(I)}\Phi(r) \frac{\partial}{\partial \theta} {}^{(I)}Y_l^m(\theta, \phi) + \frac{1}{r} {}^{(II)}\Phi(r) \frac{1}{\sin \theta} \frac{\partial}{\partial \phi} {}^{(II)}Y_l^m(\theta, \phi), \tag{12}$$

$$W = \frac{1}{r} \frac{d}{dr} {}^{(I)}\Phi(r) \frac{1}{\sin \theta} \frac{\partial}{\partial \phi} {}^{(I)}Y_l^m(\theta, \phi) - \frac{1}{r} {}^{(II)}\Phi(r) \frac{\partial}{\partial \theta} {}^{(II)}Y_l^m(\theta, \phi), \tag{13}$$

where we have used the relation

$$\Lambda^2 Y_l^m(\theta, \phi) = -l(l+1) Y_l^m(\theta, \phi). \tag{14}$$

It should be noted that the components of magnetic field can be written in similar forms regardless of the assumption concerning compressibility of the medium, and that for a compressible medium the components of velocity at a steady state can be given by replacing U, V, and W with ϱU, ϱV, and ϱW, respectively.

If we assume further that the flow has symmetry with respect to the polar axis ($\theta = 0$) and the equatorial plane ($\theta = \pi/2$), we find the following conditions for the choice of l and $|m|$, excluding $|m| \neq 0$,

$$l \geqslant |m|, \quad |m| \geqslant 2$$

$$l - |m| = \begin{cases} 0 \quad \text{or even integers for } (I) \\ \text{odd integers for } (II). \end{cases} \tag{15}$$

An example of such a solution in the presence of a strong axisymmetric toroidal magnetic field is shown in Figure 1. The unipolar regions in Figure 1 are identified with the variations of the magnetic field near the surface, so that the flux coil which appeared near the surface is identified as one polarity, while the submerged region is the opposite polarity. It may be noted that in the absence of a surface differential rotation we may choose $\mathbf{v}\|\mathbf{B}$, so that the magnetic flux could be considered to represent the flow fields. Also it should be noted that without a strong axisymmetric toroidal magnetic field the solutions with the choise of l and m given in the condition (15) represent celluar structures rather than the continuous configurations shown in Figure 1.

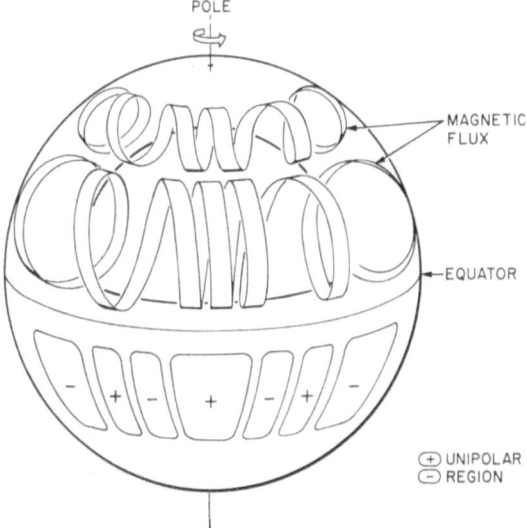

Fig. 1.   An example of schematic large scale non-axisymmetric magnetic field.

In the present problem, however, with the assumed surface differential rotation, we have $W(\theta) \neq 0$ at $r=1$ while $\mathbf{B}=0$ at $r=1$. Hence, $\mathbf{v} \nparallel \mathbf{B}$ near the surface (also in the interior), this boundary condition leads to the generation of an induced magnetic field by the flow near the surface. The subsequent interactions between the flow and magnetic fields then lead to the non-linear variations (oscillations) of the whole flow and magnetic field. It is considered that these non-linear oscillations represent the solar cycle in our model.

4. Results and Discussions

The results which remain somewhat preliminary at present indicate that the velocity field distorted strongly near the surface (to satisfy the surface boundary condition of the observed differential rotation) induces a strong distortion of magnetic fields and subsequently oscillations of both velocity and magnetic fields. The strong distortion of magnetic fields near the surface could be considered as the zone of activity. It is

found that the oscillations of the flow and magnetic fields propagate in both the longitudinal and latitudinal directions, somewhat similar to the results of linear analyses
of the Rossby-type waves by Kato (1969), and Kato and Nakagawa (1970). The
propagation of such waves could be identified with the change of pitch of the magnetic
flux shown in Figure 1, thus the migration of large unipolar regions. From the surface
differential rotation given by Newton and Nunn (1951) and the differential rotation
monotonically decreasing with depth we find that the zone of activity gradually
migrates towards the equator, while each unipolar region grows and decays in the
order of a year.

However, more detailed studies are needed since a number of parameters involved
require further examination. For example, with faster interior rotation, we may still
satisfy Spörer's law of zone of activity by increasing the number of meridional circulation cells so that near the surface the circulation is directed toward the equator.
Further, it should be noted that the strength of the interior magnetic field could be
taken somewhat arbitrary irrespective of the emerged strength of the sunspot magnetic
field. The requirement of the virial theorem (e.g. Nakagawa and Trehan, 1970) could
be satisfied by an interior magnetic field of strength well over 10^6 G.

In the analysis by Kato and Nakagawa (1970), it was shown that the most unstable
waves propagate with a sub-Alfvénic speed, and the present preliminary results support this prediction. In this connection, it may be worthwhile to note that in the state
of axisymmetric equilibria, it has been shown (Mestel, 1961; Nakagawa, 1970b) that
the poloidal convective velocity must remain either super-Alfvénic or sub-Alfvénic
throughout the circulation. This condition may provide the limit to the strength of
the interior magnetic field, if a realistic velocity of the interior convective circulation
is determined.

It is advantageous to start our discussions with the results of previous axisymmetric
incompressible study (Nakagawa and Swarztrauber, 1969), as the basic physical
mechanisms of the problem are represented in the model. Figure 2 shows the developments of the toroidal magnetic field in the axisymmetric model by differential rotation
at 2, 4, 6, 8, 10, and 12 yr after the initial state, when the surface differential rotation
is given by the formula of Newton and Nunn (1951). The migration of a strong toroidal
magnetic field developed near the surface towards the equator can be seen clearly in
Figure 2. Another representation of this migration of the toroidal magnetic field in the
form of the 'Butterfly' diagram is shown in Figure 3. It should be noted that the results
of the study also showed that the poleward migration of active regions (identified with
the maxima of the induced toroidal magnetic field) results with the formula of the
surface differential rotation given by d'Azambuja (1948).

We may now discuss the similarities and differences of the present non-axisymmetric
model and the previous axisymmetric model. In both models, the viscous dissipation
due to granulation and supergranulation are considered, thus a net angular velocity
transfer towards the equator is required, in order to maintain the faster equatorial
rotation. This angular momentum transfer was achieved by a meridional circulation
directed towards the equator near the surface. Also in both models, the zone of activity

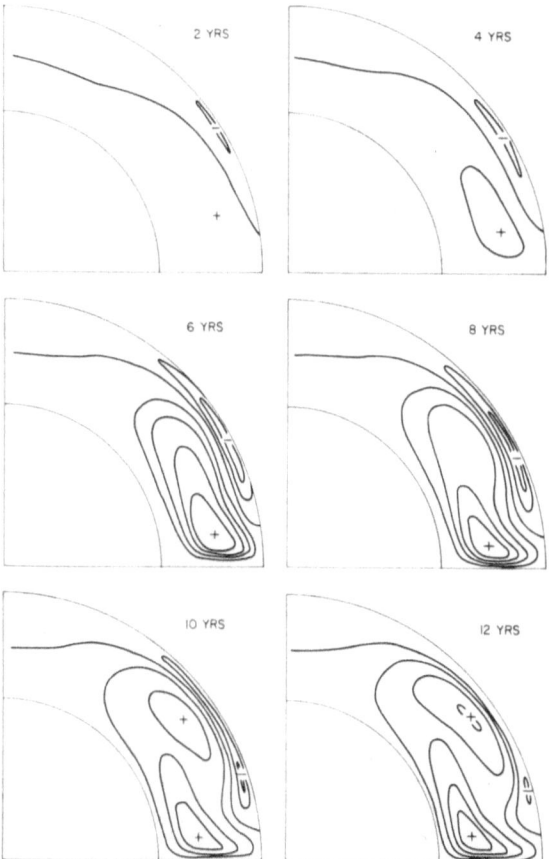

Fig. 2.   The equatorial migration of the zone of activity (identified with – sign near the surface) at 2, 4, 6, 8, 10, and 12 yr after the initial state for the axisymmetric incompressible model (Nakagawa and Swarztrauber, 1969).

is generated by a strong distortion of toroidal magnetic field near the surface. The major difference between these two models is that the present model contains the possible mechanism for the formation of sunspots.

It has been shown (Nakagawa and Malville, 1969; Nakagawa, 1970a) that the instability due to the magnetic buoyancy could be identified with the instability arising from the local magnetic shear in a stratified fluid in the proper context of hydromagnetics. In the present model such a magnetic shear can be generated locally due to dispersive nature of oscillations of the interior flow and magnetic field. Recently thermal instability due to ionization has been reported by Defouw (1970), thus this mechanism must be examined in more detail in relation to the hydrogen ionization and hopefully account for Hale's law governing the polarity of emerging sunspots.

In the present study, our attention is directed only to the kinematics of the interior primary magnetic and flow fields, and the observed surface magnetic field is con-

sidered as the secondary phenomenon. However, we may illustrate the general concept
of the ultimate model which incorporates the possible convective phenomena of the
Sun. Figure 4 shows such a model. On the surface dominant convective motions are
the granulations and supergranulations. The observed rather homogeneous characters
of these phenomena all over the solar surface suggest little connection with large scale
flow and magnetic fields. Hence we may consider the pertinent physical mechanisms
responsible for these phenomena could be confined within the shallow layer of the
depth of strong hydrogen ionization zones. Below the strong hydrogen ionization zone

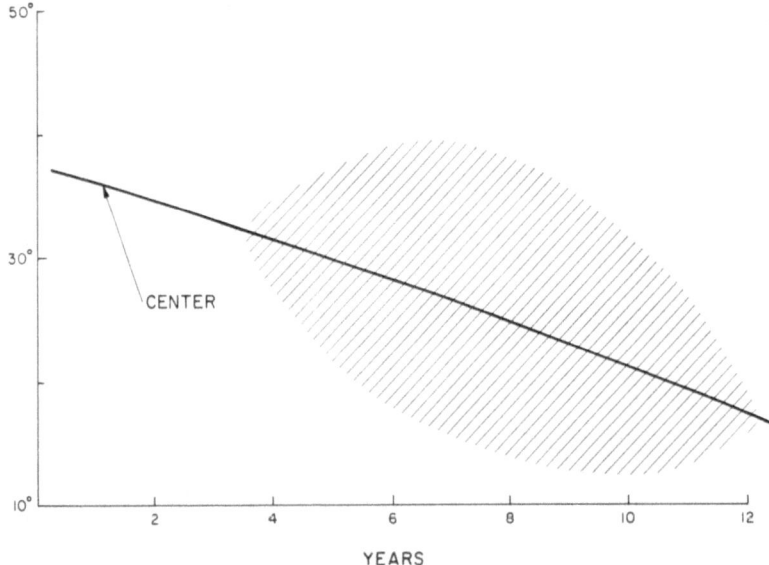

Fig. 3.   The 'Butterfly' diagram (Nakagawa and Swarztrauber, 1969). The shaded area represents the
strength of the toroidal magnetic field near the surface above 175 G.

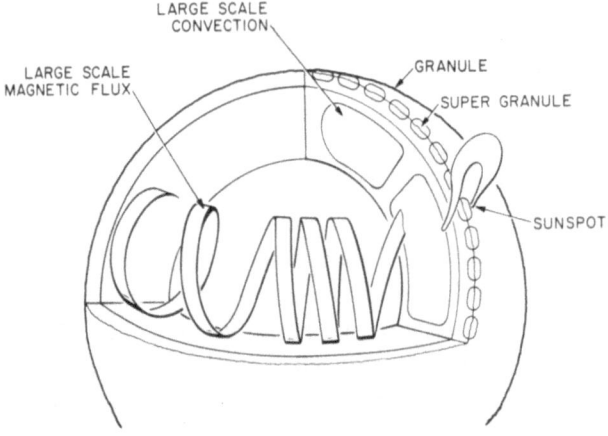

Fig. 4.   A schematic model of convective motions and large scale magnetic field of the Sun.

we may then postulate the existence of large scale convection motion together with the large scale magnetic field for which the present study is directed.

The results of the present study indicate the slow variations of the flow and magnetic field which migrate both in the latitudinal and longitudinal directions. These variations could be identified with the variations of large scale magnetic fields reported by Wilcox (1971). Further, the local strong shear in the flow and magnetic fields could be considered the eruptive zone of sunspots, then the present model does contain this mechanism, although the exact manner of formation of the bipolar sunspots still remains for future study.

5. Concluding Remarks

In this paper, models of the solar cycle pertinent to numerical studies of the problem are briefly reviewed. Then the physical characteristics of possible averaged state of axisymmetric equilibria are discussed. Some results of preliminary studies and possible non axisymmetric solutions of the hydromagnetic equations are also presented. It is suggested that the observed surface phenomena are somewhat secondary in nature with respect to the interior solutions. This argument is partially supported by the 'magneto-kinematic' model (Leighton, 1969) of the solar cycle which indicates the random nature of the surface phenomena and partially by the Krause model (1967) which emphasizes the induced effect of hydromagnetic turbulence.

It may be worthwhile to note some of the detailed physical characteristics of the solar activity summarized by Babcock (1961) since the large scale model must eventually incorporate these phenomena. The east-west asymmetry and preferred longitude of solar activities could be interpreted in the present non-axisymmetric model in terms of a local concentration of the pitch of the interior magnetic flux. The forward tilt of sunspot axes towards the direction of rotation (Minnaert, 1946) could be interpreted by a faster surface rotation. The present non-axisymmetric model, however, suggests a possible modulation of the surface rotation by the interior flow. In other words, the surface differential rotation could be a function of short-time variations closely correlated with the movements of the large scale surface unipolar magnetic regions.

In summary, it seems apparent that the preliminary task of the hydromagnetic model should be directed to the study of the convective circulation with the magnetic field in the solar convection zone. The one-dimensional model of convection zone such as given by Baker and Temesvary (1966) has to be improved to accommodate some meridional circulations. Currently, a formulation of an axisymmetric model of the solar convection zone with a meridional circulation and poloidal as well as toroidal magnetic field is in progress, and it is hoped that the present incompressible model with simulated convective circulation can be integrated into a more realistic compressible hydromagnetic model in this exceedingly complex problem of great importance.

References

d'Azambuja, M. and d'Azambuja, L.: 1948, *Ann. Obs. Meudon* **6**, fasc.7.
Babcock, H. D.: 1959, *Astrophys. J.* **130**, 364.

Babcock, H. W.: 1961, *Astrophys. J.* **133**, 572.

Baker, N. H. and Temesvary, S.: 1966, *Tables of Convective Stellar Envelope Models*, Institute for Space Studies, NASA, New York.

Busse, F.: 1970, *Astrophys. J.* **159**, 629.

Carrington, R. C.: 1858, *Monthly Notices Roy. Astron. Soc.* **19**, 1.

Chandrasekhar, S.: 1961, *Hydrodynamic and Hydromagnetic Stability*, Clarendon Press, Oxford, p. 622.

Defouw, R. J.: 1970, *Astrophys. J.* **161**, 55.

Ferraro, V. C. A.: 1937, *Monthly Notices Roy. Astron. Soc.* **97**, 458.

Hale, G. E.: 1908, *Astrophys. J.* **28**, 100, 315.

Hale, G. E.: 1913, *Astrophys. J.* **38**, 27.

Janke, E. and Emde, F.: 1945, *Tables of Functions*, Dover Publ., New York.

Kato, S.: 1969, *Astrophys. J.* **157**, 827.

Kato, S. and Nakagawa, Y.: 1969, *Solar Phys.* **10**, 476.

Kato, S. and Nakagawa, Y.: 1970, *Solar Phys.* **14** 138.

Krause, F.: 1967, Habilitationschrift, University of Jena.

Leighton, R. L.: 1964, *Astrophys. J.* **140**, 1547.

Leighton, R. L.: 1969, *Astrophys. J.* **156**, 1.

Mestel, L.: 1961, *Monthly Notices Roy. Astron. Soc.* **122**, 473.

Minnaert, M.: 1946, *Monthly Notices Roy. Astron. Soc.* **106**, 98.

Nakagawa, Y.: 1970a, *Solar Phys.* **12**, 419.

Nakagawa, Y.: 1970b, *Astrophys. Space Sci.* **8**, 327.

Nakagawa, Y. and Malville, J. M.: 1969, *Solar Phys.* **9**, 102.

Nakagawa, Y. and Swarztrauber, P.: 1969, *Astrophys. J.* **155**, 1111.

Nakagawa, Y. and Trehan, S. K.: 1970, *Astrophys. J.* **160**, 725.

Newton, H. W. and Nunn, M. L.: 1951, *Monthly Notices Roy. Astron. Soc.* **111**, 413.

Spörer, G.: 1894, *Publ. Potsdam Obs.* **10**, Part. I, 144.

Wilcox, J. M.: 1971, this volume, p. 744.

Woltjer, L.: 1959a, *Astrophys. J.* **130**, 405.

Woltjer, L.: 1959b, *Astrophys. J.* **131**, 227.

Discussion

Meyer: The equations and the boundary conditions of your model are axisymmetric. The solution you showed contains an azimuthal variation. How was the period for this variation determined, was it the outcome of a time evolution or was it introduced by the initial condition?

Nakagawa: The equations are non-axisymmetric and we chose a set of non-axisymmetric initial solutions. The period of variation varies with wave number and at present we are examining different initial conditions to be more general in the future.

Ward: Your model's conversion of kinetic energy, combined with Howard's observations of rapid changes in the velocity fields should constitute at least a partial answer to Altschuler's question earlier this morning.

DYNAMICS OF LARGE-SCALE MAGNETIC FIELDS

G. V. KUKLIN

Siberian Institute of Terrestrial Magnetism, Ionosphere and Radio Propagation, Irkutsk, U.S.S.R.

Abstract. A discussion is given of the dynamics of large-scale solar magnetic fields as a joint action of a transport by a regular velocity field, by a generalized diffusion, by coming to the surface and sinking and by sources of origin and destruction of magnetic fields. Also these processes are discussed using the observed picture of magnetic field dynamics.

The evolution and dynamics of the large-scale weak solar magnetic fields studied first by Bumba and Howard (1965) are not clear, although investigations of such kind by many authors are in progress now. Only the structure of these large-scale magnetic fields and their connection with other solar phenomena were studied relatively in detail. Because the large-scale magnetic fields are closely connected with the interplanetary magnetic field and with the theory of solar activity, some formal considerations may be of interest.

Let us consider the simplest form of the fundamental equation of hydromagnetics describing the magnetic field evolution in the case of the Sun.

$$\frac{\partial \mathbf{H}}{\partial t} = \mathrm{rot}\,[\mathbf{V} \times \mathbf{H}] - \mathrm{rot}\,(v_m\,\mathrm{rot}\,\mathbf{H}). \tag{1}$$

This equation for the z-component of the magnetic field may be written in such a way

$$\frac{\partial \mathbf{H}_z}{\partial t} = -\,(\mathbf{V}\,\nabla)\,H_z - H_z\,\mathrm{div}\,\mathbf{V} + v_m\,\Delta H_z + \nabla v_m\,\nabla H_z +$$

$$+ v_m\,\frac{\partial^2 H_z}{\partial z^2} - V_z\,\frac{\partial H_z}{\partial z} + \mathbf{h}\,\nabla V_z - \nabla v_m\,\frac{\partial \mathbf{h}}{\partial z} \tag{2}$$

where V and h are the transverse velocity and magnetic fields and div, ∇ and Δ are the differential operators in the XOY plane. If we identify H_z with the longitudinal component of the magnetic field $H_{\parallel}$ observed at a solar surface element which is approximately normal to the sight line, then the above equation of the magnetic field dynamics may be interpreted in the following way.

The term $v_m\,\Delta H_z$ describes the ohmic dissipation of the longitudinal magnetic field H_z in the XOY plane. The term $\nabla v_m\,\nabla H_z$ takes into consideration the changes of v_m along the XOY plane. The term $-(\mathbf{V}\,\nabla)H_z$ is connected with the transport of the frozen magnetic field by the velocity field. The term $-H_z \times \mathrm{div}\,\mathbf{V}$ describes the sinking or coming to the surface caused by the sources of the transverse (horizontal) velocity field. The terms $v_m(\partial^2 H_z/\partial z^2) - \nabla v_m(\partial \mathbf{h}/\partial z) + h\,\nabla v_z - v_z(\partial H_z/\partial z)$ may be combined into the term describing the sources of the magnetic field H_z. The term $v_m(\partial^2 H_z/\partial z^2)$ describes the diffusion of the longitudinal magnetic field H_z from under the surface. The

Howard (ed.), Solar Magnetic Fields, 737–743. All Rights Reserved.

term $-v_z(\partial H_z/\partial z)$ is connected with the transport of the frozen magnetic field H_z by the vertical velocity v_z from under the surface. The term $-v_m(\partial \mathbf{h}/\partial z)$ takes into consideration the changes of v_m which cause the transformation of the transverse magnetic field h into the longitudinal one during the diffusion to the surface. A similar meaning may be ascribed to the term $\mathbf{h} \nabla v_z$ connected with the deformation of the force lines during the transport to the surface. Practically speaking, the values included in these four expressions are unobservable and therefore we can consider only the general source of the magnetic field φ. The next detail is that we must replace v_m with the formal diffusion coefficient D which includes some processes existing simultaneously, for example the magnetic field diffusion caused by the ohmic dissipation, the diffusion of a turbulent type proposed by Leighton (1964) etc.

Now we can use the following generalized equation of the dynamical model instead of the initial one (2).

$$\frac{\partial A}{\partial t} = \varphi - \mathrm{div}\,(A\mathbf{V} - D\nabla A), \tag{3}$$

where A is an arbitrary physical parameter (in our case $A \equiv H_z$). The problem is to reconstruct the functions φ, D and $\mathbf{V}$ depending on the time and the coordinates on the basis of an observed time-spatial distribution of the value A, in other words to solve a reverse problem of the differential equations. Strictly speaking this problem is at least two times incorrect in Tikhonov's sense, because we need to compute derivatives using only numerical data surely burdened with errors, and to reconstruct the behavior of three apparently independent functions using only one function A. It is necessary to introduce some simplifying assumptions.

There are three ways to solve our problem.

(1) The 'global' method permits us to apply the expansions of the functions A, φ, D, v_x and v_y using some orthogonal basis and to replace Equation (3) with the system of linear equations for the expansion coefficients of φ, D, v_x and v_y. Difficulties connected with boundaries vanish if we use a spherical surface (the whole Sun), but other obstacles appear. First it is not clear to what degree the Equation (2) used is valid in the polar regions of Sun where the observed $H_{\parallel}$ does not correspond to the vertical component. Generally speaking this difficulty remains valid in other methods. Secondly if we want to get information about φ, D, v_x and v_y with good spatial resolution (for example $\sim 20°$) then using the spherical function expansion we need about 18 harmonics, which means more than 300 coefficients for every unknown function, or a system of more than 1000 equations which surely will be very ill-conditioned.

(2) The 'continuous local' method requires us to assume the functions A, φ, D, $\mathbf{V}$ to be smooth functions of coordinates in the neighborhood of the point considered and the functions φ, D and $\mathbf{V}$ to be weakly dependent on time (practically constant) close to the moment considered. Then differentiating (3) with respect to the time we can obtain the necessary number of linear equations for the functions φ, D, $\mathbf{V}$ and their derivatives with respect to coordinates (no more than of the 2nd order). In the case of 2 spatial coordinates and of anisotropic diffusion it leads to a system of 10 linear

equations. Here we run into the incorrectness of the numerical determination of derivatives of A and the solution of the linear system which as a rule is ill-conditioned. Unfortunately up to now we have had no success in working out an algorithm which provides a stable solution. It is necessary to mention that the Courant-Friedrichs-Levy condition requires the time interval to be 1 solar rotation (the Mount Wilson atlas of solar magnetic fields has such a minimal step in time) so that the spatial interval for the derivative computation is of the order of 30° or 40°. This does not permit us to use the simplest formulae of differentiation (for example 3-point formula).

(3) The 'discrete local' method differs from the previous one in the absence of the derivative computation. Instead of this we obtain the expansions of A as a function of time and of coordinates in the neighborhood of the point considered with the help of Legendre polynomials. Here we must assume the functions φ, D and V to be constant during some solar rotations necessary to write down a full system of equations – that means the obtained values of φ, D and V are averaged during these rotations. The Courant-Friedrichs-Levy condition is very difficult to fulfill in this case and it leads to an instability in the solution. Moreover the validity of using the Legendre polynomial derivatives is not confirmed if the order of polynomials is high.

For a first attempt we used the third method with the following parameters. The averaging interval was equal to 5 solar rotations and the coordinate step corresponded to 5° (nonconformity to the Courant-Friedrichs-Levy condition). The values of $H_{\parallel}$ were taken from the maps of the solar magnetic fields. The computations were made for two periods: rotations 1431–1435 (September-December 1960) and 1502–1506 (December 1965-April 1966).

The instability of the solution is expressed in deviations of the same order as the terms of Equation 3 and in disordered or unreal values of velocity. Therefore the results of the computation may be considered only as accurate up to the sign. They are presented at Figures 1–6 (maps of φ, D and div$V$). The positive values of φ describe the origin of the N-polarity magnetic fields and the destruction of the S-polarity magnetic fields. The negative values of div$V$ correspond to the magnetic field sinking. The value orders are correspondingly of 10^{-2} G/s for φ, of $10^3 + 10^4$ km²/s for D and of 10^{-3} s^{-1} for div$V$. These values seem to be real ones. Unfortunately the large averaging interval is the reason for the mixing of the dynamical picture: active regions develop and begin to decay during some months. Thus on our maps processes of increasing and decreasing of magnetic fields become superimposed and located in the same regions. This is especially noticeable for the second period. In the first period the maps reflect a complex developed system of long-living features.

The most striking fact is an appearance of negative values of the diffusion coefficient D. If the stable solution will confirm this fact, that means the negative diffusion coefficient is not caused by computation errors, then it shows a new unknown mechanism of magnetic field concentration by random walk opposite to Leighton's mechanism. We think that the existence of such a phenomenon is very unlikely because at first sight it contradicts the second law of thermodynamics and most probably it is caused by the incorrectness of the problem.

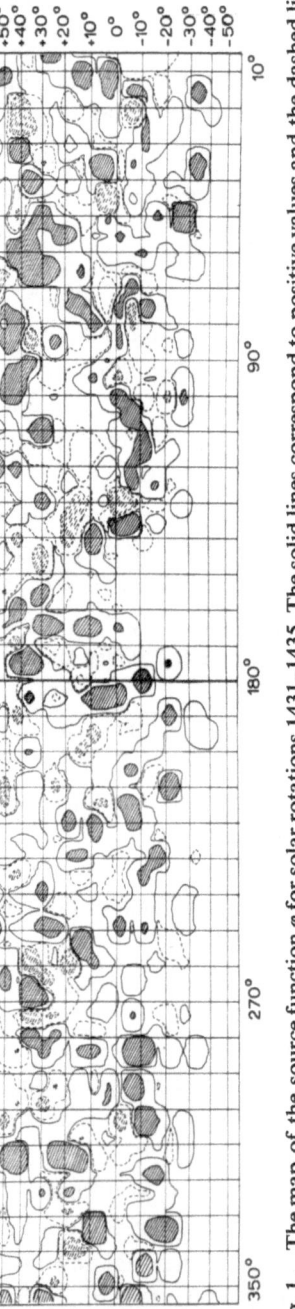

Fig. 1. The map of the source function φ for solar rotations 1431–1435. The solid lines correspond to positive values and the dashed lines to negative ones. The shaded regions correspond to $|\varphi| \geqslant 0.05$ G/s.

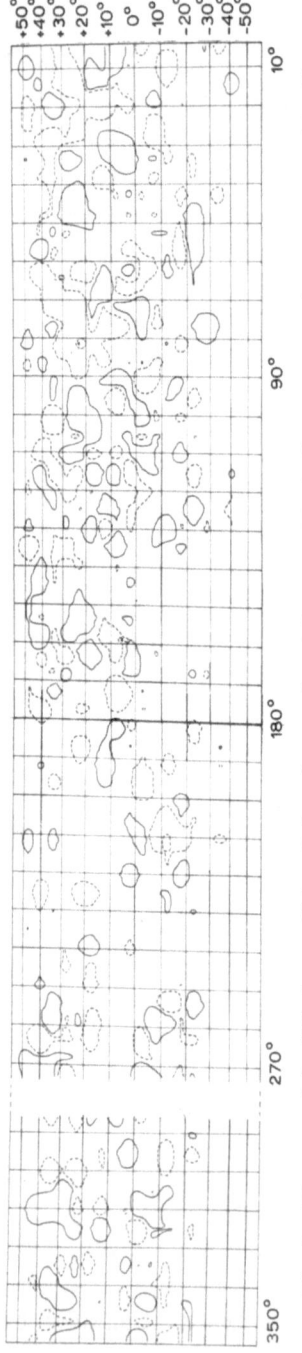

Fig. 2. The map of the diffusion coefficient D for solar rotations 1431–1435. The isolines correspond to $|D| \geqslant 5 \times 10^3$ km²/s.

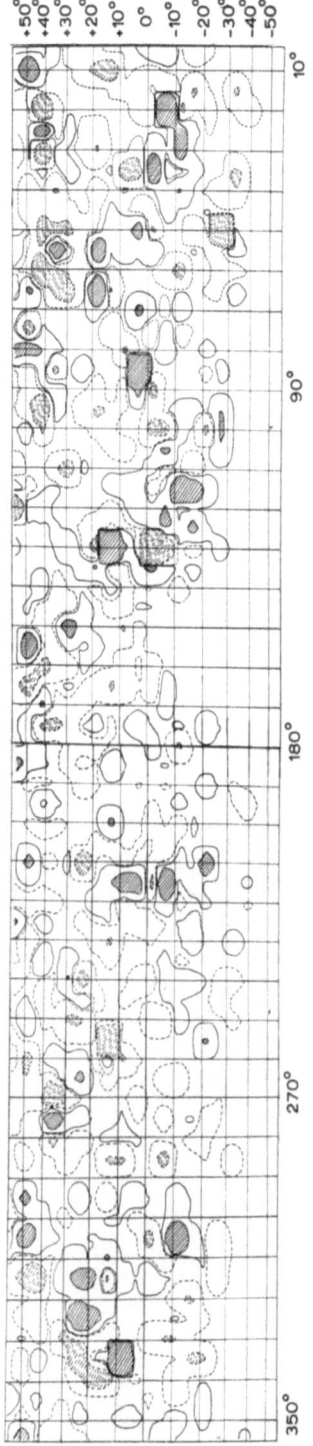

Fig. 3. The map div V for solar rotations 1431–1435. The shaded regions correspond to $|\text{div}\,\mathbf{V}| \geqslant 10^{-3}\ \text{s}^{-1}$.

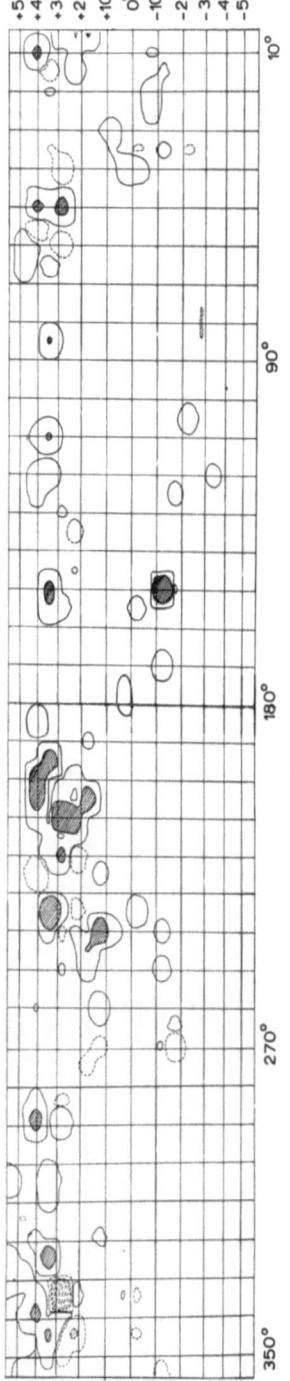

Fig. 4. The map of the source function φ for solar rotations 1502–1506. The shaded regions correspond to $|\varphi| \geqslant 0.05$ G/s.

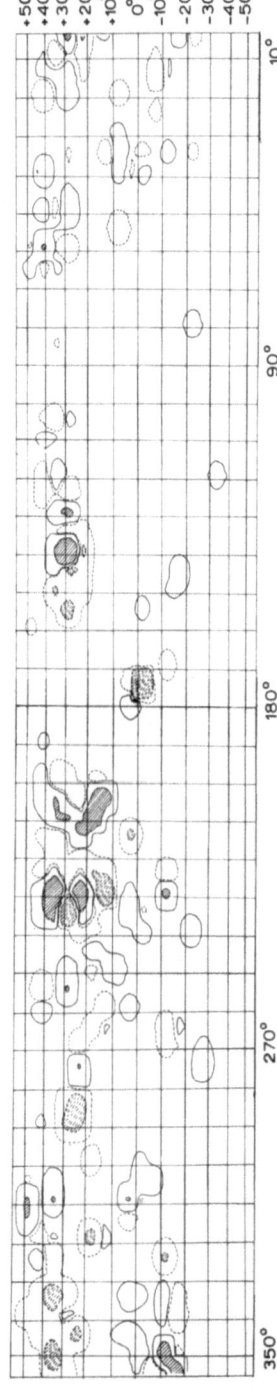

Fig. 5. The map of the diffusion coefficient D for solar rotations 1502–1506. The shaded regions correspond to $|D| \geqslant 5 \times 10^4$ km^2/s.

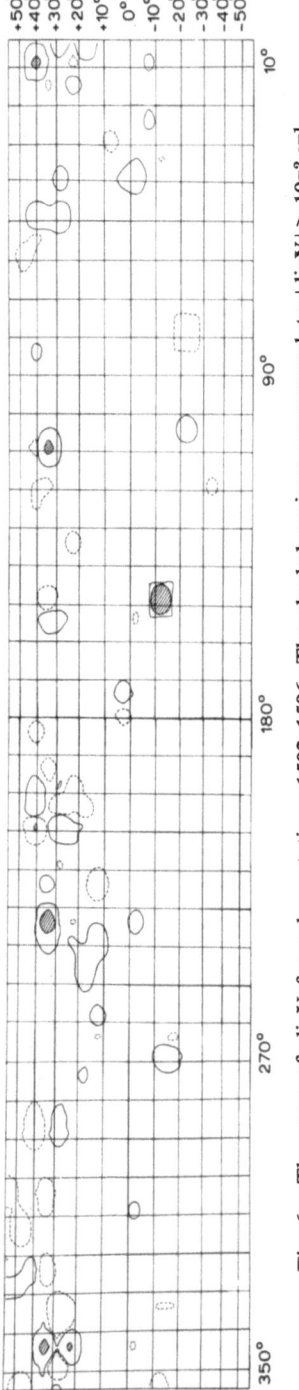

Fig. 6. The map of divV for solar rotations 1502–1506. The shaded regions correspond to $|\mathrm{div}V| \geqslant 10^{-2}$ s^{-1}.

It is very interesting that the regions of ascending or descending magnetic fields (large values of divV) practically coincide with the regions of powerful sources of magnetic fields. The sizes of the regions are about 20°–40°. Therefore the theoretical assumption about the origin of sunspots as the coming to the solar surface of the subphotospheric magnetic tubes is confirmed not for a single magnetic tube but for a whole system of tubes occupying a large region.

We intend to continue the study of the magnetic field dynamics using more perfect algorithms for the solution of incorrect problems. Perhaps our preliminary results will be confirmed and will require more detailed analysis.

Acknowledgements

The author wishes to thank Mr V. Korolev for the help in computations and Drs B. Wertlieb, S. Vainstejn, I. Krinberg, and J. Katz for valuable discussions.

References

Bumba, V. and Howard, R. F.: 1965, *Astrophys. J.* **141**, 1502.
Leighton, R. B.: 1964, *Astrophys. J.* **140**, 1547.

Discussion

Altschuler: What was the role of the transverse magnetic field?

Kuklin: As we had no data on the transversal component of the magnetic field we included all terms depending on the transverse magnetic field (unobservable values) into the source function Φ.

Gilman: I would comment that the second law of thermodynamics does not generalize to turbulent transport processes.

Kuklin: Nevertheless it is difficult to talk about discrepancies with the second law of thermodynamics loudly.

SECTOR STRUCTURE OF THE SOLAR MAGNETIC FIELD

JOHN M. WILCOX

Space Sciences Laboratory, University of California, Berkeley, Calif. 94720, U.S.A.

Abstract. The solar sector structure consists of a boundary in the north-south direction such that on one side of the boundary the large-scale weak photospheric magnetic field is predominantly directed out of the Sun, and on the other side of the boundary this field is directed into the Sun. The region westward of a solar sector boundary tends to be unusually quiet and the region eastward of a solar sector boundary tends to be unusually active. This tendency is discussed in terms of flares, coronal enhancements, plage structure and geomagnetic response.

A solar magnetic sector pattern has been inferred from comparisons of the observed photospheric magnetic field with the observed interplanetary magnetic sector pattern. A schematic of an average boundary during 1965 of the solar sector pattern is shown in Figure 1. The boundary is approximately in the north-south direction. Over a large region on both sides of the equator to the west of the boundary the large-scale weak photospheric magnetic field is predominantly into the Sun, and similarly to the east of the boundary this field is predominantly out of the Sun. The solar sector boundary rotates in an approximately rigidly rotating system since the shearing effects to be expected from differential rotation have little or no effect on the boundary. Thus the solar sector pattern differs from the classical model of solar magnetism given by Babcock (1961) in two fundamental respects: (1) The sector magnetic structure rotates in a rigid system while the Babcock model depends on differential rotation for field amplification, and (2) the solar sector pattern extends across the equator without

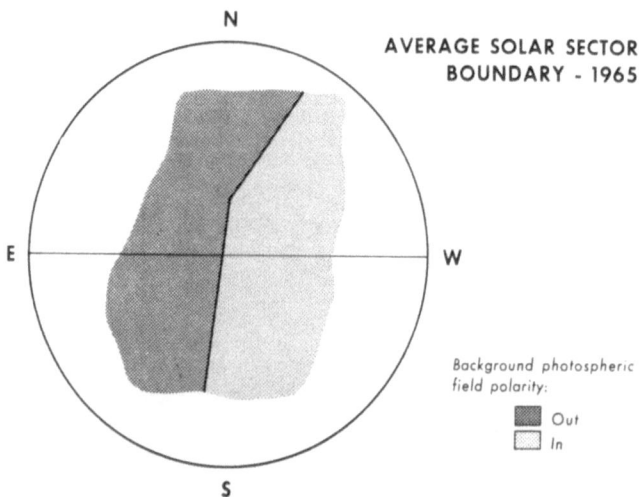

Fig. 1. A schematic of the average position of a solar sector boundary during 1965. On each side of the boundary the weak background photospheric magnetic field is predominantly of a single polarity in equatorial latitudes on both sides of the equator (after Wilcox *et al.*, 1969).

Howard (ed.), Solar Magnetic Fields, 744–753. All Rights Reserved.
Copyright © 1971 by the IAU.

change of polarity while the Babcock model has opposite polarities for bipolar regions on the two sides of the equator. In the Babcock model (and in observations of sunspots) the solar equator is a well defined position such that if sunspots are only a degree or two away from the equator they will almost always obey the polarity law of the Babcock model. By contrast the analysis that leads to the sector structure crosses the equator without any perceptible change.

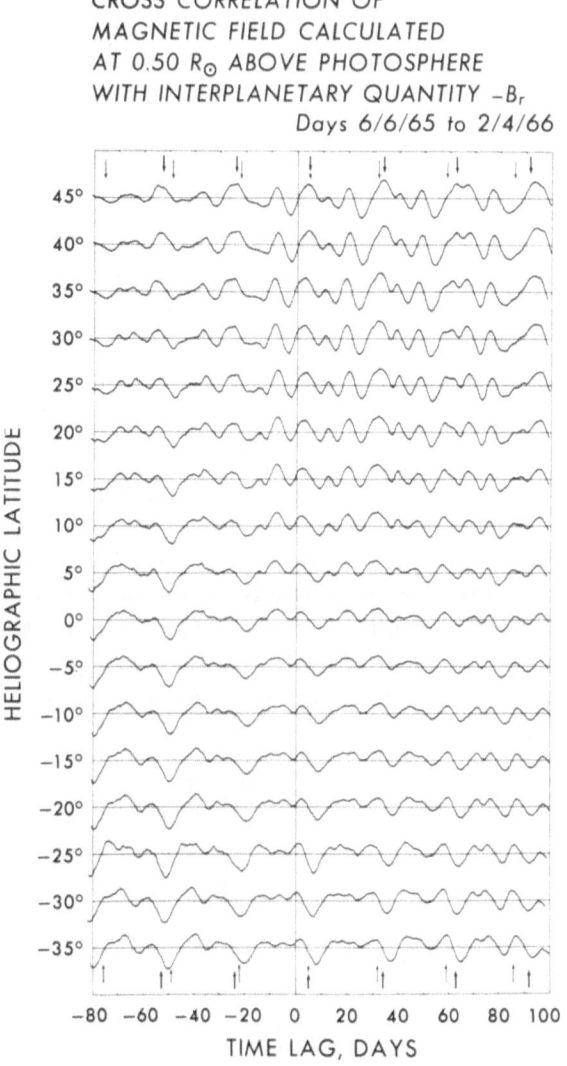

Fig. 2.   Cross correlation of the magnetic field near the Sun with the radial component of the interplanetary magnetic field as a function of time lag. Nine solar rotations of data are utilized with correlations extending from 45°N to 35°S in intervals of 5°. The horizontal line labeled 45° represents zero correlation for this latitude. The line just above represents 1.0 and the line just below represents − 1.0. Correlations for all other latitudes are displaced in the same format (after Schatten *et al.*, 1969).

The solar sector pattern is obtained from cross correlations of the interplanetary sector pattern observed with spacecraft near the ecliptic and the photospheric magnetic field observed with the solar magnetograph at Mount Wilson Observatory. The result of such a cross correlation is shown in Figure 2. Consider first a cross correlation in Figure 2 for a solar latitude near the equator. This correlation has a peak at a lag near $4\frac{1}{2}$ days, which represents the transit time of solar wind plasma from Sun to Earth. This peak may be regarded as the result of a physical transport by the solar wind plasma of field lines from the Sun to the spacecraft near the Earth. If we now examine the other cross correlations in Figure 2 we find that at each solar latitude there is a peak near a lag of 4 or 5 days. All of these peaks are interpreted not in terms of a direct connection from the solar latitude involved to the position of the spacecraft but rather as indicating that all of the solar latitudes have a common pattern in the large-scale field.

The observation of this large-scale solar pattern is a signal-to-noise problem. Two kinds of 'noise' obscure the observation of this pattern: first, the magnitude of the photospheric fields involved is comparable to the minimum fields that can be detected with the solar magnetograph, and second, small-scale features such as active regions and bipolar sunspots appear as 'noise' in the attempts to observe a large-scale region with a single predominant polarity. The solar sector boundaries schematic in Figure 1 can be inferred from the cross correlations of Figure 2 in a manner described by Severny *et al.* (1970).

During the last solar minimum the interplanetary sector pattern consisted of four sectors per rotation that existed in a quasi-stationary manner for approximately one year. Each of these sectors was a coherent entity in the sense that several observed interplanetary quantities changed in a smooth and unitary manner within each sector. In the years near solar maximum attention has been concentrated on the solar sector boundaries. In this paper we present several lines of evidence indicating that the region westward of a solar sector boundary tends to be unusually quiet and that the region eastward of a solar sector boundary tends to be unusually active.

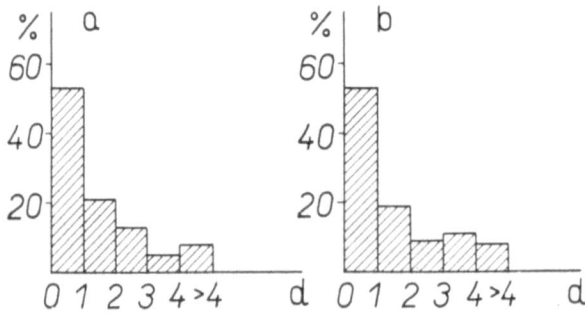

Fig. 3. Histograms of frequency distribution of the time difference between the central meridian passage of spot groups and the position of solar sector boundaries for the groups: (a) with flares of importance 1+ or greater; (b) with a number of flares equal or greater than 10 (after Bumba and Obridko, 1969).

Flares are most likely to occur near sector boundaries, and the larger the flare the more likely it will be near a boundary (Bumba and Obridko, 1969). Figure 3 shows their results for a histogram of frequency distribution of the time differences between the central meridian passage of spot groups and the position of the solar sector boundary for the groups: (a) with flares of importance 1 + or greater; (b) with a number of flares equal or greater than ten. Eleven of the fourteen proton-flare regions examined by Bumba and Obridko occurred at a distance smaller or equal to one day from the boundary of the sector structure. It is interesting to note from the work of these authors that the solar sector structure is consistent with but more quantitative and precise than the concept of active longitudes. When Bumba and Obridko constructed histograms similar to Figure 3 but referring to the center of gravity of an active longitude they found peaks in the distribution that were smaller and broader than the peaks shown in Figure 3. Vladimirsky (private communication) using a more recent and larger data sample has confirmed the results of Bumba and Obridko. Vladimirsky has found that flares tend to be concentrated in the region just eastward of the sector boundary.

Coronal enhancements have been detected just eastward of solar sector boundaries by Couturier and Leblanc (1970) using the Nançay radio interferometer at 169 MHz (1.77 m). Most of the emission detected originates at altitudes between 0.2 and 0.5 $R_\odot$

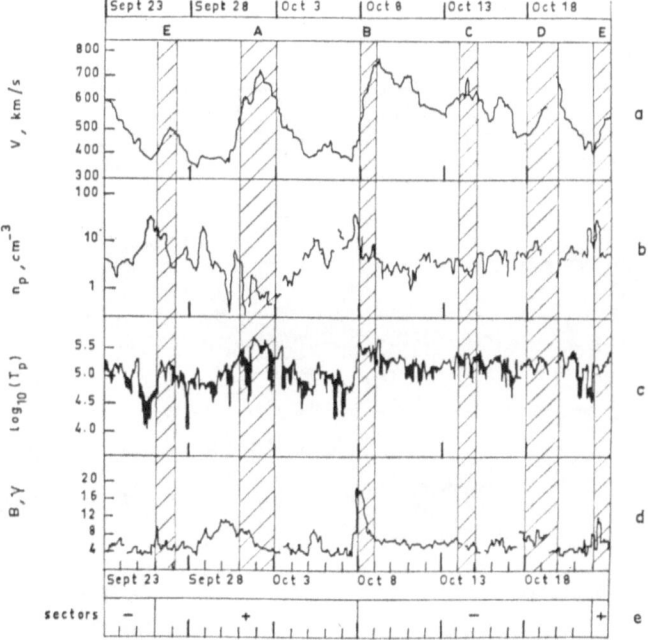

Fig. 4. Solar wind activity during solar rotation 1768. The vertical hatched regions represent CMP of coronal enhancements. a, solar wind velocity; b, proton density; c, temperature (upper and lower limits); d, interplanetary magnetic field magnitude; and e, sector polarity pattern (after Couturier and Leblanc, 1970).

above the photosphere. The electron density and temperature of these enhancements are higher than those of the 'quiet' Sun. Their results are shown in Figure 4. At the bottom of this figure there is shown the sector polarity pattern. The vertical cross hatched regions above indicate the position of the coronal enhancements. It can be seen that just after (eastward) each sector boundary there is a coronal enhancement. The opposite however is not necessarily correct. We see that within a sector there may be one or two additional coronal enhancements. These may be related to what we may term 'subsectors', i.e. a single magnetic sector may consist in some cases of two or three adjacent subsector regions on the Sun.

Evidence for the existence of subsectors has already been available from observations of the solar wind velocity. It is found that just after almost every sector boundary the solar wind velocity rises (as is shown in the top panel of Figure 4) but that again the converse is not necessarily true. Within a given magnetic sector there may be two or three large-scale increases in solar wind velocity. Near the last solar minimum each sector was a discrete entity, i.e. it did not contain any subsectors. With the increase of

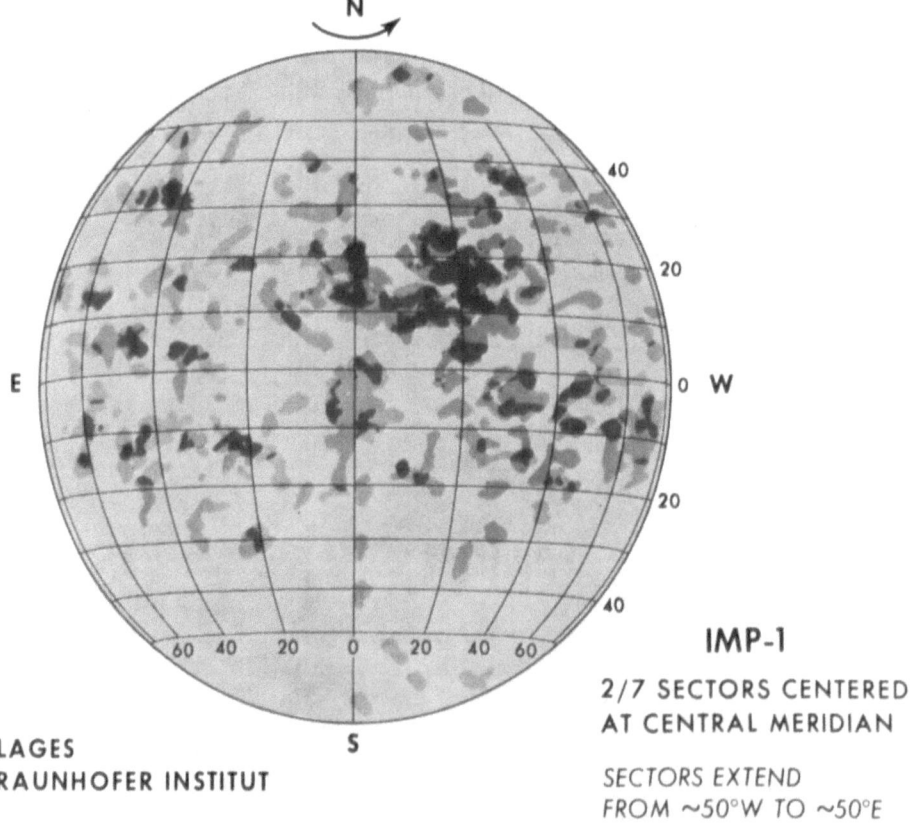

Fig. 5. Superposed-epoch analysis of calcium plage structure obtained from the daily Fraunhofer Institute maps of the Sun. The sectors are approximately centered at central meridian, so that the leading edge of the sector is at about 50° W and the trailing edge of the sector about 50° E longitude (after Wilcox and Ness, 1967).

solar activity in the rising portion of the sunspot cycle the concept of the subsector appears. Therefore much attention during the period of solar activity has been concentrated on the sector boundaries rather than on the entire sectors, which may be complicated by the presence of one or two subsectors. It remains a challenging problem in solar physics to understand the origin of both the sectors and the subsectors.

An analysis near solar minimum showed that plages were most probable in the preceding portions of solar sectors. Fraunhofer Institute daily solar charts were selected for days on which a solar sector was approximately centered at solar meridian. A number of these charts were then overlayed to form a kind of superposed epoch analysis of plage location with respect to sector position. The results are shown in Figure 5. It can be seen that plage activity is more intense to the west of central meridian (the preceding part of the sector) than to the east of central meridian (the following portion of the sector). Thus the above result that flare activity and coronal enhancements are most pronounced eastward of the sector boundary extends also to plage activity.

In the discussion of Figure 3 we showed that the position of flare-producing regions is related to the solar sector boundaries. We now show that the magnetic polarity of sunspots tends to be related to the solar sector pattern. The mean magnetic field of the Sun observed as though it were a star has been compared with the observed interplanetary sector structure (Wilcox *et al.*, 1969). The top portion of Figure 6 shows the results, in which the dots represent the observed mean solar field and the bars represent the polarity of the observed interplanetary sector pattern. The transit time of $4\frac{1}{2}$ days of solar wind plasma from the Sun to the Earth has been allowed for in this comparison in Figure 6. The agreement between the two sets of observations is seen to be very good. The bottom portion of Figure 6 shows the contribution of sunspot magnetic

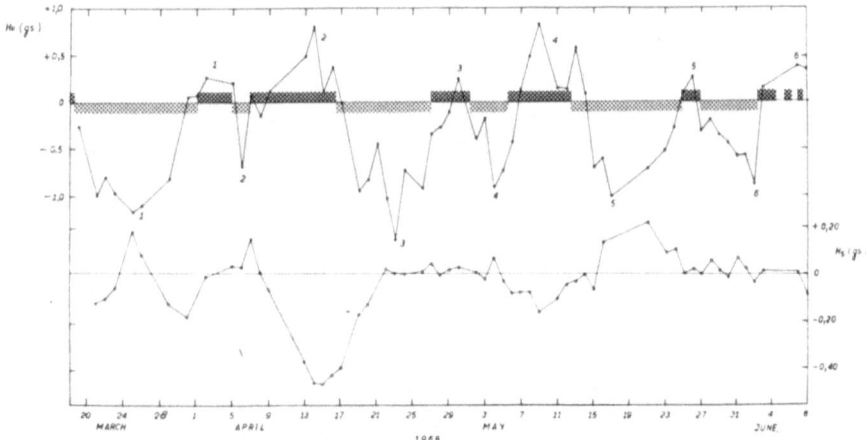

Fig. 6.   Top: mean value of the solar magnetic field (dots) and polarity of the interplanetary magnetic field (bars). The interplanetary field is displaced to allow for the transit time of solar wind plasma from the Sun to the Earth. Bottom: contribution of sunspot magnetic fields to the mean solar field shown above (after Wilcox *et al.*, 1969).

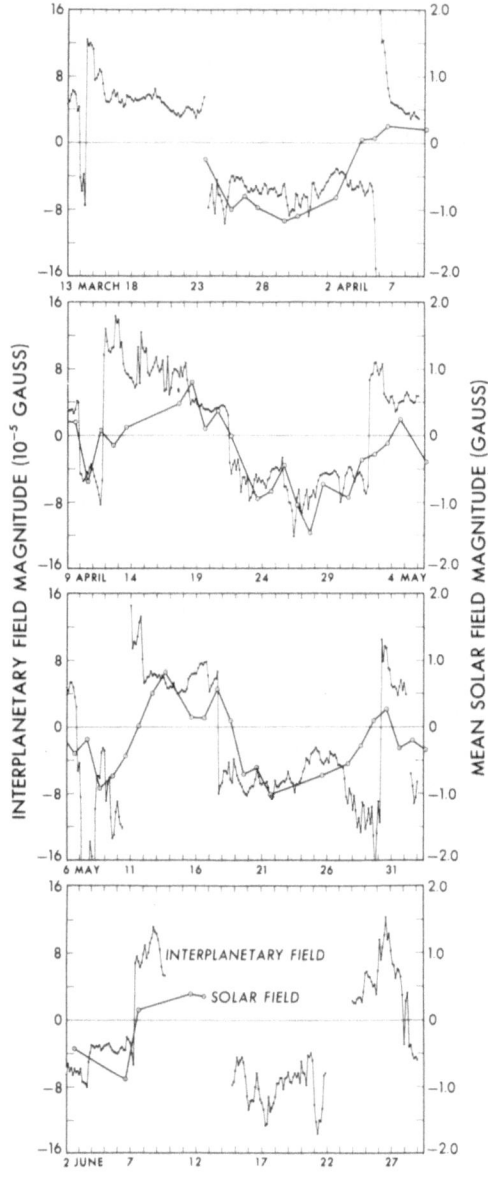

Fig. 7.   Comparison of the magnitude of the mean solar field and of the interplanetary field. The open circles are the daily observations of the mean solar field, and the dots are 3-hr average values of the interplanetary field magnitude observed near the Earth. The solar observations are displaced by $4\frac{1}{2}$ days to allow for the average Sun-Earth transit time. The abscissa is at the time of the interplanetary observations (after Severny *et al.*, 1970).

fields to the observed mean solar field. The net sunspot field contribution is often sensibly zero, but when it has an appreciable magnitude the net spot polarity is usually opposite to the polarity of the mean solar field and the interplanetary sector pattern. This suggests that the magnetic polarity of sunspots may be related to the solar sector pattern. The details of this relationship are another challenging problem in solar physics.

A comparison of the magnitude of the observed mean solar field with the magnitude of the observed interplanetary magnetic field is consistent with the solar sector pattern. A comparison of these magnitudes is shown in Figure 7. The scale of the ordinate has been chosen so that in the middle portions of large sectors the two fields approximately coincide. This then yields an experimental scaling of an average interplanetary field of 6×10^{-5} G corresponding to an average mean solar field of 0.75 G. Let us see if this scaling is reasonable. The average magnitude for the interplanetary field of 6×10^{-5} G corresponds to a radial component of 4×10^{-5} G, since the Archimedes spiral angle near the Earth is about $45°$. If we now scale this field by $1/r^2$ corresponding to the spherical expansion of the solar plasma (and field) we reach a magnitude of 1.8 G at the solar surface. The mean field observations refer to the line-of-sight component of the field and depending on specific assumptions about the average direction of the photospheric field the 1.8 G will correspond to an observation in the range of 1.3 to

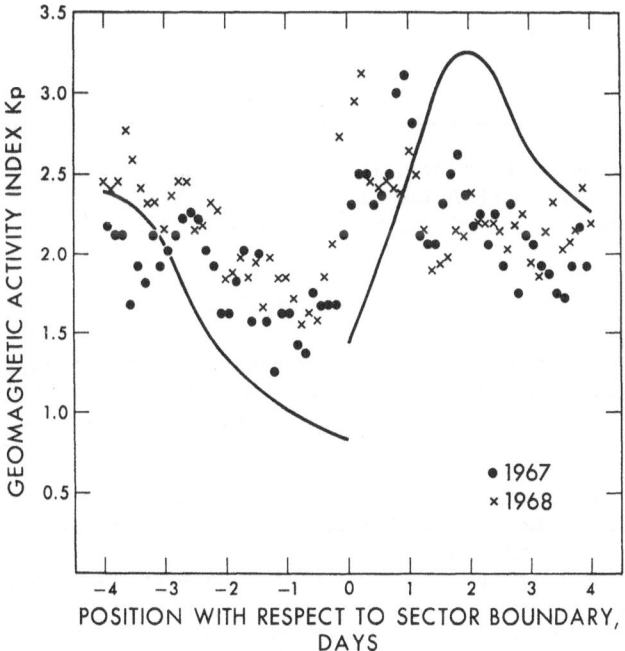

Fig. 8.   Superposed epoch analysis of the magnitude of the planetary magnetic 3-hr-range indices *Kp* as a function of position with respect to a sector boundary. The abscissa represent position with respect to the sector boundary, measured in days, as the sector pattern sweeps past the Earth. The solid line represents similar results obtained near solar minimum, the dots represent results in 1967 and the × represent results during 1968 (after Wilcox and Colburn, 1970).

0.9 G. Thus we find an agreement within less than a factor of two with the experimental scaling. This indicates that the solar sector pattern must occupy a large area on the Sun. If the solar sector pattern that is the source of the interplanetary sector pattern were confined to a small range of latitudes on the Sun the comparison of scaling discussed above would be changed by a large factor.

Finally, we may use the Earth as a detector for sector conditions. Figure 8 shows the average response of geomagnetic activity as a sector boundary rotates past the Earth. It can be seen that on the average geomagnetic activity has a monotonic decline in the following portion of a sector with an abrupt increase near the sector boundary to a peak early in the preceding portion of the sector. Again we find that the region just after (eastward) a sector boundary is the most active and that the region just before (westward) the sector boundary tends to be the most quiet.

Acknowledgements

This work was supported in part by the Office of Naval Research under contract N00014-69-A-0200-1016, by the National Aeronautics and Space Administration under grant NGL 05-003-230, and by the National Science Foundation under grant GA-1319.

References

Babcock, H. W.: 1961, *Astrophys. J.* **133**, 572.
Bumba, V. and Obridko, V. N.: 1969, *Solar Phys.* **6**, 104.
Couturier, P. and Leblanc, Y.: 1970, *Astron. Astrophys.* **7**, 254.
Schatten, K. H., Wilcox, J. M., and Ness, N. F.: 1969, *Solar Phys.* **6**, 442.
Severny, A., Wilcox, J. M., Scherrer, P. H., and Colburn, D. S.: 1970, *Solar Phys.* **15**, 3.
Wilcox, J. M. and Colburn, D. S.: 1970, *J. Geophys. Res.* **75**, 6366.
Wilcox, J. M. and Ness, N. F.: 1967, *Solar Phys.* **1**, 437.
Wilcox, J. M., Severny, A., and Colburn, D. S.: 1969, *Nature* **224**, 353.

Discussion

Krause: It seemed that the rigid rotation of an active region observed by you indicates that the toroidal field of the Sun is mainly produced by a radial variation of the angular velocity at deeper layers, and not – as it is the general opinion up till now – by the variation of the angular velocity with the latitude as observed at the surface of the Sun. Just this fact is used in the dynamo model for the Sun developed by us.

Pick: You have noted that active centers with large flare activity are located near a sector boundary. What is in this case the size of the interplanetary sector? Sometimes, we may note the existence of 'a secondary' boundary inside a large sector – this kind of boundary is probably associated directly with an active center. I would like to note that the enhancement observed on metric wavelengths is not characteristic of the activity.

Wilcox: The analysis includes the average properties of all solar sector boundaries. It is true, however, that newly-formed sectors tend to have large associated geomagnetic activity. Newly-formed sectors are quite narrow, starting with a width of perhaps 30° in longitude, and growing as they become older.

Vaiana: (1) Could you clarify the relation between the sector structures and the large scale complex of activity (the so-called active longitudes) seen for instance in the synoptic magnetic charts?
(2) You mentioned that one of the problems of establishing sector boundaries on the basis of solar

observations alone is the fact that one is dealing with a basically noisy set of maps. I would like to notice that X-ray maps may be a noise-free method of correlation with interplanetary sectors.

Wilcox: (1) The active longitudes and complexes of activity can fit within the solar sector structure. This can be seen in a quantitative way in the work of Bumba and Obridko already referred to. They made a histogram of flare occurrence as a function of active longitudes, but the resulting peak was shorter and broader as compared with the analysis using sector boundaries. It appears that solar sectors are more precisely defined than are active longitudes.

(2) I quite agree, and believe that the X-ray maps are a very promising technique.

Bumba: If we compare the 'supergiant regular structure' I showed before with Dr Wilcox's fine structure of interplanetary magnetic field sectors, we may see that individual smaller structures, in the photospheric magnetic field distribution (the internal structure of the supergiant features) nicely coincide with Dr Wilcox's 'subsectors' and other details.

Newkirk: One aspect of the sector structure appears to imply a contradiction. On the one hand we have the sector structure itself which, as you have mentioned, does not participate in the differential rotation. On the other hand there appears to be a statistically preferred position of active regions within the sector structures. The active regions, of course, display differential rotation and this one should expect that any preferred position would be quickly destroyed unless we are dealing with something like a standing wave pattern for the sector structure.

Wilcox: This is indeed an interesting problem. Part of the solution may be found in the property of the photospheric magnetic field to display both *differential* and *rigid* rotation. The field pattern near a solar sector boundary seems to display differential rotation on the time scale of a few Solar rotations and to display rigid rotation on a time scale of several rotations.

THE SUN AS A MAGNETIC ROTATOR

JAAKKO TUOMINEN

Astrophysics Laboratory, University of Helsinki, Helsinki, Finland

Severny (1969) used a technique which enables the Sun to be observed like a magnetic star. Because of rotation, different sides of the Sun are visible at different times. A curve of variable magnetic field is obtained. During nearly three periods of rotation the Sun appears to be a quadrupole magnetic rotator. Generally, the integrated field of sunspots seems to be opposite to the field of the Sun as a whole. The curve seemed to remain similar throughout the year 1968.

This observed phenomenon can be explained in the following way:

Magnetic flux through the leading part of a bipolar magnetic region, or bipolar group, must be equal (but of opposite polarity) to that through the following part. The following part disintegrates much faster than the leading one, which means that the field of the following part is distributed over a large area. When the magnetic field of the Sun as a whole is observed, the fields of the spots affect the result less, because their temperature is lower than that of the photosphere and their area smaller. Therefore the disintegrated following parts have a greater influence on the observed large scale fields. In the field of the spots the leaders are the more important.

The long life-time of the large scale magnetic regions can be explained on the basis of the phenomenon that new spot groups have a tendency to develop in close proximity to the places where spots already appeared. The phenomenon seems to continue for years, which means that the Sun will have long-lasting large-scale magnetic regions.

References

Severny, A.: 1969, *Nature* **224**, 53.
Tuominen, J.: 1962, *Z. Astrophys.* **55**, 110.
Tuominen, J.: 1970, *Nature* **228**, 1179. The above text is a summary of this article.

Discussion

Wilcox: Your discussion conserves magnetic flux in a bipolar magnetic region, the same flux coming out of one part of the region and going into the other part. As a discussion of Prof. Severny's results it is incomplete because there is no discussion of the net photospheric lines of one polarity that go out to form interplanetary magnetic sector patterns.

PART VIII

THEORIES OF LARGE SCALE FIELDS AND
THE ACTIVITY CYCLE

THEORIES OF LARGE SCALE FIELDS AND
THE MAGNETIC ACTIVITY CYCLE

N. O. WEISS

Dept. of Applied Mathematics and Theoretical Physics, University of Cambridge, England

Abstract. The magnetic field is effectively frozen into the ionized gas in the Sun and it is therefore necessary first to describe the motion in the convective zone. Large scale motion in giant cells is strongly affected by Coriolis forces, giving a radial shear in the angular velocity, while the interaction of convection and rotation leads to the equatorial acceleration. Many hydromagnetic dynamo mechanisms have been proposed in the last few years. In particular, meridional fields can be generated from azimuthal fields owing to a preferred sense of helicity in the motion $\overline{(\mathbf{u} \cdot \operatorname{curl} \mathbf{u}} \neq 0$, the '$\alpha$-effect'). Such regeneration is included in Leighton's phenomenological model, which reproduces many features of the solar cycle. More detailed models will have to treat the concentration of magnetic flux into ropes by individual convection cells.

1. Introduction

It is impossible to provide a comprehensive review in one short lecture and I shall concentrate on a few selected topics in which recent progress has been substantial. Since magnetic fields in the Sun are effectively frozen into the ionized gas it is necessary to describe the motion of this gas before the behaviour of the fields can be understood. Furthermore, the scale of active regions (comparable with the solar radius) indicates that magnetic fields penetrate deep into the convective zone. In the last few years we have gained a better understanding of the structure of this zone and also of the differential rotation, which dominates the solar dynamo. I shall first discuss the velocity field in the convective zone and then go on to recent developments in the theory of hydromagnetic dynamos.

At one time it was reasonable to doubt whether any such dynamo could work: now a host of possible mechanisms has appeared. At the same time, interest has been stimulated by discoveries of recent reversals in the Earth's field and of magnetic fields in stars and pulsars. Some dynamo models have fluid dynamical significance, some relate to the Earth and others imitate the oscillatory behaviour that we see in the Sun. Many details of the solar cycle are reproduced by a pair of remarkably faithful models and the process is becoming more comprehensible. However, these models still describe smoothed mean fields, rather than the complicated patterns that Professor Severny has shown us, and I shall conclude with some remarks about the concentration of magnetic flux into ropes by individual convection cells.

2. Motion in the Convective Zone

Models based on mixing length theory ascribe a depth of 100000 to 200000 km to the convective zone (Baker and Temesvary, 1966). It is conventionally supposed that there is a hierarchy of eddies whose scales are characterised by a mixing length equal to the local pressure or density scale height. Unless their vertical size is limited, continuity implies that all flow would be predominantly horizontal. On the other hand, observation shows only a limited number of preferred length scales. The diameter of granules is comparable with the scale height at the base of the photosphere; the formation of sunspots at junctions in the chromospheric network indicates that supergranules are associated with deep-seated convection; Bumba's (1967) suggestion of giant cells is supported by the pattern of magnetic fields (Bumba *et al.*, 1969) and by Howard's (1971) analysis of large-scale velocity fields; and now Bumba (1971) suspects that there may be supergiant cells too. There are some theoretical grounds (Simon and Weiss, 1968) for believing that cells can extend over about three scale heights and therefore that there should, in addition to granules and supergranules, be giant cells with diameters of about 300000 km and velocities around 0.1 km s^{-1}, lasting for periods of a month or more. The convective zone could then be divided into three layers, each with its own preferred scale of motion.

All magnetic features share in the differential rotation of the Sun. Therefore it must be a deep-seated phenomenon and not just a superficial wind. The angular velocity Ω is found to vary with both latitude and depth. Let us consider the radial variation first. The solar wind exerts a decelerating torque on the convective zone and this deceleration may be enhanced by turbulence (Gough and Lynden-Bell, 1968). It has been suggested (e.g. by Plaskett, 1966) that the Sun's interior rotates ten times faster than the surface, and this hypothesis is supported by measurements of the Sun's oblateness (Dicke, 1970). It is hard to accept the existence of so great a variation in Ω if any magnetic field permeates the region, and harder still to see how an ambient field could be entirely excluded from the shear zone. If the Sun indeed has a rapidly rotating core it cannot affect magnetic fields observed in the photosphere. Nevertheless, conservation of angular momentum does impose a constraint upon motion in convective cells and the importance of the Coriolis force is given by the ratio of the turnover time to the rotation period of the Sun. For granules this is negligible and for supergranules it is slight. However, for giant cells there will be a tendency to maintain an angular momentum independent of r (the distance from the centre of the Sun) so that $\Omega \propto r^{-2}$; this will be limited by turbulent or kinetic friction. Rising fluid can thus move azimuthally through about 90° with a consequent distortion of the cells. Recent observations of the Sun's rotation rate (Howard and Harvey, 1970; Wilcox and Howard, 1970) show that magnetic features rotate more rapidly than the surrounding photosphere and also that the large scale field has a shorter rotation period than sunspots, thereby supporting the hypothesis that $\partial\Omega/\partial r < 0$.

It is apparent that the observed equatorial acceleration must result from the effect of rotation upon convection in a spherical shell and various simplified models have been put forward to explain this. Plaskett (1959) suggested that a heliostrophic wind could be driven by the meridional pressure gradient resulting from cooling at the poles, where rotation inhibits convection. (It is only in axisymmetric systems that convection is suppressed at the equator (Durney, 1968, 1970).) If convection cells near the equator are elongated parallel to the rotation axis, the temperature difference on an equipotential surface might be 30°, which would suffice to drive the differential rotation (Weiss, 1965). However, there is no direct evidence for such elongation and the observations summarized by Caccin *et al.* (1970) indicate that the temperature difference is less than 20°, while Plaskett (1962, 1970) used measurements of limb darkening to find an *excess* temperature of 300° at the poles.

Biermann (1958) introduced an isotropic eddy viscosity to describe the transport of angular momentum by convective eddies; reduced diffusion in the radial direction then leads to a meridional circulation with equatorward motion at the surface. That such a motion can produce an equatorial acceleration follows from a simple argument due to Kippenhahn (1964). Let (r, θ, φ) be spherical polar co-ordinates and consider the region $r_1 < r < r_0$ of a system rotating with constant angular velocity Ω. Suppose there is an axisymmetric meridional flow with a θ-component $v(r, \theta)$ such that $v < 0$ $(r_1 < r < r_2)$ and $v > 0$ $(r_2 < r < r_0)$. Then conservation of mass implies that

$$\left| \int_{r_1}^{r_2} \varrho v r \, dr \right| = \left| \int_{r_2}^{r_0} \varrho v r \, dr \right|.$$

Hence, comparing the transport of angular momentum in the two regions, we find that

$$\left| \int_{r_1}^{r_2} \varrho v \Omega r^3 \, dr \right| < \left| \int_{r_2}^{r_0} \varrho v \Omega r^3 \, dr \right|.$$

Thus angular momentum is transported towards the equator and the flow can maintain a steady state with an equatorial acceleration. This behaviour has been demonstrated in specific models (Kippenhahn, 1963; Köhler, 1966) and Köhler (1970) has shown that anisotropic viscosity leads to an equatorial acceleration for an incompressible fluid in a spherical shell, constrained to rotate between free boundaries.

Another approach draws on meteorological analogy. In the terrestrial atmosphere transport of angular momentum is dominated by baroclinic (Rossby) waves. These waves have been studied theoretically and in laboratory experiments, using a rapidly rotating cylindrical annulus with an imposed lateral temperature gradient (Hide, 1970). A symmetrical regime is unstable and growing waves develop into asymmetric eddies. The flow is predominantly horizontal, and is dominated by the Coriolis force and horizontal pressure gradients; any unstable vertical variation in temperature must be small compared with the horizontal temperature difference. This baroclinic instability gives rise to cyclones or anticyclones in the Earth's atmosphere. Attempts have been made to identify baroclinic waves in the Sun, either from the proper motions of sunspots (Ward, 1965, 1966) or by directly measuring velocities (Plaskett, 1966), and then to show that these waves are responsible for the equatorial acceleration (Starr and Gilman, 1965; Kato, 1969; Kato and Nakagawa, 1969; Starr, 1968, has conveniently assembled most of the relevant papers). The essential feature of this process is that the acceleration is maintained not by a general meridional circulation but owing to a net non-linear transfer of angular momentum by non-axisymmetric eddies. The discussion of this effect is most illuminating but the model cannot be applied directly to the Sun. The convective zone differs significantly from the Earth's atmosphere. The super-adiabatic stratification cannot be neglected, for the transport of energy requires large scale vertical motions in regions where the scale height is an appreciable fraction of the solar radius. Observationally, Rossby waves cannot be distinguished from the giant cells described above and horizontal motions should not be treated independently of convection.

These simplified models have demonstrated that an equatorial acceleration is less remarkable than had once been supposed. Further progress requires a proper study of non-linear convection in rotating systems. Busse (1970) has formulated the problem for convection in a Boussinesq fluid contained in a slowly rotating spherical shell. Different modes are described by spherical harmonics in the form

$$ f(r)\, P_l^m(\cos\theta)\, e^{im\varphi} . $$

The first mode to become unstable has $m=l$, corresponding to a sectorial harmonic with a segmented structure like an orange and a pronounced maximum at the equator. For a thin shell of depth h, instability sets in with

$$ l \approx \frac{\pi r_0}{\sqrt{2h}} ; $$

thus motion occurs in rolls whose equatorial cross-section is rectangular with an aspect ratio of $\sqrt{2}$. Proceeding to higher order in an expansion about the critical Rayleigh number, Busse shows that there is a net equatorial acceleration whose magnitude (after inserting suitable values for parameters) is comparable with that observed in the Sun. Similar results were obtained by Davies-Jones and Gilman (1970, 1971). As a model for giant cells they considered a thin cylindrical annulus with rectangular cross-section, rotating about a vertical axis and heated from below. The relevant equations were solved to second order. When rotation is dominant the motion tends to be in rolls and produces differential rotation with an equatorial acceleration. Durney (1970, 1971) has used a computer to tackle a more complicated model: the problem resembles that of Busse, with free boundaries, but non-linear terms are represented in the mean field (Herring) approximation. This provides (as is necessary) a limited representation of non-axisymmetric flow, though there is no coupling between different values of m. The fields are expanded in vector spherical harmonics and truncated; only three poloidal and two toroidal harmonics, each with five radial functions, are retained. The qualitative features of Busse's analysis are confirmed by Durney's results. The dominant mode has the predicted value of l and the heat flux is maximized when $m=l$; the estimated effect of fluctuating interactions is to mix angular momentum and to produce an equatorial acceleration; the heat transport is a maximum at the equator (though the inhibition of convection at the poles is enhanced by the Herring approximation); and the convection pattern lags behind the rotation of the shell as a whole.

Obviously further work remains to be done but it is now possible to give a plausible model of the convective zone. The radiative core rotates with a constant angular velocity. In the convective zone, below about 15 000 km convection occurs in giant (or possibly in supergiant) cells and $\partial\Omega/\partial r < 0$. These cells are probably elongated perpendicular to the equator. Above this level are the supergranular cells in which the horizontal velocity is predominantly outward from a rising central plume. The Coriolis force therefore introduces a swirling flow in the opposite sense to the Sun's rotation (clockwise in the northern hemisphere, anti-clockwise in the southern, like terrestrial anticyclones) corresponding to a net radial vorticity. An equatorial acceleration is present throughout but drops to zero at the base of the convection zone. We might speculate that the existence of longitudinal structure in the field (Bumba et al., 1969) and of magnetic sectors with a fixed rotation period of 27 days (Wilcox and Ness, 1967) indicates the presence of convection in rolls parallel to the axis of rotation. However, preferred longitudes for solar activity might persist as a consequence of the dynamo process itself (Leighton, 1969) in which case it can only be inferred that the 27 day rotation period is somehow typical of the deep convection zone.

3. The Dynamo Problem and Models of the Solar Cycle

Despite the presence of strong local fields, the average magnetic energy density is small compared with the energy of motion. It is naturally convenient to consider smoothed magnetic fields and to treat them kinematically. The main features of the dynamo that produces the solar cycle are now generally accepted (Babcock, 1961; Leighton, 1964; Schmidt, 1968). The process runs as follows:

(i) At sunspot minimum there is an initial poloidal field and the toroidal field is small.

(ii) Differential rotation forms a strong toroidal field, first at medium and then at lower latitudes.

(iii) Some instability produces kinks in the toroidal field, which float upwards (or are borne up by rising supergranules) to the surface. The associated swirl produces tilted bipolar regions with the preceding part closer to the equator, thus generating a poloidal field opposite to that originally present.

(iv) The supergranules provide an effective eddy diffusion, which may be represented by a random walk process, and the reversed poloidal field diffuses to the poles by the next sunspot minimum.

(v) The sequence is then repeated to give the full cycle.

More detailed investigation involves the general theory of hydromagnetic dynamos. Fifteen years ago there was doubt as to whether any homogeneous dynamo mechanism was possible (Cowling, 1953). This gloom has been dispelled: indeed, G. O. Roberts (1970) has recently stated that 'almost all motions' will give dynamo action. (The phrase is given a mathematically precise meaning and obvious symmetrical models are still excluded.) Thus Lortz (1968) has devised a self-exciting dynamo with helical symmetry, while G. O. Roberts (1971), starting with an axially symmetric velocity, has produced a growing, non-axisymmetric field. The present state of dynamo theory is reviewed in detail by P. H. Roberts (1971) and discussed by Parker (1970a, b) in the context of astronomy.

A successful dynamo must avoid the effects of Cowling's theorem. The induction equation has the form

$$\frac{\partial \mathbf{B}}{\partial t} = \mathrm{curl}\,(\mathbf{u} \times \mathbf{B}) + \eta \nabla^2 \mathbf{B}, \tag{1}$$

where $\mathbf{B}$ is the magnetic field, $\mathbf{u}$ is the velocity and η the magnetic diffusivity. Suppose that $\mathbf{B}$ and $\mathbf{u}$ are axisymmetric. They can be separated into poloidal (meridional) and toroidal (azimuthal) parts:

$$\mathbf{B} = \mathbf{B}_p + B_\varphi \mathbf{i}_\varphi, \quad \mathbf{u} = \mathbf{u}_p + u_\varphi \mathbf{i}_\varphi.$$

It is convenient to introduce as an independent variable the distance from the axis,

$$\varpi = r \sin \theta$$

together with the operator

$$\Delta = \nabla^2 - \frac{1}{\varpi^2}.$$

Then the φ-component of (1) becomes

$$\frac{\partial}{\partial t}\left(\frac{B_\varphi}{\varpi}\right) + \nabla \cdot \left(\frac{B_\varphi}{\varpi}\mathbf{u}_p\right) = \mathbf{B}_p \times \nabla \left(\frac{u_\varphi}{\varpi}\right) + \frac{1}{\varpi}\,\eta\,\Delta B_\varphi. \tag{2}$$

In this equation the second term on the left hand side merely represents the advection of toroidal field by the flow. On the right hand side, the creation of toroidal flux from the poloidal field as a result of differential rotation compensates for ohmic dissipation. The poloidal field is best expressed in terms of a toroidal vector potential,

$$\mathbf{B}_p = \mathrm{curl}\,(A\mathbf{i}_\varphi).$$

Then (1) can be integrated to give

$$\frac{\partial}{\partial t}(\varpi A) + \mathbf{u}_p \cdot \nabla (\varpi A) = \eta \varpi\,\Delta A. \tag{3}$$

In contrast with (2), Equation (3) does not allow the regeneration of poloidal flux from B_φ. Consequently $\mathbf{B}_p$ ultimately decays, followed by B_φ. Thus an axisymmetric field cannot be maintained (Cowling, 1934).

A realistic dynamo model must therefore be non-axisymmetric, with an azimuthally averaged field $\mathbf{B}$, whose poloidal part is maintained by a toroidal emf

$$E_\varphi = \overline{(\mathbf{u} \times \mathbf{B})_\varphi}. \tag{4}$$

Most models incorporate two scales of motion. Averaging over φ then produces a simplified form of E_φ which can be used to solve (1) for the mean field **B**. Parker (1955, 1970b) argued heuristically that the rate of regeneration of the poloidal field in the Sun as flux tubes are brought up to the surface should be proportional to the toroidal field strength, so that

$$E_\varphi = \Gamma B_\varphi$$

and (3) becomes

$$\frac{\partial A}{\partial t} + \frac{1}{\varpi}\, \mathbf{u}_p \cdot \nabla(\varpi A) = \Gamma B_\varphi + \eta\, \Delta A. \tag{5}$$

Dynamo action is then possible. For example, if deviations from axial symmetry are small and the magnetic Reynolds number $R_m \gg 1$, it is possible to expand **B** and **u** in powers of $R_m^{-1/2}$. Then (2) and (5) hold for modified average fields B_φ and A, with Γ a function of $\mathbf{u}_p$, and dynamo action can be demonstrated (Braginsky, 1964).

In order to maintain the magnetic field it is necessary that the system should not be axially symmetric and this is certainly true of the turbulent convective zone. But "order does not arise spontaneously out of chaos" (Cowling, 1965). For homogeneous isotropic turbulence the average quantities $\bar{\mathbf{u}}$, $\overline{\text{curl}\,\mathbf{u}}$ and $\overline{\mathbf{u}\cdot\text{curl}\,\mathbf{u}}$ are all zero. However, if the random turbulence lacks reflectional symmetry it will have a preferred sense of helicity, so that

$$\overline{\mathbf{u}\cdot\text{curl}\,\mathbf{u}} \neq 0.$$

The effect of this small scale turbulent flow on the mean field **B** is given by a mean emf

$$E = \alpha\mathbf{B} - \beta\,\text{curl}\,\mathbf{B}$$

with

$$\alpha = -\tfrac{1}{3}\tau\,\overline{\mathbf{u}\cdot\text{curl}\,\mathbf{u}} \tag{6}$$

and

$$\beta = \tfrac{1}{3}\tau\,\overline{\mathbf{u}\cdot\mathbf{u}},$$

where τ is the correlation time for a turbulent eddy. Then (5) is once more obtained in the form

$$\frac{\partial A}{\partial t} + \frac{1}{\varpi}\, \mathbf{u} \cdot \nabla(\varpi A) = \alpha B_\varphi + (\eta + \beta)\, \Delta A. \tag{7}$$

Thus the helicity allows regeneration of the poloidal field by the so-called 'α-effect' in (7). That such a dynamo operates has been shown by the Potsdam group (Steenbeck et al., 1966; Steenbeck and Krause, 1966, 1969; Krause et al., 1971) on the assumption that

$$v^2\tau \ll l^2/\tau \ll \eta$$

(where v, l are typical velocities and length scales for the turbulent eddies) and also by

Moffatt (1970) on the assumption that $vl/\eta \ll 1$. As indicated below, this theory is also the basis for a model of the solar cycle.

The various models of the sun represent smoothed fields and include some form of eddy diffusion. The effect of a meridional circulation on a poloidal field has been computed by Maheswaran (1969), while Nakagawa and Swarztrauber (1969) have studied the generation of toroidal flux. They adopt an axisymmetric velocity that satisfies the equation of motion for an incompressible fluid in a spherical shell with boundary conditions such that $\partial\Omega/\partial r > 0$. Differential rotation creates B_φ from $\mathbf{B}_p$ in a zone that migrates equatorward with velocity $\mathbf{u}_p$ (Bullard, 1955) but more complicated flows are needed to reverse $\mathbf{B}_p$ (Nakagawa, 1971).

Many features of the solar cycle are reproduced when the field is distorted by baroclinic waves and Gilman (1968, 1969a, b) has argued that the Sun is indeed a Rossby wave dynamo. He considers a simplified model with waves driven by horizontal temperature gradients in a small, rapidly rotating cylindrical annulus. The fields are represented by drastically truncated expansions but the Coriolis force does allow generation of reversed poloidal flux from the toroidal field, leading to an oscillatory dynamo. A more realistic model, with spherical symmetry, has been studied by Gordon (1970). Similar processes can be expected in the convective zone, where the super-adiabatic vertical gradient is more important than temperature differences between the poles and the equator (Davies-Jones and Gilman, 1970).

A successful model has been constructed in which the interaction between convection and rotation is represented by the α-effect (Steenbeck and Krause, 1969; Krause et al., 1971). Owing to the density stratification, rising material is associated with an outward radial flow and thus with anticyclinic motion. Thus turbulent convection has a left-handed screw sense in the northern hemisphere, and right-handed in the southern. An expression for α emerges after solving the Navier-Stokes equation and Krause (1968) finds that (6) reduces to

$$\alpha = \frac{v^2\tau^2\Omega}{H}\cos\theta, \tag{8}$$

where H is the density scale height. The resistivity is dominated by the eddy diffusivity β. The model separates the convective zone into a lower region (giant cells) with differential rotation such that $\partial\Omega/\partial r < 0$, and an upper region (supergranules) in which the α-mechanism operates. With suitably chosen parameters reversals are obtained and a convincing butterfly diagram is produced.

This turbulent dynamo resembles Leighton's (1969) physical model, which reproduces the solar cycle with extraordinary accuracy. In this model the fields are projected onto a sphere of radius r_0 (by averaging radially through the convective zone) and then averaged over longitude to give mean fields that are functions only of the colatitude and time. The differential equations governing $\mathbf{B}$ are constructed from terms representing the physical effects responsible for the solar cycle. Once again, Ω is assumed to vary in the lower convective zone. In the upper region, toroidal fields erupt and form poloidal fields which are in turn dispersed by a random walk process.

Furthermore, the eruption occurs only when $|B_\varphi|$ exceeds a critical value B_c. The essential equations are then

$$\frac{\partial B_r}{\partial t} = F \delta \frac{C}{\sin \theta} \frac{\partial}{\partial \theta} (B_\varphi \sin \gamma) + \frac{1}{T_D \sin \theta} \frac{\partial}{\partial \theta} \left(\sin \theta \frac{\partial B_r}{\partial \theta} \right) \tag{9}$$

and, from (2),

$$\frac{\partial B_\varphi}{\partial t} = \sin \theta \left(B_\theta \frac{\partial \Omega}{\partial \theta} + r_0 B_r \frac{\partial \Omega}{\partial r} \right) - \delta \cdot C' |B\varphi| B_\varphi, \tag{10}$$

where

$$\delta = 0 \quad (|B_\varphi| < B_c)$$
$$= 1 \quad (|B_\varphi| \geqslant B_c)$$

and C, C' are constants depending on properties of the convective zone. The remaining field component, B_θ, is calculated from $\nabla B = 0$. In (9), the first term on the right-hand side generates new poloidal flux. The arbitrary parameter F is an adjustable efficiency factor and γ is the tilt of active regions. Now this term corresponds to an emf.

$$E_\varphi = \Gamma B_\varphi$$

with

$$\Gamma = F \delta \frac{C \sin \gamma}{\sin \theta}, \tag{11}$$

cf. Equation (8), and Leighton takes $\sin \gamma = \frac{1}{2} \cos \theta$. Thus his mechanism is similar to the α-effect described above. (Γ remains finite since $|B_\varphi| < B_c$ when θ is small.) The remaining term in (9) represents the random walk process, with a decay time $T_D \approx 22y$, corresponding to the eddy diffusivity in (7). Similarly, Equation (10), which includes the effects of differential rotation in the lower region and depletion of B_φ as a consequence of eruption into the upper zone, resembles (2). The main features of the solar cycle – Spörer's law, the butterfly diagram, the poleward drift of prominence zones and the 22 year periodicity – can be reproduced for a variety of assumptions about $\Omega(r, \theta)$. In this paper, Leighton discusses a number of experiments in which the values of parameters are altered. For example, the dynamo is found to be most efficient (i.e. the value of F that allows a 22 yr period is least, about 0.6) when the differential rotation is dominated by a radial variation such that

$$h \frac{\partial \Omega}{\partial r} = - 18 \sin^2 \theta \text{ radians } y^{-1}.$$

This heuristic model demonstrates that the observed behaviour of the Sun's magnetic field is compatible with the dynamo process outlined at the beginning of this section. Work by others, notably the Potsdam group, shows that the model can be made more rigorous. Many details of these mean field dynamos remain to be investigated. In addition it is now necessary to seek a better understanding of the effect of individual convection cells in producing the flux ropes that emerge into the photosphere.

4. Concentration of Magnetic Fields into Flux Ropes

Magnetic flux is mainly concentrated around the perimeter of solar convection cells and particularly at corners where several cells meet. For weak fields this concentration is purely kinematic and the field is limited by the magnetic Reynolds number

$$R_m = \frac{vl}{\eta}.$$

A steady solution of (1) with a balance between diffusion and advection is achieved after the initial field B_0 has been amplified locally to a magnitude

$$B \approx R_m^{1/2} B_0 \quad \text{or} \quad B \approx R_m B_0 \tag{12}$$

for two or three dimensional flow respectively (Parker, 1963; Clark, 1965; Weiss, 1966; Clark and Johnson, 1967). In the Sun R_m is so large that forces exerted by these fields could not be withstood. It is generally supposed that concentration proceeds until equipartition is achieved. Then the field is limited by dynamical effects, and, locally,

$$\tfrac{1}{2}\varrho u^2 \approx \frac{B^2}{2\mu}. \tag{13}$$

Indeed, Beckers (1971) has listed this as a means of determining the field strength. However, there has hitherto been no more precise calculation on which this assumption could be based.

The process can be studied through a simple model problem. Consider a layer of incompressible fluid, heated from below in the presence of an imposed vertical field B_0. The linearised treatment is well understood (Chandrasekhar, 1961; Danielson, 1961; Weiss, 1964) and non-linear solutions have recently been obtained on a computer. Three regimes are found to exist, depending on the value of B_0: weak fields are distorted kinematically and reach a maximum strength given by (12), moderate fields are concentrated until they are powerful enough to affect the motion, while strong fields hinder and ultimately suppress convection, as predicted by linear theory. Preliminary results for two dimensional flow within free boundaries show a maximum magnetic energy density for moderate fields that is four to ten times greater than the kinetic energy density corresponding to the greatest horizontal velocity. The strongest fields may therefore be several times greater than the values derived from (13). Moreover, the computations confirm not only that overstable linear modes grow into finite amplitude oscillations but also that certain exponentially growing perturbations develop into non-linear oscillations.

Further investigation should help to clarify the factors governing the formation of flux ropes and also, perhaps, to explain the transmission of energy in sunspots (Weiss, 1969). The magnetic field in sunspot umbrae is strong enough to inhibit steady convection, yet the energy emitted cannot be supplied by radiative transfer. As

Sweet (1971) and Wilson (1969, 1971) have already pointed out, some wave process must be present. Savage (1969) considers that overstable modes can be excited in sunspots, owing to a coupling of hydromagnetic waves in the unstable region with gravity waves in the stably stratified layer above.

5. Conclusion

It is apparent that great advances have been made in our understanding of large scale magnetic fields in the Sun. Many suggestive models illuminate various aspects of the solar cycle; but details are frequently obscure and more comprehensive calculations have still to be completed. A proper treatment of the convective zone must include the effects of compressibility as well as rotation, in three dimensional spherical geometry. The development of photospheric fields by flux concentration between individual granules and supergranules must be studied, as proposed by Kuklin (1971). And these results must be combined with the full dynamo problem in order to provide a proper model that can explain the complicated fields that Severny (1971) has described.

References

Ambrož, P., Bumba, V., Howard, R., and Sýkora, J.: 1971, this volume, p. 696.
Babcock, H. W.: 1961, *Astrophys. J.* **133**, 572.
Baker, N. and Temesvary, S.: 1966, *Tables of Convective Stellar Envelope Models*, NASA, New York.
Beckers, J. M.: 1971, this volume, p. 3.
Biermann, L.: 1958, in *Electromagnetic Phenomena in Cosmical Physics,* (ed. by B. Lehnert), Cambridge University Press.
Braginsky, S. I.: 1964, *Zh. Eks. Teor. Fiz.* **47**, 1084 *Soviet Phys. JETP* **20**, 726, 1965.
Bullard, E. C.: 1955, in *Vistas in Astronomy* **1** (ed. by A. Beer), Pergamon, London.
Bumba, V.: 1967, in *Plasma Physics* (ed. by P. A. Sturrock), Academic Press, London.
Bumba, V., Howard, R., Kopecky, M., and Kuklin, G. V.: 1969, *Bull. Astron. Inst. Czech.* **20**, 18.
Busse, F. H.: 1970, *Astrophys. J.* **159**, 629.
Caccin, B., Falciani, R., Moschi, G., and Rigutti, M.: 1970, *Solar Phys.* **13**, 33.
Chandrasekhar, S.: 1961, *Hydrodynamic and Hydromagnetic Stability*, Clarendon Press, Oxford.
Clark, A., Jr.: 1965, *Phys. Fluids* **7**, 1455.
Clark, A., Jr. and Johnson, A. C.: 1967, *Solar Phys.* **2**, 433.
Cowling, T. G.: 1934, *Monthly Notices Roy. Astron. Soc.* **94**, 39.
Cowling, T. G.: 1953, in *The Sun* (ed. by G. R. Kuiper), University of Chicago Press.
Cowling, T. G.: 1965, in *Solar and Stellar Magnetic Fields* (ed. by R. Lüst), North-Holland, Amsterdam.
Danielson, R. E.: 1961, *Astrophys. J.* **134**, 289.
Davies-Jones, R. P. and Gilman, P. A.: 1970, *Solar Phys.* **12**, 3.
Davies-Jones, R. P. and Gilman, P. A.: 1971, *J. Fluid Mech.*, to be published.
Dicke, R. H.: 1970, *Astrophys. J.* **159**, 1.
Durney, B.: 1968, *J. Atmospheric Sci.* **25**, 771.
Durney, B.: 1970, *Astrophys. J.* **161**, 1115.
Durney, B.: 1971, *Astrophys. J.* **163**, 353.
Gilman, P. A.: 1968, *Science* **160**, 760.
Gilman, P. A.: 1969a, *Solar Phys.* **8**. 316.
Gilman, P. A.: 1969b, *Solar Phys.* **9**, 3.
Gordon, C. A.: 1970, Ph.D. thesis, Mass. Inst. Tech.
Gough, D. O. and Lynden-Bell, D.: 1968, *J. Fluid Mech.* **32**, 437.

Hide, R.: 1970, in *Global Circulation of the Atmosphere* (ed. by G. A. Corby), Roy. Meteorol. Soc., London.

Howard, R.; 1971, *Solar Phys.* **16**, 21.

Howard, R. and Harvey, J.: 1970, *Solar Phys.* **12**, 23.

Kato, S.: 1969, *Astrophys. J.* **157**, 827.

Kato, S. and Nakagawa, Y.: 1969, *Solar Phys.* **10**, 476.

Kippenhahn, R.: 1963, *Astrophys. J.* **137**, 664.

Kippenhahn, R.: 1964, in *Atti del convegno sulli campi magnetici solari e la spettroscopia ad alta risoluzione* (ed. by M. Cimino), Barbera, Firenze.

Köhler, H.: 1966, *Mitt. Astron. Ges.* **21**, 91.

Köhler, H.: 1970, *Solar Phys.* **13**, 3.

Krause, F.: 1968, Habilitationsschrift, Universität Jena.

Krause, F. and Rädler, K.-H.: 1971, this volume, p. 770.

Kuklin, G. V.: 1971, this volume, p. 737.

Leighton, R. B.: 1964, *Astrophys. J.* **140**, 1559.

Leighton, R. B.: 1969, *Astrophys. J.* **156**, 1.

Lortz, D.: 1968, *Plasma Phys.* **10**, 967.

Maheswaran, M.: 1969, *Monthly Notices Roy. Astron. Soc.* **145**, 435.

Moffatt, H. K.: 1970, *J. Fluid Mech.* **41**, 435.

Nakagawa, Y.: 1971, this volume, p. 725.

Nakagawa, Y. and Swartztrauber, P.: 1969, *Astrophys. J.* **155**, 295.

Parker, E. N.: 1955, *Astrophys. J.* **122**, 293.

Parker, E. N.: 1963, *Astrophys. J.* **138**, 552.

Parker, E. N. 1970a, *Astrophys. J.* **160**, 383.

Parker, E. N.: 1970b, *Ann. Rev. Astron. Astrophys.* **8**, 1.

Plaskett, H. H.: 1959, *Monthly Notices Roy. Astron. Soc.* **119**, 197.

Plaskett, H. H.: 1962, *Monthly Notices Roy. Astron. Soc.* **123**, 541.

Plaskett, H. H.: 1966, *Monthly Notices Roy. Astron. Soc.* **131**, 407.

Plaskett, H. H.: 1970, *Monthly Notices Roy. Astron. Soc.* **148**, 149.

Roberts, G. O.: 1970, *Phil. Trans. Roy. Soc. A* **266**, 535.

Roberts, G. O.: 1971, to appear.

Roberts, P. H.: 1971, in *Lectures in Applied Mathematics* (ed. by W. H. Reid), American Mathematical Society, Washington.

Savage, B. D.: 1969, *Astrophys. J.* **156**, 707.

Schmidt, H. U.: 1968, in K. O. Kiepenheuer (ed.), 'Structure and Development of Solar Active Regions', *IAU Symp.* **35**, 95.

Severny, A. B.: 1971, this volume, p. 675.

Simon, G. W. and Weiss, N. O.: 1968, *Z. Astrophys.* **69**, 435.

Starr, V. P.: 1968, *Energetics of Eddy Actions in the Solar Atmosphere and Related Astronomical Topics*, MIT Dept. of Meteorology, Planetary Circulations Project, Report No. A3.

Starr, V. P. and Gilman, P. A.: 1965, *Astrophys. J.* **141**, 1119.

Steenbeck, M. and Krause, F.: 1966, *Z. Naturforsch.* **21a**, 1285.

Steenbeck, M. and Krause, F.: 1969, *Astron. Nachr.* **291**, 49.

Steenbeck, M., Krause, F., and Rädler, K.-H.: 1966, *Z. Naturforsch.* **21a**, 369.

Sweet, P. A.: 1971, this volume, p. 457.

Ward, F.: 1965, *Astrophys. J.* **141**, 534.

Ward, F.: 1966, *Astrophys. J.* **145**, 416.

Weiss, N. O.: 1964, *Phil. Trans. Roy. Soc. A* **256**, 99.

Weiss, N. O.: 1965, *Observatory*, **85**, 37.

Weiss, N. O.: 1966, *Proc. Roy. Soc. A* **293**, 310.

Weiss, N. O.: 1969, in *Plasma Instabilities in Astrophysics* (ed. by D. G. Wentzel and D. E. Tidman), Gordon and Breach, New York.

Wilcox, J. M. and Howard, R.: 1970, *Solar Phys.* **13**, 251.

Wilcox, J. M. and Ness, N. F.: 1967, *Solar Phys.* **1**, 437.

Wilson, P. R.: 1969, *Solar Phys.* **10**, 404.

Wilson, P. R.: 1971, this volume, p. 475.

Discussion

Sreenivasan: It is very nice to hear that you have obtained magnetic energy density an order of magnitude higher than kinetic energy density in your computations.

(1) Did you actually solve the Navier-Stokes equation with a $(\mathbf{j} \times \mathbf{B})$ term on the right hand side and integrate

$$\frac{\partial \mathbf{B}}{\partial t} = \nabla \times (\mathbf{u} \times \mathbf{B}) + \lambda \nabla^2 \mathbf{B}$$

to obtain your flux concentration at the edges of the cells?

(2) If the answer to question (1) is yes, how does one physically understand your conclusion, in the light of equipartition arguments of Batchelor?

(3) What is the reaction due to the growing magnetic field doing to the velocity field?

I have recently shown that, starting from the assumption that an initial magnetic field is force-free, there is only one class of velocity fields which permit the magnetic field to remain force-free in time. These are *Beltrami fields* and obey the relation:

$$\xi = \alpha \xi \quad \text{and} \quad \nabla \times \eta = \alpha \eta$$

where $\xi = (\mathbf{B} \cdot \nabla)\mathbf{u}$ and $\eta = (\mathbf{u} \cdot \nabla)\mathbf{B}$ and $\mathbf{B}$ is given by $\nabla \times \mathbf{B} = \alpha \mathbf{B}$ *for incompressible flow*. The compressible case is more complicated. It has also been shown that these helial motions can amplify the magnetic field $\mathbf{B}$, beyond the equipartition limit, provided α satisfies an inequality. In this picture, there is no reaction by the magnetic field since the Lorentz force is zero. The amplification limit is set by the dynamical stability of the configuration.

Weiss: Yes, the time-dependent Navier-Stokes, induction and heat flow equations were solved simultaneously. Of course, the enhanced magnetic field tends to slow down the motion, particularly where $\mathbf{B}$ is strongest. But the magnetic field is only concentrated locally, at the boundaries of a cell, so the overall energy density, averaged over a whole cell, is not necessarily larger than the average kinetic energy density. Moreover, equipartition arguments, which relate magnetic fields to velocities or vorticity, apply to three dimensional homogeneous turbulence and the numerical experiments don't really reproduce this ideal configuration.

Gilman: As far as the observational evidence from sunspot motion is concerned (in particular the correlation of longitude and latitude motion giving an equatorward transport of momentum) we can not tell the difference between a giant convective cell and a baroclinic or Rossby wave.

DYNAMO THEORY OF THE SUN'S GENERAL MAGNETIC FIELD ON THE BASIS OF A MEAN-FIELD MAGNETOHYDRODYNAMICS

F. KRAUSE and K.-H. RÄDLER

Zentralinstitut für Astrophysik, Potsdam, Germany

Abstract. An outline of the mean-field magnetohydrodynamics suggested and developed by M. Steenbeck and the authors and its application to the dynamo theory of the solar cycle is presented. Four basic requirements are formulated which have to be satisfied by any dynamo model which claims to explain the solar cycle. The models investigated allow conclusions about the differential rotation. In this connection Leighton's work is criticized.

1. Basic Ideas

The general magnetic field of the Sun is observed to be an alternating field, therefore the possibility of a relic field is almost completely excluded. In this way the idea that the Sun is a self-excited dynamo seems to be the only one promising success for the explanation of the magnetic field on the basis of classical physics.

According to our opinion any dynamo model for the Sun has to satisfy the following four requirements:

(1) The dynamo action must be provided by motions, which can be expected at the Sun;

(2) the dynamo excitation must occur for arbitrarily small seed fields;

(3) the dynamo must be an alternating field dynamo;

(4) the excited field must be antisymmetric with respect to the equatorial plane.

According to requirement (1) one has to take velocity fields, which correspond to the observations or are derived from the Navier-Stokes equation under appropriate conditions. Requirement (2) is formulated according to the conception that without dynamo action only insignificantly small fields can exist at the Sun. A first consequence of (2) is that we are allowed to neglect the Lorentz force in excitation. A second consequence of this requirement is that the feedback mechanism proposed by Babcock (1961) cannot be accepted. Babcock suggests that the desired production of poloidal field from the toroidal field is given by the systematic tilt of sunspot pairs. This observed tilt clearly gives a production of a poloidal field, however, this mechanism does not work until the field strength is greater than a certain critical value. Hence a dynamo using this mechanism cannot work at arbitrarily small fields. Leighton (1969) allowed this feedback mechanism to work also for arbitrarily small fields, but we will return to this paper later.

Requirement (3) expresses the observed fact that the Sun's general magnetic field is an alternating one with a period of 22 yr, and requirement (4) that the field shows the opposite orientation regarding space points situated symmetric to the equatorial plane. Whereas requirement (3) expresses the change of the polarity from one cycle to

Howard (ed.), Solar Magnetic Fields, 770–779. All Rights Reserved.
Copyright © 1971 by the IAU.

the following, requirement (4) expresses that the polarity also changes from one hemisphere of the Sun to the other. Requirement (4), for instance, includes the observation that the preceding spots of one hemisphere have the opposite polarity of the preceding spots on the other hemisphere.

The formulation of these four basic requirements is rather general. Obviously one can formulate more detailed requirements, because more detailed results are furnished by observations. The main reason for choosing these four basic requirements is that they can be satisfied by a linear theory as is shown in this paper. Such details of the solar field as the butterfly diagram are apparently strongly influenced by non-linear effects (action of the Lorentz force on motion). Therefore it is of less value to require from a linear theory very accurately fitted models. It is clear how to include the non-linear effects in this theory but then great mathematical difficulties will arise. That will be a matter of work to be done in future.

2. Mean-Field Magnetohydrodynamics

As is well known there exists no solution of the dynamo problem in which both magnetic and velocity fields have a simple structure. One is tempted to draw attention to the turbulence. The plasma of the convection zone shows turbulent motions and there are some hints that they are important for the dynamo action. In this connection a mean-field magnetohydrodynamics seems to be the appropriate theory, which was suggested and developed in earlier papers (Steenbeck *et al.*, 1966; Rädler, 1968; Krause, 1968, 1969; and Rädler, 1969). We give only an outline of the basic ideas in the present paper*.

The basic equations of magnetohydrodynamics are:
Maxwell's equations

$$\operatorname{curl} \mathbf{E} = - \dot{\mathbf{B}}, \quad \operatorname{curl} \mathbf{H} = \mathbf{j}, \quad \operatorname{div} \mathbf{B} = 0, \tag{1}$$

the constitutive equations

$$\mathbf{B} = \mu \mathbf{H}, \quad \mathbf{j} = \sigma (\mathbf{E} + \mathbf{v} \times \mathbf{B}), \tag{2}$$

the Navier-Stokes equation

$$\varrho \, (\partial \mathbf{v} / \partial t) + (\mathbf{v} \cdot \operatorname{grad}) \, \mathbf{v} = - \operatorname{grad} p + \mathbf{j} \times \mathbf{B} + \mathbf{F} \tag{3}$$

and further equations which, as well as the forces **F** in the Navier-Stokes equation, need not be given explicitly here. We use the notations: **E** electric field strength, **B** magnetic induction, **H** magnetic field strength, **j** current density, **v** velocity field, p pressure, ϱ density.

---

* An extended representation of the mean-field magnetohydrodynamics will appear soon in Rompe and Steenbeck, *Handbuch der Plasmaphysik und der Gaselektronik*, Band II, Akademie-Verlag Berlin.

We get the basic equations for the mean fields by averaging the foregoing equations (1), (2), (3). Averaged quantities are denoted by a bar and fluctuations by a dash. We have

$$\text{curl}\, \bar{\mathbf{E}} = - \dot{\bar{\mathbf{B}}}, \quad \text{curl}\, \bar{\mathbf{H}} = \bar{\mathbf{j}}, \quad \text{div}\, \bar{\mathbf{B}} = 0 ; \tag{4}$$

$$\bar{\mathbf{B}} = \mu \bar{\mathbf{H}}, \quad \bar{\mathbf{j}} = \sigma (\bar{\mathbf{E}} + \bar{\mathbf{v}} \times \bar{\mathbf{B}} + \overline{\mathbf{v}' \times \mathbf{B}'}) ; \tag{5}$$

$$\varrho \left(\frac{\partial \bar{\mathbf{v}}}{\partial t} + (\bar{\mathbf{v}} \cdot \text{grad})\, \bar{\mathbf{v}} \right) = - \text{grad}\, \bar{p} + \bar{\mathbf{j}} \times \bar{\mathbf{B}} + \bar{\mathbf{F}} + \overline{\mathbf{j}' \times \mathbf{B}'} - \overline{\varrho (\mathbf{v}' \cdot \text{grad})\, \mathbf{v}'} . \tag{6}$$

Maxwell's equations are valid for the mean fields as well as for the original fields. This is a consequence of their linearity. But the non-linear terms involved in Ohm's law and in the Navier-Stokes equation provide for new physical quantities like the electromotive force $\overline{\mathbf{v}' \times \mathbf{B}'}$ and the ponderomotive forces $\overline{\mathbf{j}' \times \mathbf{B}'}$ and $-\overline{\varrho (\mathbf{v}' \cdot \text{grad})\, \mathbf{v}'}$.

In order to get a closed system of differential equations again one has to express the additional terms $\overline{\mathbf{v}' \times \mathbf{B}'}$, $-\overline{\varrho (\mathbf{v}' \cdot \text{grad})\, \mathbf{v}'}$, $\overline{\mathbf{j}' \times \mathbf{B}'}$ by the mean fields. The analysis of this problem is developed to some extent for the electromotive force $\overline{\mathbf{v}' \times \mathbf{B}'}$ (Steenbeck *et al.*, 1966; Rädler, 1968; and Krause, 1969). The other two terms can be treated in an analogous manner (Krause, 1969), but more detailed work is needed.

For a treatment of the dynamo problem according to our requirements (1)–(4) we need to be concerned with the emf $\overline{\mathbf{v}' \times \mathbf{B}'}$ only. The explicit expression of this emf $\overline{\mathbf{v}' \times \mathbf{B}'}$ will depend on the structure of the turbulence. A turbulence like that performed by the plasma of the convection zone undergoes in particular the influence of some vector quantities, like the gravity force, the density gradient, the temperature gradient and finally the rotational motion represented by the pseudo-vector $\boldsymbol{\omega}$ of the angular velocity. The first three vector quantities are parallel, parallel to the radial direction, and their influence is subsumed by the general property of the turbulence that the radial direction is a preferred direction, a direction of anisotropy. We denote it by $\mathbf{g}$. In addition the direction parallel to the axis of rotation is a preferred direction, but this is represented by the pseudo-vector $\boldsymbol{\omega}$. Any average vector field can be a combination of these two vectors $\mathbf{g}$ and $\boldsymbol{\omega}$ and in addition the mean magnetic field $\bar{\mathbf{B}}$ only. It may be noticed that $\bar{\mathbf{B}}$ is also a pseudo-vector, whereas $\overline{\mathbf{v}' \times \mathbf{B}'}$ is a vector.

Taking into account the general symmetry laws of physics the emf $\overline{\mathbf{v}' \times \mathbf{B}'}$ can only be an expression like

$$\overline{\mathbf{v}' \times \mathbf{B}'} = - \alpha_1 (\mathbf{g} \cdot \boldsymbol{\omega})\, \bar{\mathbf{B}} - \alpha_2 (\mathbf{g} \cdot \bar{\mathbf{B}})\, \boldsymbol{\omega} - \alpha_3 (\boldsymbol{\omega} \cdot \bar{\mathbf{B}})\, \mathbf{g}$$
$$- \beta\, \text{curl}\, \bar{\mathbf{B}} - \beta_1 \boldsymbol{\omega} \times \text{curl}\, \bar{\mathbf{B}} - \beta_2\, \text{grad}\, (\boldsymbol{\omega} \cdot \bar{\mathbf{B}}) - \gamma \mathbf{g} \times \bar{\mathbf{B}}, \tag{7}$$

where $\alpha_1, \alpha_2, \alpha_3, \ldots, \gamma$ are scalars.

The most important result we find is an emf $\alpha \bar{\mathbf{B}}$ parallel to the mean magnetic field. We have now introduced the pseudo-scalar according to

$$\alpha = - \alpha_1 \mathbf{g} \cdot \boldsymbol{\omega} . \tag{8}$$

This effect of establishing an emf parallel to the magnetic field is a quite natural effect appearing in turbulence in rotating systems, although it is quite unknown in common electrodynamics. We call it the 'α-effect'. This effect can be obtained from general physical arguments as shown here. In order to determine the value of the scalar α_1 more detailed calculation is needed.

There is a close relationship between α and the appearance of helical motions in a conducting fluid. α is positive if there are motions according to a lefthanded screw being more probable than motions according to a righthanded screw. α is negative in the opposite case. According to Moffatt (1970) a turbulence of this kind is called 'turbulence with helicity'.* A characteristic property of a turbulence with helicity is that the average $\overline{\mathbf{v}' \cdot \mathrm{curl}\,\mathbf{v}'}$ is not equal to zero. The helicity of the turbulence within the convection zone is furnished by the action of the Coriolis forces. As a consequence of (8) a dependence proportional to $\cos\theta$ is expected, where θ is the colatitude.

A calculation of α was made by a perturbation method (Steenbeck et al., 1966; Krause, 1968). An inhomogeneous and consequently anisotropic turbulence was perturbed by a rotational motion. The following result was obtained (Krause, 1968)

$$\alpha = \alpha_0 \cos\theta, \quad \alpha_0 = \tau_{\mathrm{cor}}^2 v_{\mathrm{eff}}^2 \omega / L . \tag{9}$$

τ_{cor} denotes the correlation time, v_{eff} the rms value of the velocity and L the scale-height. With the data observed at the surface of the Sun we have

$$\alpha_0 \approx 0.3 \ \mathrm{m/s}. \tag{10}$$

The α-effect is of interest for dynamo theory (Krause, 1968; Moffatt, 1970; Steenbeck and Krause, 1966, 1968; Vainstein, 1970; and Fitremann and Frisch, 1969). This is so because toroidal currents are caused by toroidal magnetic fields by means of the α-effect. In this case one gets a closed circuit: Toroidal magnetic fields will be produced from poloidal magnetic fields by the differential rotation, and poloidal magnetic fields will be produced from toroidal magnetic fields by the action of the α-effect.

3. The Dynamo Models

For the Sun there arises a certain difficulty which is related to the observed fact that the Sun has an alternating magnetic field with the well-known period of 22 yr. The situation is depicted in Figure 3 schematically.

The field $\mathbf{B}_{\mathrm{tor}}^{(1)}$ is produced from $\mathbf{B}_{\mathrm{pol}}^{(1)}$ by the differential rotation $\mathbf{B}_{\mathrm{pol}}^{(2)}$ with the desired opposite sign appearing because of the toroidal currents caused by the field $\mathbf{B}_{\mathrm{tor}}^{(1)}$ by means of the α-effect. However, if the α-effect and the differential rotation act at the same place at the same time, the original field $\mathbf{B}_{\mathrm{pol}}^{(1)}$ is weakened immediately by the new field $\mathbf{B}_{\mathrm{pol}}^{(2)}$ and as a result there will be a damped oscillation which dies out during a few periods.

---

* In German 'Turbulenz mit bevorzugtem Schraubensinn'.

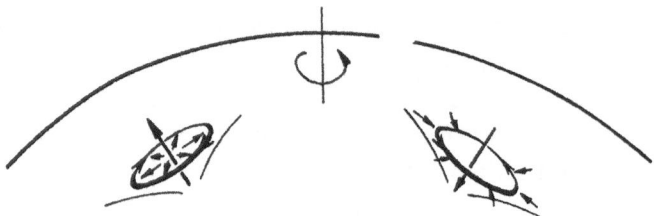

Fig. 1. This figure illustrates the helicity of the convection in the northern hemisphere of the Sun. Rising matter expands because of the density gradient and undergoes the influence of Coriolis forces. A motion according to a lefthanded screw results. For sinking matter both the radial motion and the circulation change sign; hence a motion according to a lefthanded screw appears too.

This behavior does not appear if the two induction mechanisms act at different places (Krause and Steenbeck, 1965). In Figure 4 we have two layers separated by a third layer of thickness δ. In one layer $\mathrm{grad}\,\omega$, in the other α, is not equal to zero. Now we have to regard each field component diffusing through the separating layer. We distinguish the fields at the places where they are induced and where they occur by diffusion, and mark them by the indices i and d. If we have $\mathbf{B}_{\mathrm{pol}}^{(d)}$ in the left layer we get $\mathbf{B}_{\mathrm{tor}}^{(1i)}$ by the action of the differential rotation. Now $\mathbf{B}_{\mathrm{tor}}^{(1i)}$ diffuses and after a time of about $\mu\sigma\delta^2$ we have $\mathbf{B}_{\mathrm{tor}}^{(d)}$ in the right layer, where the field $\mathbf{B}_{\mathrm{pol}}^{(2i)}$ is generated by the α-effect. Now $\mathbf{B}_{\mathrm{pol}}^{(2i)}$ also diffuses and reaches the left layer, weakening $\mathbf{B}_{\mathrm{pol}}^{(1)}$ and finally changing the sign of it. The distance δ between both layers determines the frequency Ω; there is some proportionality between Ω and $1/\mu\sigma\delta^2$, roughly speaking. Explicit numerical treatments of some models proved that this mechanism works (Steenbeck and Krause, 1968). The basic form of this model is depicted in Figure 5.

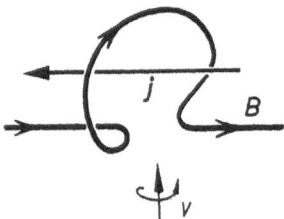

Fig. 2. A magnetic flux line undergoing a helical motion is deformed to a twisted Ω. For a motion according to a righthanded screw, as depicted here, this deformation is accompanied by a current antiparallel to the magnetic field (α negative). For lefthanded screw motions the accompanying current is parallel (α positive).

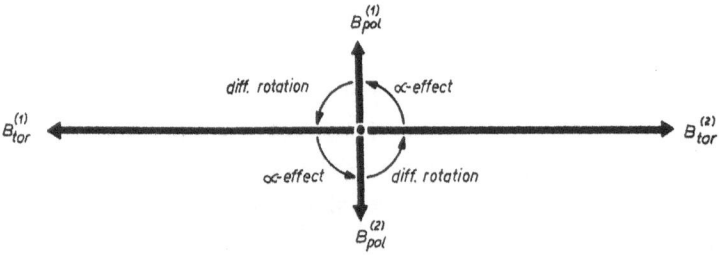

Fig. 3. Schematic representation of a model where both the induction mechanisms act at the same place. The weakening of the original field occurs immediately and damped oscillating field results.

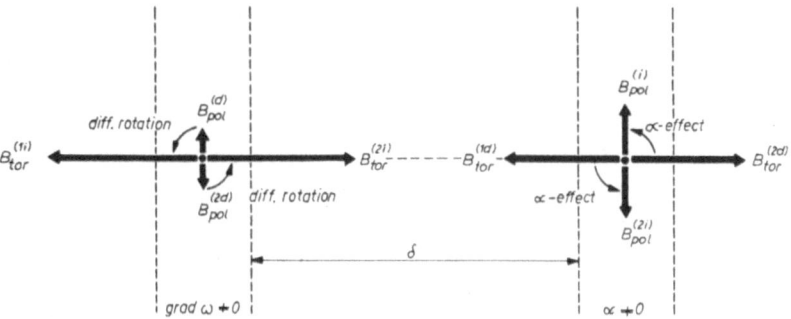

Fig. 4.   Schematic representation of a model where different induction mechanisms act at different layers separated by an additional layer. The induced field needs a certain time for diffusing through the separating layer. Hence the weakening of the original field occurs with a certain time retardation. This model is proved to maintain an oscillating field (Steenbeck and Krause, 1968).

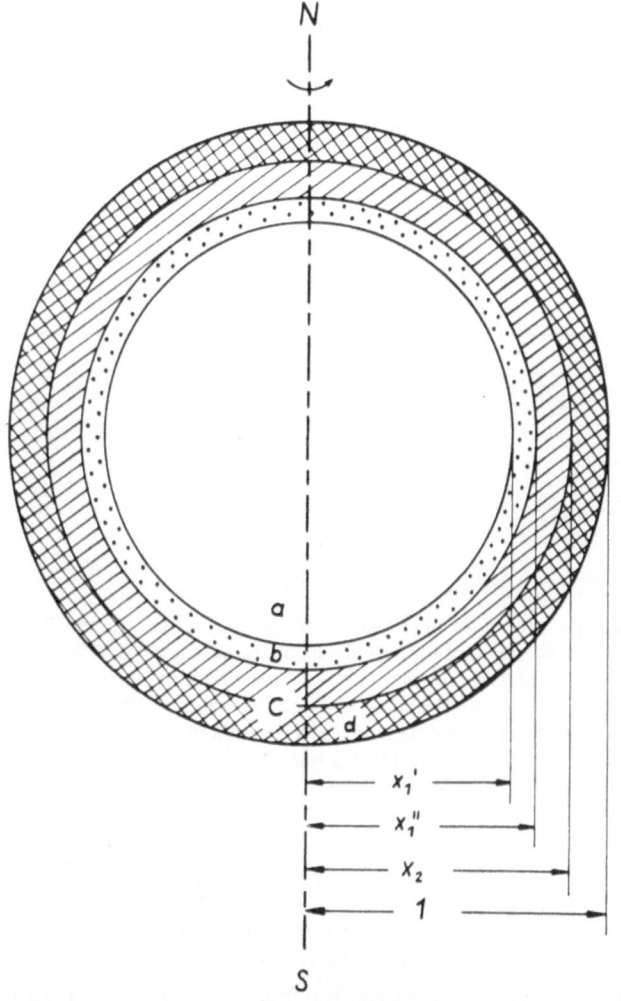

Fig. 5.   Representation of the basic type of the investigated spherical alternating field dynamos. *a* denotes a rigid rotating core and *c* a spherical shell also rotating rigidly but with a lower angular velocity. *b* denotes the transition shell and *d* a shell with α-effect.

It may be remarked that the separation of the two layers must not be complete; there can occur an overlapping.

Surprisingly a butterfly diagram is obtained quite similar to the observed one taking into account an *r* dependence of the angular velocity only. This result suggests that for the production of the toroidal field of the Sun the radial variation of the angular velocity is more important than the variation with latitude, as has been believed till now. An observation of Wilcox (1971) supports this conception. According to this observation the active regions show rigid rotation and do not take part in the differential rotation observed at the surface of the Sun.

It may be underlined that the alternating field dynamo presented here only works if the two induction mechanisms act on different layers. This is a definite statement concerning the structure of the Sun which would be of less importance if there were other alternative models. Therefore it is of interest to analyze the model of Leighton (1969). It seems that Leighton involved a certain singularity in his model. In Equation (4) of his paper a factor $\sin\theta$ is missed. Therefore a toroidal field for the production

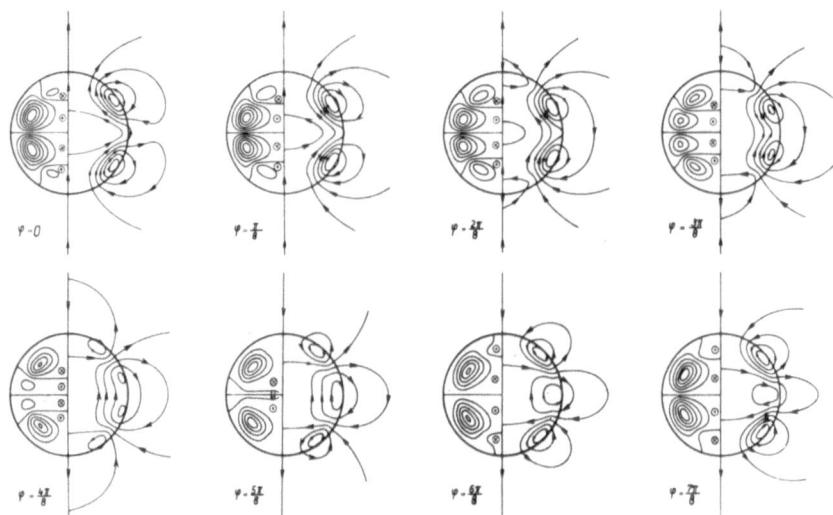

Fig. 6.   Eight phase pictures of the field maintained by a calculated dynamo. The right half of every phase picture shows the field lines of the poloidal field whereas the left half shows the lines of constant field strength of the toroidal field. The phase $\Omega t = 0$ corresponds to maximum toroidal field strength. The migration of the field towards the equator should be noted. Further attention may be drawn to the region of the phases where the dipole changes its sign (about $5\pi/8 \ldots 6\pi/8$). The field configuration is determined by the higher multipoles there.

of B_r is assumed, which is singular at the poles of the Sun. At first this singularity is excluded by the assumption of the threshold field B_c, but the results show a dying oscillatory field. Therefore Leighton for a fraction of the B_r field, takes this equation without the threshold condition. Now the oscillation is maintained but the singularity is included too. For this reason the Leighton model cannot be accepted as an alternative.

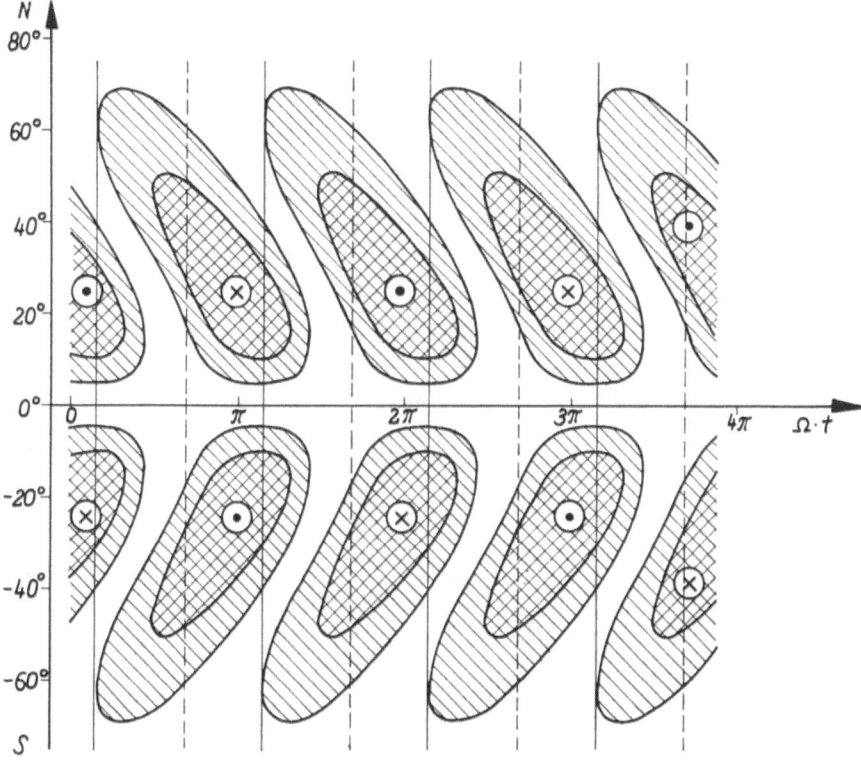

Fig. 7.   Butterfly diagram derived from a calculated model. The typical observed behavior, i.e. the migration towards the equator, can also be noticed here. The dotted line denotes the phase where the dipole changes its sign.

4. Conclusions

We promised to furnish a dynamo satisfying the requirements (1)–(4). Concerning requirement (1) we have to remark that the α-effect is a quite natural one appearing in turbulence in rotating systems as already mentioned. The second induction mechanism is provided by the differential rotation. There is no doubt about the fact that both effects will occur at the Sun, but their radial dependence is rather unknown. We can assume this dependence according to what is needed for having a dynamo because there is no completely developed theory. If a similarity appears between the calculated and the observed field the dynamo model can be considered as a probe giving information about the structure of the layers of the Sun which take part in the dynamo excitation. A first hint in this direction is the statement that there is a strong r dependence of the angular velocity at deeper layers. This statement is supported by the observation of Wilcox (1971) already mentioned. More detailed information can be expected from the non-linear theory.

Requirement (2) is satisfied because of the linearity of the equations describing the dynamo. Requirement (3) needs no further discussion.

Concerning requirement (4) it may be remarked that the field equations have eigen solutions which are symmetric in relation to the equatorial plane – the leading term for the poloidal field is a quadrupole – in addition there are antisymmetric solutions with a dipole as the leading term. A model will excite that field which makes the smallest pretensions to the induction effects, in other words, which has the smallest eigenvalue characterizing the pretension. The eigenvalue for the symmetric field was proved by calculations to be greater than that for the antisymmetric field. Hence the antisymmetric field will be excited.

As we have mentioned an accurate fitting of the models to the observations is only reasonable in the framework of a non-linear theory. For this purpose it is desirable to derive as much information as possible concerning the space-time behavior of the mean field. The butterfly diagram is one kind of evaluation of the observational data. Another one could be the determination of the dependence on time and latitude of the mean poloidal field at the surface of the Sun. A first step in this direction has been carried out by Stenflo (1971) with encouraging results.

References

Babcock, H.: 1961, *Astrophys. J.* **133**, 572.

Fitremann, M. and Frisch, U.: 1969, *Compt. Rend.* **268**, B705.

Krause, F.: 1968, Habilitationsschrift, Univ. of Jena.

Krause, F.: 1969, *Proc. 4th Consult. Solar Phys. and Hydromagn.*, Sopot 1966, Acta Univ. Wratisla-viensis No. 77, 157.

Krause, F.: 1969, *Mber. Deut. Akad. Wiss. Berlin* **11**, 188.

Krause, F. and Steenbeck, M.: 1965, *Proc. 3rd Consult. Solar Phys. and Hydromagn.*, Tatranska Lomnica 1964, *Czech. Acad. Sci. Astron. Inst. Publ.* **51**, 36.

Leighton, R. B.: 1969, *Astrophys. J.* **156**, 1.

Moffatt, H. K.: 1970, *J. Fluid Mech.* **41**, 435.

Rädler, K.-H.: 1968, *Z. Naturforsch.* **23a**, 1841, 1851.

Rädler, K.-H.: 1969, *Mber. Deut. Akad. Wiss.* **11**, 194.

Steenbeck, M. and Krause, F.: 1966, *Z. Naturforsch.* **21a**, 1285.

Steenbeck, M. and Krause, F.: 1968, *Astron. Nachr.* **291**, 49, 271.

Steenbeck, M., Krause, F., and Rädler, K.-H.: 1966, *Z. Naturforsch.* **21a**, 369.

Stenflo, J. O.: 1971, this volume, p. 714.

Weinstein, S. I.: 1970, *J. Exp. Theor. Phys.* **58**, 153.

Wilcox, J. M.: 1971, this volume, p. 744.

Discussion

Gilman: I am concerned about the separation required by the two dynamo processes (differential rotation and α-effect). It seems to me that the α-effect due to swirl motion induced by the basic rotation may be just as strong, or stronger, near the bottom of the convection zone where you are also assuming the differential rotation effect to be strongest.

Krause: Apparently α rises at first with the depth inside the convection zone, but it is zero at the bottom. The wanted radial variation of the angular velocity ω may be given in the layer just under the convection zone where we expect a smooth transition of ω to the value inside the core.

Nagarajan: My question is about the prescription of the pseudo scaler stochastic parameter α which effectively linearizes the problem. This has to be dynamically determined from the fluctuation equation, unless you neglect all fluctuations below a certain length scale (magnetic field or velocity). I do not see how the leaving out of the Lorentz term is consistent. For smaller scales even a crude testing would show that $\overline{v \cdot \nabla v}$ and $\overline{j \times B}$ would be of the same order. Secondly, in this approximation of treating α's as prescribed stochastic parameters, we run into the statistical difficulty about separation of mean and fluctuation time scales, which is rather difficult in turbulence.

Krause: α is determined by a certain disturbation method from the Navier-Stokes equations. If we include Lorentz forces, α is lowered by a term proportional to the square of the mean magnetic field. It should be added that the α-effect was also experimentally proved by an experiment carried out in the Institute of Physics in Riga using liquid sodium.

A DYNAMO MODEL FOR THE LARGE SCALE FIELDS

I. K. CSADA

Konkoly Observatory, Budapest, Hungary

Abstract. The expansion of the magnetic field in a series of force-free constituents for the internal field and that of 'semi' force-free constituents for the external field results in the skew symmetry of the coupling tensor of a magnetohydrodynamic dynamo. The periodic solution which exists in this case is deduced by the method of eigenvalues. Based on the numerical results a dynamo model is proposed to draw up the characteristic features of the magnetographic observations carried along the central meridian and at the polar areas.

The coupling tensor α_{sn} in the eigenvalue theory of the magnetohydrodynamic dynamo is transformed to skew-symmetric form. In this case the solutions to the dynamo equation give a magnetic field with periodical variation superimposed on decays. The velocity is steady.

On applying Gauss's integral theorem, the tensor α_{sn} can be written in the form

$$\alpha_{sn} = -\frac{\displaystyle\int \mathbf{v}\,(\mathrm{curl}\,\mathbf{H}_s \times \mathbf{H}_n)\,\mathrm{d}\tau}{\displaystyle\int H_s^2\,\mathrm{d}\tau}. \tag{1}$$

The surface integral involved in the complete expression goes here to zero owing to the boundary conditions imposed on the velocity and magnetic field.

Two methods are available for the transformation of α to skew-symmetric form.

(1) Let $\mathbf{H}_n$ be a force-free vector satisfying the equation

$$\mathrm{curl}\,\mathbf{H}_n \times \mathbf{H}_n = 0. \tag{2}$$

(The magnetic field is not force-free since it is defined as

$$\mathbf{H} = \sum_{n=0} B_n(t)\,\mathbf{H}_n(r, \theta, \phi) \tag{3}$$

and the sum of force-free vectors is not force-free).

Applying the dynamo to the internal field of the Sun, $\mathbf{H}_n$ is governed by the equation

$$\Delta\mathbf{H}_n + \frac{k_n^2}{\kappa}\,\mathbf{H}_n = 0 \tag{4}$$

and the coupling tensor has the form

$$\alpha_{sn} = -\frac{k_n\displaystyle\int \mathbf{v}\,(\mathbf{H}_s \times \mathbf{H}_n)\,\mathrm{d}\tau}{\displaystyle\int H_n^2\,\mathrm{d}\tau}. \tag{5}$$

Howard (ed.), Solar Magnetic Fields, 780–782. All Rights Reserved.
Copyright © 1971 by the IAU.

(2) In the other method one uses the assumption that only the r and θ components of $\mathbf{H}_n$ have force-free character and the velocity has only a ϕ component, then

$$\alpha_{sn} = - \frac{k_s \int v_\phi \begin{vmatrix} H_{sr} & H_{s\theta} \\ H_{nr} & H_{n\theta} \end{vmatrix} d\tau}{\int H_s^2 \, d\tau} \tag{6}$$

which can be easily transformed to skew-symmetric form.

Applying this formulation to the external field, $\mathbf{H}_n$ is governed by the equation

$$\Delta \mathbf{H}_n - \frac{k_n^2}{\kappa} \mathbf{H}_n = 0. \tag{7}$$

In the next step the infinite tensor α_{sn} is reduced to a finite tensor by the modification of the velocity and thus the infinite determinant of α_{sn} splits into the double product

$$|\alpha_{sn} - i\Omega\delta_{sn}| = D_1 D_2 = 0, \tag{8}$$

where D_1 is a finite and D_2 is an infinite determinant. By putting $D_1 = 0$, no higher terms than those present at the start of the dynamo will be generated.

Two models were computed in terms of the present dynamo. Both have the same internal field but in the first case the external field is assumed to be random without any average contribution to the field fluctuations while in the second no random fluctuations are supposed to occur in the external field deduced from (6) and (7).

The numerical and analytical results obtained in these models seem to account for the interpretation of the following phenomena in solar physics.

(i) The cyclical variation of the general field. In a general expansion of the theory the cyclical variation is presented by the superposition of a large number of periodic terms arising in $B_n(t)$. In the present solution the analysis is performed up to the first four terms of (3) and thus two periods were obtained. The periods evaluated with the use of the second model are about ten times shorter than the corresponding periods deduced in the first model, i.e.

$$P_1 = 27.36 \text{ yr} \quad P_2 = 137.45 \text{ yr} \quad \text{first model}$$
$$P_1 = 1.60 \text{ yr} \quad P_2 = 95.33 \text{ yr} \quad \text{second model}$$

It seems reasonable to describe the cyclical variations of the solar activity by coupling of the two models. Thus, the second model applies to the minimum since its time scale is short and negligible magnetic disturbances are observed in the photosphere, while the first model, having a longer time scale during which practically no general field is present in terms of the averages of the magnetic fluctuations, holds for the maximum.

(ii) The average surface distribution and the variation of weak photospheric fields. These phenomena are accounted for in terms of the second model. The field component of the line of sight is calculated for the central meridian as a function of time (Figure 1). These results are intended to explain the magnetogram observations averaged over rotation periods.

(iii) The field reversal observed at the poles is deduced by the first model (Figure 2). No simple variation of the polar field is obtained. As an approximate law of variation one can calculate with the superposition of two periods. The longer period has a larger amplitude at one of the poles while at the other the amplitude of the shorter period is longer.

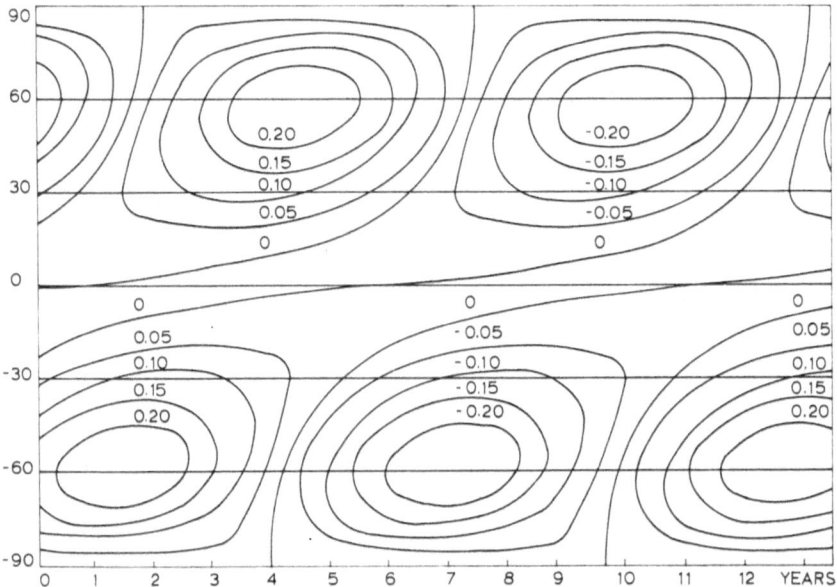

Fig. 1. Characteristic variation of the magnetic field (in arbitrary units) as it should be observed along the central meridian and averaged over rotation periods during the solar cycles. The results are based on the present dynamo model. The solar latitude is plotted on the vertical axis, the time on the horizontal axis.

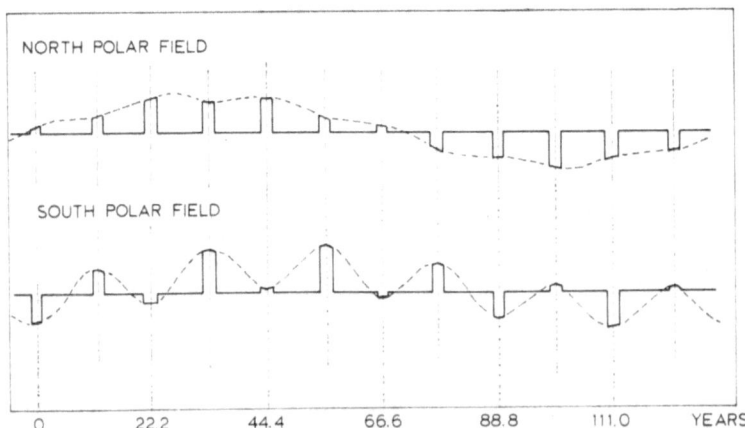

Fig. 2. The variations of the polar fields during the solar cycles as they are obtained in the present theory. The field strengths are measured in arbitrary units.